International Symposium on Food and Agro-Biodiversity (ISFA 2024)

IOP Conference Series: Earth and Environmental Science
Volume 1460

Semarang, Indonesia
2-4 October 2024

ISBN: 979-8-3313-1582-5
ISSN: 1755-1307

Printed from e-media with permission by:

Curran Associates, Inc.
57 Morehouse Lane
Red Hook, NY 12571

proceedings.com

Some format issues inherent in the e-media version may also appear in this print version.

Institute of Physics
Dirac House, Temple Back
Bristol BS1 6BE UK

Phone: 44 1 17 929 7481
Fax: 44 1 17 920 0979

techtracking@iop.org

Additional copies of this publication are available from:

Curran Associates, Inc.
57 Morehouse Lane
Red Hook, NY 12571 USA
Phone: 845-758-0400
Fax: 845-758-2633
Email: curran@proceedings.com
Web: www.proceedings.com

TABLE OF CONTENTS

IOP Conf. Series: Earth and Environmental Science **1460** (2025) 011001 doi:10.1088/1755-1315/1460/1/011001

Preface

The **4th International Symposium on Food and Agro-biodiversity (ISFA)** was successfully held from October 2–4, 2024, at the Oak Tree Hotel, Semarang, Indonesia. Organized by the Faculty of Animal and Agricultural Sciences, Universitas Diponegoro, in collaboration with the Indonesian Food Technologists (IFT), this triennial event provided an esteemed platform for researchers, academics, industry professionals, and students to engage in thoughtful discussions under the theme of **"For Healthy, Affordable, and Sustainable Food."**

We extend our heartfelt gratitude to all who contributed to the success of this symposium. Special appreciation is given to our distinguished **keynote speakers**, including **Prof. Jan Van Impe** (Belgium), **Prof. Hii Ching Lik** (Malaysia), **Prof. Daniel Rodriguez** (Australia), **Prof. Masahiro Ogawa** (Japan), **Assoc. Prof. I. H. Tekiner** (Turkey), and **Prof. Sugiharto** (Indonesia), whose expertise and insights set the stage for impactful discussions. We also thank our **invited speakers**, **Dr. Mulyana Hadipernata** and **Dr. Eko H. Purnomo**, for their valuable contributions in enriching the knowledge exchange during the symposium.

This year' symposium brought together **139 participants** from a wide range of professional backgrounds, including researchers, educators, industry experts, and students. The event also welcomed international participants from **Malaysia, Timor-Leste, Japan, Belgium, Turkey, the Philippines, Pakistan, India, and Myanmar.** Their active participation fostered a dynamic, cross-cultural dialogue and underscored the symposium's global relevance.

The ISFA Scientific Committee undertook a meticulous review and selection process, resulting in the publication of **70 high-quality articles** in the **IOP Conference Series: Earth and Environmental Science.** These articles, presented during the symposium through oral and poster sessions, reflect cutting-edge research and innovative approaches to addressing challenges in food and agro-biodiversity.

We express our profound gratitude to all individuals, institutions, and organizations whose support and dedication made this event a resounding success. The contributions of the organizing team, speakers, presenters, and participants have reaffirmed ISFA's role as a key platform for fostering collaboration and innovation in sustainable food systems.

We look forward to the next ISFA with great enthusiasm, striving to make it an even more impactful event that continues to inspire progress and global collaboration in the years to come.

Warm regards,

The Organizing Committee
4th International Symposium on Food and Agro-biodiversity (ISFA)

IOP Conf. Series: Earth and Environmental Science 1460 (2025) 011001 doi:10.1088/1755-1315/1460/1/011001

Symposium Documentation

 doi:10.1088/1755-1315/1460/1/011001

Symposium Documentation

**All participants agree to the inclusion of documentation in the following proceedings.*

IOP Conf. Series: Earth and Environmental Science **1460** (2025) 011001 doi:10.1088/1755-1315/1460/1/011001

COMMITTEE OF THE 4TH INTERNATIONAL SYMPOSIUM ON FOOD AND AGRO-BIODIVERSITY

Steering Committee:
Prof. Sugiharto, S.Pt., M.Sc., Ph.D.
Dr. Ir. Limbang K. Nuswantara, S.Pt., M.P., I.P.U.
Prof. Agus Setiadi, S.Pt., M.Si., Ph.D.

Organizing Committee:
Yoga Pratama, S.TP., M.Sc., Ph.D.
Dr. Rafli Zulfa Kamil, S.TP.
Swastika Dewi, S.TP., M.Sc.
Dr. Sri Mulyani, S.Pt., M.P.
Annisa Firdauzi, S.P., M.P.
Dr. Heni Rizqiati, S.Pt., M.Si.
Setya Budi M. Abduh, S.PT., M.Sc., Ph.D.
Prof. Ir. Anang Moh. Legowo, M.Sc., Ph.D.
M. Iqbal Fauzan, S.P., M.Si.
drh. Siti Susanti, Ph.D.
Asep Setiaji, S.Pt., M.Si., Ph.D.
Hega Bintang Pratama Putra, S.TP. M.Sc.
Dr. Ir. Bambang Dwiloka, M.S.

Reviewer:
Ir. Rudy Hartanto, S.Pt., M.P. Ph.D., IPM.
Prof. Sugiharto, S.Pt., M.Sc. Ph.D
Sutaryo, S.Pt., M.P. Ph.D
Nuruliarizki Shinta Pandupuspitasari, S.Pt., M.Si., M.Sc., Ph.D
Prof. Dr. Ir. Florentina Kusmiyati, M.Sc.
Dr. Ir. Eny Fuskhah, M.Si.
Dr. Dian Safitri, S.P., M.Si.
Prof. Dr. Ir. Titik Ekowati, M.Sc.
Ir. Joko Mariyono, M.P. Ph.D
Ir. Kustopo Budiraharjo, M.P.
Siwi Gayatri, S.Pt., M.Sc. Ph.D
Prof. Dr. Ir. Valentinus Priyo Bintoro, M.Agr.
Dr. Ir. Antonius Hintono, M.P.
Dr. Ir. Nurwantoro, M.S.
Dr. Ir. Yoyok Budi Pramono, S.Pt., M.P.
Ahmad Ni'matullah Al-Baarri, S.Pt., M.P. Ph.D

Committee affiliation:
Faculty of Animal and Agricultural Sciences, Universitas Diponegoro, Jl. Prof. Soedarto, Tembalang, Kec. Tembalang, Kota Semarang, Jawa Tengah, Indonesia

IOP Conf. Series: Earth and Environmental Science **1460** (2025) 011001 doi:10.1088/1755-1315/1460/1/011001

LIST OF PARTICIPANTS

Full Name	Affiliation	Nationality
Abib Fajri	Universitas Diponegoro	Indonesia
Adeel Shahzad Alvi	Chiang Mai University Thailand	Pakistan
Adi Magna Patriadi Nuhriawangsa	Universitas Sebelas Maret	Indonesia
Aditya Wahyudhi	Institut Teknologi Dan Sains Nahdlatul Ulama Jambi	Indonesia
Afrizal Malik	National Research and Innovation Agency	Indonesia
Agung Prabowo	National Research and Innovation Agency	Indonesia
Agung Setyarini	Universitas Veteran Bangun Nusantara	Indonesia
Ahmad Muzaki Khoiruman	Universitas Diponegoro	Indonesia
Ahmad Ni'matullah Al-Baarri	Universitas Diponegoro	Indonesia
Ahmed Hafizd Aththoriq	Universitas Brawijaya	Indonesia
Ailsa Ocha Alifa	Universitas Diponegoro	Indonesia
Aisna Rahma Tiarani	Universitas Brawijaya	Indonesia
Akhmad Arief	Universitas Diponegoro	Indonesia
Albertus Fajar Irawan	Universitas Diponegoro	Indonesia
Alif Talitha Zahra	Universitas Diponegoro	Indonesia
Alya Griya Cahyani	Universitas Sebelas Maret	Indonesia
Amaliya Putri	Universitas Diponegoro	Indonesia
Andi Agus Mumang	Universitas Hasanuddin	Indonesia
Andro Klemens Nugrahanto	Universitas Diponegoro	Indonesia
Arief Rakhman Affandi	Universitas Gadjah Mada	Indonesia
Arif Bestari	Universitas Diponegoro	Indonesia
Arif Rizqi Nurwidiyanto	Universitas Diponegoro	Indonesia
Arum Kurnia Sari	PT Mitra Asiatek Biosensor	Indonesia
Aung Thu Zin	University of Singaperbangsa Karawang	Myanmar
Bambang Dwiloka	Universitas Diponegoro	Indonesia
Bella Putri Maharani	Universitas Gadjah Mada	Indonesia
Bhakti Etza Setiani	Universitas Diponegoro	Indonesia
Binti Ma'rifah	Universitas Diponegoro	Indonesia
Dania Rahma Anandini	Universitas Diponegoro	Indonesia
Daniel Rodriguez	Queensland University	Australia
Dearestantrianto Hadits Fardana	Universitas Diponegoro	Indonesia
Deni Novia	Universitas Andalas	Indonesia
Dhany Hermansyah	IPB University	Indonesia
Dimas Rahadian Aji Muhammad	Universitas Sebelas Maret	Indonesia
Dwi Ishartani	Universitas Sebelas Maret	Indonesia
Dwi Sulistiawati	Tadulako University	Indonesia
Dwita Indrarosa	Universitas Diponegoro	Indonesia

Full Name	Affiliation	Nationality
Dwiyati Pujimulyani	University Of Mercu Buana Yogyakarta	Indonesia
Eko H Purnomo	IPB University	Indonesia
Endah Subekti Kuntariningsih	Universitas Diponegoro	Indonesia
Enny Purwati Nurlaili	Universitas 17 Agustus	Indonesia
Erlyn Grace P. Aguilar	University of Science and Technology	Philippines
Fadhil Adi Nugraha	Universitas Muhammadiyah Purwokerto	Indonesia
Faiq Yusron	Universitas Diponegoro	Indonesia
Farhan Akmala Putra	Universitas Diponegoro	Indonesia
Fariz Nurmita Aziz	Universitas Diponegoro	Indonesia
Fariz Zharfan Haris	Universitas Diponegoro	Indonesia
Fida Ghaisani Yusuf	Universitas Diponegoro	Indonesia
Fidelino Carvalho	Univesity of Tribhuwana Tunggadewi	Timor Leste
Firda Tasya Kamila	Universitas Diponegoro	Indonesia
Fitra Inna Istifara	Universitas Diponegoro	Indonesia
Fitriyah Zulfa	Universitas Sebelas Maret	Indonesia
Fredericus Enrico Satrio Bawono	Universitas Diponegoro	Indonesia
Ghanashyam Sharma	The Mountain Institute India	India
Halim Abdelaziz	Universitas Diponegoro	Indonesia
Hanis Syazwani Mat Gani	Universiti Sultan Zainal Abidin	Malaysian
Haris Rinaldi	Others	Indonesia
Hefti Salis Yufidasari	Universitas Brawijaya	Indonesia
Hega Bintang Pratama Putra	Universitas Diponegoro	Indonesia
Hii Ching Lik	Nottingham University Malaysia	Malaysia
Huda M. Elmatsani	Badan Riset Dan Inovasi Nasional	Indonesia
Ichlasia Ainul Fitri	University Of Mercu Buana Yogyakarta	Indonesia
Ilham Ramadhan	Universitas Diponegoro	Indonesia
Ilhemi Ismasari	Universitas Diponegoro	Indonesia
Indra Maulani	Universitas Diponegoro	Indonesia
Isabel E. Bihag	Caraga State University	Philippines
Ismail Hakkı Tekiner	Haliç University	Turkiye
Jabosar Ronggur Hamonangan Panjaitan	Institut Teknologi Sumatera	Indonesia
Jade Dhapnee Z. Compendio	Visayas State University	Philippines
Jan Van Impe	KU Leuven	Belgium
Khalila Virgy Violetta	Universitas Diponegoro	Indonesia
Laila Rahmawati	Research Center For Food Technology And Processing	Indonesia
Latifah Dita Cahyani	Universitas Sebelas Maret	Indonesia
Lee Ikrimah Ma'ruf	Politeknik Pariwisata Negeri Makassar	Indonesia
Lilik Krismiyanto	Universitas Diponegoro	Indonesia
Lilik Retna Kartikasari	Universitas Sebelas Maret	Indonesia

IOP Conf. Series: Earth and Environmental Science **1460** (2025) 011001 doi:10.1088/1755-1315/1460/1/011001

Full Name	Affiliation	Nationality
Lutfi Purwitasari	Universitas Diponegoro	Indonesia
Masagus Haidir Tamimi	Universitas Diponegoro	Indonesia
Masahiro Ogawa	Kagawa University	Indonesia
Mgs. Muhammad Prima Putra	Universitas Gadjah Mada	Indonesia
Michelle Wahyu Effendy	Universitas Diponegoro	Indonesia
Muhamad Ghazi Agam Sas	Universitas Diponegoro	Indonesia
Muhammad Hizbullah Al Anshary	Universitas Gadjah Mada	Indonesia
Muhammad Qasim Ali	University Malaysia Pahang Al Sultan Abdullah	Pakistan
Muhammad Rizky Ramanda	Institut Teknologi Sumatera	Indonesia
Mulia Winirsya Apriliyani	Universitas Gadjah Mada	Indonesia
Mulyana Hadipernata	National Research and Innovation Agency	Indonesia
Nafisah Zainuba Hasan	Universitas Diponegoro	Indonesia
Nailah Salma Razita	Universitas Diponegoro	Indonesia
Naurah Syifa Syahputri	Universitas Diponegoro	Indonesia
Ni Kadek Karisma Dewi	Universitas Diponegoro	Indonesia
Nida Fariha	Universitas Diponegoro	Indonesia
Nishya Afisah	Universitas Diponegoro	Indonesia
Novian Wely Asmoro	Universitas Gadjah Mada	Indonesia
Novian Wely Asmoro	Universitas Veteran Bangun Nusantara	Indonesia
Nuke Faradhila	Universitas Jenderal Soedirman	Indonesia
Nur Annisa Metya Novikasari	Universitas PGRI Semarang	Indonesia
Nurul Hasniah	Universitas Diponegoro	Indonesia
Nurwantoro	Universitas Diponegoro	Indonesia
Okky Rizkia Yustian	Bina Nusantara University	Indonesia
Premy Puspitawati Rahayu	Universitas Brawijaya	Indonesia
Putu Novia Gariri	Universitas Diponegoro	Indonesia
Rafli Zulfa Kamil	Universitas Diponegoro	Indonesia
Ratih Paramastuti	Universitas Diponegoro	Indonesia
Ria Millati	Universitas Gadjah Mada	Indonesia
Riuh Wardhani	Chungnam National University	Indonesia
Rosita Dewati	Universitas Veteran Bangun Nusantara	Indonesia
Rudy Hartanto	Universitas Diponegoro	Indonesia
Ruth Dameria Haloho	Universitas Sulawesi Barat	Indonesia
Sandi Darniadi	National Research and Innovation Agency	Indonesia
Septrial Arafat	Universitas Diponegoro	Indonesia
Shaula Virginia Bellatrix	Universitas Sebelas Maret	Indonesia
Sirly Eka Nur Intan	Universitas Diponegoro	Indonesia
Siti Susanti	Universitas Diponegoro	Indonesia
Siti Tamaroh Cahyono Murti	Universitas Mercu Buana Yogyakarta	Indonesia
Sri Sumarsih	Universitas Diponegoro	Indonesia

Full Name	Affiliation	Nationality
Sugiharto	Universitas Diponegoro	Indonesia
Sukmawati Utami	Universitas Diponegoro	Indonesia
Swastika Dewi	Universitas Diponegoro	Indonesia
Tengku Farizan Izzi Che Ku Jusoh	Universiti Sultan Zainal Abidin	Malaysia
Thryrone A. Ordanel	Visayas State University,	Philippines
Titri Siratantri Mastuti	Universitas Pelita Harapan	Indonesia
Tizar Muhammad Kautsar Bijaksana	Institut Teknologi Bandung	Indonesia
Ulil Afidah	Universitas Diponegoro	Indonesia
Usman Pato	Universitas Riau	Indonesia
Weni Audi Anjani	Universitas Diponegoro	Indonesia
Widitya Tri Nugraha	Universitas Gadjah Mada	Indonesia
Wilbur Donald R. Pokatong	Universitas Pelita Harapan	Indonesia
Yaktiworo Indriani	University of Lampung	Indonesia
Yan Ariana	Universitas Diponegoro	Indonesia
Yan Ariana	Universitas Diponegoro	Indonesia
Yani Magfiroh	Soegijapranata Catholic University	Indonesia
Yannie Asrie Widanti	Universitas Gadjah Mada	Indonesia
Yasmin Aulia Rachma	Universitas Diponegoro	Indonesia
Yoga Arif Firmansyah	Universitas Gadjah Mada	Indonesia
Yoga Pratama	Universitas Diponegoro	Indonesia
Yoyok Budi Pramono	Universitas Diponegoro	Indonesia
Yuli Perwita Sari	Universitas Mercu Buana Yogyakarta	Indonesia
Zahra Nafizha Amani	Universitas Diponegoro	Indonesia

Peer Review Statement

All papers published in this volume have been reviewed through processes administered by the Editors. Reviews were conducted by expert referees to the professional and scientific standards expected of a proceedings journal published by IOP Publishing.

- **Type of peer review:** Single Anonymous
- **Conference submission management system:** Morressier
- **Number of submissions received:** 72
- **Number of submissions sent for review:** 72
- **Number of submissions accepted:** 70
- **Acceptance Rate (Submissions Accepted / Submissions Received × 100):** 97.2
- **Average number of reviews per paper:** 2
- **Total number of reviewers involved:** 25
- **Contact person for queries:**
  Name: Yoga Pratama, PhD
  Email: yogapratama@live.undip.ac.id
  Affiliation: Department of Food Technology, Faculty of Animal and Agricultral Sciences, Universitas Diponegoro, Indonesia

IOP Conf. Series: Earth and Environmental Science **1460** (2025) 012001 doi:10.1088/1755-1315/1460/1/012001

Enhancing Production Efficiency: The Role of Inflection Points in Non-Linear Growth of Bali Cattle

Putu Novia Gariri, Sutopo Sutopo, Asep Setiaji*, Fatmawati Mustofa, Fariz Zharfan Haris, Maulida Arkaan Muhammad Da'i, Pupus Galau Prahara, Firda Tasya Kamila, Syaddad Verahry Philco, Tirta Dwi Tamanningrum, Suratman Suratman

Department of Animal Science, Faculty of Animal and Agricultural Sciences, Universitas Diponegoro, Semarang 50275, Central Java, Indonesia

*E-mail: asepsetiaji93@gmail.com

Abstract. Bali cattle (*Bos sondaicus*) are an important genetic resource for Indonesia's beef industry due to their adaptability to tropical environments, resilience, and efficient reproductive performance. Contributing approximately 26.92% to national beef production, these cattle are highly valued for their efficient feed conversion, high fertility, and early maturity. This study aimed to analyze the growth patterns of male Bali cattle using the Gompertz non-linear growth model, utilizing data from 256 cattle aged 0–730 days. The results indicated an asymptotic mature weight of 423.99 kg, higher than previously reported for this breed. The inflection point, where maximum growth occurs, was estimated at 155.98 kg at approximately 398.54 days (13 months), marking the optimal time for intensified nutritional support. The gradual growth pattern, characteristic of Bali cattle, reflects adaptation to variable feed quality in tropical regions. The Gompertz model demonstrated high accuracy in predicting growth, providing valuable insights into cattle management and feeding strategies. Identifying the inflection point offers a practical tool for optimizing production by adjusting nutrition and management during critical growth phases. This model not only helps to forecast growth but also assists in the selection of high-performing cattle and enhances the efficiency of livestock production. The findings underscore the importance of advanced growth curve models in improving Bali cattle productivity and ensuring the sustainable development of Indonesia's beef industry.

Keywords: bali cattle, growth performance, non-linear model, production efficiency

1. Introduction

Bali cattle (*Bos sondaicus*) are represent an important genetic resource for the beef cattle industry in Indonesia. These cattle, domesticated from the wild Banteng of Indonesia. Optimizing Bali cattle production is critical to meeting the global protein demand. Bali cattle, known for their adaptability and resilience, play a significant role in the agricultural economy of many regions. Acknowledging the breed's original geographic distribution across Indonesia and its significance as part of the nation's genetic livestock resources that need to be protected and preserved (Soares *et al.*, 2011). Bali cattle have adapted well and exhibit good reproductive capabilities, allowing their spread and development across the country. Bali cattle demonstrated remarkable reproductive performance, achieving around 80% pregnancy and birth rates (Wawo, 2018). Female Bali cattle gain an average of 0.6 kg per day, while males achieve 0.7 kg per day. Additionally, it has 51.5% and 59.8% meat to bone (Biscarini *et al.*, 2015; Tahuk *et al.*, 2018). Bali cattle are also known for their lean meat quality. Farmers prefer Bali cattle due to their efficient use of low-quality feed, high fertility and conception rates, as well as their early maturity (Purwantara *et al.*, 2012). Bali cattle play a significant role by contributed around 26.92% in fulfilling the national beef demand (Sutarno and Setiawan, 2015).

Continuous efforts have been made to enhance Bali cattle productivity, including breeding programs aimed at improving their genetic quality. However, traditional growth models often fail to fully account for the complexities in their growth trajectories, which may not follow a linear pattern. Utilizing advanced mathematical and statistical models, growth phases is identified significantly impact productivity, offering practical solutions for cattle management. Growth curve models assist in decision-making related to maintenance strategies, culling of low-productivity animals, and selecting the best-performing animals based solely on their growth patterns (Ghavi and Zandeh, 2017). Growth curve analysis using mathematical models serves as an effective tool to control and optimize the systems of livestock production (Vazquez *et al.*, 2012. Non-linear models based on body weight data at different age ranges can predict the growth curve in beef cattle. Several non linear mathematical models, including Brody, Gompertz, Logistic, von Bertalanffy, and Richards, have been utilized to analyze growth curves in local cattle breeds (Mokhtari *et al.*, 2023). The Gompertz growth model curve demonstrated a high level of accuracy in illustrating the growth patterns of male Turkish Holstein cattle (Putra and Fajrina, 2021). Another Study on Madura and Madrasin cattle have demonstrated that non-linear models can effectively represent growth trends with high accuracy (Widyas *et al.*, 2018)

The determination of the inflection point is biologically challenging, but with the assistance of non-linear growth curves, this issue can be resolved (Suryawan *et al.*, 2017). The equation that represents the inflection point is the second derivative of a nonlinear equation, which is used to estimate the value at which the inflection point of the growth curve occurs (Goshu and Koya, 2013). The inflection point is considered the most efficient because it corresponds to the lowest mortality and the fastest growth. Determining the inflection point biologically is challenging, so it can be addressed using nonlinear growth curves (Nahashon *et al.*, 2006). The inflection point represents the transition from acceleration to deceleration in the growth of body weight in livestock. This critical juncture typically coincides with the attainment of sexual maturity and is considered the most economically advantageous phase of growth. Identifying the inflection point serves as a valuable indicator of sexual maturity in livestock (Brody, 1974). This point delineates two distinct phases of growth: the acceleration phase, characterized by rapid growth rates, and the retardation phase, where the growth rate diminishes until it reaches a plateau, indicating cessation of further cellular or tissue development (Yasser, 2022). The growth analysis for the bali cattle is still highly uncommon. Understanding the inflection points those crucial stages where growth patterns shift can offer valuable insights into improving overall production efficiency. This paper examines the role of these inflection points in the non-linear growth of Bali cattle, with the aim of optimizing production practices and enhancing growth performance.

2. Material and Method

2.1 Material

The research utilized body weight recording data of 256 male Bali cattle from 0 days to 2 years old, with weight measurements taken at key age points: 0 days (birth), 205 days (weaning), 365 days (a year), 550 days (1.5 years), and 730 days (2 years) at the Pulukan Breeding Center (BC), BPTU-HPT Denpasar in Jembrana Regency, Bali over the past 6 years from 2016 to 2021.

Table 1. Mean (kg) and Standard Deviation of body weight for male Bali Cattle

Age (days)	Mean	Standard Deviation
0	18.26	1.9
205	91.52	18.8
365	138.59	28.6
550	1,93.51	42.6
730	273.91	59.3

2.2 Data Analysis

Body weight data were analyzed using Gompertz's nonlinear growth curves (Gompertz, 1825). The NLIN procedure from the Statistical Analysis System (SAS) On Demand for Academics (SAS, 2021) was employed to fit the Gompertz nonlinear growth curve models to the observed body weights of male Bali cattle. The model used to predict body weight is as follows::

Table 2. The Mathematical Equations of Gompertz Non-Linear Model

Model	Function	Inflection Points	
		Wi	Ai
Gompertz	y = A*(1-B*exp(-k*t))	A/e	(ln B)/K

Where: Y = Body weight at t time (kg), A = Body weight (kg) (Asymptote), B = Integral constant, e = Basic logarithm (2,71828), k = Average growth rate until adult age, t = Time unit (day), Wi = Weight of inflection, Ai = Age at an inflection point, ln = natural logarithm.

3. Result and Discussion

The growth parameters of Bali cattle, estimated using the Gompertz model, reveal key insights into their developmental stages.

3.1 Growth Parameters

The asymptotic body weight was estimated at 423.99 kg (table 3), this higher than the average body weight of male Bali cattle aged 2-3 years was 266.70 ± 56.64 kg (Latulumamina, 2013). The mature weight of Bali cattle estimated in this study was higher than that reported in a previous study, which found the mature weights of Bali cattle from Bali, Nusa Tenggara Timur, Nusa Tenggara Barat, and South Sulawesi respectively to be 303.3 kg, 221.5 kg, 241.9 kg, and 211.0 kg (Martojo, 2012). These differences could be affected by environmental factors, including the management system (extensive, semi-intensive, or intensive) and the availability of feed (Martojo, 2012). The integral constant (B = 3.75) helped shape the growth curve and indicated how growth progressed toward this final weight. The growth rate constant (K = 0.01151/year) reflected Bali cattle's slower, more gradual growth compared to other breeds. This steady growth was an adaptation to tropical environments, where resources like feed may have been limited.

Table 3. Growth Parameters estimated by Gompertz Models

Model	Parameter				
	A	B	K	Wi	Ai
Gompertz	423,99	3,75	0,01151	155,98	398,54

A = Body weight (Asymptote); B = Integral Constant, Ai = Age at inflection point, Integral costantan; K= Average growth rate until adult age; Wi = Weight of inflection.

3.2 Inflection Points

The body weight at the inflection point was the body weight at which Bali cattle experienced their maximum growth rate before it began to slow down as they approached maturity. A weight of 155.98 kg signified a critical developmental stage, and it corresponded with the time at which cattle should have received optimal nutrition to maximize growth. Previous research on other native Indonesian cattle using the Gompertz model showed lower results, with the inflection weight for male Madura cattle being 110.40 kg (Hartati and Putra 2021). Understanding this inflection weight allowed farmers to make informed decisions about feeding strategies, ensuring that Bali cattle received adequate nutrition before growth decelerated. This stage was particularly important for managing young cattle, as proper growth at the inflection point could influence overall productivity, reproductive performance, and meat quality.

The age of cattle at the inflection point indicated when the growth rate of the cattle was at its highest, at 398.54 days or approximately 1.09 years (around 13 months). Bali cattle exhibited a higher maturation rate of 0.109±0.011 compared to Brakmas cattle's 0.089±0.005, indicating that they grew faster and reached mature weight at a younger age than Brakmas. However, in other the study the inflection age of Ongole grade cattle using the Gompertz model was 15.7 months (Adinata *et al.*, 2022). This inflection point occurred later compared to that of Bali cattle, which generally reached their inflection age at an earlier stage of development. This was supported by the research which compared the growth of Bali cattle with Brakmas cattle and showed that Bali cattle reached the inflection point earlier (Hafiz *et al.*, 2018). Bali cattle reached their inflection point, marking a transition from rapid juvenile growth to slower growth as they approached their mature weight. This timing was consistent with the onset of sexual maturity in many cattle breeds, suggesting that around this age, Bali cattle were nearing reproductive maturity as well. Understanding these differences was important for managing each breed effectively.

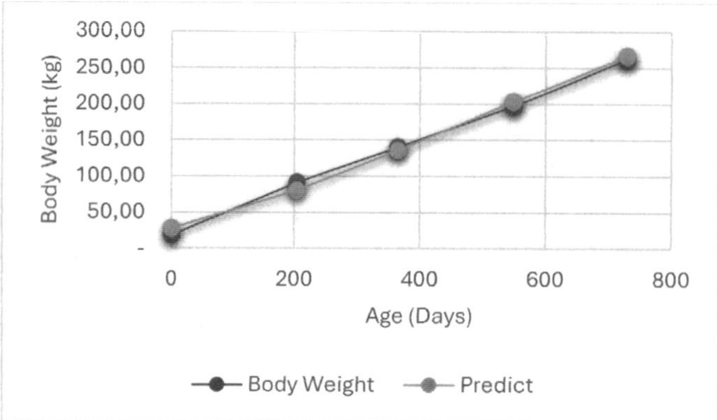

Figure 1. The observed and Predicted Body Weight for male Bali Cattle

The growth curve (figure 1) a strong correlation between the observed body weight and the predicted values across all age ranges. This alignment indicates that the Gompertz model used to predict the cattle's growth is highly accurate and reliable. The Gompertz model is a nonlinear approach used to describe growth patterns in biology, livestock, agriculture, and chemistry (Setiaji *et al.*, 2021). Previous study of the Gompertz model was shown to accurately predict body weight based on age in livestock (Ghiasi *et al.*, 2018).The lines for actual and predicted body weight follow a nearly identical trajectory, suggesting that the model effectively captures the growth pattern of Bali cattle over time. The growth curve is the steady increase in body weight as age progresses, with no major fluctuations or deviations. This linearity in the growth curve suggests that Bali cattle experience consistent growth from birth to around 700 days of age. This steady pattern is characteristic of tropical cattle breeds, which grow more gradually compared to European breeds raised in intensive farming systems. The gradual increase is likely an adaptation to the tropical environment, where feed availability may vary and resources are typically lower in quality.

Furthermore, the growth curve insights into the efficiency of the growth prediction model. The close fit between the observed and predicted data indicates that the Gompertz model can be effectively used for planning and managing cattle growth. By using this model, farmers can predict the growth rate of their cattle, adjust feeding strategies, and optimize cattle management practices to ensure that cattle reach their potential body weight in a timely manner.

4. Conclusion

This study showed that the Gompertz non-linear model effectively predicted Bali cattle growth, with an inflection point at 155.98 kg around 13 months, offering critical insights for optimizing feeding strategies and improving productivity. The model's accuracy underscored its potential for enhancing cattle management and sustainable beef production in Indonesia.

Acknowledgments
The authors would like to express their sincere gratitude to the Ministry of Agriculture of Indonesia for providing the scholarship that made this research possible. Their support was instrumental in the successful completion of this study. We also extend our appreciation to the Pulukan Breeding Center (BPTU-HPT Denpasar) for their invaluable assistance in data collection and to all individuals and institutions who contributed to this research.

Conflict of Interest
The authors declare that they have no conflict of interest

References

Adinata Y, Noor RR, Priyanto R, Cyrilla L, Sudrajad P. 2022. Comparison of growth curve models for Ongole Grade cattle. Trop Anim Health. 54(252): 1-6.

Biscarini F., E. L. Nicolazzi, A. Stella, P. J. Boettcher, and G. Gandini. 2015. Challenges and Opportunities in Genetic Improvement of Local Livestock Breeds. Frontiers in Genetics. 6(33): 1-7

Brody, S. 1945. Bioenergetics and growth, with special reference to the efficiency complex in domestic animals. Reinhold Publishing Corporation, New York

Ghavi, H. and N. Zadeh. 2017. Modelling growth curve in Moghani sheep: comparison of non-linear mixed growth models and estimation of genetic relationship between growth curve parameters. J. Agric. Sci. 155(7): 1150–1159.

Ghiasi, H., T. M. Lupi, and T. M. Mokhtari. 2018. The estimation of genetic parameters for growth curve traits in Raeini Cashmere goat described by Gompertz model. Small Ruminant Research. 165: 66–70

Gompertz, B. 1825. On the nature of the function expressive of the law of human mortality, and on a new mode of determining the value of life contingencies. Philosophical Transactions. 115: 513-585.

Goshu AT and P. R. Koya. 2013. Derivation of inflection points of nonlinear regression curves-implications to statistics. American Journal of Theoretical Applied Statistics. 2: 268-272

Hafiz, A.W, H. R. Mohamad, O. M. Ariff, A. J. Izuan Bahtiar, and A. L. Faezal Ashraff. 2018. Comparison of growth pattern for body weight in Brakmas and Bali cattle using non-linear regression models Mohd. Mal. J. Anim. Sci. 21(1): 19-28

Hartati, H and W. P. B. Putra. 2021. Predicting the growth curve of body weight in madura cattle. Kafkas Üniversitesi Veteriner Fakültesi Dergisi. 27(4): 431–437.

Latulumamina, M. 2013. Korelasi antara umur dan berat badan sapi Bali (Bos sondaicus). Agrinimal. 3(1): 35-40.

Martojo, H. 2012. Indigenous Bali cattle is most suitable for sustainable small farming in Indonesia. Reprod. Dom. Anim. 47:10-14.

Mokhtari, M., A. Esmailizadeh, R. Mirmahmoudi, J. P. Gutierrez, and E. Mohebbinejad. 2023. Comparison of non-linear models and genetic parameter estimation for growth curve traits in the Murciano-Granadina goat breed. Small Ruminant Research. 226:1-7. 107059.

Nahashon S. N, S. E. Aggrey, N. A. Adefope, A. Amenyenu and D. Wright. 2006. Growth characteristics of pearl gray guinea fowl as predicted by the Richards, Gompertz, and Logistic models. Poultry Science. 85: 359-363

Purwantara, B., R. R. Noor, G. Andersson, and H. Rodriguez-Martinez. 2012. Banteng and Bali cattle in Indonesia: Status and forecasts. Reproduction in Domestic Animals. 47(1): 2-6.

Putra, W. P. B dan N. Z. Fajrina. 2021. Model kurva pertumbuhan berat badan pada Sapi Turkish Holstein. Jurnal Peternakan Lingkungan Tropis. 3(1): 23-30.SAS (Statistical Analysis System). 2021. SAS/STAT® 9.4. User's Guide. SAS Institute Inc. Cary, North Carolina

Setiaji, A., D. Arakaki, dan T. Oikawa. 2021. Bayesian estimates of genetic parameters of non-return rate and success in first insemination in Japanese Black Cattle. Anim. Biosci. 34 (7): 1100-1104.

Soares F.S. and Dryden, G.M. 2011. A body condition scoring system for bali cattle. Asian-Aust. J. Anim Sci. 24(11): 1587-1594.

Suryawan I. M, I. P. Sampurna, and I. K. Suatha. 2017. Pola Pertumbuhan Dimensi Panjang Alat Gerak Tubuh Itik Bali Betina. Buletin Veteriner 9(2): 178-186

Sutarno S., and A. D. Setyawan. 2015. Genetic Diversity of Local and Exotic Cattle and Their Crossbreeding Impact on the Quality of Indonesian Cattle. Biodiversitas Journal of Biological Diversity. 16(2): 327-354.

Tahuk, P.K., S. P. S. Budhi, P. Panjono, and E. Baliarti, E. 2018. Carcass and Meat Characteristics of Male Bali Cattle in Indonesian Smallholder Farms Fed Ration with Different Protein Levels. Tropical Animal Science Journal. 41(3): 215–223.

Vazquez, J.A, J. M. Lorenzo, P. Fucinos, D. Franco. 2012. Evaluation of non-linear equations to model different animal growths with mono and bisigmoid profiles. J. Theor. Biol. 314: 95–105.

Wawo A.A. 2018. Effect of Bulls on birth rate and birth weight by using semi-intensive Bali Cattle maintenance. Chalaza J. Animal Husbandary. 3(1): 24–28.

Widyas, N, S. Prastowo, T. S. M. Widi, and E. Baliarti. 2018. Predicting Madura cattle growth curve using non-linear model. In IOP Conf. Ser.:Earth Envirin. Sci. 142 012006. DOI 10.1088/1755-1315/142/1/012006

Evaluating Semen Quality of Ettawa Grade and Boer Goats to Enhance Productivity in Central Java

Fariz Zharfan Haris[1], Asep Setiaji[2], Enny Tantini Setiatin[2], Putu Novia Gariri[1], Maulida Arkaan Muhammad Da'i[1], Pupus Galau Prahara[1], Tirta Dwi Tamanningrum[1], Firda Tasya Kamila[1], Syaddad Verahry Philco[1]

[1]Master of Animal Science, Animal and Agricultural Sciences Faculty, Universitas Diponegoro, Semarang, Indonesia

[2]Animal Science Departement, Animal and Agricultural Sciences Faculty, Universitas Diponegoro, Semarang, Indonesia

*E-mail: asepsetiaji93@gmail.com

Abstract. This research examines the modelling for semen production in Ettawa Grade (PE) and Boer goats to enhance productivity in Central Java. With increasing demand for high-quality breeds for meat and milk, optimizing reproductive efficiency through semen quality assessment is vital. This study focused on prediction through modelling based on three critical semen quality parameters such as semen volume (SV), individual motility (IM) and semen concentration (SC). 6 models were being used to predict the productivity throughout the production period of each breed. Cubic used as the best mode for SV and IM, exponential model used as the best model for SC. Both breeds, Ettawa grade and Boer goats tend to have constant semen productivity throughout the production period. Semen concentration has the best amount at the end of production period, while semen volume reached its peak at the beginning of production period, and individual motility has best productivity within 6 to 7 production period, with all parameters have no significant decrease or increase in every period. Incorporating these evaluations into breeding programs can improve herd performance and contribute to more sustainable goat farming practices in Central Java.

Keywords: Goat productivity, semen quality, non-linear modelling, reproductive efficiency

1. Introduction

Goat meat is a significant commodity in Indonesia's meat supply chain. Optimal meat production is achieved only from goats with superior genetic quality. The inheritance of desirable traits comes from selected male goats, chosen based on their performance in production, reproduction, and environmental adaptability. Boer and Ettawa Grade goats are the most widely developed breeds in Indonesia as superior breeding males due to their adaptability to tropical climates and their good reproductive performance, assessed by the quality and quantity of semen produced [1].

The variability in production and reproductive traits of superior goats can be assessed through genetic parameter observations. Genetic parameters can also predict traits that will be passed on to the next generation through predictions of breeding values [2]. Breeding values obtained can be used as parameters for selecting livestock to produce offspring with desirable traits while eliminating undesirable ones [3]. Male goats play a crucial role in the inheritance of reproductive quality traits based on their breeding values. The reproductive performance of male goats can serve as a basis for predicting the reproductive quality traits of their offspring [4].

Artificial Insemination Centers have been established in various regions of Indonesia to meet the demand for frozen semen used in artificial insemination. The artificial insemination program is a government initiative aimed at accelerating livestock population growth to meet national meat needs. In Indonesia, artificial insemination in goats is widely practiced due to its cost-effectiveness, simplicity, and efficiency [1]. One key indicator of successful artificial insemination is the quality of frozen semen. The quality of frozen semen is influenced by several factors, including the genetic quality of the sire producing the sperm, semen processing, and the production of frozen semen, as well as the implementation of artificial insemination in the field [5].

Common semen evaluation methods typically assess parameters such as sperm concentration, volume, motility, and the proportion of morphologically normal and abnormal cells [6]. However, these evaluations are often limited to capturing data at the time of sampling, without providing insights into the long-term trends or future seminal quality in males. There is a noticeable lack of research aimed at predicting the future progression of seminal characteristics.

Mathematical regression models offer a promising approach to addressing this gap, potentially identifying key stages in the evolution of semen quality throughout an individual's life [7]. This approach could be highly valuable not only from a conservation perspective but also in clinical and biological contexts, enhancing reproductive management in goats by optimizing timing for reproductive handling and identifying early anomalies that could impact reproductive efficiency [8].

This study aims to address the limited information regarding the modelling prediction of semen quality in superior male goats in Indonesia. Modelling based estimation is conducted to ensure the quality of these superior male goats, with the expectation that their production traits will be passed on to their offspring. The estimated values could serve as a basis for selecting males with desirable traits. In this study, the focus is particularly on observing reproductive performance parameters [9].

The novelty of this study lies in the modelling estimation performed on two goat breeds, Ettawa Grade (PE) and Boer, which are indigenous animal genetic resources in Indonesia. The estimated productivity values are also analyzed based on three parameters of semen quality (volume, motility, concentration), the age of the males, the frequency of semen collection, as the results of interaction between the two breeds and non-genetic factors.

2. Materials and Methods

2.1 Data

This research did not require any approval from the animal care and welfare committee as only secondary data was used and there are no treatments or field experiments were performed on animals. The materials used for analysis consist of 7002 ejaculates from total of 26 individuals, consisting 18 PE goats and 8 Boer goats, with average age of 6-7 years or peak production age. Semen production data, including pedigrees of Boer and PE goats, which encompassed information on ancestry and progenitor origins, as well as data on fresh semen production sourced from collection records and semen quality tests at the Production and Distribution Section of the Ungaran Artificial Insemination Center over the past five years, from January 2018 to December 2022. Semen production was being recapped every 3 months period of total semen collection, resulting in total of 20 production period for PE and 11 production period for Boer. The parameter analyzed for semen quality traits focused on volume (ml), individual motility (%), and sperm concentration (million/ml). The extracted data were tabulated and compiled using Microsoft Excel 365. This compilation process was conducted under the supervision and guidance of technical personnel with authority in the section.

2.2. Collection of semen

Semen collection was part of Artificial Insemination Center standard operation routine and being conducted before all the data was recorded. Semen collection was also performed by trained officer in Artificial Insemination Center. Semen collected regularily every twice a week usually in Monday and Thursday. Real semen collection performed after 2 to 3 false mounts before actual semen collection. Semen collected would be stored temporarily inside semen tube connected below artificial vagina. Semen was then immediately transported to laboratory for quality analyisis. Volume was observed by measuring the volume directly from the tube. Concentration was counted using spectrophotometer. Individual motility was observed using electronic microscope in 40x and 100x magnification. Fresh semen which fulfilled the minimum criteria according to Indonesian National Standard Number 4869-3:2023 will be processed furthermore into frozen semen.

2.3. Data normality test

Data subset sample was analyzed by Shapiro-Wilk's test (n<50) taken from whole population of semen production data, using SAS OnDemand for Academics. The Durbin-Watson test also performed for possible autocorrelations among residuals and it was reliable for sample sizes larger than 15. Durbin-Watson (DW) usually a good fit for time series research [9]. Another key assumption of ordinary least squares (OLS) regression is that the variance of the residuals remains constant across different levels of the predicted values, a condition known as homoscedasticity. When heteroscedasticity occurs, it is typically observed that lower values of the standardized predicted values exhibit reduced variance around zero.

2.4. Mathematical models and curve shape parameters

Outliers data were omitted from analysis process in order to get more accurate prediction and prevent the violation of the normality assumption. Outliers data elimination used Robust Regression techniques that reduce the influence of outliers without explicitly removing them and can be used to fit a curve that is not influenced by outliers. In this study, one linear model and five non-linear models were employed to describe the progression of semen production in 20 PE goats and 11 Boer goats. Table 1 presents the equations for the six models used, along with their abbreviations and corresponding model syntax equations. The syntax for these models was designed for seamless integration into the non-linear regression task within SAS OnDemand for Academics [10].

The curve estimation function within the SAS regression procedure was utilized to iteratively determine parameter bounds for each model using the Levenberg-Marquardt method which is iterative optimization algorithm used to solve non-linear least squares problems. It is specifically designed for minimizing the sum of squared differences between observed data and a model, making it useful in regression analysis and curve fitting when the relationship between variables is non-linear. The Levenberg-Marquardt method refines an initial guess of the parameters iteratively to find the values that minimize the residual sum of squares (RSS), which is the difference between observed and predicted data [11].

Table 1. Models and syntax

Model	Model syntax/equation
Linear	Y = b0 + (b1*t)
Inverse	Y = b0 + (b1/t)
Logarithmic	Y = b0 + (b1 * ln(t))
Cubic	Y = b0 + (b1 * t) + (b2 * t ** 2) + (b3 * t ** 3)
Exponential	Y = b0 * (e ** (b1 * t)) or ln(Y) = ln(b0) + (b1 * t)
Quadratic	Y = b0 + (b1 * t) + (b2 * t ** 2)

*b0 represents constant; b1, b2, b3, e represents parameters bound; t represents production period

2.5. Model selection criteria

The selection criteria for identifying the most explanatory and predictive models included the Residual Sum of Squares (RSS), Mean Squared Error (MSE), R-Squared (R^2) along with its standard deviation across subjects, Akaike Information Criterion (AIC), and Bayesian Information Criterion (BIC). The Residual Sum of Squares (RSS) is a statistical measure that quantifies the portion of variance in a dataset not accounted for by the regression model, thereby assessing the model's explanatory capacity. Mean Squared Error (MSE) was used to evaluate error variation as an alternative, which has been noted to be sensitive to the number of parameters, especially in cases where sample sizes are small [10]. R^2 could provide a reliable measure of a model's ability to predict new observations, as it accounts for and penalizes the potential overfitting that may occur with the inclusion of a large number of predictors. Based on the principles of information theory, several methods have been developed to evaluate model performance, including the Akaike Information Criterion (AIC) and Bayesian Information Criterion (BIC). AIC was used to compare models in terms of their ability to explain the variability within the dataset, while BIC was employed to assess the predictive capacity of each model [11].

In situations involving a limited number of observations (N) or complex models with many parameters, the corrected AIC is often considered more accurate. However, as the sample size increases, AIC converge in their estimates. It is recommended to use AIC when the ratio of observations to parameters (N/K) is less than 40. Similar to R^2, BIC serves as a model selection criterion that penalizes the inclusion of additional parameters in more complex models and was calculated accordingly in this study [12].

3. Results

3.1. Descriptive statistics and regression analysis

Table 2 indicates that all variables displayed non-normal distributions and heteroscedastic residuals ($p<0.05$). This, along with the relatively small sample size, necessitated the use of nonlinear regression models and Bayesian inference to compare the results. Table 3 provides the descriptive statistics for the all parameters in both breeds, with no outliers identified in any of the studied parameters. Table 2 provides a summary of the determination coefficients, curve shape parameters, and peak values for the best-fitting models across all variables in the study, while Table 3 presents the results obtained from fitting the eleven models evaluated in the study. Table 3 provides a summary of model fit and flexibility criteria, including mean square error (MSE) along with the variability explanation capacity assessed through the Akaike Information Criterion (AIC). Additionally, the predictive power of the best-fitting linear or nonlinear regression models for each trait is evaluated using the Bayesian Information Criterion (BIC).

Table 2. Summary of the results for determination coefficients, curve shape parameters, and peak for best fitting models for all variables

Traits	Model	Predict	Sig.	b0	b1	b2	b3	e	t	R^2
PE SV	Cubic	1.605	0.050	0.997	2.000	0.200	1.000	-	1.000	0.000
PE IM	Cubic	61.488	0.050	46.488	5.000	5.000	5.000	-	5.000	0.000
PE SC	Expo	2506.000	0.050	4.008	5.000	-	-	5.000	0.800	0.000
Boer SV	Cubic	1.757	0.050	0.997	3.000	3.000	5.000	-	5.000	0.000
Boer IM	Cubic	54.560	0.050	39.560	5.000	1.000	5.000	-	5.000	0.000
Boer SC	Expo	2894.000	0.050	2.071	5.000	-	-	5.000	0.900	0.000

*SV: Semen volume; IM: Individual motility; SC: Semen concentration; Expo: Exponential; Sig.: Significance; b0: constant; b1,b2,b3, e: parameter bounds of each model; t: production period

Table 3. Summary of the measures for model fitness and flexibility criteria of the best fitting linear or nonlinear regression models for each parameter

Traits	R	RSS	SS	dF	MSE	SSE	AIC	BIC
PE SV	0.000	-0.005	0.128	19	0.007	0.128	-43.981	-13.494
PE IM	0.000	5.362	34791.900	19	19.260	365.945	43.550	26.293
PE SC	0.000	-232.052	3076626.000	19	161896.000	3076020.000	142.954	71.475
Boer SV	0.000	-0.037	0.337	10	0.034	0.336	-33.360	-8.662
Boer IM	0.000	11.080	952.600	10	95.260	952.598	54.074	31.077

*SV: Semen volume; IM: Individual motility; SC: Semen concentration

3.2. Semen production modelling

The values of R (correlation coefficient) and R-squared (coefficient of determination) was zero, which indicates a specific relationships or lack thereof between all variables in a model. R-squared measures the proportion of the variance in the dependent variable that can be explained by the independent variables in the model. If R-squared = 0, it means the model does not explain any of the variability in the dependent variable. The independent variables had no predictive power over the dependent variable. In other words, changes in the independent variables do not correspond to changes in the dependent variable. R value also 0 indicates no linear correlation between the variables. The variables do not move together in a linear and could be non-linear. In nonlinear regression models, R-squared might approach zero, even when significant nonlinear relationship existed between the variables, because R-squared assumes a linear relationship when measuring explained variance. Other goodness-of-fit measures such as AIC, BIC, or residual sum of squares could provide a better evaluation of model fit [13].

A constant trend of prediction was indicated in all traits SV, IM, SC from both breeds along with the production period. PE semen volume tend to coincide with the prediction trajectory (Figure 2), indicating that semen would have the same amount of volume along the production period. PE individual motility and semen concentration has reached its peak within 15 and 16 production period (Figure 3 and 4). All Boer parameters (SV, IM, SC) tend to coincide with the predictive value of semen production. Boer SC progression has reached its peak value right in the end of production period after decreasing to its lowest value (Figure 7). Boer SV reached its peak at the beginning of production period (Figure 5), while IM has reached its peak between 6 to 7 period, with both parameters have no significant decrease or increase every period.

SV and IM parameters best fitted for using the cubic model showing the constant performance of semen productivity. However, at the end of production period, the trend for both parameters tend to decrease. SC for both breeds used exponential model for the best fitted result. Both SC has resulted in different progression, PE SC tend to decrease while Boer SC increased significantly at the end of production period.

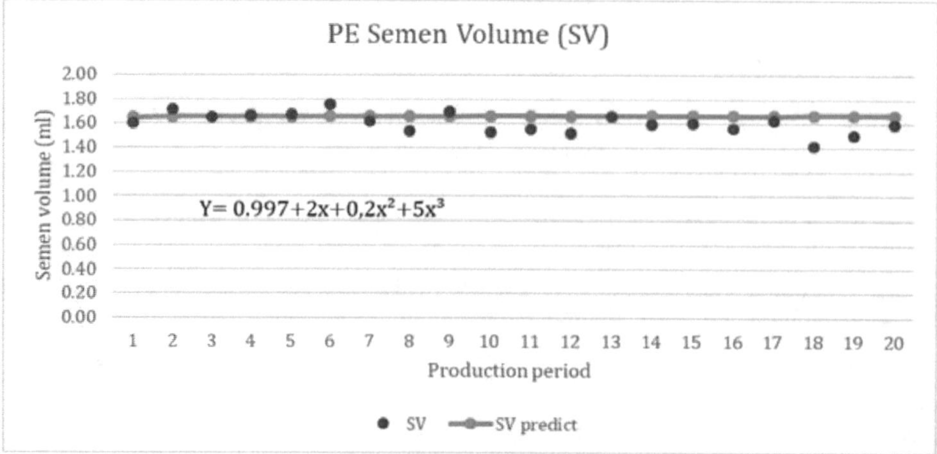

Figure 1. PE semen volume (ml) modelling using cubic function

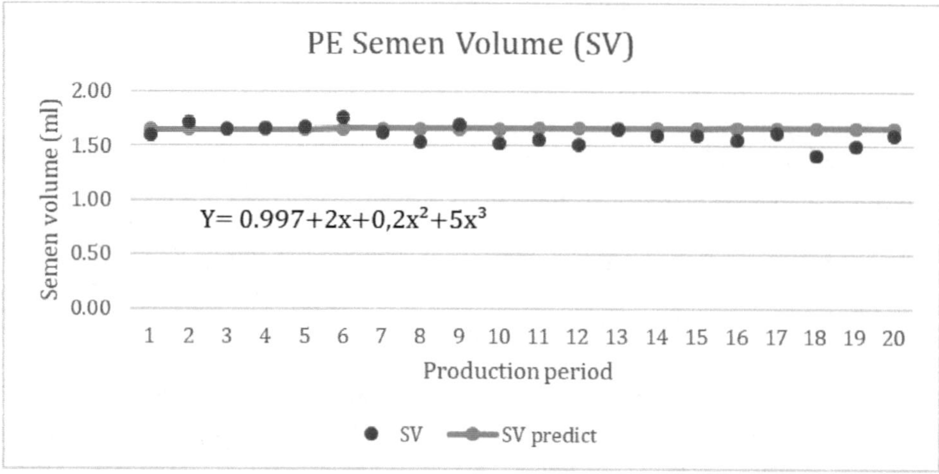

Figure 2. PE semen volume (ml) modelling using cubic function

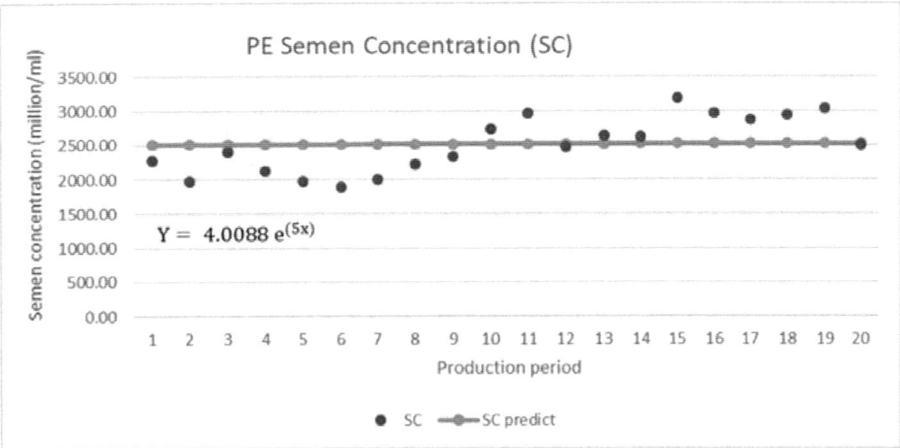

Figure 3. PE sperm individual motility (%) modelling using cubic function

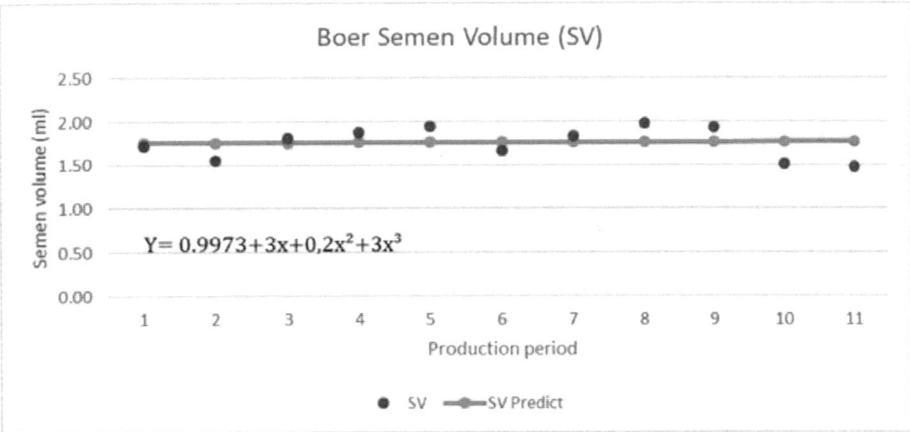

Figure 4. PE semen concentration (million/ml) modelling using exponential function

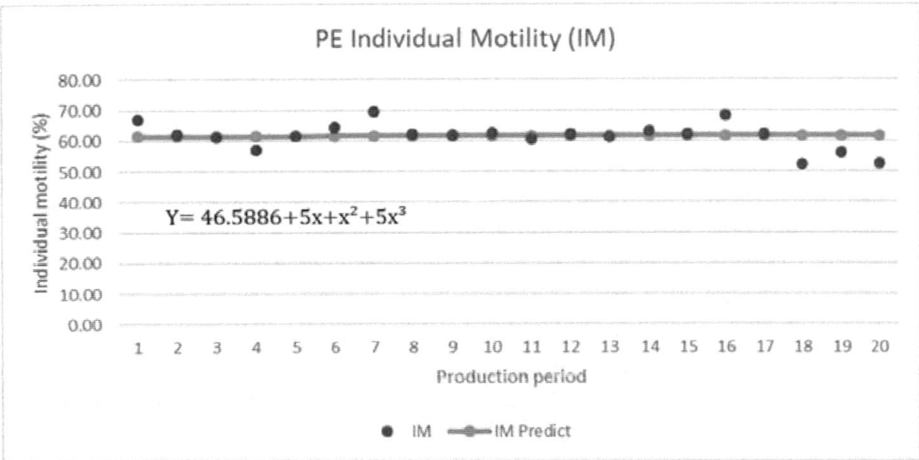

Figure 5. Boer goat semen volume (ml) modelling using cubic function

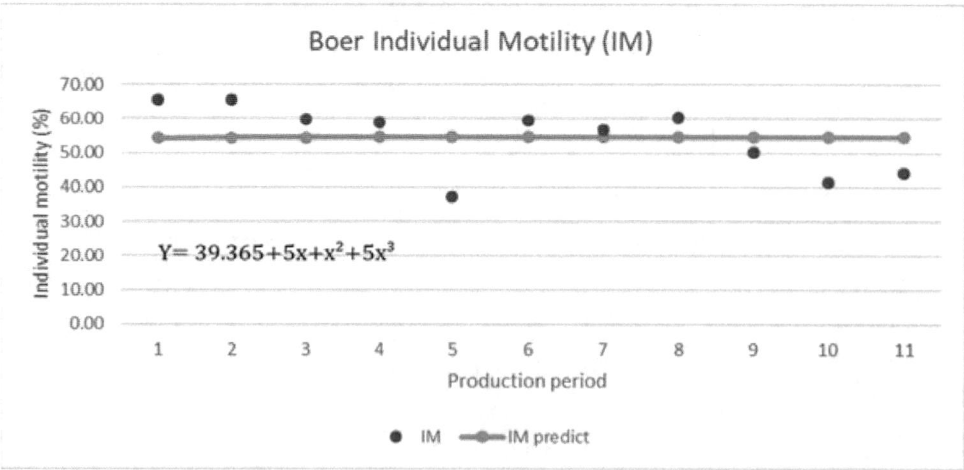

Figure 6. Boer goat sperm individual motility (%) modelling using cubic function

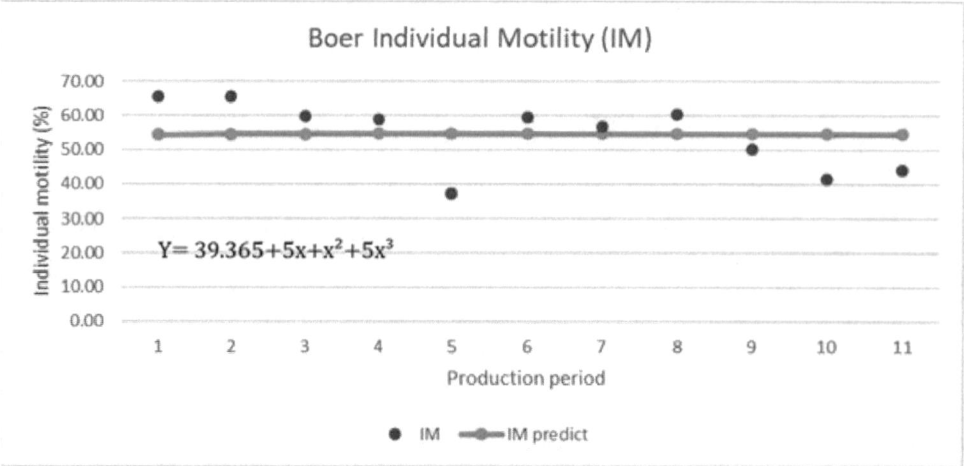

Figure 7. Boer goat semen concentration (million/ml) modelling using exponential function

4. Discussion

4.1. Semen volume and individual motility

Two goat breeds PE and Boer employed the same cubic modeling approach for traits such as semen volume and individual motility. Cubic modeling is a form of polynomial regression where the relationship between the independent variable such as production period and the dependent variable such as semen production, is expressed as a third-degree polynomial. This method is particularly useful when the relationship between variables is nonlinear and cannot be effectively captured by simpler models, such as linear or quadratic regressions. The cubic model allows for a more complex representation of semen production in goats, capturing fluctuations across different phases of life. For instance, semen production may initially increase during early maturity, reach a peak, and then decline as the goat ages. This type of relationship, characterized by multiple inflection points, is well-suited to a cubic model [14].

The progression of semen volume and motility in goats often stabilizes at peak production levels. In PE goats, peak productivity typically occurs between 2 to 4 years of age, whereas Boer goats tend to reach their peak between 5 to 6 years. Dairy breeds like PE generally mature earlier than meat breeds such as Boer, largely due to their higher metabolic rates. This accelerated metabolism supports faster growth and reproductive development, leading to earlier puberty and semen production [15]. Additionally, dairy goats are often bred to maximize reproductive efficiency, enabling them to mature earlier and align with the intensive breeding schedules required in dairy production systems. Earlier sexual maturity in PE goats facilitates early semen production, enhancing herd turnover and supporting more controlled breeding programs [16].

In contrast, meat breeds like Boer have been selected primarily for traits such as muscle growth or fiber production, rather than for reproductive efficiency or early maturation. As a result, these breeds tend to mature more slowly, as their energy is directed towards muscle or fiber development rather than rapid reproductive growth. Consequently, meat-type goats often take longer to reach their reproductive and semen production peaks [17].

4.2. Semen concentration modelling

The prediction of semen concentration traits in both goat breeds is best estimated using the exponential method. Exponential modelling is commonly employed to describe various biological processes, including semen concentration in goats, particularly when the changes follow a non-linear trajectory. This model is effective in capturing patterns where semen concentration initially increases or decreases rapidly before stabilizing over time [8]. In the context of goat semen production, semen concentration can fluctuate due to factors such as age, seasonality, and reproductive development. Typically, semen concentration increases rapidly as goats approach maturity, reaching a peak during their reproductive prime. This rise in production continues through early reproductive stages, peaking at specific points, and subsequently declining as the reproductive system ages and efficiency diminishes [4].

Dairy goats such as the PE breed generally exhibit higher semen concentration per millilitre of ejaculate compared to meat-type breeds like Boer. Even at their respective peak production periods, PE goats maintain a higher semen concentration than Boer goats, despite potentially lower semen volume. Physiologically, PE goats possess more active reproductive endocrine systems, which contribute to superior reproductive performance. Elevated hormone profiles, particularly higher testosterone levels, promote enhanced testicular development, increased sperm production, and overall improved semen quality [18]. Boer goats on the other hands tend to exhibit lower testosterone levels, as selective breeding in these animals has prioritized physical growth and meat yield over reproductive efficiency. Additionally, PE goats demonstrate superior testicular development, which is critical for achieving high sperm production [19]. Larger, well-functioning testes are directly linked to increased sperm production and enhanced semen quality. In meat-type breeds such as Boer goats, the emphasis on muscle and meat production often results in less advanced testicular development, leading to lower semen concentration per millilitre of ejaculate [20]. Boer goats typically exhibit more pronounced seasonality in their reproductive cycles, with semen concentration being significantly influenced by seasonal variations. Specifically, outside of the breeding season, decreases in testosterone levels lead to reduced sperm production, resulting in greater fluctuations in semen concentration over the course of the year [21].

5. Conclusion

Both breeds, Ettawa grade and Boer goats tend to have constant semen productivity throughout the production period. PE semen volume tend to coincide with the prediction trajectory, while individual motility and semen concentration has reached its peak within 15 and 16 production period. Boer semen volume, motility, and semen concentration also tend to coincide with the predictive value of semen production. Regression analysis indicates optimal semen concentration at production period's end, while semen volume reached its peak at the beginning of production period, and individual motility has best productivity within 6 to 7 production period, with all parameters have no significant decrease or increase in every period. Subsequently, information given from this study hopefully could help the Central Java Government to create a more effective and feasible breeding program especially for Ettawa grade and Boer goats as the indigenous goat breeds from Indonesia.

Acknowledgements

Authors thank the Central Java Province Government, Central Java Agency of Animal Health and Production, and Ungaran Artificial Insemination Center for their unlimited amount of support and eagerness to make this research possible. Also, authors thank to Animal Breeding, Genetic, and Reproduction Laboratory family, which have given all the help, motivation, and many of their two-cents for the better of this writing. Approval from the committee on the care and use of animals was not sought because no treatment and field experiments were conducted on animals.

References

[1] Mokoagow, F. E. Pudjihastuti, M. J. Hendrik, U. Paputungan. 2021. Fresh semen macroscopic of Ettawa Grade (PE), Boer and Saanen in Artificial Insemination Lembang. Zootec. **41** (1): 150 – 157.

[2] Menezes, L. M., W. H. Sousa, E. P. Cavalcanti-Filho, and L. T. Gama. 2016. Genetic parameters for reproduction and growth traits in Boer goats in Brazil. Small Ruminant Research. **136**: 247 – 256.

[3] Jembere, T., T. Dessie, B. Rischkowsky, K. Kebede, A.M. Okeyo, and A. Haile. 2017. Meta-analysis of average estimates of genetic parameters for growth, reproduction and milk production traits in goats. Small Ruminant Research. **153**: 71 – 80.

[4] Ramukhithi, F. V. T. C. Chokoe, T. Ronald, and K. C. Lehloenya. 2021. Characterisation of semen and phenotypic parameters in relation to male goat fertility. In: S. Kukovics (Ed.). Goat Science - Environment, Health and Economy. Hungarian Sheep and Goat Dairying Public Utility Association. IntechOpen, Hungary. pp. 1 – 27.

[5] Isnaini, N., A. A. Hakim, D. Amertaningtyas, H. E. Sulistyo, A. Irsyammawati, and F. Andri. 2021. Comparative study of semen quality traits between Etawah grade and Senduro bucks. IOP Conf. Series: Earth and Environmental Science. **888**: 1 – 4.

[6] Dorado, J. D. Acha, M. Galvez, L. Ortiz, J. Carrasco, B. Dıaz, V. Gomez-Arrones, R. Calero-Carretero, and M. Hidalgo. 2013. Sperm motility patterns in Andalusian donkey (Equus asinus) semen: effects of body weight, age, and semen quality. Theriogenology. **79**(7): 1100–1109.

[7] Blanchard, T. L., S. P. Brinsko, D. D. Varner, C. C. Love, and J. P. Morehead. 2013. Progression of reproductive changes accompanying testicular dysfunction in aging thoroughbred stallions: case studies. In: Proceedings of the 59th American Association of Equine Practitioners (AAEP), Nashville, TN, USA. pp. 532–536.

[8] Martins-Bessa, A., M. Quaresma, B. Leiva, A. Calado, A. Arando, C. Marín and F. J. Navas. 2021. Age-related linear and nonlinear modelling of semen quality parameters in Miranda donkeys. Italian Jo. of Anim. Sci. **20** (1): 1029-1041.

[9] Bhave, K., T. P. J. Koilpillai, V. Ragothaman, S. Sontakke, G. Joshi, and V. Ducrocq. 2020. Semen production and semen quality of indigenous buffalo breeds under hot semiarid climatic conditions in India. Trop. Anim. Health and Prod. **52**: 2529-2539.

[10] R Core Team (2018). R: A language and environment for statistical computing. R Foundation for Statistical Computing, Vienna, Austria.

[11] SAS Institute Inc. 2016. SAS/STAT® 14.2 User's Guide: The REG Procedure.

[12] Greenberg, D., S. Kerwick, T. Encheva, D. Williamson, Z. Mingyuan, K. Muthuraman, L. Moliski, and J. Murray.

2020. Time Series Analyses. Part 3. In STA 371G; Greenberg, D. (Ed.). University of Texas. Austin, TX, USA.

[13] Dakhlan, A., A. Saputra, M. D. I. Hamdani, and Sulastri. 2020. Regression models and correlation analysis for predicting body weight of female Ettawa Grade goat using its body measurements. Adv. Anim. Vet. Sci. 8 (11): 1142–1146.

[14] Karangeli, M., Z. Abas, T. Koutroumanidis, C. Malesios, and C. Giannakopoulos. 2011. Comparison of models for describing the lactation curves of Chios sheep using daily records obtained from an automatic milking system. In: Proceedings of 5th Hellenic Association for Information and Communication Technologies in Agriculture, Food and Environment (HAICTA), 8–11th September 2011, Skiathos Island, Greece. pp. 571–589.

[15] Setiaji, A., D. Arakaki, and T. Oikawa. 2021. Bayesian estimates of genetic parameters of non-return rate and success in first insemination in Japanese Black cattle. Anim. Biosci. 34 (7): 1100–1104.

[16] Brewer, A. N. 2024. A Novel Goat Model to Characterize Male Fertility Following Paternal Cannabidiol Exposure. School of Animal Sciences, Louisiana State University, Baton Rouge, Louisiana, USA. (Thesis).

[17] Mocé, E., M. L. Mocé, S. A. Lozano-Palazón, J. Bernácer, M. Martínez I. Granell, C. Esteve, F. Bernat, S. J. Contreras, I. Villalba, and E. A. Gómez. 2022. Fertility prediction in dairy goats from Murciano-Granadina breed: The role of sperm evaluation and female traits. Animal. **16** (5): 100525

[18] Slamet, N. C. K., Mudawamah, and Sumartono. 2023. Comparison productivity of PE and Boer goats based on body size and semen macroscopic. J. Ternak. **14** (2): 96-102.

[19] Akpa, G. N., A. L. Ambali, and I. O. Suleiman. 2013. Body conformation, testicular and semen characteristics as influenced by age, hair type and body condition of Red Sokoto goat. New York Sci. J. **6** (7): 44–58.

[20] Cadena-Villegas, S., J. A. Hernández-Marín, J. Gallegos-Sánchez, C. G. Germán- Alarcón, and P. Pérez-Hernández. 2021. Reproductive management of the male goat: a review. Agro Productividad. **14** (8): 2102.

[21] France, K. P. I. Monau1, U. Moreri, and E. Waugh. 2024. Assessment of sperm quality of communal goats in the hard-veld region of Botswana. South African J. Anim. Sci. **54**: 147 – 156.

IOP Conf. Series: Earth and Environmental Science **1460** (2025) 012003 doi:10.1088/1755-1315/1460/1/012003

Evaluation of qualitative and quantitative characters in F1 generation from crossing of red, orange and yellow cherry tomato

Florentina Kusmiyati*, Muhamad Ghazi Agam Sas, Endang Dwi Purbajanti, and Putri Ayu Noor Kusumaningtyas

Department of Agricultural, Faculty of Animal and Agricultural Sciences, Universitas Diponegoro, Semarang, Indonesia

*E-mail: fkusmiyati@live.undip.ac.id

Abstract. Recently, cherry tomatoes [*Solanum lycopersicum* var. *cerasiforme* (Dunnal) A. Grey] are becoming popular in Indonesia. There are not many varieties of cherry tomatoes in Indonesia. So, breeding cherry tomatoes is essential. The objective of this research was to evaluate qualitative and quantitative characters in F1 generation from crossing of red, orange and yellow cherry tomatoes. In this research, there were 3 parents of cherry tomatoes with different colours of ripe fruit (red, orange and yellow). The artificial crossing was done in all combinations. Seed from crossing was harvested and planted as F1 generation. Three parental and six genotypes of F1 were planted in monofactor completely randomized design. Parameter measured were qualitative and quantitative characters. The results showed that there were variation of qualitative and quantitative characters between parent and F1 progeny. There were variation of fruit longitudinal shape in F1 such as ovate, oblong, elliptic, cordate. There were no variation of fruit flesh color at F1. Ripe fruit colours of F1 that were difference from the parents were vivid orange, yellow and strong orange. Progeny with significantly larger fruit weight and diameter than parent were P2-4xP1-3, P2-1xP3-2 and P3-1xP2-9. Three genotypes (P3-1xP1-12, P3-5xP1-1 and P3-8xP1-2) had a higher total soluble solid than both parents. The qualitative and quantitative diversity of F1 is the key to successfully breeding cherry tomato varieties with larger fruit size and sweet taste.

Keywords : cherry tomato, breeding, F1 generation, color variation, fruit shape

1. Introduction

Lately, cherry tomatoes become popular as fresh fruit, salad, and other dishes in Indonesia and other parts of the world. The popularity of cherry tomatoes is due to their richness in glucose and fructose (sugar), which leads to higher sweetness than large-sized tomatoes. They also contain organic acids, organic volatile compounds [1], β-carotene, lycopene, and other nutrients [2]. In Indonesia, the price of cherry tomatoes is still relatively high, and they are only sold at modern markets because productivity is lower, and there are few producers.

Cherry tomatoes (*Solanum lycopersicum* var. cerasiforme (Dunnal) A. Grey) are extant accessions of *Lycopersicon esculentum* var. cerasiforme. These subspecies appears to be an admixture of cultivated and wild tomatoes, not a transitional step of domestication from wild tomatoes [3] . In Indonesia, only a few cherry tomato varieties have been registered in the Indonesian Ministry of Agriculture for the past 10 years [4]. Farmers in Indonesia planted imported seeds from various countries. Those imported seeds are costly, not always available and not certified. In other hand, increasing demand of cherry tomatoes requires the availability of certified and sufficient seeds. So, the breeding of cherry tomatoes is very necessary to fulfill those seed demand. Cherry tomatoes breeding will produce a variety that adapted to the Indonesian climate, is cheaper, always available, and certified.

The main steps in plant breeding are establishing genetic diversity improvement through artificial crossing, progeny/lines selection, preliminary yield test or selection based on production, and multi-location trial. The available genetic diversity is the key to breeders' success in obtaining the desired superior characters. In plant breeding, evaluation of qualitative and quantitative characters is essential to develop desirable characters in plants. Qualitative characters are influenced by a single or a few major genes and express distinct phenotypic variations. Some qualitative character often following Mendelian inheritance patterns. These qualitative characters are either present or absent. Dissimilar qualitative characters, quantitative characters are controlled by multiple genes and environmental factors, so the inheritance of quantitative characters are more complex.

There are many ripe fruit tomato cherry color such as red, yellow and orange. The longitudinal shapes of cherry tomato fruit vary, such as ovate, obovate, cordate, cylindric and circular. Our plant breeding program have already started doing artificial crossing in dialelic crossing of three parents (red cherry, orange cherry and yellow cherry). The objective of this research was to evaluate qualitative and quantitative characters in F1 generation from crossing of red, orange and yellow cherry tomato.

2. Materials and Methods

2.1 Materials

The research was conducted in the Agricultural and Animal Sciences Faculty greenhouse, Diponegoro University. The plant materials used in this study were 3 parents of cherry tomatoes with different colour of ripe fruit (red/P1, orange/P2 and yellow/P3) and 14 genotypes of F1, which were developed from full diallel crosses among three parents (Table 1).

Table 1. Plant materials for evaluation characters in F1

Female	F1	Number of plants
P1		15
	P1–6 x P2-11	7
	P1-11 x P3-8	7
	P1-5 x P2-14	4
	P1-10 x P2-10	10
P2		15
	P2-4 x P1-3	4
	P2-1 x P3-2	4
	P2-4 x P3-2	15
	P2-20 x P3-4	13
P3		15
	P3-1 x P1-12	15
	P3-5 x P1-1	11
	P3-8 x P1-2	6
	P3-1 x P2-9	7
	P3-6 x P2-4	11
	P3-7 x P2-11	13
	Total	172

IOP Conf. Series: Earth and Environmental Science **1460** (2025) 012003 doi:10.1088/1755-1315/1460/1/012003

2.2 Methods

The experimental design was a completely randomized design monofactor. Seeds of 3 parents and 14 genotypes were sown in a seed box for 14 days, followed by transplanting to 35 cm x 35 cm polybags containing 8 kg of latosol soil, manure and husk (1:1:1 by volume). Each polybag consisted of one plant. Recommendations for cherry tomato cultivation and plant protection were adopted. Fertilization application of 14 g urea/polybag, 8 g SP-36/polybag and 4 g KCl/polybag was applied twice i.e. at transplanting and 14 days after transplanting.

Parameter measured were qualitative and quantitative characters. Qualitative characters were longitudinal shape of fruit, fruit flesh colour and ripe fruit colour that were based on [5] (Figure 1, 2 and 3). All qualitative characters was measured at ripe fruit. Fruit flesh colour and ripe fruit colour is observed using the RHS colour chart.

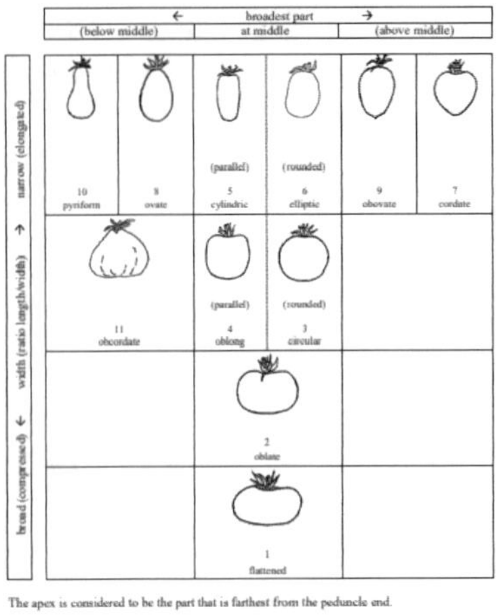

Figure 1. Longitudinal shape of fruit [5]

Ripe fruit color of cheery tomatoes are beige, yellow, orange, pink, red, brown and green (Figure 2). Ripe fruit colour was observed using RHS color chart for each fruit at harvest time

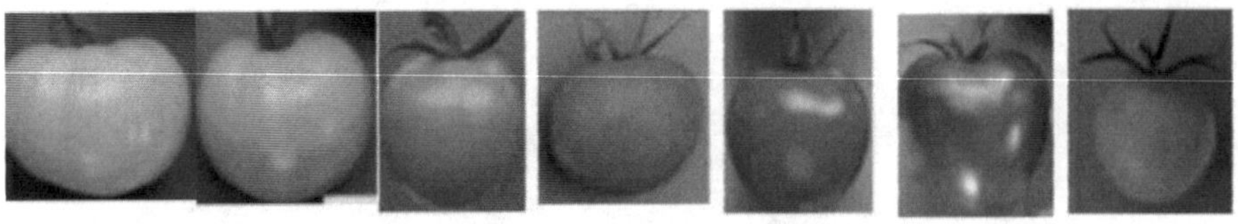

Beige Yellow Orange Pink Red Brown Green

Figure 2. Ripe fruit colour based [5]

Flesh fruit color of cheery tomatoes are beige, yellow, orange, pink, red, brown and green (Figure 3). Cherry tomatoes were cut into two pieces, then flesh fruit colour was observed using RHS coloor chart for each fruit at harvest time.

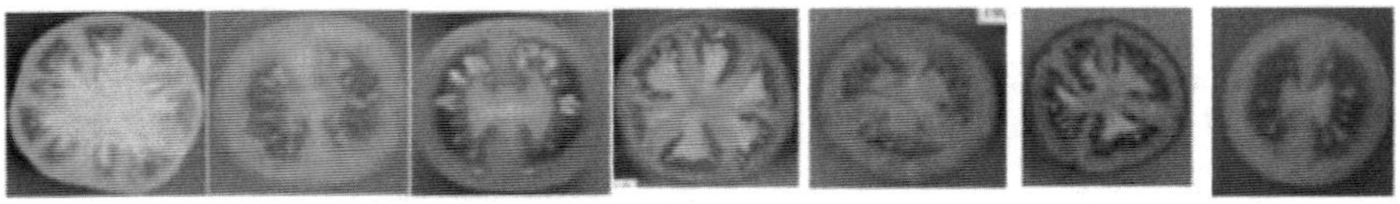

Figure 3. Flesh fruit colour based [5]

Parameter of quantitative characters were fruit weight, fruit length, fruit diameter, sugar content, and total titrated acid. All those parameters were measured at ripe fruit. Sugar content analysis is calculated based on total dissolved solids using a refractometer. Total titrated acid is measured [6].

Analysis data of qualitative character was descriptive. Meanwhile, quantitative character was analyzed using a T-test and compared to female or male parents.

3. Results and Dicussion

3.1. Qualitative Characters in F1

The longitudinal shape of the parental fruit is obovate, cylindric, and pyriform (Table 1 and Figure 1). Fruit longitudinal shape variation in F1 includes ovate, oblong, elliptic, and cordate (Figure 4). Crossing P1 (obovate) and P2 (cylindric) resulted in the ovate longitudinal shape of the fruit. At the same time, the reciprocal of P2 x P1 resulted in an oblong longitudinal shape of fruit. The longitudinal shape of the fruit progeny from P1 x P3 was elliptic and cordate, while the progeny from the reciprocal (P3 x P1) was pyriform and ovate. P2 x P3 and P3 x P2 crossing progeny had elliptic and ovate longitudinal fruit shapes. Only the progeny of P3 x P1 had longitudinal fruit shapes similar to those of female parents. In contrast, other progeny had a different longitudinal fruit shape than parental traits. The gene contribution of two parental caused the variation of fruit longitudinal shape at F1. The fruit longitudinal shape of progeny was the combination of female and male parents, although some progeny had fruit longitudinal shapes similar to either female or male parents.

Table 2. Longitudinal shape of fruit, Fruit Flesh Colour and Ripe Fruit Colour of Parental and F1

	Longitudinal shape of fruit	Fruit flesh colour	Ripe fruit colour
P1	Obovate	Red	Vivid Reddish Orange
P2	Cylindric	Orange	Vivid Reddish Orange
P3	Pyriform	Yellow	Vivid Yellow
P1 x P2	Ovate	Red	Vivid Reddish Orange
P2 x P1	Oblong	Red	Vivid Reddish Orange
P1 x P3	Elliptic, Cordate	Red, Orange	Vivid Reddish Orange
P3 x P1	Pyriform, Ovate	Orange	Vivid Reddish Orange
P2 x P3	Elliptic, Ovate	Yellow	Vivid Reddish Orange, Strong Orange
P3 x P2	Ovate	Red, Yellow	Vivid Reddish Orange, Vivid Orange Yellow, Strong Orange

IOP Conf. Series: Earth and Environmental Science **1460** (2025) 012003
doi:10.1088/1755-1315/1460/1/012003

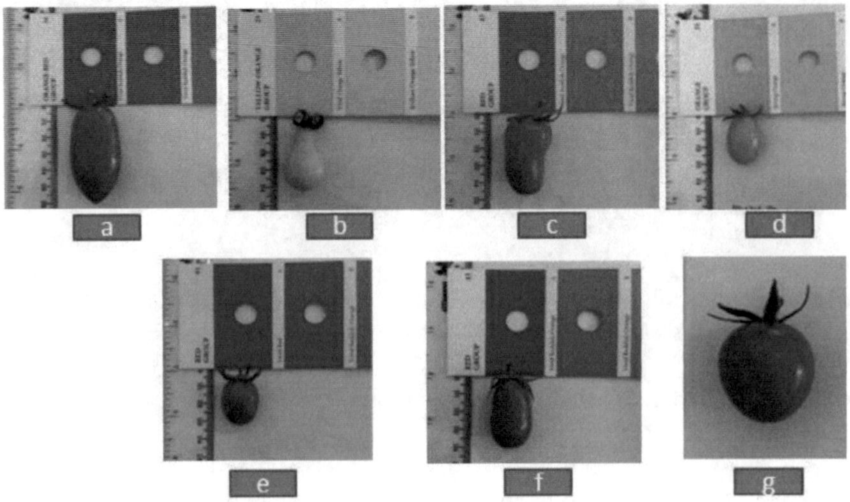

Figure 4. Diversity of longitudinal shape of fruit (a) Obovate, (b) Pyrifrom, (c) Cylindric, (d) Ovate, (e) Oblong, (f) Eliptic (g) Cordate

The fruit flesh color of the parental is red (P1), orange (P2), and yellow (P3). The progeny also had similar colors, such as red, orange, and yellow. There was no variation in fruit flesh color at F1, although some progeny had different flesh colour from that of the parental. All progeny of F1 from P1 X P2 and P2 x P1 were red, similar to the flesh color of the fruit P1. Crossing red (P1) and yellow (P3) or P3 x P1 resulted in red and orange fruit flesh color progeny. The orange fruit flesh colour of this progeny may be a result of gene combination of two parentals. The progeny of P2 x P3 had fruit flesh colour similar to that of male parents. While progeny of P3 x P2 had red and yellow fruit flesh colour.

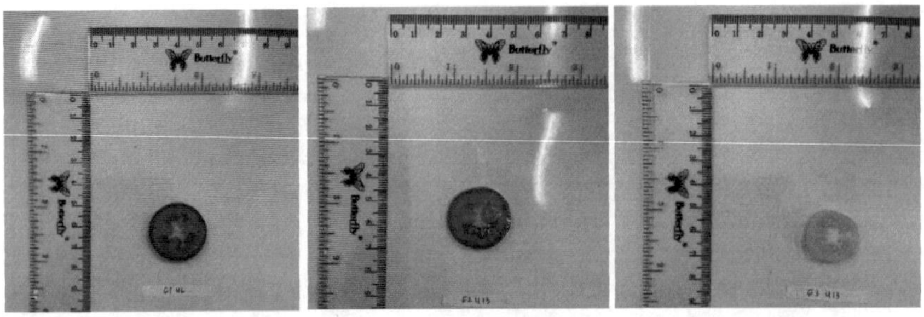

Figure 5. Fruit flesh colour (a) Red, (b) Orange, (c) Yellow

Ripe fruit colour of parental is vivid reddish orange (P1 and P2) (RHS 43A) and vivid yellow (P3). The ripe fruit colour of F1 from female parent P1 was vivid reddish orange that was similar to female parent. There were variation of ripe fruit colour from crossing P2 and P3 that were vivid reddish orange and strong orange (RHS 25 A). Variation of ripe fruit colour also was shown by F1 of crossing P3 x P2 that were vivid reddish orange, vivid orange yellow, and robust orange. The ripe fruit colors of F1 that differed from parental were vivid orange yellow and strong orange.

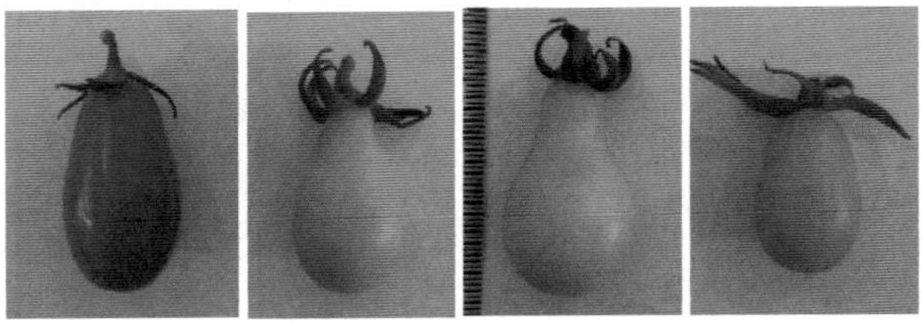

Figure 6. Ripe fruit colour : (a) Vivid Reddish Orange (b) Vivid Yellow (c) Vivid Orange Yellow, (d) Strong Orange

Fruit quality and flavor differs from the colour of cherry tomatoes. Red cherry tomatoes had the highest lycopene content, sourness amino acid contents and significantly higher Zn content than other cherry tomatoes. Yellow cherry tomatoes had the highest soluble sugar, essential amino acids, aromatic amino acids and sweetness amino acids. Yellow cherry tomatoes also had the best flavor [7].

3.2. Quantitative characters

The fruit weight, polar diameter and equatorial diameter of cherry tomato at this research were 7.28 – 17.09 g, 2.50 – 4.30 cm and 1.31 – 2.23 cm respectively (Table 3). All fruit weight of cherry tomato was heavier than four g/fruit. So, of all cherry tomatoes classified as "A" grade marketable fruit yield based on weight [8]. Fruit weight of cherry tomato was higher than earlier reported fruit weight was in the range of 3.96 – 13.45 g [9]. Fruit polar diameter of cherry tomato in this research was longer However, equatorial diameter was shorter compared to fourteen advanced breeding lines collected from locations in India [10].

One of female parent (P1) had longer polar diameter (4.21 cm) compared to five introduced cherry tomato lines in Lembang highland [11]. Total soluble solid and titratable acidity content were 5.38 – 8.47 °Brix and 0.40 – 0.78 %, respectively. The total soluble solid of cherry tomato was similar however, titratable acidity content was higher than fourteen advanced breeding lines collected from locations in India that reported total soluble solid and titratable acidity content were 6.90 °Brix and 0.37 %, respectively [10]. The total soluble solid of cherry tomato was similar to the cherry tomato cultivar Sunstream [12].

Based on the T-test, the fruit weight of some progeny at F1 was not significantly different from that of the female parent, with only six significantly different genotypes. Genotypes of P1-11 x P3-8, P2-1 x P3-2, and P2-4xP3-2 had significantly lower fruit weight compared to the female parent. Progeny of P2-4xP1-3, P3-5xP1-1, and P3-1xP2-9 had significantly higher fruit weight compared to the female parent. Compared to male parent, only four genotypes had significantly lower fruit weight. There were P1–6 x P2-11, P1-10 x P2-10 and P3-7 x P2-11. The fruit weight of cherry tomatoes can be classified as large (± 16.6 g), medium (±12.6 g), and small (±8.1 g) [13]. The fruit weight of parents in this research was classified as medium and large cherry tomatoes. Our goal of cherry tomato breeding is large fruit weight. Progeny that fit our goal are P1–6 x P2-11, P2-4xP1-3, and P3-1xP2-9. Cherry tomatoes with large fruit weight were firmest. The respiration rates and ethylene production of large cherry tomatoes are also lower than that of small fruit weight [13].

Table 3. Fruit weight, polar diameter, equatorial diameter, total soluble solid and titratable acidity content of parents and F1 progeny

Female parent	F1	Fruit Weight (g)	Polar diameter (cm)	Equatorial diameter (cm)	Total soluble solid (°Brix)	Titratable acidity content (%)
P1		17.09	4.21	1.91	6.80	0.65
	P1–6 x P2-11	14.46^{x}	2.61^{a}	1.31^{ax}	6.43	0.64^{x}
	P1-11 x P3-8	9.21^{a}	3.39^{x}	1.34^{ax}	7.43	0.59^{ax}
	P1-5 x P2-14	13.68	3.23	1.58^{ax}	7.00	0.61^{ax}
	P1-10 x P2-10	14.20^{x}	3.28^{x}	1.92	8.00	0.44^{ax}
P2		15.09	3.61	1.90	6.60	0.63
	P2-4 x P1-3	16.00^{a}	3.45	2.23^{ax}	6.75	0.63^{x}
	P2-1 x P3-2	7.28^{a}	4.30^{ax}	2.13^{ax}	6.63^{x}	0.70^{ax}
	P2-4 x P3-2	7.84^{a}	3.41^{x}	1.87^{x}	7.67^{ax}	0.66^{ax}
	P2-20 x P3-4	8.00	3.00^{ax}	1.53^{ax}	7.00^{ax}	0.63^{x}
P3		9.57	2.77	1.60	5.67	0.78
	P3-1 x P1-12	9.57	3.24^{a}	1.52^{x}	8.47^{ax}	0.67^{ax}
	P3-5 x P1-1	10.15^{a}	2.97^{ax}	1.45^{x}	8.36^{ax}	0.62^{ax}
	P3-8 x P1-2	9.67	3.02^{ax}	1.48^{x}	8.33^{ax}	0.51^{ax}
	P3-1 x P2-9	14.57^{a}	3.34^{a}	1.67^{x}	6.14^{x}	0.64^{ax}
	P3-6 x P2-4	7.70	2.50^{ax}	1.61^{x}	7.10^{x}	0.40^{ax}
	P3-7 x P2-11	7.91^{x}	2.67^{ax}	1.86^{x}	5.38	0.60^{ax}
Mean			3.23			

Note : superscript "a" in same column and female parent are significantly different ($P < 0.05$) to female parent. Supercript "x" in same column and female parent are significantly different ($P < 0.05$) to male parent based on T test.

Some progeny of cherry tomatoes had significantly different polar diameters to female and male parents. All progeny with P3 as female parent had polar diameter that was significantly different to female parent. It means that there were no effect of maternal. One genotype (P2-1xP3-2) had polar diameter significantly higher than both parent. Polar diameter of cherry tomato can be divided in large (3.1 cm), medium (2.8 cm) and small (2.4 mm) [13]. Polar diameter of parents in this research were classified as large and medium. Genotype P2-1xP3-2 and P3-1xP2-9 can be selected for next generation based on our goal of large cherry tomato. Only one genotype of F1 that had equator diameter was not significantly different to male parent. Equator diameter of the other genotype was significantly different to male parent. Genotype P2-4xP1-3 and P2-1xP3-2 could be the candidate for large cherry tomato based on equatorial diameter. Large cherry tomato had the longest shelf life compared to small one. Occurance of fungal and microbial activity also the lowest at large size of cherry tomato. So, when the cherry tomatoes needs to be exported over long distances, it may be best to grow larger cherry tomatoes [13].

Total soluble solid of all genotype with P1 female parents was not significantly different for both parents. Only one genotype of F1 progeny from the P2 female parent with total soluble solids was not substantially different from the male parent, while others were significantly different. Total soluble solids (TSS) reflect the taste and quality of the produce. They are considered an index of ripening and an indicator of the number of soluble minerals and sugar present in fresh produce

[14]. Three genotypes (P3-1xP1-12, P3-5xP1-1, and P3-8xP1-2) had higher total soluble solids than both parents. The total soluble solid of those three genotypes was 8.33 – 8.47 0Brix, which were higher than all parents. Total soluble solid of cherry tomato can be reached 10.8 ^{0}Brix [15]. Total soluble solid is the important character in cherry tomato. Consumers prefer cherry tomatoes because they taste sweeter than other types of tomatoes. Total soluble solid of other types of tomato *((Lycopersicon esculentum* L) is around 4^{0}brix [16]. Three genotypes (P3-1xP1-12, P3-5xP1-1 and P3-8xP1-2) can be good candidate genotype for next generation of our breeding program.

Titratable acidity content of all F1 progeny were significantly different to male parents. Some progeny also had titratable acidity content that were significantly different to female parents. Some progeny had lower titratable acidity content compared to both parents. Titratable acidity content closely correlates with sourness [17]. The sour taste in tomato is attributed mainly to citric and malic acids [18]. Titratable acidity content of tomato (*Lycopersicon esculentum* L.) is 4.5 % [19]. While titratable acidity content of cherry tomato is 0.6 % [20] . There were four genotype that had lower titratable acidity content (lower than 0.6 %). There were P1-11 x P3-8, P1-10 x P2-10, P3-8 x P1-2 and P3-6 x P2-4. Those four genotype can be select for next evaluation in F2 generation.

4. Conclusion

There were variation of qualitative and quantitative characters between parent and F1 progeny. There were variation of fruit longitudinal shape in F1 such as ovate, oblong, elliptic, cordate. There were no variation of fruit flesh color at F1. Ripe fruit colour of F1 that were difference to parental were vivid orange yellow and strong orange. Progeny with significantly larger fruit weight and diameter compared to parent were P2-4xP1-3, P2-1xP3-2 and P3-1xP2-9. Three genotypes (P3-1xP1-12, P3-5xP1-1 and P3-8xP1-2) had total soluble solid higher significantly different to both parents. Qualitative and quantitative diversity of F1 is the key success for breeding cherry tomato varieties with larger fruit size and sweet taste.

Acknowledgement

The research leading to these results received funding from the Faculty of Animal and Agricultural Sciences – Universitas Diponegoro 2023.

References

[1] Pérez-Marín, J. Issa-Issa, H. Clemente-Villalba, J. García-Garví, J.M. Hernández, F. Carbonell-Barrachina, Á.A. Calín-Sánchez, A. Noguera-Artiaga, L, 2021, Physicochemical, volatile, and sensory characterization of promising cherry tomato (Solanum lycopersicum L.) cultivars: Fresh market aptitudes of pear and round fruits, Agronomy 11, 618
[2] Mun H.I., Kwon, M.C., Lee N.R., Son S.Y., Song D.H., Lee C.H., 2021, Comparing metabolites and functional properties of various tomatoes using mass spectrometry based metabolomics approach. Front. Nutr 8, 659646
[3] Nesbitt, T.C. and Tanksley S.D., 2002, Comparative sequencing in the genus Lycopersicon : implication for the evolution of fruit size in the domestication of cultivated tomatoes, Genetics, 162, 365-379
[4] https://benih.pertanian.go.id/varietas-daftar. Access date November 20. 2024
[5] UPOV, 2021, Guidelines For The Conduct Of Tests For Distinctness, Uniformity And Stability Tomato (*Solanum lycopersicum* L.), Geneva.

[6] Sudarmadji, S., Haryono B, dan Suhardi, 1997, Prosedur Analisis untuk Bahan Makanan dan Pertanian. Yogyakarta, Liberty.

[7] Chang Y., Xiaodan Z., Cheng W., Ning M., Jianming X. and Jing Zhang, 2024, Fruit Quality Analysis and Flavor Comprehensive Evaluation of Cherry Tomatoes of Different Colors , Foods, 13, 1898

[8] Kusumiyati K., Farhan A., Muhamammad R.K., Mochamad A. S and Ristina S.S, 2023, Productivity of cherry tomato cultivars as influenced by watering capacities and microclimate control design, The Open Agriculture Journal, 17, e18743315280566, DOI: 10.2174/0118743315280566231119172531.

[9] Al-Mfargy, O.K.A and Juwari M.S.M, 2023, Estimation of Hybrid Vigor for Flowering and Yield Cherry Tomato Hybrids, IOP Conf. Series: Earth and Environmental Science 1158 (2023) 042066 , doi:10.1088/1755-1315/1158/4/042066.

[10] Debmala M, Praveen K , Tridip B, Swadesh B, Subhrajyoti C, Sudipa M, Ivi C, Soumitra C, Sunanda C, Anirban M and Arup C, 2020, Assessment of Breeding Potential of Cherry Tomato [Solanum lycopersicum var. Cerasiforme (Dunnal) A. Gray] Grown under Open Field to Identify Desirable Alleles , Int.J.Curr.Microbiol.App.Sci , 9(4), 2152-2171

[11] Gaswanto. R., 2021, Adaptation test of five introduced cherry tomato lines in Lembang highland, IOP Conf. Series: Earth and Environmental Science 883, 012066 IOP Publishing doi:10.1088/1755-1315/883/1/012066

[12] Rivelli, A.R.; Castronuovo, D.; Gatta, B.L.; Liberatore, M.T.; Libutti, A. Qualitative Characteristics and Functional Properties of Cherry Tomato under Soilless Culture Depending on Rootstock Variety, Harvesting Time and Bunch Portion. Foods 2024, 13, 1450. https://doi.org/ 10.3390/foods13101450

[13] Islam M.Z., Young-Tack L., Mahmuda A.M., In-Lee C., and Ho-Min K, 2019, Effect of fruit size on fruit quality, shelf life and microbial activity in cherry tomatoes , AIMS Agriculture and Food, 4, Issue 2, 340–348.

[14] Abiso E, Neela S and Addisalem H, 2015, Effect of storage methods and ripening stages on postharvest quality of tomato (Lycopersicom esculentum Mill) cv Chali, Annals. Food Science and Technology, 16, 127 -137

[15] Lone, S, Hussain K , Ajaz M , Khalid Z. M , Dar Z A , Nageena N , Zahedullah Z and Gowher A, 2021, Genetic variability studies in cherry tomato for quality traits in protected condition, Annals of Horticulture, 14 (1), 68-71 DOI : 10.5958/0976-4623.2021.00013.X

[16] Mai Al-Dairi a , Pankaj B. Pathare , Rashid Al-Yahyai. 2021. Chemical and nutritional quality changes of tomato during postharvest transportation and storage. Journal of the Saudi Society of Agricultural Sciences 20, 401–408

[17] Tandon, K.S., E.A. Baldwin, J.W. Scott and R.L. Shewfelt, 2003. Linking sensory descriptors to volatile and non-volatile components of fresh tomato flavor. J. Food Sci., 68: 2366-2371.

[18] Petro-Turza, M., 1987. Flavor of tomato and tomato products. Food Rev. Int., 2: 309-351.

[19] Majidi H., Saeid M, Morteza A, Younes M, 2011, Total Soluble Solids, Titratable Acidity and Repining Index of Tomato In Various Storage Conditions, Australian Journal of Basic and Applied Sciences, 5(12), 1723-1726.

[20] Murtic S, Rodoljub O , Mirela S, M., , Amila V, Asima A and Hamdija C, 2018, Cherry tomato productivity as influenced by liquid organic fertilizer under different growth conditions . Journal of Central European Agriculture, 19(3), 503-516

Isolation and Identification of Bacteria - Producing Cold-Adapted Lipase from *Morganella morganii* strain F6C1 Isolated from Fish Eggs

Hefti Salis Yufidasari[1,2*], Aji Sutrisno [1], Agustin Krisna Wardani [1], Teti Estiasih[1]

[1] Faculty of Agricultural Tecnology, Universitas Brawijaya, Malang, Indonesia
[2] Faculty of Fisheries and Marine Science, Universitas Brawijaya, Malang, Indonesia

*E-mail: hefti_salis@ub.ac.id

Abstract. Lipase (triacylglycerol acylhydrolase, EC3.1.1.3) is an enzyme that has an important role in modern biotechnology. Lipase is known to have high activity in hydrolysis reactions and in synthetic chemistry. Lipase can act as a biocatalyst for hydrolysis, esterification, alcoholysis, acidolysis and aminolysis reactions. Lipase hydrolyzes triglycerides into free fatty acids, partial glycerides, and glycerol. Lipase is an enzyme that is widely distributed in nature and is produced by several plants, animals and microorganisms which has quite large potential physiologically and industrially. Isolate is a bacterial isolate obtained from fish egg, samples stored at cold temperatures, where the isolate is thought to be capable of producing enzymes that have high lipolytic activity at low temperatures. The purpose of this research was to characterize the bacterial isolate that produces the lipase enzyme, namely testing enzyme activity and molecular identification to estimate bacterial species. The results showed that isolate TL2510B had the highest enzyme activity at a temperature of 25^{0}C of 141.33 U/mL, the enzyme activity based on pH showed the highest at pH 10, namely 106.00 U/mL, while in optimization the highest enzyme production was obtained at the 24th hour, of 104.69 U/mL. Identification of the isolate based on 16s rRNA analysis showed 97.02% similarity to *Morganella morganii* strain F6C1 (accession number JX999959.1).

Keywords: *Morganella morganii*, fish eggs, cold-adapted, lipase, hydrolyzed

1. Introduction

The increasing demand for sustainable and eco-friendly alternatives in industrial processes has highlighted the importance of enzyme-based reactions over traditional chemical methods. Enzymes are considered environmentally advantageous because they generate less waste, consume less energy, and pose a reduced pollution risk, leading to improved product quality (Kavitha, 2016). Among these enzymes, lipases (acyl glycerol hydrolases, EC 3.1.1.3) stand out due to their extensive application across various fields, such as biotechnology, food processing, pharmaceuticals, biodiesel production, and other industries (Jaeger & Eggert, 2002). Lipases are versatile, capable of catalyzing the hydrolysis of glycerol esters and long-chain fatty acids as well as synthesizing these esters, making them valuable tools in industrial processes (Jaeger & Eggert, 2002).

A significant area of research in enzyme development focuses on their activity under extreme conditions (Santiago et al., 2016). For instance, thermophilic enzymes, which are stable at high temperatures, have been studied extensively to enhance reaction rates and maintain stability during industrial applications (Cowan, 2010). However, reactions conducted at elevated temperatures often result in higher production costs and may lead to the denaturation of sensitive components in products like food and pharmaceuticals (Collins et al., 2012). As a response to these challenges, research has increasingly shifted towards cold-adapted enzymes, particularly cold-adapted lipases (CLPs). CLPs are specialized enzymes capable of operating effectively at low temperatures, offering significant energy savings and maintaining the integrity of temperature-sensitive compounds during processing (Feller, 2010).

Microorganisms that thrive in cold environments, such as the polar regions and deep-sea habitats, are promising sources of these cold-active enzymes. These microorganisms, classified as psychrophiles, have evolved unique metabolic adaptations that enable them to survive and function in extreme cold conditions (Georlette et al., 2004). Several bacterial genera, including *Moraxella*, *Psychrobacter*, and *Pseudomonas*, have been identified as sources of cold-adapted lipases (Feller et al., 1991; Kulakova et al., 2004). Despite the potential of these microorganisms, research on lipase-producing bacteria from cold environments remains limited, with much of the focus placed on high-temperature and mesophilic conditions (Xuezheng et al., 2010; Zhang and zeng (2008).

While cold-adapted enzymes from polar and alpine environments are valuable, their stability at moderate temperatures is often inadequate for industrial use. However, some cold-adapted lipases derived from tropical environments have shown greater thermal stability compared to those from colder regions (Kavitha & Shanthi, 2013). This characteristic makes them particularly attractive for applications requiring both cold and moderate temperature stability, expanding their industrial utility.

The purpose of this research was to characterize the bacterial isolate that produces the lipase enzyme. Isolate is a bacterial isolate obtained from fish egg, samples stored at cold temperatures, where the isolate is thought to be capable of producing enzymes that have high lipolytic activity at low temperatures. The research involves the screening, identification, and characterization of these bacterial isolates to determine their optimal conditions for lipolytic activity. By focusing on microorganisms from seafood cold storage, which mimics the low-temperature environments of psychrophilic habitats, this study hopes to discover strains capable of producing stable and efficient cold-active lipases suitable for industrial applications. The results will contribute to the broader understanding of microbial diversity in cold environments and the development of sustainable enzymatic processes for various industries.

2. Materials and Methods

2.1 Screening for Lipase-Producing Bacteria

Fish egg, clams and crab meat samples were collected from three cold storage facilities at a modern supermarket in Malang City, East Java, Indonesia. Each 1-gram sample was ground and combined with 50 mL of Basal Salt Medium (BSM), followed by incubation at 25°C for 48 hours. The mixture underwent serial dilution using 0.9% NaCl. Dilutions ranging from 10^{-5} to 10^{-9} were then spread on agar media containing 5 g/L tryptone, 2.5 g/L yeast extract, 9 g/L agar, and 1% Tween 20 or Tween 80 for initial lipolytic screening. Incubation was performed for 2 to 10 days at 16°C, 25°C, and 37°C. Lipolytic activity was indicated by the formation of white precipitation around the bacterial colonies. Colonies showing this white precipitation were collected for further screening to confirm their lipolytic activity.

Each bacterial colony was taken and separately inoculated by spotting on Sierra's Medium (10 g/L peptone, 5g/L NaCl, 20g/L agar, 0.1 g/L $CaCl_2$, and 1% Tween 80 or Tween 20 as lase substrates). According to Mukesh et al. (2012), incubation was performed at three distinct temperatures (16°C, 25°C, and 37°C) for durations ranging from 2 to 6 days. Lipolytic activity was identified by the formation of white precipitates surrounding the bacterial colonies. Three bacterial isolates with the largest precipitation zones were chosen and cultured on agar medium using the streak plate technique to isolate single colonies.

2.2 Production of Lipase Crude Extract

The selected bacterial isolates were individually inoculated into 10 mL of broth medium containing 1% tryptone, 1% sodium chloride, and 0.5% yeast extract, then incubated for 24 hours with agitation at 150 rpm. The following day, a 2% culture was transferred into 50 mL of production medium composed of 1% tryptone, 1% sodium chloride, 0.5% yeast extract, and 1% lipase substrate. Incubation was carried out with agitation at 150 rpm for 2x24 hours. The culture was centrifuged at 10,000 rpm for 15 minutes. The supernatant was transferred into a sterile tube and referred to as the lipase crude extract.

2.3 Activity of Lipase Crude Extract

The lipase activity assay followed the procedure outlined by Stoytcheva et al. (2012), with slight modifications to the reaction mixture. In the assay, 2 mL of lipase substrate was combined with a reaction mixture consisting of 2 mL of 0.1 M sodium phosphate buffer at pH 6.5, 2.5 mL of distilled water, 0.5 mL of 0.1 M $CaCl_2$, and 1 mL of crude enzyme extract. The mixture was then agitated at 150 rpm for 15 minutes. The reaction was halted by adding 96% ethanol. Enzyme activity was determined using a titration method with 0.05 M NaOH and phenolphthalein as an indicator. A single unit of lipase activity in the crude extract was defined as the quantity of enzyme required to generate 1 mole of fatty acid per minute under the specified assay conditions.

2.4 Effect of Temperature and pH on Enzyme Activity

The characterization of lipolytic activity was conducted across different temperature and pH conditions. To determine the optimal temperature, the crude extract was tested at a range of temperatures: 10°C, 20°C, 30°C, 40°C, 50°C, 60°C, and 70°C. For the optimal pH assessment, the crude extract was incubated in various buffer systems spanning a pH range from 5 to 11. The buffer systems included CH_3COONa/CH_3COOH for pH 5, KH_2PO_4/Na_2HPO_4 for pH 6 and 7, Tris-HCl for pH 8 and 9, and $Na_2HPO_4/NaOH$ for pH 10 and 11. The procedure followed the standard method, and the lipolytic activity at each temperature and pH was expressed as relative activity, using the highest observed lipase activity as the reference point (100% relative activity).

2.5 Amplification of the partial gene encoding 16s rRNA

The potential isolates underwent species identification based on the 16s rRNA sequence. Genomic DNA was extracted using the Wizard Genomic DNA Purification Kit (Promega, USA), and its concentration and purity were measured with a Nanodrop 2000 (Thermo Fisher Scientific, USA). The 16s rDNA sequence was then amplified using PCR with the forward primer 63F and the reverse primer 1387R.

The PCR reaction mixture consisted of 1 μL of genomic DNA, 5 μL of GoTaq Green Master Mix (Promega, USA), 1 μL of each primer, and nuclease-free water, making up a total volume of 10 μL. Verification of the PCR product was performed through electrophoresis in 1% agarose gel (w/v). The amplified product was purified and analyzed using the 3130 DNA analyzer (Applied Biosystems, Foster City, USA) according to the standard protocol. The contiguous sequence obtained was then compared using NCBI-BLAST (https://www.ncbi.nlm.nih.gov/) to identify the closest matching species.

3. Results and Discussions

Lipase enzymes are found in nearly all living organisms, including microorganisms (such as bacteria and yeasts), plants, and animals (Gupta et al., 2004). Lipases from microorganisms are particularly valuable in the industrial sector due to their ease of production and efficiency. Many studies have explored and utilized lipases that function at extreme temperatures since most industrial processes involve high temperatures. However, using high temperatures can be less advantageous as it increases energy consumption and production costs, and it may also compromise the quality of sensitive components in products like food and pharmaceuticals.

As a result, there has been growing interest in exploring lipases that function at lower temperatures, known as cold-adapted or cold-active lipases, due to their efficiency under mild conditions. Consequently, extensive research has been dedicated to screening and characterizing cold-adapted lipases derived from bacterial and fungal sources.

The current study focuses on isolating and characterizing lipase enzymes from fish egg, clams and crab meats samples stored under cold conditions. Bacterial colonies were cultured on non-selective Plate Count Agar (PCA) medium. To facilitate the identification of lipase-producing colonies for further analysis, Tween 20 and Tween 80 were incorporated into the medium. Incubation was conducted at three different temperatures: 16°C, 25°C, and 37°C.

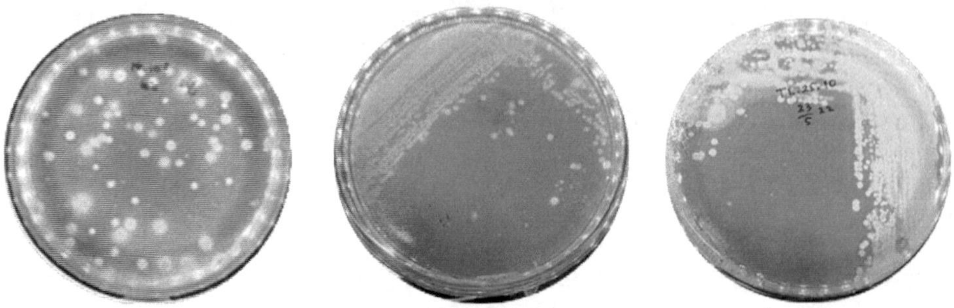

A B C

Figure 1. The growth of total bacteria on PCA medium added with Tween 80 (A), the lipolytic activity at Sierra's Medium containing Tween 20 (B) and Tween 80 (C). The white presipitation around the colony indicating lipolytic activity

The incubation time required for bacterial growth varied significantly depending on the incubation temperature: approximately 18 hours at 37°C, 24-48 hours at 25°C, and up to 7 days at 16°C. The clam, fish egg, and crab meat samples obtained from a modern market in Malang City had bacterial counts ranging from 6.27 × 10^5 to 1.23 × 10^7. Bacterial colonies that showed white precipitation were selected and re-inoculated on Sierra's Medium to validate their lipolytic activity, as illustrated in Figure 1.

Lipolytic activity assay using Tween 20 or 80 based on the principle of calcium salt precipitation. Hydrolysis of tween releases fatty acids which bind to calcium derived from

$CaCl_2.2H_2O$ in the medium. This complex will form insoluble crystals, appear as white precipitation around the colony (Figure 1), as indication that the bacterial colony produces extracellular lipase Rai et al. (2014).

Screening results from 3 different incubation temperatures (16°C, 25°C and 37°C) showed that at 16°C no colonies were found indicating that there were bacterial isolates producing lipase enzymes, whereas at 25°C and 37°C colonies were found that were suspected of producing lipase enzymes. The data is presented in Table 1.

Table 1. Kualitative Screening isolated bacteria of lipase producing bacteri from each sample and temperature of incubation

Sample	Incubation Temperature (°C)		
	16	25	37
Fish Egg (TL)	-	+++	++
Clams (K)	-	++	+
Crab meats (JL)	-	+	+

From the screening results, 48 colonies were obtained which showed an opaque zone around the colonies, which came from the 3 types of samples used. Next, 24 colonies that were thought to have the most potential to produce the lipase enzyme were taken and their enzymatic activity would be tested. The isolates are JL252A, JL252B, JL254A, JL254B, JL259A, JL259B, JL259C, JL371A, JL371B, KD2510A, KD2510B, KD2517A, KD2517B, TI2510A, TI2510B, TI2510C, TI2 518A, TI2518B, TI2519, TI3713A, TI3713B, TI3713C and TI3713D

Enzyme activity was tested using the titration method, carried out on three repetitions of the crude lipase extract enzyme, from potential bacteria and the highest lipase crude extract activity was obtained in isolate TL2510B, namely 11.3 U/mL (Figure 2). Lipase activity was indicated by a color change from colorless to pink with the addition of phenolphthalein indicator. This change occurs due to changes in pH in the reaction solution. The pink color occurs when there is no more NaOH that can bind to the fatty acid so the pH of the solution becomes alkaline, because fatty acids have a pH that is weakly acidic.

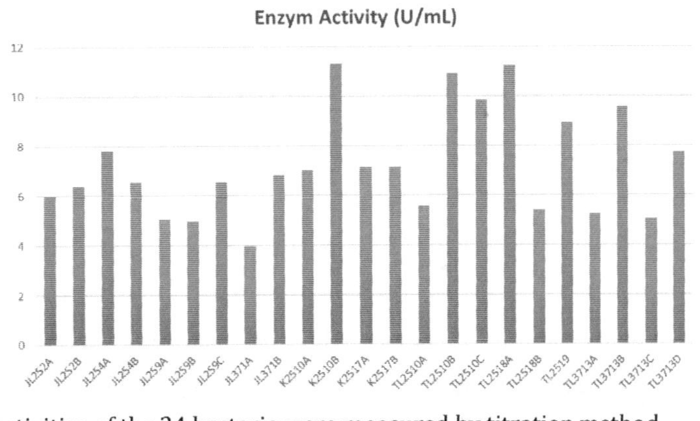

Figure 2. Lipase activities of the 24 bacteria were measured by titration method

Of the 24 isolates, 1 isolate was then selected which had the highest activity, namely isolate TL2510B, and continued by determining the enzyme activity at optimum temperature, pH and molecular identification to determine the species.

Based on the test results for the effect of temperature on lipase activity, it was found that the optimum temperature was 25°C (Figure 3). The isolate that had the highest enzyme activity was isolate TL2510B, namely 141.33 U/mL. Enzyme activity will continue to increase until the optimum temperature and will slowly break down if it passes that point.

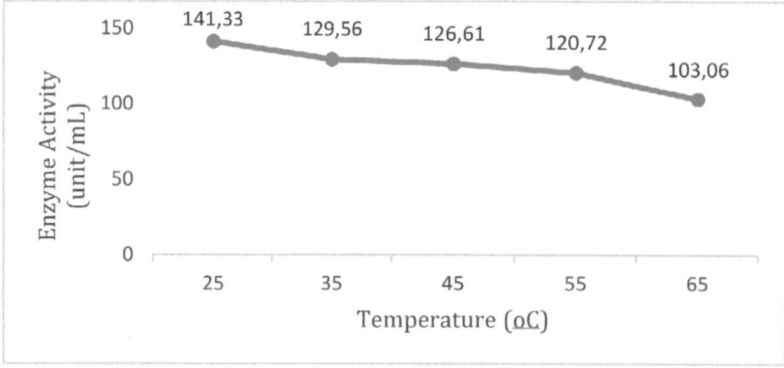

Figure 3. Lipolytic activity of lipase crude extract from potential bacteria under various temperature

Determination of the optimum pH of lipase was carried out using the pNPP method in the pH range 5-11. Test results show that lipase works optimally at pH 10. Enzyme activity at pH is influenced by the ion groups on the catalytic side of the enzyme and the charge on the three-dimensional structure of the protein. The pH level of the medium significantly influences enzymatic activities and the movement of compounds across cell membranes (Kuddus and Remteke, 2008). Thus, in general enzymes will experience denaturation at extreme pH. Usually the active site of enzymes is composed of polar amino acids which easily undergo ionization so that their catalytic activity depends on pH.

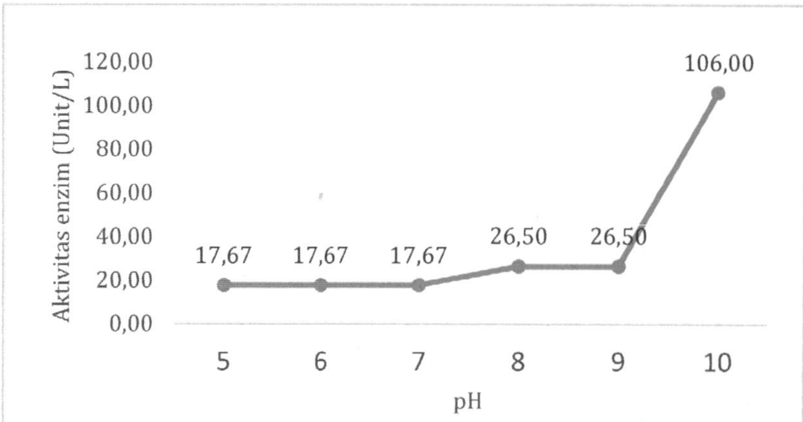

Figure 4. Lipolytic activity of lipase crude extract from potential bacteria under various pH

Optimization of the harvest period was carried out on the selected isolate. The procedure is incubation of lipolytic bacteria inoculated in Luria Bertani (LB) liquid media for 54 hours and the enzyme activity is tested at 0, 6, 12, 18, 24, 30, 36, 42, 48, and 54 hours.

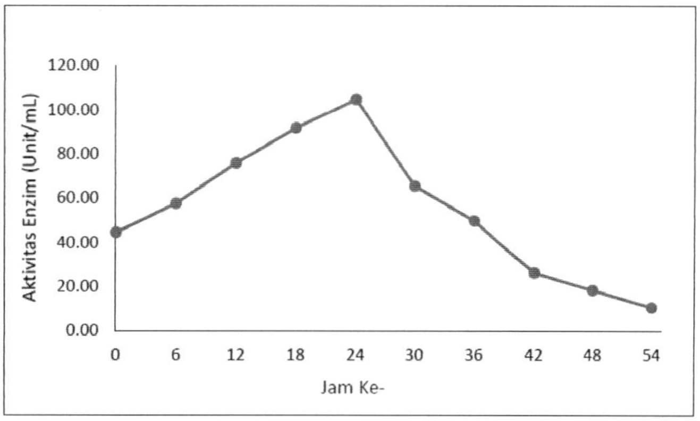

Figure 5. Optimization of Harvest Period

The Fig.5 shows that at hour 0 it is still included in the adjustment phase, which shows an enzyme activity of 44.49 Units/mL. In the following hours, namely the 6th, 12th, 18th and 24th hours, they enter the bacterial growth phase, namely the exponential phase, where the bacteria continue to multiply as indicated by the enzyme activity increasing successively, namely 57.58 units/ mL, 75.90 units/mL, 91.60 units/mL, and 104.69 units/mL. After that, at the 30th hour to the 54th hour, it showed that bacterial growth entered a stationary phase, which showed that there was no increase in enzyme activity, but continued to decrease with enzyme activity respectively 65.43 units/mL, 49.73 units. /mL, 26.17 units/mL, 18.32 units/mL, 10.47 units/mL. This phenomenon indicates that the bacteria are approaching the death phase.

Species identification was conducted by analyzing the 16s rRNA coding sequence. The 16s rRNA gene is widely used as a molecular marker in bacterial studies due to its presence and consistent function across all organisms. This molecule contains both conserved and variable regions. The conserved regions, which evolve slowly, are valuable for building universal phylogenetic trees and are essential in designing reaction primers. In contrast, the variable regions are utilized to identify diversity and distinguish bacteria at the species or strain level (Stackebrandt and Goebel, 2000).

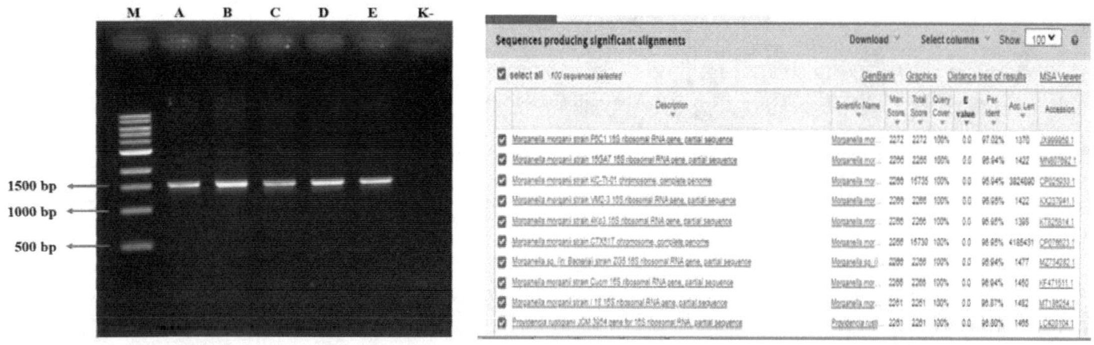

Figure 6. The results of the PCR analysis (left), The results of BLAST analysis on the gene database (www.ncbi.nlm.nih.gov)

The results of the PCR analysis showed that the DNA band was about 1500 bp in size. In accordance with the estimated results of 16s rRNA gene amplification using Marchesi et al (1998) and the results of BLAST analysis on the gene database (www.ncbi.nlm.nih.gov) 97.02% similarity to *Morganella morganii* strain F6C1 (accession number JX999959.1).

Morganella moganii is a Gram negative, facultative anaerobic and oxidase negative bacterium, with a straight rod shape between 0.6 and 0.7 mm and a diameter of 1.0 and 1.8 mm.

The colonies appear whitish and opaque when planted on agar, and can live optimally at temperatures below 30° C (Kumaran et al., 2019), so they were categorized as CLPs.

In addition, its activity is relatively stable at moderate temperatures.

4. Conclusions

Screening and isolation of lipase-producing bacteria from fish eggs in cold storage resulted in 24 isolates of bacteria that produce lipase on Sierra's media. The highest activity of crude extract enzymes was shown by bacteria TL25.10B, which was 11.3 U/mL using the titration method.

Lipase enzyme TL25.10B was isolation from fish eggs showed optimum activity in the range of 25-30 oC and pH 10, that it can be categorized as CLPs potential. Identification of isolate TL25.10B based on 16s rRNA analysis showed 97.02% similarity to *Morganella morganii* strain F6C1 (accession number JX999959.1).

References

[1] Arribas, E · H. Bisswanger A. Sotos-Lomas ,M. Garcia-Moreno F. Garcia-Canovas, J. Donoso-Pardo F. Muñoz-Izquierdo R. Varon, 2008, A method, based on statistical moments, to evaluate the kinetic parameters involved in unstable enzyme systems, J Math Chem 44:379–404

[2] Collins, T., Gerday, C., & Meuwis, M. (2012). Cold-Adapted Enzymes.

[3] Cowan, D. (2010). Concentration of Omega-3 Fatty Acids Using Enzymes. Agro FOOD Industry Hi-Tech.

[4] Feller, G. (2010). Protein Stability and Enzyme Activity at Extreme Biological Temperatures. Journal of Physics Condensed Matter, 22(32), 1-17.

[5] Feller, G., Thiry, M., Arpigny, J. L., & Gerday, C. (1991). Cloning and Expression in Escherichia coli of Three Lipase-Encoding Genes from the Psychrotrophic Antarctic Strain Moraxella TA144. Gene, 102(1), 111-115.

[6] Georlette, D., et al. (2004). Some Like It Cold: Biocatalysis at Low Temperatures. FEMS Microbiology Reviews, 28(1), 25-42.

[7] Gupta R, Gupta N, Rathi P. 2004. Bacterial lipases: an overview of production, purification and biochemical properties. Appl. Microbiology and Biotechnology. 64:763-781. https://doi.org/10.1007/s00253-004-1568-8

[8] Jaeger, K. E., & Eggert, T. (2002). Lipases for Biotechnology. Current Opinion in Biotechnology, 13(4), 390-397.

[9] Joseph B, Ramteke PW, Thomas G. 2008. Cold active microbial lipases: Some hot issues and recent developments. Biotechnol Advances 26:457-470. https://doi.org/10.1016/j.biotechadv.2008.05.003

[10] Kavitha, M. (2016). Cold Active Lipases - An Update. Frontiers in Life Science, 9(3), 226-238.

[11] Kavitha, M., & Shanthi, C. (2013). Isolation and Characterization of Cold-Active Lipase Producing Pseudomonas sp. 4 from Marine Samples of Tamilnadu Coast.

[12] Kuddus, M., & Ramteke, P. W. (2008) Purification and properties of cold-active metalloprotease from Curtobacterium luteum and effect of culture conditions on production, Shengwu Gongcheng Xuebao/Chinese Journal of Biotechnology, Vol. 24, No. 12, 2074-2080. https://www.sciencedirect.com/science/article/abs/pii/S1872207509600121

[13] Kulakova, L., et al. (2004). Cold-Active Esterase from Psychrobacter sp. Ant300: Gene Cloning, Characterization, and Effects of Gly→Pro Substitution Near the Active Site on Its Catalytic Activity and Stability. Biochimica et Biophysica Acta - Proteins and Proteomics, 1696(1), 59-65.

[14] Kumaran, V. B. Sandeep, T. Ramesh, D. Annadurai, M.Thangaraj, R. Rajasekaran, 2019, Screening and Preliminary Optimization of Protease in Morganella Morganii VBS1 Strain Isolated from a Gastropod, Cerithidea cingulate. 6 (2), IJRAR- International Journal of Research and Analytical Reviews

[15] Maayer P.D., Dominique A, Craig C, Don A.C. 2014. Some like it cold: understanding the survival strategies of psychrophiles. Embo reports: 1-10. DOI 10.1002/embr.201338170

[16] Marchesi, J. R., Sato, T., Weightman, A. J., Martin, T. A., Fry, J. C., Hiom, S. J., & Wade, W. G. (1998). Design and Evaluation of Useful Bacterium-Specific PCR Primers That Amplify Genes Coding for Bacterial 16S rRNA. Applied and Environmental Microbiology, 64(2), 795–799. doi:10.1128/aem.64.2.795-799.1998
[17] Mukesh, K., Rejitha, R., Devika, S., Balakumaran, M. D., Rebecca, I. N., & Kalaichelvan, P. T. (2012) Production, optimization and purification of lipase from Bacillus sp. MPTK 912 isolated from oil mill effluent, Advances in Applied Science Research, Vol. 3, No. 2, 930-938.
[18] Rai, B., Shrestha, A., Sharma, S., & Joshi, J. (2014) Screening, Optimization and Process Scale up for Pilot Scale Production of Lipase by Aspergillus niger, Biomedicine and Biotechnology, Vol. 2, No. 3, 54-59. Retrieved from https://doi.org/10.12691/bb-2-3-3
[19] Santiago, M., Ramírez-Sarmiento, C. A., Zamora, R. A., & Parra, L. P. (2016) Discovery, molecular mechanisms, and industrial applications of cold-active enzymes, Frontiers in Microbiology, 7, 1408. Retrieved from https://doi.org/10.3389/fmicb.2016.01408
[20] Stackebrandt and Goebel, 2000, Nonculturable Microorganisms in the Environment, Springer Nature Link
[21] Xuezheng, L., Shuoshuo, C., Guoying, X., Shuai, W., Ning, D., & Jihong, S. (2010) Cloning and heterologous expression of two cold-active lipases from the Antarctic bacterium Psychrobacter sp. G, Polar Research, Vol. 29, No. 3, 421-429. Retrieved from https://doi.org/10.3402/polar.v29i3.6087
[22] Zhang, J. W., & Zeng, R. Y. (2008) Molecular cloning and expression of a cold-adapted lipase gene from an antarctic deep sea psychrotrophic bacterium Pseudomonas sp. 7323, Marine Biotechnology, Vol. 10, No. 5, 2008, 612-621. Retrieved from https://doi.org/10.1007/s10126-008-9099-4

IOP Conf. Series: Earth and Environmental Science **1460** (2025) 012005 doi:10.1088/1755-1315/1460/1/012005

Analysis of Dairy Business Development Problems in Cigugur District, Kuningan Regency

Sukmawati Utami[1*], Siwi Gayatri[2] and Limbang Kustiawan Nuswantara[1]

[1] Animal Sciences Study Program, Faculty of Animal and Agricultural Science Diponegoro University, Semarang, Indonesia
[2] Agribusiness Study Program, Faculty of Animal and Agricultural Science Diponegoro University, Semarang, Indonesia

*E-mail: sukmawaa1206@gmail.com

Abstract. Cigugur District, Kuningan Regency, holds potential for dairy business development in West Java Province, contributing 6.93% of the total milk production in West Java. However, the development of the dairy business in the region has not reached its full potential due to limited access to production inputs and farmers' low levels of education and skills. This study aims to identify the problems and solutions for the dairy cattle business in Cigugur District, Kuningan Regency. The research used a survey method, with a sample of 115 dairy farmers selected through a stratified sampling method. The lack of community empowerment programs and the ineffective and inefficient use of production inputs are significant issues faced in the development of the dairy business. Moreover, coordination and collaboration among various stakeholders are essential to build and developed a sustainable dairy farming business in Cigugur District, Kuningan Regency.

Keywords: dairy farming, development, problems, sustainable

1. Introduction

West Java Province is one of the national centers for milk production, ranking second in milk production after East Java Province. In 2021, the population of dairy cattle reached 119.92 thousand, producing 28.33 tons of milk annually and contributing 29.44% to the national milk supply [1]. There are three major milk-producing regions in West Java Province, namely West Bandung Regency (98,175 tons/year), Bandung Regency (54,446 tons/year), and Garut Regency (34,148 tons/year) [2]. Kuningan Regency is also one of the milk-producing regions in West Java Province, with a production of 19.64 tons/year or 6.93% annually [3].

Cigugur, Kuningan Regency, holds great potential for developing dairy farming businesses. Several opportunities exist, including, first, the location of Cigugur District at the foot of Mount Ciremai, which provides ideal geographical and agroclimatic conditions for dairy farming. Second, the increasing national milk consumption, which is not yet fully met by Domestic Fresh Milk Production (SSDN), presents a market opportunity for developing dairy farming. Third, the development of Cigugur District as a tourist area opens up opportunities for product diversification in milk processing and expands the marketing segments of dairy products as souvenirs for both local residents and tourists.

However, dairy farmers in Cigugur District, Kuningan Regency, West Java, have not fully optimized these opportunities. This is evident from the low daily milk production [4] and the small-scale ownership of livestock, with almost 90% of existing farms focusing solely on on-farm activities. Consequently, dairy processing and post-harvest activities remain limited to Micro, Small, and Medium Enterprises (MSMEs), which are still home-based businesses with relatively few numbers.

The underutilization of opportunities for dairy farming development is caused by several issues, both directly and indirectly, hindering the growth and development of the dairy farming business in Cigugur District, Kuningan Regency. Therefore, this study aims to analyse the problems that hinder business development and provide solutions for the dairy farming business in Cigugur District, Kuningan Regency, by examining issues related to farmers' human resources as key business actors, business input utilization, ecological, economic, and socio-cultural factors, as well as the institutional conditions of farmers and non-farmers.

2. Material and methods

Data collection was conducted through a survey method in three villages in Cigugur District, Kuningan Regency, West Java: Cisantana Village, Cigugur Village, and Cipari Village. These villages were selected based on their status as central areas for dairy farming in Cigugur District, Kuningan Regency with statistical analysis. The study used a sample of 115 dairy farmers from Cisantana Village, Cigugur Village, and Cipari Village. The sample was determined by randomly selecting respondents from these three villages using a stratified sampling method. The groups were divided based on administrative regions, namely the subdistricts. The sample size for each group was determined using the following formula:

$$\frac{\mathbf{Ni}}{\mathbf{N}}\,\mathbf{X\ n}$$

Explanation:
Ni: number of farmers in the village; N: total population; n: total sample size
The sample size for each village/subdistrict is shown in Table 1.

Table 1. Sample Size by Location

Location	Population (People)	Sample (People)
Cisantana Village	614	46
Cigugur Subdistrict	530	39
Cipari Subdistrict	408	30
Total	1552	115

The sample distribution was as follows: 46 respondents from Cisantana Village, 39 respondents from Cigugur Subdistrict, and 30 respondents from Cipari Subdistrict. Each respondent was given a questionnaire related to the development process of dairy farming, covering aspects such as farmers' human resources, business input usage, ecological, economic, socio-cultural factors, as well as the institutional conditions of farmers and non-farmers, along with the challenges and issues encountered in the development of dairy farming businesses.

3. Results and Discussion

3.1. Dairy Farming Conditions in Cigugur District

Cigugur District is one of the regions with significant potential in the livestock sector. Specifically, Cigugur District serves as the center for dairy farming in Kuningan Regency. Based on observations and interviews, the farming businesses developed in the area are generally small-scale, family-owned farms with limited livestock holdings [5]. Approximately 90% of family farms are small-scale, with livestock ownership ranging from 1 to 5 heads. The farming system used is traditional, where all activities are performed manually without modern machinery. About 95% of farms in the area focus on on-farm cultivation, selling their dairy products in the form of fresh milk to dairy cooperatives. Most of the milk marketing is done individually, with farmers selling directly to the cooperatives. This individual marketing is contingent on the farmers' membership in the cooperative. However, some farmers are part of a livestock group, which collectively sells milk through the group, enabling them to market the milk under the group's name to the cooperative. The advantage of this system is that farmers have greater bargaining power when negotiating milk prices with the cooperative.

3.2. Opportunities for Dairy Farming Business Development

First, the geographical conditions and agroclimatic suitability in Cigugur District support the development of dairy farming. The location of Cigugur District at the foot of Mount Ciremai places it in a highland area with an ideal temperature and humidity for dairy farming. Cigugur District is situated at an elevation of 400-1,100 meters above sea level, with temperatures ranging from 18 to 32°C and an average humidity of 60-70% [4]. These conditions are considered ideal for dairy farming. FH cows are optimally raised at an elevation of 561-750 meters above sea level, with a temperature range of 22-31°C [6].

Second, public awareness and knowledge about nutrition is a key factor driving the increasing demand and consumption of fresh milk in Indonesia. However, this condition is not yet supported by the stability and continuity of domestic fresh milk production. The still wide-open demand and market present an important aspect to be explored in developing dairy farming businesses in a region [7].

Third, the development of the tourism sector in Cigugur District creates opportunities for diversification in dairy products. The influx of tourists visiting the area's attractions presents a market opportunity to introduce various types of dairy products from Cigugur. Moreover, processed dairy products can be marketed as souvenirs for tourists to purchase after visiting Cigugur District.

IOP Conf. Series: Earth and Environmental Science **1460** (2025) 012005 doi:10.1088/1755-1315/1460/1/012005

3.3. Problems in the Development of Dairy Farming Businesses

Based on the results of observations and interviews, several technical and non-technical issues were identified in the development of dairy farming businesses in Cigugur District, Kuningan Regency, as presented in Table 2.

Table 2. Issues in Dairy Farming in Cigugur District, Kuningan Regency

No.	Aspect	Issue
1	Human Resources	• Low farmer regeneration rates. Most farmers are elderly and face physical limitations in managing their farms. • Lack of knowledge and information among farmers regarding population control, especially related to replacement stock.
2	Production Input	• Ineffective use of production inputs in small-scale businesses, resulting in low milk production, income, and profit from dairy farming.
3	Environment	• High rates of land conversion, leading to difficulties in obtaining sufficient forage during the dry season. • Unprocessed farm waste causing environmental issues and complaints from the community, especially during the rainy season.
4	Economy	• Lack of markets to select breeding stock or sell cull bulls and cows, leading to high dependence on intermediaries and lengthy market chains that disadvantage farmers.
5	agricultural institution	• Insufficient empowerment of the Women Farmers Group (WFG) in milk processing.
6	non-agricultural institution	• Limited access to capital after the Foot-and-Mouth Disease outbreak.

Based on the issues identified through observation and interviews, community empowerment is needed to enhance farmers' capacities as business actors. Farmer empowerment can be achieved by providing proper understanding through extension services or farmer-to-farmer communication, enabling them to improve farm management systems and achieve desired goals [8]. Other empowerment efforts can involve collaboration between government and private sectors to conduct training aimed at developing farmers' human resource capacities, particularly by introducing farmers to appropriate innovations and technologies that can increase the effectiveness and efficiency of dairy farming. This approach transforms dairy farming into an agribusiness model, with a focus not only on the on-farm sector but also on upstream activities and post-harvest processing.

The results of this study indicate that the use of business inputs in family-owned dairy farms with ownership of 1-5 dairy cattle have an R/C value of 0,89-0,93 or R/C below 1, means that the dairy cattle business activity was less profitable and inefficient. Business inputs referred to in this

IOP Conf. Series: Earth and Environmental Science **1460** (2025) 012005 doi:10.1088/1755-1315/1460/1/012005

study include livestock, barns and equipment, forage, concentrate feed, and labour. The low number of livestock owned, small business scale, and limited capital are the reasons for the suboptimal use of business inputs.

The main problem related to business inputs in this study is the provision of breeding stock and the limited availability of forage feed. The difficulty in obtaining good-quality breeding stock at affordable prices remains a challenge for farmers in selecting breeding stock for their businesses. Most of the market chain for breeding stock passes through middlemen, leading to high prices price rate upset IDR 20.000.000 for Heifer with middle quality not high quality that are disadvantageous for farmers. This issue is exacerbated by the fact that not all farmers have replacement stock, with many only keeping lactating cows. Government support in the form of facilities such as regular livestock markets and fairs can be implemented as a measure to provide high-quality breeding stock and shorten the lengthy market chain that harms farmers.

Feed is a crucial input in dairy farming. Most farmers do not face difficulties with forage during the rainy season; however, forage shortages become a problem during the dry season. Farmers incur higher costs for forage feed during the dry season [9]. Farmers need to be introduced to superior varieties of forage that have high productivity and nutrient content for cultivation. The utilization and processing of agricultural waste should also be optimized to address the forage shortage during the dry season.

The high rate of agricultural land conversion in Cigugur District into housing, tourist attractions, lodges, and restaurants has led to a reduction in available forage land for livestock. One factor contributing to the limited availability of forage for ruminants is the decreasing availability of land [10]. The socio-economic transformation of Cigugur District from an agricultural area to a tourist destination has also contributed to the reduction of forage land. The demand for land continues to rise with the increasing population and economic development [11]. The trend of land conversion for various purposes [12], including the shift from agricultural to non-agricultural use [13], has become a major issue. Policies on land management reform based on agrarian reform in highland areas involving local communities, private investors, and the government as policymakers are needed to ensure development in the region does not disadvantage any party. Agrarian reform is expected to create an agro-tourism area that fosters collaboration between the tourism sector and local agriculture and livestock sectors.

Another finding relates to the provision of business capital. Limited capital remains a major obstacle for most smallholder dairy farms. Based on field observations and interviews, there was a significant change in access to capital before and after the Foot-and-Mouth Disease (FMD) outbreak. Before the outbreak, farmers had relatively easy access to loans from cooperatives and banks. The high level of trust that financial institutions had in dairy farmers made borrowing easier, in part because this sector generates regular income through milk sales, which can be verified through monthly sales receipts from the cooperative. Trust is a key factor that has a significant positive impact on dairy farmers' loyalty [14]. After the FMD outbreak, however, farmers faced difficulties in borrowing capital due to business losses and their inability to repay previous loans. Not all farmers have the capacity to access these funding sources [15]. Enhancing the capacity of farmer groups to establish partnerships with various stakeholders, including investors who can support the provision of production facilities and capital for farmers, is essential. Additionally, low-interest credit access through the government's People's Business Credit (PBC) remains an alternative source of capital for farmers.

This study also found a lack of awareness among farmers regarding the waste produced by their farms, as well as insufficient performance in terms of coordination, oversight, and communication between the government and the community. These factors have hindered the enforcement of regulations on sustainable farming systems and waste management in Cigugur District. The existing communal Wastewater Treatment Plant (IPAL) is insufficient to handle water and waste during the rainy season, causing overflows and contamination of residents' water channels. One of the main environmental issues commonly faced in farming areas is the odor and waste pollution, especially during the rainy season, as waste piles become breeding grounds for mosquitoes and disease sources [16]. Conflicts between community members can also arise due to dairy farm waste [17]. The development of an integrated waste processing center for compost and biogas production should be pursued as a long-term solution for farm waste management. Additionally, community education and empowerment are necessary for promoting crop-livestock integration systems in Kuningan Regency.

4. Conclusion

Several problems hinder the development of dairy farming in Cigugur District, Kuningan Regency, which need to be addressed appropriately to benefit farmers as key business actors in the future. Cooperation and coordination among various stakeholders, including businesses, the government, and the private sector, are essential for developing a sustainable dairy farming business in Cigugur District. The government needs to review regulations and policies related to regional development to ensure that no parties feel disadvantaged by these policies.

References

[1] Direktorat Jendral Peternakan dan Kesehatan Hewan. 2022. Statistik Peternakan dan Kesehatan Hewan 2022. Direktorat Jendral Peternakan dan Kesejahteraan Hewan, Jakarta

[2] Badan Pusat Statistik. 2022. Provinsi Jawa Barat Dalam Angka 2022. Badan Pusat Statistik, Bandung

[3] Badan Pusat Statistik. 2023. Kabupaten Kuningan Dalam Angka 2023. Badan Pusat Statistik, Kuningan

[4] Balai Penyuluhan Pertanian. 2023. Programa Kecamatan Kuningan 2023. Balai Penyuluhan Pertanian, Kuningan

[5] Matondang, R.H., dan S. Rusdiana. 2013, Langkah langkah strategis dalam mencapai swasembada daging sapi/kerbau 2014. Jurnal Litbang Pertanian. **32** (3): 131-139

[6] Sudrajad, P. dan Adiarto. 2011. Pengaruh stress panas terhadap performa produksi susu sapi frisien Holstein di balai besar pembibitan ternak unggul sapi perah baturaden. Prosiding Seminar Nasional Teknologi Peternakan dan Veteriner 2011, Bogor. Hal: 341-346

[7] Siregar. S.B. dan Kusnadi. Peluang usaha sapi perah di daerah dataran rendah Kabupaten Cirebon. Media Peternakan. **27** (2): 77-87

[8] Sondakh, I.Th.P., J.A. Malingkas., J. Lainawa., G.D. Lenzun. 2019. Analisis kinerja penyuluh terhadap pemberdayaan kelompok usaha peternakan sapi di Desa Tondegesan Kecamatan Kawangkoan Kabupaten Minahasa. Zootec. **39**(1): 101-111

[9] Rahayu. E.T. 2013. Analisis pendapatan usaha ternak sapi perah di Kecamatan Cepogo Kabupaten Boyolali. Sains Peternakan. **11**(2). 99-105

[10] Zailzar, L., Sujono., Suyanto., Yani, A. 2011. Peningkatan kualitas dan ketersediaan pakan untuk mengatasi kesulitan di musim kemarau pada kelompok peternak sapi perah. Jurnal Dedikasi. **8**: 16-28

[11] Al-Vatia, T.V dan P.N. Djojomartono. 2019. Analysis of the effect of land use planning and land value in Gamping Subsdistrict, Sleman, D.I Yogyakarta from 2013 to 2018. J of Geospatial Information Sci and Engineering. **2**(2): 245-254

[12] Adiyaksa, F. dan P.N. Djojomartono. 2020. Evaluation of land use change of agricultural land into industrial area in Kendal Regency in period 2014-2018. J of Geospatial Information Sci and Enginering. **3**(1): 71-78

[13] Syafitri, A.K.N dan P.B Santosa. 2020. Spatial analysis of Kulon Progo District development from 2007-2030 with cellular automata markov model. Proseding The 1st International Conference on Geodesy, Geomatics and Land Administration: Knowledge E. 269-277

[14] Susanty, A., A. Bakhtiar, N.B. Puspitasari, dan N. Susanto. 2019. The performance of dairy supply chain in Indonesia: a system dynamics approach. International Journal of Productivity and Performance Management. **68** (6): 1141–1163.

[15] Satiati, E., S. Andarwati., T.A. Kusumastuti. 2022. The role of women in livestock farmers' Groups in Samiran, Boyolali, Central Java. The Journal of Social Sciences and Humanities. **12**(1): 78-98

[16] Santosa, A. Sumari., S. Marfuah., Muntholib., R. Retnosari. 2020. Pemanfaatan limbah sapi perah untuk biogas sebagai energi terbarukan pada kelompok peternak. Jurnal Graha Pengabdian. **2**(2): 144-123

[17] Indri A, S. Marina., M.M. Ali. 2015. Persepsi masyarakat terhadap manfaat dan dampak negative limbah peternakan sapi perah (kasus di Desa Rancamulya Kecamatan Sumedang Utara Kabupaten Sumedang. Students E-journal. **4**(3).

IOP Conf. Series: Earth and Environmental Science **1460** (2025) 012006 doi:10.1088/1755-1315/1460/1/012006

Sensitivity Analysis of Biodiesel Production from Crude Palm Oil (CPO) in Indonesia: Evaluation of CPO and Biodiesel Prices in 2024

Jabosar Ronggur Hamonangan Panjaitan*

Chemical Engineering Program, Institut Teknologi Sumatera, Lampung 35365, Indonesia

*E-mail: jabosar.panjaitan@tk.itera.ac.id

Abstract

Indonesia, as a palm oil producer in the world, can produce biodiesel using palm oil as raw material. Biodiesel production is increasing every year due to the need for alternative fuels to replace fossil fuels. In this study, biodiesel production in Indonesia from Crude Palm Oil (CPO) as raw material was studied and its sensitivity to the influence of raw material and biodiesel prices was analyzed. Based on research, it was known that high CPO prices and low biodiesel prices result in biodiesel production factories experiencing losses. The increase in CPO prices in Indonesia can be caused by several factors, including climate change which disrupts CPO productivity, increased demand for CPO, increased production of other vegetable oils, Indonesian government policies towards CPO, global geopolitical influences and economic changes. Apart from that, the increase in biodiesel prices can be caused by several factors, including the increase in CPO prices, increased demand for biodiesel, and the Indonesian government's policy regarding B35. The profits of a palm oil biodiesel factory depend on CPO price. The increase in biodiesel prices as a result of B35 policy requires support such as evaluating market acceptance of biodiesel and adequate facilities to support B35 implementation.

Keywords: biodiesel, crude palm oil, sensitivity

1. Introduction

Indonesia as the largest palm oil producer in the world can produced 45 million tons of palm oil with a total plantation area of 14 million hectares in 2020 [1]. Therefore, Indonesia has the potential to utilize palm oil biorefineries that can convert palm oil biomass into bioelectricity, biodiesel and bioethanol [2].

The main product from oil palm plantations is crude palm oil (CPO). In its designation, some CPO was exported while the rest was consumed domestically to produce various CPO derivative products such as cooking oil and margarine. In energy sector, CPO was widely applied to produce biodiesel to maintain energy needs and reduce CO_2 emissions [3].

Diesel from fossil fuels can be replaced with biodiesel. Biodiesel comes from fat which is a mono alkyl ester with a carbon chain of C12 to C24 [4]. The commonly used biodiesel production method is transesterification [5]. Several advantages of biodiesel such as non-toxic, can reduce emissions, biodegradable and sulfur-free are attractive for the widespread use of biodiesel. As for the cetane value, biodiesel has a higher value than petroleum diesel [6].

The feasibility of biodiesel production can be determined by conducting an economic evaluation. An important factor that is a source of costs in the factory is raw materials. Biodiesel production using various alternative raw materials and evaluating their economics has also been studied [7] [8] [9]. In addition, the use of catalysts also affects the cost of biodiesel production such as biochar [10] and hydroxyapatite [11] catalysts. Several studies have also been conducted to evaluate biodiesel production in several countries such as Iran [12], Hong Kong [13], Egypt [14], and Indonesia [15].

Economic sensitivity analysis of biodiesel production is part of biodiesel economic analysis which showed the effect of changes in certain price parameters in biodiesel production. Economic sensitivity analysis research on biodiesel production had been conducted previously. Tang et al (2015) studied life cycle cost and unit cost in biodiesel production. Based on their research, it was found that raw material prices and interest rates affect economic analysis [16]. Ong et al (2012) studied life cycle cost and payback period for biodiesel production in Malaysia. Based on their research, it was known that CPO price dominates biodiesel production price so that changes in raw material price can affect biodiesel price [17]. Therefore, this study examined the sensitivity analysis of biodiesel production in Indonesia with CPO as raw material which parameters to be tested for sensitivity were CPO price and biodiesel price in 2024.

2. Method

2.1 Biodiesel economic evaluation

The economic evaluation of biodiesel production in this study refers to Panjaitan et al (2023) [18]. Based on Panjaitan et al (2023), biodiesel was produced using CPO with two reactors then purified by washing, sedimentation, and evaporation according to Figure 1 [18]. The results of the economic evaluation and profit analysis can be seen in Table 1 and Table 2.

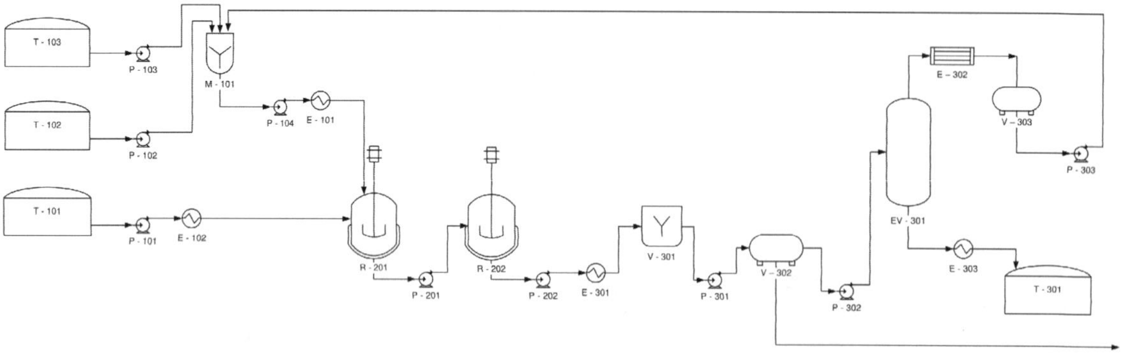

Figure 1. Process flow diagram [18]

Table 1. Economic evaluation [18]

No.	Economic Parameters			Total Price
1	Total Capital Investment			12,635,223
2	Total Production Cost			50,367,616
3	Raw Material Cost			41,756,987
	Raw Materials	Demand (Ton/Year)	Cost (US$/Ton)	
	CPO	68,462	600	
	Methanol	971	300	
	Sodium Hydroxide	971	400	
4	Total Revenue			57,801,714
	Products	Production (Ton/Year)	Cost (US$/Ton)	
	Biodiesel	68,000	788	
	Glycerol	7,030	600	

Table 2. Profit analysis [18]

Parameters	Value
Return on investment (ROI)	31.94%
Payback period (PBP)	2.35 years
Net present value (NPV)	US$ 22,237,387
Internal rate of return (IRR)	33.5%

2.2 Sensitivity Analysis Calculation

Determination of deviation in sensitivity analysis refers to the maximum price of CPO and biodiesel. Based on Indonesia ministry of trade in January 2024, there was an increase in CPO price around US$ 774.93/ton [19]. Meanwhile, the selling price of biodiesel based on Indonesia ministry of energy and mineral resources showed the selling value of biodiesel for January 1, 2024 was IDR 10,896/liter or US$ 1912/ton [20]. Based on the information, if referring to the price of CPO and biodiesel in the study of Panjaitan et al (2023) [18], the deviation of CPO and biodiesel price was greater than 30%. Therefore, in this study, the deviation value was determined at -60%, -30%, +30%, and +60%. The deviation at 0% was US$ 600/ton and US$ 788/ton for CPO and biodiesel price according to Panjaitan et al (2023) [18]. The magnitude of this deviation was used to determine CPO and biodiesel price used in the sensitivity analysis according to Table 3.

Table 3. CPO and biodiesel price deviation

Deviation	CPO Price (US$/Ton)	Biodiesel Price (US$/Ton)
+60%	960	1260.4
+30%	780	1024.4
0%	600	788
-30%	420	551.6
-60%	240	315.2

The change in deviation value was used in two sensitivity parameters, namely CPO and biodiesel price to obtain economic profitability values such as NPV, IRR, PBP and ROI.

3. Results and Discussion

3.1 Sensitivity analysis

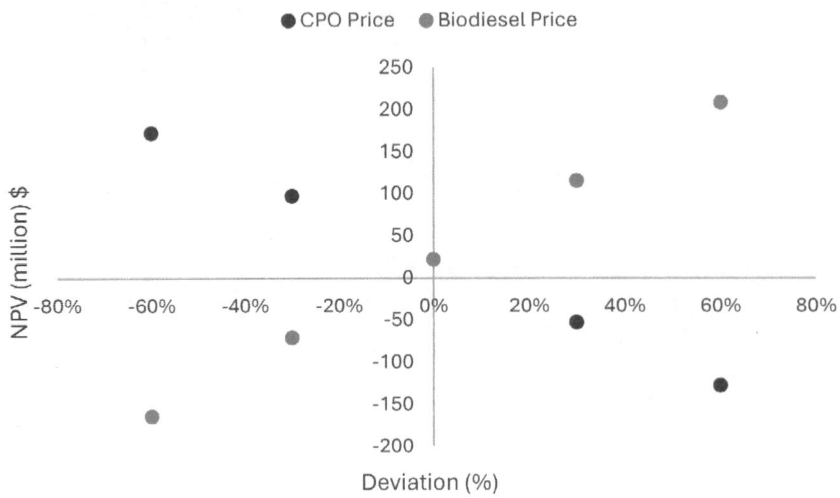

Figure 2. Sensitivity analysis of NPV

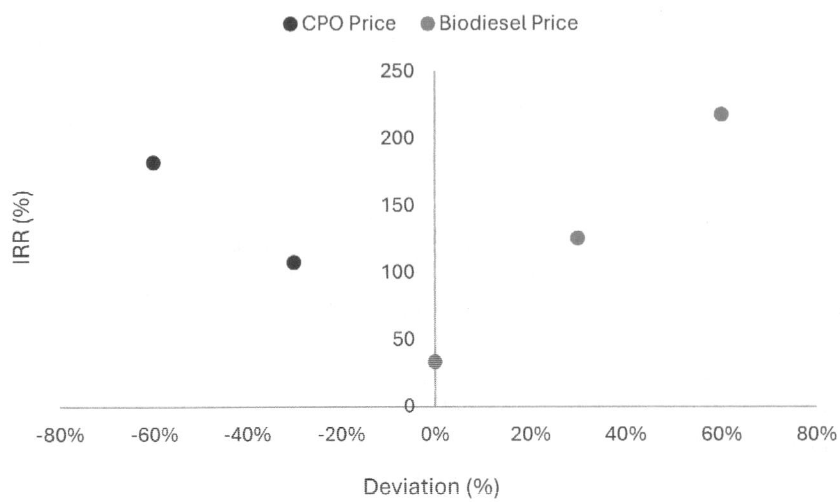

Figure 3. Sensitivity analysis of IRR

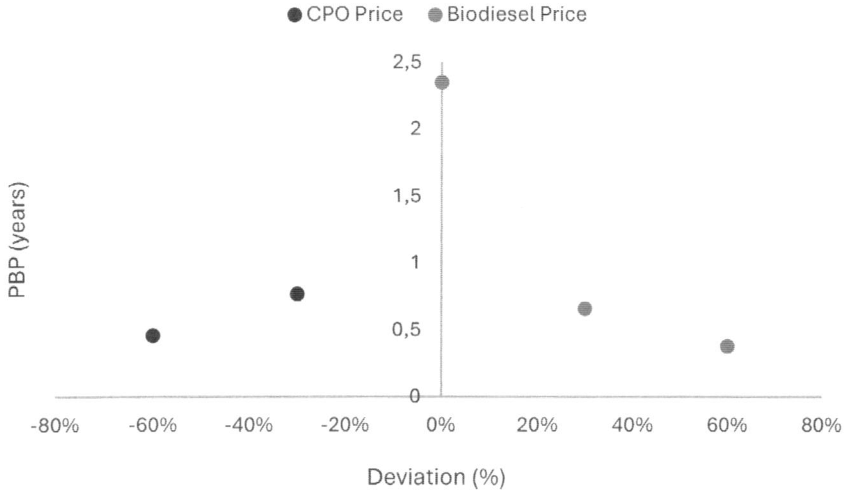

Figure 4. Sensitivity analysis of PBP

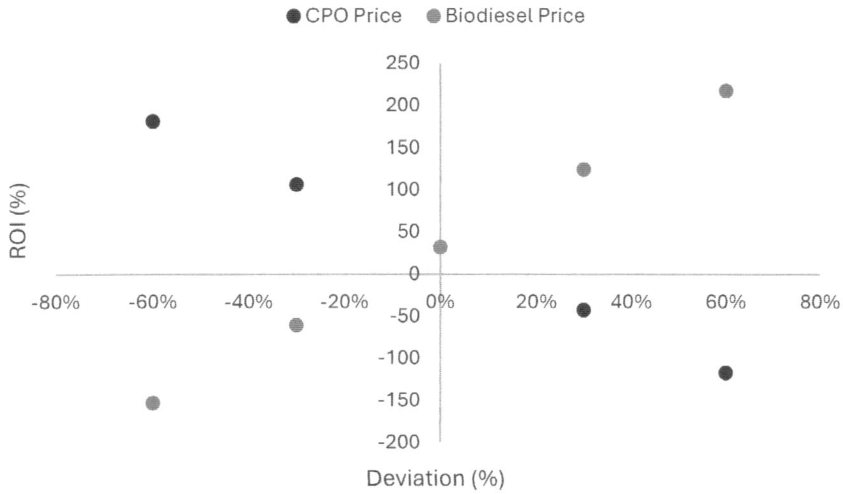

Figure 5. Sensitivity analysis of ROI

NPV, IRR, PBP and ROI values based on the sensitivity analysis calculation can be seen in Figures 2 to 5. Based on Figures 2 to 5, it can be seen that changes in CPO and biodiesel prices greatly affect the NPV, IRR, PBP and ROI values. In addition, large deviation values in this research had major impact on profitability value. CPO dan biodiesel price at deviations of +60% and +30% resulted in significant losses to the factory so that the IRR and PBP values cannot be evaluated.

Based on the results, the increase in CPO prices and the decrease in biodiesel selling prices had caused significant losses to the factory and made the factory unfit to be built. This was happened because the prices of CPO and biodiesel were the highest prices in raw materials and products. According to Table 1, CPO had the highest price compared to other raw materials, namely US$ 600/ton. While biodiesel in Table 1 was the main product with a higher price than glycerol, namely US$ 788/ton.

3.2 CPO Price Increase in 2024

Crude Palm Oil (CPO) is the main raw material in biodiesel production. The increase in CPO prices can be caused by several reasons such as:

1. The existence of El Nino or La Nina climate change which disrupts CPO productivity [21] [22].
2. Increased demand and limited CPO stock [22].
3. Increased production of other commodity vegetable oils (sunflower seed oil, rapeseed, soybean oil) [22].
4. Indonesian government policy on CPO (B35 policy, monetary policy, Indonesian export policy, and changes in biodiesel policy) [22] [23].
5. The influence of global geopolitics and economic changes [22][24].

Aprina (2014) stated that Indonesia as the largest CPO producer in the world can affect world CPO prices [25]. In January 2024, there was an increase in CPO prices to the value of CPO around US$ 774.93 per ton. Several factors caused the CPO price to increase in January 2024 such as an increase in world crude oil prices, an increase in the price of soybean oil, and tightening of palm oil supplies from Malaysia. In addition, Indonesia also reduced palm oil exports in 2024 due to stagnation production in Indonesia, increasing domestic demand for palm oil (especially for the production of oleochemical products), and B35 fuel policy [19].

3.3 Biodiesel price increase in 2024

The production and application of biodiesel in Indonesia was a step to accelerate the use of clean and sustainable energy in Indonesia. Determining the biodiesel fuel mixture was one way to reduce greenhouse gas emissions [26]. The increase in biodiesel prices can be caused by several reasons, including:

1. Increase in Crude Palm Oil (CPO) price.
2. Increased demand for biodiesel.
3. The Indonesian government's policy in producing B35 fuel according to the Decree of the Minister of Energy and Mineral Resources Number 208.K/EK.05/DJE/2022 dated December 28, 2022 concerning the implementation of B35 [27][28]. B35 triggers an increase in biodiesel consumption [29].

The B35 fuel policy has an impact on CPO, including reduced export value of palm oil in 2023, reduced sales, and increased domestic consumption [30]. On the other hand, research by Sahara et al (2022) shows that the mixed fuel policy can increase the selling value of the chemical product sector related to palm oil such as cleaning agents, synthetic resins and cosmetics [31].

Based on the website of the Ministry of Energy and Mineral Resources, the selling value of biodiesel in January 1, 2024 was IDR 10,896/liter [20]. The price of biodiesel was still quite high. One way that can be used to reduce the price of biodiesel was to use the biorefinery concept in a biodiesel production plant. Wirawan et al (2024) research showed that the economic value of biodiesel can be increased in several ways, including finding sources of oil other than palm oil so that it does not have to compete with food industry, finding technology that can produce cheaper and better quality biodiesel, and doing research about the use of biodiesel on environmental impacts [32]. Harahap et al (2018) stated that the concept of an integrated biorefinery for the palm oil biodiesel industry can simultaneously produce biodiesel, electricity, heat, and biofertilizers [33]. These additional products can increase profits compared to only producing biodiesel. On the other hand, although the B35 fuel implementation policy had been issued, the implementation of B35 had not been implemented comprehensively with various obstacles such as the existence of B30 stock and limited facilities or infrastructure [27]. In addition, although biofuel is environmentally friendly, the escalation of oil palm planting must still consider deforestation and loss of biodiversity [34].

4. Conclusion

This research examined the sensitivity analysis of biodiesel production in Indonesia from Crude Palm Oil (CPO) by evaluating the influence of CPO and biodiesel prices. Based on the research that had been done, it was known that higher CPO price and the lower biodiesel price made factory financial loss. The increase in CPO price in Indonesia can be caused by several factors, including climate change that disrupts CPO productivity, increased demand for CPO, increased production of other commodity vegetable oils, Indonesian government policies on CPO, global geopolitical influences and economic changes. In addition, the increase in biodiesel price can be caused by several factors, including the increase in CPO prices, increased demand for biodiesel, and Indonesian government policies regarding B35. The increase in CPO prices was an important parameter for determining the profit of palm oil biodiesel factory due to the higher raw material price. On the other hand, the B35 fuel policy that increases the price of biodiesel required some support such as an evaluation of market acceptance about biodiesel and adequate facilities to support the B35 fuel implementation.

References

[1] Mayandi and Suharjito, "Palm oil-Based Biodiesel Industry Sustainability Model using Dynamic Systems to Balance Food, Energy, and Export Allocations," *Smart Agric. Technol.*, vol. 7, no. February, p. 100421, 2024, doi: 10.1016/j.atech.2024.100421.

[2] F. Harahap, S. Leduc, S. Mesfun, D. Khatiwada, and F. Kraxner, "Meeting the bioenergy targets from palm oil based biorefineries : An optimal configuration in Indonesia," *Appl. Energy*, vol. 278, no. May, p. 115749, 2020, doi: 10.1016/j.apenergy.2020.115749.

[3] R. Nabila *et al.*, "Oil palm biomass in Indonesia : Thermochemical upgrading and its utilization," *Renew. Sustain. Energy Rev.*, vol. 176, no. January, p. 113193, 2023, doi: 10.1016/j.rser.2023.113193.

[4] I. A. Pratama, I. Kurniaty, U. H. Hasyim, and G. Fitriyano, "Review: Pemanfaatan Biji Nyamplung (Calophyllym Inophyllum) Sebagai Bahan Baku Biodiesel Berdasarkan Proses Produksi dan Penambahan Katalis," *J. Konversi*, vol. 10, no. 01, 2021.

[5] E. S. Aktaş, Ö. Demİr, and D. Uçar, "A review of the biodiesel sources and production methods," vol. 5, no. 1, pp. 1–10, 2020.

[6] L. Aisyah and C. S. Wibowo, "A Review of Biodiesel Development in Indonesia: Current Status, Future Potential and Its Impact on The Environment," vol. 2010, no. June 2010, pp. 177–188, 2025.

[7] T. Tipsri and V. Lersbamrungsuk, "Techno-economic analysis of biodiesel production from waste cooking oil using supercritical and subcritical processes," vol. 13, pp. 153–162, 2019.

[8] A. Mohammadi and B. Ghobadian, "Iranian Journal of Chemical Engineering Techno-Economic Optimization of a Continuous Process for the Biodiesel Production from Waste Vegetable Oils," vol. 18, no. 4, pp. 20–35, 2022, doi: 10.22034/ijche.2022.318805.1419.

[9] E. G. Al-sakkari, M. G. Mohammed, A. A. Elozeiri, I. Ismail, and I. Ashour, "Comparative Technoeconomic Analysis of Using Waste and Virgin Cooking Oils for Biodiesel Production Summary of Process Designs," vol. 8, no. December, pp. 1–13, 2020, doi: 10.3389/fenrg.2020.583357.

[10] J. Lee, B. Lee, Y. Sik, and H. Lim, "Preliminary techno-economic analysis of biodiesel production over solid- biochar," *Bioresour. Technol.*, vol. 306, no. January, p. 123086, 2020, doi: 10.1016/j.biortech.2020.123086.

[11] R. V. Bôas, "Techno-economic evaluation of biodiesel production using by-product as raw material and hydrotalcite-hydroxyapatite as catalyst," vol. 2022, pp. 1–18, 2022.

[12] B. Ruhani, P. Movahedi, S. Saadi, A. Ghasemi, A. Kheradmand, and M. Dibaj, *Comprehensive Techno-Economic Analysis of a Multi-Feedstock Biorefinery Plant in Oil-Rich Country : A Case Study of Iran*. 2022.

[13] S. K. Karmee, R. D. Patria, C. Sze, and K. Lin, "Techno-Economic Evaluation of Biodiesel Production from Waste Cooking Oil — A Case Study of Hong Kong," pp. 4362–4371, 2015, doi: 10.3390/ijms16034362.

[14] H. I. El Shimi, A. S. Fawzy, N. K. Attia, G. I. El Diwani, and S. T. El Sheltawy, "Evaluation of Biodiesel Production from Spent Cooking Oils : A Techno-Economic Case Study of Egypt," vol. 11, no. 17, pp. 10280–10290, 2016.
[15] M. Y. Abduh, R. Manurung, and H. J. Heeres, "Techno-Economic Analysis for Small Scale Production of Rubber Seed Oil and Biodiesel in Palangkaraya , Indonesia," vol. 5, no. 4, pp. 1–6, 2017, doi: 10.18178/jocet.2017.5.4.381.
[16] Z. Tang, L. Zhenzhou, L. Zhiwen, and X. Ningcong, "Uncertainty analysis and global sensitivity analysis of techno-economic assessments for biodiesel production," *Bioresour. Technol.*, vol. 175, pp. 502–508, 2015, doi: 10.1016/j.biortech.2014.10.162.
[17] H. C. Ong, T. M. I. Mahlia, H. H. Masjuki, and D. Honnery, "Life cycle cost and sensitivity analysis of palm biodiesel production," *Fuel*, vol. 98, pp. 131–139, 2012, doi: 10.1016/j.fuel.2012.03.031.
[18] J. R. H. Panjaitan, D. F. Nury, V. V. Suswanto, and L. D. Putri, "Economic evaluation of biodiesel plant design in Economic evaluation of biodiesel plant design in Bontang , East Kalimantan , Indonesia," in *6th International Symposium on Sustainable Urban Development 2023*, 2023. doi: 10.1088/1755-1315/1263/1/012045.
[19] Kementrian Perdagangan RI, "Harga Referensi CPO Januari 2024 Menguat, Pungutan Ekspor Dipatok US$75 per Ton." [Online]. Available: https://www.kemendag.go.id/berita/pojok-media/
[20] ESDM, "Harga Indeks Pasar (HIP) Bahan Bakar Nabati (BBN) Jenis Biodiesel Bulan Januari 2024." [Online]. Available: https://ebtke.esdm.go.id/post/2024/01/05/3698/
[21] CNBCa, "Konglomerat RI ini Bakal Ketiban Berkah dari Sawit." [Online]. Available: https://www.cnbcindonesia.com/market/20231202002804-17-493938/
[22] CNBCb, "Harga CPO Bisa Tembus US$ 1.000, RI Siap-Siap Pesta Pora?" [Online]. Available: https://www.cnbcindonesia.com/market/20231104061141-17-486329/
[23] G. Chandrarin, K. Sohag, D. S. Cahyaningsih, D. Yuniawan, and H. Herdhayinta, "The response of exchange rate to coal price, palm oil price, and inflation in Indonesia: Tail dependence analysis," *Resour. Policy*, vol. 77, 2022.
[24] Sawit Indonesia, "Dorab Mistry: Biodiesel Akan Dongkrak Harga CPO RM 4.500/Ton." [Online]. Available: https://sawitindonesia.com/
[25] H. Aprina, "The Impact of Crude Palm Oil Price on Rupiah's Rate," *Bull. Monet. Econ. Bank.*, vol. 16, no. 4, pp. 295–314, 2014.
[26] Indonesia Palm Oil, "Fact Sheet: Indonesian Biodiesel And Palm Oil." [Online]. Available: https://www.indonesiapalmoilfacts.com/
[27] Liputan6, "Harga Indeks Pasar Biodiesel Turun Jadi Rp 10.234 per Liter di Juni 2023." [Online]. Available: https://www.liputan6.com/bisnis/read/5315118/
[28] Indonesia Palm Oil, "Prospects For Indonesian Biodiesel In 2024 And Beyond." [Online]. Available: https://www.indonesiapalmoilfacts.com/
[29] Ofimagazine, "Indonesia's new B35 blending rate expected to raise biodiesel consumption by 25% to 13bn litres." [Online]. Available: https://www.ofimagazine.com/news/
[30] J. and Chu, "Indonesia's biodiesel policy, dry weather to keep palm oil prices elevated." [Online]. Available: https://www.reuters.com/world/asia-pacific/
[31] Sahara., A. Dermawan, S. Amaliah, T. Irawan, and S. Dilla, "Economic impacts of biodiesel policy in Indonesia : a computable general equilibrium approach," *J. Econ. Struct.*, 2022, doi: 10.1186/s40008-022-00281-9.
[32] S. S. Wirawan, M. D. Solikhah, H. Setiapraja, and A. Sugiyono, "Biodiesel implementation in Indonesia : Experiences and future perspectives," *Renew. Sustain. Energy Rev.*, vol. 189, no. PA, p. 113911, 2024, doi: 10.1016/j.rser.2023.113911.
[33] F. Harahap, S. Silveira, and D. Khatiwada, "Cost competitiveness of palm oil biodiesel production in Indonesia," *Energy*, 2018, doi: 10.1016/j.energy.2018.12.115.
[34] J. Sen, "Indonesia Eyes a Biodiesel Consumption Figure of 13.4 Million Cubic Meters in 2024." [Online]. Available: https://www.chemanalyst.com/NewsAndDeals/NewsDetails/

IOP Conf. Series: Earth and Environmental Science **1460** (2025) 012007 doi:10.1088/1755-1315/1460/1/012007

Income determination of rice traders and how they perceived the rice import policy in Indonesia: an economic and ecological impacts

Fadhil Adi Nugraha[1*] and Ernoiz Antriyandarti[2]

[1]Study Program of Agribusiness, Universitas Muhammadiyah Purwokerto, Purwokerto, Indonesia
[2]Study Program of Agribusiness, Universitas Sebelas Maret, Surakarta, Indonesia

*E-mail: fadhiladi@ump.ac.id

Abstract. Rice is the principal source of carbohydrates for most people in Indonesia. Over time, the government implemented some measures, including importing rice, to meet the population's daily need for rice. Traders have a vital role in the marketing chain before the community finally purchases and consumes local and imported rice. The welfare of rice traders must be considered, which becomes the basis for researching the factors determining the traders' incomes. Besides, this study aimed to observe the traders' perception of the rice import policy. The sampling method was conveniently executed, utilizing a sample of forty rice traders from four regions on Indonesia's Java Island. Interviews were carried out using a questionnaire to collect the data. The results show that the respondent's age and rice stock influenced the trader's income. Most rice traders agreed that the import policy increased rice availability, decreased rice prices, and reduced competition among traders. Although most traders confirmed that imports were still needed, the government must control and oversee the execution to maximize economic benefits and reduce negative environmental impacts.

Keywords: import policy, income determination, perception, rice trader.

1. Introduction

As one of the top global rice consumers, Indonesia made rice a strategic commodity because it contributes significant income to the country's economy and employs many people [1]. Rice must also be accessible and large enough to meet the demand [2]. Since rice is a common source of carbohydrates that Indonesians eat before engaging in daily activities, such as working to support themselves, rice impacts the Indonesian people's welfare [3]. In addition to being a source of energy, rice may also help reduce the risk of human diseases like diabetes and heart disease [4].

An inelastic demand characterizes rice as a commodity. Demand is steady despite price swings because individuals must eat a certain amount of rice daily [5]. However, due to its tendency to rise in price, consumers (especially those with low incomes) may find rice costly if it is scarce [6]. Climate change and population growth can negatively affect the domestic rice supply; therefore, the Indonesian government imported rice to prevent that problem [7,8]. Indonesia imports rice from countries like India, Thailand, and Vietnam (Statistics Indonesia, 2024). The

IOP Conf. Series: Earth and Environmental Science **1460** (2025) 012007 doi:10.1088/1755-1315/1460/1/012007

government's efforts to provide rice as a nutritious and safe food for all residents align with Sustainable Development Goal (SDG) Number 2: Zero Hunger [9].

The provision of local or imported rice involves various groups, including rice traders. Sujatmiko [10] defines a trader or seller as someone who engages in commerce by purchasing and reselling items that he does not manufacture to turn a profit. Meanwhile, Mariyono et al. [11] found that traders play a vital role in creating high marketing channels. However, many traditional traders in Indonesia depend on the rice supply in the market for their livelihood. The amount of rice that can be sold to customers and the earnings will undoubtedly fluctuate if the rice stock is low. Importing rice from overseas will surely boost the availability of goods and services, ultimately supporting SDG Number 8: Decent Work and Economic Growth [12].

The importance of the role of rice traders in Indonesia's economic activities, especially after the implementation of the rice import policy, is the basis for this research. Similar studies have been conducted previously. For example, Sondakh et al. [13] inferred that capital, price, stock, and output significantly influence rice traders' income when price fluctuations occur in Manado City, Indonesia. Meanwhile, in Ebonyi State in Nigeria, the performance of local rice entrepreneurs was influenced by factors such as age, sex, farm gate price, marital status, transportation and marketing costs, household size, distance to market, and volume size [14].

Lançon et al. [15] concluded that traders in Nigeria's urban centers need to consider the price and availability of rice before importing rice because they need to maintain good relations with their customers. In the meantime, traders in Sierra Leone prefer imported rice over local rice because it is more profitable and favored by consumers [16]. It must be noted that importing countries are experiencing environmental problems because the price of foodstuffs on the market has become cheaper, and local farmers have decided to change the commodities they plant, which require more fertilizer and water [17]. However, the amount of previous research is still limited, and none have specifically discussed Indonesian traders' perspectives on the economic and ecological aspects of rice imports.

Java was chosen for the study site because this island is the center of Indonesia's economic activity. With 145 million people living there [18], there is a significant demand for rice, which is being met by selling local and imported rice. This study noted characteristics related to age, education, business experience, stock, and price, along with the hypothesis that these factors affect the income of rice merchants. Rice traders' opinions about the economic and ecological impacts of the import policy for rice were also evaluated to understand the rice market's state. The results of this study should help develop policies by the Indonesian government to boost the income of rice traders and control the distribution of foreign rice per local demands on Java Island. Additionally, these policies must consider the environmental implications of rice importation, such as carbon emissions from transportation and the potential ecological benefits of local rice production. The study's findings will also benefit future research on Indonesian rice traders' income, import policy, and the environmental sustainability of these practices.

2. Methods

The primary methodologies used in this study are descriptive and analytical methods. This study employs a practical sampling technique by selecting random respondents who were in a location relevant to the research subject [19]. The research respondents numbered forty people on Java Island, with details of ten rice traders of the Special Region of Yogyakarta Province (five from Sleman District and five from Yogyakarta City), ten rice traders of Central Java Province (five from Grobogan District and five from Solo City), ten rice traders of West Java Province (five from

Bandung City and five from Indramayu District), and ten rice traders of East Java Province (five from Surabaya City and five from Jember District). These provinces are all located on the island of Java, where rice is a staple food for the vast majority of people.

The hypothesis formed from previous research is that age [14], [20], [21], education [22], [23], business experience [22], [23], rice stock [13], [24], and rice price [13], [14] impacted the income of Indonesian rice traders. The regression model is formulated as below:

$$Y = C + D_1X_1 + D_2X_2 + D_3X_3 + D_4X_4 + D_5X_5 + E$$

Where Y is the trader's income (IDR/month), C is the constant, X1 is the trader's age (years), X2 is the trader's education (years), X3 is the trader's business experience (years), X4 is rice stock (kg/month), X5 is rice price (IDR/kg), D1–D5 are regression coefficients, and E is the standard error.

3. Result and Discussion

3.1 Summary of Descriptive Statistics

A summary of the statistical measures used to collect the data for the study is shown in Table 1. The average income of rice traders on Java Island was IDR 4,026,500/month. The trader's average age was 47.5 years, which means they were still in productive age. They had an average of 10.9 years of educational background (high school level). They have already run their business for an average of 13.6 years. Traders stocked the rice at an average of 8,333.4 kg/month while selling the rice at an average price of IDR 10,035/kg. The types of rice sold were varied, but primarily domesticated rice, such as C4, IR 64, Mentik Wangi, and Rojo Lele. Rice was sold in bulk or sack packages ranging from 2 kg to 50 kg. The rice sacks used were made of plastic, so they were not environmentally friendly because they did not quickly decompose in nature. The locations traders chose to sell rice in this study were traditional markets, kiosks, and private stalls.

Table 1. Descriptive Statistics of Rice Traders

Variable	Description	Unit	Min	Max	Mean
Y	Income	IDR/month	50,000	25,000,000	4,026,500
X1	Age	Years	24	69	47.5
X2	Education	Years	6	16	10.9
X3	Business experience	Years	1	35	13.6
X4	Stock	Kg/month	25	150,000	8,333.4
X5	Price	IDR/kg	7,000	12,500	10,035

Source: Primary Data Processed, 2019.

3.2 Factors Determining the Rice Trader's Income

In this study, a validity and reliability test was performed to make sure the questionnaire worked. All the questionnaire's statements were valid, according to the results of the data analysis. Furthermore, the instrument that will be used is trustworthy. Next, Table 2 displays the multiple linear regression results to assist in determining the factors affecting the income of rice traders.

IOP Conf. Series: Earth and Environmental Science **1460** (2025) 012007
doi:10.1088/1755-1315/1460/1/012007

Table 2. Estimation Result of Income Determination

Variable	Coefficients	Standard Error
Constant	11,963,864.27^{ns}	9,910,491.378
Age (X1)	−211,016.387**	88,954.659
Education (X2)	−173,385.659^{ns}	315,003.191
Business experience (X3)	651,406.510^{ns}	93,488.186
Stock (X4)	75.734*	37,645
Price (X5)	264.002^{ns}	880.980
F value	2.031*	
R square	0.230	

Note: ** and * indicate significance at the 5% and 10%, respectively. Meanwhile, ^{ns} means not significant.
Source: Primary Data Processed, 2019.

The R Square value was 0.230, meaning that the trader's age, education, business experience, rice stock, and price influence 23.0% of rice traders' income in this study. Other factors outside the variables studied influence the remaining 77.0%. Although the proportion of the impact of the independent variables studied on the dependent variables is not large, it still needs to be considered so that traders' income can be maximized and in line with SDG Number 8, which promotes equitable, long-term, and sustainable economic growth, full-time, productive employment, and good work for everybody.

Separately, only the trader's age and rice stock significantly affected their monthly income. A year increase in the trader's age could reduce income by IDR 211,016.387/month. This study's findings are consistent with those of Yang et al. [20], who found a substantial inverse relationship between flexible workers' age and income. As people age, their health declines, especially in lower-income countries [21]. Rice traders with poor health certainly cannot maximize their potential in managing their business and making profits. The ability to adapt to sustainable practices and new environmental policies regarding the rice trade may also vary with age.

In the meantime, if the trader increases their rice stock by 1 kg/month, their monthly income could rise by IDR 75.734. The more rice stock they had, the more rice available for consumers to purchase. The source of this rice stock, whether locally produced or imported, has environmental implications. If grown sustainably, locally produced rice can reduce the carbon footprint and appeal to consumers who prioritize eco-friendly products. However, the trader's income increase is not much, and they must pay attention to their rice storage capacity. Traders also need to consider the durability of rice so that it does not rot while in storage or when sold to consumers; this is consistent with the findings of Lestari & Saidah [25], who claimed that consumers prefer rice with a one-month shelf life because they worry that the rice may go rotten if it is kept for an extended period before being consumed.

3.3 Traders' Perceptions of the Rice Import Policy

The importation of rice profoundly impacts the local economy and environment. While it can stabilize rice prices and ensure food security, it also has environmental implications. For example, the transportation of imported rice releases more greenhouse gases into the atmosphere than does the cultivation itself [26]. Therefore, understanding the perspective of rice traders on import policies is very important because traders are a crucial part of the rice marketing chain in Indonesia. Consumers will have difficulty buying rice to meet their daily food needs without traders. Environmental considerations add another layer of complexity to these perceptions.

Table 3. Traders' Perceptions of the Rice Import Policy's Impact

No.	Question	Responses (%)				
		VNI	NI	LI	I	VI
1.	Did the rice import policy impact the increasing market availability of rice?	0.00	12.50	7.50	60.00	20.00
2.	Did the rice import policy impact the decrease of rice prices?	0.00	20.00	2.50	42.50	35.00
3.	Did the rice import policy impact the decreasing competition of rice traders?	0.00	17.50	10.00	50.00	22.50

Note: VNI = Very Not Impactful, NI = Not Impactful, LI = Less Impactful, I = Impactful, VI = Very Impactful.
Source: Primary Data Processed, 2019.

Table 3 shows that most traders in this study argued that the rice import policy increased the rice availability in the market (60.00%), while 20.00% opined that the policy was very impactful. Before import, local rice was already sold in the market. The addition of imported rice increased the quantity of rice that was available for consumers to buy. However, this increased availability must be weighed against the environmental impact of importing rice, which includes carbon emissions from transportation and potential strain on local ecosystems due to increased competition for resources. Some traders thought that imports had less impact (7.50%) and no impact (12.50%) on rice availability. These traders perceived that the quantity of rice in the market was stable or unchanged.

Traders mostly said that the rice import policy impacted the decrease in rice prices in the market (42.50%). Next, 35.00% of traders argued it was very impactful. If rice availability in the market increases, the price will decrease. Consumers with lower economic conditions will have more substantial purchasing power to buy rice to meet their daily food needs. Other traders thought the policy was not impactful (20.00%) and less impactful (2.50%). They reasoned that they did not know the answer or that the rice price decrease was due to factors like season change or government law.

The rice import policy was considered impactful and very impactful in decreasing the competition among 50.00% and 22.50% of traders, respectively. The reason was that traders could sell other types of rice, namely imported rice, which could compete with local rice. In the meantime, 17.50% of traders opined that it was not impactful, and only 10.00% thought it was less impactful. They assumed that local and imported rice have different market segments, so there would be no competition.

Table 4. Traders' Perceptions of the Rice Import Implementation

No.	Question	Responses (%)				
		VI	I	LE	E	VE
1.	Does the government require rice imports?	9.00	34.75	18.50	36.75	0.00
2.	Does the government have to control how rice imports are implemented?	0.00	0.00	5.00	52.50	42.50
3.	Does the government have to oversee how the import policy for rice is being applied?	0.00	0.00	5.00	60.00	35.00

Notes: VI = Very Inessential, I = Inessential, LE = Less Essential, E = Essential, VE = Very Essential.
Source: Primary Data Processed, 2019.

The following evaluation will be predicated on how traders view Indonesia's import policy for rice (Table 4). Most vendors said that rice imports were required by the Indonesian government (36.75%). When the lean season occurs, local rice stocks can drop, and imports are necessary to keep rice available and satisfy customer demand. Some traders (34.75%) said that since Indonesia still has many rice fields and local farmers can produce enough rice to prevent hunger and meet people's carbohydrate needs, there is no need to import rice. The environmental benefits of supporting local rice production include reduced transportation emissions and enhanced sustainable agricultural practices. In the meantime, 18.50% and 9.00% of vendors contended that imports are less essential and very inessential, respectively.

Most respondents concurred that import regulations for rice are essential (52.50%) and very essential (42.50%) for several reasons, including preventing fraud, managing the amount of rice in the market, and safeguarding local farmers. According to Suasih & Yasa [27], most rice farmers in Indonesia are considered small farmers with low incomes. Farmers may experience financial losses and perhaps become impoverished if their domestic rice is not as competitive as imported rice and is not sold well in the market. The remaining traders, unsure of the impact of restricting rice imports, believed it was less necessary (5.00%).

When asked if they thought it was necessary to monitor the import policy for rice, the majority of vendors said that it was (60.00%), followed by "very essential" (35.00%). They maintained that during import, oversight is necessary to stop fraud (corruption and collusion) and the smuggling of other items. Routine monitoring also attempts to keep the price of rice reasonable and ensure that environmental regulations are upheld, minimizing the ecological footprint of imported rice. Meanwhile, according to 5.00% of vendors, overseeing the import policy was less essential because there would not be any significant changes.

4. Conclusion

Based on the study's results, it can be concluded that the trader's age and rice stock partially influence the trader's income; meanwhile, the trader's education, business experience, and rice price have no significant influence. Most rice traders agreed that the import policy is impactful in increasing the rice's availability, decreasing rice prices, and reducing the trader's competition in the market. Regarding the implementation of the import policy, most of them agreed that imports are still required, but the Indonesian government must control and oversee the execution. Besides, it is essential to consider the environmental implications of these policies. By doing so, Indonesia can ensure a resilient rice market supporting economic growth and ecological conservation, benefiting consumers, traders, and the broader ecosystem.

Acknowledgement

The authors thank Universitas Sebelas Maret for funding the research.

References

[1] B. Arifin, N. Nuryartono, S. H. Pasaribu, F. Yasmin, M. A. Rifai, and R. Kurniadi, "Profitability and Labor Productivity in Indonesian Agriculture," Washington, DC, USA, 2019.

[2] E. Siswanto, B. M. Sinaga, and H. Harianto, "The Impact of Rice Policy on Rice Market and the Welfare of Rice Producers and Consumers in Indonesia," *J. Ilmu Pertan. Indones.*, vol. 23, no. 2, pp. 93–100, 2018, doi: 10.18343/jipi.23.2.93.

[3] D. Septiadi, "Dampak Kebijakan Perberasan terhadap Pengentasan Kemiskinan di Indonesia," Institut Pertanian Bogor, 2016.

[4] N. K. Fukagawa and L. H. Ziska, "Rice: Importance for Global Nutrition," *J. Nutr. Sci. Vitaminol. (Tokyo).*, vol. 65, pp. S2–S3, 2019, doi: 10.3177/jnsv.65.S2.

[5] W. Tarigan, Z. Lubis, and Z. Zein, "Analisis Permintaan dan Penawaran Beras di Provinsi Sumatera Utara," *Agrica (Jurnal Agribisnis Sumatera Utara)*, vol. 4, no. 1, pp. 18–30, 2011.

[6] Y. Haryati and I. M. Hendrati, "Ekonomi Perberasan: Keterkaitan Pasar Beras Dunia dengan Pasar Indonesia,"

J. Mitra Ekon. dan Manaj. Bisnis, vol. 1, no. 2, pp. 194–201, 2010.

[7] S. Meylinah, "El Nino Conditions Compel Indonesia to Import Additional 2 MMT of Rice," Washington, DC, USA, 2023.

[8] F. Widadie and A. Sutanto, "Model Ekonomi Perberasan: Analisis Integrasi Pasar dan Simulasi Kebijakan Harga," *SEPA*, vol. 8, no. 2, pp. 125–136, 2012.

[9] United Nations, "Goal 2: End hunger, achieve food security and improved nutrition and promote sustainable agriculture," 2024..

[10] E. Sujatmiko, *Kamus IPS, Cetakan 1*. Surakarta: Aksara Sinergi Media, 2014.

[11] J. Mariyono, J. Waskito, A. Kuntariningsih, G. Gunistiyo, and S. Sumarno, "Distribution channels of vegetable industry in Indonesia: impact on business performance," *Int. J. Product. Perform. Manag.*, vol. 69, no. 5, pp. 963–987, 2020, doi: 10.1108/IJPPM-11-2018-0382.

[12] United Nations, "Goal 8: Promote sustained, inclusive and sustainable economic growth, full and productive employment and decent work for all," 2024..

[13] E. Sondakh, R. Kaunang, and P. A. Pangemanan, "Faktor-Faktor Yang Mempengaruhi Pendapatan Pedagang Beras Dalam Menghadapi Fluktuasi Harga di Kota Manado," *ASE*, vol. 12, no. 1A, pp. 103–120, 2016.

[14] A. A. Emmanuel, "Influence of Middlemen Roles on the Performance of Local Rice," Michael Okpara University of Agriculture Umudike, 2017.

[15] F. Lançon, O. Erenstein, S. O. Akande, S. O. Titilola, G. Akpokodje, and O. O. Ogundele, "Imported Rice Retailing and Purchasing in Nigeria: A Survey," Abidjan, Cote d'Ivoire, 2003.

[16] J. S. Kamara, S. M. Bockari-Gevao, P. J. Luseni, A. U. Leigh, and R. A. Cooke, "A National Survey of Rice (Oryza sativa L.) Grain Quality in Sierra Leone I: Perception of Traders and Consumers," *African J. Food, Agric. Nutr. Dev.*, vol. 14, no. 4, pp. 9117–9135, 2014.

[17] J. Sun *et al.*, "Importing food damages domestic environment: Evidence from global soybean trade," *Proc. Natl. Acad. Sci. U. S. A.*, vol. 115, no. 21, pp. 5415–5419, 2018, doi: 10.1073/pnas.1718153115.

[18] Statistics Indonesia, "Population of Indonesia: Result of the 2015 Intercensal Population Census," Jakarta, Indonesia, 2015.

[19] M. Musa, N. Othman, and F. A. Fatah, "Determinants of Consumers Purchasing Behavior for Rice in Malaysia," *Am. Int. J. Contemp. Res.*, vol. 1, no. 3, pp. 159–167, 2011.

[20] H. Yang, C. Li, and Z. Sun, "The Impact Mechanism of Work Experience on the Income of Flexible Workers: Evidence from China," *Sustainability*, vol. 15, p. 16422, 2023, doi: 10.3390/su152316422.

[21] C. Li and S. Managi, "Income raises human well-being indefinitely, but age consistently slashes it," *Sci. Rep.*, vol. 13, p. 5905, 2023, doi: 10.1038/s41598-023-33235-7.

[22] B. Wahyono, "Analisis Faktor-Faktor yang Mempengaruhi Pendapatan Pedagang di Pasar Bantul Kabupaten Bantul," *J. Pendidik. Ekon.*, vol. 6, no. 4, pp. 388–399, 2017.

[23] M. Jomi, S. Widodo, and E. Hariani, "Analisis Faktor-Faktor yang Mempengaruhi Pendapatan Pedagang di Pasar Reok Kabupaten Manggarai Nusa Tenggara Timur Tahun 2020," *Econ. J. Ilmu Ekon.*, vol. 2, no. 2, pp. 104–119, 2020, doi: 10.30742/economie.v2i1.1151.

[24] Ferawati, K. D. Fersiartha, Yusmalina, and I. Yuliana, "Analisis Pengaruh Persediaan Barang dan Penjualan Terhadap Laba Perusahaan (Studi Kasus CV Davin Jaya Karimun)," *J. Cafe.*, vol. 1, no. 2, pp. 33–44, 2020, doi: 10.51742/akuntansi.v1i2.146.

[25] A. P. Lestari and Z. Saidah, "Analysis of Consumer Preferences for Rice Attributes in Cibeunying Kidul District, Bandung City," *Agrikultura*, vol. 34, no. 1, pp. 28–36, 2023, doi: 10.24198/agrikultura.v34i1.40305.

[26] M. Li *et al.*, "Global food-miles account for nearly 20% of total food-systems emissions," *Nat. Food*, vol. 3, pp. 445–453, 2022, doi: 10.1038/s43016-022-00531-w.

[27] N. N. R. Suasih and I. N. M. Yasa, "Indonesian eat rice, but why farmers are poor?," *Sci. Pap. Manag. Econ. Eng. Agric. Rural Dev.*, vol. 17, no. 3, pp. 403–410, 2017.

IOP Conf. Series: Earth and Environmental Science **1460** (2025) 012008 doi:10.1088/1755-1315/1460/1/012008

Digital transformation, change management, and organizational resilience in the culinary sector in West Java

Okky Rizkia Yustian*, Chyntia Ika Ratnapuri

Entrepreneurship Departement, Binus Business School Undergraduate Program
Bina Nusantara University, Bandung, Indonesia

*E-mail: okky.yustian@binus.ac.id

Abstract. The study explores the organizational resilience in small cafes of the culinary sector in West Java influenced by digital transformation and change management. It reveals the impact of digital transformation initiatives and effective change management practices on the development of organizational resilience in this specific context. The authors surveyed 162 small café managers/owners in West Java to explore variable relationships. Structured questionnaires were used with a numerical scale (1-5) for data consistency. PLS statistical technique analyzed the quantitative data. The findings of the study reveal the significant role of digital transformation in shaping the organizational resilience of small cafes in West Java. The integration of digital technologies into various aspects of cafe operations emerges as a critical driver of organizational resilience. The study highlights the crucial mediating role of change management in the relationship between digital transformation and organizational resilience. The findings provide valuable insights for cafe owners, managers, and policymakers, offering guidance on effective strategies and practices to foster organizational resilience and ensure long-term success in the dynamic and evolving culinary sector in West Java. The novelty of this study lies in its exploration of the interrelationships between digital transformation, change management, and organizational resilience, specifically within the context of small cafes in the culinary sector of West Java, providing unique insights into the factors that drive resilience in this specific industry and region.

Keywords: change management, culinary sector, digital transformation, organizational resilience.

1. Introduction

Like many other industries, culinary sector businesses, including Indonesia, experienced setbacks during the global Covid-19 pandemic. The pandemic has forced small cafes in West Java, Indonesia, to navigate through unprecedented challenges, including temporary closures, reduced customer demand, supply chain disruptions, and changing health and safety regulations. In the post-pandemic period, some businesses in this sector have started to crawl again to continue their

IOP Conf. Series: Earth and Environmental Science **1460** (2025) 012008 doi:10.1088/1755-1315/1460/1/012008

business. In response to these challenges, many cafes have turned to digital transformation and digital marketing to adapt, survive, and thrive in the post-pandemic areas of [1].

Digital transformation in the culinary sector involves integrating digital technologies, such as online ordering platforms, contactless payment systems, and delivery services, into the core operations of small cafes. The adoption of these digital solutions enables cafes to enhance their operational efficiency, improve customer experience, and explore new revenue streams [2]. However, the successful implementation of digital transformation initiatives relies on technological aspects and effective change management practices [3].

Change management plays a crucial role in guiding the transition from traditional operations to a digitally transformed state. It encompasses a range of activities, including communication, training, stakeholder engagement, and addressing resistance to change. By leveraging change management strategies, small cafes can guide the complexities associated with digital transformation, facilitate employee acceptance and adoption of new technologies, and drive organizational resilience [4–6].

Organizational resilience is essential to successful digital transformation and effective change management. In the face of disruptions, such as the COVID-19 pandemic, resilient cafes can quickly adapt, respond, and recover, positioning themselves for long-term sustainability and growth. Resilience allows cafes to withstand uncertainties, mitigate risks, and identify opportunities in a dynamic and ever-changing market [7].

Understanding the relationship between digital transformation, change management, and organizational resilience is crucial for the culinary sector in West Java as it seeks to recover and rebuild in the post-pandemic era. By exploring how digital transformation initiatives will impact change management and organizational resilience, this research aims to provide insights into the strategies and practices that small cafes can employ to navigate the challenges of the pandemic and beyond.

While the relationship between digital transformation, change management, and organizational resilience is recognized, several theoretical gaps in the current literature warrant further exploration. Firstly, limited research focuses explicitly on the culinary sector in West Java, leaving a gap in understanding the unique challenges and opportunities faced by small cafes in this region. By investigating the interplay between digital transformation, change management, and organizational resilience in the context of West Java's culinary sector, this study aims to bridge this gap and provide valuable insights for local cafe owners and stakeholders.

Secondly, although the impact of digital transformation on organizational resilience has been examined in various industries, the specific role of change management as a mediator in this relationship still needs to be explored. This study seeks to address this gap by examining the mediating effect of change management on the relationship between digital transformation and organizational resilience within the culinary sector. The study will shed light on how change management practices contribute to developing organizational resilience in digital transformation by delving into this mediation process.

Lastly, given the recent COVID-19 pandemic and its profound impact on the culinary sector, there is a need to investigate the role of digital transformation and change management in facilitating post-pandemic recovery and future-proofing small cafes. This study seeks to contribute novel insights by exploring how small cafes in West Java can leverage digital transformation and effective change management practices to navigate the challenges of the pandemic and position themselves for long-term sustainability and resilience in the evolving culinary sector.

The concept of Digital Transformation involves the incorporation of digital technologies in an organization's operations, business processes, and customer interactions [8]. This process aims to utilize technology to change how organizations operate fundamentally, provide value to their customers, and remain competitive in a digital-first world. Digital Transformation is not just about adopting specific technologies. However, it involves a comprehensive strategy that requires rethinking business models, strategies, and organizational culture in order to take advantage of the opportunities presented by digital advancements [9].

Change management is a methodical approach and a series of processes that aim to facilitate and manage the shift from one state to another within an organization. It entails leading individuals, teams, and the entire organization through planned changes to achieve desired results efficiently. Change Management involves communication, engaging with stakeholders, training, and addressing any resistance to change [10]. The ultimate goal of Change Management is to minimize disruption, maximize employee acceptance, and ensure the successful adoption and implementation of changes within an organization [11].

The term Organizational Resilience refers to an organization's ability to prepare for, adapt to, and overcome disruptions, challenges, or unexpected events while still fulfilling its primary functions and providing value to stakeholders [5]. It includes being able to endure shocks, recover from setbacks, and even emerge stronger from adversity. Achieving Organizational Resilience requires strategic, operational, and cultural elements that allow an organization to navigate uncertainties, mitigate risks, and recognize opportunities for growth and innovation. It involves proactive planning, robust risk management, agility in decision-making, and the effective utilization of resources to respond to and recover from disruptions [7].

2. Methods

The authors conducted a cross-sectional survey to explore the relationship between variables. The survey was completed by 162 managers/owners of small cafes in West Java, selected from a population of 278. Structured questionnaires were used to gather data on the study's constructs, using a numerical scale (1 to 5) for ease and consistency, specifically an online survey through Google Forms. The quantitative data were analyzed using the Partial Least Square (PLS) statistical technique. The researchers also conducted validity and reliability tests on the questionnaire to ensure accurate and consistent results.

Data collection was based on the development of a questionnaire that refers to the operationalization of variables:

1. Digital Transformation is an antecedent variable with four indicators: online presence and visibility, adoption of digital tools and technologies, integration of data analytics, and e-commerce and delivery capabilities.
2. Change Management is a mediation variable with four indicators: communication and stakeholder engagement, training and development initiatives, change readiness and acceptance, and change implementation and monitoring.
3. Organizational Resilience is a subsequent variable also with four indicators: adaptive capacity, resource flexibility, innovation and diversification, and collaborative networks and partnerships.

The proposed research model can be described in Figure 1.

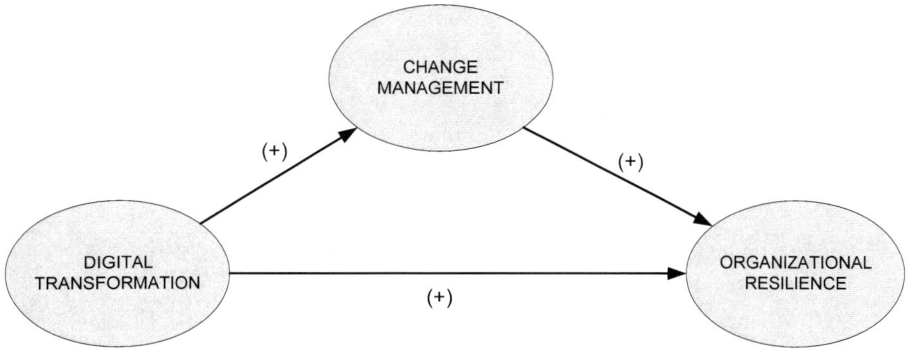

Figure 1. Proposed Research Model

Based on the model, here are the hypotheses:

H1: Digital transformation has a significant positive effect on change management. It implies that there is an expectation that implementing digital transformation initiatives within small cafes in the culinary sector of West Java will lead to a positive impact on change management practices.

H2: Change management has a significant positive effect on organizational resilience. Effective change management practices implemented within small cafes in the culinary sector of West Java will lead to a positive impact on organizational resilience.

H3: Digital transformation has a direct and indirect effect on organizational resilience through change management. This hypothesis recognizes the dual influence of digital transformation, with both direct and mediated effects on organizational resilience through the lens of change management.

Referring to the research framework and paradigm, operationalization of the variables from Digital Transformation (X1) as the antecedent variable, Change Management (X2) as the mediating variable, and Organizational Resilience (Y) as the subsequent variable, the variables can be operationalized in Table 1.

Table 1. Operationalization of Variables

Latent Variables	Manifests (Dimensions)	Indicators	Item Number
Digital Transformation (X1)	Online presence and visibility	1.01. Website traffic 1.02. Social media engagement 1.03. Online customer reviews	1-3
	Adoption of digital tools and technologies	1.04. Online ordering system 1.05. Point-of-sale (POS) software 1.06. Digital payment options	4-6
	Integration of data analytics	1.07. Customer data analysis 1.08. Operational efficiency metrics 1.09. Decision-making based on data	7-9
	E-commerce and delivery capabilities	1.10. Online sales revenue 1.11. Delivery fulfillment rate 1.12. Menu customization for online platforms	10-12
Change Management (X2)	Communication and stakeholder engagement	2.01.Frequency of communication 2.02.Clarity of communication 2.03.Stakeholder feedback mechanisms	1-3

Latent Variables	Manifests (Dimensions)	Indicators	Item Number
	Training and development initiatives	2.04.Training program participation 2.05.Training effectiveness 2.06.Employee feedback on training	4-6
	Change readiness and acceptance	2.07.Attitudes towards change 2.08.Perceived organizational support 2.09.Level of resistance to change	7-9
	Change implementation and monitoring	2.10.Change management frameworks 2.11.Key performance indicators (KPIs) 2.12.Alignment between intended and actual changes	10-12
Organizational Resilience (Y)	Adaptive capacity	3.01.Agility in decision-making 3.02.Agile supply chain management 3.03.Proactive risk management	1-3
	Resource flexibility	3.04.Scalability of operations 3.05.Adaptability of workforce 3.06.Financial resilience	4-6
	Innovation and diversification	3.07.New product or service development 3.08.Market expansion initiatives 3.09.Continuous improvement practices	7-9
	Collaborative networks and partnerships	3.10.Alliances and collaborations 3.11.Information sharing and knowledge exchange 3.12.Disruption response coordination	10-12

Scale: Interval

3. Results and Discussion

Based on the profiles of the respondents, the sample consisted of 162 respondents, with a majority of males (98 respondents, 60.5%) and a smaller proportion of females (64 respondents, 39.5%). The gender distribution was skewed towards males, with males representing a larger percentage of the respondents. Level of education is distributed across different levels. Combining all educational levels, the respondents reflect a diverse range of educational backgrounds, including high school graduates, Bachelor's degree holders, Master's degree holders, and individuals with a Doctoral degree. The majority of respondents had attained at least a Bachelor's degree, as they accounted for more than half of the sample (50.6%). In the length of work distribution, the sample reflects a range of lengths of work experience, including respondents with various levels of professional experience. The most significant proportion of respondents had worked for more than four years, making up the majority of the sample (41.4%). Respondents with 3-4 years of work experience represented the second largest group (29.6%), followed by those with 2-3 years of work experience (19.8%). The smallest group consisted of respondents with 1-2 years of work experience (9.3%). Table 2. represents the profiles of the respondents in terms of gender, level of education, and length of work.

Table 2. Profiles of the Respondents

Gender	Frequency	Percentage
Male	98	60.5
Female	64	39.5
Total	162	100.0

Level of Education:	Frequency	Percentage
High School	62	38.3
Bachelor	82	50.6
Master	14	8.6
Doctoral	4	2.5
Total	162	100.0

Length of work:	Frequency	Percentage
1 - 2 years	15	9.3
2 - 3 years	32	19.8
3 - 4 years	48	29.6
more than 4 years	67	41.4
Total	162	100.0

Source: primary data (2023)

Based on the data in Table 3, we can conclude that the studied context has effectively undergone digital transformation, developed robust organizational resilience, and invested in staff development. Although all three variables received high mean scores, staff development and organizational resilience received slightly higher scores than digital transformation. These mean scores also align with the overall average. The descriptive statistics of each variable can be shown in Table 3.

Table 3. Descriptive Statistics

Dimension	**Mean**	**Category**
Online presence and visibility	4.08	Very High
Adoption of digital tools and technologies	3.75	High
Integration of data analytics	3.87	High
E-commerce and delivery capabilities	4.04	Very High
Digital Transformation (X1)	3.93	High

Dimension	**Mean**	**Category**
Communication and stakeholder engagement	4.01	Very High
Training and development initiatives	4.05	Very High
Change readiness and acceptance	3.98	High
Change implementation and monitoring	3.98	High
Staff Development (X2)	4.01	Very High

Table 4. Descriptive Statistics (continue)

Dimension	**Mean**	**Category**
Adaptive capacity	3.92	High
Resource flexibility	4.09	Very High
Innovation and diversification	4.10	Very High
Collaborative networks and partnerships	3.86	High
Organizational Resilience (Y)	3.99	High

Source: primary data (2023)

In the measurement model, we assess the reliability and validity of measurement scales using Cronbach's Alpha (CA), Composite Reliability (CR), and Average Variance Extracted (AVE), as presented in Table 4.

Table 5. Descriptive Statistics

	CA	CR	AVE
X1: Digital Transformation	0.786	0.861	0.609
X2: Staff Development	0.841	0.893	0.677
Y: Organizational Resilience	0.807	0.874	0.635

Source: SmartPLS output (2024)

Cronbach's Alpha is a way to gauge how consistent the items in a variable or scale are with one another. Essentially, it looks at how much the items are interrelated or correlated. The higher the value of Cronbach's Alpha, the stronger the internal consistency among the items in the scale. In the table provided, X1, X2, and Y have respective CA values of 0.786, 0.841, and 0.807. These values suggest that the items within each variable are consistent.

Composite Reliability (CR) measures how consistent a group of items is within a construct in a structural equation model (SEM) framework. It is similar to Cronbach's Alpha in that higher values indicate a firmer internal consistency among the items. In the table provided, the CR values for X1, X2, and Y are 0.861, 0.893, and 0.874, respectively, indicating a high level of internal consistency for these constructs.

AVE measures convergent validity by evaluating how much variance a construct captures in relation to measurement error. A higher AVE value indicates that the construct explains a larger proportion of the variance in its observed indicators. In the table provided, X1, X2, and Y have AVE values of 0.609, 0.677, and 0.635, respectively, implying that these constructs considerably impact their respective indicators. Figure 2 depicts the overall model of Digital Transformation, Staff Development, and Organizational Resilience in the Culinary Sector in West Java.

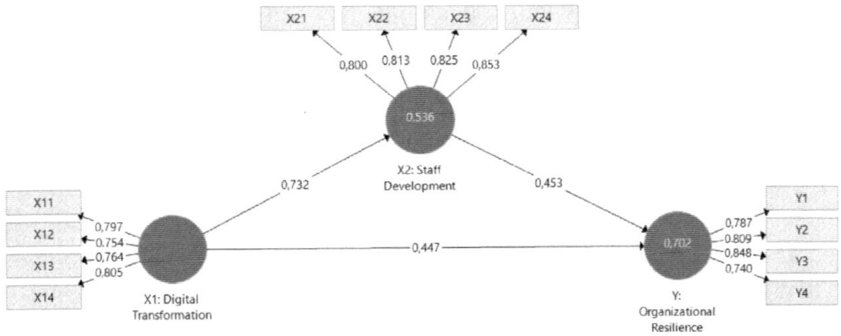

Source: SmartPLS outputs (2023)
Figure 2. Overall Model

Based on the provided outer loadings, all the indicators (Online presence and visibility, Adoption of digital tools and technologies, Integration of data analytics, E-commerce and delivery capabilities) exhibit a positive and significant relationship with the Digital Transformation construct. "Online presence and visibility" and "E-commerce and delivery capabilities" have relatively higher outer loadings, indicating stronger associations with Digital Transformation than the other indicators.

In the culinary sector, particularly for a small cafe in West Java, the concepts of "Online presence and visibility" and "E-commerce and delivery capabilities" are strongly related to digital transformation. Having an online presence means establishing a digital footprint for the cafe, including a website, social media profiles, or online listings. Visibility is crucial because it determines how easily potential customers can discover the cafe online. To reach a wider audience, increase brand awareness, and attract customers, cafes need to enhance their online presence and visibility [12]. It can be achieved through various strategies, such as creating an engaging website, maintaining active social media accounts, optimizing online search presence, and utilizing digital marketing techniques [13]. Digital transformation in the culinary industry often involves leveraging technology to expand reach, engage customers, and improve business operations [13]. A solid online presence is a crucial aspect of this transformation.

In Staff Development, all indicators (communication and stakeholder engagement, training and development initiatives, change readiness and acceptance, and change implementation and monitoring) also display a positive and significant relationship with the construct. "Change readiness and acceptance" and "Change implementation and monitoring" have relatively higher outer loadings than the other indicators, indicating stronger associations with Staff Development.

In a cafe setting, change readiness is the willingness and preparedness of individuals and teams to adapt to changes. On the other hand, acceptance is the extent to which staff members support changes in the workplace. Both play a crucial role in staff development by facilitating the implementation of new practices, processes, or technologies that enhance skills and capabilities [14]. To achieve this, staff must be open to new learning opportunities, training programs, and skill development initiatives. By fostering change readiness and acceptance, the cafe can create a positive environment that encourages staff members to embrace growth, acquire new skills, and adapt to changes in the culinary sector. It promotes staff development by enabling individuals to expand their knowledge, improve their performance, and contribute to the overall success of the cafe. It may implement new practices, processes, or initiatives to improve the cafe's operations and work environment [15]. The ongoing evaluation and assessment of these changes is called monitoring. It is crucial for the successful skill-building and professional growth of the staff [16]. The cafe may offer training programs, workshops, or mentorship opportunities to enhance the skills of its staff. Effective implementation and monitoring of these initiatives ensure staff members receive the necessary support, resources, and guidance to develop their skills. The cafe can track the progress of staff development efforts, identify areas for improvement, and make necessary adjustments to maximize the impact of the initiatives.

All indicators in Organizational Resilience (Adaptive capacity, Resource flexibility, Innovation and diversification, and Collaborative networks and partnerships) show a positive and significant relationship with the construct. "Resource flexibility" and "Innovation and diversification" have relatively higher outer loadings than the other indicators, indicating stronger associations with the Organizational Resilience of small cafes in West Java.

Resource flexibility is an essential component of organizational resilience in small cafes located in West Java. It refers to the ability of an organization to adapt and allocate its resources effectively in response to changing circumstances or challenges [17]. This flexibility is crucial for ensuring operational agility and resilience in cafes, especially since they often face dynamic market conditions, seasonal fluctuations, and unforeseen events. Small cafes must adjust their resources, such as staff, ingredients, and equipment, to meet customer demands and optimize efficiency. The "Resource flexibility" indicator is strongly associated with organizational resilience in small cafes. It suggests that cafes that exhibit higher levels of resource flexibility are more likely to withstand disruptions, adapt to changes, and maintain their operations effectively [18].

The R-squared values indicate the proportion of variance in Staff Development (53.3%) and Organizational Resilience (69.8%), which are explained by the other variables in the model. The f-square values provide information about the effect sizes and the amount of variance explained by the independent variables (Digital Transformation and Staff Development) on the dependent variables (Staff Development and Organizational Resilience) while controlling for other factors.

The relationship between Digital Transformation and Staff Development is relatively strong, as indicated by the f-square value 1.155. This value represents the effect size for X1's impact on X2. The moderate relationship between Digital Transformation and Organizational Resilience is reflected in the f-square value of 0.311, indicating that X1 explains a moderate amount of variance in X3. Additionally, the f-square value of 0.319 suggests a moderate relationship between Staff Development and Organizational Resilience, with X2, explaining a moderate variance in Y.

From the overall model, we can observe that the direct effect of X1 on X2 is 0.732; X1 on Y is 0.447, and X2 on Y is 0.453. In this model, the indirect effect from X1 on Y through X2 is 0.332. The results suggest that both Digital Transformation and Staff Development directly affect Organizational Resilience. However, Digital Transformation has a more substantial direct effect on Staff Development compared to its direct effect on Organizational Resilience. Meanwhile, Staff Development has a relatively comparable direct effect on Organizational Resilience as Digital Transformation. The significant indirect effect suggests that Staff Development is mediating the relationship between Digital Transformation and Organizational Resilience, contributing to the overall impact of Digital Transformation on Organizational Resilience.

Organizational resilience refers to a company's ability to adapt and recover from unexpected challenges and disruptions. Digital transformation involves leveraging technology to improve business operations and processes. Meanwhile, staff development concentrates on enhancing the skills and knowledge of employees. Both factors play a crucial role in determining a company's level of resilience [5]. Investing in digital transformation and staff development can better equip

IOP Conf. Series: Earth and Environmental Science **1460** (2025) 012008 doi:10.1088/1755-1315/1460/1/012008

a company to handle unexpected challenges and disruptions and emerge more robust and agile in the face of adversity.

A small cafe's absence of digital transformation and staff development can limit its adaptability, hinder operational efficiency, restrict growth opportunities, stagnate employee skills, and increase vulnerability to disruptions [19]. Without digital transformation, the cafe may struggle to keep up with the changing business landscape. It may miss opportunities to streamline operations, improve customer experience, and stay competitive. This lack of adaptability can hinder the cafe's ability to respond to evolving customer preferences and market trends. Digital tools and technologies are crucial in enhancing operational efficiency and effectiveness. Without leveraging these tools, the cafe may encounter inefficiencies in various areas, such as order processing, inventory management, and customer communication. It can lead to increased costs, lower productivity, and decreased customer satisfaction.

Digital transformation can open doors to new growth opportunities for small cafes. It enables online presence and visibility, expands customer reach through e-commerce capabilities, and supports data analytics for informed decision-making [20]. The cafe may miss potential growth and revenue generation avenues without embracing digital transformation. Neglecting staff development can lead to stagnant employee skills and limited professional growth. Failure to invest in staff training and development can decrease employee motivation and productivity, resulting in a higher turnover rate. The cafe should consider supporting training programs to attract and retain talented staff to improve service quality and customer satisfaction. In today's unpredictable business climate, organizational resilience is crucial. Without digital transformation and staff development, the cafe may struggle to navigate unforeseen challenges, making it more susceptible to market fluctuations, operational disruptions, and changes in customer preferences.

4. Conclusion

Organizational Resilience is impacted by Digital Transformation and Staff Development. However, Staff Development has a more direct effect on Organizational Resilience than Digital Transformation. Meanwhile, Digital Transformation has a relatively equal direct effect on Organizational Resilience and Staff Development. The significant indirect effect means that Staff Development mediates the relationship between Digital Transformation and Organizational Resilience, ultimately contributing to the overall impact of Digital Transformation on Organizational Resilience.

Investing in digital transformation and staff development is crucial for a company's organizational resilience. Digital transformation enhances business operations and processes, while staff development enhances employee skills and knowledge. Both factors directly affect organizational resilience, with staff development mediating the relationship between digital transformation and organizational resilience. Neglecting digital transformation and staff development can limit adaptability, hinder operational efficiency, restrict growth opportunities, stagnate employee skills, and increase vulnerability to disruptions. Small cafes can benefit from digital transformation and staff development by improving their online presence, expanding customer reach, supporting data analytics, improving service quality, and customer satisfaction.

To foster organizational resilience, businesses should adopt practical policies such as allocating budgets for digital tools and training programs, incentivizing continuous learning through professional development plans, and integrating digital technology into daily operations. For example, small cafes can implement a customer relationship management (CRM) system to streamline operations and gather customer insights while providing staff with training on digital tools and customer engagement strategies. These policies enable cafes to anticipate market trends, personalize customer experiences, and increase operational efficiency. The implications of such investments include enhanced competitive advantage, improved employee retention due to skill enrichment, and long-term sustainability by building adaptability to future disruptions.

References

[1] Rini Rahayu Kurniati and Fitri Hidayat 2022 Adaptive and Digital Marketing: A Sustainable Strategy for Developing Culinary SME Business (Law of The Republic of Indonesia, Number 20 of 2008) Post-pandemic *International Journal Of Artificial Intelligence Research* 6

[2] Kraft C, Lindeque J P and Peter M K 2022 The digital transformation of Swiss small and medium-sized enterprises: insights from digital tool adoption *Journal of Strategy and Management* 15 468–94

[3] Bellantuono N, Nuzzi A, Pontrandolfo P and Scozzi B 2021 Digital Transformation Models for the I4.0 Transition: Lessons from the Change Management Literature *Sustainability* 13 12941

[4] Fleron B, Pries-Heje J and Baskerville R 2022 Becoming a Most Digitalized Country: A History of Digital Organizational Resilience in Denmark *Communications of the Association for Information Systems* 51 120–39

[5] He Z, Huang H, Choi H and Bilgihan A 2023 Building organizational resilience with digital transformation *Journal of Service Management* 34 147–71

[6] Sobczak A 2022 Robotic Process Automation as a Digital Transformation Tool for Increasing Organizational Resilience in Polish Enterprises *Sustainability* 14 1333

[7] Lisdiono P, Said J, Yusoff H and Hermawan A A 2022 Examining Leadership Capabilities, Risk Management Practices, and Organizational Resilience: The Case of State-Owned Enterprises in Indonesia *Sustainability* 14 6268

[8] Hanelt A, Bohnsack R, Marz D and Antunes Marante C 2021 A Systematic Review of the Literature on Digital Transformation: Insights and Implications for Strategy and Organizational Change *Journal of Management Studies* 58 1159–97

[9] Holzinger A, Weippl E, Tjoa A M and Kieseberg P 2021 Digital Transformation for Sustainable Development Goals (SDGs) - A Security, Safety and Privacy Perspective on AI pp 1–20

[10] Jami Pour M and Hosseinzadeh M 2021 An integrated framework of change management for social CRM implementation *Information Systems and e-Business Management* 19 43–75

[11] Kho J, Gillespie N and Martin-Khan M 2020 A systematic scoping review of change management practices used for telemedicine service implementations *BMC Health Serv Res* 20 815

[12] Rina Hansen and Siew Kien Sia 2015 Hummel's Digital Transformation Toward Omnichannel Retailing: Key Lessons Learned *MIS Quarterly Executive* 14 51–66

[13] Matt C, Hess T and Benlian A 2015 Digital Transformation Strategies *Business & Information Systems Engineering* 57 339–43

[14] Lyndon Pugh 2016 *Change Management in Information Services*

[15] Durmaz O, Hawrami S S and Hamasaeed A M 2022 The suitable leadership for industry 4.0 *Journal of Global Economics and Business* 3 113–24

[16] Burlea-Schiopoiu A and Ferhati K 2020 The Managerial Implications of the Key Performance Indicators in Healthcare Sector: A Cluster Analysis *Healthcare* 9 19

[17] Senivongse C, Bennet A and Mariano S 2019 Clarifying absorptive capacity and dynamic capabilities dilemma in high dynamic market IT SMEs *VINE Journal of Information and Knowledge Management Systems* ahead-of-print

[18] Lin H-F 2017 Antecedents and consequences of electronic supply chain management diffusion *The International Journal of Logistics Management* 28 699–718

[19] Margiono A 2021 Digital transformation: setting the pace *Journal of Business Strategy* 42 315–22

[20] Li L, Lin J, Turel O, Liu P and Luo X (Robert) 2020 The impact of e-commerce capabilities on agricultural firms' performance gains: the mediating role of organizational agility *Industrial Management & Data Systems* 120 1265–86

Analysing the Impact of Special Corn Programs on Farmers Income in Gunungkidul Regency

Endah Subekti Kuntariningsih[1*], Bambang, WHEP[2] and Agus Setiadi[3]

[1]Department of Master of Agribusiness, Universitas Diponegoro, Semarang, Indonesia
[2]Department of Animal Science, Universitas Diponegoro, Semarang, Indonesia
[3]Department of Agribusiness, Universitas Diponegoro, Semarang Indonesia

*E-mail: endahsubektikuntarin@students.undip.ac.id/eskgunungkidul@gmail.com

Abstract. Although traditionally sidelined in favour of rice, corn is recognized as a crucial local commodity for food security. In 2022, corn production from Gunungkidul farmers decreased by 1.20% to around 228,861 tons, while in 2023 it reached 276,589 tons, an increase of 1.20% from 2022. Therefore, Gunungkidul Regency Government recognized the importance of their role, along with relevant institutions, in providing support through Special Program for Corn to continuously strengthen sustainable and innovative agricultural strategies. This study analysed the impact of a special corn program on farmers' income in Gunungkidul Regency, Indonesia. The study surveyed 200 corn farmers across three sub-districts: Saptosari, Tanjungsari, and Semin, using interviews and questionnaires. Data were analysed using multiple linear regression to determine the impact of independent variables on income, and the Wilcoxon Signed-Rank Test to assess income differences before and after the program. The results showed that farm size had the most significant impact on income, with a regression coefficient of 0.412, followed by production costs (0.410), both indicating strong positive relationships. Training and extension services (0.083), seed quality (0.059), technology adoption (0.044), and access to fertilizer (0.039) also positively influenced income, although to a lesser degree. The Wilcoxon test revealed a significant income increase after the program, with average income rising from IDR 2,081,982.54 to IDR 3,100,108.13. The study concluded that the special program effectively enhanced corn farmers' incomes in Gunungkidul Regency.

Keywords: Farmers Income, Maize, Special Programs

1. Introduction

Corn (*Zea mays L.*) is one of the world's major crops with high economic value, serving as a source of food, livestock feed, and industrial raw material [1]. Corn was first cultivated about 10,000 years ago in Mexico and has since spread worldwide, including to Indonesia. In Indonesia, corn is an important commodity grown in various regions, including Gunungkidul Regency in Yogyakarta. This regency is known for its fertile and diverse agricultural land, where corn is one of the leading crops after rice. With fertile agricultural land and ongoing support programs, Gunungkidul Regency has the potential to become one of Indonesia's highly competitive corn farming centers.

The corn harvest in Gunungkidul Regency is not only used for local consumption but also as livestock feed and raw material for industry, contributing to increased farmers' income [2]. The corn agribusiness in Gunungkidul Regency faces challenges such as pest attacks, including stem borers, and diseases like downy mildew, which are major issues requiring careful management and appropriate pesticide use. Additionally, limited access to modern technology and production inputs, such as high-quality seeds and fertilizers, as well as a lack of technical extension services, slow down the adoption of more efficient farming practices. Unpredictable climate changes, including unstable rainfall patterns, also pose a serious threat, affecting the productivity and sustainability of corn farming in this region [3].

The Central Bureau of Statistics of Gunungkidul Regency recorded that corn production in 2021 reached 275,913 tons, an increase of 31.7% from the target of 209,501 tons of dry shelled corn in 2020. In 2022, corn production from Gunungkidul farmers decreased by 1.20% to around 228,861 tons, while in 2023 it reached 276,589 tons, an increase of 1.20% from 2022 [4]. Observing this fluctuating trend, it is essential for the Gunungkidul Government to continuously strengthen sustainable and innovative agricultural strategies. The Gunungkidul Regency Government recognized the importance of their role, along with relevant institutions, in providing support through counselling, training, and access to modern agricultural technology. Therefore, the Gunungkidul Government has launched a Special Efforts (UPSUS) program from the Ministry of Agriculture to enhance the production and productivity of corn as a strategic local commodity, as well as to increase farmers' incomes. Such programs included adoption of agricultural technologies by providing modern farming tools, the supply of high-quality seeds, and easier access to fertilizers.

In view of the above, the income of corn farmers in Gunungkidul Regency is heavily influenced by harvest yields, access to fertilizers, the quality of corn seeds, the price of corn seeds, and the selling price of corn. With the application of appropriate agricultural technology, it is expected that corn productivity will increase, leading to more stable and higher farmer incomes. The local government and related institutions also play a crucial role in providing support through extension services, training, and access to modern agricultural technology. These efforts aim to improve the welfare of corn farmers in Gunungkidul Regency and ensure the sustainability of corn production in the future. Therefore, an in-depth analysis of the impact of agricultural special programs on the income of corn farmers in Gunungkidul Regency is necessary.

2. Research Methods

The study applied a quantitative approach. The interview method with questionnaires and observation was used for collecting data. The locations in this study were determined purposively based on the criteria of sub-districts that benefited from the Special Program for Corn and had been running for 1 year, resulting in 3 sub-districts: Saptosari Sub-district, Tanjungsari Sub-district, and Semin Sub-district. The research sample consisted of farmers who are beneficiaries of the special corn program in three sub-districts (Saptosari, Tanjungsari, and Semin), with a total population of 8,986 individuals. A sample of 200 respondents was obtained from the total population using the Slovin method, and the distribution of samples across the three sub-districts was determined using proportionate random sampling. The results yielded 108 respondents from Saptosari sub-district, 47 respondents from Tanjungsari sub-district, and 45 respondents from Semin sub-district. The data obtained were presented in the form of statistics, including tables, diagrams, and measurable figures, and the results are then explained descriptively in a quantitative manner. The data were analysed quantitatively using multiple linear regression analysis to test the hypothesis regarding the causal relationships between the variables presented in this study. The Wilcoxon Signed-Rank Test is used to evaluate whether there is a statistically significant difference in farmers income levels before and after the corn special program is implemented.

IOP Conf. Series: Earth and Environmental Science **1460** (2025) 012009 doi:10.1088/1755-1315/1460/1/012009

3. Results and Discussion

3.1 Safety Nets for Agriculture

Social safety nets play a crucial role in enhancing agricultural productivity by providing support to vulnerable groups, particularly in rural areas where agriculture is the main livelihood. Social safety net programs such as cash transfers, food assistance, and labor-intensive work programs are designed to protect households from economic shocks and improve food security. When these programs are integrated with agriculture-focused initiatives, they not only shield farmers from immediate risks but also promote long-term productivity by providing the resources and stability needed to invest in agricultural activities. This integration is key to reducing poverty and ensuring sustainable agricultural development [5].

An effective example of an agriculture-focused safety net program is the Productive Safety Net Programme (PSNP) in Ethiopia. This program combines cash transfers with agricultural inputs such as fertilizers, improved seeds, and access to credit, enabling farmers to invest in their fields. Additionally, PSNP provides training on sustainable farming techniques, such as soil conservation, which can enhance agricultural yields. A study indicated that beneficiaries of this program have experienced increased crop yields and livestock ownership, ultimately improving household incomes and food security. These outcomes highlight the importance of linking safety nets with innovations in agriculture[6]. Despite the success of programs like PSNP, challenges remain in fully integrating social protection with agricultural development. One barrier is the accessibility of land, credit, and inputs for the poorest farmers, who often cannot afford formal market options.

Given the potential of social safety nets to enhance agricultural productivity and improve food security, it becomes essential to explore the specific impacts of such programs on local farming communities. This is particularly important in regions like Gunungkidul, where agriculture, especially corn farming, plays a pivotal role in the livelihoods of many families. By conducting research on the impact of corn-specific programs on farmers' incomes, we can gain valuable insights into how targeted interventions can address local challenges and improve economic outcomes.

3.2 Characteristics of Research Locations

From the collected data, for the Semin subdistrict, there was significant variation in terms of agricultural land size, with the largest farm size being 5,000 m^2 and the smallest being only 500 m^2. In the Saptosari subdistrict, there was also significant variation in land size, with the largest agricultural land measuring 25,000 m^2 and the smallest being 2,000 m^2. Field research in the Tanjungsari subdistrict showed that the average land size ranges from 2,000 to 5,000 m^2, with the largest land owned by farmers reaching 10,000 m^2 and the smallest only 500 m^2. This variation indicated diversity in the scale of agricultural businesses in the region.

In terms of age, the youngest farmer was from Tanjungsari subdistrict, who was 28 years old and has 8 years of farming experience. On the other hand, the oldest farmer was from Saptosari subdistrict, who was 81 years old with 60 years of farming experience. Based on the education level of farmers in the three subdistricts, farmers in Tanjungsari were predominantly those who completed primary education (elementary school). Meanwhile, most farmers in Semin and Saptosari have completed junior high school as their highest level of education.

IOP Conf. Series: Earth and Environmental Science **1460** (2025) 012009 doi:10.1088/1755-1315/1460/1/012009

3.3 Classical Assumption Tests

3.3.1 Normality Test

Figure 1 shows the results of the normality test on the residual histogram with the dependent variable of Farmers' Income, indicating that the regression model used meets the assumption of residual normality. The histogram displays a symmetrical residual distribution resembling a normal curve, suggesting that there is no significant skewness in the data. This is further supported by the curved line that follows the theoretical normal distribution, indicating that the residual distribution of this model aligns with the expected normal distribution. Additionally, the mean residual value close to zero (-6.38E-15) shows that the average prediction error is unbiased, while a standard deviation of 0.985 suggests that the variation in prediction errors is within reasonable limits. With a total of 209 observations, these results can be considered valid and representative. Therefore, it can be concluded that the regression model used was reliable, and the statistical inferences made from this model can be considered valid, providing a strong foundation for further analysis and interpretation related to farmers' income. Based on the results of the One-Sample Kolmogorov-Smirnov test, it can be concluded that the residuals of this regression model are normally distributed. The high significance value (0.200) indicated that there was no significant deviation from normality.

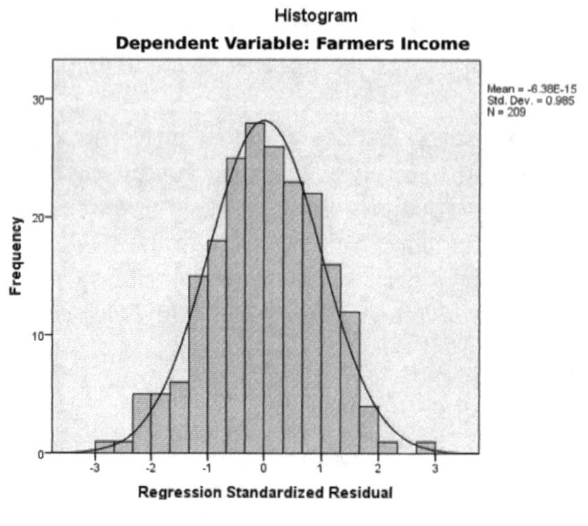

Figure 1. The Histogram of the Residual

3.3.2 The Coefficient of Determination (R^2) Test

The R Square value of .652 indicates that 65.2% of the variation in the dependent variable can be explained by the independent variable included in the model. This suggested that the model has a reasonably good predictive ability, considering the complexity of factors that may influence the dependent variable in the socio-economic research context [7]. The Adjusted R Square, which is a more conservative version of R Square, showed a value of .642. This value accounts for the number of independent variables in the model and the sample size, providing a more accurate estimate of the model's predictive strength in the population [8].

3.3.3 The Collinearity and Heteroscedasticity Test

The collinearity test results (**Table 1**) show that all variables have a Tolerance value > 0.1 and VIF <10, which are the general thresholds for determining the presence or absence of multicollinearity. The lowest tolerance value is 0.668 for farm size, which is still far above the 0.1 threshold. The highest VIF value is 1.497 for farm size, well below the threshold of 10. This indicates that there is no excessively strong correlation between independent variables, so each variable can be considered to contribute uniquely to the model.

Table 1. Collinearity and Heteroscedasticity Test.

Variable	Collinearity Statistics		Heteroscedasticity (Spearman's Rank)
	Tolerance	VIF	Sig. (2-Tailed)
(Constant)			
Technology Adoption	.871	1.148	.732
Training and extension services	.971	1.030	.735
Seed Quality	.868	1.153	.855
Farm Size	.668	1.497	.785
Production Costs	.686	1.457	.792
Access to Fertilizers	.923	1.084	.911

The heteroscedasticity test (**Table 1**) shows that all significance values are > 0.05, indicating the absence of heteroscedasticity. The highest value is 0.911 for Access to Fertilizers, indicating that this variable has the most consistent residual variance. The lowest value is 0.732 for Technology Adoption, but it is still far above the 0.05 threshold. In addition, the observation of the distribution of points on the scatterplot (**Figure 2**) does not show a clear systematic pattern. The points are scattered randomly above and below the value of 0 on the Y-axis. This indicates that the assumption of homoscedasticity is met, where the residual variance remains relatively constant for each predicted value [9].

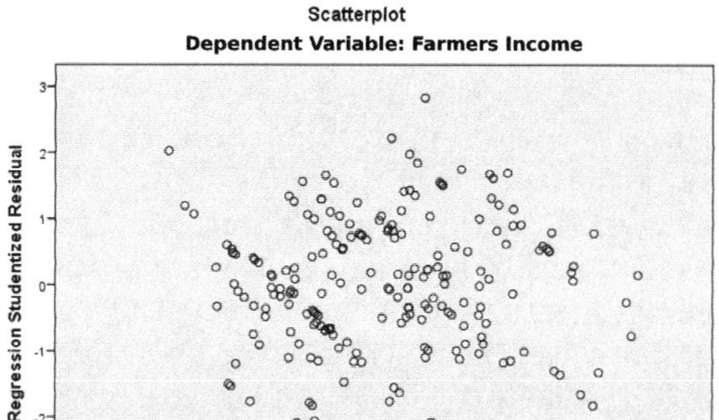

Figure 2. Scatterplot Diagram

3.4 The Multiple Regression Analysis on Farmers' Income

The regression results (**Table 2**) show that the constant of 1.750 ($p < 0.001$) indicates the value of the dependent variable when all independent variables are zero. Technology adoption (X_1) with a regression coefficient of 0.044 ($p = 0.047$) indicates a significant positive relationship; for every 1% increase in agricultural technology adoption, farmers' income will increase by 0.044%, assuming other variables remain constant. This result is consistent with the study of [10] in Kenya who identified the adoption of improved feed technologies. The study estimated the adoption of improved feed technologies to increase the dairy income from US$2.87 to US$3.39. The literature study of [11] concluded that most studies regarding the influence of technology adoption on farmers' well-being indicated that technology adoption improved farmers' well-being as measured by productivity and income, although these measures excluded value judgments by farmers, such as happiness.

Table 2. Multiple Regression Analysis

Variable	Unstandardized Coefficients		Sig.
	β	Std. Error	
(Constant)	1.750	.269	.000
Technology Adoption	.044	.022	.047
Training and extension services	.083	.023	.000
Seed Quality	.059	.024	.016
Farm Size	.412	.046	.000
Production Costs	.410	.048	.000
Access to Fertilizers	.039	.018	.030

Training and extension services (X_2) have a regression coefficient of 0.083 ($p < 0.001$), indicating a highly significant positive relationship; every increase in training and extension services is associated with a 0.083% increase in farmers' income. This result is in line with a study conducted by [12] in Ethiopia that assessed the impact of training on crop productivity and households' income. The study applied propensity score matching (PSM) to analyse the impact of training on the aforesaid outcome variables. The results found that trainees increased their wheat and maize yield, respectively as the outcome of the conducted training and extension services. It has been expressed as well by [13] that agricultural training might possess potential effective methods to disseminate relevant new technologies to increase productivity and alleviate rural poverty. In addition, the development and growth of training and extension networks for both men and women are crucial for enhancing skills, boosting productivity, and improving income-generating potential [14].

The variable seed quality (X_3), with a regression coefficient of 0.059 ($p = 0.016$), also shows a significant positive relationship; for every 1% increase in the use of quality seeds, there will be a 0.059% increase in income. This result contributes to the argument of [15] that enhancing the quality of seeds from any chosen variety is fundamental to increasing agricultural productivity. In addition, higher yields, increased productivity, and farmers' income are closely linked to the availability and use of enhanced crop varieties [16].

Farm size (X_4) shows the largest impact among the independent variables with a regression coefficient of 0.412 ($p < 0.001$), indicating a highly significant positive relationship; for every 1% increase in farm size, farmers' income will increase by 0.412%. This result is consistent with the study of [17] that found a significant effect of farm size as an independent variable on income of corn farmers in East Lampung Regency. [17] also reported that the farm size owned by individual farmers significantly affects the income of corn farmers; a larger farm size leads to increased production, which in turn enhances their revenue. In addition, a system dynamics model proposed by [18] also underlined the significant role of farm size in improving the production and income of corn farmers.

For Production costs (X_5) with a regression coefficient of 0.410 ($p < 0.001$), it has the second-largest effect, showing a highly significant positive relationship; every 1% increase in production costs will lead to a 0.410% increase in farmers' income. The impact of production costs on corn farmers; income is significant, as it affects profitability and sustainability [19]. For example, a study on U.S. corn production found that rising costs for inputs such as seeds, fertilizers, and labour reduce farmers' net returns, even when yields are high. This underscores the importance of managing costs to maintain profitability. Additionally, small-scale farmers are more sensitive to input price fluctuations than larger farmers, which can directly impact their income levels [20].

Lastly, for the variable Access to fertilizer (X_6) with a regression coefficient of 0.039 ($p = 0.030$), it shows a significant positive relationship, though its effect is relatively small. However, as access to fertilizer increases, farmers' income will rise by 0.039%. This result is consistent with [21] who reported that access to subsidized fertilizer was associated with a reduction in the likelihood of using manure of 5% and an increase in farmers' total gross margin of 11 %. The fertilizer subsidy policy is seen as having a beneficial effect by boosting agricultural productivity and increasing farmers' incomes, particularly for food crops [22].

3.5 The Wilcoxon Signed-Rank Test

The results of the descriptive analysis in the Wilcoxon model (**Table 3**) show that the average income of farmers before receiving the Corn special program was IDR 2,081,982.54 with a maximum value of IDR 5,745,000.00, and after receiving the program, an increase in the average income to IDR 3,100,108.13 with a maximum income after receiving the Corn special program of IDR 9,685,000 was observed. The ranks and statistics analysis found the following

Table 3. Wilcoxon's Ranks and Statistics Analysis

		N	Mean Rank	Sum of Ranks	Post - Pre
Post - Pre	Negative Ranks	8[a]	109.25	874.00	
	Positive Ranks	200[b]	104.31	20862.00	
	Ties	1[c]			
Asymp. Sig. (2-tailed)					.000

The analysis revealed that 8 farmers experienced a decrease in income after receiving the Corn special program with an average rank of 109.25 and a total rank of 874.00. However, 200 farmers experienced an increase in income after receiving the Corn special program, with an average rank of 104.31 and a total rank of 20,862. In addition, one person showed that their income remained the same both before and after receiving the program. The statistical analysis result is indicated by the Asymp. Sig. (2-tailed) value of 0.000 or < 0.05, showing a significant difference in farmers' income after receiving the Corn special program.

4. Conclusion

Agricultural programs focused on corn, including government efforts to promote better seed varieties, fertilizers, and pest control methods, have a notable effect on productivity. The adoption of technologies like drought-resistant and pest-tolerant seeds has helped farmers boost yields by allowing for denser planting and cultivation in previously less viable areas. These advances also reduce dependence on resources such as irrigation and pesticides, fostering more sustainable farming practices. Such programs, therefore, support productivity growth by improving both technological adoption and farm management practices [18, 21]. The results of this study contribute to these arguments by proving that the special programs effectively enhanced corn farmers' income in Gunungkidul Regency. The Multiple regression analysis results indicated that farm size had the most significant impact on income, with a regression coefficient of 0.412, followed by production costs (0.410), both indicating strong positive relationships. Training and extension services (0.083), seed quality (0.059), technology adoption (0.044), and access to fertilizer (0.039) also positively influenced income, although to a lesser degree. In addition, the Wilcoxon test revealed a significant income increase after the program, with average income rising from IDR 2,081,982.54 to IDR 3,100,108.13.

References

[1] P. Ainun, M. Sayuthi, and N. Pramayudi, "Kelimpahan Serangga Hama Pada Tanaman Jagung (Zea mays) Varietas Hibrida Di Lahan Perkebunan Badan Standardisasi Instrumen Pertanian (BSIP) Aceh (Abundance of Insect Pests on Hybrid Corn (Zea mays) Varieties on Agricultural Instrument Standardization Agency (BSIP) Aceh," *Jurnal Ilmiah Mahasiswa Pertanian*, vol. 8, no. 4, 2023, [Online]. Available: www.jim.usk.ac.id/JFP

[2] S. Eka Pramono, R. Widyaningsih, and D. Sulistianingsih, "Pemberdayaan Masyarakat Melalui Produk Olahan Untuk Meningkatkan Perekonomian Masyarakat Desa Dlisen," *Jurnal Pengabdian Hukum Indonesia*, vol. 02, no. 2, pp. 192–198, 2020, [Online]. Available: https://journal.unnes.ac.id/sju/index.php/JPHI/index

[3] Ahmad Ni'matullah Al-Baarri, *Pembangunan Bidang Pertanian dalam Rangka Meningkatkan Ketahanan Pangan Nasional*, vol. 1. Semarang: Penerbit Indonesian Food Technologies, 2020.

[4] BPS Kabupaten Gunungkidul, "Luas Panen, Produksi, dan Produktivitas Jagung Menurut Provinsi, 2022-2023," 2023.

[5] D. O. Gilligan, "Safety Nets for Agriculture and Nutrition," in *Agriculture for Improved Nutrition: Seizing the Momentum*, International Food Policy Research Institute (IFPRI), 2019, ch. 10, pp. 104–112.

[6] G. Berhane, D. O. Gilligan, J. Hoddinott, N. Kumar, and A. Seyoum Taffesse, "Can Social Protection Work in Africa? The Impact of Ethiopia's Productive Safety Net Programme," *Econ Dev Cult Change*, vol. 63, no. 1, 2014.

[7] J. F. Hair JR, W. C. Black, B. J. Babin, and R. E. Anderson, *Multivariate Data Analysis*, 7th ed., vol. 7th. Pearson Prentice Hall, 2010.

[8] B. G. Tabachnick and L. S. Fidell, *Using Multivariate Statistics*, 6th ed. Boston: Pearson, 2013.

[9] I. Ghozali, *Aplikasi Analisis Multivariate Dengan Program SPSS versi 19*, vol. 19. Semarang: Universitas Diponegoro Publishing, 2011.
[10] V. Ngeno, "Adoption of dairy feed technology and its impact on smallholder farmers' income and poverty in Kenya's south-western region," *Sci Afr*, vol. 23, p. e02123, Mar. 2024, doi: 10.1016/J.SCIAF.2024.E02123.
[11] M. Abdul-Majid, S. A. Zahari, N. Othman, and S. Nadzri, "Influence of technology adoption on farmers' well-being: Systematic literature review and bibliometric analysis," *Heliyon*, vol. 10, no. 2, p. e24316, Jan. 2024, doi: 10.1016/J.HELIYON.2024.E24316.
[12] K. M. Wonde, A. S. Tsehay, and S. E. Lemma, "Training at farmers training centers and its impact on crop productivity and households' income in Ethiopia: A propensity score matching (PSM) analysis," *Heliyon*, vol. 8, no. 7, p. e09837, Jul. 2022, doi: 10.1016/J.HELIYON.2022.E09837.
[13] Y. Nakano, T. W. Tsusaka, T. Aida, and V. O. Pede, "Is farmer-to-farmer extension effective? The impact of training on technology adoption and rice farming productivity in Tanzania," *World Dev*, vol. 105, pp. 336–351, May 2018, doi: 10.1016/J.WORLDDEV.2017.12.013.
[14] A. Gaaya, "Extension education in agricultural and rural development: Role of international organizations - the FAO experience," *CIHEAM*, vol. 4, no. 2, pp. 33–45, 1994.
[15] D. Nabuuma, C. Reimers, K. T. Hoang, T. J. Stomph, K. Swaans, and J. E. Raneri, "Impact of seed system interventions on food and nutrition security in low- and middle-income countries: A scoping review," *Glob Food Sec*, vol. 33, p. 100638, Jun. 2022, doi: 10.1016/J.GFS.2022.100638.
[16] S. Buah, S. K. Nutsugah, and A. Karikari, "Enhancing farmers' access to technology for increased rice productivity in Ghana," *Afr J Agric Res*, vol. 6, no. 6, pp. 4455–4466, 2011, [Online]. Available: http://www.academicjournals.org/AJAR
[17] A. R. Wardani, E. S. Rahayu, and Agustono, "Analysis of social economic factors affecting maize farming income in Marga Tiga district of East Lampung Regency," in *IOP Conference Series: Materials Science and Engineering*, Institute of Physics Publishing, Oct. 2019. doi: 10.1088/1757-899X/633/1/012054.
[18] E. Suryani, U. E. Rahmawati, and A. A. Zahra, "Improving Maize Production and Farmers' Income Using System Dynamics Model," *Journal of Agricultural Science*, vol. 14, no. 6, p. 68, May 2022, doi: 10.5539/jas.v14n6p68.
[19] Q. Yang, N. Zhang, Q. Lu, and X. Han, "Family Net Income, Input Factor Prices and Agriculture Services Selection Behavior of Maize Farmers," *Agriculture (Switzerland)*, vol. 14, no. 1, Jan. 2024, doi: 10.3390/agriculture14010062.
[20] M. Saavoss, T. Capehart, W. Mcbride, and A. Effland, "Trends in Production Practices and Costs of the U.S. Corn Sector," 2021. [Online]. Available: www.ers.usda.gov
[21] A. Ricome, J. Barreiro-Hurle, and C. Sadibou Fall, "Government fertilizer subsidies, input use, and income: The case of Senegal," *Food Policy*, vol. 124, Apr. 2024, doi: 10.1016/j.foodpol.2024.102623.
[22] W. R. Susila, "Fertilizer Subsidy Policy: Revisited," *Jurnal Litbang Pertanian*, vol. 29, no. 2, p. 2010, 2010.

Analyzing the agricultural sector's value-added contribution to Timor-Leste's gross domestic product and the affecting factors

Fidelino Carvalho, Eri Yusnita Alvianti, Dyanasari

Agricultural Economics, University of Tribhuwana Tunggadewi, Malang, Indonesia
*E-mail: fidelinocarvalho13@gmail.com

Abstract. This research analyzes the contribution of Agriculture Value Added (AV) income to Timor-Leste's GDP and the factors influencing it. Using time series data from the Word Bank data center, the annual Timor-Leste budget allocation report, and the Ministry of Agriculture and Forestry, from 2002 to 2022. the WarpPLS version 7.0 analysis tool was used to analyze the relationship between the Government Budget Allocated to the Ministry of Agriculture (GBAMA), Agricultural Infrastructure Investment (AII), and Found From Agricultural Development Partners (FADP). The findings show that FADP and GBAMA significantly contribute to AV income and GDP over 20 years. AII to GDP is also significant, but the impact of AII's relationship to AV is not significant. Furthermore, the impact of the direct relationship of AV on GDP insignificant relationship. In addition, the influence of indirect relationships from GBAMA, AII, and FADP on GDP through AV for 20 years also did not have a significant impact on GDP or the economic performance of Timor-Leste in general. The study suggests that adequate infrastructure investment in the agricultural sector can increase AV income and GDP. The government must develop independent funding capacity in the agricultural industry, monitor and evaluate its effectiveness, and ensure the success and sustainability of each project or program in the GBAMA and FADP funds. Considering that climate change and extreme weather are becoming world problems that will result in hunger and poverty in the future being inevitable, the Timor-Leste government needs to consider how important it is to invest in the agricultural sector for food security, poverty, and sustainable economic diversification.

Keywords: agriculture, affecting factors, GDP, Timor-Leste, value-added

1. Introduction

Regaining its independence on May 20, 2002, Timor-Leste heavily relies on oil production worldwide. The oil fund is the state's primary revenue source, with savings totaling over USD 12 billion. However, the government of Timor-Leste is more focused on growing the non-oil sector of the national economy due to the fluctuating price of oil. Additionally, Timor-Leste announced the National Strategic Development Plan in July 2011 in collaboration with Development Partners, who have created several initiatives to enhance the non-oil economy, this established the nation's long-term objectives for poverty alleviation and economic growth (1).

The National Strategic Development Plan for Timor-Leste 2011-2030 aims to build social capital, develop core infrastructure, and increase economic development. It aligns with UN millennium development goals and aims to achieve prosperity over different timeframes. However, fluctuating oil prices pose challenges for long-term planning in mineral-rich countries (2). Nonetheless, Timor-Leste's population is projected to increase by 13% between 2015 and 2022, reaching almost one million, a 16% increase from the first post-independence census in 2004, and 156 thousand more people between the 2015 and 2022 censuses. Competition in business activities is fueled by population growth, and as a nation's population increases, so does the size of its potential market, which motivates entrepreneurs to launch new businesses. Many population economists highlight the benefits of population expansion and identify people as the most important factor in economic success (3).

The World Food and Agriculture Organization's (FAO) production data estimates Timor-Leste's GDP, which heavily relies on petroleum. Agriculture, despite declining from 28.3% in 2008 to 17.1% in 2016, remains a significant contributor to the non-oil and gas GDP. Livestock experienced the highest increase, contributing between 2% and 3% of non-oil GDP (4). The Ministry of Agriculture and Fisheries in Timor-Leste has seen a significant reduction in funding for fisheries and aquaculture, with the sector receiving only 0.1% of the national budget allocation and 5% of the Ministry's budget between 2008 and 2016. The Ministry of Agriculture and Fisheries' budget was reduced to over USD 7 million in 2018, the lowest in ten years. To boost growth and diversification, the government and stakeholders must improve productivity and sustainability in the agricultural sector, encourage environmentally friendly farming, invest in infrastructure, R&D, and capacity building, and enhance the value of agricultural and animal products. The application of high-yielding crop varieties, improved irrigation and water management techniques, efficient soil management practices, and the use of fertilizers can all lead to increased agricultural productivity, which in turn has a large multiplier effect on the economy. This is especially true in lower-income countries where agriculture accounts for the majority of national income, employment, and foreign trade (5).

The background information provided means that to carry out this study, researchers must analyze the relationship between variables both directly and indirectly. These variables include the Ministry of Agriculture's government budget allocation, investments in agricultural infrastructure, financial support from development partners, and the effects these relationships have on the agricultural sector's added value and the Gross Domestic Product. To satisfy the National Strategic Development Plan, it is anticipated that this research will also be used as a resource by stakeholders creating strategic policies related to the sector's development.

Growth in the agricultural sector is two to four times more effective in increasing the incomes of the poorest people than in other sectors. Agriculture is also important for economic growth as agriculture accounts for 4% of the global Gross Domestic Product (GDP), in some lower-income countries, agriculture accounts for more than 25% of GDP *(World Bank, 2023*). The share of agriculture in global GDP is around 4%, which has been almost stable since 2000 (FAO, 2022). In addition, the agricultural sector has an important role in the national economy by increasing Gross Domestic Product (GDP), state foreign exchange, providing food and raw materials, alleviating poverty, and providing jobs, It has a multiplier effect. the relationship between input and output between industry, consumption, and investment. It is another indirect contribution made by the agricultural sector and its direct contribution. The agriculture sector is the primary one deserving of development for the national economy due to the comparatively significant multiplier impact (6) and (7). A nation's economy is driven by a multitude of sectors and

IOP Conf. Series: Earth and Environmental Science **1460** (2025) 012010 doi:10.1088/1755-1315/1460/1/012010

subsectors. One of the most significant indicators of a nation's economic activity is its gross domestic product, or GDP, which particularly shows the size and performance of an economy (8). However, in the context of the contribution of the agricultural sector to the GDP of Timor-Leste, it shows that from 2017 to 2020, Timor-Leste's GDP grew negatively, but in 2021, it slightly increased, this could be because the economy, which is heavily reliant on the public sector, is impacted by factors like El Niño, COVID-19, population growth, and political uncertainty, 20% of Timor-Leste's non-oil GDP comes from the agricultural sector, which has decreased from 32% in 2006 to 17.6% in 2015 and is still declining due to several issues, including a drop in average per capita production (9). In Timor-Leste, almost 80% of impoverished households reside in rural areas and rely on agriculture as their primary source of income. As a result, it is generally accepted that expanding the agriculture industry is a key way to lower poverty in the nation. Despite this, the nation's agricultural production system is among the least developed in the world, and subsistence farming and extremely risk-averse farmers dominate all subsectors (10-11).

The total value of all products and services generated over a certain period is known as the gross domestic product or GDP. Typically, GDP can be computed annually. GDP might be made up of all government spending, investments, exports less imports that take place inside a certain area, as well as all consumption private and public. GDP is typically used to gauge the expansion and general health of the economy (12). Timor-Leste's yearly periodic GDP income data, which is the market worth of all finished products and services produced in a nation and is usually expressed in dollars, is used in this analysis. GDP acts as a standard to measure economic progress and make comparisons. Agriculture Value Added (AV) is the total value added from agricultural products produced in a region or country annually, including grains, fruits, vegetables, meat, dairy, and animal products. It assesses productivity and performance and is used to evaluate the sector's economic health. The entire government budget allotted to the Ministry of Agriculture is known as the Government Budget Allotted to the Ministry of Agriculture (GBAMA). It is computed by figuring out how much money is given to the Ministry of Agriculture throughout a specific time frame. The Ministry of Finance's annual report typically calculates this fund allocation using the value of the US dollar. The Ministry of Public Development receives funding for projects related to the agricultural sector through the Agricultural Infrastructure Investment (AII) program, which is funded by the State Budget. The distribution of infrastructure funding to the development of the agricultural sector is used in this study. The Ministry of Agriculture acquires funding for Agriculture Development Partners (FADP) by procuring development project programs from foreign organizations. The annual state budget reports use this aid, which is valued in USD.

1.1 Research Hypothesis Statement

H1 It is suspected that GBAMA, AII, and FADP for 20 years have impacted gross domestic product (GDP) income.

H2 It is suspected that VA for 20 years has impacted GDP income.

H3 It is suspected that GBAMA, AII, and FADP for 20 years have impacted AV income.

H4 It is suspected that GBAMA, AII, and FADP through VA for 20 years have impacted gross domestic product (GDP) income.

2. Methodology

2.1 Methodology and research design

The research design aims to structure a study and determine the data collection method. This study used secondary data from various sources, including journal articles, official documents, books, and newspapers. Data was collected from reports of Timor-Leste's Ministry of Finance (TLMF) and World Development Indicators (WDI) for 20 years in the form of time series data with the consideration that the series data for the past 20 years is data available starting from Timor-Leste's independence in 2022, with using Microsoft Word, Excel, and the WarpPLS version 7.0.

2.2 Framework

In this research, data for the measuring variables used were obtained from the annual report of Timor-Leste's Ministry of Finance, as well as WDI data for the years 2002–2022. The availability of data is the reason this time was selected, Gross Domestic Product (GDP) is the dependent variable, and agriculture income value added (AV) serves as a mediating variable, the independent variables are the Government Budget Allotted to the Ministry of Agriculture (GBAMA), Agricultural Infrastructure Investment (AII), and Funds from Agricultural Development Partners (FADP). The 20 years of yearly data collected for a specific purpose will be the basis for the data sample used in the econometric analysis.

The model of the conceptual framework for this research can be seen in Figure 1 as follows:

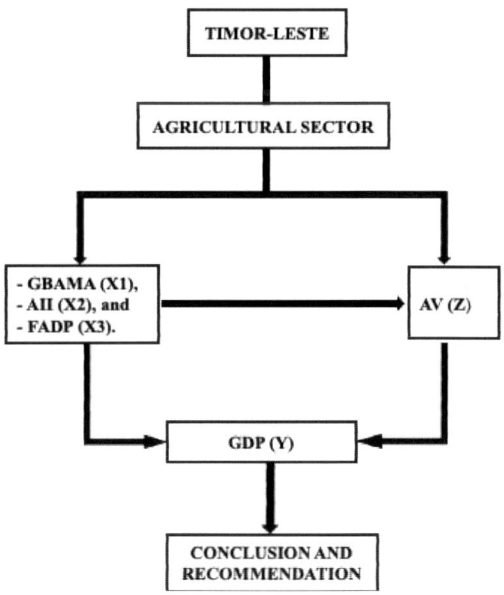

Figure 1. Illustration model of conceptual framework

2.3 Data Collection Method

Data was collected from the annual report of Timor-Leste's Ministry of Finance (TLMF) and Word Bank Development Indicators (WDI), and their period is 2002–2022. The details of the data are given in Table 1.

Table 1. Functional relationship and source of data

Name of variable	Type of variable	Abbreviation	Source of data	Time
Gross Domestic Product	Dependent	GDP	WDI	2002-2022
Agriculture Value Added	Mediation	AV	WDI	2002-2022
Government Budget Allocated to Ministry of Agriculture	Independent	GBAMA	TLMF	2002-2022
Agricultural Infrastructure Investment	Independent	AII	TLMF	2002-2022
Funds from Agriculture Development Partners	Independent	FADP	TLMF	2002-2022

3. Results

3.1 Results from the Influence of direct effect relationships

The SmartPLS analysis's findings, which indicate a direct effect link between factors, are displayed in Table 2. Accordingly, if the P-value is less than 0.05, there is an influence from a direct relationship. If the P-value is greater than 0.05, there is no influence from a direct relationship.

Table 2. Direct effect test results

No.	Relationship between variables		Path Coefficient	P-value	Description
1.	GBAMA	GDP	-0.561	<0.001	Significant
2.	AII	GDP	-0.278	0.039	Significant
3.	FADP	GDP	0.453	<0.001	Significant
4.	AV	GDP	-0.257	0.053	Insignificant
5.	GBAMA	AV	0.282	0.036	Significant
6.	AII	AV	-0.151	0.178	Insignificant
7.	FADP	AV	-0.297	0.029	Significant

According to the test results, which may then be further interpreted to address the assumptions that have been developed in the past, the following conjectures can be answered:

1. The GBAMA (X1), to GDP (Y), has a significant relationship because with a p-value of less than 0.05, namely (<0.001), this hypothesis is accepted.
2. The allocation of funding for AII funds (X2) to GDP (Y) has a significant relationship because it has a p-value of less than 0.05, which is (0.039), so this hypothesis is accepted.

3. The FADP fund assistance (X3) to GDP (Y) has a significant relationship because with a p-value of less than 0.05, namely (<0.001), this hypothesis is accepted.
4. The AV income (Z) to GDP (Y) has an insignificant relationship because, with a p-value greater than 0.05, namely (0.053), this hypothesis is rejected.
5. The GBAMA (X1) to AV (Z), has a significant relationship because with a p-value of less than 0.05, namely (0.036), this hypothesis is accepted.
6. The allocation of funding for AII (X2) to AV (Z) has an insignificant relationship because, with a p-value greater than 0.05, namely (0.178), this hypothesis is rejected.
7. The FADP (X3) to AV (Z) has a significant relationship because, with a p-value of less than 0.05, namely (0.029), this hypothesis is accepted.

3.2 Results from the influence of indirect effect relationships

In addition to examining direct correlations, WarPLS allows for the analysis of indirect correlations as well. Specifically, this includes the indirect impact of the relationship between exogenous variables (GBAMA, AII, and FADP) and endogenous variables (GDP) via mediation variable (AV). See Table 3 below for further information on the outcomes of evaluating this indirect effect.

Table 3. Results of test indirect effect relationship

No.	Variables Relationship between			Path Coefficient	*P-value*	Description
	Exogen	Mediation	Endogen			
1.	GBAMA	AV	PDB	0.144	0.110	Insignificant
2.	AII	AV	PDB	0.071	0.276	Insignificant
3.	FADP	AV	PDB	-0.116	0.163	Insignificant

Based on the results of the analysis of the indirect influence test above, it can be further interpreted to answer the conjecture of the hypothesis that has been built previously, which is as follows:

1. The allocation of GBAMA (X1) to GDP (Y) through AV (Z) has an insignificant relationship effect because, with a p-value greater than 0.05, namely (0.110), this hypothesis is rejected.
2. The allocation of AII (X2) to GDP (Y) through AV (Z) has an insignificant relationship because, with a p-value greater than 0.05, namely (0.276), this hypothesis is rejected.
3. The FADP (X3) fund assistance to GDP (Y) through AV (Z) has an insignificant relationship because, with a p-value greater than 0.05, namely (0.163), this hypothesis is rejected.

4. Discussion

4.1 The influence of the direct relationship between the GBAMA, AII, and FADP on GDP.

Based on the results of this study to answer the hypothesis and research objectives, the findings in this study reveal that:

1. Hypothesis 1: It is suspected that GBAMA, AII, and FADP for 20 years have impacted gross domestic product (GDP) income.
   GBAMA has had an impact on GDP over the past 20 years because it has a significant relationship in the strong category, showing that the GBAMA has had a significant impact on Timor-Leste's GDP income over the past 20 years. These results are consistent with a study

(13) that reported non-oil GDP growth of about 2 percent in 2019, showing a rebound from two years of negative growth, -3.8 percent in 2017 and -0.8 percent in 2018. The relationship between AII and GDP has a significant impact but is in the weak category, this shows that AII has had an impact on Timor-Leste's GDP income over the last 20 years. The analysis's findings are consistent with the report (14). According to (15), the infrequent use of irrigation will affect improving agricultural productivity and decreasing agricultural yields, given the significance of irrigation for crop production. The lowest percentages of irrigation use are seen in the eastern region (2%), central region (6%), and western region (12%). The relationship of FADP to GDP also has a significant impact in the strong category, this shows that for 20 years FADP has had an impact on the income of Timor-Leste's Gross Domestic Product. The results of this analysis are in line with the research report (16), which notes that even if the GDP of agriculture is growing at an average yearly rate of 5%, less money was allocated to this sector in 2019. This is a result of the agricultural sector's comparatively low government investment and heavy reliance on foreign development assistance funds, which account for twice as much of the sector's total spending.

2. Hypothesis 2: It is suspected that VA for 20 years has impacted GDP income.
   The relationship between AV value-added income and GDP has an insignificant influence, so the H2 hypothesis is rejected. This shows that for 20 years AV revenue has no impact on GDP income. The results of this analysis follow a report (Ministry of Finance of Timor-Leste, 2021) which states that in 2019, agriculture contributed 14.2% to Timor-Leste's non-oil GDP, but the value-added income from this agricultural sector has been on a downward trend since 2006, until now. In addition, the research report (17)also adds that during three of the previous five years 2017, 2018, and 2020 the actual non-oil GDP growth of the economy of Timor-Leste was at a negative rate. GDP per capita fell in 2021 despite a small increase in GDP due to an average annual population growth of 1.9 percent.
3. Hypothesis 3: It is suspected that GBAMA, AII, and FADP for 20 years have impacted AV income.
   The relationship between GBAMA and AV has a significant relationship, so this hypothesis is accepted, this shows that for 20 years GBAMA has had a significant impact on the value-added income of AV Timor-Leste. This result is in line with a report by *The Ministry of Agriculture and Fisheries Timor-Leste* which reports that Timor-Leste's GDP contribution from the agricultural sector is around 20%. In addition, since 2010, non-oil and gas GDP has received an average contribution from the agricultural sector of 19.2% (18). The relationship between AII and AV has a non-significant impact, so this hypothesis is rejected. This shows that AII's funding for 20 years has not had an impact on AV Timor-Leste. In addition, the relationship between FADP and AV has a significant relationship, so this hypothesis is accepted. This shows that over the past 20 years of FADP funding has had an impact on Timor-Leste's AV revenues, these findings indicate that the government's 20-year reliance on dynamic FADP funding may have been followed by Timor-Leste's lack of value-added (AV) results. The results of this analysis are in line with the research report (19) which says that although agricultural GDP has an average annual growth rate of 5%, but the budget allocation for the agricultural sector decreased in 2019, public spending for the agricultural sector is very low and greatly exceeds the foreign development assistance funds for the agricultural sector, which amounts to more than double the public expenditure on the agricultural sector.

IOP Conf. Series: Earth and Environmental Science **1460** (2025) 012010 doi:10.1088/1755-1315/1460/1/012010

4.2 The influence of the indirect relationship between the GBAMA, AII, and FADP through VA on GDP.

1. Hypothesis 4: It is suspected that GBAMA, AII, and FADP through VA for 20 years have impacted gross domestic product (GDP) income.
   The impact of the indirect relationship of GBAMA on GDP through AV shows an insignificant influence, so this hypothesis is rejected. It can be concluded that there is no GDP impact from GBAMA through AV after 20 years. This demonstrates that there is no discernible effect on GDP from the 20-year GBAMA allocated to boost gross domestic product income through agricultural AV. The subsequent AII to GDP through AV is also insignificant, so this hypothesis is rejected, this shows that AII funding allocated from the government budget to increase gross domestic product income through agricultural value-added income AV for the last 20 years the impact on GDP is not significant. In addition, FADP on GDP through AV is also insignificant, so this hypothesis is rejected. This shows that FADP, which funds the programs and projects of the Ministry of Agriculture to increase gross domestic product income through agricultural value-added income AV, has not had a significant impact on GDP over the past 20 years.

4.3 The implication and novelty of this research

Timor-Leste's agricultural value-added (AV) income has not increased much due to the distribution of AII funds for the past 20 years. Furthermore, the value-added income (AV) performance during the past 20 years has not had a major impact on GDP or gross domestic product income, or Timor-Leste's economic performance. The novelty of this study is that this study analyses the significance of the relationship between variables that have been determined both in direct and indirect relationships, with the parameters of all the measurement variables used being all funding buckets that have been allocated for 20 years by the government of Timor-Leste.

5. Conclusion

Based on the results of this study using the WrapPLS in analysing the impact of direct and indirect relationships of independent variables (GBAMA, AII, FADP), variable mediation (AV), and endogenous or dependent variables (GDP), the results of the analysis in this study can be concluded that:

The impact of the direct relationship between the GBAMA and FADP over the past 20 years has played an important role in contributing to the AV and GDP or the overall economic performance of Timor-Leste. The influence of AII's direct relationship to GDP is also significant, but the impact of AII's relationship to AV is not significant. This shows that the funding allocated for AII for 20 years has no impact on AV revenue.

Furthermore, the impact of the direct relationship of AV on GDP also shows an insignificant relationship in the context of the analysis of this study, this means that perhaps for 20 years the income from AV-added products does not have a significant impact on the income of the Gross Domestic Product (GDP) of Timor-Leste.

In addition, the influence of indirect relationships from GBAMA, AII, and FADP on GDP through AV for 20 years also did not have a significant impact on GDP or the economic performance of Timor-Leste in general.

Recommendation

An adequate government budget for the Ministry of Agriculture and the Ministry of Development can guarantee the production and productivity of the agricultural sector which can ultimately increase the added value of agricultural commodities and contribute to Timor-Leste's GDP income. Therefore, the government of Timor-Leste in the future needs to increase the development of irrigation infrastructure, modernize agricultural equipment, and provide effective farmer training programs, the government must also evaluate the projects and programs of the ministry that have been implemented so far and change the marketing structure of agricultural products, and also increase training for farmers on the added value of agricultural products to encourage the growth of agricultural SMEs, and it is also necessary to build a market information portal that is transparent, accessible and accessible to market participants.

Acknowledgments

I would like to express my deepest gratitude to the lecturers from the postgraduate of Tribhuwana Tunggadewi University who have given me their direction and support in this research process. I would also like to thank for all of my family's support given during my research process.

References

[1] Harmadi SHB, Gomes RA. Developing Timor-Leste's Non-Oil Economy: Challenges and Prospects. Southeast Asian Economies. 2013;30(3):309.
[2] John S, Papyrakis E, Tasciotti L. Is there a resource curse in Timor-Leste? A critical review of recent evidence. Development Studies Research. 2020 Jan 1;7(1):141–52.
[3] Timor-Leste National Institute of Statistics 2023, Main Report Population and Housing Census 2022.
[4] Efuntade AO, Omotayo EO. Effect Of Population Growth On Economic Growth In Nigeria The impact of taxation on transfer pricing in Nigeria Economy View project. Available from: https://www.researchgate.net/publication/364669894
[5] López Angarita J, Hunnam K, Pereira M, Jonathan Mills D, Pant J, Shwu Jiau T, et al. General Fisheries Directorate, Ministry of Agriculture and Fisheries, Government of Timor-Leste, Dili, Timor-Leste Citation Creative Commons License Photo credits [Internet]. 2019. Available from: www.worldfishcenter.org
[7] Amoussouga Gero A, Egbendewe AYG. Macroeconomic effects of semi-subsistence agricultural productivity growth: Evidence from Benin and extension to the WAEMU countries. Sci Afr. 2020 Mar 1;7.
[8] Khairiyakh R, Irham I, Mulyo JH. Contribution of Agricultural Sector and Sub Sectors on Indonesian Economy. Ilmu Pertanian (Agricultural Science). 2016 Apr 11;18(3):150.
[9] Albetris A. Kontribusi Sektor Pertanian dan Pengaruhnya Terhadap Perekonomian Daerah Provinsi Jambi. J-MAS (Jurnal Manajemen dan Sains). 2019 Apr 29;4(1):96.
[10] Chowhan S, Rahman MdM, Sultana R, Abdur Rouf Md, Islam M. Contribution Of Agriculture Sector In The GDP Growth Of Bangladesh. Food and Agri Economics Review. 2023;3(2):51–4.
[11] Scheiner C. Timor-Leste economic survey: The end of petroleum income. Asia Pac Policy Stud. 2021 May 1;8(2):253–79.
[12] Timor-Leste Timor-Leste Poverty Making Agriculture Work for the Poor in Timor-Leste [Internet]. 2019. Available from: www.worldbank.org
[13] Inder^ B, Brown A, Datt^ G. Poverty and the Agricultural Household in Timor-Leste: Some Patterns and Puzzles.
[14] Adb. Sector Assesment (Summary): Agriculture, Natural Resources, And Rural Development 1.
[15] Đokić A, Jović S. Evaluation of agriculture and industry effect on economic health by ANFIS approach. Physica A: Statistical Mechanics and its Applications. 2017 Aug 1;479:396–9.
[16] Lopes H. Macroeconomic Policy And Financing For Development Division Timor-Leste to Graduate from LDC Category and Beyond through Structural Transformation and Economic Diversification [Internet]. Available from: www.unescap.org/knowledge-
[17] Timor-Leste Timor-Leste Poverty Making Agriculture Work for the Poor in Timor-Leste [Internet]. 2019. Available from: www.worldbank.org
[18] Fanzo J, Bonis-Profumo G. The Ministry of Agriculture and Fisheries The Democratic Republic of Timor-Leste The Sustainable Agriculture Productivity Improvement Project Timor-Leste Agriculture Sector Review. 2019.

[19] Fanzo J. Timor-Leste Agriculture Sector Review Process Planning and Program Generation for STEP- Compliant NC Manufacture of Complex Surface Components View project Improving evidence synthesis and decision-making in environment and health View project. 2019; Available from: https://www.researchgate.net/publication/354156322

The Role of Field Agricultural Extension Officers in Foot and Mouth Disease Control on Smallholder Cattle Farms in Semarang City, Central Java

Yan Ariana*, Joko Mariyono and Enny Tantini Setiatin

Department of Animal Science Master's Program, Diponegoro University, Semarang City, Indonesia

*E-mail: januarimaniez87@gmail.com

Abstract. Foot and mouth disease (FMD) is acute and contagious in cloven-hoofed animals caused by infection from RNA viruses belonging to the Picornaviridae family and genus characterised by erosion and blisters on the mouth, tongue, gums, nostril, nipples, and on the skin around the nails. The purpose of the study was to examine the incidence of the FMD pandemic in Semarang City, analyse the participation of field agricultural extension workers in FMD management for farmers in Semarang City, and analyse the field actions of field agricultural extension workers in the FMD Control team in Semarang City. Research uses qualitative methods by collecting information through questionnaires, interviews, observations and documentation of actual circumstances. The subject of the study was a field agricultural extension worker in Semarang City. The data obtained were processed using qualitative descriptive analysis. Data processing is carried out by objectively describing the nature of a situation or symptom at the research time using numbers from the beginning of data collection to data interpretation and presentation. The results of the study show that the involvement of extension workers in the extension process with roles as motivators, educators, facilitators and communicators is considered to have a positive and efficient impact

Keywords: Field agricultural extension, foot and mouth disesase (FMD), smallholder cattle farms

1. Introduction

Cattle farming is divided into two sectors, namely breeding and fattening, breeding aims to obtain puppies or milk as the main result while fattening aims primarily as a source of meat or a source of animal protein [1]. Local cattle in Indonesia are a Bali Cattle, Madurese Cattle and Peranakan Ongole [2]. Since 2020 there have been many diseases that attack cattle both bacteria and viruses, diseases in cattle caused by viruses are Foot and mouth disease (FMD) and Lumpy Skin Disease (LSD) [3]. Foot and mouth disease (FMD) is an acute and highly contagious viral infection in cloven-hoofed animals. The formation of vesicles/blisters and erosion in the mouth, tongue, gums, nostrils, nipples, and the skin around the nails characterises this disease. Foot and mouth disease can cause significant economic losses due to decreased meat/milk production and become an obstacle in the trade of animals and their products. The rapid spread occurs through the traffic of animals and their products, vehicles, and objects that are contaminated with the FMD virus. FMD cases in the city of Semarang in cattle, both beef and dairy cattle, reached 1,385 cases, with 88.73% recovered, 2.1% dead, and 8.47% forced to slaughter. The ability to detect and diagnose FMD quickly and accurately and control the traffic of susceptible animals and their products to other areas that are still free are actions to minimise the spread of FMD. FMD surveillance programs, both clinically and serologically, are routinely carried out in high-risk areas with appropriate epidemiological designs. Thus, suspected FMD cases can be detected early, and control and eradication programs in the event of an FMD outbreak can be implemented quickly and accurately. The increasing cases of FMD outbreaks have become a complicated problem for livestock farmers after 36 years of the Republic of Indonesia being free from the outbreak [4]. FMD control requires various parties, including the Semarang City Agriculture Service, and, in this case, an extension. The role of extension workers, including facilitators, motivators, educators, and communicators, is vital. In carrying out their roles, extension workers carry out many actions relevant to the problems, in this case, FMD control. The actions carried out include FMD counselling, intensive assistance, livestock mobility permits, and the launch of livestock vaccinations. The role and actions of Field Agricultural Extension Workers (PPL) need to be studied to determine their performance in controlling FMD outbreaks. The results of this study can be used to evaluate the FMD control program so that it will be faster and more efficient in the future.

2. Research Method

The study used qualitative methods by collecting information through questionnaires, interviews, observations and documentation of actual conditions. The study subjects were agricultural extension workers in the field of Semarang City. The data obtained were processed using qualitative descriptive analysis. Data processing was carried out by objectively describing the nature of a condition or symptom that occurred during the study using numbers from the beginning of data collection to data interpretation and presentation.

3. Result and Discussion

3.1. Livestock Conditions in Semarang City

Semarang City, which has a unique topography and consists of upper and lower city morphological parts, creates its advantages for Semarang City in terms of business diversity. One of the businesses supported by this diverse topography is livestock farming. Livestock farming is generally located in the Upper Semarang City morphological part, namely Ngaliyan District, Gunungpati District, Banyumanik District, Mijen District and part of Tembalang District. Currently, the livestock area of Semarang City has a total of around 1,767 hectares spread across several districts in Upper Semarang City. In Semarang City, three districts with great potential for developing cattle farming are Mijen District, Gunungpati District, and Ngaliyan District. Detailed data on the number of livestock and livestock groups in Semarang City can be seen in Table 1.

Cattle farming is one of the mainstay sub-sectors of livestock in agro-industrial activities. The development of this livestock business has a very positive impact on job creation and promises cash income so it can motivate the community to play an active role in developing superior livestock agribusiness to increase income and fulfil community welfare. The condition of livestock in Semarang City is mostly dominated by small-scale livestock farming, where all management activities are still carried out traditionally. 90% of livestock farming is a small-scale business with ownership ranging from 1-5 cattle [5]. The condition of livestock farming with minimal application of biosecurity in the cage environment increases the speed of FMD spread in the affected area. This is exacerbated by the farmers' need for knowledge in efforts to prevent and handle FMD

Table 1. Number of farmer groups and livestock in Semarang Municipality

Sub-district	Number of farmer groups	Slaughter Cattles (head)	Dairy cattles (head)
Mijen	15	2.690	67
Gunungpati	27	522	879
Banyumanik	6	87	119
Tembalang	5	228	119
Tugu	1	30	-
Semarang Barat	1	9	-
Ngaliyan	4	72	37
Pedurungan	1	6	-
Gayamsari	1	9	-
Total	61	3.653	1.221

Source: Dinas Pertanian Kota Semarang (2024)

3.2. FMD Outbreak in Semarang City

FMD outbreak is a disease that must be reported to the World Organization for Animal Health (OIE). It is the most critical disease for ruminant animals. It is the biggest threat to livestock development in Indonesia. Indonesia was declared an FMD-free country in 1986 through the Decree of the Minister of Agriculture 260/Kpts/TN.510/5/1986, and in 1990, Indonesia succeeded in gaining world recognition for its FMD-free status without vaccination as stated in OIE Resolution Number XI of 1990 after OIE, FAO/APHC and ASEAN sent teams to evaluate the FMD status in Indonesia. The first confirmed FMD case in Semarang City was in the Gunungpati District. The FMD case in Gunungpati District infected beef cattle originating from livestock trading activities from Semarang Regency. The absence of strict traffic control at this time has caused control of FMD-infected livestock to not be optimally implemented in the Semarang City area. This has caused livestock infected with FMD to enter the Semarang City area.

3.3. Efforts to Prevent and Control FMD in Semarang City

Prevention and control of FMD in Semarang City refers to the Circular of the Minister of Agriculture Number 04/SE/PK.300/M/6/2022 concerning the control of Foot and Mouth Disease that there are 5 main steps in controlling FMD outbreaks, namely:

1. Optimisation of the Agricultural Extension Agency (BPP)

Control of the spread of FMD is carried out in all administrative areas down to the sub-district level. BPP plays a direct role in controlling FMD, including a) Coordination and synergy of FMD control activities with the Regency/City, Provincial, and National Veterinary Authorities; b) Data collection related to livestock profiles in each region, including livestock populations at risk and their locations (by name by address) for contingency measures; Strengthening village extension posts in the implementation of public communication by agreed posts; c) Assistance and supervision of FMD control activities.

2. Increased supervision of traffic, animal products, and FMD carriers.
3. Animal immunisation through vaccination
4. Training and reporting
5. Communication, Information and Education.
   Communication, Information, Education (KIE) is one strategy in achieving increased public understanding. Increasing public awareness in the form of communication, information, and education (KIE) activities so that the public knows all the symptoms of the disease and helps secure infected animals and does not move animals. This needs to be done so that farmers participate in the prevention, control, and early detection of FMD in cattle so that control is more effective.

3.4. The Role and Action of Agricultural Extension Workers in FMD Handling Efforts

The Agricultural Extension Agency (BPP), as an extension officer at the sub-district level, has the closest access to farmers who are threatened by the FMD outbreak and has a great responsibility in the government's efforts to overcome and control this outbreak. Extension workers are the spearhead of agricultural development in Indonesia, one of the successes of agriculture is in the hands of extension workers because extension workers interact directly with farmers so that programs can be directly delivered and implemented to farmers [6]. Extension workers' role in handling FMD outbreaks is divided into five roles: facilitator, motivator, educator, communicator and action.

The first role of extension workers is as a facilitator. Extension workers as facilitators provide supporting infrastructure for activities, increasing learning motivation and expertise [7]. In handling FMD outbreaks in Semarang City, the role of extension workers as facilitators includes helping farmers to get services from the FMD task force team, helping farmers monitor access to FMD-prone areas, providing assistance with vaccinations for cattle, helping farmers verify cattle infected with FMD, helping farmers isolate areas exposed to FMD, providing counselling on FMD and its management, helping groups to get access to the Department of Agriculture, helping farmers in data collection at IKSIKNAS, and facilitating compensation for cattle that died of FMD infection. Facilitation of feeding and vaccination has also been carried out well.

The second role of the extension worker is as a motivator. Extension workers, as motivators, provide motivation and encouragement for farmers and livestock breeders to develop their businesses and encourage them to increase their production results. In the extension process, the role of extension workers as motivators is to encourage farmers to maintain the cleanliness of the pens, encouraging farmers to maintain livestock health, encouraging farmers to maintain livestock that has been infected with FMD, encouraging farmers to innovate to overcome FMD, encouraging farmers to want to use new technology, encouraging farmers to follow technical guidance held by extension workers/agricultural services, and encouraging strict biosecurity activities in livestock groups. Extension activities for livestock breeders require the role of a motivator because it is very important to support farmers in adopting good innovations and

IOP Conf. Series: Earth and Environmental Science **1460** (2025) 012011 doi:10.1088/1755-1315/1460/1/012011

developing protection against diseased livestock. Extension workers' motivators motivate farmers to develop their livestock businesses or livestock groups, motivate farmers about good livestock farming methods, and support livestock behaviour to raise livestock better to achieve optimal results [8].

The third role of extension workers as educators in livestock extension activities is to educate farmers and help provide solutions to problems faced by farmers. In handling FMD outbreaks, extension workers help provide education to increase public awareness through communication, information, and education (KIE) activities so that the public knows all the disease symptoms and help secure infected animals and not move animals. Extension workers carry out their role as educators by providing training to livestock groups, providing ideas or concepts to farmers, demonstrating directly after providing ideas or concepts, demonstrating how to deal with FMD, demonstrating how to care for infected cattle, excellent and correct sanitation methods, providing information about the impact of FMD infection. This follows the statement that Extension workers carry out activities that can increase farmers' insight, provide examples of good cultivation, and implement one of the innovations in the livestock field to farmers [9].

The fourth role of the extension worker is as a communicator. In acting as a communicator for farmers, the extension worker plays a role in communicating well with farmers, helping to accelerate the flow of information received by farmers, and assisting farmers in decision-making [10]. The role of a communicator by the extension worker has been carried out by being able to communicate well with farmers, being able to guide farmers well, using print media in extension activities, having extensive knowledge about FMD so that they can carry out knowledge transfer well, preparing materials before delivering information about FMD before extension activities, having good technical knowledge and practices during extension activities, and delivering information that is easy for farmers to understand.

The last role is action; in this role, it describes the ability of the extension worker to carry out direct action in the field. The direct action carried out by this extension worker aims to ensure that the extension that has been carried out has an impact and is implemented and applied correctly by farmers. This action also avoids misunderstandings between the extension material and the capture of farmers so that the final goal of the extension can be achieved, which is impactful and effective. The impact of the outreach and actions that have been carried out is to ensure that livestock farmers receive services from the FMD task force team, livestock farmers are aware of and aware of access to FMD-prone areas, vaccination assistance for cattle to be carried out properly and correctly, ensuring that farmers verify cattle infected with FMD, ensuring that farmers isolate areas exposed to FMD, ensuring that counselling regarding FMD and its management is carried out properly and correctly, ensuring that farmer groups get access to assistance from the Agriculture Service, ensuring that farmers are recorded in IKSIKNAS, and facilitating compensation for cattle that died of FMD infection.

4. Discussion

The success of the extension can be measured by the extent to which the extension worker carries out his role as a facilitator, motivator, educator, communicator and action in handling the FMD outbreak problem in Semarang City. The success of the extension worker from each indicator can be seen in how effective the extension worker is in organising and moderating this activity, how well the extension worker is able to promote cooperation between farmers and other agencies, how well the extension worker can adjust the communication style with various audiences, either using technical language or language that is easier for farmers to understand, how well the extension worker can improve farmers' knowledge and practical skills, and the extent to which farmers apply these new techniques in their agricultural practices, how much influence it has in inspiring positive changes in farmers' attitudes and behaviour, real changes in agricultural yields, increased farmer income, or improved their overall quality of life. The ratio of respondents' answers to the indicators of the success of the role of extension workers is presented in full in Table 2.

Based on the results in Table 2, it is known that the role of PPL has been carried out well, as can be seen from the high value of each indicator. However, in the facilitator indicator, the success value still needs to be higher due to limited facilities, infrastructure, and costs in handling the FMD outbreak in Semarang City. In addition to the facilities provided in the form of feed and vaccine assistance, facilities are also needed to update the procedures for recording livestock data, which are more advanced to facilitate management and ensure that livestock traffic is carried out and appropriately supervised. In the next few years, it is hoped that there will be digitalisation of livestock data collection, this can be done with an application or website that can be accessed by farmers who have previously been socialised about the existence of the application or website. In the communicator indicator, four extension workers have never played a role in handling the FMD outbreak, this is because the respondents are new extension workers who have only started working as extension workers in 2023 after the FMD outbreak is over.

Table 2. The success indicator of the PPL role

PPL Role	Often	Seldom	Never
Facilitator	17	11	0
Motivator	20	8	0
Educator	19	9	0
Communicator	19	5	4
Action	18	9	1

Source : Primer Data (2024)

5. Summary and Recommendations

The involvement of extension workers in the extension process has fulfilled the role of facilitators, educators, motivators, and communicators. The extension officials' actions are considered to positively and efficiently impact handling the FMD outbreak in Semarang City. It is necessary to optimise BPP as the government's last hand in overcoming and controlling the threat of epidemics that may occur in the future, as well as improving facilities and digitalising BPP in prevention efforts and long-term programs in handling possible epidemics and diseases in the livestock sector to minimise losses for farmers as the main actors in the livestock business and increasing the capacity of agricultural-human resources, both extension workers and farmers through routine training programs so that they can overcome technical and non-technical obstacles in the livestock business being carried out.

IOP Conf. Series: Earth and Environmental Science **1460** (2025) 012011 doi:10.1088/1755-1315/1460/1/012011

References

[1] Firman, A., I. Trisman and R.H. Puradireja. 2022. Dampak ekonomi akibat outbreak penyakit mulut dan kuku pada ternak sapi dan kerbau di Indonesia. Mimbar Agribisnis: Jurnal Pemikiran Masyarakat Ilmiah Berwawasan Agribisnis. 8 (2):1123-1129.
[2] Sudarmono A. S. dan Y. B. Sugeng. 2008. Sapi Potong. Penebar Swadaya, Jakarta.
[3] Akbar, W. 2023. Analisis struktur pasar ternak sapi kurban berdasarkan konsentrasi penjual di kabupaten Batang hari pada masa pandemi covid-19. Universitas Jambi. (THESIS)
[4] Silitonga. R. J., A, Caraka D., 2016. Ancaman masuknya virus penyakit mulut dan kuku melalui daging ilegal di Entikong, Perbatasan Darat Indonesia dan Malaysia. Jurnal Sain Veteriner. 34 (2); 147–154
[5] Matondang, R.H., dan S. Rusdiana. 2013, Langkah langkah strategis dalam mencapai swasembada daging sapi/kerbau 2014. Jurnal Litbang Pertanian. **32** (3): 131-13
[6] Purwatiningsih, N. A., A. Fatchiya, dan R. S. H. Mulyandari. 2018. Pemanfaatan internet dalam meningkatkan kinerja penyuluh pertanian di Kabupaten Cianjur. Jurnal Penyuluhan. 14 (1).
[7] Wibowo. H.S., N. Sutjipta, I.W india. 2018 'Peranan penyuluh pertanian lapangan (PPL) sebagai fasilitator dalam penggunaan metode belajar pendidikan orang dewasa (andragogi) (Kasus di Gapoktan Madani, Desa Sampalan Klod, Kecamatan Dawan, Kabupaten Klungkung, Provinsi Bali)', Jurnal Agribisnis dan Agrowisata (Journal of Agribusiness and Agritourism).
[8] Marbun. D.N., V.D., S. Satmoko dan S. Gayatri. 2019. Peran penyuluh pertanian dalam pengembangan kelompok tani tanaman hortikultura di Kecamatan Siborongborong, Kabupaten Tapanuli. Jurnal Ekonomi Pertanian dan Agribisnis.
[9] Tumengkol, T., Dahar2, D., & Ashari3, U. 2021. Respon Petani Cengkeh Terhadap Peranan Penyuluh Pertanian Lapangan di Desa Momalia II Kecamatan Posigadan Kabupaten Boolang Mongondow Selatan. 5(4), 1157–1168
[10] Khairunnisa, N. F., Saidah, Z., Hapsari, H., & Wulandri, E. (2021). Presepsi petani tentang peran penyuluh pertanian dalam peningkatan pendapatan petani jagung hibrida. Jurnal Pemikiran Masyarakat Iimiah Berwawasan Agribisnis, 7(1), 486–498.

Rearing Management and Profitability Analysis of Buffalo Farming in Mamasa Regency

Ruth Dameria Haloho[1*], Marsudi[1], Siti Nuraliah[1], Taufik Dunialam Khaliq[1], Agus Setiadi[2], Edy Rianto[2], Nadlirotun Luthfi[3], Irmayanti[1]
[1]Faculty of Animal Science and Fisheries, Universitas Sulawesi Barat, Indonesia;
[2]Faculty of Animal and Agricultural Sciences Universitas Diponegoro, Indonesia
[3]Faculty of Animal Husbandry, University of Darul Ulum Islamic Centre, Sudirman, Indonesia

*) Corresponding author: ruthdameria.haloho@unsulbar.ac.id

Abstract. This study was conducted to evaluate the rearing system and profitability of a smallholder buffalo farming in Mamasa Regency, West Sulawesi Province, Indonesia. This study used a survey method. The number of respondents were 60 buffalo farmers randomly. The data analysis were quantitatively and descriptively. The parameters were income analysis, profitability using One Sample T-Test using SPSS. The results showed that buffalo farming in Mamasa Regency was still very traditional. The feed provided consisted of forages such as napier, setaria, and pasture grass. Additional feed was only given to the dappled buffalo (the most expensive buffalo). The mating method was natural mating. The majority of rearing system was still extensive, i.e. the buffaloes were grazed on the open field. The total of revenue was IDR 34,991,667 per year; production cost was IDR 21,022,372.17 per year. The amount of income was IDR 13,969,294.5 per year, the profitability value was 66.44 %. It indicated that the profitability was greater than bank interest, i.e 4%, BEP was IDR 11,457,530.42 equal to a Pudu Buffalo heifer price. It can be concluded that the buffalo farming business was profitable and feasible to be developed.

Keywords: buffalo, rearing, profitablitiy, analysis, agribusiness

1. Introduction

In Indonesia, buffalo have a very important role in supporting the income of small farmers and contributing to increasing economic development as a source of food, working animals, and traditional ceremonies [1,2]. This can be seen from the increase in buffalo meat production from 2021 by 18,941,730 kg to 2023 by 22,110,800 kg [3]. Previous study showed that buffalo farming was still traditional and only being a side business and savings. Buffalo farming is currently not fully able to provide prosperity for farmers and the community [2]. Buffalo farming with a traditional system causes a lack of role for farmers in managing their livestock. An analysis of buffalo business has to be considered for the progress and development of buffalo farming businesses because the success and failure of a livestock business can be seen from the profit and cost ratio [4].

IOP Conf. Series: Earth and Environmental Science **1460** (2025) 012012 doi:10.1088/1755-1315/1460/1/012012

Some data showed that the population of buffalo in West Sulawesi in 2020 was 8,948, and decreased in 2021 to 7,761. Mamasa Regency, is the area with the largest buffalo population in West Sulawesi Province. The buffalo population in Mamasa Regency in 2020 was 7,392 heads and in 2023, it was 7,195 heads. In terms of area, Mamasa Regency is large enough to support the buffalo livestock business development program [5,6]. Problems that often arise in buffalo farming are lack of knowledge of rearing management, how to gain profit and business analysis to evaluate when the business is not developed. However, studies on buffalo business opportunities in Indonesia are still very limited. Its potential to improve the economy of farmers has not been adequately studied or discussed [1] especially in Mamasa Regency. Therefore, before developing buffalo business and its potential for improving the economics of small farmers, it is necessary to evaluate a rearing management and buffalo business analysis in Mamasa Regency.

2. Material and Method

The material used in the study was a buffalo farming business unit. The study was conducted using a case study and survey method. The study was located in Mamasa Regency, West Sulawesi Province, Indonesia. This area was chosen for the highest buffalo population in West Sulawesi Province, Indonesia. The respondents were randomly selected and the number of buffalo farmers chosen were 60.

Data collection was conducted through interviews with parties who have functions and duties according to the field with a questionnaire guide. The data consisted of primary and secondary data. Primary data were obtained from direct surveys in the field and interviews according to questionnaires regarding technical and financial aspects. Secondary data were obtained from books or literature, data from the Animal Husbandry Service and the Central Bureau of Statistics. Data analysis was conducted using descriptive and quantitative methods, namely to determine profitability, BEP.

Total revenue (TR) was the amount of money received by the farmers in a certain period of time, for example one year [7]. The income (μ) was obtained by subtracting the total cost (TC) from TR, where TC were calculated using the formula:

$TC = TFC + TVC$

Where:

TC = Total Cost

TFC = Total Fixed Cost

TVC = Total Variabel Cost

then income was calculated using formula:

$\mu = TR - TC$

$TC = TVC + TFC$

$TR = Y.Py$

Where:

μ = income

TR = Total Revenue

TC = Total Cost

TVC = Total Variabel Cost

TFC = Total Fixed Cost

Y = the number of livestock sold

Py = the selling price

Profitability ratio was the comparison between TR and TC in percentage (%).

Description:

1. If profitability > prevailing bank interest rate then the business was profitable.

2. If profitability < the prevailing bank interest rate then the business was not profitable.

Break Even Point analysis was used to examine the relationship between fixed costs (FC), variable costs (VC), profits and activity volume. If the income from total activity was the same as the TC, the business did not make a profit or loss , then this was called the break-even point (BEP), which was calculated using the following formula:

$$BEP\ in\ a\ Unit\ (Q\ BEP)\ = \frac{Total\ Fix\ Cost}{Selling\ Price\ Per\ Unit - Variable\ Cost\ Per\ unit}$$

$$BEP\ in\ IDR\ (Q\ Sales)\ = \frac{Total\ Fix\ Cost}{1 - \frac{Total\ Variable}{Sales\ Volume}}$$

3.Results and Discussion

3.1 *Rearing management*

Buffalo farming in Mamasa Regency was carried out extensively. Buffalo were grazed from morning to evening by considering the availability of grass on the land. The feed available in the pasture was grass. The grass consisted of napier, setaria and pasture grass. Additional feeding was only carried out by a few farmers who have dappled buffalo (the most expensive buffalo). This provided benefits to the farmers because there were no additional costs for feed. The farmers did not pay attention to the quality and quantity of the feed that was able to support the growth of buffalo. Farmers in Mamasa Regency only provided grass to the buffalo due to limited capital and lack of knowledge of farmers regarding the importance nutrition for their livestock. It indicated that the farmers in Mamasa Regency were still traditional. Therefore, routine counseling was required to introduce the provision of high quality of feed to livestock. [1] claimed that in Indonesia, the grazed buffaloes utilized natural grass as a source of feed and there was tendency to be no improvement because the diversity of grass in the grasslands still depended on natural vegetation. This study also showed that the grassland for buffaloes provided by natural pastures and some were planted by the buffalo farmers. The source of feed was still very possible to be developed both in the form of pasture, especially in range land, agricultural land that was not cultivated by farmers. Traditional farming practices contributed to low productivity of the buffaloes, because they were grazed on land with low quality grass. The most of farmers relied on the natural grass and only very few farmers used grass of high quality or additional feed such as concentrates for their buffalo [8]. The poor quality feed availability and inefficient feeding also resulted in poor reproductive performance and production of the buffaloes due to delayed oestrus cycles, increased calving intervals, and reduced growth rates [9]

Not only the traditional rearing system, but also the mating of buffalo in Mamasa Regency used natural mating. The buffaloes in Mamasa Regency were mated by bringing in male buffaloes when the females were in oestrous. In general, farmers did not mated their female buffaloes by artificial insemination (AI). This indicated that the mating process had not used technology to be more efficient. It led to inbreeding, low growth rate and reduced the birth rate of calves. So that, it could decrease buffalo production in Mamasa Regency. The low in the buffalo population was caused by not enough attention of rearing systems to animal nutrients requirements, relying on natural mating, and did not use a good recording [1]. A little attention and knowledge about

IOP Conf. Series: Earth and Environmental Science **1460** (2025) 012012 doi:10.1088/1755-1315/1460/1/012012

breeding programs also leads to high inbreeding rates in buffalo rearing systems [10].. Inbreeding occured randomly over several generations, so that the inbreeding rate increases by 10%–30% in each generation caused by natural mating in buffalo population. Some physical characteristics of inbreeding buffalo were albinoid skin colour, either a few or many white lines on the skin, one or two horns pointing downwards, short tails, and blue eyes [11].

3.2. Business analysis

The revenue of buffalo farmers is presented in Table 1. The study showed that the revenue of buffalo farmers was obtained from the sale of buffaloes that consisted of the sale of males and female buffalo, male and female calves, and culls. A total sales of buffalo per year was IDR 2,099,500,000 and an average of buffalo sales was IDR 34,991,666. This study showed that buffalo sales were carried out at any time, not only to fulfil the needs of farmers but also to traditional ceremonies requirements. In this study, the selling price of buffalo was influenced by several factors including age, gender, and body posture of the buffalo. [2] claimed that Buffalo sales are based on their characteristics including type, color, eye shape, body posture and buffalo horns. The revenue in this study was higher than the finding [2] that revenue of buffalo farmers in West Sulawesi per year in 2023 was IDR 30,200,000. It was might be due to the difference total of buffalo sold. The revenue is greatly influenced by the number of buffaloes sold by the farmer. The more buffaloes sold, the higher the total revenue received by the farmer.

The amount of buffalo production in this study was IDR 21,022,372.28 presented in Table 2, so the average income in this study was IDR 13,969,293.72. The income in this study was greatly affected by production costs. Production costs consisted of fix and variable costs. Fix costs in this study consisted of buffalo, cage and facility depreciation. Variable costs consisted of feed, labour and medicine costs. The amount of fix costs was IDR 3,751,157.00 and variable cost was IDR 17,271,214.28. [12] stated that the total production cost might increased in every year, depended on the increase in business scale.

Table 1. The revenue of buffalo business of farmers in Mamasa Regency

No	Kind of buffalo	The amount of revenue (IDR/ year)
1	Buffalo bull selling	1,454,000,000
2	Buffalo female calf selling	5,000,000
3	Buffalo male calf selling	192,000,000
4	Buffalo heifer selling	25,000,000
5	Buffalo young bull selling	123,000,000
6	Buffalo cow selling	228,500,000
7	Culled buffalo selling	20,000,000
	Total	**2,099,500,000**
	Average revenue of buffalo selling	**34,991,666**

Table 2. Analyses of Production Cost in Mamasa Regency

No	Variables	IDR / year
	Fix Costs	
1	Buffalo depreciation	2,800,000
2	Cage Depreciation	450,000
3	Facility depreciation	501.157
	Total Fix Costs	3,751.157
	Variable Costs	
1	Feed	11,516,081.28
2	Labor	5,629,333
3	Medicine	125,800

Total Variable Costs	17. 271.214.28
Total of Production Cost	21,022,372.28

Table 3. The Profitability of buffalo business in Mamasa Regency

No	Variables	IDR / year
1	Revenue	34,991,666.67
2	Production cost	21,022,372.28
3	Income	13,969,293.72
	Profitability (%)	66.44

The profitability of buffalo business in Mamasa Regency is presented in Table 3. The profitability obtained in study was 66.44% per year. This figure was higher than the bank interest (4% per year in 2024). It indicated that the buffalo farming business in Mamasa Regency was still profitable. [13] found that the livestock business was profitable if the profitability percentage was higher than the bank interest rate. Therefore, the buffalo farming business in Mamasa Regency was feasible to be run because it resulted in quite high profits. [14,15] claimed that the number of animals may lead the profit and the livestock business efficiency. The higher the number of buffalo, the greater the profitability of buffalo farming.

Table 4. Break even poin of buffalo business in Mamasa Regency

No	Variables	IDR /year
1	Fix cost	3.751.158,00
2	Variable cost	17.271. 214,28
3	Revenue	34.991.666,67
4	**BEP**	**11.457.530,42**
5	**BEP (head)**	**A pudu buffalo heifer**

The BEP of buffalo business in Mamasa Regency is presented in Table 4. This study showed that the BEP of buffalo units in this study was IDR 11,457,530.42. This value was equivalent to 0.56 ST or the price of a buffalo heifer in the Mamasa Regency. It indicated that farmers was at the break-even point (no profit and no loss) if they raised a heifer buffalo. The BEP in this study was greatly influenced by fix cost, variable cost and revenue. The lower the production cost and the higher the revenue obtained, the lower the BEP achieved. This showed that BEP in the buffalo business in Mamasa Regency could be achieved at low number of buffalo being raised. The study [16] found that BEP of the Molan Beef Cattle Fattening Business was IDR 267,858,142 and was equivalent to raising 21 cattle. According to [8], the production cost in the buffalo business could be offset by the selling price of buffalo. Buffalo business conditions in breeders can provide optimal benefits, if the production cost was lower than the revenue.

Analisis *One Sampel T-test*

The analysis used to determine data observations with the assumption of an average variable with a constant or hypothesis value using one sample t-test analysis [17]. One Sample T-Test Analysis was used to determine whether the buffalo business was profitable by comparing the profitability value with bank interest of 4%. Based on the results of the calculation of the sig value of the profitability of the buffalo farming of (0.000) meant that the profitability of the buffalo farming was very significantly different from the bank interest, so that the business was profitable and feasible to develop.

Conclusion

The management of buffalo farming in Mamasa Regency was run in a traditional way. Nevertheless, this business was profitable. The profit was 66.44 percent of the cost, that was greater than bank interest rate of 4 percent, and the BEP was achieved when the farmer raised one head buffalo heifer with the value of IDR 11,457,530.42. Therefore, buffalo farming business is feasible to be developed in Mamasa Regency.

Acknowledgments

We would like to thank the Directorate of Research, Technology and Community Service (DRTPM) Ministry of Education, Culture, Research and Technology for providing funding through national collaboration research program in 2024 with contract agreements 192/UN55.C/PG/2024

References

[1] Prihandini PW, Tribudi YA, Hariyono DNH, Sari APZNL, Praharani L, Handiwirawan E, Tiesnamurti B, Romjali E, Matitaputty PR, and Wiyono A (2023) Biodiversity of Indonesian indigenous buffalo: First review of the status, challenges, and development opportunities, Veterinary World, 16(11): 2217–2229

[2] Haloho RD, Palayukan J, Setiadi A, Rianto E, Luthfi N (2024). The feasibility of business of buffalo used in the traditional funeral ceremony (Rambu solo) in West Sulawesi, Indonesia. Adv. Anim. Vet. Sci., 12(3):523-531

[3] Central Bureau of Statistics. 2024. Diakses 1 Agustus 2024. https://www.bps.go.id/id/statistics-table/2/NDgxIzI=/produksi-daging-kerbau-menurut-provinsi.html

[4] Anis SD, E. Wantasen., S. Dalie., D. A Kaligis and U. Paputungan. 2015. Beef cattle feasibility study of house hold farm in Bolmong Regency, North Sulawesi Province of Indonesia. International

[5] Rompis JEG, Paat JF, Kawatutu MM, Demmalona (2013). Tatalaksana pemeliharaan ternak kerbau belang di kecamatan mamasa kabupaten mamasa provinsi Sulawesi barat. Jurnal Zootek (Zootek J.), 33(1): 68-79

[6] Suryaalim AM, Fausiah A (2023). Community perceptions of comparison of buffalo price criteria in Mamasa Regency, West Sulawesi Province. AGRIMOR, 8(2): 68-76

[7] Soekardono. 2009. Ekonomi Agribisnis Peternakan Teori dan Aplikasinya, Akademika Pressindo, Jakarta

[8] Rusdiana, S., Praharani, L. and Sianturi, R.G. (2020) The economic efficiency of buffalo business (Bubalus bubalis) with the shepherd. SOCA J. Sos. Ekon. Pertanian, 14(2): 275–287

[9] Deb, G.K., Nahar, T.N., Duran, P.G. and Presicce, G.A. (2016) Safe and sustainable traditional production: The water buffalo in Asia. Front. Environ. Sci., 4(38): 1–7

[10] Rusli, R. and Syahidin, S. (2021) Characteristics of breeders and development strategies for Gayo buffalo livestock in the Peruweren system [Karakteristik peternak dan strategi pengembangan ternak kerbau Gayo sistem peruweren]. J. Ilmu Teknol. Peternakan, 9(2): 81–89

[11] Talib, C., Herawati, T. and Hastono, H. (2014) Strategies for increasing buffalo productivity through improvement in feed and genetic. Wartazoa, 24(2): 83–96.

[12] Nafiu LO, Ode L, Sani A, Salam I (2020). Smallholder farmbased buffalo breed-ing in bombana regency reviewed from financial feasibility aspect. Int. J. Sci. Eng. Res., 11(1): 1228–1232

[13] Haloho RD. (2020). Analisis Usaha Penggemukan Sapi Potong Dengan Menggunakan Paradigma Agribisnis (Studi Kasus Pada Peternakan Sapi Potong Molan) di Kecamatan Binjai Barat Kota Binjai Provinsi Sumatera Utara. Agrimor., 5(1): 17-19.

[14] Bahta S, Baker TSM (2015). Determinants of profit efficiency among smallholder beef producers in botswana. Int. Food Agribus. Manage. Rev., 18(3): 107-129.

[15] Sugiarto M, Wakhidati YN, Einstein A, Khaerudin (2017). The competitiveness of beef cattle business on various agroecological zones in tegal regency. Anim. Prod., 19(2): 127- 134.

[16] Haloho RD, Elvany T (2021). Manajemen Pakan dan Analisis Profitabilitas Usaha Peternakan Sapi Potong Rakyat di Masa Pandemi Covid 19 di Kabupaten Langkat. Agrimor., 6(4):180-185.

[17] Trihendradi, C. 2019. Langkah Mudah Melakukan Analisis Statistik Menggunakan SPSS 19. Penerbit Andi, Yogyakarta.

IOP Conf. Series: Earth and Environmental Science **1460** (2025) 012013 doi:10.1088/1755-1315/1460/1/012013

Rice-Beef Cattle Integration on Irrigated Rice Fields to Increase Farmers' Income in Central Java, Indonesia

Afrizal Malik[1*], Agus Setiadi[2], Joko Mariyono[2] and Sutaryo[2]

[1] Research Center for Behavioral and Circular Economics, National Research and Innovation Agency, Jakarta, Indonesia
[2] Department of Agriculture, Faculty of Animal and Agricultural Sciences, Diponegoro University, Semarang, Indonesia

*E-mail: afri010@brin.go.id

Abstract. This study aims to analyze the income level of farmers who integrating lowland rice and beef cattle on irrigated land in Central Java. The study was conducted using a survey method with a questionnaire in Sukoharjo and Karanganyar Regency, Central Java Province. The selection of this location was determined purposively on the basis that these two regencies are centers of rice fields and beef cattle. The number of respondents was 140 farmers who cultivate rice fields and have beef cattle. Primary data collected included education, age, experience in cultivating rice fields and beef cattle, as well as input and output of rice field production. For beef cattle farming, input and output were analyzed. Data were analyzed using descriptive qualitative and qualitative analysis. The results of the study showed that the age of farmers was at a productive age (55.05 years), education (8.3 years) with 25.1 years of experience in rice farming and 21.6 years of cattle farming. Farmers in rice farming used the Inpari-32 variety as much as 88% and other varieties 12%. Farmers use inorganic fertilizers below the recommendation because they are substituted with organic fertilizers. The productivity of the paddy fields is 6,054 tons/ha. From the analysis of the integrated rice-beef cattle business, farmers' income is IDR 23,782,163 /planting season. Farmers' income is greater than the value of the Regency/City Minimum Wage (R/CMW) of the research location. This means that the integrated rice and beef cattle farming business is promising.

Keywords: paddy fields, beef cattle, integration

1. Introduction

The national rice requirement in 2024 is 31.2 million tons, while production is estimated at 31.9 million tons. From this data, there is a surplus of around 700 thousand tons as rice reserves, but to strengthen domestic rice reserves, 2.4 million tons are imported [1]. The decline in rice production is due to the productivity of lowland rice in Indonesia in the last 10 years stagnating at around 5.10-5.25 tons/ha [2]. Increasing the productivity of lowland rice continues to be carried out by implementing productivity-increasing technologies such as the use of inorganic fertilizers. This effort has a negative impact on decreasing soil fertility and environmental degradation.

The target for meat consumption in 2022 is 14.1 kg/capita/year and has only been met by 12.8 kg/capita/year [3]. The target is from ruminant livestock and poultry. The level of beef/buffalo meat consumption is still at 2.93 kg/capita/year, to meet national beef needs, some are still met through imports [4].

To reduce imports, domestic beef production needs to be increased. One effort that can be made is to increase the application of cattle farming technology, including the integration of paddy fields with cattle. The program aims to improve the condition of paddy fields sustainably through the use of organic fertilizers from cattle while increasing paddy productivity, encouraging the development of cattle populations, and increasing farmers' incomes.

Integrated rice-cattle farming can increase farmers' income by 69.88% with a rice planting area of one hectare and 11 cows [5]. The implementation of sustainable rice-cattle integration can increase the availability of nutrients in the soil and reduce the need for large amounts of feed [6]. The rice-cattle integration model is sustainable because it reduces the use of external inputs. Therefore, rice-cattle integration is an alternative strategy to increase rice and cattle productivity sustainably and farmers can produce biogas [7].

The application of crop-cattle integration in the utilization of local resources through the use of straw as animal feed and cow dung as organic fertilizer can increase productivity, reduce production costs and increase farmers' income [8]. Integrated rice and cattle integration was developed to overcome the problem of declining rice productivity, although intensification efforts have been made [9]. The integration of rice plants with cattle financially and technically prevents land degradation, so it is an alternative that can increase organic matter content, input utilization efficiency, and agricultural sustainability [10;11].

There are three main technological components in the crop and livestock integration system, especially rice-cattle integration, namely: (a). livestock cultivation technology, (b) rice cultivation technology, and (c) straw and compost processing technology [10;12]. These three technological components can be integrated synergistically, so the development of crop-livestock integration is carried out with an institutional approach [13]. In addition, the crop-livestock system can also be considered as a diversification of income sources for farmers [14]. There is a relationship and cycle of nutrients in the crop-livestock integration model [7]. The rice-cattle integration model can provide feed of around 33.3% of the total grass needed [15].

The implementation of this integration can reduce the use of inorganic fertilizers so that it is worth developing. Raising beef cattle using feed in the form of agricultural waste resulted in an average daily weight gain of 0.7 kg/head/day, and manure production of 0.45 tons/head/period for 4 months [16].

The benefits of rice-beef cattle integration are 27% compared to no integration [17]. Farmer income, land area and number of livestock have a very significant effect on the implementation of rice-beef cattle integration [18]. This study aims to analyze the income level of farmers who implement rice and beef cattle integration in irrigated rice fields in Central Java.

2. Research methods

The study was conducted in Sukoharjo and Karanganyar Regencies, Central Java Province. The selection of these locations was determined purposively on the basis that these two regencies are centers of paddy fields and beef cattle and have implemented a paddy field-beef cattle integration program. From the selected regencies, sub-districts and villages were then determined (Table 1).

Table 1. Research locations with samples of lowland rice and beef cattle farmers, 2024.

No.	Regency	Subdistrict	Number of farmer samples (farmers)	Village
1.	Karanganyar	Gedong Sari	70	Kragan and Karangturi
2.	Sukoharjo	Polokarto	70	Genengsari, Polokarto, and Kemasaan
	Total		140	

Primary data collected include education, age, experience in rice farming and livestock farming, inputs for rice production (seeds, fertilizers, pesticides, herbicides, and labor) and output. All data are converted per hectare. For beef cattle farming, input and output are analyzed.

The data was analyzed descriptively qualitatively and quantitatively to obtain a holistic picture, both input and output, while qualitative analysis to measure quantitative variables used simple statistical parameters such as percentage, maximum and minimum values, average values, and average growth which were carried out using the revenue and cost balance analysis method (R/C) and . benefit cost ratio (B/C)

After the data is tabulated, a descriptive analysis is carried out and the analysis of rice farming is calculated using the formula [19; 20; 21], namely:

$$\text{Income} = QPq - (\Sigma XiPXi + \Sigma yiPYi) \quad (1)$$

where:

Q = Amount of rice production (kg/ha)
Pq = Price of grain production (Rp/kg)
Xi = Type of variable input for rice farming Xi (i = 1,2,3,...,x)
PXi = Price of variable input for rice farming Xi (i = 1,2,3,...p)
Yi = Type of fixed input in rice farming Yi (i = 1,2,3,...,y)
PYi = Price of fixed input in rice farming Yi (i=1,2,p)

3. Results and discussion

3.1 Farmer characteristics

Farmer characteristics are indicators that measure farmers' ability to make decisions related to their farming activities [23]. The characteristics of farmers studied include age, education, and farming experience. The age of farmers is at a productive age. Age can also be a benchmark for the success of farming activities. Farmers who are of productive age will usually work better and more optimally compared to farmers who are no longer of productive age [24].

Farmers' experience in cultivating rice fields and cattle farming is quite long and experienced. Older age has longer farming experience so that it can compensate for the decrease in physical energy with deeper knowledge and skills when managing farming, especially traditional farming [25].

Table 2. Age, education level, experience in rice and cattle farming in the research area, 2024.

No.	Description	Average (years)	Range (years)
1.	Age	55.05	33-73
2.	Education	8.3	6-15
3.	Experience in farming rice fields	25.1	8-65
4.	Experience in cattle farming	21.6	3-50

Source: Primary data (processed), 2024

Experience in rice farming is one of the factors that plays an important role, because it can encourage and support the achievement of expected production. Farming experience is closely related to the age of farmers [26]. Age, education level and experience in rice farming and cattle farming can be seen in Table 2.

3.2 Performance and analysis of lowland rice farming

Farmers in cultivating paddy farming use several new superior varieties (NSV) and are dominated by Inpari-32 as much as 88%, while 12% of them use NSV, such as Sintanur, Ciherang, Ir-64, Mentik, Mekongga, M.70 D, and Cakra Buana. Farmers use these NSV because of their high productivity, and resistance to pests and diseases (P/D). New superior varietas (NSV) is a technological innovation that has a great opportunity to be adopted by farmers because it is technically easy to do, has high yields, and is resistant to certain P/D [27]. Farmers use more seeds than recommended (20-25 kg/ha). The high use of seeds by farmers is due to their lack of confidence in the quality of the seeds to be planted.

The use of inorganic fertilizers by farmers is lower than the fertilizer recommendations according to Permentan Number 13/2022 at the research location in both Karanganyar and Sukoharjo Regency, namely urea 350 kg/hectare, SP-36 50 kg/hectare, and KCl 50 kg/hectare or the use of NPK compound fertilizer 250 - 275 kg/hectare. The low and diverse use of fertilizers by farmers is more due to purchasing power due to the scarcity of capital by farmers. However, what is encouraging is that farmers tend to use organic fertilizers sourced from cow dung as a substitute for the use of inorganic fertilizers. The use of this organic fertilizer will improve soil structure.

Farmers use several types of pesticide brands. Farmers carry out P/D control economically, meaning that farmers use pesticides selectively. Farmers spray pre-emergence herbicides with the brands DMA, Gramaxone, and Run Up to reduce weeds in rice fields. DMA is used for broad-leaf weeds, while Gramaxone and Run Up are used for pre-emergence plants. Farmers also use ZPT/PPC as a substitute for inorganic fertilizers. Farmers process the land using hand tractors with a rental system. For planting, weeding, and harvesting, farmers use more contract systems, using either threshers or combine harvesters. Details of the use of seeds, fertilizers, pesticides, herbicides, and labor can be seen in Table 3.

The productivity of paddy fields produced is higher than the results in the application of paddy field integration with beef cattle, which is 5.34 tons/ha [28]. This is due to the use of inputs such as various fertilizers. In addition, it is thought to be due to the high use of manure (organic) which improves soil structure.

Table 3. Analysis of lowland rice farming in the research area (MT-II/March/April-June/July), 2024.

No.	Description	Physique	Value (IDR)
A	**Output**		**37,019,454**
B	**Input**		**15,965,379**
1.	Seed (kg)	35	548,326
2.	Fertilizer (Kg)		3,322,290
	a. NPK Phoska	315	724,500
	b. Urea	325	731,250
	c. ZA	12	32,400
	d. SP-26	16	44,800
	e. KCl	3	22,500
	f. Manure (kg)	1,595	797,500
	g. ZPT/PPC (IDR)	-	27,662
3.	Pesticide (IDR)	-	562,211
4.	Herbicide (IDR)	-	379,465
5.	Labor		12,094,761
	Male (Working People's Day/WPD)	40,5	4,050,000
	Women (Working People's Day/WPD)	30,6	2,488,000
	Contract (land cultivation, planting, harvesting) (IDR)		5,201,761
C	**Profit (A-B)**		**21,054,075**
	B/C (C/B)		**1.39**
	R/C (A/B)		**2.39**

* Source: primary data (2024)

3.3 Performance and analysis of beef cattle business

From the research results, the types of cattle raised by farmers are Simmental 80% and PO 20%. Farmers raise livestock for the purpose of (1) a source of family savings and (2) a source of manure. From the analysis of beef cattle farming with an average ownership of 1.8 Animal Units (AU) it is profitable. If the outpouring of labor is dominated by the family, then the profit is greater (Table 4).

From the analysis of integrated business (combining paddy fields and cattle farming), the profit from paddy fields farming is IDR 21,054,075 plus from beef cattle farming of IDR 2,728,088, so the profit from integrating paddy fields and beef cattle at the research location is IDR 23,782,163 per planting season. From this analysis, the value obtained is higher than the R/CMW of Sukoharjo Regency in 2024 of IDR 2,215,482/month and IDR 2,288,366/month for Karanganyar Regency. This means that the integrated paddy fields and beef cattle farming promises increased income, if managed properly according to productivity improvement technology.

Table 4. Analysis of beef cattle farming business in Sukoharjo and Karanganyar Regencies, 2024 (4 months).

No.	Description		Value (IDR)
A.	**Output**		**7,455,438**
1.	Adult cattle income		5,275,356
2.	Young cattle income		1,282,142
3.	Income from calves		601,190
4.	Income from manure		296,750
B.	**Input**		**4,727,350**
1.	Grass		578,571
2.	Bran		555,828
3.	Straw		73,142
4.	Artificial insemination (AI)		64,143
5.	drugs		27,666
6.	Labor @ Rp. 100,000		3,428,000
C.	**Profit (A-B)**		**2,728,088**
		B/C (C/B)	**0.58**
		R/C (A/B)	**1.58**

* Source: primary data (2024)

4. Conclusion and suggestions

Rice-cattle integration increases rice and cattle productivity, as well as farmer income. Rice-cattle integration provides income of Rp 23,782,163/planting season or equivalent to Rp 5,945,549/month. Farmer income is greater than the R/CMW value of the research location. Integrated rice and beef cattle farming is very promising, if managed properly according to productivity improvement technology. The contribution of cattle farming profits to lowland rice income is 12.95% (Rp 2,728.088)/lowland rice planting season. Rice-cattle integration is highly recommended to be carried out and developed, both at the research location and in other districts.

References

[1] Bapanas. 2024. Ketersedian beras nasional. Badan Pangan Nasional.

[2] Sudarmo A. 2021. Produktivitas Padi Nasional Stagnan, Kementan Genjot Penyuluh Pertanian. Liputan6.com.2021. Available from: https://www.liputan6.com/bisnis/read/4775878/ produktivitas-padi-nasional-stagnan-kementan-genjot-penyuluh-pertanian

[3] Direktorat Ketersediaan Pangan. 2022. Situasi Konsumsi Pangan Nasional Tahun 2022. Deputi Bidang Ketersediaan dan Stabilitas Pangan. Badan Pangan Nasional. http://badanpangannasional.go.id

[4] BPS Indonesia. 2023. Peternakan dalam angka di Indonesia 2022. Jakarta.

[5] Tiro, B. M.W dan P. Beding. 2019. Jurnal Pertanian Agros 21(1): 1 – 8.

[6] Carlos, F. S., de Oliveira Denardin, L. G., Martins, A. P., Anghinoni, I., de Faccio Carvalho, P. C., Rossi, I., M.P. Buchain & de Oliveira Camargo, F. A. (2020). Integrated crop–livestock systems in lowlands increase the availability of nutrients to irrigated rice. Land Degradation & Development, 31(18), 2962-2972. https://doi.org/10.1002/ldr.3653.

[7] Swastika, D. K. S. , A.Priyanti. , A. M. Hasibuan., D. Sahara., N.N. Arya., A. Malik., N..Ilham. , A. L. Sayekti., J. Triastono., R. Asnawi., D. Sugandi., N. Q.Hayati and Atman. 2024. Pursuing circular economics through the integrated crop-livestock systems: An integrative review on practices, strategies and challenges post Green Revolution in Indonesia. Journal of Agriculture and Food Research 18: 101269. https://doi.org/10.1016/j.jafr.2024.101269.

[8] Utami, S dan K. Rangkuti. 2021. Sistem pertanian terpadu tanaman ternak untuk peningkatan produktivitas lahan: *A Review.* AGRILAND Jurnal Ilmu Pertanian 9 (1) Januari-April 2021 1-6

[9] Priyanti, A. 2007. Dampak Program Sistem Integrasi Tanaman-Ternak Terhadap Alokasi Waktu Kerja, Pendapatan Dan Pengeluaran Rumahtangga Petani. Disertasi. Sekolah Pascasarjana Institut Pertanian Bogor (tidak dipublikasikan)

[10] Allen, V. G., M. T. Baker, E. Segarra, and C. P. Brown. 2007. Agron. J. 99:346–360.

[11] Ilham N. 2021. Reformulasi Kebijakan Pengembangan Sentra Produksi Sapi Potong Berbasis Sumber Daya Pakan. Jakarta, IAARD Press.

[12] Diwyanto K., A. Priyanti dan R.A. Saptati. 2007. Prospek Pengembangan Usaha Peternakan Pola Integrasi. Sains Peternakan 5 (2): 26-33.

[13] Haryanto. B., I. Inounu., Arsana. B dan K. Diwyanto. 2002. Sistem Integrasi Padi-Ternak. Badan litbang Pertanian. Departemen Pertanian. Jakarta.

[14] Hasibuan AM., Wulandari S., Ardana IK., Saefudin and Wahyudi A. 2023. Understanding Climate Adaptation Practices Among Small-Scale Sugarcane Farmers In Indonesia: The Role Of Climate Risk Behaviors, Farmers' Support Systems, and Crop-Cattle Integration. Resources, Environment and Sustainability 13:100129 (2023). https://doi.org/10.1016/j.resenv.2023.100129.

[15] Kariyasa, K. 2005. Sistem Integrasi Tanaman-Ternak Dalam Perspektif Reorientasi Kebijakan Subsidi Pupuk Dan Peningkatan Pendapatan Petani. Analisis Kebijakan Pertanian. Volume 3 No. 1, Maret 2005 : 68-80. Pusat Penelitian Sosial Ekonomi dan Kebijakan Pertanian.

[16] Sugandi, D., Y. Surdianto. dan A. Gunawan. 2002. Sistem Usahatani Tanaman-Ternak di Lahan Sawah. BPTP Jawa Barat. Badan Litbang Pertanian.

[17] Muklis., M. Noer., Nofialdi dan Mahdi. 2015. Jurnal Penelitian Lumbung 14(1).

[18] Novitri, K., K. Sukiyono., Purmini and M.h O. Sativa. 2021. Adoption Level of Integrated Farming System Based on Rice–Cattle and Its Determinants Related to Sustainable Agriculture. E3S Web of Conferences 226, 00034. https://doi.org/10.1051/e3sconf/202122600034 ICoN BEAT 2019.

[19] Hernanto, F. 1989. Ilmu Usahatani. Penebar Swadaya Jakarta.

[20] FAO. 1993. Farming System Development: A. General Guideline. FAO. Roma.

[21] Manti, I dan R. Hendayana, 2005. Kajian Kelayakan Ekonomi Rakitan Teknologi Usahatani Jagung di Lahan Kering. JPPTP. Badan Litbang Pertanian. Vol 8 (1) Maret 2005. Hal 55-66.

[22] Sasmi, M., Susanto, H., Mula, E. P. K., & Alatas, A. (2022). Farmer Characteristics' Effect on The Decision of Farmers in Choosing Local Variety Rice Farming Business. Business Review and Case Studies, 3(2), 104–112. https://doi.org/10.17358/brcs.3.2.104.

[23] Gusti, I. M., Gayatri, S., & Prasetyo, A. S. (2021). Pengaruh umur, tingkat pendidikan dan lama bertani terhadap pengetahuan petani mengenai manfaat dan cara penggunaan kartu tani di Kecamatan Parakan. Jurnal Litbang Provinsi Jawa Tengah, 19 (2), 209–221. https://doi.org/10.36762/jurnaljateng.v19i2.926.

[24] Kurmi, A., Kaushik, S., Pandey, S. K., Nagre, S., Shweta, S., & Thomas, M. (2022). Traditional knowledge-based agricultural practices in Tribal dominated District Anuppur, Madhya Pradesh. Plant Science Today, 9(sp3), 52-62.

[25] Suardana, P.A., M. Antara., & M.N. Alam. 2013. Jurnal E-Agrotekbis 1(5): 447-484.

[26] Taufik, M., Raliman dan R. Hermawan. 2011. Analisis Produktivitas Padi Sawah di Kupang Timur Nusa Tenggara Timur. Jurnal Ilmu Ilmu Pertanian 8(2):105-114.

[27] Basuni R, Muladno, Kusmana C, dan Suryahadi. 2010. Sistem Integrasi Padi-Sapi Potong di Lahan Sawah. Iptek Tanaman Pangan Vol. 5 No. 1 – 2010.

Soil water availability in coffee plantations during dry season at various altitudes

Septrial Arafat[1], Muhammad Iqbal Fauzan[1*], Sugeng Prijono[2], Sudarto[2], Soemarno[2]

[1] Agroecotechnology Study Program, Agriculture Department, Faculty of Animal and Agricultural Sciences, Semarang City, Central Java, Indonesia.
[2]Soil Sciences Department, Faculty of Agriculture, Brawijaya University.

*E-mail: muhammadiqbalfauzan@lecturer.undip.ac.id.

Abstract. Water availability to plants is scarce in the dry season due to intensive solar radiation, high temperatures, increased wind speed, and also due to reduced humidity. The minimal availability of water for plants causes the plants to wilt and the productivity of coffee plants decreases. However, from observations in South Malang, water can still be accessed by coffee plants in several locations. This research aims to determine the fundamental factors that cause water to remain available at the peak of the dry season using a stratified random sampling research method with 9 observations points in a 30m x 30m plot, repeated four times, and observed three times. The parameters examined include the microclimate of coffee plants (air temperature, air humidity, intensity of solar radiation, and wind speed), soil physical properties (soil bulk density, soil weight density, soil porosity, soil texture, and soil water content at various levels), altitude, and number of coffee plants and their shade. The analysis used in the research uses correlation displayed with a scatter plot of two simple variables.

Keywords: drought, micro climate, shading plant, soil physic

1. Introduction

There is great concern about the impact of climate change on global agricultural production. Much of the world's plant is grown in the tropics, and climate change is expected to change the frequency, intensity and duration of extreme climate events there. Drought is defined in a variety of ways. 'Agricultural drought' refers to lack of water in the top three feet of soil (root zone) affecting crops (1,2).

Soil water content has an important role in the availability of groundwater for plants (3). Knowledge of soil moisture content is important as the main parameter in designing irrigation programs as a solution to increase crop production (4). Soil water content influences the hydrological cycle, infiltration, and evapotranspiration. Lack of water in agricultural crops adversely affects their physiological aspects. water shortages reduce world agricultural output by more than 50%. In coffee plants, lack of water has an impact on reduced flower and fruit production (5). The results of the study concluded that the productivity of Arabica and Robusta coffee in 2050 will decrease by 20% and 40%, respectively (6).

The results of crop simulations using climate projection scenarios based on the Global Circulation Model (GCM) conclude that the decline in agricultural production will be more severe in the tropics (7). Climate change in the tropics causes crop damage, decreased production, soil erosion, and failure of tillage due to heavy rains and land degradation due to landslides and drought (8). Climate change causes an increase in temperature which will reduce the rate of growth, flowering, and fruiting of coffee plants (9). Coffee farmers in Nicaragua reported that changes in rainfall patterns in the last 20 years had an impact on irregular flowering, incomplete fruit ripening, and frequent fruit fall (10).

Coffee is a very sensitive plant to water and requires water during its critical growth phases, especially during stages such as pre-flowering, post-flowering, fruiting and ripening to achieve maximum yields. Water availability in soil is greatly influenced by plant microclimate, including plant shade, soil physics, temperature, humidity and insolation intensity (11). Canopies buffer against extreme climates and promotes a functional microclimate. As a refuge for undergrowth species in changing climates. The plant canopy can reduce the intensity of solar radiation, so that the temperature received at the ground surface becomes lower (12).

Soil properties are the properties of soil that are responsible for the circulation of air, heat, water, and minerals dissolved in the soil, which can support plant growth and increase land productivity. Some of the various physical properties such as soil texture, soil structure, soil porosity, soil volume weight, and soil particle density. These physical properties are different for each soil type and can change with different tillage methods. Soil quality is one of the physical properties of soil that expresses the roughness and fineness of the soil (13,14).

Soil texture is often related to permeability, water retention resistance, air and cation exchange capacity, and soil fertility. The ability of soils to hold water varies. Based on soil permeability (porosity), it can be estimated that planting areas with specific soil textures have different water requirements for soil to be used as a good planting site (15).

Differences in altitude are closely related to microclimates such as temperature, humidity and solar radiation. The coffee tree planting pattern is close to square height (16).

Long wet months often occur in La Nia climates which cause a decrease in coffee production at the Sumber Asin Experimental Garden, Malang, by up to 98% (17). The wet month that occurred throughout the 1996 La Nina year reduced coffee production in KP Jollong, Pati, by 48% (18). Coffee quality is very sensitive to temperature and rainfall. An increase in temperature and a decrease in rainfall which is projected to occur in the south of the equator has an impact on a decrease in coffee production. For example, coffee productivity which is only 150kg/ha in Sulawesi is partly due to the lack of a dry month period (19-20).

The ambient temperature exhibits a decline concomitant with the elevation increment. This phenomenon is attributed to adiabatic cooling, wherein ascending air undergoes a temperature reduction of roughly 0.6 °C per 100 meters of ascent. Concurrently, alterations in air humidity, solar radiation intensity, and wind velocity transpire in accordance with the altitude, consequently instigating fluctuations in the microclimatic conditions pertinent to coffee cultivation (21,22).

Knowing the soil moisture content characteristics of coffee plantations in South Malang will help overcome future droughts and facilitate rural irrigation decisions for sustainable coffee production.

2. Materials and methods

This study used a descriptive method on various land uses of coffee plants in 4 sub-districts, namely Ampelgading, Sumbermanjing Wetan, Tirtoyudo, and Dampit sub-districts with layered sampling with 9 observation points in a plot measuring 30 m x 30 m with 4 observation points and 3 times replication observation. Observations included the microclimate of coffee plants, namely: air temperature, humidity, intensity of solar radiation, and wind speed; soil physical properties in the form of: soil density, soil density, soil porosity, soil texture, soil moisture content at pF 0, pF 2.5, pF 4.2, and real water content in the field; altitude and number of coffee plants and their shade. The measuring instrument for measuring temperature, humidity, solar radiation intensity, and wind speed is the Mastech MS6300 environment tester. Fill weight was measured using a sample ring which was dried in an oven, specific gravity was measured using a pycnometer method, pF 0, pF 2.5, and pF 4.2 were measured using a soil moisture pressure plate apparatus, height was measured using software 123, and the number of plants was counted manually by researchers.

3. Results

3.1. Microclimate in South Malang

A microclimate refers to the climate conditions within a small, localized area that may differ from the surrounding larger climate. Various factors contribute to the creation of microclimates, including geographical features, topography, and vegetation.

Table 1. Microclimate in South Malang.

Location	temperature (°C)	humidity (%)	max radiation	sunlight intensity (%)	wind speed m/s
Purwodadi	28.25	70.65	61009.29	84.25	3.48
Purwoharjo	25.01	76.71	58998.67	82.25	2.66
Sri Mulyo	26.39	78.11	54463.01	74.25	2.62
Sukodono	26.14	79.14	52009.55	67.75	2.74
Sumbermanjing Wetan	26.73	77.69	62390.25	91.50	2.70
Tirtoyudo	26.69	79.27	57799.14	79.00	2.41
Taman Ksatriyan	22.12	84.71	49862.04	49.86	5.19
Taman Sari	22.60	84.26	42779.58	42.78	5.63

Based on the provided table describing microclimates in South Malang, we can make several observations. The temperatures across the locations range from 22.12°C (Taman Ksatriyan) to 28.25°C (Purwodadi). Taman Ksatriyan and Taman Sari have notably lower temperatures compared to other locations, indicating a relatively cooler microclimate. Humidity levels vary from 70.65% (Purwodadi) to 84.71% (Taman Ksatriyan). Taman Ksatriyan and Taman Sari have the highest humidity levels, suggesting a more humid microclimate.

Purwodadi has the highest maximum radiation (61009.29) among the locations, indicating a higher exposure to solar radiation. Taman Ksatriyan and Taman Sari have lower maximum radiation values, suggesting less exposure to solar radiation. Sunlight intensity ranges from 42.78% (Taman Sari) to 91.50% (Sumbermanjing Wetan). Taman Sari has the lowest sunlight intensity, indicating lower sunlight exposure compared to other locations.

Wind speeds vary from 2.41 m/s (Tirtoyudo) to 5.63 m/s (Taman Sari). Taman Sari has the highest wind speed. Purwodadi stands out with the highest temperature and maximum radiation, indicating a warmer and sunnier microclimate. Taman Ksatriyan and Taman Sari exhibit lower temperatures, higher humidity, and lower sunlight intensity, suggesting cooler and more humid microclimates with less sunlight exposure. Wind speeds vary across locations, with Taman Sari having the highest wind speed.

This study used a descriptive method on various land uses of coffee plants in 4 sub-districts, namely Ampelgading, Sumbermanjing Wetan, Tirtoyudo, and Dampit sub-districts with layered sampling with 9 observation points in a plot measuring 30 m x 30 m with 4 observation points and 3 times replication observation. Observations included the microclimate of coffee plants, namely: air temperature, humidity, intensity of solar radiation, and wind speed; soil physical properties in the form of: soil density, soil density, soil porosity, soil texture, soil moisture content at pF 0, pF 2.5, pF 4.2, and real water content in the field; altitude and number of coffee plants and their shade. The measuring instrument for measuring temperature, humidity, solar radiation intensity, and wind speed is the Mastech MS6300 environment tester. Fill weight was measured using a sample ring which was dried in an oven, specific gravity was measured using a pycnometer method, pF 0, pF 2.5, and pF 4.2 were measured using a soil moisture pressure plate apparatus, height was measured using software 123, and the number of plants was counted manually by researchers.

3.2. *Soil Physics in South Malang*

Soil physics is a branch of soil science that focuses on the physical properties and behavior of soils. It involves the study of how soils function as a medium for plant growth, as well as the interaction between soils and water, air, and heat.

Table 2. Soil physics in South Malang.

Location	bulk density	particle density	porosity (%)	pF 0 (%)	pF 2.5 (%)	pF 4.2 (%)	sand (%)	silt (%)	clay (%)
Purwodadi	1.32	2.41	45.31	77.81	28.62	13.66	51.33	30.84	17.83
Purwoharjo	1.53	2.54	39.79	73.06	52.68	23.57	14.53	81.09	4.39
Sri Mulyo	1.20	2.35	48.75	71.05	35.65	24.32	40.27	28.68	31.05
Sukodono	1.28	2.46	47.86	78.96	46.17	30.21	49.45	27.43	23.13
Sumbermanjing	1.26	2.39	47.19	75.02	40.94	27.37	31.19	31.43	37.38
Tirtoyudo	1.18	2.37	49.97	61.61	34.94	18.64	45.77	37.12	17.11
Taman Ksatriyan	1.13	2.64	57.09	54.67	17.16	10.02	90.00	7.26	2.74
Taman Sari	1.17	2.64	55.73	52.33	15.96	10.50	94.32	3.43	2.25

Based on the provided table describing soil physics in South Malang, we can make several observations: Bulk density varies across locations, ranging from 1.13 (Taman Ksatriyan) to 1.53 (Purwoharjo). Lower bulk density generally indicates a more porous soil. Particle density ranges from 2.35 (Sri Mulyo) to 2.64 (Taman Ksatriyan and Taman Sari). Particle density represents the mass per unit volume of soil particles.

Porosity, which is the percentage of pore space in the soil, varies from 39.79% (Purwoharjo) to 57.09% (Taman Ksatriyan). Higher porosity is indicative of more space available for air and water in the soil. pF 0 (%), pF 2.5 (%), and pF 4.2 (%) represent the soil moisture tension at different levels. Lower pF values generally indicate wetter conditions. Locations like Purwoharjo and Taman Ksatriyan show lower pF values at 0% and 2.5%, suggesting higher moisture availability.

IOP Conf. Series: Earth and Environmental Science **1460** (2025) 012014 doi:10.1088/1755-1315/1460/1/012014

Soil texture is described by the proportions of sand, silt, and clay. Different locations exhibit variations in these components. Taman Ksatriyan and Taman Sari have high sand percentages, indicating a sandy texture. Purwoharjo has a high clay percentage. Taman Ksatriyan stands out with the lowest bulk density, highest porosity, and high sand content, suggesting a well-structured, porous, and sandy soil. Purwoharjo has relatively higher bulk density, lower porosity, and a higher clay content, indicating a denser and potentially less well-drained soil. Locations like Purwodadi, Sri Mulyo, and Sukodono show moderate values for various soil physics parameters.

3.3. *Altitude, amount coffee plantation, and water soil availability*

Altitude, coffee plantation, and water soil availability are interconnected factors that play a crucial role in the cultivation of coffee. The relationship between these elements can significantly impact the quality and yield of coffee crops.

Table 3. Altitude, amount coffee plantation, and water soil availability.

Location	amount of shade plant varieties	altitude (mdpl)	coffee plantation	Water soil availability (%)
Purwodadi	3	515	1466	14.90
Purwoharjo	3	630	1506	16.32
Sri Mulyo	2	531	1104	14.10
Sukodono	3	695	1493	15.46
Sumbermanjing Wetan	2	647	1318	12.23
Tirtoyudo	4	653	1404	17.71
Taman Ksatriyan	3	1255	2562	2.26
Taman Sari	2	1028	1862	1.61

Based on the provided table describing altitude, amount of coffee plantation, and water soil availability in different locations, we can make several observations: The amount of shade plant varieties varies across locations, ranging from 2 to 4. Taman Ksatriyan has the highest amount of shade plant varieties (4), while Sri Mulyo has the lowest (2).

Altitude differs among locations, ranging from 515 meters (Purwodadi) to 1255 meters (Taman Ksatriyan). Taman Ksatriyan is situated at the highest altitude, indicating that it is likely to have cooler temperatures and potentially different microclimatic conditions compared to lower-altitude locations. The extent of coffee plantation, measured in terms of the number of coffee plantations, varies from 1104 to 2562. Taman Ksatriyan has the largest coffee plantation, while Sri Mulyo has the smallest.

Water soil availability is given as a percentage and varies from 1.61% (Taman Sari) to 17.71% (Tirtoyudo). Taman Sari has the lowest water soil availability, suggesting potential water scarcity, while Tirtoyudo has the highest, indicating better water availability.

Taman Ksatriyan stands out with the highest altitude and the largest coffee plantation, potentially indicating different growing conditions and the possibility of a unique microclimate. Tirtoyudo has the highest water soil availability, which could positively impact the water supply for the coffee plantation. Taman Sari and Taman Ksatriyan have the lowest water soil availability percentages, which might suggest the need for additional irrigation or water management practices.

4. Discussion

4.1. Microclimate insight

The provided data on microclimates in South Malang highlights the diverse weather conditions across different locations. Purwodadi exhibits the warmest temperature and highest maximum radiation, indicating a sunnier and warmer microclimate. Conversely, Taman Ksatriyan and Taman Sari showcase cooler temperatures, higher humidity, and lower sunlight intensity, suggesting a more temperate and humid microclimate. Wind speeds also vary, with Taman Sari experiencing the highest speed. This information is valuable for understanding local climate variations, aiding agricultural planning and resource management.

4.2. Soil physics overview

Soil physics data reveals variations in soil properties across South Malang. Taman Ksatriyan stands out with the lowest bulk density, highest porosity, and sand content, suggesting well-structured, porous, and sandy soil. In contrast, Purwoharjo has higher bulk density, lower porosity, and a higher clay content, indicating denser and potentially less well-drained soil. Understanding soil physics is crucial for optimizing agricultural practices, as it influences water retention, aeration, and nutrient availability.

4.3. Altitude, coffee plantation, and water soil availability

Altitude plays a pivotal role in shaping microclimates and, consequently, impacting coffee cultivation. Taman Ksatriyan, situated at the highest altitude, may experience cooler temperatures, influencing its microclimate. The size of the coffee plantation also varies, with Taman Ksatriyan having the largest and Sri Mulyo the smallest. Water soil availability percentages indicate potential challenges, with Taman Sari facing water scarcity, emphasizing the need for efficient irrigation or water management practices.

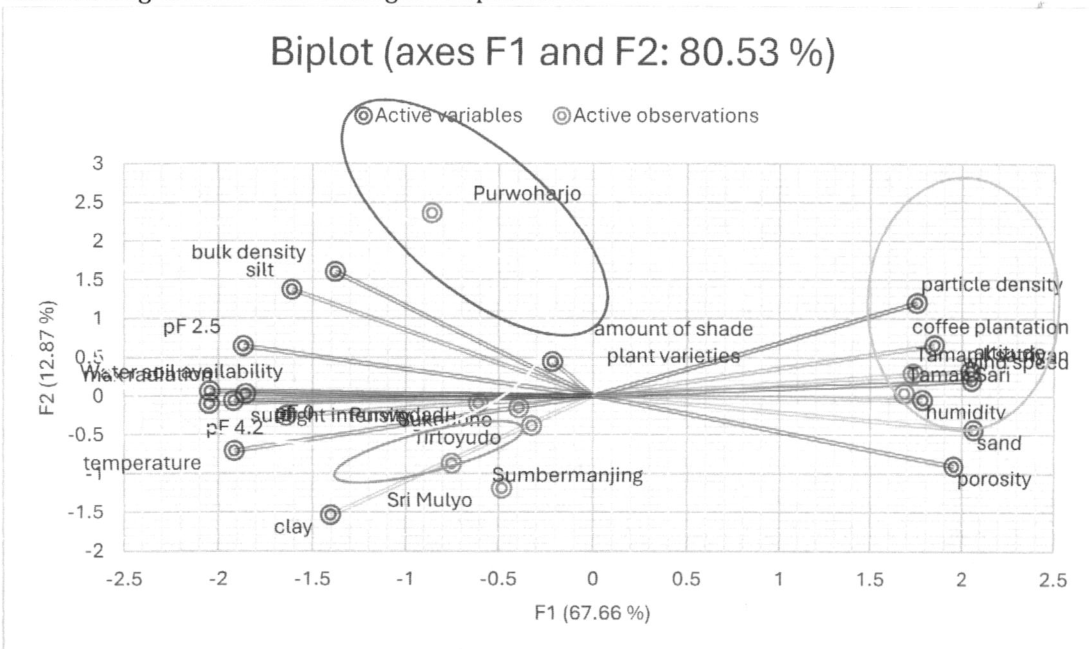

Figure 1. Biplot diagram of South Malang coffee land plantation

From **Figure 1.** Biplot diagram, an inverse correlation between water content and altitude is observed. In higher altitudes, water content tends to decrease significantly due to high light intensity and wind speed, leading to a reduction in water content.

The relationship between groundwater availability and sand texture in the soil is directly proportional, where higher porosity and a greater presence of sand texture result in reduced water content. The quantity of plants and shading in coffee plants in South Malang does not exert a significant impact on groundwater availability.

5. Conclusion

Altitude plays a significant role in the characteristics of the locations. Taman Ksatriyan, situated at the highest altitude (1255 meters), is likely to have cooler temperatures and potentially different microclimatic conditions compared to lower-altitude locations. This could influence the type of coffee plants that thrive in these areas.

Water soil availability, indicated as a percentage, ranges from 1.61% (Taman Sari) to 17.71% (Tirtoyudo). Tirtoyudo has the highest water soil availability, suggesting better water supply for the coffee plantation, while Taman Sari has the lowest, indicating potential water scarcity. Taman Sari and Taman Ksatriyan, with the lowest water soil availability percentages, may require additional irrigation or water management practices to sustain the coffee plantation. In contrast, Tirtoyudo's high water soil availability could positively impact its water supply for coffee cultivation.

References

[1] Dai, M., Huang, S., Huang, Q., Leng, G., Guo, Y., Wang, L., & Zheng, X. (2020). Assessing agricultural drought risk and its dynamic evolution characteristics. *Agricultural Water Management, 231*, 106003.

[2] Gupta, A., Rico-Medina, A., & Caño-Delgado, A. I. (2020). The physiology of plant responses to drought. *Science, 368*(6488), 266-269.

[3] Brocca, L., Ciabatta, L., Massari, C., Moramarco, T., Hahn, S., Hasenauer, S., Kidd, R., Dorigo, W., Wagner, W., Levizzani, V., 2014a. Soil as a natural rain gauge: estimating global rainfall from satellite soil moisture data. J. Geophys. Res. 119 (9), 5128–5141.

[4] Brocca, L., Camici, S., Melone, F., Moramarco, T., Martínez-Fernández, J., Didon-Lescot, J.F., Morbidelli, R. 2014b. Improving the representation of soil moisture by using a semiíanalytical infiltration model. Hydrol. Processes 28 (4), 2103–2115.

[5] Osakabe, Y., Osakabe, K., Shinozaki, K., Tran, L. S. P. 2014. Response of plants to water stress. *Frontiers in plant* science, p. 5.

[6] Lemma, D. T., & Megersa, H. G. (2021). Impact of climate change on East African coffee production and its mitigation strategies. *World Journal of Agricultural Sciences, 17*(2), 81-89.

[7] Cerri, C. E. P., Sparovek, G., Bernoux, M., Easterling, W. E., Melillo, J. M., Cerri, C. C. 2007. Tropical agriculture and global warming: impacts and mitigation options. Scientia Agricola 64: 83–99.

[8] Solomon, S., Qin, D., Manning, M. 2007. Climate Change 2007: The Physical Science Basis. Contribution of Working Group I for the Fourth Assessment Report of the Intergovernmental Panel on Climate Change. Cambridge, UK: Cambridge University Press.

[9] DaMatta, F. M., Avila, R. T., Cardoso, A. A., Martins, S. C., & Ramalho, J. C. (2018). Physiological and agronomic performance of the coffee crop in the context of climate change and global warming: A review. *Journal of agricultural and food chemistry, 66*(21), 5264-5274.

[10] Jaramillo, J., Chabi-Olaye, A., Kamonjo, C., Jaramillo, A., Vega, F. E., Poehling, H. M., Borgemeister, C. 2009. Thermal tolerance of the coffee berry borer hypothenemus hampei: predictions of climate change impact on a tropical insect pest. PLoS One, 4(8): e6487.

[11] Kath, J., Byrareddy, V. M., Craparo, A., Nguyen-Huy, T., Mushtaq, S., Cao, L., & Bossolasco, L. (2020). Not so robust: Robusta coffee production is highly sensitive to temperature. *Global Change Biology, 26*(6), 3677-3688.

[12] Bosco, L. C., Bergamaschi, H., & Marodin, G. A. B. (2020). Solar radiation effects on growth, anatomy, and physiology of apple trees in a temperate climate of Brazil. *International Journal of Biometeorology*, *64*(11), 1969-1980.

[13] Lal, R. (2020). Soil quality and sustainability. In *Methods for assessment of soil degradation* (pp. 17-30). CRC press.

[14] Lehmann, J., Bossio, D. A., Kögel-Knabner, I., & Rillig, M. C. (2020). The concept and future prospects of soil health. *Nature Reviews Earth & Environment*, *1*(10), 544-553.

[15] Cardoso, R., Vieira, J., & Borges, I. (2022). Water Retention Curve of Biocemented Sands Using MIP Results. *Applied Sciences*, *12*(20), 10447.

[16] Klinges, D. H., & Scheffers, B. R. (2021). Microgeography, not just latitude, drives climate overlap on mountains from tropical to polar ecosystems. The American Naturalist, 197(1), 75-92.

[17] Sugianto, P., Sudarsono, S., Sukma, D., & Sumirat, U. (2022). Adaptation of introduced Robusta coffee clones in some agroclimate types in East Java. *Pelita Perkebunan (a Coffee and Cocoa Research Journal)*, *38*(2), 98-107.

[18] Rokhmah, D. N., Supriadi, H., & Heryana, N. (2023, September). Sustainable specialty coffee production: An agronomy perspective (A review). In *IOP Conference Series: Earth and Environmental Science* (Vol. 1230, No. 1, p. 012067). IOP Publishing.

[19] Panggabean, Y. B. S., & Arsyad, M. (2022, December). Sustainability agricultural supply chain in improving the welfare of North Toraja Arabica coffee farmers. In *IOP Conference Series: Earth and Environmental Science* (Vol. 1107, No. 1, p. 012065). IOP Publishing.

[20] Neilson, J., Hartari, D. S. F., Lagerqvist, Y. F. 2013. Coffee-based livelihoods in South Sulawesi, Indonesia. Appendix 8 to the final report for ACIAR Project SMAR/2007/063. Australian Centre for International Agricultural Research, Canberra.

[21] Aryani, R. D., Basuki, I. F., Budisantoso, I., & Widyastuti, A. (2022). Pengaruh Ketinggian Tempat terhadap Pertumbuhan dan Hasil Tanam Cabai Rawit (Capsicum frutescens L.). *Journal of Applied Agricultural Sciences AGRIPRIMA*, *6*(2), 202-211.

[22] Widiyani, D. P., Gusta, A. R., Aziz, A., Usodri, K. S., Hartono, J. S., & Hamdani, H. (2022). Identifikasi Hubungan Iklim Mikro pada Tanaman Kopi Robusta (Coffea chanepora) Dataran Rendah. *Jurnal Agrinika: Jurnal Agroteknologi dan Agribisnis*, *6*(2), 151-160.

IOP Conf. Series: Earth and Environmental Science **1460** (2025) 012015 doi:10.1088/1755-1315/1460/1/012015

The Effect of Bread Waste Utilization in Feed on Kampung Unggul Balitbangtan (KUB) Chicken Growth

Subiharta[1], Dwinta Prasetianti[2], Rini Nur Hayati[1] and Agung Prabowo[3*]

[1] Research Center for Animal Husbandry, National Research and Innovation Agency, Cibinong, Indonesia
[2] Central Java Agricultural Instrument Standardization Agency, Semarang, Indonesia
[3] Research Center for Sustainable Production System and Life Cycle Assessment, National Research and Innovation Agency, Tangerang Selatan, Indonesia

*E-mail: agungsuwarjo@gmail.com

Abstract. Feed is an important factor and the biggest cost component in the poultry industry, including Kampung Unggul Balitbangtan (KUB) chickens. We proposed that bread waste could be a partial substitute for commercial feed. The aim of the study was to identify the feed formulation that made from local feed for KUB chickens in the growth period. This study was carried out in Karanganyar Regency. Ninety (90) chickens aged 45 days were used in this study and grouped into 3 feed treatments (T1, T2, T3) with 3 repetitions, each repetition consisting of 10 chickens. T1: 100% commercial feed, T2: 70% commercial feed, 3% Distillers Dried Grains with Soluble (DDGS) Maize, 25.449% bread waste, 1.5% fish meal, 0.006% molasses, and 0.045% micro-enzyme, and T3: 50% commercial feed, 5% Distillers Dried Grains with Soluble (DDGS) Maize, 42.415% bread waste, 2.5% fish meal, 0.01% molasses, and 0.075% micro-enzyme. The parameters observed include body weight, body weight gain, crude protein consumption, correlation crude protein consumption on body weight gain, and feed conversion ratio (FCR). Data was analyzed using the SPSS program, with a completely randomized design (CRD). The results showed that body weight, body weight gain, crude protein consumption, and FCR were not significantly different ($P>0.05$) for three treatments. The body weight of three treatments were T1 1,823.31 kg/chicken, T2 1,840.03 kg/chicken, and T3 1,796.34 kg/chicken. The bread waste could be used as KUB chicken feed to reduce feed cost and has no negative effect on growth.

Keywords: bread waste, feed, growth, KUB chicken

1. Introduction

Livestock business will be influenced by 3 important things known as the golden triangle, namely: breeds, feed and breeding management [1]. These three factors mutually influence each other, if all factors are run properly then the business will be sustainable.

Kampung Unggul Balitbangtan (KUB) chickens is one of the local chickens because of the Indonesian Agency for Agricultural Research and Development (IAARD) research and which has been released by the Minister of Agriculture. KUB chicken is the result of selection of native chickens for 6 generations. The results of the selection of these chickens are hereinafter referred

IOP Conf. Series: Earth and Environmental Science **1460** (2025) 012015 doi:10.1088/1755-1315/1460/1/012015

to as the KUB-1 chicken [2]. Based on its description, KUB -1 chicken is known as local layers whose production reaches 160-180 eggs/chicken/year, with an average Hen Day Production (HDP) of 50%, peak HDP of 65-70% [3].

Indonesian Agency for Agricultural Research and Development (IAARD) has also released the KUB-2 chicken which is an improvement from the KUB-1 chicken. The average HDP is 60%, production is 200 eggs/chicken/year, and the peak HDP is 70-75% [4]. KUB -1 and KUB -2 chickens, apart from being known for their high egg production, the two KUB chickens are also grow fast. At the age of 2.5 months, KUB -1 female body weight reached an average of 0.8 kg and 1.2 kg for the male chicken [5]. The slaughter weight desired by local chicken culinary entrepreneurs is between 0.8 – 1.2 kg [6]. KUB chicken farmers in Central Java cultivate KUB chickens for breeding and fattening purposes, arguing that two businesses can complement each other, and provide the highest profit compared to individual businesses.

As previously mentioned, apart from breed, feed is an important factor and is the highest cost component in the chicken business [7, 8, 9]. Including KUB chickens, to produce optimally, feed is needed according to their nutritional needs and in sufficient quantity. Good quality of feed is obligate for KUB chickens to produce according to their genetic capacity [10]. Feed adapted to its physiological phases, namely starter, grower and layer. We should bear in mind that each physiological phase has different levels of growth, production and reproductive activity [11]. During the growth period, the protein requirement of KUB chickens is higher compared to other physiological phases, this occurs because the formation of muscle tissue is highest in this phase [12].

An alternative way of increasing the supply of poultry product at cheaper prices is by reducing the cost of production with cheaper, locally available sources of energy source feed. Bread waste is a by-product of the bakery industry that is made from flour, and it is not consumed by human, it has always been disposed. This paper discusses the performance of KUB chickens in the growth period given rations made from local ingredients such as bread waste.

2. Materials and Methods

A study of the use of local feed in KUB chickens was carried out in the box pen owned by farmers in Karanganyar Regency from January to June 2019. The study used 90 KUB chickens aged 45 days with an observation period of 4 months. The chickens were kept in a 180x90x60 cm box with a base height of 40 cm from the floor. For 1 – 45 days, the chickens are fed commercial feed. The total cages used were 18 cages, 9 cages for chicken aged 1-45 days and 9 cages for chicken aged 1,5-4 months. Chickens were grouped into 3 feed treatments with 3 repetitions, each repetition consisting of 10 chickens. The feed treatment is presented in Table 1 below.

Table 1. The Feed treatment of KUB chickens in the grower period.

Treatments	Description
T1	100% commercial feed
T2	70% commercial feed, 3% DDGS Maize, 25.449% bread waste, 1.5% fish meal, 0.006% molasses, 0.045% micro-enzyme
T3	50% commercial feed, 5% DDGS Maize, 42.415% bread waste, 2.5% fish meal, 0.01% molasses, 0.075% micro-enzyme

Amount of feed given can be seen in Table 2. The amount of feed given is according to the age of the chicken. Feed is given twice, namely in the morning and evening. At the age of 1 – 45 days, the chickens were given ad libitum commercial feed. Feed was given twice a day with the amount of ration given according to the growth of the chicken (Table 2). Drinking water is provided ad libitum.

Table 2. Amount of feed given based on age (weeks).

KUB chicken age (weeks)	Amount of feed (g/chicken/day)
0-3	5-15
>3-6	20-35
>6-10	35-60
>10-14	>60-80
>14-18	>80-90
>18-22	>90-100

The amount of crude protein consumption is calculated by multiplying the percentage of crude protein in the feed by the amount of feed consumption. Body weight (BW) and Body Weight Gain (BWG) was measured and recorded. Weighing was carried out every 2 weeks during the 4 weeks of the study. The remaining weighing in the study was carried out every 4 weeks. Feed Conversion Ratio was calculated by dividing the amount of feed consumed by the increase in body weight over the same period. The economic value of feed is calculated by multiplying the feed price per kg by the FCR. This value showed feed cost to produce 1 kg body weight.

Data was analysed using the SPSS program, with a Completely Randomized Design (CRD) in one direction testing three feed treatments and three replications. If the results of the analysis provide significant differences between treatments, then proceed with Duncan's New Multiple Range Test (DMRT) [13].

3. Results and Discussion

3.1 Consumption of crude protein

Feed is an important factor in the success of KUB chicken cultivation, considering that the nutritional needs of KUB chickens are supplied from the feed provided, so that KUB chickens need to receive feed intake with nutritional content according to their nutritional needs [14]. Realizing this, it is necessary to pay attention to the formulation of free-range chicken feed, namely the content of crude protein (CP), energy metabolism (EM), calcium (Ca), available phosphorus (P av), methionine and lysine [15]. Crude protein is useful for the process of growth and tissue formation [16], a substance for building and replacing damaged cells [17]. In the formulation of free-range chicken feed, it is prioritized to use local feed ingredients because they are relatively cheap, easy to obtain according to the specific location, do not compete with human needs and are in the form of by-products or industrial waste [18]. Several techniques are needed in mixing feed, namely: 1) to lend the feed ingredient with the highest amount at the bottom, followed by other ingredients with a smaller amount; 2) to mix the small amount of feed ingredients with other ingredients and then stirred until evenly distributed (homogeneous) [19].

Crude protein consumption can be seen in Table 3. This consumption is calculated from the start of feed treatment (6 weeks of age). The crude protein consumption of T1 treatment tended to be higher than T2 and T3. Likewise, T2 treatment was higher than T3. This was because the crude protein of T1 treatment feed was higher than T2 and T3. Likewise, the crude protein of T2 treatment feed was higher than T3. This difference was caused using bread waste in the feed.

Table 3. Crude protein consumption (g/chicken).

KUB chicken age (weeks)	Crude protein consumption (g/chicken)		
	T1	T2	T3
10	302.84	293.13	282.63
14	749.13	710.81	669.33
18	1,291.06	1,217.99	1,138.91
22	1,896.74	1,784.83	1,663.73

3.2 *KUB chicken growth rate*

The growth rate of KUB chickens is shown by the regression of body weight gain (Figure 1). Figure 1 shows that the regression lines of treatments T1 and T2 overlap and are above the regression line of treatment T3. This indicates that the rate of weight gain of KUB chickens in treatments T1 and T2 is faster than treatment T3, as shown by the regression coefficient in Table 4, namely the regression coefficient of treatments T1 and T2 is higher than treatment T3.

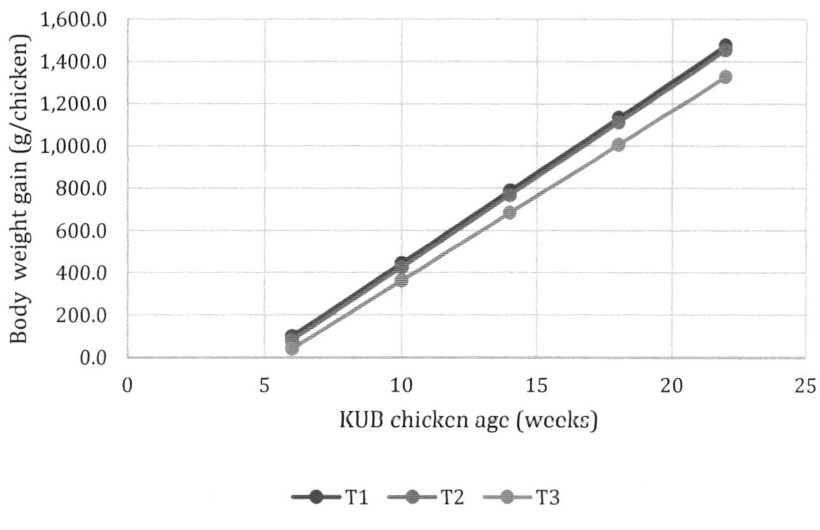

Figure 1. Body weight gain regression.

The regression coefficient of KUB chicken weight gain can be seen in Table 4. This regression coefficient can describe the growth rate of KUB chicken. The higher the regression coefficient, the faster the growth.

Table 4. Regression the equation and coefficient.

Treatment	Regression equation	Regression coefficient	R^2
T1	y = 85.994x – 415.57	85.994	0.9748
T2	y = 85.755x – 433.68	85.755	0.9713
T3	y = 80.399x – 441.96	80.399	0.9889

3.3 Productivity performance of KUB chicken

Feed analysis was carried out at the Nutrition Laboratory of the Faculty of Animal Husbandry, Gadjah Mada University. The composition and nutritional content of feed was shown in Table 5. Commercial feed for starter phase was given at aged 0-3 weeks with 20% crude protein in all treatments. Meanwhile, at 4-6 weeks of age, commercial feed for growth phase was given in all treatments.

Table 5. Feed ingredients and nutritional content.

Description	T1	T2	T3
Feed Ingredients			
Commercial Feed, %	100	70	50
Distillers Dried Grains with Soluble (DDGS) Maize, %	0	3	5
Bread Waste, %	0	25.449	42.415
Fish Meal, %	0	1.5	2.5
Molasses, %	0	0.006	0.01
Micro – Enzyme, %	0	0.045	0.075
Nutritional Content*			
Dry Matter, %	91.48	91.47	91.23
Ash, %	8.26	25.68	32.85
Crude Protein, %	22.77	21.31	19.73
Crude Fat, %	6.46	5.09	4.8
Crude Fiber, %	6.52	6.61	6.67

*Analysis Results of UGM Faculty of Animal Husbandry Nutritional Biochemistry Laboratory, 2019

After 8-9 weeks of age, commercial feed is reduced the same as at 10-14 and 15-22 weeks of age that became 70% and 50% respectively. The inclusion of the bread waste level of T1, T2 and T3 were 0%; 25. 449%; and 42. 415% respectively. To produce optimal weight, KUB chickens need nutritional content such as protein, metabolism energy, calcium, phosphorus, amino acid lysine and methionine found in feed [20]. The crude protein content after the reduction in commercial feed T2 and T3 was still quite high, namely 21.31% and 19.73% above the recommended minimum limit of 14% [21].

Feed consumption for males and females was different, males consume feed higher than females [15]. Some of the advantages of KUB chickens compared to ordinary free-range chickens are: 1). being able to produce 160-180 eggs/year; and 2) a slaughter weight of 800-900 grams at 10 weeks of age [22]. KUB chickens had the highest male body weight at T3, namely 1,055.20 g/head for 8-10 weeks of age (Table 6). In addition, KUB chickens can digest feed more effectively than ordinary native chickens, thereby reducing production costs [23]. Feed for KUB chickens aged >10-14 weeks is 60-80 g/head/day and at the age of >14-22 weeks, namely 80-100 g/head/day (Table 2). The KUB chicken in layer phase requirements feed during production range from 80-85 g/head/day. Meanwhile, the standard for feeding local chickens aged 2-5 months ranges from 55-100 g/head/day [24].

Table 6. The average body weight (BW) of the KUB chicken (g/chicken).

Description	Average body weight (g/chicken)			p-value
	T1	T2	T3	
Male				
8-10 weeks of age	975.00	1,045.66	1,055.20	-
>10-14 weeks of age	1,410.40	1,515.20	1,553.14	-
>14-22 weeks of age	1,992	2,131	2,011	-
Female				
8-10 weeks of age	924.10	911.33	842.82	-
>10-14 weeks of age	1,295.17	1,344.36	1,207.21	-
>14-22 weeks of age	1,717	1,771	1,691	-
Unsexing				
6 weeks of age	427.80^{c}	492.10^{b}	525.46^{a}	<0.001
10 weeks of age	932.58	926.25	880.75	0.176
14 weeks of age	1,315.75	1,376.00	1,287.93	0.228
18 weeks of age	1,581.33	1,660.60	1,554.96	0.210
22 weeks of age	1,823.31	1,840.03	1,796.34	0.815

Comparison of body weight of KUB chickens at 6-22 weeks of age in each treatment did not show a significant difference, except at 6 weeks of age (Table 6). This means that alternative feed treatments of T2 and T3 can be used when commercial feed is more expensive. The recommended protein content in KUB chickens is 16-17% [20]. The crude protein content in the alternative feed treatment (T3) is 19.73% (2% higher) so it has a high level of protein efficiency [25]. It is suspected that the nutritional adequacy of KUB chickens is fulfilled from alternative feed (T3) so that it affects the performance of KUB chickens and is almost the same as the performance of KUB chickens that are fed commercial feed (T1).

Table 7. The body weight gain of KUB chickens aged 22 weeks.

No.	Treatment	Feed price (IDR/kg)	Average body weight (g/chicken)		Initial BW	BW gain (g/chicken)	
			Male	Female		Male	Female
1.	T1	7,200	1,992	1,717	427.8	1,564.2	1,289.2
2.	T2	6,600	2,131	1,771	492.1	1,638.9	1,278.9
3.	T3	6,100	2,011	1,691	525.5	1,485.5	1,165.5

3.4. Crude protein consumption with the body weight gain correlation

The regression of correlation crude protein consumption with body weight gain can be seen in Figure 2.

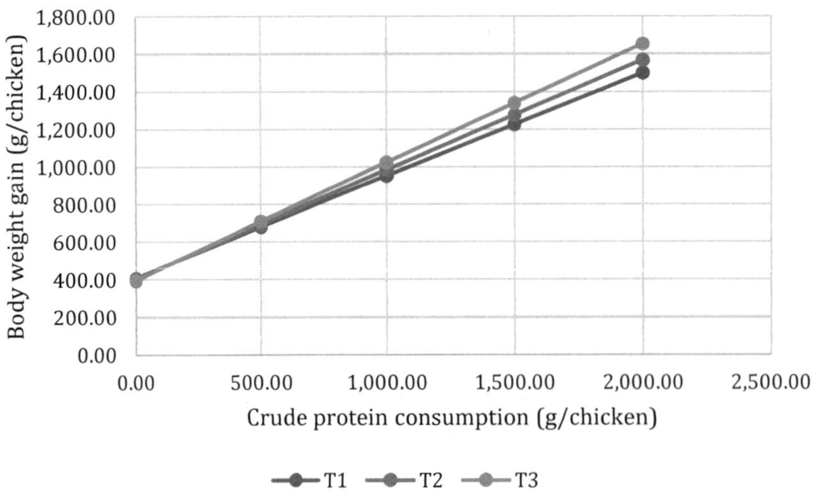

Figure 2. Crude protein consumption with body weight gain correlation.

The regression lines of the three treatments overlap for crude protein consumption of 0-500 g/head. This indicates that the efficiency of feed crude protein from the three treatments is the same. The efficiency of feed crude protein for crude protein consumption of >500-2000 g/head from the three treatments began to differ (Figure 2), treatment T3 was more efficient than T1 and T2, and treatment T2 was more efficient than T1, as shown by the regression coefficient in Table 8, the regression coefficient of treatment T3 was higher than treatments T1 and T2, and T2 was higher than T1.

Table 8. Regression equation and coefficient of crude protein consumption with body weight gain correlation.

Treatment	Regression equation	R^2
T1	y = 0.5453x + 407.46	0.9687
T2	y = 0.5827x + 401.80	0.9687
T3	y = 0.6293x + 394.74	0.9889

The regression coefficient of the relationship between crude protein consumption on weight gain of KUB chickens can be seen in Table 8. This regression coefficient shows the level of efficiency of crude protein consumption. The higher the regression coefficient, the more efficient crude protein consumption. Table 8 shows that the crude protein consumption of T3 treatment is more efficient than T1 and T2 treatments, and T2 treatment is more efficient than T1.

3.5. The feed conversion ratio

The results of FCR calculation can be seen in Table 9. This FCR illustrates the amount of feed needed to produce one kg of KUB chicken body weight. FCR figure increases with the age of the chicken (Table 9).

Table 9. Feed conversion ratio.

KUB chicken age (weeks)	T1	T2	T3
10	2.6	3.1	3.7
14	3.7	3.7	4.3
18	4.9	4.9	5.5
22	6.0	6.2	6.6

The results showed that the FCR of treatment T1 was lower than treatments T2 and T3, and treatment T2 was lower than T3. These results indicate that the T1 treatment feed is more efficient than treatments T2 and T3, and the T2 treatment feed is more efficient than treatment T3. Feed costs are the largest production cost in livestock business that is approximately 70%.

Efficiency in the use of feed is expected to increase income of smallholder livestock businesses so that they can be sustainable and develop according to the desired results. The results of this study showed that the FCR for chickens up to 10 weeks of age with T3 treatment was the same as that carried out by other researchers in Bendan Village, Boyolali Regency, namely 3.7 [26].

3.6. The economic value of feed

The feed cost to produce one kg of KUB chicken body weight raised from 6 to 22 weeks of age can be seen in Table 10. The calculation results show that the T2 and T3 treatment feeds are more economical than the T1 treatment feed.

Table 10. Feed Cost to produce 1 kg body weight.

Treatment	Feed price (IDR/kg)	FCR	Feed cost to produce 1 kg body weight (IDR)
T1	7,200	6	43,200
T2	6,600	6.2	40,920
T3	6,100	6.6	40,260

4. Conclusion

Bread waste can be used as KUB chicken feed. The use of bread wastes up to 42.415%: 1) does not interfere with the growth of KUB chickens, 2) increases the efficiency of feed crude protein consumption, 3) reduces the price of KUB chicken feed, and 4) makes KUB chicken business more economical.

References

[1] Amam A and Harsita PA 2019 Tiga Pilar Usaha Ternak : Breeding, Feeding, and Management. J Sain Peternak Indones. 14(4):431–439. doi:10.31186/jspi.id. 14.4.431-439
[2] Sartika T, Desmayati, Iskandar S, Resnawati H, Setioko AR, Sumanto, Sinurat AP, Isbandi, Tiesnamurti B Romjali E 2013 Ayam KUB-1. Jakarta (ID): IAARD Press
[3] Keputusan Menteri Pertanian RI Nomor 274/2014 2014 Keputusan Menteri Pertanian Republik Indonesia Nomor 274/Kpts/SR.120/2/2014 tentang Pelepasan Galur Ayam KUB-1 Jakarta 24 February 2014
[4] Sartika T, Iskandar S, Komarudin, Zainal H, Pratiwi N, Sopiyana S, Kostaman T, Ishak ABL 2021 Proposal Pelepasan Galur Ayam KUB-2 Janaka Agrinak Ciawi-Bogor (ID): Balai Penelitian Ternak
[5] Balai Penelitian Ternak 2011 Ayam Kampung Unggul Balitnak hhtp://balitnak.litbang deptan.co.id
[6] Subiharta, Komalawati, Pita S and Rini N H 2022 Overview of KUB chicken rearing at the small farmer level in Central Java. ICESAI 2022 ABSR 28 pp 132 - 134
[7] Thirumalaisamy G, Muralidharan J, Senthilkumar S, Sayee R H and Priyadharsini M 2016 Cost-effective feeding of poultry. Int. J. Sci. Environ. Technol. 5 (6) 3997-4005
[8] Afandi R, Hartono B, Djunaidi I 2020 The Analysis of Production Costs of Laying Hen Farms Using Semi Self-Mixing and Total Self-Mixing Feeds in Blitar Regency, East Java Trop. Anim. Sci J 43(1):70-76
[9] Noblet J, Wu S B, and Choct M 2022 Methodologies for energy evaluation of pig and poultry feeds: A review. Anim. Nutr. 8 185-203
[10] Hidayat C, Iskandar S, Sartika T, Wardhani T 2016 Growth Response of Improved Breeds of Native Chicken to Diets Differed in Energy and Protein Content IJAVS 21(3):174-181

[11] Hidayat C, Iskandar S, Sartika T 2011 Respon Kinerja Perteluran Ayam Kampung Unggul Balitnak (KUB) Terhadap Perlakuan Protein Ransum Pada Masa Pertumbuhan JITV 16(2):83-89

[12] Liu J, Fu R, Liu R, Zhao G, Zheng M, Cui H, Li Q 2016 Protein Profiles for Muscle Development and Intramuscular Fat Accumulation at Different Post-Hatching Ages in Chickens. 1–19.doi:10.1371/journal.pone.0159722

[13] Aqil M and Efendi B 2015 Aplikasi SPSS dan SAS untuk Perancangan Percobaan.Absolute Media Yogyakarta p 226

[14] Iskandar S, Hidayat C and Cahyaningsih T 2014 Pengaruh pemberian ransum pre- starter pada efisiensi kinerja ayam lokal KUB JITV 19(3):203-209

[15] Iskandar S, Hidayat C, Sartika T, Resnawati H and Kadiran 2010 Optimasi energi dan protein ransum untuk pertumbuhan maksimum umur 0-18 minggu pada ayam KUB Laporan hasil penelitian Balai Penelitian Ternak Ciawi Bogor

[16] Munir I M 2009Pengaruh suplementasi DL-metionin dan L-lisin kadaluarsa dalam pakan ayam broiler Skripsi. Yogyakarta (Indonesia) : Universitas Gadjah Mada.

[17] Tillman A D, Reksohadiprodjo S, Prawirokusumo S, Hartadi H, Lebdosoekojo S 1991 Ilmu Makanan ternak Dasar Gadjah Mada university Press Yogyakarta

[18] Zainuddin D 2005 Strategi Pemanfaatan pakan sumberdaya Lokal dan Perbaikan Manajemen ayam Lokal Lokakarya nasional Inovasi teknologi Pengembangan Ayam lokal Bogor Balitnak pp 32-41

[19] Purba M 2014 Kebutuhan Gizi dan Formulasi Ransum Ayam KUB Pusat Penelitian dan Pengembangan Peternakan Jakarta pp 41-53

[20] Sartika T 2017 Panen ayam kampung 70 hari Penebar Swadaya Jakarta p 116

[21] Nawawi N T and Nurrohmah 2002 Ransum Ayam Kampung PT Trubus Agrisarana Surabaya

[22] Sartika T, Resnawati H, Iskandar S, Purba M, Zaenuddin D and Unadi A 2014 Teknik formulasi ransum ayam KUB berbasis bahan pakan lokal Bogor Puslitbangnak

[23] Tirajoh S, Usman and Baliadi Y 2017 Kelayakan UsahaTani Ayam KUB melalui Pemanfaatan Daun Lamtoro sebagai Pakan Lokal. Prosiding Seminar Nasional Teknologi Peternakan dan Veteriner pp 500-505

[24] Iskandar S and Sinurat A P 2017 Petunjuk Teknis Produksi Ayam Lokal Pedaging Unggul Pusat Penelitian Pengembangan Peternakan Badan Penelitian dan Pengembangan Pertanian. Kementerian Pertanian Jakarta p 61

[25] Sidadolog J H P and Yuwanta T 2009 Pengaruh Konsentrasi Protein-energi Pakan terhadap Pertambahan Berat badan, Efisiensi Energi dan efisiensi Protein pada Masa Pertumbuhan Ayam Merawang Journal Animal production 11(1) : 15-22

[26] Prasetyo A, Prabowo A, Sudrajad P and Hayati RN 2021 Effectiveness of using local ingredients ration to increase KUB chicken productivity in the grower period. The 3rd International Conference of Animal Science and Technology. IOP Conf. Series: Earth and Environmental Science 788 (2021) 012037. IOP Publishing: p1-5. doi:10.1088/1755-1315/788/1/012037

Morphology Traits and Yield Components of F8 Yardlong Bean Genotypes (Vigna unguiculata L. Ssp. Sesquipedalis) in Jambi Province

Florentina Kusmiyati[1], Fajrin Pramana Putra[1], Muhammad Ghazi Agam Sas[1], Aditya Wahyudhi[2]*, Agus Arianto2, Muhammad Mukhlis2

[1] Agroecotechnology Program, Departement Agriculture, Faculty of Animal and Agriculture, Diponegoro University, Semarang, Indonesia.
[2] Agronomy Program, Institut Teknologi dan Sains Nahdlatul Ulama Jambi, Jambi.

*E-mail: adityawahyudhi2@gmail.com

Abstract. Red yardlong beans are a horticultural product with significant potential for development. Hybridization of yardlong bean has resulted in genotypes with diverse traits. Evaluation of the offspring is needed to see the potential superior characters. The aim of this research was to assess the morphological characters and yield components of the red yardlong beans F8 generation. The research was conducted in Muaro Jambi, Jambi, Ibru village. This study used 15 genotypes of red yardlong beans as treatments, this experiment used a randomized complete block design (RCBD). The morphological characters observed include stem color, leaf color, petiole color, leaf shape, leaf tip shape, leaf edge shape, calyx color, flower color, pod color, seed shape, main seed color and pattern second seed color. Yield component variables include plant height, flowering time, harvesting time, pod length, pod width, pod weight, number of pods per plant and pod weight per plant. The results of the research showed that the genotypes of the F8 generation of red yardlong beans had diverse morphological characters, especially for the leaf color, leaf shape, calyx color, flower color, primary seed color and pattern second seed color. The genotype FA 1-1-10-1-6 showed better yield components than the check cultivar in flowering time, harvesting time, number of pods per plant, and pod weight per plant. The morphological characters and yield components that have been characterized can be used as potential donors for future hybridization programs to develop superior yardlong bean varieties that have a short harvesting time with high production.

Keywords : Red yardlong beans, hybridization, progeny, morphology, yield components

1. Introduction

Lifestyle changes, especially dietary patterns related to food nutrition, cause various degenerative diseases such as stroke, heart disease, diabetes, obesity and several types of cancer that have an impact on morbidity and mortality [1]. Efforts to improve the health and nutritional status of the community can be encouraged by increasing access to healthy food consumed by many people [2]. Vegetables are a food ingredient that is widely consumed by the community because they are cheap and easy to obtain [3]. In Indonesia, yardlong beans are vegetables that are widely consumed by the wider community, and are known to have crispy and soft pods so that they can be consumed fresh or cooked [4]. One of the qualities that is considered in yardlong beans is the nutritional value such as anthocyanins as antioxidants related to human health. In general, anthocyanins are found in purplish red vegetables [5]. Young/ready-to-eat yardlong bean pods are usually green or whitish green which do not contain anthocyanins. Breeding red long bean pods is important to increase their nutritional value, especially antioxidant levels. Plant breeding or genetic modification through artificial crossing methods has been reported to successfully engineer antioxidant content, fruit flavor, and fruit color in several plant species, including blueberries [6], strawberries [7], papaya [8], peaches [9], apples [10], and tomatoes [11]. However, efforts to genetically modify yardlong beans with red/purple color and high antioxidants are reported to be still separate from efforts to assemble yardlong bean varieties with high production. In addition, there have been no reports of superior lines of red-purple yardlong beans containing antioxidants and high production.

The main steps in plant breeding are the formation of genetic diversity through artificial crossing, progeny selection/line selection, preliminary yield test/advanced yield test, and multilocation tests [12]. The available genetic diversity is the key to the success of breeders in obtaining the desired superior characters. Increasing genetic diversity has been carried out by researchers with artificial crossing on diallelic crosses of three parents. Diversity of agronomic markers was found in the F2 and F3 generations [13, 14]. Selection based on agronomic markers including pod color was also carried out in the F3, F4, F5 and F7 generations [15, 16, 17, 18, 19, 20]. Selection based on agronomic characters in the F7 generation also still showed diversity in agronomic characters including production components [19]. Another very significant thing is that the agronomic characters of the lines are different from the characters of their parents so that they can be continued to obtain new varieties [18, 20]. The test results in the F7 generation have produced 14 red-podded yardlong bean lines that have the potential to be further tested and selected to obtain lines with red pods and stable high production.

2. Material and method

2.1.Study area and materials genetic

The experiment was conducted in Muaro Jambi, Jambi, Ibru village, between November 2023 and February 2024. The field site was located at an altitude of 28 m asl. Fifteen genotypes used for this experiment consist of thirteen genotypes of F8 generations and two cultivar checks (Table 1.)

2.2. Procedure

The experiment was laid out in a randomized complete block design (RCBD) with yardlong beans as the single factor treatment. Each experiment unit is a 5 x 1 m^2 plot with 20 plants per plot, repeated three times. A bamboo pole used to support each plant and tied up every four bamboo at the top. The manure was added to the soil with 20 tons ha-1 two weeks before planting. The succeeding fertilizations were carried out once in two weeks using NPK Mutiara 16:16:16 100 kg.ha-1. Pest and plant diseases were controlled mechanically and chemically using insecticide (profenofos 2 ml/L and carbofuran 17 kg.ha- 1). The parameters morphology characteristic observed consist of stem color, leaf color, petiole color, leaflet shape, leaf ap. Shape, leaf bs. Shape, calyx color, flower color, pod color, seed shape, first seed color, pattern second seed color, The parameters yield component observed consist of plant height, flowering time, harvesting time, pod length, pod width, pod weight, number of pod per plant, pods weight per plant.

Table 1. Material genetics in this study

Genotype	Notes
FG (Cultivar Check 1)	Cultivar check (Purple-podded)
AH (Cultivar Check 2)	Cultivar check (Green-podded)
FA 1-1-9-15-B	FG x AH
FA 1-1-5-3-18-B	FG x AH
FA 3-2-13-14-B	FG x AH
FA 1-1-7-3-12-B	FG x AH
FA 1-1-3-3	FG x AH
FA 1-1-2-1-14-B	FG x AH
FA 1-1-4-3	FG x AH
FA 1-1-10-1-6	FG x AH
FS 3-4-10-3-B	FG x SP
FS 3-4-7-1-B	FG x SP
FS 3-4-8-2-B	FG x SP
FS 3-4-14-2-1	FG x SP
FS 3-4-10-1-B	FG x SP

2.3. Data analysis

Data from this study was analyzed with analysis of variance (ANOVA) with a significance level of 5% followed by Duncan Multiple Range Test (DMRT) was performed using STAR software.

3. Results and Discussions

3.1. Leaf characters of yardlong beans genotype

According to the study, several leaf traits exist, as indicated in Table 2. The stem color, petiol color, and leaf base shape are not different for all genotypes. The leaf color of FA 1-1-7-3-12-B, FA 1-1-3-3, FA 1-1-10-1-6, FS 3-4-10-3-B, and FS 3-4-8-2-B have different leaf colors from both cultivar checks. The leaflet shape FA 1-1-9-15-B, FA 1-1-7-3-12-B, and FA 1-1-3-3 have different leaf colors from both cultivar checks. The leaf apical shape for all genotype are similar with cultivar check 1.

The changes in leaf color can be caused by effect of crossing between parentals (Fagiola with Aura Hijau and Fagiola with Super Putih. According to the research of [21], major genes from the green color of the leaves is very strong, so it can be influenced by minor genes and can be affects the visible character. It supported by [22] that reported crop yield and quality are controlled from genetic factors and influenced by environmental factor so that differences can occur. The apical leaf from all genotypes are influenced by female parent (Fagiola).

Table 2. Stem and leaf characters of yardlong beans genotype

Genotypes	Stem Color	Leaf Color	Petiole Color	Leaflet Shape	Leaf Ap. Shape	Leaf Bs. Shape
FG (Cultivar Check 1)	144 B	147 A	187 A	lenceolate	Acute	Cuneate
AH (Cultivar Check 2)	144 B	139 A	137 A	ovate-lanceolate	Acuminate	Cuneate
FA 1-1-9-15-B	144 B	139 A	187 A	long lanceolate	Acute	Cuneate
FA 1-1-5-3-18-B	144 B	139 A	187 A	lenceolate	Acute	Cuneate
FA 3-2-13-14-B	144 B	139 A	187 A	lenceolate	Acute	Cuneate
FA 1-1-7-3-12-B	144 B	137 A	187 A	long lanceolate	Acute	Cuneate
FA 1-1-3-3	144 B	NN137 A	187 A	long lanceolate	Acute	Cuneate
FA 1-1-2-1-14-B	144 B	139 A	187 A	lenceolate	Acute	Cuneate
FA 1-1-4-3	144 B	147 A	187 A	lenceolate	Acute	Cuneate
FA 1-1-10-1-6	144 B	137 A	187 A	lenceolate	Acute	Cuneate
FS 3-4-10-3-B	144 B	137 A	187 A	lenceolate	Acute	Cuneate
FS 3-4-7-1-B	144 B	139 A	187 A	ovate-lanceolate	Acute	Cuneate
FS 3-4-8-2-B	144 B	NN137 A	187 A	lenceolate	Acute	Cuneate
FS 3-4-14-2-1	144 B	147 A	187 A	lenceolate	Acute	Cuneate
FS 3-4-10-1-B	144 B	139 A	187 A	lenceolate	Acute	Cuneate

3.2. Flower, pod and seed characters of yardlong beans genotype

Table 3 shows the variability of flower characters on Yardlong bean genotypes. Based on Table 3, the calyx color of genotype and cultivar check are yellow green group. The petal color of FA 1-1-9-15-B, FA 3-2-13-14-B, FA 1-1-7-3-12-B,FA 1-1-3-3, FA 1-1-2-1-14-B, FA 1-1-4-3, and FA 1-1-10-1-6 have different color from both cultivar checks (purple-violet group), while other genotypes have a similar color with both cultivar checks (violet group). The pod color for all genotype are similar with cultivar check 1. The seed shape of FA 1-1-9-15-B, FA 1-1-5-3-18-B, FA 3-2-13-14-B, FA 1-1-7-3-12-B, FA 1-1-3-3, FA 1-1-2-1-14-B, FA 1-1-4-3, and FA 1-1-10-1-6 have a similar shaped with both cultivar checks, but the other genotypes are different.

The first seed color of FA 1-1-5-3-18-B, FA 1-1-3-3, and FA 1-1-2-1-14-B have a similar shape with both cultivar checks 2 (white) and FA 1-1-9-15-B, FA 3-2-13-14-B, FA 1-1-7-3-12-B, FA 1-1-4-3, FA 1-1-10-1-6, and FS 3-4-8-2-B have a similar shape with both cultivar checks 1 but the other genotypes are different. The pattern second seed color of FA 1-1-9-15-B, FA 1-1-7-3-12-B, FA 1-1-4-3, and FA 1-1-10-1-6 6 have different pattern from both cultivar checks. FA 3-2-13-14-B have same patern with Cultivar Check 1, while other genotypes have a similar color with cultivar checks 2.

The color difference that occurs on the petals due to the appearance of streaks maybe the result of cross between Fagiola with Aura Hijau and Fagiola with Super Putih so there ate anthocyanin pigments in the flowers. According to [23] anthocyanins are red or purple pigments found in all plant tissues such as leaves, stems, roots, flowers, and fruit. According to the research of [24] gene that regulates the purple color of yardlong beans are dominant to green gene so the purple color can be visible. This supported by [25] research which show that the yardlong bean genotype which has green petals produces green pods. Based on the research of [26] stated that

IOP Conf. Series: Earth and Environmental Science **1460** (2025) 012016 doi:10.1088/1755-1315/1460/1/012016

the color of the red bean pods is caused by the accumulation of anthocyanin pigments. The red color difference that occurs in the crosses is thought to be caused by modified genes. The research of [21] stated that modified genes can affect the appearance of other genes so that they change the character of the major genes.

Table 3. flower, pod and seed characters of yardlong beans genotype

Genotypes	Calyx Color	Petal Color	Pod Color	Seed Shape	first Seed Color	Pattern second Seed Color
FG (Cultivar Check 1)	149 A	N88 A	187 A	kidney shaped	Brown Red	in streak/veins
AH (Cultivar Check 2)	149 A	86 C	143 B	kidney shaped	White	mottled on part of seed
FA 1-1-9-15-B	149 A	N82 C	187 B	kidney shaped	Brown Red	in patches at tip
FA 1-1-5-3-18-B	149 C	N87 A	187 B	kidney shaped	White	mottled on part of seed
FA 3-2-13-14-B	149 C	N82 A	187 A	kidney shaped	Brown Red	in streak/veins
FA 1-1-7-3-12-B	149 B	N81 B	187 A	kidney shaped	Brown Red	in patches at tip
FA 1-1-3-3	149 A	N81 B	187 B	kidney shaped	White	mottled on part of seed
FA 1-1-2-1-14-B	149 C	N81 C	187 A	kidney shaped	White	mottled on part of seed
FA 1-1-4-3	149 A	N81 C	187 B	kidney shaped	Brown Red	in patches at tip
FA 1-1-10-1-6	149 B	N81 C	187 A	kidney shaped	Brown Red	in patches at tip
FS 3-4-10-3-B	149 D	86 A	187 B	curved	Dark Red	mottled on whole seed
FS 3-4-7-1-B	149 A	N88 A	187 B	curved	Dark Red	mottled on whole seed
FS 3-4-8-2-B	149 D	86 A	187 B	curved	Brown Red	mottled on whole seed
FS 3-4-14-2-1	149 A	N87 A	187 B	curved	Dark Red	mottled on whole seed
FS 3-4-10-1-B	149 D	86 A	187 B	curved	Dark Red	mottled on whole seed

3.3 plant height, flowering and harvesting time of Yardlong Bean Genotypes

The plant height of FA 1-1-9-15-B, FA 3-2-13-14-B, FA 1-1-2-1-14-B, FA 1-1-4-3, and FS 3-4-10-1-B was lower significantly different from Cultivar Check 2. The genotypes FA 1-1-5-3-18-B, FA 1-1-7-3-12-B, FA 1-1-10-1-6, FS 3-4-10-3-B, and FS 3-4-10-1-B have a shorter time to harvest than other genotypes and cultivar check with an average ± 48 DAP (Table 4.). The variation of time to harvest can be affected by flowering time, because flowering time influenced maturity duration, which is early flowering lines also mature early [27]. The study about flowering time have strong pleiotropic effects on yield traits in rice. Flowering is a prerequisite for crop production whenever seeds or fruits are harvested. In general, flowering time is a complex phenological trait influenced by a variety of physiological and environmental factor [27], there are two factor that can regulating flowering time in crop, that are differential endogenous and environmental [28]

Table 4. plant height, flowering and harvesting time of Yardlong Bean Genotypes

Genotypes	Plant Height	Flowering Time	Harvesting Time
FG (Cultivar Check 1)	185,83 b	44 b	51 b
AH (Cultivar Check 2)	191,80 a	46 a	53 a
FA 1-1-9-15-B	186,02 b	41 cd	48 cd
FA 1-1-5-3-18-B	186,75 ab	41 d	48 d
FA 3-2-13-14-B	185,93 b	43 bc	50 bc
FA 1-1-7-3-12-B	187,00 ab	41 d	48 d
FA 1-1-3-3	187,05 ab	42 bcd	49 bcd
FA 1-1-2-1-14-B	185,38 b	42 bcd	49 bcd
FA 1-1-4-3	185,92 b	41 cd	48 cd
FA 1-1-10-1-6	187,33 ab	41 d	48 d
FS 3-4-10-3-B	187,97 ab	41 d	48 d
FS 3-4-7-1-B	186,92 ab	42 bcd	49 bcd
FS 3-4-8-2-B	186,82 ab	43 bc	50 bc
FS 3-4-14-2-1	186,47 ab	42 cd	49 cd
FS 3-4-10-1-B	185,55 b	41 d	48 d

Notes : values followed by the same letter within the same column were not significantly different to DMRT α =5%.

Table 5. Yield Components of Yardlong Bean Genotypes

Genotypes	Pod Length	Pod Width	Pod Weight	Number of Pod per Plant	Pods Weight per Plant
FG (Cultivar Check 1)	45,02^{d}	5,64^{abc}	11,86^{bc}	19,24^{abc}	153,92^{bc}
AH (Cultivar Check 2)	58,63^{a}	5,29^{abc}	15,91^{a}	13,54^{c}	108,34^{c}
FA 1-1-9-15-B	51,51^{bc}	5,22^{abc}	13,59^{abc}	24,29^{abc}	194,31^{abc}
FA 1-1-5-3-18-B	50,44^{bcd}	5,76^{abc}	12,64^{abc}	24,07^{abc}	192,55^{abc}
FA 3-2-13-14-B	44,87^{d}	5,66^{abc}	12,49^{abc}	19,19^{abc}	153,52^{bc}
FA 1-1-7-3-12-B	48,10^{bcd}	5,39^{abc}	12,74^{abc}	17,34^{bc}	200,94^{abc}
FA 1-1-3-3	47,97^{bcd}	5,10^{abc}	11,78^{bc}	32,19^{a}	257,57^{ab}
FA 1-1-2-1-14-B	50,43^{bcd}	5,58^{abc}	12,28^{bc}	23,37^{abc}	186,93^{abc}
FA 1-1-4-3	49,73$_{bcd}$	6,03^{a}	13,24^{abc}	27,21^{abc}	217,71^{abc}
FA 1-1-10-1-6	51,25^{bc}	5,04^{bc}	12,62^{abc}	33,55^{a}	268,37^{a}
FS 3-4-10-3-B	52,76^{b}	4,96^{c}	15,20^{ab}	19,09^{abc}	152,72^{bc}
FS 3-4-7-1-B	46,20^{cd}	5,91^{ab}	11,24^{c}	20,60^{abc}	164,77^{abc}
FS 3-4-8-2-B	52,52^{b}	4,92^{c}	12,65^{abc}	22,45^{abc}	179,56^{abc}
FS 3-4-14-2-1	49,90^{bcd}	5,09^{abc}	13,05^{abc}	28,33^{abc}	226,61^{ab}
FS 3-4-10-1-B	49,07^{bcd}	5,09^{abc}	11,96^{bc}	31,20^{ab}	249,63^{ab}

Notes : values followed by the same letter within the same column were not significantly different to DMRT α =5%.

The genotypes with the highest number of pods than other genotypes and cultivar check were FA 1-1-10-1-6 (± 33.55 pods per plant). However, FA 1-1-10-1-6 was not significantly different from FA 1-1-3-3. The number of pods per plant observed was the accumulation of the number of pods from the first harvest to the sixth harvest in each plant tested. The weight pods per plant range from 108.34 g to 268.37 g (Table 5). Genotypes FA 1-1-10-1-6 had the highest pod weight per plant than other genotypes and cultivar check (check 1 and check 2), with an average weight of ±268.37 g pods per plant.

The pod width of FA 1-1-4-3 is the wider than FA 1-1-10-1-6, FS 3-4-10-3-B, FS 3-4-7-1-B, FS 3-4-8-2-B. The cultivar check 2 still had the best pod weight, with an average weight of ±15.91 g. However, genotypes FA 1-1-9-15-B, FA 1-1-5-3-18-B, FA 3-2-13-14-B, FA 1-1-7-3-12-B, FA 1-1-4-3, FA 1-1-10-1-6, FS 3-4-10-3-B, FS 3-4-8-2-B, and FS 3-4-14-2-1 had pod weight better than the cultivar check 1.

The cultivar check 2 still had the best pod length, with an average length of ±58.63 cm. However, genotypes FA 1-1-9-15-B, FA 1-1-10-1-6, FS 3-4-10-3-B and FS 3-4-8-2-B had pod length significantly better than the cultivar check 1. Genotypes with the shortest pod length were FA 3-2-13-14-B with an average length of ±44.87 cm was not significantly different from cultivar check 1. Table 5 showed that cultivar check 2 still had a superior character in pod length. Pod length is one of the main criteria for selecting a superior yardlong bean, because consumers prefer yardlong beans with long pod [29].

4. Conclusion

The genotypes of the F8 generation of red yardlong beans had diverse morphological characters, especially for the leaf color, leaf shape, calyx color, flower color, primary seed color and pattern second seed color. The genotype FA 1-1-10-1-6 showed better yield components than the check cultivar in flowering time, harvesting time, number of pods per plant, and pod weight per plant. The morphological characters and yield components that have been characterized can be used as potential donors for future hybridization programs to develop superior yardlong bean varieties that have a short harvesting time with high production.

Acknowledgement

This research is funded by the Ministry of Education, Culture, Research, and Technology of the Republic of Indonesia through the Research Grant Program, and Community Service, basic research scheme, domestic collaborative research in 2023.

References

[1] Trovato, G. M. 2012. Behavior, nutrition and lifestyle in a comprehensive health and disease paradigm: skills and knowledge for a predictive, preventive and personalized medicine. EPMA Journal 3(1): 8

[2] Evans, A., K. Banks, R. Jennings, E. Nehme, C. Nemec, S. Sharma, A. Hussaini, dan A. Yaroch. 2015. Increasing access to healthful foods: a qualitative study with residents of low-income communities. International Journal of Behavioral Nutrition and Physical Activity 12 (Suppl 1): S5

[3] Angelia, I.O. 2017. Kandungan pH, total asam tertitrasi, padatan terlarut, dan vitamin C pada beberapa komoditas hortikultura. Journal of Aritech Science 1(2): 68-74

[4] Alam, S.S., R. Tasmin, I. Jahan, M.A. Habib, dan S.S. Sultana. 2013. Fluorescent banding dan RAPD analysis of five cultivars in Vigna unguiculata ssp. Sesquipedalis (L.) Verc. Cytologia 78 (1): 73-79.

[5] T.J. Ha, M.H. Lee, C.H. Park, S.B. Pae, K.B. Shim, J.M. Ko, S.O. Shin, I.Y. Baek and K.Y. Park. Identification and characterization of anthocyanins in Yard-Long Beans (Vignaunguiculata ssp. sesquipedalis L) by high-performance liquid chromatography with diode array detection and electrospray ionization/ mass spectrometry, J. Agric. And Food Chem, 58, 2571- 2576, 2010.

[6] Ribera, A.E., M. Reyes-Diaz, M. Alberdi, G.E. Zuniga, dan M.L. Mora. 2010. Antioxidant compounds in skin and pulp of fruits change among genotypes and maturity stages in highbush blueberry (Vaccinium corymbosum L) grown in Southern Chile. Journal of Soil Science and Plant Nutrition 10 (4): 509-536.

[7] Diamanti, J., F. Capocasa, F. Balducci, M. Battino, J. Hancock, dan B. Mezzetti. 2012. Increasing strawberry fruit sensorial and nutritional quality using wild and cultivated germplasm. PloS One 7(10): e46470

[8] Lamjud, K., S. Srimat, P. Sangwanangkul, S. Wasee, K. Thaipong. 2016. Antioxidant properties and fruit quality of selected papaya breeding lines. Science Asia 42 (5): 323- 339

[9] Cirilli, M., D. Bassi, dan A. Ciacciulli. 2016. Sugars in peach fruits: a breeding perspective. Horticulture Research 3: 15067[10] Aprea, E., M. Charles, I. Endrizzi, M.L. Corollaro, E. Betta, F. Biasioli, dan F. Gasperi. 2017. Sweet taste in apple: the role of sorbitol, individual sugars, organic acids and volatile compounds. Scientific Reports 2017 (7): 44950

[11] Manoharan, R.K., H.J. Jung, I. Hwang, N. Jeong, K. H. Kho, M.Y. Chung, dan I.S. Nou. 2017. Molecular breeding of novel orange-brown tomato fruit with enhanced beta carotene and chlorophyll accumulation. Hereditas 154: 1

[12] Reynolds, M., dan P. Langridge. 2016. Physiological breeding. Current Opinion in Plant Biology 31:162-171

[13] Kusmiyati, F., S. Anwar, U. Jamiati, B. Herwibawa. 2021. Segregation ratio and heritability of agronomic character in yardlong bean F2 population. Prosiding International terindeks Scopus. IOP Conf. Ser.: Earth Environ. Sci. 803 012027. https://iopscience.iop.org/article/10.1088/1755-1315/803/1/012027

[14] Kusmiyati, F., S. Anwar, B. Herwibawa. 2021. Agronomic character evaluation of F3 yardlong bean progenies. Prosiding International terindeks Scopus. AIP Conference Proceedings 2353, 030019 (2021). https://aip.scitation.org/toc/apc/2353/1?size=all

[15] Bakhtiar, I.F., S. Anwar, F. Kusmiyati. 2020. Analisis keragaman dan heritabilitas karakter agronomi generasi F4 kacang panjang hasil persilangan Fagiola IPB x Super Putih. J. of Tropical Crop Sci. Technol 2(1) : 13 - 20. https://ejournal.umm.ac.id/index.php/jtcst/article/view/11455

[16] F. Kusmiyati, S. Anwar, and B. Herwibawa. 2022. Study on genetic variability and heritability in F5 segregating generation for yield and its compenents in yardlong bean. Advances in Biological Sciences Research, volume 17. https://www.atlantispress.com/proceedings/ictaff-21/125968309

[17] S. Anwar, F. Kusmiyati, and D. R. Lukiwati. 2022. Estimation of varian component, heritability and correlation to determine selection criteria in F5 population of yardlong bean. Prosiding International terindeks Scopus. AIP Conference Proceeding 2586, 020004. https://doi.org/10.1063/5.0109912

[18] F. P. Putra, F. Kusmiyati, S. Anwar, Widaryati and M.G.A. Sas. 2022. Variation in leaf characters and agronomical traits of yard-long bean genotypes between Dimocarpus longan and Psidium guajava stand. Biodiversitas 23 (11) : 5752 -5758. DOI: 10.13057/biodiv/d231127

[19] Widaryati, F. Kusmiyati, S. Anwar, M.G.A Sas and F.P. Putra. 2023. Yield component of red yardlong bean genotypes ((Vigna unguiculata (L.) ssp. sesquipedalis). IOP Conference Series (In Press)

[20] T D Ihsanto, F Kusmiyati, S Anwar, M G A Sas, and F P Putra . 2023. Agro-morphological characteristics, anthocyanin, and sweetness level of Yardlong bean genotypes (*Vigna unguiculata* (L.) ssp. sesquipedalis). IOP Conf. Series: Earth and Environmental Science 1230 (2023) 012126 doi:10.1088/1755-1315/1230/1/012126

[21] Sriwidarti S, Sa'diyah N, Utomo SD, and Barmawi M 2020 Inheritance of qualitative character on yardlong bean (Vigna sinensis var. sesquipedalis L.) zuriat brown x black seed skin J. Agrotropika 15 43-47 (http://dx.doi.org/10.23960/ja.v15i1.4247)

[22] Hua S, Bianco MD, and Chen ZH 2022 Crop yield and quality response to the interaction between environment and genetic factors Front. Genet. 13 1-3 (https://doi.org/10.3389/fgene.2022.823279)

[23] Bondre S, Patil P, Kulkarni A, and Pillai MM 2012 Study on isolation and purification of anthocyanins and its application as ph indicator Int. j. adv. biotechnol. res. 3 698-702

[24] Kuswanto, Waluyo B, and Hardianingsih P 2013 Segregation and selection of observed yardlong bean (Vigna sesquipedalis L. Fruwirth) to get expected lines of purple pod Int. Res. J. Agric. Sci. Soil Sci. 3 88-92

[25] Khadijah N, Kuswanto K, and Damanhuri D 2012 Distinctness assessment on yardlong bean (Vigna sesquipedalis (L.) Fruhw.) varieties (case study for five yard long bean varieties in PVP right application) AGRIVITA J Agric Sci. 34 198-206 (https://doi.org/10.17503/Agrivita-2012-34-2- p198-206)

[26] Lisbona FJ, Gonzales AM, Capel C, Alcazar MG, Capel J, Ron AMD, Santalla M, and Lozano R 2014 Genetic variation underlying pod size and color traits of common bean depends on quantitative trait loci with epistatic effects Mol. Breeding. 33 939-952 (https://doi.org/10.1007/s11032-013-0008-9)

[27] Zhang, Z. Z., K. Wang, L. Guo,Y. J. Zhu,Y. Y. Fan, S. H. Cheng, and J. Y. Zhuang. 2012. Pleiotropism of the photoperiod-insensitive allele of Hd1 on heading date, plant height and yield traits in Rice. PLoS ONE 7(12): e52538. https://doi.org/10.1371/journal.pone.0052538

[28] Vivalapali, S. K., V.A. Celine, and A. M. Vahab. 2014. Assessment of genetic divergence in among Yard long bean (Vigna unguiculata subsp. sesquipedalis[L.]) Genotypes. Legume Genomics Genet. 5(1) : 1 – 3. http://lgg.biopublisher.ca

[29] Kuswanto, B. Waluyo, and P. Hardianingsih. 2013. Segregation and selection of observed yardlong bean (Vigna sesquipedalis L. Fruwirth) to get expected lines of purple pod. Journal of Agricultural Science and Soil Science. 3(3) : 88-8.

The effect of differences in energy and protein consumption on the quality of dairy cow's milk in smallholder farms

Nida Fariha*, Rudy Hartanto, Anis Muktiani

Department of Animal Sciences, Universitas Diponegoro, Semarang, Indonesia

*E-mail: nidafariha08@gmail.com

Abstract. The study aimed to examine the effect of energy and protein consumption on lactose, fat, and milk protein production in dairy cows in smallholder farms fed potluck. This research was conducted on an observational basis. Fifty-six cows in the lactation phase in community farms in the Semarang City area were sampled from two sub-districts with the largest dairy cow population. The parameters measured were crude protein consumption and energy consumption in the form of Total Digestible Nutrients (TDN) as independent variables and lactose, protein, and milk fat production as fixed variables. The data were tested using correlation and linear regression tests. The results showed that the average production of lactose, protein, and milk fat was 0.53±0.17, 0.35±0.11, and 0.36±0.17 kg/day. The average TDN and crude protein content in rations was 61.64±5.31% and 12.12±4.52%. TDN had an effect on lactose production of 13.9% with the equation Y1 = 0.337 + 0.026X1 (R2 = 0.139, p = 0.005) and on milk protein production by 13.9% with the equation Y2 = 0.224 + 0.017X1 (R2 = 0.139, p = 0.005), while on milk fat production had no real effect. Consumption of crude protein has no significant effect on milk nutrient production. The conclusion is that TDN consumption is closely related to lactose and protein production but not milk fat in dairy cows. Crude protein consumption does not correlate with lactose, protein, and milk fat production.

Keywords: energy consumption, protein consumption, lactose, milk protein, milk fat.

1. Introduction

The nutrient needs of lactated dairy cows are influenced by basic living needs and the need for production. The basic living needs of dairy cows are determined by the weight, health, and activity of cows, while the need for milk production is influenced by milk production and fat content. The nutrient needs of lactated dairy cows must be met by the feed consumed, which can be seen in the amount of dry matter intake (DMI) in the feed. The dry matter (DM) contains several important nutrients, such as fat, protein, calcium, phosphorus, vitamins, also energy, which can be measured in Total Digestible Nutrient (TDN). If the nutrient needs of lactating cows are not met from the feed consumed, milk production will be less than optimal [1]. Lactating dairy cows with a body weight (BW) of 450 kg, which has a milk fat content of 4% and milk protein of 3-4%

in milk production of 10 kilograms, require feed containing 12.4 kg of DM (2.7% of BW) with a TDN content of 68% and crude protein (CP) of 11.9-13.4%. If the milk production is 15 kg, then the cow requires feed with DM content of 9.4 kg (2% of BW) with a TDN 78% and CP of 16-19.4% [2].

The most important components that must be met in the ration for lactating dairy cows are energy and protein [3]. DMI is not the main thing that must be considered in rations for lactating dairy cows. This is because, during the first phase of lactation, there is a rapid increase in milk production to the peak of lactation. However, feed consumption is low due to the involution or shrinkage of the digestive tract in cows. This condition will result in cows experiencing a nutrient deficit if the feed provided does not have high energy. Physiologically, the increase in DMI in lactating cow will occur when entering the second phase, when the peak of lactation has passed [4].

Energy adequacy is very important to pay attention to because a lack of energy causes protein overhaul to produce energy [5]. The energy needs of lactating dairy cows are determined by the need for basic livelihood, which is influenced by body weight. In contrast, the need for milk production is influenced by the amount of milk secreted and the fat content contained in the milk [6]. An imbalance in nutrient content to meet the energy needed from feed will result in cows utilizing food reserves in the body to meet their needs, causing a decrease in body weight [7].

The quality of nutrients influences the nutrient components in milk in the feed given to cows. This is because most of the precursors for milk synthesis in the mammary gland come from blood, whose number depends on the quality of feed and the process of absorption of nutrients in the body [8]. Protein and TDN can affect the production and quality of milk, so nutrient intake is very important for the milk synthesis process. TDN derived from coarse fibre feed will be used by cows as a basic ingredient for synthesizing lactose and milk fat. At the same time, protein will be broken down into amino acids and used by cows in milk biosynthesis, especially milk protein. If the consumption of TDN in lactating cows cannot meet their needs, it will decrease cows' body weight and low milk production [1].

Semarang City is one of the highest milk-producing cities in Central Java province. Dairy cattle in Semarang City are managed in the people's livestock business, most of whom are already members of the Livestock Farmers Group/ Kelompok Tani Ternak (KTT). The dairy cows raised are Friesian Holstein obtained from animal markets or existing broods. Based on data from the Semarang City Agriculture Service, it is known that the number of dairy cattle in Semarang City in 2022 was 1,633. The dairy cattle population is spread across several sub-districts, namely North Semarang 2, Pedurungan 6, Tembalang 35, Banyumanik 150, Gunungpati 1,156, Mijen 211, and Ngaliyan 73. In 2022, dairy cow milk production in Semarang City reached 3,040,350.66 liters. However, in mid to late 2022, Semarang City was affected by the spread of Foot-and-mouth Disease (FMD and Lumpy Skin Disease (LSD) so that the dairy cattle population experienced a drastic decline.

Feeding small-scale smallholder farms still needs to pay attention to the adequacy of nutrients in cows. So far, the farm feed is only aimed at increasing production, but the milk quality has not been paid much attention. Farmers usually provide only potluck feed that is easy to get from the surrounding environment. As a result, the milk produced by dairy cows is not too high, and the quality is not very good. Feed adequacy, both quantity and quality, is crucial to support high milk production. Quality feed must contain the nutrients needed by dairy cows. Based on the description above, it is necessary to conduct a study to determine the relationship between energy and protein consumption in dairy farms and the milk quality produced. This study aims to determine the effect of protein and energy consumption from feed on milk quality which can be seen from lactose production, milk protein production, and milk fat production.

2. Materials and methods

The research was conducted on January 23 – February 28, 2023, at a farm/ KTT in Semarang City, Central Java. Milk and feed analysis was conducted at Meet and Dairy Production Laboratory and Nutrition and Feed Science Laboratory, Faculty of Animal Husbandry and Agriculture, Universitas Diponegoro, Semarang. Ethical approval for this study was obtained from the Animal Ethics Committee at Universitas Diponegoro (Semarang, Indonesia) with approval number 61-01/A-02/KEP-FPP (January, 2023). Briefly, Friesian Holstein cows in lactation months 1-10 were directly observed for their feed consumption and milk production for 7 days. There were no changes in feed provision or any treatment carried out in this study, the cows continued to carry out their normal daily activities.

The material used in this study was 56 lactation phase cows in the farm/summit in the Semarang City area with an average BW of 383.11±42.92 kg. The equipment used is books and stationery to record data, a 50 kg hanging digital scale to weigh feed, a bottle as a milk sample container, a cooling box for temporary storage of milk, plastic clips as a container for concentrated feed samples, plastic bags as a forage feed sample container, and a camera for documentation. The parameters measured were crude protein (CP) consumption and energy consumption in the form of Total Digestible Nutrients (TDN) as independent variables and lactose, milk protein, and milk fat production as fixed variables.

The research was conducted in an observational manner. The location of the study was determined by purposive sampling, which is a sample determination technique with certain considerations based on the sub-district area with the largest dairy cattle population, namely in Gunungpati and Mijen Districts. The samples were taken from 8 KTT with details of 5 from Gunungpati District and three from Mijen District. The selection of cows that used as a sample is based on the following criteria: the cows used is FH cattle and it is ensured that the breeder always mates pure FH male seeds, lactation months in the range of 1-10 months, healthy cows and not sick in any way, especially FMD, Mastitis, and LSD, cows are milked two times a day completely, cows are given forage and concentrate feed regularly with relatively the same type of feed, and the farmer allows his cows to be observed.

2.1. Sampling and testing of nutrient content

Forage feed and concentrated feed samples, as well as milk samples, were taken from each farm to be tested for nutrient content and quality. Observation was carried out for one week at each farm/ KTT.

Feed sampling, both forage and concentrate, is carried out 3-4 times weekly. Feed samples were taken from forage, and concentrates were used for morning and evening feeding. Each sample was aerated to dry the air, weighed, and homogenized into a feed sample. Samples of leftover feed were taken from leftover feed in the morning and leftovers in the afternoon. Each sample is aired so that the air is dry, weighed, and then homogenized into a sample of feed residues. DMI is measured by reducing the amount of feed given with the rest on the next day multiplied by the feed DM level. CP and TDN consumption is calculated by multiplying DMI by CP and TDN feed levels.

Testing the nutrient content of the feed is carried out by proximate analysis (Weende method). The TDN level is calculated using a formula developed by Sutardi [3], namely:

1. Feed with a CP content of < 20% and crude fiber > 18%, namely 70.6 + (0.259 × %CP) + (1.01 × %ether extract) - (0.760 × %crude fiber) + (0.0991 × %nitrogen free extract)
2. Feed with a CP content < 20% and crude fiber < 18%, namely 2.79 + (1.17 × %CP) + (1.74 × % ether extract) - (0.295 × % crude fiber) + (0.810 × % nitrogen free extract)
3. Feed with a CP content of > 20% and crude fiber < 18%, namely 25.6 + (0.5 × %CP) + (1.7 × %ether extract) - (0.474 × %crude fiber) + (0.732 × %nitrogen free extract)
4. Feed with a CP content of > 20% and crude fiber > 18%, namely 3.17 + (0.64 × %CP) + (2.08 × %ether extract) - (0.0675 × %crude fiber) + (0.94 × %nitrogen free extract)

At the end of the observation week, milk production was measured from morning and evening milking and added up to get daily milk production in kg/day. The nutrient content of the feed is tested using the Lactoscan Milk Analyzers.

2.2. Data analysis

The data was tested using a linear regression test to determine whether there is a relationship between the consumption of energy nutrients and feed protein with milk quality, including the production of lactose, milk protein, and milk fat. The form of a simple linear regression equation is as follows:

$$Y = a + bx$$

Information:
Y = lactose production, milk protein production, milk fat production
a = constant
b = regression coefficient
x = TDN consumption, CP consumption

The significance test at the level of $\alpha = 5\%$ was carried out with ANOVA and continued with a hypothesis test with a t-test to find out whether the regression coefficient of the independent variable (X) partially had a significant effect on the dependent variable (Y).

3. Results and discussion

3.1. Consumption of TDN with milk quality

Table 1. Results of simple regression analysis between TDN consumption and milk nutrient production

X	Y	R^2	Equation
TDN Consumption	Lactose Production	0,139	Y = 0.337 + 0.026X
	Milk Protein Production	0,139	Y = 0.224 + 0.017X
	Milk Fat Production	0,059	Y = 0.236 + 0.017X

Table 1 presents the results of the analysis and regression equation between TDN consumption and milk nutrient production. TDN consumption is measured from the amount of TDN consumption in a cow in one day, as is milk nutrient production.

A simple linear regression analysis on milk lactose production with TDN consumption yielded a significant value (P<0.05) with the equation Y = 0.337 + 0.026X (r2 = 0.139, p = 0.005). This equation shows that every 1 kg of TDN consumption affects lactose production by 0.363 kg. The value of the determination coefficient of the equation is 0.139, which means that milk lactose production is influenced by TDN consumption by 13.9%, while other factors influence the remaining 86.1%. A simple linear regression analysis on milk protein production with TDN consumption yielded a significant value (P<0.05) with the equation Y = 0.224 + 0.017X (r2 =

Table 2. Average TDN Consumption and Milk Nutrient Production

TDN Consumption		n	Milk Nutrient Production		
			Lactose	Protein	Fat
----- (kg/cow)-----			-------------------- (kg/cow)--------------------		
	<5.0	10	0.44±0.12	0.29±0.08	0.32±0.16
	5,0-6,0	6	0.53±0.22	0.35±0.15	0.41±0.30
	6,0-7,0	6	0.38±0.11	0.25±0.07	0.28±0.11
	7,0-8,0	7	0.68±0.19	0.45±0.13	0.24±0.06
	8,0-9,0	6	0.51±0.14	0.34±0.09	0.32±0.16
	9,0-10	9	0.57±0.12	0.38±0.08	0.48±0.17
	> 10	11	0.61±0.16	0.40±0.11	0.43±0.11
Average	7.54±2.47		0.53±0.17	0.35±0.11	0.36±0.17

0.139, p = 0.005). This equation shows that every 1 kg of TDN consumption will affect 0.139 kg of milk protein production. The value of the determination coefficient of the equation is 0.139, which means that milk protein production is influenced by TDN consumption by 13.9%, while the remaining 86.1% is influenced by Other factors.

In addition to TDN consumption, other factors affecting lactose and milk protein production are easily digestible carbohydrates (Nitrogen Free Extract), protein consumption, and digestibility. Easily digestible carbohydrates (Nitrogen Free Extract) are a source of glucose as a precursor to milk lactose, while protein is a source of amino acids as a precursor to milk protein. Glucose, which results from the digestion of carbohydrates, will be a precursor to lactose synthesis, while in protein synthesis, the raw material is an amino acid from CP [9]. Low feed quality can also affect the production of milk nutrients. Low ration quality, indicated by a low TDN content even though the dry ingredients are as needed, will affect milk quality [10].

A simple linear regression analysis between milk fat production and TDN consumption had an insignificant value (P>0.05), with the equation Y = 0.236 + 0.017X (r2 = 0.059, p = 0.072). These results show that milk protein production in dairy cows in Semarang City people's farms is not directly affected by TDN consumption. The average TDN consumption and milk nutrient production per cow per day are presented in Table 2.

TDN consumption is closely related to lactose production because TDN is an energy source for cows. TDN is generally used to express energy needs but can also express the adequacy of other nutrients such as carbohydrates, fats, and proteins. TDN adequacy can be interpreted as the adequacy of carbohydrates, proteins, and fats, which are precursors of the synthesis of milk components [11]. Lactose synthesis is formed from glucose and galactose, resulting from glucose absorption into the udder secretory cells. The glucose inside the udder cells is converted to glucose 6-phosphate and can eventually be converted back to glucose, whereas galactose is converted to UDP-galactose. All of these processes require energy in the form of ATP. ATP is required to change glucose to glucose 6-phosphate and, subsequently, the process of lactose synthesis [12]. Apart from being an energy source, TDN consumption can affect lactose production because energy source feed contains much glucose, a precursor to lactose formation. The feed TDN derived from the energy source feed will produce glucose, channelled to the udder cells and used as a precursor to lactose formation [1].

TDN consumption is closely related to milk protein production as an energy contributor. The process of protein synthesis in milk requires energy in the form of ATP to transport amino acid components and DNA information from the nucleus to the ribosome. Transport of amino acids through the cell membrane will require up to 2 ATP, and the synthesis of mRNA, tRNA, and DNA also requires a certain amount of ATP [13]. Peptide bond synthesis is an endergonic process that requires energy to start the reaction, so it needs to be supported by energy donation [14]. The higher the TDN consumed by cows, the higher the energy that can be used for protein synthesis. Energy supply for lactation and metabolic energy can increase the production of milk components, including milk protein because an increased energy supply will increase the absorption of glucose and amino acids towards the udder [15].

TDN consumption has no real relationship with milk fat production. TDN as an energy source has no direct effect on milk fat production. This can be caused by the fact that the quality of the feed consumed by cows is still not good. The consumption of TDN by cows in Semarang City people's farms is 61.64±5.309% of the total rations given. In contrast, according to NRC (1989), the TDN content in rations for lactated dairy cows with milk production between 10 – 15 litres/day is at least 63% [16]. This low TDN content makes the energy supply consumed by cows less than what can be used to support production. The quality of the ration can be seen from the high content of TDN as energy; if the energy from the ration is still insufficient to meet the needs of cows, it will result in milk metabolism and biosynthesis becoming suboptimal [17]. Other factors that can affect this are environmental conditions and feed management. Smallholder farmers in Semarang City do not have good feed management, as evidenced by the provision of forage and concentrate, which is still unbalanced. Good feeding of dairy cows is the ratio of forage 60% and concentrate 40% BK or the provision of fresh forage as much as 10% BB and concentrate as much as 1-2% BB [18]. Meanwhile, in smallholder farms, some farmers provide forage in very small quantities, up to below 40% of the total feed given.

3.2. Consume CP with milk quality

Table 3 presents the analysis results and the regression equation between CP consumption and milk nutrient production. Protein consumption is measured from the amount of CP consumption in a cow in one day, as is milk nutrient production.

A simple linear regression analysis on lactose, protein, and milk production with CP consumption yielded insignificant values ($P>0.05$) with the equations $Y = 0.466 + 0.044X$ ($r2 = 0.044$, $p = 0.121$), $Y = 0.309 + 0.030X$ ($r2 = 0.045$, $p = 0.116$), and $Y = 0.312 + 0.034X$ ($r2 = 0.026$, $p = 0.239$). This means that milk lactose production in lactated dairy cows in Semarang City is not

Table 3. Results of simple regression analysis between CP consumption and milk nutrient production

X	Y	R^2	Equation
CP Consumption	Lactose Production	0,044	Y = 0.466 + 0.044X
	Milk Protein Production	0,045	Y = 0.309 + 0.030X

directly affected by CP consumption. The average CP consumption and milk nutrient production per cow per day are presented in Table 4.

CP consumption does not directly affect the production of milk nutrients in the form of lactose, protein, and milk fat. This can be caused by the consumption of CP in cows, which is still quite low, with an average of 1.52±0.81 kg/cow/day. The CP requirement in lactated cattle is 1.47 – 1.66 kg for BB 454 kg with milk production of 10 kg and 2.26 – 2.59 kg for cows with milk production of 20 kg [2]. The quality of feed provided to cows is also still not good enough. Based on the results of the analysis, the average CP content of rations given to cows by smallholder farmers in Semarang City was 12.12±4.52%. Proteins or amino acids that are glucogenic can be

Table 4. Average CP Consumption and Milk Nutrient Production

CP Consumption		n	Milk Nutrient Production		
			Lactose	Protein	Fat
----- (kg/cow)-----			------------------- (kg/cow)-------------------		
	> 0.5	2	0.43±0.02	0.29±0.01	0.28±0.18
	0,5 – 1,0	15	0.46±0.17	0.31±0.12	0.37±0.21
	1,0 – 1,5	18	0.55±0.19	0.36±0.12	0.33±0.19
	1,5 – 2,0	7	0.55±0.18	0.36±0.11	0.31±0.13
	2,0 – 2,5	6	0.68±0.02	0.48±0.12	0.44±0.14
	2,5 – 3,0	3	0.58±0.01	0.38±0.01	0.46±0.05
	3,0 – 3,5	5	0.50±0.07	0.33±0.05	0.42±0.09
Average	1.52±0.81		0.53±0.17	0.35±0.11	0.36±0.17

converted into glucose for milk lactose synthesis or glycerol for milk fat synthesis. However, because the protein content of the feed is low, it cannot be used for this purpose.

In addition, insignificant results can also be caused by poor protein quality from feed. The quality of protein can be seen from the balance of essential amino acids in it and the absence of deficient amino acids. The quality of protein is determined by its array of amino acids. High-grade protein contains all important amino acids in sufficient quantities for the body's metabolic needs [19]. Methionine is an amino acid that has an important role in synthesizing milk proteins. Methionine initiates protein formation, so if there is a deficiency of the amino acid methionine, the formation of milk protein will decrease, resulting in low milk protein production. Methionine is very Methionine binds to the UAG codon to initiate protein synthesis within the cell. It starts on the mRNA at the P ribosome site, the first amino acid formed on the polypeptide chain [20]. Tofu pulp, the most widely used protein source feed ingredient by cows, is a derivative product of soybean processing and has a high protein content but low content of the amino acid methionine. Beans, pods, or other legumes contain a complete protein content but low methionine [19]. Soybean meal is one of the feed ingredients for legumes, and it has high protein but is deficient in the amino acid methionine [21]. Thus, although CP consumption is high, more than low protein quality is needed to directly encourage milk nutrient production.

However, lactose production has increased numerically along with CP consumption, which has increased to 2.5 kg/day, although statistically insignificant (P>0.05). The numerical increase in lactose production indicates that CP consumption has an indirect relationship with lactose synthesis. Protein has an important role in the biosynthesis process of milk lactose, namely as an enzyme. The lactose synthetase enzyme is an enzyme composed of 1,4-galactosyltransferase 1 and alpha-lactalbumin. Lactose synthesis occurs within the Golgi apparatus, where lactose synthase, which is a unique enzyme consisting of β-1,4-galactosyltransferase-1 and α-lactalbumin joins glucose and UDP-galactose through glycosidic bonds β-1-4 [22]. In addition, protein acts as a glucose transporter that absorbs glucose from the blood circulation and as other supporting hormones such as prolactin, glucocorticoids, insulin, thyroid hormones, and epidermal growth factors. Glucose is absorbed into the mammary gland via insulin-independent glucose-transporter-1 (GLUT-1) [23]. The enzyme lactose synthetase is a determining factor in lactose synthesis, where the expression of both in udder cells is strictly regulated by important hormones such as prolactin, glucocorticoids, insulin, progesterone, thyroid hormone, estrogen, and epidermal growth factor [24].

Milk protein production also increased numerically to 2.5 kg/day consumption. This indicates that CP consumption affects milk protein production even indirectly. CP consumption should have a close relationship with milk protein production because the main precursors in milk protein synthesis are amino acids obtained from CP digestion. Milk protein synthesis combines several amino acids to form proteins that occur in alveoli epithelial cells and are controlled by genes containing DNA [25]. Milk protein production decreased again with more than 2.5 kg/day CP consumption. This can be caused by the consumption of CP that exceeds cows needs. The CP requirement for lactated dairy cows at 454 kg is 1.47 – 1.66 kg/cow with a milk production of 10 kg and 2.26 – 2.59 kg/cow for cows with a milk production of 20 kg [2]. CP consumption that is too high in cows causes protein not to be used properly. Protein consumption from feed that exceeds the standard has little effect on protein production and milk production because protein will be deamination [26]. The deamination process will convert amino acids into α-keto and ammonia, which will then undergo a conversion process that requires much energy. The deamination activity will form α-keto acid, which will enter the Krebs cycle and combine with acetyl-CoA to produce energy, as well as form ammonia, which will be converted to carbonyl phosphate with 2 ATP and then converted into urea, which requires 3 ATP [27].

Milk fat production is not directly affected by CP consumption. This can be caused by the fact that amino acids, which result from protein digestion from feed, are not the main precursor to milk fat synthesis. VFA is the ultimate precursor to milk fat synthesis [28]. Milk fat synthesized in the udder is derived from short-chain milk fatty acids from the synthesis of de novo acetate and β-hydroxy butyrate in the udder glands, long-chain fatty acids absorbed from chylomicron and LDL plasma triglycerides of blood, and medium-chain fatty acids from both pathways [29]. Protein's role in synthesising milk fat is to help transport fat in the blood. Triglycerides are distributed throughout the body through the lymphatic system in the form of chylomicron, a combination of triglycerides, cholesterol, phospholipids, fat-soluble vitamins, and proteins [27].

References

[1] Adi, D. S., D. W. Harjanti, and R. Hartanto. 2020. Evaluation of protein and energy consumption on milk production of lactation early dairy cows. J. Animal Husbandry in Indonesia. 22 (3): 292 – 305.

[2] National Research Council. 2001. Nutrient Requirement of Dairy Cattle. 7th Revised Edition. National Academy Press, Washington D. C.

[3] Sutardi, T. 2001. Revitalization of dairy farms through the use of plantation waste-based rations and organic mineral supplementation. Final report of RUT VIII 1. Office of the State Ministry of Research and Technology and LIPI

[4] Phillips, C. J. C. 2018. Principles of Cattle Production, 3rd Edition. CSIRO Publishing. Clayton South, Australia.

[5] Sunaryati, A. Muktiani and J. Achmadi. 2013. Supplementation of Curcuma xanthoriza and zn proteinate on milk energy consumption and production in dairy cows. J. Anim. Agric. 2. (1): 168 – 174.

[6] Novianti, J., B.P. Purwanto, and A. Atabany. 2014. Milk production efficiency and digestibility of elephant grass (Pennisetum purpureum) in FH dairy cows with different cut sizes. J. Production Science and Technology of Animal Husbandry Products. 2 (1): 224 – 230.

[7] Putra, N. G. W., D. N. Ramadani, A. Ardiansyah, F. Syaifudin, R. I. Yulinar, and H. Khasanah. 2022. Review: strategies for the prevention and treatment of metabolic disorders in ruminant livestock. J. Animal Husbandry in Indonesia. 24 (2): 150 – 159.

[8] Adriani, A. Latif, S. Fachri and I. Sulaksana. 2014. Improvement of milk production and quality of Peranakan Etawah goats as a response to feed quality improvement. J. Scientific Animal Science. 17 (1): 15 – 21.

[9] Suhendra, D., G. T. Anggiati, S. Sarah, A. F. Nasrullah, A. Thimoty, and D. W. C. Utama. 2015. Display of milk quality of dairy cows due to different balance of concentrates and forage. J. Animal Science 25 (1): 42 – 46.

[10] Mariana, A., D. N. Hadi, and N. Q. Agustin. 2016. Physiological response and quality of Friesian Holstein dairy cows in the long dry season in the highlands. J. Agripet. 16 (2): 131 – 139.

[11] Muktiani. A. 2017. Correlation between protein consumption, energy, and lactation month with dairy cow milk production in Semarang Regency. J. Research and Development of Central Java Province. 15 (2): 153 – 160.

[12] Chandel, N. S. 2021. Glycolysis. J. Cold Spring Harb Perspect Biol. 13 (5): 1 – 12.

[13] Chuzaemi, S. 2012. Physiology of Ruminant Nutrition.UB Press. Malang.

[14] Erastova, V., M. T. Degiacomi, D. G. Fraser, and H. C. Greenwell. 2017. Mineral surface chemistry control for origin of prebiotic peptides. J. Nat Commun. 8 (1): 1 – 9.

[15] Anger, J. C., C. Loncke, C. Omphalius, M. Boutinaud, J. G. Flament, H. Lapierre, and S. Lemosquet. 2023. Synthesis of milk components involves different mammary metabolism adaptions in response to net energy and protein supplies in dairy cows. J. Dairy Sci. 107 (5): 2883 – 2899.

[16] National Research Council. 1989. Nutrient Requirements of Dairy Cattle. 6th Edition, National Academy Press, Washington D. C.

[17] Anggiati, G. T., Sudjatmogo, and T. H. Suprayogi. 2015. The efficiency and persistenceency of milk production on Friesian Holstein dairy cows fed at different forage and concentrate feeding ratio. J. Anim. Agric. 4 (2): 234 – 238.

[18] Wadjdi, M. F. and U. Ali. 2021. Assistance for independent dairy farming business in Pesanggrahan Village, Batu City, East Java. J. Progressive Community Service. 5 (1): 387 – 393.

[19] Yatno, Nelson, R. Murni, Suparjo, and H. L. Syarifa. 2018. Protein Isolation and Analysis of Amino Acids of Kaliandra Leaf Protein Concentrate as an Effort to Provide Animal Feed Supplements: 1. Quality of Protein Concentrated Amino Acids from Kaliandra Leaf Extraction. Proceedings of the National Seminar of the Faculty of Agriculture, University of Jambi in 2018. Jambi. Page : 699 – 70.

[20] Widyobroto, B. P., Rochijan, Ismaya, Adiarto, and Y. Y. Suranindyah. 2016. The impact of balanced energy and protein supplementation to milk production and quality in early lactating dairy cows. J. Indonesian Trop. Anim. Agric. 41(2): 83 – 90.

[21] Malianti, L., E. Sulistiyowati, and Y. Fenita. 2019. Amino acid and nutrient profile of durian seed waste (Durio zibethinus Murr) fermented with tape yeast (Saccharomyces cerevisiae) and tempeh yeast (Rhizopus oligosporus). J. Naturalist - Research on Natural Resources and Environmental Management. 8 (1): 59 – 66.

[22] Sadovnikova, A., S. C. Garcia, and R. C. Hovey. 2021. A comparative review of the cell biology, biochemistry, and genetics of lactose synthesis. J. Mammary Gland Biol. Neoplasia 26 (2): 181 – 196.

[23] Gross, J. J. 2023. Dairy cow physiology and production limits. J. Anim Front. 13 (3): 44 – 50.

[24] Sadovnikova, A., S. C. Garcia, and R. C. Hovey. 2021. A comparative review of the extrinsic and intrinsic factors regulating lactose synthesis. J. Mammary Gland Biol. Neoplasia 26 (2): 197 – 215.

[25] Akhdiat, T., N. Widjaya, H. Permana, R. F. Christi, and A. Suherna. 2021. Effect of premix administration in rations on milk production and quality of Friesian Holstein dairy cows. J. Zootec. 41 (2): 355 – 363

[26] Indriani, A.P., A. Muktiani, and E. Pangestu. 2013. Consumption and production of lactated dairy cow milk protein supplemented with curcumin (Curcuma xanthorrhiza) and zinc proteinate. J. Anim. Agric. 2 (1): 128 – 135.

[27] Adiwimarta, K. I. S. 2021. Ruminant Nutrition, Energy and Protein Levels. Gajah Mada University Press. Sleman.

[28] Ali, U., B. Muwakhid, and N. Sholikah. 2021. Effect of the Use of Concentrate in Tmr Feed on Fat, Lactose and Tsnf Levels in Lactated Dairy Cow's Milk. Seminar "Research Innovation and Community Service for Strengthening Freedom of Learning during the Pandemic". Item : 716 – 722.

[29] Tanuwiria, U. H., R. Hidayat, R. F. Christi, and A. M. Rizki. 2021. Effect of adding ruminer in dairy cattle ration on the production of 4% fcm and milk nutrients. J. Agripet. 21 (2): 200 – 206.

IOP Conf. Series: Earth and Environmental Science **1460** (2025) 012018 doi:10.1088/1755-1315/1460/1/012018

Potential of *Averrhoa carambola* L. as Natural Antioxidant of Broiler Chicken

B Ma' rifah*, E Suprijatna, D Sunarti, S Sugiharto, L D Mahfudz, S Kismiati, T A Sarjana, R Muryani, H D Shihah, N M Wahyuni

Department of Animal Science, Faculty of Animal and Agricultural Sciences, Universitas Diponegoro, Semarang

*E-mail: marifahbinti2908@gmail.com

Abstract. Feed additives from natural sources are much sought after to replace synthetic antibiotics. One of the natural ingredients that is often found around us is sweet star fruit leaves. This study aims to determine the potential of sweet star fruit leaves (*Averrhoa carambola* L.) as a Source of Natural Antioxidants". Parameters that will be taken include: antioxidant activity; phenolic compound content includes flavonoids, tannins, saponins, alkaloids and anthocyanins; fiber, fat, and protein content; amino acid content; and essential and non-essential fatty acid content. *Averrhoa carambola* leaves has a great potential as natural antioxidant. Its has 9.32 ppm for antioxidant activity; 79.278 mg/100g Flavonoid; 4399.69 mg/100g phenol; 4998.276 mg/100g total tannin; 215.517 mg/100g saponin; 5501.557 mg/100g alkaloid; 13.46% terpenoid; 13.36 mg/100g antosianin. *Averrhoa carambola* leaves also has a complete amino acid and essential fatty acid content that will give a big benefit as natural antioxidant to broiler chicken. *Averrhoa carambola* containt several essensial amino acid like alanin, arginin, aspartat, glisin, glutamic acid, histidin, isoleusin, leusin, lisin, valin, fenilalanin, prolin, serin, treonin. This leaves also content several essensial fatty acid like oleic acid, linoleat acid, linolenat acid, omega 3, 6, 9 fatty acid. The complete content of active compounds, amino acids and essential fatty acids makes *Averrhoa carambola* leaves have strong potential as a natural antioxidant.

Keywords: *Averrhoa carambola* l., antioxidant, amino acid, fatty acid, broiler chicken

1. Introduction

A recent development in the poultry industry is the search for safe substitute antibiotics for livestock. Probiotics, prebiotics, organic acids, enzymes, phytobiotics such essential oils, and so forth are a few substitute sources of antibiotics [1]. Due to their numerous advantages, Indonesian herbal plants, which are extensively distributed, are very concerning. The word "herbal plants" refers to plants that have medicinal properties. A herbal plant that is widely available in Indonesia and have a medical properties is the sweet starfruit plant (*Averrhoa carambola* L). Star fruit plants grow in almost all regions of Indonesia. *Averrhoa carambola* is also known by the names starfruit, carambola, Chinese star fruit, five-angled fruit, and star apple. *Averrhoa carambola* is a little cerama tree with a short stem that develops slowly, according to [2]. The wood is white, the branches hanging. The delicate, medium-green compound leaves are alternately arranged in a spiral around the branches. Pinnate leaves are composed of one terminal leaf blade, five to eleven almost opposite leaflets, and one oblong or oval-ovate leaf blade that is 3.8–9 cm long. The leaf has a shape 15 to 20 cm long. The leaves have smooth upper surfaces and smooth, pale beneath surfaces. The leaves are sensitive to abrupt shocks and are light-sensitive, folding at night [2].

This plant is widely used for its fruit and leaves. People use the leaves of the star fruit plant as herbal medicine. Sweet starfruit leaves have anti-inflammatory properties, and are widely used as a fever-reducing medicine [3]. Sweet starfruit leaves also contain large amounts of flavonoids, especially the compounds apigenin and quercetin. Quercetin is a flavonoid compound with high antioxidant activity. According to [4] sweet starfruit leaves extract has the potential as an antioxidant compound because it contains large amounts of flavonoids, saponins, steroids and tannins. Antioxidants play an important role in preventing the oxidation of long chain unsaturated fatty acids. According to [5], antioxidants are compounds that can stop the oxidation process by stopping the oxidation chain reaction. This shields the body from oxidative damage. Oxidative stress and oxidative damage are closely related to disease [6].

A part from containing high antioxidants, sweet star fruit leaves also contain large amounts of essential fatty acids. The fatty acids contained in star fruit leaves include linolenic acid, MUFA (stearic acid and oleic acid), and PUFA (linolenic acid). Linolenic acid was present in large quantities. This fatty acid is an essential fatty acid that is not produced by the body but is very necessary to help metabolic processes in the livestock's body. Double chain unsaturated fatty acids (PUFA) are also found in large amounts [7]. The availability of fatty acids in sweet starfruit leaves is expected to have a positive influence on the broiler chicken meat produced. Several ingredients contained in starfruit leaves are expected to have a positive influence on the performance and productivity of broiler chickens. This complete bioactive compound in this leaves hope will give a beneficial effect as antioxdiant to broiler chicken.

2. Material And Methods

2.1. Materials

This research used a leaves of *Averrhoa carambola* L. as a main material, water to clean the leaves, knive, and grinder.

2.2. Methods

2.2.1. Extraction

The first step in the process was gathering starfruit leaves from farmers in the Demak region. The leaves was cleaned under running water, allowed to dry for 72 hours (until the water content drops to 10%), separated from the stems, and pulverised using a grinder. 32% of the weight of new leaves is generated as leaves meal. The leaves meal was analized for antioxidant activity, thotal phenol, flavonoid, tanin, saponin, fatty acid content, amino acid content, and proximate analisys.

2.2.2. Measurment of Antioxidant Activity

The methanol extract's capacity to scavenge free radicals was used to gauge its antioxidant potential (DPPH). A 1:25 methanol to leaves flour ratio is used to soak the leaves (1 g of leaves in 25 ml of methanol). For 72 hours, the mixture was macerated. Whattman filter paper was used to filter the filtrate after 72 hours. 3 ml of the resultant filtrate should be purified with n-hexane until a clear solution is obtained. A spectrophotometer with a wavelength of 517 nm was utilised for the measurements.

2.2.3. Measurment of Active Compound

Using gallic acid as a standard, the Folin-Ciocalteu technique was used to determine the total phenol concentration [7]. One tube had 0.3 mL of aquadest steril, to which 0.2 mL of extract was added. The solution was homogenised after being combined with 2.5 mL of 20% Na_2CO_3 solution and 0.5 mL of Foline reagent. At a wavelength of 725 nm, the absorbance of the sample and blank solution was measured during a 40-minute incubation period at room temperature. Using (+)-catechin as a reference, the total flavonoid concentration was determined. Pure leaves were used as a standard in the butanol-HCl method to quantify the content of tannin. The amount of anthocyanin was determined by weighing 600 g of freshly crushed leaves, combining them with 60 g of NaCl for 60 minutes, macerating the combination in 1500 mL of distilled water for 20 hours, and then storing the mixture at 4°C until HPLC testing [8].

2.2.4. Measurment of Fatty Acid Content

Fatty acid analysis using the gas chromatography method. Following the freeze-drying of the flour leaves samples, 0.5 g of the sample was weighed and put into analysis tube along with 1 mL of hexane, 4 mL of acetylchloride-methanol solution (1:10), and 1 mL of C11:0 standard solution (1 mL, 1 mg/mL). The mixture was then mixed and the tube was placed in water at 80°C for 4 hours. Five millilitres of potassium carbonate (7%) was added to the tube once it had reached room temperature. The combination was then centrifuged for five minutes at 180 bar pressure, filtered through filter paper and then subjected to gas chromatography analysis.

2.2.5. Measurment of Amino Acid Content

Analysis of amino acids : 0.1 g of dry, of the leaves were hydrolysed for 24 hours at 110°C in sealed tubes using 10 ml of 6 N HCL. Following hydrolysis, the extra HCL was eliminated by vacuum-assisted evaporation with sporadic water additions. Any insoluble material was removed by filtering the residue after it had been dissolved in sodium citrate buffer (PH 2.2), the optically clear solution (30 μl) is chromatographically tested in an amino acid analyser.

3. Results and Discussion

3.1. Active compound

The antioxidant activity of sweet starfruit leaves as a result of research was 9.32 ppm. This indicates that star fruit leaves have strong antioxidant activity. According to [5], sweet starfruit leaves have strong antioxidant content, with a high polyphenol content, especially proanthocyanidin. The fraction ethyl acetate of star fruit leaves exhibited strong antioxidant activity with an IC50 value of 96 mg/ml in the DPPH. The leaves of *Averrhoa carambola* are rich in various phytochemicals, including flavonoids, phenols, tannins, saponins, alkaloids, terpenoids and antosianin. Sweet starfruit leaf extract has high antioxidant activity related to phenol content [9]. [4] claim that sweet starfruit leaf extract has a high concentration of flavonoids, saponins, steroids, and tannins, it may have antioxidant properties. Antioxidants are crucial in stopping long-chain unsaturated fatty acids from oxidising. Tannins are a component of polyphenols that blocking the enzymatic system, can lower free radicals. By taking electrons from free radicals and stopping ring processes that create new free radicals, flavonoids can stabilise free radicals. Additionally, flavonoids help prevent and treat oxidative diseases like inflammation and atherosclerosis. Because flavonoids have a part in controlling brain enzymes and receptors that have a big impact on the central nervous system, they can act as medications.

Table 1. Active compound of sweet star fruit leaves powder (*Averrhoa carambola* L.)

Active compound	*Averrhoa carambola* L leave powder
Antioxidant activity (ppm)	9.32
Phenol (mg GAE/100g)	4399.69
Flavonoid mg QE/100g)	79.27
Tanin (mg GAE/100g)	4998.27
Saponin (mg DE/100g)	215.51
Alkaloid (mg Caffein/100g)	5501.55
Therpenoid (%)	13.46
Antosianin (mg/100g)	13.36

Saponins, a class of naturally occurring compounds found in various plants, have garnered attention in the field of animal nutrition, particularly as feed additives in poultry rations. These compounds are known for their ability to enhance nutrient absorption through the intestinal epithelium, which is crucial for optimal growth and health in poultry. Research indicates that saponins can positively influence gut morphology, notably increasing villus height and crypt depth, which are important indicators of intestinal health and nutrient absorption efficiency [10]. Saponins improve the permeability of the intestinal mucosa, facilitating the absorption of nutrients. They can enhance the uptake of essential nutrients by disrupting the gut barrier in a controlled manner, allowing for greater nutrient availability to the enterocytes. The inclusion of saponins in poultry diets has been associated with increased villus height and crypt depth. A higher villus height correlates with an expanded surface area for nutrient absorption, while deeper crypts indicate enhanced intestinal cell turnover and health. This morphological improvement is vital for maintaining efficient digestion and nutrient assimilation. Other studies suggest that the antioxidant compounds in *Averrhoa carambola* leaves play a crucial role in neutralizing reactive oxygen species (ROS) and preventing oxidative stress [9]. Tannin is a potential natural antioxidant and is widely used as a feed additive. Tannin can ward off free radicals and inhibit prooxidative enzymes.

3.2. Nutrien content

From this result sweet star leaves have complete nutrition that will give a beneficial to broiler chicken. some plant-based diets contain carbohydrates in the form of fiber. The fiber is usually a nondigestible polysaccharide. fiber has the benefit of helping intestinal peristaltic movement, preventing rations from clumping, accelerating the rate of digestion, and stimulating the development of digestive organs [11]. In plants, proteins are not evenly distributed; they are particularly concentrated in the leaves, where species-specific variations in protein content may be observed. On a dry mass basis, values ranging from 5 to 20% are typical; in certain tropical plants, they may even be higher [12].

Table 2. Proximate analysis of sweet star fruit leaves powder (*Averrhoa carambola* L.)

Proximate analysis	*Averrhoa carambola* L leave powder (%)
Water	13.05
Ash	5.58
Crude Fat	1.79
Crude Fiber	8.66
Crude Protein	12.54

The insoluble fibers inhibit the activity of α-amylase and delays the release ofglucose from starch [13]. Potent hypoglycemic activity has been demonstrated in vitro. A study performed on male Wister rats found a decrease in blood sugar level when they were fed with hydroalcoholic extract of leaves of Averrhoa carambola [14].

3.3. Amino acid content

According to the Table 3, sweet star fruit leaves contain several amino acid. Up to 14 amino acid were identified and quantified. The highest amino acid content is glutamic acid 12817.86 mg/kg, followed by prolin 10840.06 mg/kg and aspartic acid 10769.20 mg/kg. Natural glutamic acid can be found in various foods, such as tomatoes, seaweed, sugar cane, and certain vegetables. Glutamic acid is an amino acid with numerous health advantages, glutamic acid aids in the formation of proteins, Assisting the brain in sending information, Aiding in concentration and attention, regulating blood sugar levels. The quality of protein is determined by the completeness and balance of the essential amino acids. High-quality protein contains complete essential amino acids, sufficient and balanced amounts.

Table 3. Amino acid content of sweet star fruit leaves powder (*Averrhoa carambola* L.)

Amino acid content	*Averrhoa carambola* L leave powder (mg/kg)
Alanin	6703.34
Arginin	4885.47
Asam Aspartat	10769.20
Glisin	5508.39
Asam Glutamat	12817.86
Histidin	1818.95
Isoleusin	2642.08
Leusin	6579.20
Lisin	6314.04
Valin	3928.58
Fenilalanin	3837.59
Prolin	10840.06
Serin	4643.02
Treonin	3925.27
Tirosin	2767.88

The physiological functions of plants, including those of the leaves, depend heavily on amino acids. They provide a number of essential roles that support plant growth, development, and stress response in addition to being the building blocks of proteins. Specifically, the amino acids produced in leaves take part in a variety of metabolic processes that improve the plant's capacity to adjust to environmental stressors. Some amino acids increase the bioavailability of minerals by building complexes with them that aid in the absorption of water and nutrients. Glutamic acid, for instance, is known to enhance the efficiency of food absorption and nitrogen metabolism [15].

3.4. Total Fatty Acid content

According to the Table 4, sweet starfruit leaves contain several fatty acid. Sweet starfruit is one of the herbal plants that are rich in essensial and non essensial fatty acid. Results indicated that the fatty acids extracted from the leaves of A. carambola were found to be rich in essential fatty acids. The amounts of fatty acids in the leaves were in the following order of MUFA > PUFA, and Omega-9 fatty acid > Omega-3 fatty acid > Omega-6 fatty acid. Essential fatty acids are needed by the body, but the body is unable to produce them (Wei et al., 2014). The most abundant fatty acid found in sweet starfruit leaves is lignoseric acid. Lignoceric acid, also known as tetracosanoic acid, is a saturated fatty acid with the chemical formula $C_{24}H_{48}O_2$. It is characterized by its long carbon chain, consisting of 24 carbon atoms. Lignoceric acid can be extracted from wood tar and is also found in certain cerebrosides. It occurs naturally in small quantities in various fats, where it constitutes about 1.1% to 2.2% of the fatty acid composition. Additionally, lignoceric acid is a byproduct of lignin production, linking it to the decomposition of plant materials. Lignoceric acid plays a role in various biological processes. Due to its unique properties, lignoceric acid has potential applications in the pharmaceutical industries. Its fatty acid profile may contribute to the formulation of products aimed at improving health [16]. several fatty acid have a role as antifungal and antibacterial [17].

Table 4. Total fatty acid content of sweet star fruit leaves powder (*Averrhoa carambola* L.)

Fatty Acid Content	*Averrhoa carambola* L leave powder (%)
Capric Acid	0.0561
Lauric Acid	0.0227
Myristic Acid	0.0390
Palmitic Acid	0.6259
Stearic Acid	0.3306
Oleic Acid	0.5846
Linoleic Acid	0.2922
Linolenic Acid	0.4562
Lignoseric Acid	2.8549
Nervonic Acid	0.2220
Caprylic Acid	0.0958
Polyunsaturated Fat	0.7485
Monounsaturated Acid	0.8066
Omega 3 Fatty Acid	0.4562
Omega 6 Fatty Acid	0.2922
Omega 9 Fatty Acid	0.5846

Sweet star fruit leaves contain lauric acid, which may have antiviral, antibacterial, antioxidant, and hypercholesterolemic properties [18]. By rupturing the bacterial cell wall and cell membrane, palmitic acid, a member of the fatty acid group, has antibacterial properties. [19] discovered this chemical to be a great anti-inflammatory and antioxidant, with the capacity to improve its antibacterial action by working in concert with other active compounds.According to reports, stearic acid has flavouring, antibacterial, and hypercholesterolemic properties. Oleic acid is an unsaturated fatty acidand have a potential effect as anti-inflammatory agents [18].

4. Conclusion

Averrhoa carambola leaves have a great potential as natural antioxidant, primarily consisting of flavonoids and phenolic compounds. With a hight and complete source of amino acid and essensial fatty acid, this leave can be used to replace sintetic antibiotic to broiler chicken.

References

[1] Gadde, U., W.H. Kim, S.T. Oh, and H.S. Lillehoj. 2017. Alternatives to antibiotics for maximizing growth performance and feed efficiency in poultry: a review. Animal Health Research Review. 18 (1) : 26-45. https://doi.org/10.1017/S1466252316000207

[2] Morton, J.F. 2013. Fruits of Warm Climate. Echo Point Books and Media. ISBN 0961018410, 9780961018412. https://ntbg.org/database/plants/detail/averrhoa-carambola

[3] Yunarto, N., and N. Sulistyaningrum. 2017. Quantitative Analysis of Bioactive Compounds In Extract and Fraction of Star Fruit (*Averrhoa carambola* L.) Leaves Using High Performance Liquid Chromatography. Jurnal Kefarmasian Indonesia 7. https://doi.org/10.22435/jki.v7i1.5605.26-33

[4] Hossain, T., A.K. Barman, U. Karmakar, B. Bokshi, J. Homepage, K.U. Kumar S. Dev, and N.N. Biswas. 2017. Bioscience and Bioengineering Communications Phytochemical and Pharmacological Evaluation of Leaves of *Averrhoa carambola* Linn. (Family: Oxalidaceae) Phytochemical and pharmacological screening of plants View project Bioscience and Bioengineering Communications Phytochemical and Pharmacological Evaluation of Leaves of *Averrhoa carambola* Linn. (Family: Oxalidaceae). https://www.researchgate.net/publication/319187104.

[5] Wei, S. D., H. Chen, T. Yan, Y. M. Lin, and H.C. Zhou. 2014. Identification of antioxidant components and fatty acid profiles of the leaves and fruits from *Averrhoa carambola*. LWT 55, 278–285. https://doi.org/10.1016/j.lwt.2013.08.013

[6] Balsano, C., and A. Alisi. 2009. Antioxidant effects of natural bioactive compounds. Currents Pharmaeutical Design. 15 (26) : 3063-3073. https://doi.10.2174/138161209789058084

[7] Wei, S. D., Y.M. Lin, M.M. Liao, H.C. Zhou, and Y.Y. Li. 2012. Characterization and antioxidative properties of condensed tannins from the mangrove plant Aegiceras corniculatum. Journal of Applied Polymer Science 124, 2463–2472.

[8] Alaysuy O., R. Snari, A. Alfi, A. Aldawsari, S. Abu-Melha, M. Khalifa, N. El-Metwaly. 2022.. Development of green and sustainable smart biochromic and therapeutic bandage using red cabbage (Brassica oleracea L. Var. capitata) extract encapsulated into alginate nanoparticles. International Journal of Biological Macromolecules 211: 390–399. http://dx.doi.org/10.1016/j.ijbiomac.2022.05.062

[9] Lakmal, K., P. Yasawardene, U. Jayarajah and S.L. Seneviratne. 2021. Nutritional and medicinal properties of Star fruit (*Averrhoa carambola*): A review. Food Science and Nutrition vol. 9 1810–1823 at https://doi.org/10.1002/fsn3.2135

[10] Luan, F., L. Peng, Z. Lei, X. Jia, J. Zou, Y. Yang, X. He, N. Zeng. 2021. Traditional Uses, Phytochemical Constituents and Pharmacological Properties of *Averrhoa carambola* L.: A Review. Frontiers in Pharmacology vol. 12 at https://doi.org/10.3389/fphar.2021.699899

[11] Khalid, W., M.S. Arshad, A. Jabeen, F.M. Anjum, T.B. Qaisrani, H.A.R. Suleria. 2022. Fiber-enriched botanical : A therapeutic tool agains certain metabolic ailment. Food Science and Nutrition. 10 : 3203-3218. https://doi:10.1002/fsn3.2920

[12] Chiesa, S. and E. Gnansounou. 2011. Protein extraction from biomass in a bioethanol refifinery-possible dietary applications : uses as animal feed and potential extension to human consumption. Bioresource Technology. 102:427-436. https://doi:10.1016/j.biotech.2010.07.125

[13] Chau, C.F., C.H. Chen, and M.H. Lee. 2004. Characterization and physi-cochemical properties of some potential fibres derived from *Averrhoa carambola*. Nahrung/Food. 48:43–46. https://doi.org/10.1002/food.200300354

[14] Ferreira, E. B., Fernandes, L. C., Galende, S. B., Cortez, D. A. G. & Bazotte, R. B. 2008. Hypoglycemic effect of the hydroalcoholic extract of leaves of *Averrhoa carambola* L. (Oxalidaceae). Revista Brasileira de Farmacognosia 18 : 339–343. http://dx.doi.org/10.1590/S0102-695X2008000300005

[15] Kawade K., H. Tabeta, A. Ferjani, M.Y. Hirai. 2023. The roles of functional amino acid in plant growth and development. Plant and Cell Physiology. 64 : 1482-1493. https://doi.org/10.1093/pcp/pcad071

[16] Altinoz, M.A. and A. Ozpinar. 2019. Ppar-Õ and erucic acid in multiple sclerosis and alzheimer's disease. Likely benefits in terms of immunity and metabolism. International Immunopharmacology. 69 : 245-256. https://doi.org/10.1016/j.intimp.2019.01.057

[17] Astiti, N. and Ramona, Y. 2021. Gc-ms analysis of active and applicable compounds in methanol extract of sweet

IOP Conf. Series: Earth and Environmental Science **1460** (2025) 012018 doi:10.1088/1755-1315/1460/1/012018

star fruit (*Averrhoa carambola* l.) leaves. Hayati Journal of Biosciences 28, 12–22. http://dx.doi.org/10.4308/hjb.28.1.12

[18] Gideon V.A. 2015. GC-MS analysis of phytochemical components of Pseudoglochidion anamalayanum gamble : an endangered medical tree. Asian Journal of Plant Science and Research. 5: 36-41. https://www.semanticscholar.org/paper/GC-MS-analysis-of-phytochemical-components-of-An-An-Gideon/164fa76bcfc6043c41218e9d9832b86ace28564b

[19] Hema R., S. Kumarafel, K. Aagusundaram. 2011. GC/MS determination of bioactive components of Murraya koenigii. Journal of American Science 7:80-83. http://www.americanscience.org.

IOP Conf. Series: Earth and Environmental Science **1460** (2025) 012019 doi:10.1088/1755-1315/1460/1/012019

The Effectiveness of Fallen Ketapang Leaf Extract (*Terminalia catappa* L.) in Drinking Water to Improve Broiler Performance

E Suprijatna, B Ma'rifah*, D Sunarti, L D Mahfudz, S Kismiati, T A sarjana, R Muryani, H D Shihah, N M Wahyuni

Animal Production Laboratory, Department of Animal Science , Faculty of Animal and Agricultural Sciences, Universitas Diponegoro, Semarang

*E-mail: marifahbinti2908@gmail.com

Abstract. The purpose of this study is to assess the effect fallen ketapang leaf extract in drinking water affects broiler productivity. This study used 200 unsex broiler with average weight of 546.85±20.28 g. A completely randomised design was employed in the experiment, with five treatments and five replications. The treatment are drinking water without leaf extract (T0), drinking water with 1% fallen ketapang leaf extract (T1), drinking water with 1.5% fallen ketapang leaf extract (T2), drinking water with 2% fallen ketapang leaf extract (T3), and drinking water with 2.5% fallen ketapang leaf extract (T4). Feed convertion ratio (FCR), Body Weight Gain (BWG), feed consumption, and drinking water consumption were measured. According to the study, performance of broiler was not significantly affected (P>0.05) by administration of fallen ketapang leaf extract. Drinking water up to 2.5 % of fallen Ketapang leaf extract not effective to improve broiler performance.

Keywords: Ketapang leaf, *Terminalia catappa*, performance, broiler

1. Introduction

Because of their rapid growth, broilers are a popular variety of poultry. High temperatures can lead to oxidative damage owing to heat stress, which is a challenge that breeders in Indonesia frequently confront. The use of an antibiotic growth promoter (AGP) is one way to prevent this. AGP is a class of chemicals with potential applications in medicine, growth promotion, and production enhancement. Due to a rise in the usage of antibiotics that do not adhere to dosing instructions or consider the drug's downtime, hence increasing dangers, the use of antibiotics is currently prohibited in Indonesia. Beginning on January 1, 2018 Indonesia prohibits the use of antibiotics as growth promoters. This ban is outlined in Law No. 18/2009 juncto No. 41/2014 [1]. Antibiotic residues in animal feed, such as meat, eggs, and milk, pose a risk to human health due to the development of bacterial resistance, allergies, and poisoning. One potential answer is to employ herbal plants, such ketapang leaves, as substitute antibiotics.

Ketapang (*Terminalia catappa*) leaves is one of the plant-based phytogenic feed additives, which are natural materials made from herbs, spices, other plants and their extract, and essential oils and utilised in animal nutrition. The performance of chickens is positively impacted by phytogenic feed additives, mostly because of variations in the origin, processing, and botanical makeup. Many different types of chemical compounds can be found in herbs, spices, plant extracts, and essential oils. *Terminalia catappa* L. is a plant that grows in various tropical regions with fruit production ranging up to 700.000 tons per year. fallen ketapang leaves are known to contain various phytochemical compounds, namely flavonoids and saponins and the dominant compound, namely tannin [2]. Tannin when given in low concentrations is able to inhibit bacterial growth, while at high concentrations according to the dose given it can act as an antibacterial, namely by coagulating protoplasm in bacteria because it forms stable bonds with bacterial proteins [2].

Ketapang leaves also have antioxidants derived from flavonoid and tannin compounds which are useful for warding off and inhibiting free radicals which cause oxidative stress [3]. The antibacterial and antioxidant qualities of the T. catappa L leaf extract are enhanced by the synergy of several components. The antibacterial properties of T. catappa L leaf extract are anticipated to preserve the digestive tract's ecology and optimise nutrition utilisation. T. catappa L leaf extract has antioxidant properties that can reduce stress and promote productivity and growth. The chemicals present in T. catappa L leaves both in-vivo and in-vitro must be identified in order to assess the utility and efficacy of the leaves as growth-promoting agents in poultry.

2. Material And Methods

2.1. Materials

This research used 200 broiler chicken, Ketapang leave extract, floor pen, water, and commercial feed.

2.2. Methods

2.2.1. Animal Ethic

Chickens' care and experimental procedures were performed under approval from the Animal Ethical Committee (No. 56-09/A-1/KEP-FPP). Faculty of Animal and Agricultural Science, Universitas Diponegoro.

2.2.2. T. catappa Leaves Collection And Liquid Extract Preparation (Maseration)

T. catappa leaves were obtained from around Diponegoro University campus, Semarang, Central Java, Indonesia. The leaves were washed, cut, shade fallen at room temperature for 3-4 days. And crushed for further processing. Aqueous extract preparation was carried out using modified method with distilled water in the ratio of 20 g/100 mL. left for 72 h, mixed periodically, and filtered with Whatman paper no. 42. The filtrate was stored in the freezer until used

2.2.3. Experimental Management

A total of 200 broilers were obtained from a local breeder farm, in Central Java, Indonesia. The chicks were brooded together until 14 days in a pen with a bohlam lamp for heating. On the 14th day, the chicks were randomly split into the 5 treatment groups, with the average body weight are 546.85 ± 20.28 g. The experimental diets were prepared to be isoprotein (21.5%) and isoenergy (2,900). The diet was based on maize, soybean meal, vegetable oil, rice bran, fish meal, CaCO3. premix, methionine, and lysin. The nutrient compositions of the diets are presented in Table 1. Drinking water and feed were given with ad libitum.

Table 1. Nutrient composition of basal diets

Material (%)	**Finisher**
Calculated composition	
ME (Kcal kg^{-1})*	2,900
Crude Protein	21.5
Crude Fat	5.00
Crude Fiber	5.00
Total Ash	7.00
Ca	0.90
P	0.60

Metabolizable energy was calculated according to Bolton (1967) formula: 40.81 {0.87 [crude protein + 2.25 crude fat + nitrogen-free extract] + 2.5}

2.2.4. Sampling and Analysis
Broiler rearing is carried out for 6 weeks. At the age of 2 weeks, a brooding period is carried out during which the animals adapt to the research environment. The implementation of the research was carried out from the beginning of the 3rd week to the 6th week. The feed given until the end of the rearing period was commercial broiler feed in the starter phase CP 511. Feed and drink were given ad libitum / always available. The treatment is carried out every week on Friday, Saturday and Sunday by adding liquid extract of fallen ketapang leaves to drinking water according to the concentration determined for each treatment. Drinking water is made just before being given to livestock with 10 liters of water, namely 1% for 100 ml, 1.5% for 150 ml, 2% for 200 ml and 2.5% for 250 ml for each treatment. Weighing of remaining feed and drinking water is carried out at night, and body weight is carried out every week.

2.2.5. Experimental Design And Sample Size
Each chick was randomly allocated in the pens. Accordingly, a total of 200 chickens were randomly assigned to 5 treatments, each treatment consists of 4 replicates consisting. The treatment are T0 : Drinking water without *Terminalia catappa* extract, T1: Drinking water with 1% of *Terminalia catappa* extract, T2: Drinking water with 1.5% of *Terminalia catappa* extract, T3: Drinking water with 2% of *Terminalia catappa* extract, T4: Drinking water with 2.5% of *Terminalia catappa* extract.

2.2.6. Statistical Analyses
The data were analyzed based on a completely randomized experimental design using the Analysis of Variance (Anova). Duncan's multiple range test was applied if treatment were significant ($P<0.05$).

3. Discussion

Feed intake, water intake, body weight gain, and FCR broiler treated with fallen ketapang leaves extract is presented in Table 2.

The administration of fallen ketapang leaves extract in drinking water with a level of 1.0-2.5% showed no significant effect on feed intake. The average feed intake was obtained between 125.06-127.77 g/ bird. These results are lower than the standard of [4] that the average feed intake of broiler in 3-6 weeks is 159.40 g/bird. This is probably because the level of fallen ketapang leaves extract are still relatively low. The administration of fallen ketapang leaves extract to broiler aims to increase palatability and improve health. This statement is in line with the statement of [5], Broiler wich given herbal ingredients in feed and water to increase palatability and stamina and endurance of livestock. Ketapang leaves have a bitter taste due to the tannins contained in ketapang leaves. According to [6], that applied ketapang leaves to armyworms, tannins contained in ketapang leaves can inhibit and reduce appetite in the armyworms that are maintained. According to [7], one of the effects caused by tannins is to reduce appetite, causing ration consumption and protein consumption to decrease. Based on the active substance calculation table, the tannin content is the highest bioactive substance in the liquid extract of fallen ketapang leaves. According to [8], the percentage of tannin use in broiler rations is 0.33%, if it exceeds this limit it can reduce ration consumption. Protein complex bonds created by tannins bonding with proteins make it difficult for protease enzymes to break down these proteins.

Table 2. Feed intake, water intake, body weight gain, and FCR

Variabel	Treatment					P value
	T0	T1	T2	T3	T4	
Feed intake	127.77±4.1	125.06±3.0	126.60±4.1	125.64±2.4	126.25±4.9	0.83
Water intake	326.37±35.5	297.47±25.5	330.30±17.9	357.37±67.5	311.63±14.3	0.17
Body weight gain	60.40±4.26	61.74±3.64	65.38±3.19	61.76±2.04	62.40±5.19	0.34
FCR	2.13±0.16	2.03±0.07	1.94±0.09	2.04±0.05	2.03±0.09	0.13

Value represent no significant (P>0.05)

This research in line with the research of [9] which gives cashew leaf extract to broiler in drinking water at level of 0-20% did not significanly to increase broiler feed intake because bioactive substances such as tannins, saponins, flavonoids, curcumin and volatile oils did not affect broiler feed intake for 6 weeks. Making fallen ketapang leaves extract in this study using a ratio of extract concentration of 20 grams of fallen ketapang leaves powder and 100ml of water is still relatively low so that it has not been able to improve the performance of broiler. According to [10] which provides liquid extract of African leaves in drinking water with a concentration of 50 g/150 ml can increase broiler ration consumption where African leaves contain bioactive substances such as flavonoids, saponins, tannins that can function as antibacterials so as to result in improvements in the digestive process of broiler so that they can digest feed better due to an increase in digestive enzymes so that the digestibility of feed will increase and result in the digestive tract of chickens being emptier and can ultimately increase. faster emptying and ultimately can increase feed consumption.

The average consumption of drinking water in this study was 326.37 - 357.37 ml/day. This result is slightly lower than the Cobb (2003) standard that the water intake of broiler reared at ambient temperatures above 30°C - 32°C is 341 - 350 ml/day. The results of the analysis of variance showed that giving fallen ketapang leaves extract in water up to the level of 2.5% showed no significant (P>0.05) on broiler water intake, because the level of provision was still relatively low. Giving fallen ketapang leaves extract to broiler aims to increase palatability and improve health. This statement is in line with [5], broiler are given herbal ingredients in feed and water to increase palatability and stamina and endurance of livestock. One compound that an important role in water intake is polyphenols, because polyphenols work as antioxidants to fight free radicals that cause oxidative stress. According to [11], antioxidants can reduce free radicals, minimize DNA damage and antioxidant enzymes. This is reinforced by the opinion of [12], giving herbal ingredients to livestock that contain antioxidants can reduce the effects of free radicals which can increase water intake. However, the polyphenol content in this research is still relatively low, so it has no effect on improving broiler health.

The total polyphenol content that can act as an antioxidant is between 25 and 35%, while the total polyphenol content of fallen ketapang leaves in this study was 4%. Too low polyphenol content can be caused by dry leaves that fall from trees that are too long exposed to sunlight, so that the polyphenol content in the leaves can decrease. According to [13], the volatility and heat tolerance of polyphenol compounds make the polyphenol content in leaves easily reduced if the sun exposure time is too long or the drying process is not appropriate. Previous studies that utilized fermented ginger extract as an additive such as in [14] which added fermented ginger extract in drinking water to a level of 8% showed no significant on water intake, which is around 153.03-164.07 ml / day. The use of fermented ginger extract in this study did not cause an increase or decrease water intake. According to [15], extreme temperature changes are what cause stress in broilers that lead to dehydration, decreased appetite, stunted growth, and increased water intake.

The average daily body weight gain of broiler ranged from 60.40 - 65.38 g, while the average results of body weight gain from the beginning of the treatment period to the end of the treatment period amounted to 1510.60 - 1954.14 grams. The results of this study are still below the standard. According to [4], the weight gain of broiler aged 3-6 weeks is 2456 grams. The factor that causes the low body weight compared to the standard is because the feed consumption in this treatment is also low so that the body weight gain is not optimal. Analysis of variance showed that add fallen ketapang leaves extract in drinking water showed no significant effect ($p>0.05$) on the daily weight gain of broiler because the amount of extraction content used was still low. The content of bioactive substances in fallen ketapang leaves extract has not been able to affect body weight gain, because the levels of active substances tannins and polyphenols are still below the levels of active compounds that affect the digestibility of feed nutrients are not optimal and the protein used for protein deposition is low.

Tannins are bioactive compounds that can have a positive or negative impact depending on the amount of tannins present in the material. According to [16] ideal tannin levels can reduce pathogen activity in the digestive tract, promote growth, improve immunity, and increase feed palatability. High tannins can be antinutrients and can form protein-tannin complexes, due to their association with proteins in enzyme or nonenzyme form. This inhibits enzymes in the digestive tract, such as the enzyme trypsin, which inhibits digestibility, leading to decreased body weight gain. According to [17], if tannins are added to animal feed more than 5%, tannins function as antinutrients and function as binders of proteins, enzymes, and minerals. This interferes with the metabolism of proteins, carbohydrates, and minerals in livestock.

Polyphenolic compounds contained in fallen ketapang leaves extract also act as antioxidants, counteracting free radicals that can trigger oxidation processes that damage body cells. According to [11], antioxidants can reduce free radicals, minimize DNA damage and antioxidant enzymes. If cell damage in the body is reduced, more protein is deposited into meat rather than used to replace damaged cells. Protein deposition is used by the body for muscle tissue formation and carcass formation, so low protein deposition reduces live weight and affects carcass weight and carcass percentage. According to [18], a decrease in protein digestibility can reduce the body's nitrogen supply, thereby reducing nitrogen retention. According to [19], nitrogen is one of the main components in the animal body, which functions as a constituent of amino acids in protein synthesis, where the results of protein synthesis in the form of protein are used to form protein into muscle tissue and meat and replace damaged tissue. Nitrogen is one of the most important elements on earth because it is very important in the composition of amino acids and proteins, which play a role in enzyme formation, tissue formation, antibody formation, and hormone production, among other processes [20]. The results of other studies that utilize herbs in drinking water are the opinion of [9], using cashew leaves at the level of administration up to 20% did not change the rate of weight gain of broiler. Polyphenolic compounds contained in fallen ketepeng leaves also act as antioxidants, antidotes to free radicals that can trigger oxidation processes that damage body cells. According to [11], antioxidants can reduce free radicals, minimize DNA damage and antioxidant enzymes. If cell damage in the body is reduced, more protein is stored into meat rather than used to replace damaged cells.

The average feed conversion ratio was 1.94 - 2.13. The results of this study are relatively high compared to the standard. According to [4], the feed conversion ratio of broiler aged 3 - 6 weeks ranged from 1.27 - 1.72. The results of the analysis of variance showed that the provision of fallen ketapang leaves extract in broiler had no significant effect ($p>0.05$) on feed conversion ratio. This is because the addition of ketapang leaf extract to water has no effect on feed utilization efficiency. Giving ketapang leaf extract to broiler aims to increase palatability and improve the health of the broiler's digestive tract. According to [20], tannin compounds act as antibacterials by inhibiting DNA topoisomerase and reverse transcriptase enzymes that are important for the formation and multiplication of bacterial cells. According to [21], tannins will bind to proteins and form polypeptide-protein complexes that can degrade bacterial cell walls and membranes, causing disruption of bacterial cell permeability and eventually bacterial lysis. The next mechanism is flavonoids which function to prevent bacterial growth by preventing the synthesis of nucleic acids, disrupting the function of the cytoplasmic membrane, and disrupting energy metabolism so that bacteria cannot multiply.

The use of tannins in fallen ketapang leaves extract for a long time will cause the accumulation of tannin compounds, the higher the concentration of antimicrobials used will increase the inhibition or death of bacterial cells. According to [22], the use of tannin compounds exceeding 1 gram, with prolonged use, causes accumulation of tannins in the body of livestock, which can reduce the digestibility of feed components, because tannins bind strongly to proteins and form complex bonds between phenolic groups tannins and ketoproteins, making some proteins insoluble. degraded by non-pathogenic microbes. According to [23], low digestibility will affect the metabolic process of protein and starch, so that the body's metabolic energy for meat production will be reduced, resulting in a decrease in live weight. Tannin content that is too low cannot affect feed palatability. According to [24], adding tannins to feed in amounts up to 1g/100g can reduce feed consumption so that animals grow more slowly and produce lower live weights, but in this study at the 2.5% level of 0.00024 g/l drinking water has not reached the level that can affect feed palatability and feed consumption.

Other studies that utilize herbal ingredients as additives in the research of [25], which use liquid extract of beluntas leaves up to a level of 10% have not been able to provide a significant difference and produce feed conversion values ranging from 2.54 - 2.60. The results showed no significant difference in each treatment, this is thought to be due to the low level of feed consumption and the inefficiency of the feed given due to scattered or wasted feed. According to [26], bioactive substances play an important role in providing health effects and high antimicrobial power so that the health of the digestive tract can improve. According to [26] feed conversion rate is highly dependent on the balance between metabolic energy and nutrients, especially protein and amino acids.

4. Conclusion

Extract of fallen ketapang leaves in drinking water has no significant effect in feed intake, water intake, body wight, and FCR. Further research on the provision of more varied liquid extracts of fallen ketapang leaves to improve broiler performance is needed

References

[1] Silitonga, L., S. Wibowo dan E. B. Bangun. 2020. Pengaruh pemberian tepung bawang dayak (*Eleutherine palmifolia* MERR.) terhadap morfometrik organdalam dan kadar lemak ayam broiler. Ziraa'ah Majalah Ilmiah Pertanian. 45 (1) : 10-20.

[2] Akpanpan, A.E., and U.D. Akpubio. 2012. Evaluation of proximate composition, mineral element and anti-nutrient in Almond (Terminalia catappa) seeds. Research Journal of Applied Science. 7 (9-12) : 489-493.

[3] Apata. 2010. Effect of Terminalia catappa fruit meal fermented by Aspergilus niger as replacement of maize on growth performance, nutrient digestibility and serum biochemical profile of broiler chickens. Biotechnology Research International V. http://dx.doi.org/10.4061/2011/907546

[4] Aviagen, P. O. 2014. Ross 308 broiler performance objectives. *Aviagen Limited Newbridge.*

[5] Wakhid, A. 2013. *Beternak Itik.* Jakarta: Cetakan Pertama. Agromedia. Jakarta.

[6] Batubara, R. N. S. 2020. Uji Efektivitas Beberapa Konsentrasi Ekstrak Daun Ketapang (*Terminalia catappa L.)* Terhadap Ulat Grayak (*Spodoptera litura*) Secara In Vitro (Doctoral dissertation, Uin Sultan Syarif Kasim Riau).

[7] Widodo, T. S., Sulistiyanto, B., dan Utama, C. S. 2015. Jumlah bakteri asam laktat(BAL) dalam digesta usus halus dan sekum ayam broiler yang diberi pakanceceran pabrik pakan yang difermentasi. Jurnal Agripet. 15(2): 98-103.

[8] Sitorus, T. F., dan Telambanua, A. P. 2021. Pengaruh pemberian kulit buah kopifermentasi terhadap performans, bobot hati, panjang dan persentase bobot usus halus ayam broiler. Jurnal Visi Eksakta, 2(1), 51-71.

[9] Khothijah, S., Erwan, E., dan Irawati, E. 2021. Performa Ayam Broiler Yang Diberi Ekstrak Daun Jambu Mete (Anacardium Occidentale Linn) Dalam Air Minum. *Jurnal Ilmu Nutrisi Dan Teknologi Pakan*, *19* (1), 19-23.

[10] Damayanti, P., Mihrani., Surung, M,Y. 2019. Pemanfaatan ekstrak daun afrika (*vernonia amydalina*) terhadap performa broiler. J. Agrisistem. 15 (1): 23- 28.

[11] Hu, R., Y. He, M. A. Arowolo, S. Wu dan J. He. 2019. Polyphenol as potential attenuators of heat stress in poultry production. 8(67): 1-11.

[12] Weiss W.P. and J.S. Hogan. 2007. Effect sof dietary vitamin c on neutrophil function and responses to intra mammary infusion of *lipopolysaccharide* inperiparturient dairy cows. *Journal of DairyScience*, 90(2): 731-739.

[13] Rauf, A., U. Pato dan D. F. Ayu. 2017. Aktivitas antoksidan dan penerimaan panelis teh bubuk daun alpukat (*Persea American Mill.*) berdasarkan letak daun pada ranting. J. Faperta. 4(2): 1–12.

[14] Haroen, U., dan Budiansyah, A. 2018. Penggunaan ekstrak fermentasi jahe(*Zingiber officinale*) dalam air minum terhadap kualitas karkas ayam broiler. J. Ilmiah Ilmu-Ilmu Peternakan. *21* (2): 86-97.

[15] Zurriyati Y, Dahono. 2013. Respon Fisiologis dan Evaluasi Karkas Ayam Broiler Terhadap Suhu Pemeliharaan Dingin. Seminar Nasional Teknologi Peternakan dan Veteriner. 586-591.

[16] Redondo, L.M., P.A. Chacana., J.E. Dominguez and M.E. Fern´andez Miyakawa.2014. Perspectives in the use of tannins as alternative to antimicrobialgrowth promoter factors in poultry. Journal Front. Microbiol.5:18.

[17] Kurniadi, M., M. Andriani, F. Faturohman dan E. Damayanti. 2013. Karakteristik fisikokimia tepung biji sorghum (*Sorghum bicolor L.*) terfermentasi bakteriasam laktat Lactobacillus acidophilus. J. Agritech. 33(3): 288–295.

[18] Mandey, J. S., Y. H. S. Kowel dan M. N. Regar. 2017. Nilai retensi nitrgogen dan enrgi metabolis broiler yang diberi ransum tepung limbah sawi putih. J. Zootek. 37(1): 41-49.

[19] Julianto, T. S. 2013. Biokimia: Biomolekul dalam Perspektif Al-Qur'an.

[20] Mukhriani, F. Y. Nonci, dan Mumang. 2014. Penetapan kadar tanin total ekstrak biji jintan hitam (*Nigella sativa*)

IOP Conf. Series: Earth and Environmental Science **1460** (2025) 012019 doi:10.1088/1755-1315/1460/1/012019

secara spektrofotometri UV-VIS. J. Farmasi, 4(4): 154-158

[21] Bulu, S., Rejeki, I. G. A. S., dan Mardewi, N. K. 2018. Pemakaian sorgum (*Sorghum bicolor l.*) sebagai bahan substitusi jagung (*Zea mays l.*) pada ransum terhadap berat bagian bagian karkas ayam broiler umur 6 minggu. Gema Agro. 23(2): 124-128.

[22] Hidayah, N. 2016. Pemanfaatan senyawa metabolit sekunder tanaman (tanin dan saponin) dalam mengurangi emisi metan ternak ruminansia. J. Sains Peternakan Indonesia. 11(2): 89–98.

[23] Ebrahim, R., J. B. Liang, M. F. Jahromi, P. Shokryazdan, M. Ebrahimi, W. K. Chendan Y. M. Goh. 2015. Effect of tannic acid on performance and fatty acid composition of breast muscle in broiler chickens under heat stress. Italian J. Anim. Sci. 14(4): 572-577.

[24] Hanif, I., dan Manullang, J. R. 2022. Pemanfaatan ekstrak daun beluntas (*Plucheaindica Less*) melalui air minum terhadap performa ayam broiler. J.Peternakan Lingkungan Tropis. 3(2): 64-69.

[25] Nuningtyas, Y. F. 2014. Pengaruh penambahan tepung bawang putih (*Allium sativum*) sebagai aditif terhadap penampilan produksi ayam pedaging. Journal of Tropical Animal Production. 15(1): 65-73.

[26] Zulfaidha, M. 2012. Efektifitas Kombinasi Jumlah dan Bentuk Ramuan Herbal sebagai Imbuhan Pakan terhadap Performa Broiler. J. Nutrisi dan MakananTernak. 9 (1): 1411 - 4577.

The combination effect of *Caulerpa sp.*, coconut meal, and palm kernel meal on performance and blood profile of male laying hens

Dwi Sulistiawati[1*], Sri Sarjuni[1], Saifullah[1] and Nur Afriana[2]

[1] Animal Husbandry, Animal Husbandry and Fisheries Faculty Tadulako University, Palu, Indonesia Pos code 94228
[2] Student of Animal Husbandry, Animal Husbandry and Fisheries Faculty Tadulako University, Palu Indonesia, Pos code 94228

*E-mail: dwisulist@untad.ac.id

Abstract. This study aims to evaluate the effect of dietary Caulerpa sp (Csp), Coconut/Copra Meal (CM), and Palm Kernel Meal (PKM) on blood profile, body weight of male laying hens. A total of 160 male laying hens DOC were used in this study until the age of 10 weeks and were randomly divided into ten treatments with four replications. The treatments given include: Control ration using Fishmeal, vegetable feed without Csp (TO), Ration without Fishmeal, Vegetable feed and Csp (T0C), Ration without Fishmeal, 10% CM, 10% PKM without Csp (T1), Ration without Fishmeal, 10% CM, 10% PKM and Csp (T1C), Ration without Fishmeal-20% CM without Csp (T2), Ration without Fishmeal, 20% CM and Csp (T2C), Ration Without Fishmeal, 20% PKM without Csp (T3), Ration without Fishmeal-20% PKM and Csp (T3C), Ration without Fishmeal, 15% CM, 15% PKM without Csp (T4), Ration without Fishmeal, 15% CM, 15% PKM and Csp (T4C). The results showed that the addition of Csp, CM, and PKM in the ration improved the blood profile of chickens, including haemoglobin, haematocrit, Leukocyte, Erythrocyte, platelet, total blood protein levels, meat antioxidant activity compared to the control group. Chickens given a ration containing Caulerpa produced the same body weight compared to the control group, while the use of vegetable feed combined with copra meal and palm kernel meal gave varying results on the performance of male laying hens. The use of 20% CM was better than 20% PKM in the ration. Chickens with low performance tend to have higher value of blood profile IC_{50} in meat. The use of Caulerpa, copra meal, and palm kernel meal in the ration can affect the production performance of male laying hens, with varying effects depending on the type of additives used.

Keywords : Blood Profile, Caulerpa sp., Coconut Meal, Palm Kernel Meal, Male Laying Hens

1. Introduction

The poultry industry, especially male laying hens, faces the challenge of maximising growth and productivity with efficient feed costs. Conventional feed formulations generally rely on relatively expensive feed ingredients such as corn, soybeans, and fishmeal. Some alternative feeds that can be used include Caulerpa meal, copra meal, and palm kernel meal.

Caulerpa is a type of marine algae that is rich in nutrients such as protein, essential fatty acids, fibre, minerals, and various bioactive compounds such as chlorophyll and polyphenols. Its use in poultry rations is believed to have various health benefits, including the ability to improve blood profile parameters such as haemoglobin levels, haematocrit, and erythrocyte counts, which are associated with increased immunity and general health of poultry [1]. In addition, Caulerpa also plays a role in increasing feed conversion efficiency, thus supporting better body weight growth in chickens. According to research by [2], the use of microalgae and macroalgae in poultry feed can increase the nutritional value of feed and improve the balance of essential amino acids. Bioactive compounds in algae have also been shown to contribute to circulatory health through their role as antioxidants and anti-inflammatories.

Coconut/Copra meal is a by-product of coconut oil processing that is rich in protein and fat, although with a fairly high crude fibre content. As a source of vegetable protein, copra meal has a good amino acid content to support chicken growth. Research by [3] showed that copra meal can be used as an alternative feed ingredient that reduces dependence on soybeans, without reducing chicken growth performance. The use of copra meal in male laying hen rations is expected to maximise body weight growth by increasing the availability of protein and energy.

Palm kernel meal is a by-product of the palm oil industry that is rich in crude fibre and fat, and contains moderate amounts of protein. Palm kernel meal is generally used as a source of energy and fibre in animal feed, but its use in poultry is often limited due to the high crude fibre content which can reduce digestibility. However, in limited quantities and with proper processing, palm kernel meal can be an economical feed component. According to [4], use local feed containing 7.5% palm kernel meal can be used in Chickens rations up to a certain level without reducing growth performance or feed conversion efficiency. The use of palm kernel meal in controlled amounts can help provide additional energy and increase muscle mass and body weight of chickens. This meal also has the potential to affect lipid metabolism in the body of poultry, which in turn can affect blood profiles, especially cholesterol and triglyceride levels. Dietary PKM Fermented 30% + PKM 10% was able to increase slaughter weight, carcass weight, breast weight, lower thigh weight, my weight and reduce abdominal fat SenSi-1 Agrinak chicken [5]. Therefore, there is a growing interest in using more affordable alternative materials, such as plant-based feed sources, which can replace or complement the main feed sources. Some alternative materials considered are Caulerpa meal, Copra meal, and Palm kernel meal, which have the potential to increase growth, body weight, and affect the blood profile of male laying hens.

2. Material and Methods

Measurement of body weight, feed consumption is carried out every week and blood profile testing is carried out when the chicken is 10 weeks old.

2.1 Animals and Experimental Diets

This study used 160 male DOC chickens and will be maintained for 10 weeks divided into a series of experiments. The experimental design used in this study was a Completely Randomised Design (CRD), with 10 treatments and 4 replications that there were 40 experimental units, each unit had 4 chickens. The treatments used were: Control ration using Fishmeal, vegetable feed without Csp (TO), Ration without Fishmeal, Vegetable feed and Csp (T0C), Ration without Fishmeal, 10%

Table 1. Ingredients composition and nutrient content of dietary treatment

Feed Ingredients	Treatment									
	T0	TOC	T1	T1C	T2	T2C	T3	T3C	T4	T4C
Fishmeal	0,00	10,00	-	-	-	-	-	-	-	-
Corn	55,00	55,00	50,00	50,00	50,00	50,00	47,00	47,00	40,00	40,00
Soybeans	12,00	12,00	27,00	27,00	27,00	27,00	30,00	30,00	27,00	27,00
Copra meal (CM)	10,00	10,00	10,00	10,00	20,00	20,00	-	-	15,00	15,00
Palm kernel meal (PKM)	10,00	10,00	10,00	10,00	-	-	20,00	20,00	15,00	15,00
Caulerpa Meal	-	0,10	-	0,10	-	0,10	-	0,10	-	0,10
Dicalcium phosphate	2,70	2,60	2,70	2,60	2,70	2,60	2,70	2,60	2,70	2,60
Lysine	0,20	0,20	0,20	0,20	0,20	0,20	0,20	0,20	0,20	0,20
Methionine	0,10	0,10	0,10	0,10	0,10	0,10	0,10	0,10	0,10	0,10
Total	**100,00**	**100,00**	**100,00**	**100,00**	**100,00**	**100,00**	**100,00**	**100,00**	**100,00**	**100,00**
ME (kkal/kg)	2.988	2.989	3.063	3.064	3.075	3.076	3.053	3.054	3.063	3.064
Crude Protein (%)	18,91	18,92	18,08	18,09	18,76	18,78	18,00	18,00	18,08	18,09
Crude Fiber (%)	6,46	6,47	6,83	6,84	6,33	6,34	7,34	7,36	6,83	6,84
Crude Lipid (%)	7,64	7,65	9,51	9,51	9,22	9,22	10,10	10,00	9,51	9,51
Ca (%)	1,23	1,21	0,92	0,90	0,86	0,84	0,99	0,97	0,92	0,90
P (%)	0,88	0,86	0,65	0,63	0,67	0,65	0,63	0,61	0,65	0,63
Lysin (%)	1,99	2,00	2,48	2,48	2,72	2,72	2,38	2,38	2,48	2,48

Based on calculating from fed proximate analyses in Feed Nutrition Laboratory, Tadulako University

CM, 10% PKM without Csp (T1), Ration without Fishmeal, 10% CM, 10% PKM and Csp (T1C), Ration without Fishmeal-20% CM without Csp (T2), Ration without Fishmeal, 20% CM and Csp (T2C), Ration Without Fishmeal, 20% PKM without Csp (T3), Ration without Fishmeal-20% PKM and Csp (T3C), Ration without Fishmeal, 15% CM, 15% PKM without Csp (T4), Ration without Fishmeal, 15% CM, 15% PKM and Csp (T4C).

The ingredients of the ration with the treatment formula are listed in Table 1. The ingredients of the vegetable feed used are corn, soybeans, copra meal and palm kernel meal. Fish meal was added to the control ration. All feed ingredients will be analysed proximately in the Feed Nutrition Laboratory of Tadulako University.

2.2 *Data Collection*

This study followed the animal ethics protocol submitted and approved with approval number 2414/UN28.16/AL.04/2024. Blood samples were collected in accordance with standard animal ethics procedures with standard sampling without harming animals.

The implementation stage of the research was carried out by maintenance with treatment for 70 days. The data collection and analysis stage were carried out by taking 10 chickens as samples, 1 chicken for each treatment. Blood samples were taken from the wing vein of each chicken to determine the complete blood count calculation. The blood taken as much as ± 1 cc is put into a vacuum container containing EDTA anticoagulant (Ethylene Diamine Tetra Acetic Acid) and shaken slowly until mixed so that it does not clot. After that, it is put into a container containing ice before being analysed. Haematology measurements using an automatic haematology analyser and total blood protein using a biosystem A15 Automatic Analyzer were carried out from the Health Laboratory Central Sulawesi Province.

A total of 80 chickens were slaughtered for carcass data collection, with 1 sample taken from each treatment for measuring antioxidant activity. The meat from the breast and thighs was collected to analyse antioxidant activity in breast and thigh meat carried out at the Chemistry laboratory of Tadulako University.

2.3 Data Analysis

The parameters observed were feed intake, weight gain, Feed Conversion Ratio (FCR). Body Weight Gain refers to the increase in body weight of the bird over 10 week that weighting birds at the start and end of the experimental period (weekly measurements) using a calibrated digital weighing scale for accuracy that expressed in grams (g) per bird.

$$\text{Body Weight Gain (BWG)} = \text{Final Body Weight (FBW)} - \text{Initial Body Weight (IBW)} \quad (1.1)$$

Feed Intake (FI) (g/bird) is the total amount of feed consumed by a bird over a specific period that record the total feed offered to the birds, Weight and record feed refusals or left overs, Calculate feed consumption for the entire experimental group, and expressed in grams (g) per bird.

$$\text{Feed Intake (FI)} = \text{Feed Offered (g)} - \text{Feed Refused (g)} \quad (1.2)$$

Feed Conversion Ratio indicates the efficiency with which birds convert feed into body weight gain that using the total feed intake and total body weight gain over the same period. FCR is dimensionless (a ratio). A lower FCR indicates better feed efficiency.

$$FCR = \frac{\text{Feed Intake (g)}}{\text{Body Weight Gain (g)}} \quad (1.3)$$

Meanwhile, the results of feed intake, body weight gain, Feed Conversion Ratio (FCR) were analysed using Completely Randomised Design and data that had significant differences were continued with Tukey test analysis using SPSS program.

At the age of 10 weeks, 1 chicken blood sample was taken for each treatment to measure blood profile, total blood protein. At the age of 12 weeks, chickens were slaughtered to take 1 sample of breast and thigh meat from each treatment to measure antioxidant activity (IC_{50}). Identification of blood profile, total blood protein, meat antioxidant activity (IC_{50}t was done using a descriptive method.

3. Result and Discussion

Table 2 shows data on we're feed Intake, weight gain, Feed Conversion Ratio (FCR), blood profile, total blood protein, antioxidant activity in meat (breast and thighs). The growth performance and productivity of male laying hens can be influenced by their diet which must ensure the right balance of energy and protein. Studies on the combination of Caulerpa sp., copra meal and palm kernel meal can explore the potential for synergy between these feed ingredients. Table 2 shows that the provision of feed ingredients sourced from completely plant-based feeds resulted in lower performance compared to control feeds containing fish meal. However, the use of 0.1% Caulerpa sp. (T0C) provided better performance than the control (T0).

The lowest weight gain was achieved using 20% CM (T2), 20% PKM (T3), or a combination of 10% CM-10% PKM (T1) and 15% CM-15% PKM (T4). The addition of 0.1% Caulerpa sp can increase body weight gain. Weight gain is used as the main indicator in measuring growth as a basis for measuring the relative speed of weight gain per unit of time or absolute size after reaching a certain period of time.

The use of full plant-based feed in rations increases Leukocytes in the blood of male laying hens (T1-T4C = 13,53-14,97 $10^3 mm^{-3}$). Leukocytes of male laying hens in the T0 and T0C treatments are in accordance with the normal range of 11-13 ($10^3 mm^{-3}$) according to [6]. Leukocytes or white blood cells are part of the body's defence system that can move. Factors that determine [7]. In this study, chickens that received a control ration containing animal feed in the form of fish meal (T0) and containing 0.1% Caulerpa sp. (T0C) were still within normal limits and those with full plant-based feed containing CM, PKM or a combination of both experienced an increase in Leukocytes (Table 3). These results indicate that there is a significant physiological disorder in the body of male laying hens due to the provision of full plant-based feed. The increase and decrease in the number of leukocytes in the blood is a mechanism of the body's response to the feed given. Allergic reactions in livestock are mainly related to photosensitivity, because nutrients contain allergens [8]. The results of this study are relatively the same according to [9] in the number of leukocytes in Commercial Laying hens and Kampung chickens ranging from 2.61-2.65 ($10^3/mm^3$) in open house of tropical zone. An increase in the number of leukocytes indicates an increase in the body's defence ability [10]. The concentration of erythrocytes, haemoglobin, and haematocrit in this study were within the normal range in all treatments, except for the T0 treatment, the values of the three parameters were less than the normal range. The haematocrit value indicates the percentage of erythrocyte volume in 100 ml of blood. The haematocrit value is positively correlated with the size of erythrocytes, remaining Total erythrocytes in this study ranged from 1.46 - 3.10 ($10^6/mm^3$). The normal number of broiler chicken erythrocytes is 2.5-3.5 ($10^6/mm^3$) [6]. Haemoglobin concentration ranges from 6.4-14.3 g/dL^{-1} where the normal amount of broiler chicken haemoglobin is 7-13 ($10^6/mm^3$) [6]. The haematocrit value in this study was obtained 20.3-43.0% where the normal amount of broiler

IOP Conf. Series: Earth and Environmental Science **1460** (2025) 012020 doi:10.1088/1755-1315/1460/1/012020

Table 2. Main effects of treatment on performance, blood profile and meat antioxidant activity male laying hens

Feed Ingredients	Treatment									
	T0	TOC	T1	T1C	T2	T2C	T3	T3C	T4	T4C
Body Weight Gain (g/bird)	854.74±77,60^{c}	1,023.58±52.78^{d}	253.76±29.60^{a}	458.29±79.21^{b}	347.93±29.95^{ab}	420.70+51.12^{b}	246.46±13.33^{a}	344.13±36.53^{ab}	306.29±50.66^{ab}	414.50±69,42^{b}
Feed Intake (g/bird)	2,888.56±124,70^{b}	3,255.69±66.05^{c}	2,164.63±94.04c	2,375.29±105.79^{ab}	2,039.19±223.93^{a}	2,163.44±146.14^{a}	1,813.1±1 55.6^{a}	2,292.81±389.12^{a}	2,069.75±267.47^{ab}	2,206.44±325,57^{a}
FCR	3.40±0.34^{a}	3.19±0.20^{a}	8.48±1.26^{c}	5.32±1.04^{abc}	5.92±1.01^{abc}	5.22±0.89^{ab}	7.37±0.73^{bc}	6.76±1.57^{bc}	6.98±1,82^{bc}	5,44±1.24^{abc}
Leukocyte ($10^{3}/mm^{-3}$)	8.6	13.19	13.76	13.90	14.41	13.80	14.39	14.97	14.39	13.53
Erythrocyte ($million/mm^{3}$)	1.46	2.33	2.76	2.55	2.91	2.45	3.10	2,68	2.90	2,54
Haemoglobin (g/dL^{-1})	6.4	10.9	12.9	11.8	12.4	11.3	14.3	13.0	12.7	12,5
Haematocrit (%)	20.3	30.3	38.7	32.9	38.3	32,6	43,0	35,6	38.8	33.0
Platelets (mm^{3})	53	47	46	29	42	54	38	25	27	21
Total protein (g/dL^{-1})	4.23	3.17	5.01	3.82	3.55	3.15	3.97	3.72	4.37	3.49
Breast Antioxidant (ppm)	147.70	156.11	132.41	146.10	91.97	136.35	115.61	114.58	121.06	117.64
Thigh Antioxidant (ppm)	175.49	205.88	152.72	191.73	110.76	160.17	124.82	118.58	133.32	125.83

abc Least squares means in the same row, marked with different superscripts, were significantly different as determined by the Tukey means comparison test ($p < 0.05$)

chicken haematocrit is 35-55% [6]. The decrease in total erythrocytes in this study is thought to be due to hot environmental conditions [7]. This haematocrit value is negatively correlated with the concentration of fluid in the chicken's body. Chickens in conditions of lack of blood fluids will cause an increase in the haematocrit value. Conversely, if the chicken is in a condition of excess fluid, such as excess addition of anticoagulants, it can cause a decrease in the value haematocrit. [7] stated that the factor that can affect the formation of erythrocytes is adequate nutrition. In this study, male laying hens received nutrition containing elements that support the formation of red blood cells in sufficient quantities. These nutrients include protein, iron, vitamin B9 and vitamin B12. Protein and iron are involved in the formation of haemoglobin, while vitamin B9 and vitamin B12 play a role in the maturation of erythrocytes. This is supported by total protein in the blood ranging from 3.15-5.01 g/dL, the normal range of total protein in blood broiler chickens is 3.0-4.9 g/dL [11]. The number of platelets in the blood ranges from 21-54 g/dL, the normal range of platelets in broiler chickens is 3-33 g/dL [12]. This is in line with [13] study which used 0.020-0.035% green seaweed Ulva sp which provided all blood values obtained in this study were within the normal range for healthy broiler chickens except for lymphocytes.

The antioxidant activity of breast meat ranges from strong to weak (91-156 ppm), Strong category in 20% CM treatment and weak category in control treatment using fish meal (TO). The antioxidant activity of thigh meat ranges from moderate to very weak (119-205 ppm), Medium category in 20% CM treatment and weak category in control treatment using fish meal (TO). These results are in line with research by [14] that antioxidant enzyme activity (SOD and catalase) was high and lipid oxidation products were low in chicken breast meat that consumed 3% coconut oil and Areca vestiaria Giseke seed flour supplementation at a dose of 1250 mg kg^{-1} describes a healthy food product.

The use of dietary full plant-based feed ingredients significantly decreased the performance of male laying hens compared to the control ration (TO) containing fish meal. However, the use of 0.1% Caulerpa sp. (TOC) gave better results than the T0 control. The lowest weight gain was in the use of 20% CM (T2), 20% PKM (T3), or a combination of 10% CM-10% PKM (T1) and 15% CM-15% PKM (T4). The addition of 0.1% Caulerpa sp to rations containing CM, PKM or their combinations can increase body weight gain.

The combination of Caulerpa sp., copra meal, and palm kernel meal promises to improve the performance and blood profile of male laying hens. Each ingredient provides unique benefits, ranging from increased nutrient utilization to improved antioxidant activity. However, proper formulation is key to optimize these benefits while minimizing potential challenges, such as high fibre content and the presence of antinutrients

4. Conclusion

Feeds composed entirely of plant-based ingredients showed reduced performance compared to control diets that included fish meal. The use of 0.1% Caulerpa sp. in rations can improve the performance of chickens given CM, PKM or their combination in rations. Chickens with low performance tend to have higher Leukocyte, Erythrocyte, Haemoglobin, Haematocrit, IC_{50} in meat and low platelets. The best chicken performance was obtained from a combination of 0.1% Caulerpa sp., 10% CM, 10% PKM, 10% fish meal. Future research should focus on optimizing the inclusion levels and exploring the long-term impacts of these feed ingredients on hen health and productivity.

Acknowledgement

The Project is finances by DIPA Fund Tadulako University (0930/UN28.16/AL.04/2024), and thank you to all parties that help completed the research process.

References

[1] Coudert , E. Elisabeth & C. Berri. 2020. Use of algae in poultry production: a review, World's Poultry Science Journal To link to this article: https://doi.org/10.1080/00439339.2020.1830012
[2] Uguz, S.; Sozcu, A. Pollutant Gases to Algal Animal Feed: Impacts of Poultry House Exhaust Air on Amino Acid Profile of Algae. Animals 2024, 14, 754. https://doi.org/10.3390/ani14050754
[3] Sundu, B. U Hatta, S Mozin, N Toana, Hafsah, Marhaeni and S Sarjuni2020 IOP Conf. Ser.: Earth Environ. Sci. 492 012126 IOP Conf. Series: Earth and Environmental Science IOP Publishing doi:10.1088/1755-1315/492/1/012126
[4] Nuraini, Z.Hidayat, F. Oktria, and D. Saphira, 2023. Performance Study and Analysis of KUB Chicken Business Starter Period was Treated with Feed Local Resource Based. JPI Vol. 25 (1): 37-46 DOI: 10.25077/jpi.25.1.37-46.2023
[5] Hanafi, N.D., M. Tafsin, S.H. Sitindaon, A. Sadeli, and K. Simanungkalit, 2022 The effect of use of palm kernel oil at the level of 40% in the ration on slaughter weight, carcass commercial cuts and meat quality SenSi-1 Agrinak chicken). Jurnal Agripet, Vol 22 (1): 62-71 DOI: https://doi.org/10.17969/agripet.v22i1.21099
[6] Adamu, M., C. Boonkaewwan, N. Gongruttananun, and M. Vongpakorn. 2013. Hematological, Biochemical and Histopathological Changes Caused by Coccidiosis in Chickens. 238–246 pp
[7] Ifititah, D., B. Arisandi, R.R.R. Widyani and Juniah, 2022. Physiological Conditions of Broiler Chickens During Transportation with Vitamin Treatment and Distance Difference. Jurnal Ilmu-Ilmu Peternakan 32(3): 313 – 327
[8] Pilarczyk R, J Wójcik, P Czerniak, P Sablik, B Pilarczyk and AT Marciniak. 2013. Concentrations of Toxic Heavy Metals and Trace Elements in Raw Milk of Simmenta and Holstein-Friesian Cows from Organic Farm. Environmental Monitoring and Assessment. 185: 8383-839
[9] Ulupi, N., and T.T. Ihwantoro. 2014. Blood Profile of Kampung Chicken and Commercial Laying Hens in Open House of Tropical Zone. Jurnal Ilmu Produksi Hasil Peternakan 2(1) 219-223
[10] Salam, S., D. Sunarti and Isroli, 2013. Physiological Responses of Blood and Immune Organs Chicken Fed Dietary Black Cumin Powder (*Nigella satifa*) During Dry Seasons. J.Indonesian Trop.Anim.Agric. 38(3) : 185-191
[11] Rafis, A.N., A. Sudarman, R. Mutiah, 2023. Metabolite and Blood Profile of Late Laying Period Supplemented with Bromelain Enzyme in The Ration. Jurnal Ilmu Nutrisi dan Teknologi Pakan 21 (3) : 143-149
[12] Fanmira, A.A., F.U. Datta, A.I.R. Detha, N.D.F.K. Foeh, N.A. Ndaon, 2022. Gambaran Ertrosit dan Tombosit Ayam Broiler. Jurnal Veteriner Nusantara, 5 (18) : 1-11
[13] Matshogo, T.B., C. M. Mnisi, and V. Mlambo, 2020. Dietary Green Seaweed Compromises Overall Feed Conversion Efficiency but not Blood Parameters and Meat Quality and Stability in Broiler Chickens . Agriculture, 10, 547; doi:10.3390/agriculture10110547

, [14] Londok, J. J. M. R., Sumiati , Wiryawan, I.K.G., Manalu, W.. 2018. Antioxidants Enzyme Activity and Malondialdehyde Concentration on Broiler Fed Contain Lauric Acid and Areca vestiaria Giseke. Bulletin of Animal Science 42(2): 109-114 Doi: 10.21059/buletinpeternak.v42i2.31767

The Effect of Adding Molasses Levels to Starter Feed on Indicators of Rumen Development In Jawarandu Goat Kids (Hematocrit, Glucose and Blood Urea Levels)

Andro K[1], S. Mukodiningsih[1], B.I. Tampubolon[1], and Achmadi[1]

[1] Department of Animal Science Faculty of Animal and Agricultural Scinces, Diponegoro University, Semarang, Central Java 50275, Indonesia

*E-mail: androklemens1@gmail.com

Abstract. The study aimed to evaluate the effect of different levels of molasses addition in the starter feed on rumen development in Jawarandu goat kids, as assessed by blood hematocrit, glucose, and urea levels. The study involved 15 Jawarandu goat kids aged 7-14 days and 4.25 ± 1.2 kg initial body weight. The with a split-plot in time randomized design applied to analyze differences was employed, with three treatments: T0 (0% molasses), T1 (2% molasses), and T2 (4% molasses), each with five replications. The analysis of variance showed that the interaction between the addition of molasses level to the starter feed and the age of the livestock had no significant effect (P>0.05) on hematocrit, glucose and blood urea levels. The addition of molasses levels to the starter feed and the age of the goats also had no significant effect on blood hematocrit, glucose and blood urea levels. The hematocrit levels in week 3 and week 6 ranged between 22.6%-28.2%, and 25.6%-27.25%, glucose levels in week 3 and week 6 ranged from 58.5 mg/dL-75.76 mg/dL, and 58.08 mg/dL-75.34 mg/dL, and urea levels in week 3 and week 6 ranged from 44.39 mg/dL-57.91 mg/dL, and 21.6 mg/dL-28.18 mg/dL. Both values fall within the normal range, in conclusion, while the addition of molasses up to 4% to the starter feed given to Jawarandu goat kids up to 6 weeks old, produce the same hematocrit, glucose, and blood levels and achieve the standard, in which shows a developed rumen.

Keywords: blood glucose, blood urea, hematocrit, Jawarandu goat kids, starter feed

1. Introduction

Goats are a popular livestock choice in Indonesia. Goat meat consumption in Indonesia has increased over the last three years, with data from the National Central Bureau of Statistics indicating 59,730 tons in 2021, 60,768 tons in 2022, and 61,315 tons in 2023 [2]. Farmers prefer raising goats, particularly the Jawarandu breed, known for its resilience to Indonesia's climate and its ability to produce meat with relatively simple management practices [16]. However, smallholder farmers often rely on goat kids born from their existing stock, although the productivity of these offspring tends to be slow. Weaning for goat kids generally occurs at 90 days of age, as they rely heavily on their mother's milk for nutrition [12]. The delay in accepting solid feed makes the rearing period less efficient. To accelerate the growth of Jawarandu goat kids and their ability to digest roughage, a starter feed is needed to stimulate the development of the reticulo-rumen. Delays in receiving solid feed make the rearing period less efficient. To accelerate the growth of Jawarandu goat kids and their ability to digest fiber, starter feed is needed which can stimulate reticulo-rumen development. This is because the reticulo-rumen has not yet developed at birth, and its development is influenced by rough feed given immediately after birth [4], with a ratio of 60% coming from milk and 40% coming from solid feed, especially starter feed [13].

The reticulo-rumen is a part of the ruminant digestive system responsible for digesting high-fiber feed such as grass and leaves, which are difficult for non-ruminant animals to break down. Fermentation in the rumen breaks down fibrous components into simpler ones, providing energy for the livestock [8]. The primary function of the reticulo-rumen is to serve as a fermentation chamber for microorganisms, producing volatile fatty acids (VFAs) like propionate and butyrate, which stimulate rumen development [17]. Faster reticulo-rumen development is essential to promote papillae growth in the rumen, aiding in food degradation [6].

Starter feed is a nutritionally formulated feed that is easy to digest and is provided to young livestock during the early growth phase after weaning. Based on this need, a proper feed formulation for pre-weaning goat kids is crucial. Pre-weaning starter feed, as per the National Research Council (NRC) 2001 [13], consists of 60% milk and 40% solid feed, with 20% crude protein (CP) and 75% total digestible nutrients (TDN). However, starter feed in practice sometimes lacks palatability. One of the ingredients that can enhance palatability is molasses. Molasses is a by-product of sugar refining, characterized by its thick, dark brown liquid form. It serves as an energy source due to its high carbohydrate content and minerals that fulfil the nutritional requirements of livestock [18]. The advantages of molasses include improving palatability and acting as an easily absorbable energy source for livestock. Molasses also enhances microbial activity in the rumen, which is crucial for the fermentation process [15]. It is expected that adding molasses to starter feed can increase palatability and the productivity of Jawarandu goat kids. The productivity of goat kids can be assessed through blood profiles. Haematocrit, glucose, and urea levels are blood parameters that can serve as indicators of livestock productivity. Therefore, it is necessary to study the effects of molasses supplementation in starter feed on the blood profiles of Jawarandu goat kids. The objective of this study is to evaluate the addition of molasses in starter feed on blood parameters, including haematocrit, glucose, and urea levels. The study aims to provide information on the appropriate level of molasses supplementation in starter feed for pre-weaning goat kids, improving farm efficiency and profitability.

2. Material and Methods

This study was approved by the Ethical Clearance Commission of the Faculty of Animal and Agricultural Sciences Diponegoro University was release at February, 2024 with number 61-01/A-03/KEP-FPP. The materials used in this research included 15 pre-weaning Jawarandu young goats (both male and female) aged between 7 to 14 days, with an average initial body weight of 4.25 ± 1.2 kg. They were housed at the Teaching Farm of the Faculty of Animal and Agricultural Sciences. This study used a completely randomized design, with blood samples taken at different ages of goat kids (split plot in time). The treatment is the addition of molasses levels to the starter feed, namely T0 (starter feed without molasses), T1 (starter feed with the addition of 2% molasses) and T3 (addition of 4% molasses). Each treatment consisted of 5 replications. The starter feed was composed of several ingredients, including ground corn, rice bran, soybean meal, molasses, and a mineral mix, formulated according to the nutritional needs of young goats [11,13] with 20% crude protein (CP) and 75% total digestible nutrients (TDN). The treatment stage lasted six weeks, with the first week serving as an adaptation period and the following five weeks being the actual treatment phase. Nutritional requirements were calculated based on body weight and weekly weight gain, as outlined by [13]. The goat kids were fed milk replacer and starter feed in a 60:40 ratio. Milk was administered twice a day (morning and afternoon), and starter feed was offered 30 minutes after milk feeding, with water provided ad libitum. The parameters observed were hematocrit, glucose and blood urea levels at 3 and 6 weeks of age in all treatments. The data obtained were analyzed using analysis of variance, and if there was an effect of treatment on the observed parameters, it was continued with the Duncan's difference test.

3. Results and Discussion

The hematocrit test results in weeks 3 and 6 for Jawarandu goat kids are presented in Table 1. The results of analysis of variance showed that the interaction between the addition of molasses level to the starter feed and the age of the goats kids had no significant effect (P>0.05) on hematocrit, glucose and blood urea levels.

The addition of molasses levels to the starter feed and the age of the goats also had no significant effect on blood hematocrit levels. However, blood hematocrit levels at all ages and levels with molasses addition are still included in the standard blood hematocrit levels. The hematocrit levels in week 3 ranged between 22.6% and 28.2%, and in week 6, they ranged between 25.6% and 27.25%. Both values fall within the normal range, which is 18–38% [4]. The highest hematocrit level in week 3 was observed in the T2 treatment, while in week 6, the highest value was found in the T1 treatment. Table 2 also shows an increase in hematocrit for T0 and T1, with increases of 4.65% and 5%, respectively, while T2 showed a decrease of -2.6%. The increase in T0 and T1 is likely due to increased feed consumption [19]. Goats receiving adequate quality and quantity of feed tend to have more efficient metabolism, stimulating the production of red blood cells and hematocrit levels [1]. The decrease in T2 between weeks 3 and 6 is thought to be due to variation in blood sampling times and environmental conditions in the barn, indicating that homeostasis processes in T2 had begun. Homeostasis is the body's natural process for regulating the number of red blood cells, influenced by environmental factors and kidney function [7].

Table 1. Average hematocrit, glucose, blood urea in young goats Jawarandu

	Sampling Time (Week)	Ratio Molasses Addition			Mean
		T0 (0%)	T1 (2%)	T2 (4%)	
		(%)			
Hematocrit	3	22.6	23.6	28.2	24.8
	6	27.3	28.6	25.6	27.2
		(mg/dL)			
Glucose	3	58.5	70.04	75.76	68.1
	6	58.08	74.80	75.34	69.4
Blood Urea	3	57.91	50.02	44.39	50.77
	6	21.6	28.02	28.18	25.93

[a] All results showed no significant difference (P > 0.05)

The glucose levels in week 3 ranged from 58.5 mg/dL to 75.76 mg/dL, and in week 6, they ranged from 58.08 mg/dL to 75.34 mg/dL. These values are above and below the normal range of 62.88–69.26 mg/dL [9]. Glucose levels in T0 were below the normal range, while those in T1 and T2 were above the normal range. The lower glucose levels in T0 are likely due to the absence of molasses in the starter feed, which delays the breakdown of high-fiber feed into glucose through rumen fermentation. In ruminants, the main carbohydrate source comes from added feed and fiber [5]. In T1 and T2, with 2% and 4% molasses added, glucose levels were higher, indicating the initiation of glycostasis. Glycostasis helps maintain stable blood glucose levels during rumen fermentation, converting carbohydrates into volatile fatty acids (VFAs), such as acetate, propionate, and butyrate [14].

Blood urea levels ranged from 44.39 mg/dL to 57.91 mg/dL, and in week 6, they ranged from 21.6 mg/dL to 28.18 mg/dL. These values fall both above and below the normal range of 26.6–56.7 mg/dL [3]. In week 3, blood urea levels were above the upper limit of the normal range, while in week 6, the levels in T0, T1, and T2 were below the lower limit of the normal range. Overall, there was a decrease in blood urea levels between weeks 3 and 6, indicating that rumen fermentation had occurred. During fermentation, ammonia from protein breakdown can be utilized for microbial synthesis. Ammonia that is not used for synthesis is converted into urea by the liver and excreted as urine. This process helps control blood ammonia levels and prevents ammonia toxicity [10].

4. Conslusion

From this research, it can be concluded that the addition of molasses up to 4% to the starter feed given to Jawarandu goat kids up to 6 weeks old, produce the same hematocrit, glucose, and blood levels and achieve the standard, in which shows a developed rumen.

References

[1] Araújo, J. P., Vasconcelos, A. M., Almeida, J. R., And Costa, R. G, 2021, Livestock Science, 243, 104398.
[2] Badan Pusat Statistik Jakarta Pusat, 2023, Statistik Konsumsi Daging Kambing Di Indonesia, Jakarta Pusat : Badan Pusat Statistik.
[3] Bendryman, M. L., Schneider, J. E., And Gibbons, J. M, 2000, *Small Ruminant Research*, 38(2), 123-128.
[4] Cunningham, J. G. (2002). *Textbook Of Veterinary Physiology.*
[5] Huntington, G. B. 1997, Journal Of Animal Science, 75(3), 852-867.
[6] Jiao, J., Li, X., Beauchemin, K. A., Tan, Z., Tang, S., And Zhou, C, 2015, British Journal Of Nutrition, 113(6), 888-900.
[7] Lallo, C. H. O., Campbell, J., Garcia, G. W., And Garraway, E, 2018, Journal Of Animal Science, 96(3), 203-207.
[8] Mayulu, H., Rahayu, F., Christiyanto, M., Haris, M. I., Daru, T. P., And Rahmatullah, S. N, 2020, European Journal Of Molecular And Clinical Medicine, 7(8), 3703-3711.
[9] Mohammed, S. A., Razzaque, M. A., Omar, A. E., Albert, S., And Al-Gallaf, W. M. 2016, African Journal Of Biotechnology, 15(24), 1253-1257.
[10] Morris, S. M. Jr, 2016, The Journal Of Nutrition, 146(12), 2579S-2586S.
[11] Mukodiningsih, S., J. Achmadi, F. Wahyono, S. J. Ohh And S. K. Ill, 2015, Biological Quality Of Complete Calf Stater On Rumen Development Of Friesian Holstein Calf (Reviewed From VFA And NH3 Concetration). Proceeding International Seminar The 16th AAAP Animal Science.
[12] Muttaqin, C, 2017. Pertumbuhan Cempe Prasapih Berdasarkan Tipe Kelahiran, Fakultas Peternakan Dan Pertanian Universitas Diponegoro, Semarang (Unpublished, Bsc Thesis).
[13] National Research Council (NRC). 2001. *Nutrient Requirements Of Small Ruminants: Sheep, Goats, Cervids, And New World Camelids.* Washington, DC: The National Academies Press.
[14] Owens, F. N., Zinn, R. A., And Kim, Y. K, 1986, Journal Of Animal Science, 63(5), 1634-1648.
[15] Palmonari, A., Federiconi, A., Cavallini, D., Sniffen, C. J., Mammi, L., Turroni, S., ... And Formigoni, A, 2023, Animals, 13(4), 728.
[16] Rahmatullah, S. N., Swuandana, R., And Sulaiman, A, 2022, Jurnal Ilmiah Fillia Cendekia, 7(2), 91-97.
[17] Tilman, A. D., H. Hartadi., S. Prawirokusumo., S. Reksohadiprodjo And S. Lebdosoekojo. 1991. Ilmu Makanan Ternak Dasar. Gadjah Mada University Press, Yogyakarta.
[18] Torres-Cavazos, Z., Rico-Costilla, D. S., Moreno-Degollado, G., Hernández-Martínez, S. P., Mendez-Zamora, G., Ramos-Zayas, Y., And Kawas, J. R, 2023, Animals, 13(21), 3370.
[19] Wardhani, W. D. 2024. Evaluasi Biologi Pakan Starter Pada Cempe Jawarandu Ditinjau Dari Pertambahan Bobot Badan, Konsumsi, Pakan, Dan Efesiensi Pakan. Fakultas Peternakan Dan Pertanian Universitas Diponegoro, Semarang. (Unpublished, Bsc Thesis)

IOP Conf. Series: Earth and Environmental Science **1460** (2025) 012022 doi:10.1088/1755-1315/1460/1/012022

Extraction of Nanocrystalline Cellulose from Oil Palm Empty Fruit Bunches Using Sulfuric Acid Hydrolysis Method

Novian Wely Asmoro [1,2], Chusnul Hidayat [1], Teguh Ariyanto [3], Ria Millati [1*]
[1] Department of Food and Agricultural Product Technology, Universitas Gadjah Mada, Flora Bulaksumur No. 1 Street, Yogyakarta 55281, Indonesia.
[2] Agricultural Faculty, Universitas Veteran Bangun Nusantara, Letjen S. Humardhani No. 1 Street, Jombor, Sukoharjo 57521, Indonesia.
[3] Department of Chemical Engineering, Universitas Gadjah Mada, Grafika No. 2 Street, Yogyakarta 55281, Indonesia

E-mail: ria_millati@ugm.ac.id

Abstract. OPEFB is a cellulose-rich biomass waste from the palm oil industry that can be further utilized as nanocrystalline cellulose (CNC). One of the methods used for the isolation of crystalline cellulose is acid hydrolysis. Improper hydrolysis process conditions will cause low crystalline yield, crystallinity index, and unsuitable CNC characteristics. This study aimed to determine the influence of temperature, time, and sulfuric acid concentration on the hydrolysis process of OPEFB cellulose, especially on changes in crystalline yield and crystallinity index of CNC. In this study, the hydrolysis process conditions studied were the hydrolysis temperature, hydrolysis time, and sulfuric acid concentration used. The temperature variation used was 30-80°C, hydrolysis time variation was 30-150 minutes, and the sulfuric acid concentration variation was 10-50%. The analyses included crystalline cellulose yield, crystallinity index, and FTIR. High temperature in hydrolysis resulted in high crystallinity index but low crystalline cellulose yield. The longer hydrolysis time caused a decrease in the yield and crystallinity index of crystalline cellulose. High concentration of sulfuric acid resulted in low crystalline yield, but high crystallinity index. The highest yield of 93.53% was obtained at a low concentration of 20% at 40°C for 30 minutes.

Keywords: Acid hydrolysis, crystallinity index, crystalline cellulose, OPEFB, yield

1. Introduction

Crystalline cellulose is one of the cellulose components that can be isolated from different kinds of biomass waste, such as oil palm empty fruit bunch (OPEFB). OPEFB is solid waste generated from the palm oil industry. Crystalline cellulose can be applied in the form of cellulose nanocrystals (CNC) for the pharmaceutical and food industries [1], composite materials have superior mechanical properties and high chemical stability [2]. CNCs are cellulose particles that have a needle-like shape, with at least one dimension measuring 100 nm or less, and exhibit high crystallinity [3]. Crystalline cellulose isolation process is generally carried out using acid hydrolysis method. Acid hydrolysis uses strong acids, such as sulfuric, hydrochloric, oxalic, hydrobromic (HBr), HNO_3 and phosphoric acid [4]. The general function of acids is mainly their ability to liberate hydronium ions (H^+) during the hydrolytic cleavage of glycosidic linkages inside the amorphous structure of cellulose molecules along cellulose fibrils [5]. Sulfuric acid has the capacity to disrupt the amorphous structure of cellulose, produces crystalline cellulose with a higher crystallinity index [6], [7].

Several conditions such as temperature, time and concentration of the acid used strongly influence the acid hydrolysis process [8]. Variations in these parameters can result in significant differences in terms of cellulose yield and crystallinity index. For example, pistachio shells yielded a crystalline cellulose yield of 50±14%, crystallinity index 66%, under acid hydrolysis conditions at 50°C, for 90 min [4]. Compared to the other methods, acid hydrolysis produces nanocellulose with a higher crystallinity index., and also has a smaller size [9]. Crystalline cellulose can have a perfect crystal structure (having a crystallinity index of about 65% ~ 95%) [5].

Studies related to acid hydrolysis conditions are important to obtain high yields of crystalline cellulose with desired physicochemical characteristics. Thus, the purpose of this study was to investigate how the hydrolysis process in the isolation of crystalline cellulose from OPEFB was affected by condition parameters for acid hydrolysis. The optimal acid hydrolysis conditions for the isolation of crystalline cellulose from OPEFB biomass waste were determined.

2. Methods

2.1 Material

OPEFB originated from an oil palm plantation in Central Kalimantan, Indonesia. The cellulose of OPEFB was isolated through a pretreatment process according to the previous work [10].

2.2 Hydrolysis Process

The hydrolysis was conducted using sulfuric acid following the description of the method provided by [11] with modifications. This research was conducted in 3 stages to assess the conditions of the chemical hydrolysis process of OPEFB crystalline cellulose. **First**, determining the optimum temperature for chemical hydrolysis using sulfuric acid with a concentration of 30% for 60 minutes. The temperature variations used were 30°C, 40°C, 50°C, 60°C, 70°C, and 80°C. **Second**, determination of the optimum hydrolysis time using 30% sulfuric acid concentration at the optimum temperature obtained at the first stage. The time variations used were 30, 60, 90, 120 and 150 minutes. **Third**, determination of the optimum concentration of sulfuric acid in chemical hydrolysis. The temperature and time conditions used were from the first and the second stage. The concentration variations used were 10%, 20%, 30%, 40% and 50%.

The amount of OPEFB cellulose used was one gram at a ratio of 1:20 (w/v) to sulfuric acid solution. The hydrolysis reaction was terminated by the addition of cold distilled water at a ratio of 1:10 (v/v) and subsequent centrifugation at 4000 rpm for 15 minutes. The supernatant and natant were separated. An equal amount of cold distilled water was again added to the natant and then centrifuged. This process was repeated 4 to 5 times until the solution reached pH 5. Finally, the separated precipitate was dried at 60°C for 24 hours.

2.3 Analysis

2.3.1 The yield of crystalline cellulose

The following equation (1) was used to determine the yield of OPEFB crystalline cellulose:

$$\text{Yield of crystalin cellulose} = \frac{B}{A} \times 100\ \% \quad \ldots\ldots\ldots\ldots\ldots\ldots\ldots\ldots\ldots\ldots\ldots (1)$$

Where, A is the amount of crystalline cellulose of OPEFB and B is the amount of crystalline cellulose after hydrolysis process.

2.3.2 Crystallinity index (CrI)
The crystallinity index was determined using X-Ray Diffraction (XRD) spectra (XPERT-PRO D8 Bruker, Germany). The analysis was conducted at the integrated laboratory of the Islamic University of Indonesia, Yogyakarta, Indonesia. Analytical and instrument conditions were set at a wavelength of 1,540°A (CuKα radiation), with a scanning speed of 2° per second and a range of 2θ = 2-80°. The CrI was determined using Equation (2):

$$CrI=[\frac{I_{200}-I_{am}}{I_{200}}]x\ 100 \qquad (2)$$

Where, I_{200} is the highest intensity peak 2θ=22-23° of the crystalline fraction and I_{am} is the low intensity peak around 2θ=18° of the amorphous region [12].

2.3.3 Fourier Transmission Infra-Red spectroscopy (FTIR)
Functional group analysis of cellulose and crystalline cellulose was performed to identify the functional groups present in cellulose using the Fourier transform infrared spectroscopy method (Perkin Elmer Spectrum IR 10.7.2, USA). OPEFB crystalline cellulose samples in powder form, were scanned in the 4000-450 cm^{-1} wave number range, resulting in FTIR spectra [13].

3. Results and Discussion

3.1 Effect of Temperature on Acid Hydrolysis Method
The first evaluation was to study the influence of acid hydrolysis temperature on crystalline cellulose yield and crystallinity index. Increasing the temperature in the acid hydrolysis process (30°C - 80°C) caused a significant decrease in crystalline cellulose yield. The lowest yield of 55.40% was obtained at 80°C. And the highest yield of 91.00% was obtained at 40°C. Referring to some research results showed that the yield of crystalline cellulose derived from barley, wheat, and rice straw was 69, 75, 64%, respectively [6].

The crystallinity index of OPEFB cellulose following acid hydrolysis with different temperature conditions was determined through XRD. The crystallinity index of raw OPEFB, OPEFB cellulose, crystalline cellulose after acid hydrolysis was calculated from the XRD spectra in Figure 2. The 2θ peaks at 34.58°, 22.74° and 16.02° represented crystalline cellulose, where the crystalline part had narrow and sharp diffraction peaks, while the amorphous had wide peaks.

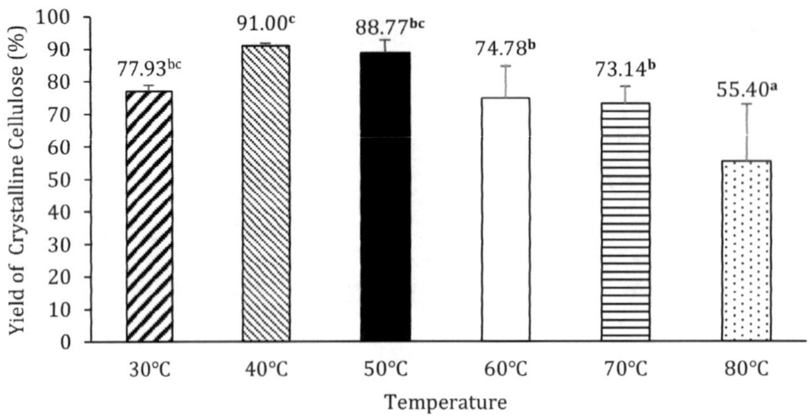

Figure 1. Acid Hydrolysis Temperature Effect on OPEFB Crystalline Cellulose Yield

Based on the data in Table 1, the lowest crystallinity index was 65.53% from hydrolysis at 30°C and the highest crystallinity index was 85.23% from hydrolysis at 80°C. The increase in crystallinity index is caused by the release of the amorphous region of cellulose due to acid hydrolysis process [14]. Hydrolysis of cellulose using sulfuric acid leads to the breaking of cellulose β-glycosidic bonds and at higher temperatures, the rate of hydrolysis reaction increases [5], [7]. The results showed that increasing the temperature of acid hydrolysis resulted in an enhancement of the crystallinity index but decreased the yield of crystalline cellulose.

Table 1. CrI of Crystalline Cellulose at Different Hydrolysis Temperatures

OPEFB crystalline cellulose (OPEFB CC)	Crystallinity index (%)
OPEFB CC-30	65.53
OPEFB CC-40	75.17
OPEFB CC-50	75.51
OPEFB CC-60	72.25
OPEFB CC-70	73.19
OPEFB CC-80	85.23

Note: the number 30-80 indicates the hydrolysis temperature

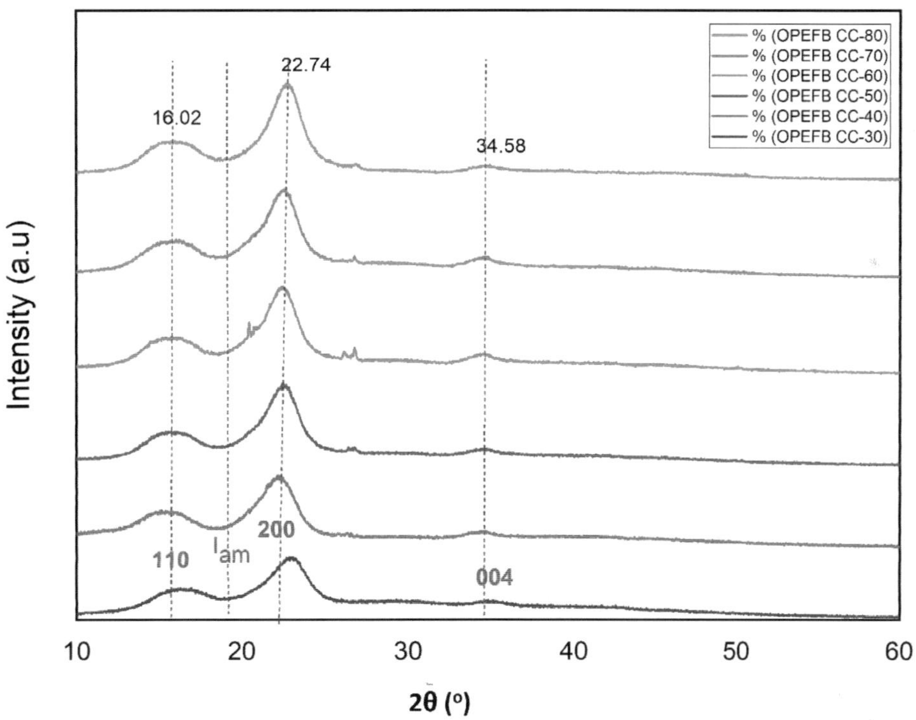

Figure 2. XRD Spectrum of OPEFB Crystalline Cellulose from Variation of Acid Hydrolysis Temperature

3.2 Effect of Time on the Acid Hydrolysis Method

The second stage of this study examined the influence of hydrolysis time on the yield and crystallinity index of crystalline cellulose. The time range selected for investigation was from 30 to 150 minutes, in accordance with the parameters of previous related study [8]. For this experiment, the hydrolysis process was conducted using sulfuric acid at a concentration of 30%

and a temperature of 40°C, based on the highest crystalline cellulose yield obtained in the first stage of the study (Section 3.1.).

Figure 3 presents the influence of hydrolysis time on the yield of crystalline cellulose. It was shown that increasing the hydrolysis time from 30 to 150 minutes caused a reduction in the yield of OPEFB crystalline cellulose. The highest yield of 94.48% was obtained at 30 minutes of hydrolysis. The decrease in yield of crystalline cellulose in the acid hydrolysis process was a consequence of the fact that sulfuric acid being able to attack the crystalline parts over extended hydrolysis times. This resulted in depolymerization of the crystalline and amorphous parts of the cellulose, which then broke down and released the individual crystals. As a result, both yield and crystallinity index decreased [15], [16]. Some studies indicated that the hydrolysis time was about 30-90 minutes [3], [17], [18]. Prolonged acid hydrolysis process could cause damage and removal of some crystal domains [19].

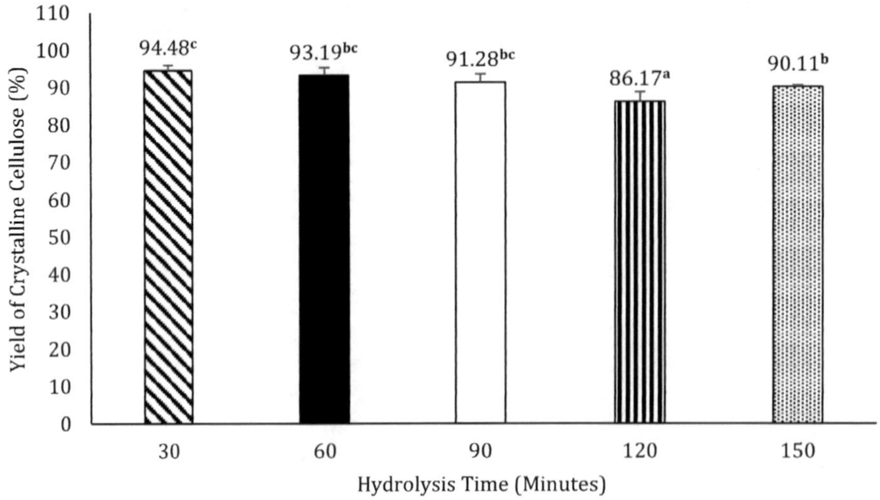

Figure 3. Acid Hydrolysis Time's Effect on the Yield of Crystalline Cellulose OPEFB

The crystallinity index of OPEFB crystalline cellulose based on different hydrolysis times was presented in Figure 4 and Table 2. Acid hydrolysis for 30 minutes resulted in the highest crystallinity index of 75.31%. Increasing the acid hydrolysis time from 60 minutes to 150 minutes led to a decrease in the crystallinity index. In the early stages of hydrolysis, sulfuric acid was more readily available and effectively broke down the cellulose structure in the amorphous region due to its irregular and looser structure. This resulted in the disruption of the amorphous structure of cellulose and a relative increase in the crystalline region. The next stage of the hydrolysis process was the attack of the crystalline structure of cellulose by sulfuric acid. The crystalline regions exhibited substantial resistance to acid hydrolysis throughout the course of the hydrolysis process; however, prolonged hydrolysis time could lead to a decrease in cellulose crystallinity [5], [9]. OPEFB CC-150 had a higher crystallinity index of 74.52% compared to OPEFB CC-120 with a crystallinity index of 71.44%. This difference can be attributed to several factors. The longer hydrolysis process of OPEFB CC-150 may have resulted in a slightly higher crystallinity index. However, it is important to note that longer hydrolysis times usually result in some damage to the crystalline regions. In the case of OPEFB CC-150, an optimal balance may have been reached where some amorphous cellulose was removed without significantly damaging the crystalline structure [3]. In addition, longer treatment times may have allowed recrystallization to occur, where partially degraded cellulose chains rearrange into a more ordered crystalline structure. This phenomenon has been observed in previous cellulose hydrolysis studies, which have reported higher crystallinity indices after long treatment times [18][17]. The sulfuric acid hydrolysis method was used to isolate CNCs from pistachio nut shells for 90 minutes at 50°C, resulting in a yield of 50.14% with a crystallinity index of 66% [4]. Comparison of the findings of this study with the results of previous research shows that the yield of crystalline OPEFB cellulose and the crystallinity index obtained are still higher when the hydrolysis method was carried out at 40°C, 30% sulfuric acid concentration, and 30 minutes of hydrolysis time.

Table 2. CrI of Crystalline Cellulose at Different Hydrolysis Times

OPEFB crystalline cellulose (OPEFB CC)	Crystallinity index (%)
OPEFB CC-30	75.31
OPEFB CC-60	75.17
OPEFB CC-90	73.52
OPEFB CC-120	71.44
OPEFB CC-150	74.52

Note: the number 30-150 indicates the hydrolysis time

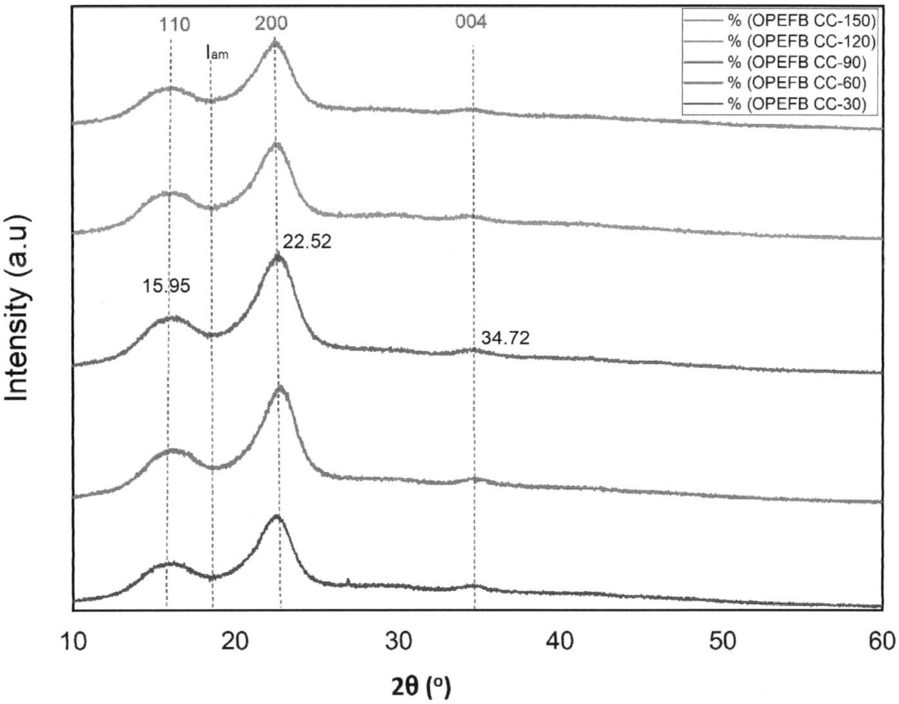

Figure 4. XRD Spectrum of Crystalline Cellulose OPEFB Variation of Acid Hydrolysis Time

3.3. Effect of Sulfuric Acid Concentration on the Acid Hydrolysis Method

The third stage of this research examined the influence of sulfuric acid concentration. The acid concentration used in some previous studies was 60% [18], [20], which is quite harmful to the environment. Thus, the purpose of this research was to find a lower acid concentration suitable for the hydrolysis process. Hence, concentrations in the range of 10-50% were evaluated. Based on the findings of the research conducted in Section 3.1. and 3.2., the hydrolysis process was conducted at 40°C for 30 minutes.

Figure 5 presented the influence of acid concentration on crystalline cellulose yield. The yield was significantly affected by the sulfuric acid concentration. The highest crystalline cellulose yield of 93.53% was obtained from hydrolysis using 20% acid concentration. While the lowest yield of 54.10% was obtained from hydrolysis using 50% acid concentration. No significant difference was observed between the results from 30% and 40% acid concentration It was found that an increase in the concentration of sulfuric acid resulted in a reduction in yield of crystalline cellulose.

Acids released hydronium ions (H+), which facilitated the hydrolysis of glycosidic linkages in the cellulose molecular chains along the cellulose fibres in amorphous regions. This process caused the hierarchical structure of the nanofibril assembly to break down into nanocrystals [5]. Amorphous regions of cellulose were more vulnerable to acid hydrolysis than crystalline regions due to the imperfections of the amorphous structure. This results in glycosidic bond breaks occurring faster than in crystalline regions [21]. As the acid concentration increases, so does the amount of hydrogen ions (H+) present, thus facilitating a better hydrolysis process. In a related study, the use of 63.5% (w/w) sulfuric acid concentration in the hydrolysis process of spruce for two hours produced a crystalline cellulose yield of 30% [8]. The use of 64% sulfuric acid concentration on rice straw cellulose produced 64% crystalline cellulose yield [6]. The crystalline cellulose yield of OPEFB in this study was identified to be relatively higher than that reported in several other studies examining various agricultural wastes.

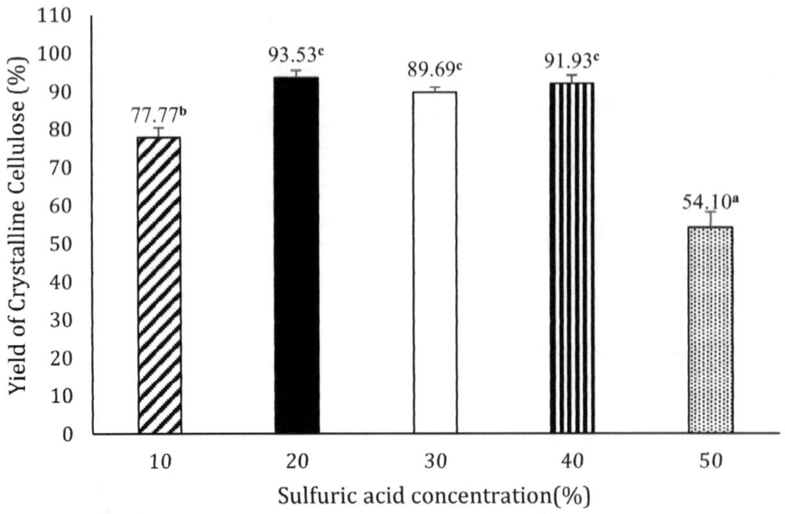

Figure 5. Effect of Sulfuric Acid Concentration on Yield of Crystalline Cellulose

As shown in Figure 6 and Table 3, the crystallinity index ranged from 62.10% to 87.07%. As the sulfuric acid concentration of increased, the crystallinity index increased. At 20% sulfuric acid concentration, the index value was 73.36%. While the highest index value observed was 87.07%, which was achieved at 50% sulfuric acid concentration. Compared with the results of other studies, the crystallinity index of this study was higher. The CNC crystallinity index of the acid hydrolysis process on sugarcane bagasse was 72.5%, using sulfuric acid concentration (64% (w/w), 1:10 g/ml (cellulose: acid solution) at 45°C for 60 minutes [3]. Crystalline cellulose from palm trunks in the hydrolysis process using 64% sulfuric acid solution, temperature 45°C for 1 hour had a crystallinity index of 69.61% [11]. The crystallinity index of crystalline cellulose obtained from sugarcane bagasse through sulfuric acid hydrolysis process with 60% concentration was 67.83%, lower than the crystallinity index obtained from hydrolysis process with 50% sulfuric acid concentration (76.01%).

Table 3. CrI of Crystalline Cellulose at Different Sulfuric Acid Concentration

OPEFB crystalline cellulose (OPEFB CC)	Crystallinity index (%)
OPEFB Cellulose	69.35
OPEFB CC-10	62,10
OPEFB CC-20	73.36
OPEFB CC-30	73.31
OPEFB CC-40	81.07
OPEFB CC-50	87.07

Note: the number 10-50 indicates the sulphuric acid concentration

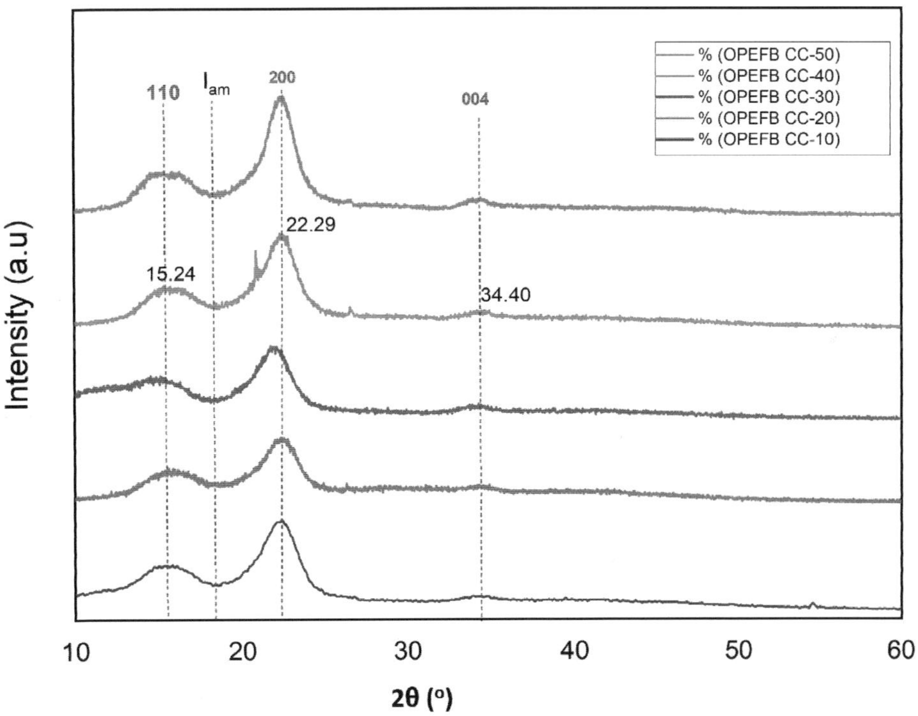

Figure 6. XRD Spectrum of Crystalline Cellulose OPEFB Variation of Sulfuric Acid Concentration

3.4 FTIR

FTIR was used to verify functional groups in OPEFB, cellulose of OPEFB, crystalline cellulose of OPEFB (Figure 6). Type I cellulose had FTIR spectra showing similar transmittance peaks in the range of 3400, 2900, 1345, 1045 cm^{-1} [22], [23]. The wavelengths shown on the FTIR graph indicate cellulose functional groups, namely C=O stretching at 1029 cm^{-1}, -O-pyranose at 1313-1642 cm^{-1}, CH_2 stretching at 2891 cm^{-1}, and OH functional groups at 3330 cm^{-1}. The transmittance peaks observed in the range of 1029-1160 cm^{-1} were connected with the pyranose ring framework's C=O stretching and C-H rocking vibrations. The transmittance peaks in the 1324 and 1634 cm^{-1} wavelength range of the spectra were associated with CH_2 symmetric bending and C-H and C=O bending vibrations of aromatic ring groups in polysaccharides [22]. The peak transmittance band around the wavelength of 3330 cm^{-1} exhibits a change in peak intensity that associated with a rise in the quantity of OH groups for CNC. This wavelength was associated with bending and stretching vibrations of OH functional groups, as well as with intra- and intermolecular hydrogen bonding. These observations suggested that the crystal domains in all samples have a type I lattice.

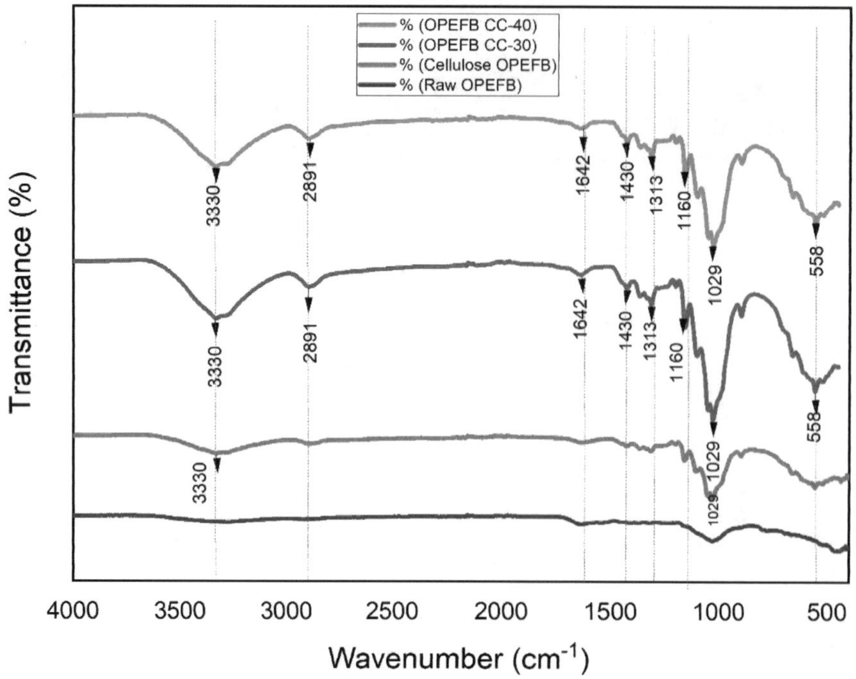

Figure 7. FTIR Spectrum of Raw OPEFB, Cellulose of OPEFB, Crystalline Cellulose of OPEFB

Conclusion

The chemical hydrolysis method using sulfuric acid in the crystalline cellulose isolation process of OPEFB is significantly affected by concentration acid, temperature and time, based on the parameters of crystalline cellulose yield and crystallinity index. This study found that high yield can be obtained from acid hydrolysis conditions at low concentration. Using 20% acid concentration of 20%, the highest crystalline cellulose yield of 93.53% was obtained.

Acknowledgement:

The authors would like to acknowledge the support of the Doctoral Dissertation Research Grant (1753/UN1/DITLIT/Dit-Lit/PT.01.03/2022) for providing financial support for this project. from the Directorate of Research, Technology, and Community Service, Directorate General of Higher Education, Research, and Technology, Ministry of Education, Culture, Research, and Technology of the Republic of Indonesia.

References

[1] R. K. Mishra, A. Sabu, and S. K. Tiwari, "Materials chemistry and the futurist eco-friendly applications of nanocellulose: Status and prospect," *Journal of Saudi Chemical Society*, vol. 22, no. 8, pp. 949–978, 2018, doi: 10.1016/j.jscs.2018.02.005.

[2] N. Auclair, A. Kaboorani, B. Riedl, V. Landry, O. Hosseinaei, and S. Wang, "Influence of modified cellulose nanocrystals (CNC) on performance of bionanocomposite coatings," *Progress in Organic Coatings*, vol. 123, no. June 2017, pp. 27–34, 2018, doi: 10.1016/j.porgcoat.2018.05.027.

[3] A. Kumar, Y. Singh Negi, V. Choudhary, and N. Kant Bhardwaj, "Characterization of Cellulose Nanocrystals Produced by Acid-Hydrolysis from Sugarcane Bagasse as Agro-Waste," *Journal of Materials Physics and Chemistry*, vol. 2, no. 1, pp. 1–8, 2020, doi: 10.12691/jmpc-2-1-1.

[4] J. Marett, A. Aning, and E. J. Foster, "The isolation of cellulose nanocrystals from pistachio shells via acid hydrolysis," *Industrial Crops and Products*, vol. 109, no. September, pp. 869–874, 2017, doi: 10.1016/j.indcrop.2017.09.039.

[5] H. M. Ng *et al.*, "Extraction of cellulose nanocrystals from plant sources for application as reinforcing agent in polymers," *Composites Part B: Engineering*, vol. 75, pp. 176–200, 2015, doi: 10.1016/j.compositesb.2015.01.008.

[6] A. A. Oun and J. Rhim, "Isolation of cellulose nanocrystals from grain straws and their use for the preparation of carboxymethyl cellulose-based nanocomposite films," *Carbohydrate Polymers*, 2016, doi: 10.1016/j.carbpol.2016.05.020.

[7] Z. Wang, Z. Yao, J. Zhou, and Y. Zhang, "Reuse of waste cotton cloth for the extraction of cellulose nanocrystals," *Carbohydrate Polymers*, vol. 157, pp. 945–952, 2017.

[8] D. Bondeson, A. Mathew, and K. Oksman, "Optimization of the isolation of nanocrystals from microcrystalline cellulose by acid hydrolysis," *Cellulose*, vol. 13, no. 2, pp. 171–180, 2006, doi: 10.1007/s10570-006-9061-4.

[9] W. T. Wulandari, A. Rochliadi, and I. M. Arcana, "Nanocellulose prepared by acid hydrolysis of isolated cellulose from sugarcane bagasse," *IOP Conference Series: Materials Science and Engineering*, vol. 107, no. 1, 2016, doi: 10.1088/1757-899X/107/1/012045.

[10] N. W. Asmoro, C. Hidayat, T. Ariyanto, and R. Millati, "Cellulose Isolation From Oil Palm Empty Fruit Bunch Using Different Pretreatment Processes," *Journal ofApplied Science and Engineering*, vol. 26, no. 11, pp. 1513–1520, 2022.

[11] J. Lamaming, R. Hashim, O. Sulaiman, C. P. Leh, T. Sugimoto, and N. A. Nordin, "Cellulose nanocrystals isolated from oil palm trunk," *Carbohydrate Polymers*, vol. 127, pp. 202–208, 2015, doi: 10.1016/j.carbpol.2015.03.043.

[12] V. Barlianti, D. Dahnum, H. Hendarsyah, and H. Abimanyu, "Effect of Alkaline Pretreatment on Properties of Lignocellulosic Oil Palm Waste," *Procedia Chemistry*, vol. 16, pp. 195–201, 2015, doi: 10.1016/j.proche.2015.12.036.

[13] N. H. Sari, E. Syafri, W. Fatriasari, and A. Karimah, "Comprehensive Characterization Of Novel Cellulose Fiber From Paederia Foetida and Its Modification For Sustainable Composites Application," vol. 26, no. 10, pp. 1399–1408, 2022.

[14] S. Collazo-Bigliardi, R. Ortega-Toro, and A. Chiralt Boix, "Isolation and Characterisation of Microcrystalline Cellulose and Cellulose Nanocrystals from Coffee Husk and Comparative Study with Rice Husk," *Carbohydrate Polymers*, vol. 191, pp. 205–215, 2018, doi: 10.1016/j.carbpol.2018.03.022.

[15] B. L. Peng, N. Dhar, H. L. Liu, and K. C. Tam, "Chemistry and Applications of Nanocrystalline Cellulose and its Derivatives : a Nanotechnology Perspective," *THE CANADIAN JOURNAL OF CHEMICAL ENGINEERING*, vol. 9999, pp. 1–16, 2011, doi: 10.1002/cjce.20554.

[16] R. J. Moon, A. Martini, J. Nairn, J. Simonsen, and J. Youngblood, *Cellulose nanomaterials review: Structure, properties and nanocomposites*, vol. 40, no. 7. 2011. doi: 10.1039/c0cs00108b.

[17] S. Elazzouzi-Hafraoui, Y. Nishiyama, J. L. Putaux, L. Heux, F. Dubreuil, and C. Rochas, "The shape and size distribution of crystalline nanoparticles prepared by acid hydrolysis of native cellulose," *Biomacromolecules*, vol. 9, no. 1, pp. 57–65, 2008, doi: 10.1021/bm700769p.

[18] F. Fahma, S. Iwamoto, N. Hori, T. Iwata, and A. Takemura, "Isolation , preparation , and characterization of nanofibers from oil palm empty-fruit-bunch (OPEFB)," *Cellulose*, vol. 17, pp. 977–985, 2010, doi: 10.1007/s10570-010-9436-4.

[19] K. C. C. de C. Benini, H. J. C. Voorwald, M. O. H. Cioffi, M. C. Rezende, and V. Arantes, "Preparation of nanocellulose from Imperata brasiliensis grass using Taguchi method," *Carbohydrate Polymers*, vol. 192, no. February, pp. 337–346, 2018, doi: 10.1016/j.carbpol.2018.03.055.

[20] Y. Habibi, L. A. Lucia, and O. J. Rojas, "Cellulose nanocrystals: Chemistry, self-assembly, and applications," *Chemical Reviews*, vol. 110, no. 6, pp. 3479–3500, 2010, doi: 10.1021/cr900339w.

[21] A. Poulose *et al.*, "Nanocellulose: A Fundamental Material for Science and Technology Applications," *Molecules*, vol. 27, no. 22, pp. 1–27, 2022, doi: 10.3390/molecules27228032.

[22] F. T. Seta *et al.*, "Preparation and characterization of high yield cellulose nanocrystals (CNC) derived from ball mill pretreatment and maleic acid hydrolysis," *Carbohydrate Polymers*, vol. 234, no. December 2019, p. 115942, 2020, doi: 10.1016/j.carbpol.2020.115942.

[23] X. An, Y. Wen, D. Cheng, X. Zhu, and Y. Ni, "Preparation of cellulose nano-crystals through a sequential process of cellulase pretreatment and acid hydrolysis," *Cellulose*, 2016, doi: 10.1007/s10570-016-0964-4.

Characteristics of Pectin Extracted from Pacitan Sweet Orange Peel by Ultrasound Assisted Extraction Method

Sifra Kusuma Dewi, Anisa Bella Mahardika Wikaningtyas, and Dian Anggraini Suroto, Ria Millati[1]*

Department of Food and Agricultural Products Technology, Faculty of Agricultural Technology, Universitas Gadjah Mada, Yogyakarta, Indonesia

*Email: ria_millati@ugm.ac.id

Abstract. The waste of sweet orange peel can be developed into pectin to increase the economic value. Ultrasonic-assisted extraction can be utilized to separate pectin from orange peel. The ultrasonic-assisted extraction method is an alternative to increase the efficiency of the pectin extraction process and improve the quality of the pectin. The purpose of this study was to determine how extraction conditions affected the yield of pectin. The variations were temperature (30 °C, 45 °C, and 60 °C); duty cycle (60%, 80%, 100%); time (20 min, 30 min, 40 min); and amplitude (60%, 80%, 100%). The pectin produced had an esterification degree of 79.52%; a moisture content of 8.69%; an ash content of 2.82%; L* 38.33; a* 1.04; b* 9.71. Temperature 60°C, duty cycle 100%, duration 40 minutes, and amplitude 100% produced the maximum yield.

Keywords: esterification degree, orange peel, pectin, ultrasound-assisted extraction

1. Introduction

Citrus (*Citrus* sp) is one of the most popular horticultural commodities in Indonesia. The Central Bureau of Statistics of Indonesia reported that citrus production in Indonesia tends to increase at 2.4, 2.5, and 2.8 million tons for 2021, 2022, 2023, respectively [1]. Furthermore, oranges are one of the fruits that are always available throughout the year at the Gemah Ripah Sleman Fruit Market in Yogyakarta Special Province with an average amount of 3694 tons every month [2]. Most of the use of oranges is only to extract the juice to be processed into squeezed orange drinks, so many parts of the orange are wasted, one of which is the orange peel. Orange peel accounts for 40% of the total weight of oranges [3]. With such a large amount, unutilized orange peels will cause environmental problems. Therefore, efforts need to be made to add value to orange peels by processing the peels into for example pectin. In dicotyledonous plants, pectin is a complex polysaccharide component that is present between the primary cell wall and the intercellular spaces. Pectin plays a role in providing flexibility and strength to plants [4]. Pectin is used extensively in the pharmaceutical, cosmetic, and food sectors as a thickener, stabilizer, and gelling agent in the production of fruit juices, jams, jellies, and sweets [4, 5].

Pectin can be extracted from plant tissues using an extraction process [6]. Conventionally, pectin can be extracted using strong acid solvents, such as nitric acid or sulfuric acid. The use of strong acids in the extraction process can produce liquid waste that can damage the environment. In addition, conventional extraction requires heat and a long time, which can cause thermal damage to pectin and reduce pectin quality [4]. Therefore, organic acids have been used to replace strong acids in environmentally friendly extraction techniques, such as acetic and citric acids [4] and using other extraction methods such as ultrasound-assisted extraction [7]. When pectin was extracted from banana peels, previous study revealed that the yield from ultrasound-assisted extraction was greater than that from the conventional method [8].

Ultrasound extraction is the process of extracting from within the plant tissue matrix using acoustic energy and solvent by diffusion and leaching mechanisms of the cell contents after the cell wall is broken by ultrasonic irradiation [9]. Ultrasound extraction methods are proven to be eco-friendly and can improve efficiency and product quality through the cavitation effects of the ultrasound mechanism as compounds are obtained rapidly with less degradation [10]. The use of citric acid in the ultrasound extraction method can reduce the environmental impact [7].

Fruit or vegetable plant tissues can contain a variety of chemicals, including polyphenols, carotenoids, aroma compounds, and polysaccharides, which can be extracted via ultrasonic-assisted extraction [11]. Pectin has frequently been extracted using ultrasound-assisted extraction from a variety of sources, including banana waste [7], passion fruit peel [12], grapefruit [13], tomato waste [14], and sour orange [15]. The extraction process with ultrasonic assistance is influenced by a number of factors including frequency, amplitude, duty cycle, power, time, temperature, solvent type, and sample-solvent ratio [11]. This study evaluated how temperature, duty cycle, time, and amplitude affected the ultrasonic assisted extraction method to extract pectin from sweet orange peel.

2. Materials and Methods

2.1 Sample collection

The material used in this research was Pacitan sweet orange (*Citrus sinensis* L. Osbeck cv. Pacitan) obtained from a supermarket in Yogyakarta. The chemicals used were citric acid, 37% HCl, NaOH, 96% ethanol, phenolphthalein indicator and distilled water.

2.2. Sample preparation

Sweet orange peels were washed to remove impurities, then cut into pieces with a size of ± 0.5 cm and dried for 24 hours at 50 °C in a cabinet dryer. The dried orange peels were pulverized using a blender and then sieved using a 20 mesh sieve and then a 60 mesh sieve. Orange peel powder that passed the 20 mesh sieve but did not pass the 60 mesh sieve was then stored in a plastic clip and given silica gel.

2.3. Pectin extraction

The device used for pectin extraction is a probe-type sonicator with the brand and type Hielscher ultrasonic Germany UP200st. The orange peel sample was weighed 10 g and put into a glass beaker, and 200 ml of 0.15 N citric acid was added. The mixture was then heated in the ultrasound probe filled with water at the specified conditions. The temperature ranges used for pectin extraction were 30 °C, 45 °C, and 60 °C. After obtaining the optimum temperature, continued with testing the duty cycle variations, namely 60%, 80%, and 100%. Then proceed with examining variations in extraction time, namely 20 minutes, 30 minutes, and 60 minutes and finally examining variations in amplitude, *i.e.* 60%, 80%, and 100%.

A vacuum filter was used to filter the extraction results. The filtered filtrate was then precipitated by adding 96% ethanol in the ratio of filtrate:ethanol (1:1) for 24 hours. After that, a vacuum filter was used to separate the pectin precipitate from the solution. The precipitate obtained was washed using 30 ml of 96% ethanol 3 times with stirring. After washing, wet pectin was obtained. An oven operated at 105 °C was then used to dry the wet pectin until its weight remained consistent. The dried pectin was then weighed to record its weight. The percentage yield was determined using the following equation:

$$Yield\ (\%) = \frac{weight\ of\ dried\ pectin\ (g)}{weight\ of\ orange\ peel\ powder\ (g)} \times 100\%$$

The pectin obtained with the highest yield was then analyzed for its characteristics, such as degree of esterification, water content, ash content, and color. The degree of esterification was calculated using the titration method, calculating the free carboxyl group first and then calculating the number of esterified carboxyl groups. The ratio of esterified carboxyl groups to total carboxyl groups is known as the degree of esterification. Color analysis of pectin was conducted using a chromameter (Konica Minolta CR-410). In addition, the functional groups of pectin were analyzed using FTIR spectroscopy.

3. Result and Discussion

Yield is calculated by dividing the final product weight after drying by the material's initial weight prior to drying. Based on the weight of the dried pectin divided by the weight of the orange peel powder used for extraction, the yield in this study was determined. The yield value is highly considered as a parameter of the efficiency of the extraction process. A higher yield value indicates a better extraction process.

Table 1. Effect of extraction process in yield value

Extraction Conditions	Yield (%)		
Temperature	2.20 ± 0.31 (30 °C)	2.85 ± 0.28 (45 °C)	3.53 ± 0.09 (60 °C)
Duty cycle	1.57 ± 0.34 (60%)	2.27 ± 0.23 (80%)	3.53 ± 0.09 (100%)
Time	1.97 ± 0.01 (20 min)	3.29 ± 0.01 (30 min)	4.23 ± 0.01 (40 min)
Amplitude	1.81 ± 0.00 (60%)	2.78 ± 0.00 (80%)	4.23 ± 0.01 (100%)

3.1 Effect of temperatures on yield value

As the extraction temperature rose in this study, the yield value increased as well. A higher temperature during extraction results in a higher pectin yield value. At 60 °C, the maximum yield value was reached during the extraction process (Table 1). The extraction temperature influences the solubility of pectin during the extraction process. At higher the extraction temperature, the solubility of pectin will rise due to the higher diffusion coefficient, which causes the extraction rate to accelerate (16). In addition, the breakdown of plant cells at high temperatures occurs maximally (17). This results in the release of pectin from plant tissues so that more pectin is produced.

3.2 Effect of duty cycles on yield value

In this study, the yield value increased as the extraction duty cycle increased. A higher pectin yield value is associated with a higher extraction duty cycle. The highest yield was obtained at 100% duty cycle. Duty cycle is the ratio of the time when the sonicator is active to the entire extraction time. According to Kobus *et al.* [18], ultrasonic vibrations of short duration will result in the formation of fewer cavitation bubbles. This will lead to stronger shock waves, resulting in greater damage to the plant tissue. The solute may diffuse into the surrounding solvent during the pause that occurs after the cavitation bubbles collapse. In addition, the active and inactive times can prevent saturation effect so that the extraction process runs optimally [11].

3.3 Effect of times on yield value

The yield value in this study rose as the extraction duration increased. The pectin yield value increases with the length of the extraction process. The highest yield was obtained at an extraction time of 40 minutes. The yield will increase with the length of the extraction time. Longer extraction time causes greater contact opportunities between the solvent and the material so that it can provide a greater opportunity for the solvent to hydrolyze protopectin [19]. However, a reduction in extraction yield may result from extraction time that above optimum limit. Extraction can be done if the solvent is not saturated [20, 21].

3.4 Effect of amplitudes on yield value

In this study, the yield value increased as the extraction amplitude increased. The higher the amplitude, the higher the pectin yield value. The highest yield was obtained at 100% amplitude. Higher ultrasonic amplitudes results in greater yields. Cavitation bubbles will form in the material solution as a result of the ultrasonic waves. When the cavitation bubbles break near the cell wall, shock waves and liquid jets are formed that may result in the cell wall breaking. When the cell wall ruptures, the components within the cell will come out and combine with the solution [22]. Increasing the amplitude will increase the cavitation energy, solvent diffusion into the material, and mass transfer rate, which increased the yield [22].

3.5 Characteristics of extracted pectin

Pectin characteristics were analyzed using pectin that had the highest yield value, which was extracted under conditions of 60 °C, 100% duty cycle, 40 minutes, and 100% amplitude (Table 2). The degree of esterification is the percentage of D-galacturonic acid residues whose carboxyl groups are esterified with ethanol. The degree of esterification is a parameter that affects the quality of pectin and its application. The degree of esterification affects pectin's solubility and gel formation, among other characteristics [24]. The pectin extracted in this study had an esterification degree of 79.52%. This shows that Pacitan sweet orange peel pectin is included in high methoxyl pectin because the degree of esterification value is more than 50%. High methoxyl pectin can form gels in environments with high sugar content, thus, in the food sectors, it is utilized as a thickener and gelling agent [25].

Table 2. Characteristics of the extracted pectin

Parameters	The Pectin (T=60; D=100%; t=40; A=100%)	IPPA Standard [23]
Degree of Esterification (%)	79.52 ± 0.09	HMP min. 50% LMP max. 50%
Water Content (%)	8.20 ± 0.00	Max. 12%
Ash Content (%)	2.82 ± 0.00	Max. 10%
Color		
L* (*lightness*)	38.33	-
a* (*redness*)	1.04	-
b* (*yellowness*)	9.78	-

HMP (High Methoxyl Pectin), LMP (Low Methoxyl Pectin)

The moisture content of pectin affects the shelf life of pectin. Materials with low moisture content are relatively more stable during storage than materials with high moisture content [26]. The maximum allowable moisture content for commercial pectin in IPPA is 12% [23]. The moisture content of the pectin produced in this research is 8.20%, thus meeting the IPPA standard.

The ash content of pectin indicates whether inorganic materials remain in the pectin after combustion. Ash content impacts the pectin's purity. The higher the ash content of pectin, the lower the purity level of pectin [27]. The maximum allowable ash content for commercial pectin in IPPA is 10% [23]. The resulting pectin has an ash content of 2.20%. The ash content has met the IPPA standard. The final pectin is displayed in Figure 1.

Figure 1. Visual of pectin

The extracted pectin in this study has an L* value of 38.33; an a* value of 1.04 that shows a reddish color; and a b* value of 9.78, which shows a yellow color. According to Rahmanda K.W et al. [28] the difference in color of orange peel's pectin is related to the presence of pigments in the orange peel raw material. During extraction, the pigments contained in the orange peel are extracted into the pectin. The ethanol is used to determine the color of the pectin. Ethanol can dissolve color during the pectin washing process, so maximum washing will increase the brightness of the resulting pectin. Meanwhile, the yellow color of the pectin is influenced by the drying process.

FTIR spectrum measurement serves to identify the functional groups present in pectin that was extracted from sweet orange peel (Figure 2). The presence of hydroxyl groups (-OH) is shown at wave number 3350.01 cm^{-1}. The aliphatic C-H group can be detected at wave number 2923.61 cm^{-1}. In addition, the typical absorption of pectin is also shown at wave number 1735.01 cm^{-1}

which is the absorption of carbonyl group (C=O). The wave number 1439.60 cm^{-1} shows the presence of methyl group (CH_3). The absorption at wave number 1144.34 cm^{-1} indicates the presence of C-O bonds in alcohols, carboxylic acids, and esters. The measurement results of the FTIR spectrum of pectin in this study are similar to the measurement results of the FTIR spectrum of pectin conducted by Méndez *et al.* [29].

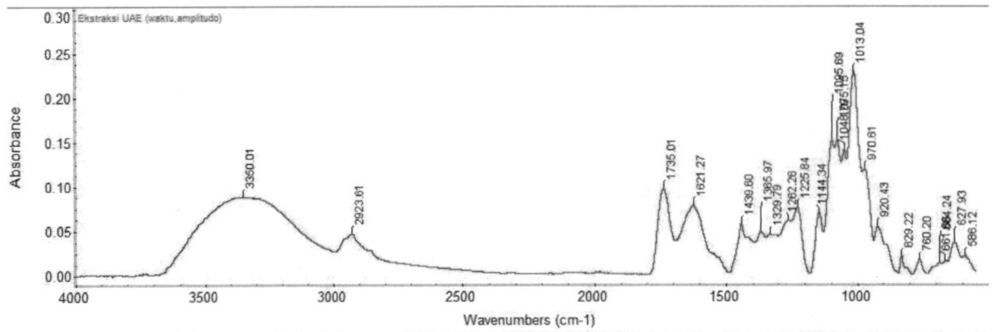

Figure 2. FTIR spectrum of the extracted pectin

4. Conclusion

The yield of pectin produced by the ultrasonic-assisted extraction method of extracting it from sweet orange peel depends on the operating conditions. Extraction at 60 °C for 40 minutes, duty cycle 100%, and amplitude 100% resulted in the highest pectin yield of 4.23%. The IPPA quality requirement was met by the pectin that was extracted from Pacitan sweet orange peel using an ultrasound-assisted extraction process.

Acknowledgements

The authors would like to thank the Innovative Research Grant of the Faculty of Agricultural Technology, Universitas Gadjah Mada (2730/UN1/FTP.1.3/SET-D/KU/2021) for funding this research work.

References

[1] Badan Pusat Statistik. Produksi Tanaman Buah-buahan, 2021-2023 [Internet]. 2024 [cited 2024 Sep 16]. Available from: https://www.bps.go.id/id/statistics-table/2/NjIjMg==/produksi-tanaman-buah-buahan.html

[2] Millati R, Nurrihadini OD, Suroto DA, Cahyari K. Waste Refinery Program in Indonesia: Characterization of Waste from "Gemah Ripah" Fruit Market as Feedstock for Biogas Production. Proceedings of International Symposium on Sustainable Energy and Environmental Protection (ISSEEP). 2009;1–6.

[3] M'hiri N, Ioannou I, Ghoul M, Boudhrioua MN. Proximate Chemical Composition of Orange Peel and Variation of Phenols and Antioxidant Activity During Convective Air Drying. Journal of New Science. 2015;9:881–90.

[4] Belkheiri A, Forouhar A, Ursu AV, Dubessay P, Pierre G, Delattre C, et al. Extraction, Characterization, and Applications of Pectins from Plant By-products. Applied Sciences . 2021;11(14):1–25.

[5] Wang W, Ma X, Xu Y, Cao Y, Jiang Z, Ding T, et al. Ultrasound-assisted Heating Extraction of Pectin from Grapefruit Peel: Optimization and Comparison with the Conventional Method. Food Chem. 2015;178:106–14.

[6] Tuhuloula A, Budiyarti L, Fitriana EN. Karakterisasi Pektin dengan Memanfaatkan Limbah Kulit Pisang Menggunakan Metode Ekstraksi. Konversi. 2013;2(1):21–7.

[7] Prakash Maran J, Sivakumar V, Thirugnanasambandham K, Sridhar R. Optimization of Microwave Assisted Extraction of Pectin from Orange Peel. Carbohydr Polym. 2013;97(2):703–9.

[8] Adhiksana A. Perbandingan Metode Konvensional Ekstraksi Pektin dari Kulit Buah Pisang dengan Metode Ultrasonik. Journal of Research and Technology. 2017;3(2):80–8.

[9] Vinatoru M. An Overview of the Ultrasonically Assisted Extraction of Bioactive Principles from Herbs. Ultrasonics Sonochemistry 8. 2001;303–13.

[10] Al-Dhabi NA, Ponmurugan K, Maran Jeganathan P. Development and Validation of Ultrasound-assisted Solid-liquid Extraction of Phenolic Compounds from Waste Spent Coffee Grounds. Ultrason Sonochem. 2017;34:206–13.

[11] Kumar K, Srivastav S, Sharanagat VS. Ultrasound Assisted Extraction (UAE) of Bioactive Compounds from Fruit and Vegetable Processing By-products: A Review. Ultrason Sonochem. 2021;70:1–11.

[12] Freitas de Oliveira C, Giordani D, Lutckemier R, Gurak PD, Cladera-Olivera F, Ferreira Marczak LD. Extraction of Pectin from Passion Fruit Peel Assisted by Ultrasound. LWT. 2016;71:110–5.

[13] Xu Y, Zhang L, Bailina Y, Ge Z, Ding T, Ye X, et al. Effects of Ultrasound and/or Heating on the Extraction of Pectin from Grapefruit Peel. J Food Eng. 2014;126:72–81.

[14] Grassino AN, Brnčić M, Vikić-Topić D, Roca S, Dent M, Brnčić SR. Ultrasound Assisted Extraction and Characterization of Pectin from Tomato Waste. Food Chem. 2016;198:93–100.

[15] Hosseini SS, Khodaiyan F, Kazemi M, Najari Z. Optimization and Characterization of Pectin Extracted from Sour Orange Peel by Ultrasound Assisted Method. Int J Biol Macromol. 2019;125:621–9.

[16] Nurhayati N, Maryanto M, Tafrikhah R. Ekstraksi Pektin dari Kulit dan Tandan Pisang dengan Variasi Suhu dan Metode. Jurnal Agritech. 2016;36(03):327–34.

[17] Yu M, Xia Y, Zhou M, Guo Y, Zheng J, Zhang Y. Effects of Different Extraction Methods on Structural and Physicochemical Properties of Pectins from Finger Citron Pomace. Carbohydr Polym. 2021;258:1–9.

[18] Kobus Z, Krzywicka M, Starek-Wójcicka A, Sagan A. Effect of the Duty Cycle of the Ultrasonic Processor on the Efficiency of Extraction of Phenolic Compounds from Sorbus Intermedia. Sci Rep. 2022;12(1).

[19] Kataren S, Suastawan I. Effect of Vanilla Fruit Quality Level and the Ratio of Material to Solvent on the Yield and Quality of Oleoresin produced. Department of Agricultural Industrial Technology. 1995;3:161–71.

[20] Wusnah W, Zulnazri Z, Sulastri S. Effect of pH and Extraction Time on the Characteristics of Pectin from Brown Skin. Journal of Unimal Chemical Technology. 2015;4(2):27–35.

[21] Perina I, Satiruiani, Soetaredjo FE, Hindarso H. Ekstraksi Pektin dari Berbagai Macam Kulit Jeruk. Widya Teknik. 2007;6(1):1–10.

[22] Budiastra IW, Mardjan SS, Azis AA. Pengaruh Amplitudo Ultrasonik dan Waktu Ekstraksi Terhadap Rendemen dan Mutu Oleoresin Pala. Jurnal Keteknikan Pertanian. 2021;8(2):45–52.

[23] Lubis N, Nuraeni ARF, Mahmudah R, Soni D, Abdillah MN. Penetapan Jenis Ekstraksi Terhadap Karakteristik Pektin Buah Pandan Laut (Pandanus Ordorifer) dari Pantai Santolo Garut. J Sains dan Teknologi Pangan. 2022;7(5):5499–509.

[24] Suwoto, Septiana A, Puspa G. Extraction of Pectin from Super Red Dragon Fruit Peel (Hylocereus costaricencis) with Variation of Extraction Temperature and Solvent Type. Scientific Journal of Chemical Engineering UNPAM. 2017;1(2).

[25] Mellinas C, Ramos M, Jiménez A, Garrigós MC. Recent Trends in the Use of Pectin from Agro-waste Residues as a Natural-based Biopolymer for Food Packaging Applications. Materials. 2020;13(3):1–17.

[26] Pardede A, Ratnawati D, Martono HP A. Ekstraksi dan Karakterisasi Pektin dari Kulit Kemiri (Alleurites Mollucana Willd). Media Sains. 2013;5(1):1–6.

[27] Sulihono A, Tarihoran B, Agustina TE. Effect of Time, Temperature, and Solvent Type on Pectin Extraction from Bali Orange Peel (Citrus maxima). Journal of Chemical Engineering. 2012;18(4):1–8.

[28] Rahmanda K.W AF, Sukardi S, Warkoyo W. Karakterisasi Sifat Fisikokimia Pektin Kulit Jeruk Keprok Batu 55 (Citrus reticulata B), Jeruk Siam (Citrus nobilis var. microcarpa), Jeruk Manis Pacitan (Citrus sinensis L, Jeruk Nipis (Citrus aurantifolia swigle), dan Jeruk Lemon (Citrus limon L) yang Tumbuh di Kota Batu. Food Technology and Halal Science Journal. 2021;4(2):124–41.

[29] Méndez DA, Fabra MJ, Gómez-Mascaraque L, López-Rubio A, Martinez-Abad A. Modelling the Extraction of Pectin Towards the Valorisation of Watermelon Rind Waste. Foods. 2021;10(4):1–16.

IOP Conf. Series: Earth and Environmental Science **1460** (2025) 012024 doi:10.1088/1755-1315/1460/1/012024

Characteristic of Goat Milk Soap Formulation with Addition Betel Leaves as a Natural Antiseptic

Dearestantrianto Hadits Fardana*, Rudy Hartanto, and Ahmad N. Al-Baari

Master of Program Animal Science, Faculty of Animal and Agriculture Science, Universitas Diponegoro, Semarang, Indonesian

*E-mail: dhfardana.aris@gmail.com

Abstract. This research aims to examine the interaction between the amount of added goat's milk and betel leaves with different concentrations on the characteristics of goat's milk soap. The process of making goat milk solid soap uses the cold process method and is cured for 2 weeks. The soap making method uses unconsumed goat's milk and is enriched with betel leaves as a natural antiseptic. The research was conducted using a Completely Randomized Design (CRD) with a 4 x 3 factorial pattern and 3 replications, so there were 12 treatment combinations. The research results were data analyzed using Analysis of Variance (ANOVA). The treatment begins by adding 50 mL (S1), 75 mL (S2), 100 mL (S3), 125 mL (S4) of goat's milk to 700 mL of soap base solution. Then add betel leaf extract according to the treatment, namely 0 g (D1), 25 g (D2) and 50 g (D3). The research results showed from the parameters of the water content test, foam stability test, that there was no significant interaction ($P>0.05$) between the level of milk addition and betel leaf extract, but there was a significant interaction ($P \leq 0.05$) with the test pH, free alkali and fatty acid test. Based on the overall results, it can be concluded that from the soap formulation that has been made, goat's milk soap with the addition of 50 g betel leaf extract is better and can be combined with the addition of 100 mL of goat's milk.

Keywords : goat milk soap, betel leaves, natural antiseptic

1. Introduction

Goat milk is a national commodity that always gets the attention of the national dairy industry because it is believed to have many advantages, such as nutritional advantages, which have an impact on increasing the effectiveness of absorption in the intestines and also to help the body's immunity. However, behind these advantages, there are factors that affect the quality of goat milk, including the type of livestock, type of feed, age of livestock, and milking process. The quality of fresh goat milk is also determined by the method and duration of storage. The shelf life of fresh goat milk tends to be short, around 5-6 hours at room temperature. Processing fresh goat milk into other food products will extend the storage time, for example into powdered milk, Ultra High Temperature (UHT) milk, cheese, yogurt, and butter. Another alternative process that can be done to utilize fresh goat milk is to process it into bar milk soap for body cleansing. Milk in fresh condition can be a potential for processing bar soap.

Commercial body cleansing soap is actually a synthetic detergent product. Mass-produced bath soap usually has a high pH because it contains chemicals that may be harmful to the skin if used excessively. Compared to goat milk soap, soap containing detergent is more popular because it foams and is effective as a dirt remover. However, detergent is also alkaline. Alkali can change the pH balance of the skin, causing the skin to become dry. The skin has a layer that acts as a protector against bacteria, viruses, toxins, irritants, and other things that have the potential to penetrate the skin. Anything that disrupts the pH of this protective layer can cause skin inflammation. The pH value of balanced skin is 4.7 - 5.75.

Goat's milk is rich in natural beta hydroxyl acid content and can be used as a mixture for body scrubs. This mixture can function as a peeling that erodes dirt and dead skin cells. The protein content in milk is also useful as a supply of nutrients that function to moisturize and coat the surface of the skin to make it smoother and healthier [1]. Milk soap generally has advantages in its use for the body, such as the fat content in milk is quite a lot to be used on the skin.

Goat's milk is an ingredient that can be used in making soap because it has natural antibacterial properties. The natural antibacterial properties in milk are safe if used continuously. It stated that fermented goat's milk concentrate has an inhibitory power equivalent about 60% of chloramphenicol (chemical antibacterial) against *Staphylococcus aureus* bacteria [2]. Therefore, the addition of goat milk with the addition of herbs such as green betel leaves has the potential to help increase production value in the economic world. In this study, a goat milk soap formula will be prepared made from pure goat milk and other natural ingredients such as coconut oil and olive oil. Betel leaves can be added as a natural antiseptic and pH lowering agent (pH value according to soap quality requirements based is 9 – 11 [3]). In addition, green betel leaves are known to have antibacterial effects against several types of bacteria such as *Escherichia coli*, *Salmonella sp*, *Staphylococcus aureus*, *Klebsiella*, *pasteurella*, and can kill *Candida albicans*. Until now, there is only a little literature on goat milk and betel leaves, therefore this study is very important to carry out.

Based on the background and problems that have been presented, the purpose of this study is to examine the interaction between the amount of goat milk and betel leaves added with different concentrations on the characteristics of goat milk soap. This research is expected to contribute to the community environment to be able to utilize goat milk and betel leaves used for making soap.

2. Materials and Methods

2.1. Materials

Making soap is used by utilizing unconsumed goat's milk and enriched with betel leaves as a natural antiseptic. The study was conducted using a Completely Randomized Design (CRD) with a factorial pattern of 4 x 3 and 3 replications, so that there were 12 treatment combinations. The study was divided into 3 stages, namely the preparation stage, the treatment and collection stage, and the final stage, namely statistical data processing.

2.2. Methods

The first stage of preparation begins with the provision of tools for making soap, namely goat's milk, betel leaves, coconut oil, palm oil, NaOH solution, and soap making equipment. Preparation in the form of betel leaves is weighed as much as 1000 g. Making betel leaf extract, namely first the betel leaves are washed and then finely chopped, after which the fine betel leaves are extracted with 96% ethanol solvent as much as 500 g. The maceration process was carried out using 96% technical ethanol for 48 hours with solvent replacement every 24 hours accompanied by stirring, then the maceration filtrate was filtered with filter paper to obtain pure filtrate which was then evaporated using a rotary vacuum evaporator to obtain a thick extract [4].

The second stage begins with the manufacture of goat milk soap with different amounts of goat milk with the aim of knowing the best amount of goat milk addition. Treat the addition of goat milk as much as 50 mL (S1), 75 mL (S2), 100 mL (S3), 125 mL (S4) in 700 mL of soap base solution with the cold method (cold process). Making soap using the cold method is done by first making a 75 g NaOH solution mixed into a 250 mL aquades solution in a measuring cup, dissolved until homogeneous then transferred again into an empty container provided and aired for 5 minutes. Then mix 150 mL of palm oil and 300 mL of CCO oil into a container.

The next treatment is carried out by adding betel leaf extract according to the treatment, namely 0 g (D1), 25 g (D2) and 50 g (D3), then add goat milk according to the treatment and mix until homogeneous, the NaOH solution is poured and mixed again until the texture thickens within 10 - 15 minutes, if it has thickened, it is transferred into a soap mold container that is already available and the curing period is waited for approximately 2 weeks.

The characteristic variables of soap measured in the results of making natural soap are: Water content test (maximum 15% [5]) using the ASTM D460-91 method, namely the test sample is weighed 5 g into a petri dish that has been dried in an oven at a temperature of (105 ± 2.0) ° C for 30 minutes. Then the test sample is heated in an oven at a temperature of (105 ± 2.0) ° C for 6 hours. After that, it is cooled in a desiccator and weighed. Measurements are taken until a constant weight is obtained [3]. Then calculated using the formula that concerning bath soap [5]. The formula for calculating the water content value is:

$$\text{Water content} = \frac{\text{W1-W2 x 100\%}}{\text{W}}$$

W1 = soap weight + petri dish weighing (g),

W2 = soap weight + petri dish weighing after dry (g)

W = soap weight (g)

The acidity level (optimal pH 9-11 [5]) was measured with a universal pH indicator on all formulations on the 14th day. A 1g soap sample was dissolved in a container and the measured pH was observed.

Foam stability (optimal 60-70% [5]) with a 1g bar soap sample was put into a graduated tube containing 10 mL of distilled water and then closed. The tube was shaken for 20 seconds and the height of the foam formed was immediately measured.

Free alkali (maximum 0.1% [5]) with a soap sample weighed about 5 g, then put into a 250 mL beaker. Next, 100 mL of 96% alcohol and a few drops of phenolphthalein indicator solution were added. Then heated on a water bath for 30 minutes until boiling. After the solution is purple, it is titrated with 0.1 N HCl solution in alcohol until the purple color disappears. The formula used to calculate free alkali is:

$$\text{Free alkaline rate (\%)} = \frac{\text{v x N x 0,0561}}{\text{w}} \text{ x 100 \%}$$

V = Volume of HCL in titration (mL)
N = Normality of HCL (N)
W = Sample Weight (g)

Free Fatty Acid Test was conducted with 10 mL of 0.1 N potassium biphthalate primary standard solution pipetted into an Erlenmeyer flask with the addition of 25 mL of distilled water and 3 drops of 1% phenolphthalein indicator. The solution was titrated with 0.1 N NaOH secondary standard solution until a pink color change occurred. Determination of Fatty Acid Level was conducted with a 10 g sample dissolved in distilled water, 25 mL of 96% ethanol and 3 drops of 1% phenolphthalein indicator were added, the sample was titrated with 0.1N NaOH solution which was indicated by a pink color change. The formula used to calculate free fatty acids is:

$$\text{Level of free fatty acids (\%)} = \frac{v \times N \times 282}{b} \times 100\ \%$$

V = Volume of KOH used (mL)
N = Normality of KOH used
b = Weight of test sample (mg)
282 = Equivalent weight of oleic acid ($C_{18}H_{34}O_2$)

3. Results

3.1. Data Analysis of Solid Soap Water Content Test with Betel Leaf Extract

Based on the data obtained from the results of the analysis of the water content of solid soap, it is known that there is no significant interaction ($P>0.05$) between the levels of milk and betel leaf extract addition, but each main factor has a significant effect ($P\leq0.05$) on the value of the soap water content. These results indicate that all treatments are in accordance with the Indonesian National Standard [5] because they do not exceed 15%. Water content testing needs to be done because water content will affect the quality of the soap. The amount of water content can affect the solubility of soap in water when used. In addition, water content also affects the hardness of solid soap. The higher the water content of the soap, the softer the hardness of the soap will be, conversely, the lower the water content of the soap, the harder the hardness of the soap will be (6) .

Table 1. Analysis results of water content test

Betel Leaf use (g)	50 mL	75 mL	100 mL	125 mL
D1	6.36 ± 0.58	5.16 ± 0.62	6.95 ± 0.58	5.59 ± 0.13
D2	7.67 ± 0.81	8.51 ±1.17	11.37 ± 4.00	7.14 ± 2.20
D3	6.67 ± 0.03	5.97 ± 1.17	7.86 ± 0.44	7.06 ± 0.50

*) Different superscripts in the same row and column indicate a significant difference $P<0.05$.

The addition of 25 g of betel leaves shows a higher number compared to without the addition of betel leaves and the addition of 50 g. Meanwhile, the use of 100 mL of goat's milk has a higher water content than the others. The water content contained in the bar soap itself is influenced by several factors, including the water content of the raw materials and the concentration of NaOH used in making soap [7]. The high water content in the treatment of adding 25 g of betel leaf extract can be due to the different concentrations of goat's milk and its water-absorbing properties. The addition of materials that have hygroscopic properties can cause the soap to easily absorb water vapor from the air, resulting in a water content that is too high [8]. The storage period of milk soap can also affect the water content in the soap because during storage, the soap will also experience evaporation so that the water content cannot be uniform. The evaporation process can affect the water content of the soap (9).

The addition of milk to the soap shows that the addition of 100 mL is higher than other treatments. The storage period of soap can also affect the water content of the soap due to the evaporation process [9]. Soap that has a fairly low water content compared to others shows that the water content in the soap begins to decrease during the curing process. Fairly low soap water content will indicate the hardness of the solid soap produced, because the more water content added to the soap will make the soap solubility when used, so that the soap will quickly experience weight and dimension shrinkage [10].

3.2. Data Analysis of Soap Foam Stability Test

Based on the data obtained from the results of the solid soap foam stability test analysis, it is known that there is no significant interaction (P>0.05) between the levels of goat's milk and betel leaf extract addition, but each main factor has a significant effect (P≤0.05) on the soap water content value. These results indicate that all treatments are in accordance with the Indonesian National Standard [5] because they are in the range of 60 - 70%. The foam stability in all treatments of solid soap is quite high in the betel leaf treatment with the addition of 25 g of betel leaves when using 75 mL of goat's milk. The soap foam produced in solid soap is influenced by several factors of the ingredients used such as the presence of active soap ingredients or surfactants, foam stabilizers such as betaine, and other soap ingredients such as the type of oil used [11].

Table 2. Analysis results of foam stability test

Betel leaf use (g)	50 mL	75 mL	100 mL	125 mL
D1	68.00 ± 7.21	75.00 ± 4.00	69.33 ± 4.04	69.33 ± 5.13
D2	71.67 ± 5.77	67.33 ± 6.42	70.67 ± 0.57	70.67 ± 5.13
D3	64.67± 1.52	68.33 ± 2.88	64.33 ± 3.05	67.67 ± 2.52

*) Different superscripts in the same row and column indicate a significant difference P<0.05.

Foam stability is influenced by the free fatty acid content contained in the soap making process which can inhibit the cleaning power of the soap which is indicated by the lack of foam produced [12]. The foam stability show that is influenced by long stirring and the amount of water ratio and the small content of active ingredients contained in making solid milk soap [13]. The addition of oil also affects the stability of soap foam which states cause that coconut and palm oil contain high levels of palmitic acid and lauric acid so that the foam in solid soap becomes stable [14].

The addition of betel leaf levels shows a decrease in the results of the soap foam stability test. Foam stability is influenced by the material factors used, namely the acid content in the ingredients in the form of stearic acid and palmitic acid. Coconut oil, namely lauric acid and casein contained in milk help in the formation of stable foam [15]. The low foam stability when giving 50 g of betel leaf extract at a milk level of 125 mL indicates that the saponification reaction in the soap is starting to decrease. This is also show that the higher the foam produced, the better the soap will form a saponification reaction [16]. Increasing the amount of milk used does not affect the results of the soap foam stability test. This shows that the amount of goat milk added to the soap is quite stable.

Milk that contains casein which functions as a foam stabilizer. The addition of oil can affect the stability of soap foam [10]. This is also supported the research that states oil will affect the stability of the foam, namely coconut and palm oil [14]. Coconut and palm oil contain high levels of palmitic acid and lauric acid, making the foam in solid soap stable.

3.3 Data Analysis of pH
Based on the data obtained from the results of the solid soap pH test analysis, the difference in betel leaf extract and the level of goat's milk is known to provide a significant interaction ($P < 0.05$) and each main factor has a significant effect ($P \leq 0.05$) on the pH value of the soap. The results show that the higher the addition of betel leaves with the amount of milk, the pH increases. The ideal pH is seen in the combination of using 100 mL of milk with 50 g of betel leaf extract, namely pH 11. This happens because the amount of alkali in the formulation is in accordance with the standard requirements. The greater the amount of alkali in the soap, the higher the pH value of the soap [17].

Table 3. Analysis results of pH

Betel leaf use (g)	50 mL	75 mL	100 mL	125 mL
D1	11	11.67	11	11
D2	11	12	12	11.33
D3	11	11	11	12

*) Different superscripts in the same row and column indicate a significant difference $P<0.05$.

The pH test obtained shows that the formulation of solid soap preparations from betel leaf extract is in accordance between 9-11. In this study, the more betel leaf extract given to the soap, the more alkaline the soap pH. The high pH value in the 75 mL and 100 mL milk formulations with the addition of 25 g of betel leaf extract and the addition of 125 mL of milk with the addition of 50 g of betel leaf extract is because when making soap, the NaOH given is likely more than the soap formulation that should be made. Soap that has a high pH can increase the growth of *Propiombacterium* bacteria and make the skin potentially dry. This happens because at a high pH the nature of the soap can swell keratin which will make it easier for bacteria to enter and cause the skin to become dry and cracked, while soap with a low pH can cause skin irritation [18].

3.4. Data Analysis of Free Fatty Acid Test
Based on the data obtained from the results of the analysis of the free fatty acid test of solid soap, it shows that the difference in betel leaf extract and the level of goat's milk is known to provide a significant interaction ($P \leq 0.05$) and each main factor has a significant effect ($P \leq 0.05$) on the fatty acid value of the soap. Variations in the addition of green betel leaf extract to goat's milk show that the higher the addition of green betel leaf extract and the amount of goat's milk added, the lower the free fatty acid levels. The use of 50 g of betel leaves with 50 mL of goat's milk showed results that were no different from the addition of 75 mL, 100 mL and 125 mL. Although there tends to be a decrease in fatty acids, the use of high betel leaves can cause fatty acid levels to

decrease, this can happen because the ethanol extract of betel leaves positively contains secondary metabolite compounds such as flavonoids, phenols, and saponins which function as natural antioxidants [19]. The decreasing fatty acid levels indicate that the hydrolysis process between triglycerides and water that can produce glycerol and free fatty acids has not occurred properly. High levels of free fatty acids occur because of the hydrolysis process between triglycerides and water which produces glycerol and free fatty acids [20].

Table 4. Analysis result of free fatty acid test

Betel leaf use (g)	50 mL	75 mL	100 mL	125 mL
D1	0.31 ± 0.31^{b}	0.85 ± 0.20^{a}	0.34 ± 0.18^{b}	0.25 ± 0.08^{b}
D2	0.18 ± 0.15^{b}	0.18 ± 0.06^{b}	0.24 ± 0.08^{b}	1.04 ± 1.00^{a}
D3	$0.18^{b} \pm 0.15^{b}$	0.18 ± 0.06^{b}	0^{b}	0.12 ± 0.20^{b}

*) Different superscripts in the same row and column indicate a significant difference P<0.05.

The fatty acid results are in accordance standard because the value is below 2.5%. The presence of free fatty acids in soap is because the fatty acids do not undergo a saponification reaction or saponification by NaOH or KOH bases. The results show that by adding extract to milk, the soap's free fatty acid content decreases. It also stated that the higher the addition of green betel leaf extract, the higher the free fatty acid content [20]. The treatment of adding goat's milk and betel leaf extract also showed that the higher the level given, the lower the fatty acids. This happens because of the hydrolysis process between triglycerides and water which produces glycerol and free fatty acids. Low fatty acids indicate a reaction with alkali during the saponification process. This is in accordance with the opinion that free fatty acids in oil and the amount of base used in soap making will affect the levels of free fatty acids produced by the soap [18].

Table 5. Analysis result of free alkaline test

Betel leaf use (g)	50 mL	75 mL	100 mL	125 mL
D1	0.05 ± 0.08^{b}	0^{b}	0^{b}	0^{b}
D2	0^{ab}	0.10 ± 0.10^{ab}	0^{b}	0^{b}
D3	12 ± 0.19^{ab}	0^{b}	0.16 ± 0.03^{a}	0.11 ± 0^{ab}

*) Different superscripts in the same row and column indicate a significant difference P<0.05.

3.5. Data Analysis of Alkaline Test

Based on the data obtained from the results of the solid soap free alkaline test analysis, it shows that the difference in betel leaf extract and the level of goat's milk is known to provide a significant interaction ($P < 0.05$) and each main factor has a significant effect ($P \leq 0.05$) on the soap alkaline value. Betel leaf extract mixed with goat's milk shows quite high alkali compared to other parameters. The combination of adding 125 mL of milk with 50 g of betel leaf extract gives an alkali result of 0.11 and is no different from the addition of 50, 75, 100 mL of milk to 50 g of betel leaf extract. High alkaline in making soap indicates that betel leaves in mixing goat's milk do not dissolve well. The addition of green betel leaf extract to making solid soap showing that phenol is soluble in alkaline hydroxide which can cause a decrease in alkali because it can dissolve well [20].

The evaluation of the free alkali content of goat's milk solid soap with betel leaf extract obtained results from 12 samples, only 3 soap samples had a free alkaline content that did not meet the requirements, namely a maximum of 0.1% in 50 g betel leaf extract solid soap with the addition of 50 mL, 100 mL and 125 mL. The cause of the high amount of free alkaline in soap is because the amount of alkali that reacts in the saponification process is still small compared to the amount of alkali added so that there is excess alkali in the preparation [6]. The interaction between oil and alkaline will increase with the stirring time, so that the reaction shows equilibrium will reduce the free alkaline content in the soap [10]. The variation in the addition of green betel leaf extract shows that the higher the addition of green betel leaf extract, the lower the alkaline content. This is because betel leaf extract contains phenol, phenol in green betel leaf extract is an alcohol compound that functions as a natural antioxidant that is soluble in acetone, alcohol, glacial acetic acid and alkali hydroxide [20].

4. Conclusion

Based on the overall results, it can be concluded that from the soap formulation that has been made, goat's milk soap with the addition of 50 g of betel leaf extract is better and can be combined with the addition of 100 mL of goat's milk. This composition also help the soap for killed bacteria since its combine with the betel leaf that has antibacteria.

References

[1] Endang, P., E. Vebriyanti and E. Latifa Sri Suharto. 2012. Virgin Coconut Oil Goat Milk Soap Can Improve Skin Health Through pH and Good Bacteria (Lactic Acid Bacteria) and Increase Community Income. Health Proceedings. Competitive Advantage II Seminar 1(2): 1-7.

[2] Wulansari, P. D. and D. Wijayanti. 2018. Physical, Chemical and Microbiological Characteristics of Goat Milk Soap. Journal of Food and Agroindustry. 8 (3): 145-153.

[3] SNI. 2016. SNI 3532:2016 About Bath Soap Congested. Jakarta: Standardization Council National.

[4] Pakpahan, R. A., S. Khotimah and M. Turnip. 2015. Effectiveness of Ethanol Extract of Betel Leaf (Piper betle L.) and Noni Fruit (Morinda citrifolia L.) as an Alternative Tofu Preservative. Journal of Protobiont. 4 (1): 115 – 119.

[5] SNI. 1994. Bath Soap Quality Standards. SNI 06-3532-1994. Jakarta: National Standardization Council. https://jdih.bsn.go.id/public_assets/file/59d303117735a0c718e4874099f44151.pdf.

[6] Pande, P. A. L. D. and E. I. Setyawan. 2021. Effect of NaOH Concentration and Stirring Time on Soap Characteristics in Opaque Aloe Vera (Aloe vera L.). Proceedings of the 2022 National Pharmacy Workshop and Seminar. 1 (1): 1-12.

[7] Khuzaimah, S., Tritisari, A., & Fertiasari, R. 2021. Purification of Used Cooking Oil in the Solid Soap Making Process. Journal of Agrofood, 3(2): 36–42.

[8] Marpaung, J.J.A., F.A. Dewi and E. Raswen. 2019. Transparent Soap Based on Pure Coconut Oil with the Addition of Papaya Fruit Flesh Extract. Journal of Agoindustry Halal. 5(2):161-170.

[9] Sukawati, Y., Warnida, H., and Artha, A. V. 2016. Formulation of Solid Bath Soap Preparation from Ethanol Extract of Tiwai Onion Bulbs (Eleutherine Bulbosa (Mill .) Urb .) Bulbs Ethanol Extract. Media Farmasi. 13(1): 14–22.

[10] Febriani, A , V Syafriana , H Afriyando , and Y. S. Djuhariah. 2020. The Utilization of Oil Palm Leaves (Elaeis guineensis Jacq.) Waste as an Antibacterial Solid Bar Soap. IOP Conf. Series: Earth and Environmental Science 572 (2020) 012038. doi:10.1088/1755-1315/572/1/012038

[11] Gianty, H. L., A. Kusmayadi and P. D. Wulansari. 2022. Chemical Characteristics of Bar Soap from Cow's Milk and Goat's Milk Using the Cold Process Method. Journal of Animal Husbandry Study Program, Ma'arif Nahdlatul Ulama University, Kebumen. 2(2): 89- 97.

[12] Dewiarum, S. and D. A. Wijayanti. 2024. Physical Quality of Goat Milk Solid Soap with the Addition of Cascara Coffee Skin Extract. Journal of Animal Husbandry. 21(1): 66-73.

[13] Damayanti, E., S. A. Sari, and S. A. Semeru. 2021. Liquid Soap from Red Onion Skin Extract. Proceedings of the National Chemistry Seminar (SNK). Department of Chemistry, Faculty of Mathematics and Natural Sciences, Surabaya State University.
[14] Saputri, K. S., Albari, A., and Nisak, S. C 2022. Effect of Oil Base on Characteristics and Cleaning Power of Transparent Soap from Salak Skin Extract (Salacca zalacca). Medical Science: Scientific Journal of Pharmacy. 7(2): 245-253.
[15] Lugiana, G. H., A. Kusmayadi and P. D. Wulansari. 2022. Chemical Characteristics of Bar Soap from Cow's Milk and Goat's Milk Using the Cold Process Method. Journal of Nusantara Animal Science. 2 (2): 89 – 96.
[16] Rahman, S., I. K. Nugraheni1 and Y. Ningsih. 2023. Characterization of Solid Soap Made from Waste Cooking Oil with the Addition of Green Betel Leaf Powder (Piper Betle L). Journal of Argo-Industrial Technology. 10(2): 27 – 39.
[17] Salatin, W. A., T. Sukmaningsih and F. D. Evadew. 2022. pH Value and Foaming Power of Solid Soap with the Addition of Goat Milk. Journal of Livestock Media. 24 (1): 1 – 6.
[18] Wasitaatmadja, S.M. 2007. Dermatology and Venereal Diseases. Fifth edition. Second printing. Jakarta: Faculty of Medicine, University of Indonesia.
[19] Kopong, M. V. U. and N. K. Warditiani. 2022. Potential of Green Betel Leaves (piper betle l.) and Red Betel Leaves (piper crocatum) as Antioxidants. Journal of Indonesian Multidisciplinary Scientific. 2 (3): 710 – 729.
[20] Martha, A., I. Hajar and A. Serlina. 2021. Optimization of Natural Antiseptic Solid Soap Making Formula with the Addition of Green Betel Leaf Extract (Piper betle L). Journal of Kinetics. 12(1): 51-59.

Food Waste Management in Food Stalls: Anticipation and Management

Alya Griya Cahyani, Karin Maula Werdani, Juna Fatimatuz Az-Zahra, NurTyas Anggraini, and Eksa Rusdiyana*

Agricultural Extension and Communication Study Program, Faculty of Agriculture, Sebelas Maret University (UNS), Indonesia

*E-mail: eksarusdiyana@staff.uns.ac.id

Abstract. Anticipatory steps are needed at the household and business unit levels to reduce food waste production. This research aims to: (1) Identify the efforts of eateries (small, medium, large scale) in determining their menus and preventing food waste, (2) Analyze the efforts in managing food waste from these eateries. This research uses a qualitative descriptive approach, with data collected through interviews and field observations. The informants consist of consumers, small-scale food stalls operators (serving 50-100 portions/day), medium-scale (100-250 portions/day), and large stalls (>250 portions/day) around the campus in Solo City. The validity of the data is determined using source and method triangulation, while data analysis employs an interactive model. Research results show that: (1) There is no difference between small, medium, and large stores in determining the daily menu they sell because every day they serve the same menu that has been their characteristic. Food waste prevention is carried out by regulating the number of menu items. (2) Food waste occurs due to unprocessed food materials and leftovers from consumers. Leftover food that is still good will be consumed by the owner in small stalls, while in medium and large stalls, the food is reheated to be sold the next day or given to employees. Food waste from consumers that is no longer fit for consumption is collected and utilized by gatherers for animal feed.

Keywords: Consumption, food, food waste, food stalls

1. Introduction

The problem of food waste is now a major issue in Indonesia as the first ranked Southeast Asian food waste producing country [1]. Food waste itself is a problem of food waste that occursbecause it is wasted during the distribution and consumption process. Indonesia produces 20.93million tons of food waste per year [2]. In addition, the National Waste Management Information System (SIPSN) also noted that in 2022, the amount of food waste in Indonesia reached 40.5% ofthe total waste volume. Based on the data from SIPSN, it would appear that food waste dominates over all other types of waste. According to Hasanah (2022) [3], the large amount of food waste isdriven by population density. In 2024, Indonesia's population reached 282 million people, an increase of 1.7 million people compared to the previous year. According to [4], Indonesia's population of 200 million people causes Indonesia to produce a very large food waste, even each resident produces 300 kg of food waste per year [5]. Food waste sourced from households accounted for 61% of all global food waste in 2019 [6]. Food waste can also come from restaurantssuch as restaurants, cafes, stalls, canteens, and the like [7]. The distribution of centralized food stalls such as food court complexes, markets, shopping centers, schools and campuses contributes significant food waste [8]. Most students choose to eat at food stalls or canteens compared to cooking themselves because it is considered more efficient and practical. Its proximity to the place of study and low prices are certainly the reasons students choose to eat at food stalls [9].

Solo City has a state campus and more than 10 private campuses with a student population of more than 37,459 students, this opportunity raises the potential for so many food stalls and boarding houses including the potential for foodwaste. According to Sari (2022) [10], the biggestchallenge in processing food waste is changing the perspective of business owners and the community about the importance of anticipating and processing food waste. Portion discrepancy is the most common reason for food waste behavior. In addition, the storage of food ingredients also needs to be considered, if there are leftover food ingredients, these materials can be reused if the storage is done properly and correctly. Students tend to try to take as many portions of food as possible at low prices, as well as food stalls trying to provide extra menus and portions [11]. Based on this background, this study aims to identify the efforts of eateries (small, medium, largescale) in determining their menus and preventing food waste, analyze the efforts in managing food waste from these eateries.

2. Methodology

This research uses a descriptive qualitative approach. Descriptive qualitative research method is a research method based on descriptive data processing. Descriptive qualitative research is conducted to explain existing research without providing manipulation of the variable data under study by conducting direct interviews (12). The research location was chosen purposively, namely stalls around the campus in Solo City, Central Java, with a focus on stalls that are approximately 1 kilometer from the campus area. Purposive sampling is the selection of research locations based on certain considerations such as the characteristics of respondents or characteristics that are in accordance with the objectives of the researcher (13). Data collection techniques are used to collect data according to the research procedures so that the required datais obtained. Data collection techniques are the most strategic step in research, because the main purpose of research is to collect data for analysis (14). Data collection was carried out through in- depth interviews, documentation, and involved observation. The main informants in this study are managers of small stalls, medium stalls, and large stalls around the Solo City campus. Solo City was chosen with the character of the campus environment and food stalls that tend to be affordable by all groups. The grouping of stalls is divided into 3 based on the scale of the business as well as daily food production which is determined based on the portion scale. In this study we took 3 stalls at each business scale and each stall we took one informant to interview, so the total informants we used were 9 main informants. Small-scale food stalls are food stalls that provide 50-100 servings per day, medium-scale food stalls are food stalls that provide 100-250 servings per day, while large-scale food stalls are food stalls that provide more than 250 servings per day. Supporting informants include students as consumers of food stalls around campus. The data that has been collected is then subjected to source and method triangulation tests to obtain data validity.

Triangulation test which is a test of reliability by checking data from various sources in various ways, and various times (15). Data analysis uses interactive analysis, where data collected through interviews, observations and documentation are arranged and organized into categories,descriptions, or patterns.

3. Results and Discussion

3.1 Menu Management in Food Stalls: Anticipating Food waste

Menu management in food stalls is a challenge, especially in anticipating and managing food waste. Shop owners need to adjust the amount and type of food served to suit consumer demand. Without good management, many food ingredients have the potential to be wasted, thus affecting profits and impacting the environment. Menu management is one of the anticipatory steps that can be taken by stalls owners to reduce the amount of food waste, the following data on menu management in stalls around the solo campus. This data is data on the total amount of leftover food in the stalls.

Table 1. Menu management in food stalls (small, medium, large) around Solo Campus.

Aspect	Small stalls	Medium stalls	Large stalls
Defining the menu	• Serving a different menu every day	• Serving the same menu every • day.	• Serving the same menu every day • Provide menus taken from suppliers.
Strategies for cooking up a storm	• The type of menu served always varies every day so that it is not boring.	• Served again the next day if the menu is still in good condition. • Leftover menu that does not last long is given to employees	• Served again the next day if the menu is still in good condition. • Leftover non-perishable menu items are given to employees for consumption

[a] Source: Primary Data Analysis, 2024

Based on Table 1, menu management carried out in small, medium and large stalls has several differences. In small stalls, the menu served is different every day which aims to make the menu served not boring and more varied. Menu variety is one of the factors that can attract consumers. Menu variation is defined as a variety offered by entrepreneurs or shop owners with the aim of satisfying consumers and being a factor that influences the purchase of a product [16]. The existence of a variety of menus that are served and different every day is also a strategy carried out by small stalls so that the menu provided does not cause leftovers. If it causes leftovers, the menu will be consumed by the shop owner and his family.

In medium and large stalls, the menu served every day is always the same with several typesof consistent diversity. This is because the food products served are typical or iconic food from the stalls itself and have been recognized by consumers. Although the diversity of the menu served every day is the same or consistent, menu diversity is one of the factors that has a significant role in business success. One of the purchasing decisions by consumers is determined by the menu diversity factor [17]. The strategy applied in medium and large stalls so that the menu does not cause leftovers is also the same, namely re-serving menus that are

IOP Conf. Series: Earth and Environmental Science **1460** (2025) 012025 doi:10.1088/1755-1315/1460/1/012025

still feasible or have long consumption durability the next day. Before being resold, the food has been confirmed in advancethe level of safety for consumption. In addition, menus that do not last long will be given to employees for personal consumption.

3.2 Food Waste Management in Food Stalls

Menu management carried out by stalls owners around campus can run consistently and sustainably. However, it is often unavoidable that food waste arises from consumers. For this reason, it is important to map the food waste management of each food stalls to reduce the amount of food waste that arises, both food waste due to unprocessed food ingredients, leftover food that is not consumed by consumers, and leftover food because it is not sold out. Table 2 belowpresents the food waste management of food stalls around the campus. This data is data on the total amount of leftover food in the stalls.

Table 2. Food waste management in food stalls around Solo Campus.

Aspect	Small stalls	Medium stalls	Large stalls
Leftover materials that are not processed	• Collected in trash bins, Cooperation with pig farmers	• Disposed of in a trash bin where someone will pick it up as animal feed.	• Stalls dispose of food waste • The stalls stores unprocessed ingredients in the frezer.
Consumer leftovers	• Collected in regular trash bins, cooperation with farmers for feed	• Collected in regular trash bins, cooperation with farmers for feed	• Collected in special food waste bins that will be taken by people as cattle or pig feed..
Unpurchased food	• Put together in a special food waste bin that will be picked up by the farmer.	• Leftover food is thrown in the trash, picked up by the farmer • Perishable food is used for the next day • Non-perishable food is distributed to employees • Stale food is thrown away.	• Some food is reheated or fried again. • Some are given to the employees. • There are stalls that dispose of leftover food from customers in the garbage that is put together, which will later be taken by people for animal feed.
Food Waste Management	• Important, because food waste can still be used for people in need, for example for animal feed	• It is important not to cause unpleasant odors so as not to disturb customers.	• It is important not to cause unpleasant odors so as not to disturb customers.

[a] Source: Primary Data Analysis, 2024

Based on Table 2, it can be seen that food waste management in small, medium and large stalls has various comparisons. The amount of organic food waste from small stalls has an average of 1-2 kg per day, from medium stalls has an average of 5-1 kg, and from large stalls has an averageof 30-15 kg. In small stalls, unprocessed ingredients, consumer leftovers, and unbought food are generally collected in bins and later harvested for animal feed. This approach not only helps reduce food waste, but also increases the economic value of the waste. Food waste management through utilization as animal feed can improve the family economy by up to 60%, as it reduces feed costs and maximizes existing resources. This initiative not only has a positive impact on the environment, but also has the potential to improve the economic welfare of the stalls owners [18].Evaluation of food waste in small stalls is minimal, with some stalls changing their daily menu toreduce food waste. Modification of side dishes can reduce the amount of food left over by consumers, so it can be an effort to reduce the amount of food waste generated [19].

In moderate stalls, unprocessed leftover ingredients are thrown into the trash with the assumption that they will be taken for animal feed. Some stalls also used leftover perishable ingredients to be used the next day. The evaluation was conducted by reducing food portions to reduce food waste. According to Handojo (2024) [20], reducing food waste is a priority, as the entire population and food supply chain will constantly generate waste if no real action is taken to address it. In large food stalls, food waste management involves more complex strategies. Leftover unprocessed food items can be stored in freezers, re-cooked, or discarded if they are nolonger viable. One effective way to prevent food waste is to store it in the freezer. Storage at low temperatures aims to prevent spoilage, extend the shelf life of foodstuffs, maintain their quality, and ensure stock availability at all times, so that foodstuffs such as vegetables, meat, and other processed products can last longer without losing their nutritional value or quality [21,22,23]. This reduces the amount of disposal of foodstuffs that have not been used in time. Temperature and storage time affect the growth of bacteria and fungi, especially in milkfish which was the object of the study [23]. Freezers also help avoid over-purchasing, which is often one of the main causes of food waste in culinary businesses. Consumer leftovers and unbought food are usually thrown away or recycled, with some given to employees. Evaluation is done in a more structuredmanner, such as cross-checking at night to determine whether to increase food portions or reduce food portions. Determining the right portion of food should also be considered. Excessive portions tend to trigger food waste [24]. It is necessary to consider in serving food to provide portions that are in accordance with consumer needs, so that there are no unused leftovers.

4. Conclusion

Small, medium and large-scale shopkeepers have the awareness to prevent food waste through good menu management. The emergence of food waste means reducing business profits. Small- scale stalls tend to change the daily menu to reduce food waste. In medium and large stalls, the menu served every day is always the same with several types of diversity that are consistent with the characteristics of the stalls. Small, medium and large-scale stalls managers have not managed food waste independently, food waste disposal is carried out in collaboration with pig farmers. In medium-scale stalls, the remaining unprocessed ingredients are thrown into the trash with the assumption that they will be taken for animal feed. Large-scale stalls tend to be more complex in handling food waste, where the remaining food ingredients that are not processed will be storedin the freezer, re-cooked, or discarded if they are no longer suitable. The hope is that small, medium, and large-scale stalls, in addition to reducing the amount of organic waste, can also utilize organic waste, such as making it into compost for plants.

IOP Conf. Series: Earth and Environmental Science **1460** (2025) 012025 doi:10.1088/1755-1315/1460/1/012025

References

[1] Maemunah, S., Agustin, D., & Carindra, R. R. (2024). Strategi Pengelolaan Sampah Makanan untuk Meningkatkan Kesadaran Publik Tentang Food Waste. *Jurnal Agrifoodtech*, *3*(1), 1-8.

[2] United Nations Environment Programme (UNEP). (2021). *Food Waste Index 2021.* Di akses pada : 14 September 2024. https://www-unep-org.translate.goog/resources/report/unep-food-waste-index-report-2021?_x_tr_sl=en&_x_tr_tl=id&_x_tr_hl=id&_x_tr_pto=tc

[3] Hasanah, A., Putri, E. I. K., & Ekayani, M. (2022). Kerugian ekonomi dari sisa makanan konsumen di rumah makan dan potensi upaya pengurangan sampah makanan. *Jurnal Pengelolaan Lingkungan Berkelanjutan (Journal of Environmental Sustainability Management)*, 45-58.

[4] Nasional, K. P. P. (2021). Laporan Kajian: Food Loss and Waste di Indonesia Dalam Rangka Mendukung Penerapan Ekonomi Sirkular dan Pembangunan Rendah Karbon. *Kementerian Perencanaan Pembangunan Nasional, Jakarta.*

[5] Evan, D. M., Natanael, I. N., & Aryani, D. I. (2022). Perancangan Konten Digital Edukasi Tentang Anti "Food Waste" Melalui Video Animasi. *DEKAVE*, *12*(3), 235.

[6] Swamilaksita, P. D., Harna, H., Suryani, F., Sa'pang, M., & Kusumaningtiar, D. A. (2023). Sosialisasi untuk Mengurangi Food Waste Rumah Tangga di Wilayah Kerja Puskesmas Parung Panjang Kabupaten Bogor. *IdeaPengabdian Masyarakat*, *3*(02), 39-43.

[7] Secondi, L., Principato, L., & Mattia, G. (2020). Can digital solutions help in the minimization of out-of-home waste? An analysis from the client and businessperspective. British Food Journal, 122(5), 1341–1359. https://doi.org/10.1108/BFJ-03-2019-0205

[8] Fajri, T. N., & Shauki, E. R. (2023). Potensi Food Loss dan Food Waste Pada UMKM: Mfca, Nudging Dan Neutralization Theory. *Jurnal Aplikasi Akuntansi*, *7*(2), 328-345.

[9] Rachmatina, L. D., & Windi Wulandari, S. K. M. (2018). *Analisis Hygiene Sanitasi Rumah Makan Di Sekitar Kampus Universitas Muhammadiyah Surakarta* (Doctoral dissertation, Universitas Muhammadiyah Surakarta).

[10] Sari, D. S., Bainus, A., Yulianti, D., & aditiany, S. (2022). Sosialisasi *From Waste To Food* Bersama Komunitas Wahana Usaha Jaya. *COMSEP: Jurnal Pengabdian Kepada Masyarakat*, *3*(3), 326-331. https://doi.org/10.54951/comsep.v3i3.330

[11] Timoty, N. S., & Yuliati, L. N. (2022). Peningkatan Kesadaran Masyarakat Dalam Pengurangan Food Waste Untuk Membentuk Cinta Lingkungan. *Policy Brief Pertanian, Kelautan, dan Biosains Tropika*, *4*(2), 245-248.

[12] Hanyfah, S. Fernandes, R.G., Budiarso, I. (2022). Application Of Descriptive Qualitative Method for Customer Data Processing Application in Car Wash. Seminar Nasional Riset dan Inovasi Teknologi (SEMNAS RISTEK), 1(1), 339-344.

[13] Ibrahim, T.J. (2020). Agricultural Socio-Economic Research Methods. Malang: UMM Press.

[14] Prasetyo, A. D., & Abduh, M. (2021). Increasing student learning activeness through discovery learning model in elementary school. Jurnal basicedu, 5(4), 1717-1724.

[15] Prasetyo, T., & Sumiyati, S. (2022). Pendampingan Me Widya, W., Arwin, A., Yuliana, Y., Ciamas, E. S., & Siahaan, R. F. B. 2021. Analysis of Employee Motivation at PT Citra Kesawan Selaras Medan. In Seminar Nasional Sains Dan Teknologi Informasi (SENSASI), 3(1), 491-495.

[16] Prasetyo, T. and Sumiyati, S., (2022). Pendampingan Menu Engineering di Warung Sate Mpok Siti (H. Saih) Dalam Meningkatkan Daya Saing Usaha Kuliner. ABDI MOESTOPO: Jurnal Pengabdian Pada Masyarakat, 5(1), 91-99. https://doi.org/10.32509/abdimoestopo.v5i1.1829

[17] Harahap, A. N. (2023). Pengaruh keragaman menu dan lokasi terhadap keputusan pembelian pada warung dapur khenan medan. Skripsi. STIE Eka Prasetya

[18] Khusniyah, K., Muis, Z. Z., & Kumalasari, D. A. (2022). Food waste study to improve family economy. *Jurnal Agrotek Ummat*, *9*(3), 209-221. https://doi.org/10.31764/jau.v9i3.8649

[19] Widyasari, A. L. (2021). Pengaruh modifikasi resep lauk nabati terhadap sisa makanan pasien diruang Rawat inap kelas III RSU dr. H. Koesnadi Bondowoso di Kota Bondowoso. *Darussalam Nutrition Journal*, *5*(1), 20-27. https://doi.org/10.21111/dnj.v5i1.4780

[20] Handojo, M. A. P., & Asri, N. P. (2024). Kajian Tentang Food Loss Dan Food Waste: Kondisi, Dampak, Dan Solusinya. https://dspace.uc.ac.id/handle/123456789/6975

[21] Sari, D. A., & Hadiyanto, H. (2013). Teknologi dan metode penyimpanan makanan sebagai upaya memperpanjang shelf life. *Jurnal Aplikasi Teknologi Pangan*, *2*(2).http://www.jatp.ift.or.id/index.php/jatp/article/view/109

[22] Wiradharma, S., DW, I. K. K., Baskara, W. P., Soimun, A., & Gautama, N. W. (2023). URGENSI PENERAPAN COLDSTORAGE DALAM PENYIMPANAN PASOKAN BAHAN PANGAN BAGI RESTORAN DI BALI. In *Berkala Forum Studi Transportasi antar Perguruan Tinggi* (Vol. 1, No. 1, pp. 57-66). https://journal.unej.ac.id/BerkalaFSTPT/article/view/563

[23] Siburian, E. T., Dewi, P., & Martuti, N. K. T. (2012). Pengaruh suhu dan waktu penyimpanan terhadap pertumbuhan bakteri dan fungi ikan bandeng. *Life Science, 1*(2).https://journal.unnes.ac.id/sju/UnnesJLifeSci/article/view/994

[24] Luthfi, J., & Elvandari, M. (2024). Implementasi Strategi Manajemen Bahan Baku untuk Mengurangi Food Waste di Grand Mercure Jakarta Kemayoran. *Ranah Research: Journal of Multidisciplinary Research and Development, 6*(5), 1486-1493. https://doi.org/10.38035/rrj.v6i5.953

Comparative study of Indigenous microorganisms (IMO) from various livestock manures in fresh cow manure-based organic fertilizer production

Deni Novia*, **Afriani Sandra, Ely Vebriyanti, Aronal Arief Putra, and Ade Rakhmadi**

Department of Livestock Products Technology, Animal Science Faculty, Universitas Andalas, Padang, Indonesia

*E-mail: dnovia@ansci.unand.ac.id

Abstract. Raw manures contain nutrients that can be used as raw materials to manufacture potential indigenous microorganisms (IMO). IMO, it can be applied to manufacture organic fertilizer using dried cow manure, but using fresh cow manure in manufacturing has yet to be studied. This study aimed to see the potential of fresh cow manure as raw material in manufacturing organic fertilizers by utilizing IMO produced from several manures from livestock species. This study used an experimental method with a completely randomized design consisting of four treatments and four replications. The treatments were raw manures: A: cow manure, B: goat manure, C: quail manure, and D: rabbit manure. Physicochemical (ash, nitrogen (N), phosphorus (P), potassium (K) content, C/N ratio, pH, moisture content) and microbial-related parameters (total plate count, total lactic acid bacterial count, total fungal count) of organic fertilizer produced were determined. The results exhibited that Fresh cow manure has the potency to be used as raw material for organic fertilizer production by utilizing cow manure IMO with characteristics; 27.98±0.01% ash content, 2.398±0.04% N content, 1.078±0.06% P content, 0.866±0.05% K content, 15.68±0.31 C/N ratio, 5.75±0.06 pH, 40.99±1.11% moisture content, 12.34±0.01 log CFU/ml total plate count, 11.25±0.03 log CFU/ml total lactic acid bacteria count, and 12.01±0.01 log CFU/ml total fungal.

Keywords: raw cow manure, indigenous microorganism (IMO), microbial content, chemical content, organic fertilizer

1. Introduction

Livestock manure is a common by-product that often creates environmental problems such as air and water pollution. Those might be associated with its accumulation, less utilization, and global warming due to methane gases released into the air. However, livestock manure still contains nutrients to produce indigenous microorganisms (IMO), so its presence is more practiced, low cost, abundant, and available. IMO, it is usually applied to produce organic manure using initial dried-cow manure. The Drying process produces air pollution and requires time and favorable weather, thus sometimes being a problem during manufacturing.

Livestock manure is a common waste problem because of its high quantity accumulation and standing for long periods. Besides causing water and air pollution, it also triggered global warming. Cow manure that is available in abundant quantities has the potential to be utilized as organic fertilizer, with a more significant opportunity to enter the export market. According to BPS, the beef cattle population in West Sumatra in 2022 will reach 400,033 heads [1], in which one cattle produces solid manure of 23.6 kg/day. If one farmer had 6 - 7 cattle, they could obtain a high quantity of manure, i.e., 56.64 million kg/day or 1.70 billion kg/month. The farmers still pay more for inorganic fertilizers.

Natural composting of livestock manure was faster, taking 2–3 months [2]. However, the organic fertilizer production process can be significantly expedited to 2-3 weeks through activators or decomposing agents [3]. Unlike commercial decomposers such as EM4 or other brands need cost, natural decomposers from livestock manure are known as natural, cheap, and practice applications. Some livestock manure from cows, goats, rabbits, chickens, and quail have the potential to be processed into decomposers (or, in another term, known as IIMO). Livestock manure might vary in nutrients, including nitrogen, phosphor, calcium, and water. Rabbit manure is known for its high nitrogen and phosphor content; goat manure is considered to have rich calcium manure, while its high moisture content characterizes cow manure.

IMO, it is a liquid that contains microorganisms that are beneficial for plants and soil fertility. The ability of IMO to decompose organic matter is very high, so it has the potential to be used as a good decomposer [4]. Microorganism inoculation is very helpful in facilitating the composting process [5]. Indigenous cow dung contains beneficial bacteria such as *Lactobacillus plantarum*, *Lactobacillus casei*, and *Bacillus subtilis*, which enhance plant growth by promoting nutrient solubilization and pest resistance [6]. Pathogenic bacteria such as *Salmonella spp.* and *Escherichia coli* are also found in cow manure, posing potential risks for contamination [7, 8]. Goat manure, such as *Pseudomonas* and *Bacillus*, effectively degrades pollutants like *Petroleum hydrocarbons* and *atrazine*, enhancing soil health and reducing environmental contamination [9, 10]. This genus is frequently isolated from goat manure and is involved in hydrocarbon degradation, making it useful in bioremediation [9, 11].

Indigenous strains such as *Lactobacillus murinus*, *Streptococcus thermophilus*, and *Pediococcus acidilactiti* have been identified in quail, contributing to improved growth performance and short-chain fatty acid production in the ileum [12]. According to our preliminary study, IMO contains bacteria, such as lactic acid bacteria. Cow manure composting using IMO developed from various livestock manure for 21 days could improve N in organic fertilizer produced. These bacteria effectively reduce ammonia emissions during composting, improving the compost quality; *Pseudomonas sp.* and *Candida sp.* show higher potential for ammonia mitigation [13].

This research aims to produce cheap, easily applied, high-quality decomposers from various livestock manure and produce the required organic fertilizer standards. This research is expected to be applied by farmers to increase income, replace expensive chemical fertilizers, and improve soil structure.

2. Materials and Methods

2.1 Research materials

Materials used in this research were 24 hours of various manure from cow, goat, rabbit, chicken, and quail as primary material for IMO production. At the same time, the primary material for organic fertilizer was raw cow manure obtained from Edufarm, Animal Science Faculty,

IOP Conf. Series: Earth and Environmental Science **1460** (2025) 012026 doi:10.1088/1755-1315/1460/1/012026

Universitas Andalas, Padang, Indonesia. Other materials, such as coconut water and sugar, were bought from the local market (Pasar Raya Padang).

The equipment used: analytical scales, a stainless steel spoon, an electric oven, a desiccator, a porcelain cup, a mortar and pestle, a cup clasp, a funnel, a distillation flask, a distiller, beaker glass, a sieve 30 mesh, UV-Vis spectrophotometry, and a set tool for IMO fermentation and black plastic for fertilizer fermentation. This research was conducted at the Animal Product Technology Laboratory, Animal Science Faculty, Universitas Andalas, Padang, Indonesia.

2.2 Research methods

The experimental method applied for IMO-treated organic fertilizer was conducted using a completely randomized design, with various animal manure sources for IMO production, i.e., A: cow manure, B: goat manure, C: quail manure, and D: rabbit manure. Organic fertilizers treated with various IMO were then analyzed for their microbial and chemical characteristics. Mean comparisons were determined using Duncan's.

IMO and organic fertilizer manufacturing were prepared using the patent method SID201804983 [14]. Then, organic fertilizers were analyzed for chemical content (ash, N, P, K content, C/N ratio, pH, moisture content) and microbial profiles (total plate count, total lactic acid bacterial count, total fungal count). Ash content was tested using the gravimetric method with ashing in a furnace at a temperature of 550°C and moisture content using the oven method[15]. Testing of C, N, P, K using UV-Vis spectrophotometry based [16, 17]. Total plate count testing using nutrient agar media, total lactic acid bacterial count using De Man Rogosa and Sharpe (MRS) agar media, and total fungal count using Potato Dextrose Agar (PDA) media [18].

3. Results and Discussion

Ash, N, P, K content, C/N ratio, pH, and moisture content of various IMO-treated organic fertilizers could be seen in Table 1.

Table 1. Result in nutrient content of organic fertilizers compared to Minister of Agriculture Regulation (MAR) and Indonesian National Standard (SNI)

Treatments	A	B	C	D	MAR No 70*	SNI**	Quality
Ash content (%)	27.98±0.10^a	24.31±0.44^c	26.73±0.07^b	27.26±0.66^b	-	-	-
N content (%)	2.398±0.04^a	1.909±0.03^c	2.047±0.03^b	1.995±0.04^b	Macronutrients	0.4	good
P content (%)	1.078±0.06^a	0.973±0.04^b	0.777±0.07^c	0.990±0.05^b	($N + P_2O_5 + K_2O$) Min 4	0.1	good
K content (%)	0.866±0.05^b	1.064±0.04^a	0.862±0.04^b	0.681±0.10^c		0.2	good
C/N ratio(%)	15.68±0.31^c	20.90±0.52^a	18.94±0.36^b	18.77±0.39^b	15-25	10-20	good
pH	5.75±0.06^b	6.03±0.10^a	6.10±0.12^a	6.98±0.05^a	4-9	6.8-7.49	good
Moisture content (%)	40.99±0.11	42.13±0.70	41.64±0.86	42.39±0.46	15-25	Max. 50	good

*MAR No 70 [19], **SNI 19-7030-2004 [20], Means with different lowercase superscripts show significantly different effects (P<0.05). A: cow manure, B: goat manure, C: quail manure and D: rabbit manure

The nutrient content of organic fertilizer with different manure IMO compared to MAR and SNI was shown in Table 1. Treatment A, using cow manure IMO, meets the quality criteria according to MAR No. 70 of 2011 except for water content, while according to SNI 19-7030-2004, all meet the criteria except carbon. The nutrient content of organic fertilizer with different manure IMO, compared to the MAR and SNI, is shown in Table 1. Treatment A, using cow manure IMO, meets the quality criteria according to MAR No. 70 of 2011 except for moisture content, while according to SNI 19-7030-2004, everything meets the criteria.

3.1 Ash content

The ash content of organic fertilizer from fresh manure produced was 27.98 ± 0.10 - 24.31 ± 0.44%. The high ash content in treatment A (cow manure IMO) was influenced by the mineral content contained in cow manure. The ash content results from this study were still lower than the results of Pisa et al. on making compost from cow manure with an ash content of 53.1% and a composting period of 76 days [21].

The high ash content in the IMO cow manure treatment compared to the others was influenced by the relatively high ash content of the cow manure. Halhale dairy manure demonstrated an ash constituent of 23.3% within the total solid fraction [22]. Analyses revealed lower ash content (20.2%) in manure from paved facilities compared to unpaved environments (58.7%) [23].

Composting cow manure with IMO promotes greater microbial diversity, which is essential for nutrient degradation and ash formation. Research indicates that specific microorganisms in cow manure enhance carbon and nitrogen metabolism, leading to increased ash content [24]. Additionally, diverse microbial groups in cow manure, such as *Proteobacteria*, facilitate efficient decomposition, resulting in a higher mineral concentration in the final compost [25].

Cow manure typically has a higher initial nutrient density than goat manure, influencing the final ash content after composting. This nutrient-rich environment fosters microbial growth that aids in converting organic matter into ash [26]. Studies on compost and biochar characteristics show that cow manure affects soil properties differently than goat manure, suggesting that the composition of the raw material significantly shapes composting outcomes [27].

3.2 N content

N content of organic fertilizer produced from wet cow manure treated with various IMO ranged from 1.909±0.03 - 2.398±0.04%. The lowest N content was found in treatment B (goat manure IMO), while the highest was in treatment A (cow manure IMO). During the natural composting of livestock manure, cow manure decomposes faster than sheep or goat manure, thus affecting the increase in N levels [2]

N content from dry cow manure before fermentation was 0.85% [3] and then increased to 1.05% after being transformed into organic fertilizer using goat manure IMO and 1.548% used cow manure IMO. The lowest N content of organic fertilizer produced by goat manure IMO was influenced by goat manure texture and degradation. By making organic fertilizer using dry cow manure with the addition of goat manure IMO also produced the lowest organic fertilizer N content, namely 1.1129 ± 0.15% [3].

Research making compost from cow manure for 72 days obtained a total nitrogen of 1.07%. N content of organic fertilizer [21] from cow manure with the addition of IM4, i.e., 0.99-0.87%, with 18-21 days fermentation time [28]. This data was lower 2.4 times compared to the result using fresh cow manure with IMO activated with cow manure. The higher nitrogen content of the research results was greatly influenced by the type of dominant microbes from each IMO in the decomposition, the characteristics of the raw materials, and the composting conditions [29].

3.3 P content

P content of organic fertilizer treated with various IMO ranged from 0.777±0.07 - 1.078±0.06%. The lowest P content was obtained in organic fertilizer treated with quail manure IMO, and the highest in cow manure IMO. The obtained P content was lower than that of the report of Novia et al. [3]. This was influenced by the sources of manure used in this present study, as organic material was from fresh manure, while Novia's is from dry manure. Studies have shown that composting can increase the phosphorus content in cow manure by approximately 25% compared to raw manure [30].

The high P levels, with the addition of cow manure IMO, were due to the composition of the microbial community, the availability of nutrients, and the inherent properties of the manure. Cow manure hosts a diverse microbiota, including species such as *Lactobacillus* and *Bacillus*, known for their ability to solubilize phosphorus [6]. The microbial diversity in cow manure supports various metabolic activities that enhance nutrient bioavailability, such as phosphorus solubilization, with its high carbon and nitrogen content creating a favorable environment for microbial activity and promoting increased phosphorus levels [31].

The composting process of cow manure with IMO had significantly increased phosphorus content, whereas quail manure does not show a similar enhancement [26, 32]. The synergistic effects of specific microorganisms in cow manure stimulate enzymatic activities, further enhancing phosphorus availability during composting [33]. Cow manure, rich in various beneficial microorganisms, was generally more effective in promoting phosphorus solubilization than quail manure. In contrast, quail manure may contain lower nutrient levels and exhibit less microbial diversity, which limits its phosphorus solubilization capacity [26].

3.4 K content

Based on Table 2, it was found that the K content of organic fertilizers was 0.681 ± 0.10 - 1.064 ± 0.04%. The low TPC and total LAB count influenced the high K content in the goat manure IMO treatment. Loss of nutrients is caused by microbial activity that utilizes K for growth and development [34].

Goat manure supports a more diverse microbial community that promotes nutrient cycling, especially for potassium, as specific microorganisms enhance potassium solubilization efficiency during composting; thus, even under similar conditions, goat manure is more likely to yield compost with higher potassium levels [35]. Goat manure undergoes significant biochemical changes, increasing potassium concentration in the final compost product [24].

In comparison, rabbit manure may not support the same degree of microbial diversity or activity, resulting in lower potassium availability [31]. Adding IMO can further boost potassium levels in goat manure due to heightened microbial activity, a less evident benefit in rabbit manure [26].

3.5 C/N ratio

The lowest C/N ratio of organic fertilizer in this study was obtained in treatment A (cow manure IMO), and the highest ratio was obtained in treatment B (goat manure IMO) with 15.68±0.31 and 20.90±0.52, respectively. Cow manure typically exhibits a lower C/N ratio, mainly due to its higher nitrogen content and the role of ammonifying microorganisms in promoting nitrogen metabolism during composting. In cow manure composting, the introduction of novel ammonifying microorganism cultures significantly enhances the structure and functionality of the microbial community [24]. Specific microorganisms in cow manure foster a cooperative relationship between bacteria and fungi, improving nitrogen and carbon metabolism [33]. Cow manure compost can achieve a C/N ratio as low as 36.97% with effective microorganism treatments [36].

A high C/N ratio of organic fertilizer in treatment B using goat manure is due to its lower nitrogen content and less robust microbial interactions during composting, which limit nitrogen availability and affect compost quality. Enhanced enzymatic activities and metabolic processes in cow manure composting enable more efficient nutrient cycling, reducing the C/N ratio compared to goat manure, which may undergo a different level of microbial enhancement [24]. This result aligns with the C/N ratio of organic fertilizer produced using dry cow manure with the highest C/N ratio from goat manure, i.e., 25.68±4.56 [3].

The C/N ratio of organic fertilizer produced using cow manure IMO was 1.47, which is lower than the report of Murni et al. [28], i.e., 18.41-17.25. However, it is almost identical to organic fertilizer treated with quail manure, IMO. According to SNI 19-7030-2004 [20], the C/N ratio of compost is 10-20. The C/N ratio of organic fertilizer from treatments A, C, and D were met the SNI standard, while treatment B near the standard.

3.6 pH

Organic fertilizer from cow manure IMO had the lowest pH (5.75±0.06), while the highest pH from rabbit manure IMO (6.98±0.05) treatment was obtained. Cow manure supports a diverse microbial community, including ammonifying microorganisms that boost nitrogen metabolism, resulting in lower pH levels due to increased ammonium production [24]. By contrast, rabbit manure may have a less diverse microbial community, leading to metabolic pathways that are less effective in reducing pH. With a higher organic matter content, cow manure undergoes more intensive microbial activity, producing more acids during composting [36].

The result aligns with the report of Novia et al. [3], which highlighted the lowest pH in cow manure IMO (3.83±0.03) compared to goat-, quail, and rabbit manure. Besides, low pH was also influenced by higher total LAB count because bacterial activity would produce lactic acid. Moreover, Novia et al. [3] also emphasized that a high total BAL could be treated with cow manure IMO in fertilizer.

pH was decreased during fermentation of IMO production due to their microbial activity. pH manure before fermentation ranged from 7.75±0.13 (cow feses) to 9.73±0.04 (rabbit feses), while the pH of IMO ranged from 3.83±0.03 (cow feses IMO) to 4.28±0.04 (rabbit feses IMO), that represented decreasing in pH. The study on the chemical characteristics of compost made from cassava peel waste and cow manure showed a pH of 6.50, slightly lower than typical cow manure compost. This may be due to the specific combination of materials [37]. Adding cow dung as a compost enhancer can affect pH by boosting microbial activity and increasing nutrient availability, which may result in gradual pH changes over time [38].

3.7 Moisture content

The moisture content of organic fertilizer from various IMO treatments ranged from 40.99±0.11 - 42.39±0.46%. No significant difference was obtained among samples, meaning adding various IMO organic fertilizer sources statistically resulting equal moisture. Furthermore, the moisture content of samples in this present study already met the standard minimum SNI of compost moisture content.

Cow manure supports a diverse microbial population that efficiently decomposes organic matter, potentially causing moisture loss through evaporation during the thermophilic phase [24, 31]. Studies show that specific microorganisms enhance carbon and nitrogen metabolism, contributing to stable moisture levels during composting [24, 33]. Adding indigenous microorganisms supports nutrient cycling, indirectly impacting moisture retention by fostering a balanced microbial ecosystem [26].

The moisture content of organic fertilizers from the research results was lower than that of Novia *et al.* [3], namely 66.68-68.78%, with an IMO moisture content in the range of 88.51-92.21%. The moisture content of the organic fertilizer produced is almost the same as that of chicken manure composting with corn straw, which is 42.0% [39].

The total plate count, total LAB count, and total fungal count of various IMO-treated organic fertilizers can be seen in Table 2.

Table 2. Total plate count, total lactic acid bacteria count, and total fungal count of various IMO-treated organic fertilizer

Treatments	A	B	C	D
Total Plate Count/TPC (log CFU/ml)	$12.34{\pm}0.01^{b}$	$12.32{\pm}0.01^{c}$	$12.36{\pm}0.01^{a}$	$12.33{\pm}0.01^{bc}$
Total LAB Count/TLABC (log CFU/ml)	$11.25{\pm}0.03^{a}$	$10.98{\pm}0.06^{b}$	$11.24{\pm}0.06^{a}$	$11.20{\pm}0.04^{a}$
Total Fungal Count/TFC (log CFU/ml)	$12.01{\pm}0.01^{c}$	$12.08{\pm}0.01^{b}$	$12.11{\pm}0.01^{a}$	$12.08{\pm}0.01^{b}$

Note: Means with different lowercase superscripts show significantly different effects ($P<0.05$). A: cow manure, B: goat manure, C: quail manure and D: rabbit manure

The result showed that IMO treatment significantly affects TPC, Total LAB, and TFC. Organic fertilizer from cow manure IMO (treatment A) manufacturing is significantly lower in pH and total fungal than others. The lower pH result was linear with TFC, but the highest LAB was noticed. Organic fertilizer from rabbit manure IMO manufacturing had the highest TPC, total LAB count, and TFC. Nevertheless, the results were similar in pH and moisture compared to treatments B and D and in total LAB count compared to treatments A and D.

3.8 Total Plate Count (TPC)

The total plate count (TPC) of organic fertilizers produced from cow manure using several kinds of IMO from manures, based on Table 1, was very high, namely 12.32 ± 0.01 to 12.36 ± 0.01 log CFU/ml. The high TPC was affected by the TPC from IMO, the raw material used for making fertilizers and reforming during the fermentation process, so there was an increase in the number of microorganisms.

Higher TPC of samples (12.32±0.01 - 12.36±0.01 log CFU/ml) compared the report of Novia et al. [3] that found $28.66{\pm}10.10 \times 10^{4}$ - $55.25{\pm}10.21 \times 10^{4}$ CFU/ml in dry fertilizer from cow manure treated with various IMO sources. Significant differences in TPC might relate to its high moisture content due to the wet condition of cow manure used in this study. Research indicates that the effective viable count of beneficial microorganisms in cow manure compost can range from 2.5 billion to 4.5 billion per gram [40]. The total heterotrophic bacterial count (THBC) in bovine manure compost demonstrates an increase from 2.4×10^{7} CFU/g to 2.5×10^{8} CFU/g during the initial composting phase [41].

The highest TPC of quail manure IMO-treated organic fertilizer might be related to the high moisture content of its quail manure IMO. Quail manures contain 65.50% of the bacteria of the *Leuconostocaceae* family, which act as pathogens and lactic acid. The dominant bacterial species in quail manures is *Lactobacillus salivarius*, which belongs to lactic acid bacteria with the highest number, namely 38.75%, followed by *Burkholderia fungorum* bacteria (21.30%) and *Escherichia coli* (15%) as pathogenic bacteria [38]. Specific ammonifying microorganisms in quail manure may also promote microbial interactions, increasing total plate counts [20]. Quail manure contains more nitrogen and phosphorus than goat manure, which supports microbial growth and increases total plate counts during composting. The differences in organic matter also influence microbial community structure and abundance [31].

3.9 Total Lactic Acid Bacteria Count (Total LAB Count)
The highest total lactic acid bacteria (LAB) count of organic fertilizer samples was found in treatment A (cow manure IMO), i.e., 11.25±0.03 log CFU/ml. In contrast, the lowest total LAB count was found in treatment B (goat manure IMO), i.e., 10.98±0.06 log CFU/ml. Cow manure is a rich source of LAB count; the research results have a higher total LAB count because they use fresh cow manure. Biodynamic preparations from cow manure exhibited elevated lactobacilli counts, reaching up to 10^9 ml−1 [39]. Fermentation of cow manure can produce lactic acid concentrations around 13.65 g·L−1, showcasing LAB proliferation potential during composting [44].

The high lactic acid bacteria produced comes from the raw materials used in IMO. Quail manures already contain the dominant LAB, namely *Lactobacillus salivarius*, as much as 38.75% [38]. Cow manure hosts various microorganisms, including *Lactobacillus plantarum* and *Bifidobacterium*, which thrive in anaerobic conditions and contribute to higher LAB counts [6]. During the thermophilic composting phase, which involves higher temperatures, cow manure's microbial community adapts well, boosting LAB counts [24, 45].

Particular LAB strains, including *Lactobacillus gasseri*, *Lactobacillus reuteri*, and *Lactobacillus salivarius*, were extracted from dairy cow manure, signifying a varied LAB population in bovine excrement [46]. At the same time, *Bacillus licheniformis* aids in the breakdown of organic materials, leading to faster compost maturation [47] and significantly lowering nitrogen loss rates, preserving essential nutrients in the compost [48]. Research shows that cow manure enhances carbon and nitrogen metabolism, encouraging LAB proliferation [24, 33]. The nutrient composition of cow manure, particularly its higher organic matter content, supports a robust microbial community, promoting LAB growth during composting [36].

3.10 Total Fungal Count (TFC)
The total fungal count (TFC) of organic fertilizer treated with various IMO sources ranged from 12.08±0.01 to 12.11±0.01 log CFU/ml. High TFC in treatment C (organic fertilizer treated with quail manure IMO caused quail manure, derived from an omnivorous source, may support a wider variety of fungi than herbivorous cow manure, which typically favors bacterial communities [31]. Fungal metabolic activity is vital in composting. Quail manure can create a more favorable environment for fungal enzyme activity, leading to higher fungal counts [49]. The fungal flora in quail-breeding environments is dominated by genera such as Aspergillus, Penicillium, *Scopulariopsis*, and *Aureobasidium*, with Aspergillus fumigatus being particularly prevalent [50].

The low total fungal colonies in treatment B (goat manure IMO) and treatment D (rabbit manure IMO) were also influenced by ammonifying microorganisms in the manure IMO. Research shows that introducing certain ammonifying microorganisms can increase fungal populations, indicating that microbial inoculation strategies may vary in effectiveness between quail and cow manure [24].

IMOd communities are more able to utilize carbon sources in thermophilic and mature phases than in other phases. *Micothermus* dominated thermophilic and cooling phases in composting. Genus IMOd thermophilic has the potential ability to degrade carboxylate acid and polymer. In contrast, the IMOd community has a higher ability to utilize Amina, asam amino, and carbohydrate in the mature phase. Composting aerobics is an effective biotreatment to minimize pathogen IMOd and enrich beneficial microbes in cow manure [51].

As reported by Novia et al. [3], rabbit manure IMO showed the lowest total fungal count, i.e., 27.93±4.14x10^{11} CFU/ml or 12.44607 logs CFU/ml and this is higher than the results of this study. This is because Novia et al. [3] was an IMO sample. At the same time, in this study, the data from organic fertilizer is made from fresh cow manure without drying so that the microbes come from IMO and also from manure as the primary material for fertilizer. Fungal populations can vary from 2.0 × 10^4 to 8.0 × 10^6 CFU/ml, depending on the specific composting conditions and stages [52]. Fungal populations peak at different temperature stages: the highest counts (1.38 × 10^4 CFU/g) are seen between 20-40°C, while counts decrease (1.68 × 10^3 CFU/g) at 65°C [53].

The composting process reveals a variety of fungi, with common genera such as *Aspergillus*, *Thermomyces*, and *Alternaria* being prominent [54]. In one study, 44 fungal species were identified in cow manure compost, including 20 thermophilic species that were especially abundant [55]. *Candida* species persist throughout the composting process, raising concerns about microbiological safety [53].

4. Conclusion

Fresh cow manure has potency to be used as raw material for organic fertilizer production by utilizing raw cow manure IMO with characteristics; 27.98±0.01% ash content, 2.398±0.04% N content, 1.078±0.06% P content, 0.866±0.05% K content, 15.68±0.31 C/N ratio, 5.75±0.06 pH, 40.99±1.11% moisture content, 12.34±0.01 log CFU/ml total plate count, 11.25±0.03 log CFU/ml total lactic acid bacteria count, and 12.01±0.01 log CFU/ml total fungal.

Acknowledgments

The award and acknowledgments were addressed to those who have supported the implementation of this Indexed Publication Research, especially to LPPM as a funder through BOPTN contract No. 333/UN.16.19/PT.01.03/PUJK/2024, then Rector, Chairman of LPPM, Dean of Animal Science Faculty, Universitas Andalas.

References

[1]. BPS-Statistics Indonesia, 2024. Beef cattle population by province (heads), 2021-2023.

[2]. Zhu, Y., et al., 2020. The effects of climate on cattle, sheep and goat manure decomposition in Kenyan tropical pastures. *Plant and Soil.*

[3]. Novia, D., Rakhmadi, A., Purwati, E., et al., 2019. *IOP Conf. Series: Earth and Environmental Science*, vol. 287.

[4]. Mansen, E. and Irawan, 2008. Efektivitas dan Efisiensi Mikroba Decomposer Komersial dan Lokal dalam Pembuatan Kompos Jerami.

[5]. Wan, L. et al., 2020. *Bioresource Technology*, vol. 301, pp. 1–7.

[6]. Banerjee, S., Kanaujia, A. and Sharma, S., 2023. *International Journal of Current Microbiology and Applied Sciences.*

[7]. Liu, G., et al., 2021. A study on the effectiveness of a defined microbial consortium to enhance the microbiological safety of cattle manure. *Journal of the Science of Food and Agriculture.*

[8]. Zhang, Y.L., et al., 2023. Effects of cow manure application on soil microbial community in Farmland. *Environmental Science.*

[9]. Nwogu, T.P., Azubuike, C.C. and Ogugbue, C.J., 2015. Enhanced bioremediation of soil artificially contaminated with petroleum hydrocarbons after amendment with *Capra aegagrus hircus* (goat) manure. *Biotechnology Research International.*

[10]. Luo, S., et al., 2021. Accelerated atrazine degradation and altered metabolic pathways in goat manure assisted soil bioremediation. *Ecotoxicology and Environmental Safety.*

[11]. Pathma, J. and Sakthivel, N., 2013. Molecular and functional characterization of bacteria isolated from straw and goat manure-based vermicompost. *Applied Soil Ecology.*

[12]. Harimurti, S., Sudaryati, S. and Ariyadi, B., 2015. The dynamics of indigenous probiotics lactic acid bacteria on growth performance, total adherence bacteria, and short-chain fatty acids production in the ileum of male quail.

[13]. Fitriyanto, N.A., et al., 2021. *IOP Conf. Series: Earth and Environmental Science*, vol. 662.

[14]. Novia, D., Rakhmadi, A. and Purwati, E., 2022. Proses pembuatan pupuk organik dari feses sapi 0 hari memanfaatkan mikroorganisme lokal feses sapi. *Patent No. IDS000004918.*

[15]. Association of Official Analytical Chemists [AOAC]. (2016). Official Method of Analysis of The Association of Official Analytical of Chemists. Arlington, Virginia, USA: Association of Official Analytical Chemists, Inc.

[16]. SNI 1998. SNI 13-4720-1998 Tata penentuan kadar karbon organik dalam tanah. Jakarta. Standarisasi Nasional.

[17]. SNI. 2012. SNI 2803 2012 Pupuk NPK Padat. Jakarta. Badan Standarisasi Nasional.

[18]. SNI. 2009. SNI 7388:2009 Batas maksimum cemaran mikroba pada pangan. Jakarta: Badan Standar Nasional.

[19]. Minister of Agriculture Regulation (MAR), 2011. Nomor 70/Permentan/SR.140/10/2011.

[20]. SNI. 2004. SNI 19-7030-2004 about Compost specification from organic domestic waste.

[21]. Pisa, C., Wuta, M., & Muchaonyerwa, P., 2020. *Bioresource Technology Reports*, vol. 9, pp. 1-10. https://doi.org/10.1016/j.biteb.2020.100383

[22]. Aremanda, R.B., and Debretsion, S., 2023. Competence of cow manure as a sustainable feedstock for bioenergy and biofertilizer production. *International Journal of Food, Agriculture and Natural Resources.*

[23]. Sweeten, J.M., Heflin, K., Annamalai, K., and Auvermann, B.W., 2006. Combustion-fuel properties of manure or compost from paved vs. unpaved cattle feedlots. *ASABE.*

[24]. Xu, Z., et al., 2024. Applying Ammonifying microorganism cultures markedly improved the efficiency of cow manure composting. *SSRN.*

[25]. Liu, S., et al., 2024. Microbial community succession of cow manure and tobacco straw composting. *Frontiers in Microbiology.*

[26]. Novia, D., et al., 2021. *IOP Conf. Series: Earth and Environmental Science*, vol. 888.

[27]. Sudita, D.N., Situmeang, Y.P., and Suarta, M., 2021. Compost and biochar characteristics test of some animal manure waste. *International Journal of Advanced Science, Engineering and Information Technology.*

[28]. Murni, S.W., Widayati, T.W., Pratama, B.A., and Oktavianto, S., 2017. In: *Proceedings of the National Seminar on Chemical Engineering "Kejuangan,"* Yogyakarta, pp. F07-1-F07-6.

[29]. Wang, F., et al., 2024. Pilot-scale membrane-covered composting of food waste: Initial moisture, mature compost addition, aeration time and rate. *Science of The Total Environment.*

[30]. Alege, F.P., Blake, C.D., Donohoe, S.P., and Thomas, J.W., 2024. Nutrient Properties of Fresh and Composted Cotton Gin By-products and Cattle Manure for Soil Amendment. *Journal of the ASABE.*

[31]. Wan, J., et al., 2021. Livestock manure type affects microbial community composition and assembly during composting. *Frontiers in Microbiology.*

[32]. Susilowati, L.E., Mahrup, M., Arifin, Z., and Sutriono, R., 2023. Effectiveness of bio activator of phosphate solubilizing bacteria consortium on composting bag-log waste incorporated with cow manure. *Jurnal Penelitian Pendidikan IPA.*

[33]. Xu, Z., et al., 2023. A bioprocess for rapid compost production from rice straw using C/N ratio improvers and indigenous microorganisms at the lab and pilot scales. *Research Square.*

[34]. Stella, M., Theeba, M., and Illani, Z.I., 2019. *Biocatalysis and Agricultural Biotechnology*, vol. 20, pp. 1-13.

[35]. Colín-Navarro, V., et al., 2019. Propiedades químicas y microbiológicas del estiércol de caprino durante el compostaje y vermicompostaje. *Agrociencia.*

[36]. Hutapea, A.R. and Siregar, D.J.S., 2023. Quality of cow manure compost using effective microorganism (em4) and black soldier fly (BSF) Fly Larvae (Maggot). *International Journal of Research and Review.*

[37]. Sebayang, N.U.W., Akasah, W., Manurung, V.R., and Sinamo, K.N., 2023. *Abdimas Talenta: Jurnal Pengabdian kepada Masyarakt,* vol. 8(1). Pp. 604-610

[38]. Njoku, N.O., Mbachu, I.A.C., and Uba, B.O., 2019. Influence of physicochemical and microbiological properties on the composting of agro wastes using cow manure as a booster. *Animal Research International*, vol. 16, no. 1. https://doi.org/10.4314/ARI.V16I1

[39]. Wan, L., Wang, X., Cong, C., Lia, J., Xu, Y., and Lia, X., 2020. *Bioresource technology*, vol. 301, pp. 1 7.

[40]. Zhu, L., Wang, S., Zhang, J., Kou, X., Xu, R., Han, G., Wang, G., Bi, J., and Tang, H., 2017. Cow dung composted rotting promoting and nitrogen preserving bacterium agent and application method. *Patent.*

[41]. Chinakwe, E.C., et al., 2019. Changes in microbial population numbers during composting of some organic wastes in greenhouse. *Journal of Applied Microbiology and Biochemistry.*

[42]. Dalle Zotte, A., et al., 2021. *Animal*, vol. 15, pp. 1-7.

[43]. Radha, T.K. and Rao, D.L.N., 2014. Plant growth promoting bacteria from cow manure based biodynamic preparations. *Indian Journal of Microbiology*.
[44]. Garrido, R., et al., 2023. Lactic acid production from cow manure: experimental process conditions analysis. *Fermentation*.
[45]. Dutta, S., Yesankar, P.J., and Kumar, M., 2023. Potential of thermophilic bacteria isolated from cow manure-grass compost for bioethanol production using floral waste. *Biofuels*.
[46]. Lin, W.C., et al., 2020. Autochthonous lactic acid bacteria isolated from dairy cow manure exhibiting promising probiotic properties and in vitro antibacterial activity against foodborne pathogens in cattle. *Frontiers in Veterinary Science*.
[47]. Shen, Q., Li, R., Xu, X., Liu, C., He, Z., and Liu, D., 2019. High-temperature-resistant *Bacillus licheniformis* strain NJAU-ND8 for acceleration of compost maturity and application of strain.
[48]. Zhao, X., Wang, W., and Li, J., 2018. Compound bacterium agent for rapidly composting livestock and poultry manure and preparation method thereof. *China Patent CN109182228B*.
[49]. Zhou, Y., Sun, Y., Liu, J., Ren, X., Zhang, Z., and Wang, Q., 2022. Effects of microplastics on humification and fungal community during cow manure composting. *Science of the Total Environment*, vol. 803. https://doi.org/10.1016/j.scitotenv.2021.150029
[50]. Franco, A.S., Gutiérrez, J., Muizquiz Moracho, J.L., and Alonso Martinez, J.L., 1981. Environmental fungus flora in quail-breeding farms. *Avian Diseases*.
[51]. Wang, K., Yin, X., Mao, H., Chu, C., and Tian, Y., 2018. *Bioresource Technology*, vol. 255, pp. 123-130.
[52]. Adegunloye, D.V., and Adetuyi, F.C., 2009. *African Journal of Basic & Applied Sciences*, vol. 1, no. 3-4, pp. 70-75.
[53]. Glushakova, A.M., and Kachalkin, A.V., 2023. *Fungal Biology*, vol. 127, no. 6, pp. 1075-1083.
[54]. Wang, R., et al., 2022. Aerosolization behaviour of fungi and its potential health effects during the composting of animal manure. *International Journal of Environmental Research and Public Health*.
[55]. Noreen, N., Ramzan, N., Parveen, Z., and Shahzad, S., 2019. A comparative study of cow manure compost, goat pellets, poultry waste manure, and plant debris for thermophilic, thermotolerant, and mesophilic microflora with some new reports from Pakistan. *Pakistan Journal of Botany*.

Effect of drying temperatures on properties of sweet potato flour

Tengku Farizan Izzi Che Ku Jusoh[1*], Siti Nur Anis Husna Mohd Rosli[1], Hanis Syazwani Mat Gani[1], Nurul Zaizuliana Rois Anwar[1], and Mohd Adzim Khalili Rohin[2]

[1] Faculty of Bioresources and Food Industry, Universiti Sultan Zainal Abidin, Besut Campus, 22200 Besut, Terengganu, Malaysia
[2] Faculty of Health Sciences, Universiti Sultan Zainal Abidin, Gong Badak Campus, 21300 Kuala Nerus, Terengganu, Malaysia

*E-mail: farizanizzi@unisza.edu.my

Abstract. VitAto is an orange-fleshed sweet potato variety introduced by the Malaysian Agricultural Research and Development Institute (MARDI) in 2017. Due to the susceptibility of fresh sweet potatoes to deterioration during storage, they are frequently processed into flour. The drying process is crucial in preserving the quality of flour. Thus, this study investigated the impact of oven drying at three different temperatures (50, 70 and 90°C) on the physical and functional properties of VitAto sweet potato flour. The range of values for physical properties of VitAto flour were: moisture (2.09 to 6.59%), water activity (0.38 to 0.40), pH (5.79 to 6.38), bulk density (0.22 to 0.25 g/ml), and tapped density (6.94 to 7.21 g/ml). Flour dried at 50°C best preserved the orange colour (L*=84.45±0.33; a*=16.64±0.19; b*=25.65±0.42), whereas flour dried at 90°C exhibited a yellowish colour (L*=89.56±0.32; a*=2.99±0.06; b*=32.45±0.10). Increasing the drying temperature led to a gradual increase in water absorption capacity, wettability, and swelling power, but oil absorption capacity decreased. The results of the pasting profile indicated significant variations in viscosity properties due to drying temperatures. Flour dried at 50°C had the lowest peak viscosity (393.67 mPa.S), breakdown viscosity (219.33 mPa.S), setback viscosity (59.33 mPa.S) and final viscosity (233.67 mPa.S). In conclusion, the drying temperatures distinctly influenced the quality characteristics of VitAto sweet potato flours, thereby altering their functionality.

Keywords : drying temperature, flour, sweet potato, VitAto

1. Introduction

In tropical countries, roots and tubers serve as staple food crops and primary energy sources, including potatoes, cassava, sweet potatoes, and yams from various botanical families [1]. The sweet potato (*Ipomoea batatas* L.), also known as camote or batata, originates from the Americas. Its native range includes the Caribbean and regions from Guatemala to Peru, including Colombia and Ecuador [2]. Sweet potatoes exhibit diverse skin and flesh colours, alongside varying chemical compositions. These tubers vary in shape, being fusiform, globular, round or oval with surface textures that are smooth, ridged or rough. The flesh colour spans from white to yellow, orange, reddish-orange, and purple, while the skin colour ranges from white to yellow, orange, red, purple, and brown [3].

In Malaysia, the VitAto sweet potato variety is particularly prominent, especially in Terengganu, Kelantan and Perak. VitAto, an orange sweet potato cultivar, is favoured for its ease of cultivation, high yield, and enhanced nutrient content [4]. This cultivar typically reaches maturity approximately four months after planting. Crop maturity is indicated by yellowing leaves and mature tubers. Sweet potatoes are abundantly available during the monsoon season, despite being non-seasonal tubers. Harvesting usually occurs before the rainy season to prevent tuber damage or discolouration from prolonged water exposure.

In general, sweet potatoes are a rich source of vitamins, minerals, antioxidants, fibre, and protein [5]. However, their nutritional value varies depending on the species and variety. Different cultivars exhibit variations in the content of carbohydrates, lipids, proteins, vitamins, dietary fibers, and other bioactive compounds, such as anthocyanin and beta-carotene [6]. The tubers, known for their sweet and pleasant flavour, are commonly prepared by boiling and roasting. The VitAto variety contains carotenoids, which impart an orange colour to its flesh. Carotenoids are recognized for their health-promoting properties, functioning as provitamin A and antioxidants that protect against various chronic diseases. However, carotenoids are characterized by low stability and are highly susceptible to oxidation and degradation during storage, which limits their effectiveness as beneficial nutritional components [7].

In tropical climates, sweet potatoes have a short shelf life of approximately two months when stored under shade, making them susceptible to deterioration. To preserve their nutritional content and functionality, one method is to process sweet potatoes into flour through peeling, slicing, drying and milling. However, processing variables can significantly impact the characteristics of the resulting flour. Thus, this study aims to investigate the effect of drying temperatures on the quality of VitAto sweet potato flour.

2. Materials and Methods

2.1 Preparation of VitAto flour

Mature VitAto sweet potatoes were purchased from local market in District of Besut, Terengganu, Malaysia. To prepare the flour, the tubers were initially cleaned by rinsing with tap water to eliminate any dirt and other contaminants. To restrict microbial growth in the finished product, the tubers were peeled to remove any damaged parts and then uniformly be cut into 2 to 3 mm thick using a slicing machine. The slices were dried in an oven dryer (MMT-UF110, Memmert, Germany) at 50, 70 and 90°C. The samples were taken out at every 30 min and weighed until equilibrium (no weight change) was reached. After drying, the sample was cooled to room temperature, ground using a grinder (MX-AC210SWUA, Panasonic, Shah Alam, Malaysia) and sieved through 250-micron sieve. The resulting VitAto flour was vacuumed-packed to prevent moisture absorption and stored at room temperature until further analysis.

2.2 Physical properties of VitAto flours

The moisture of VitAto flours was determined by AOAC method [8]. The water activity was analyzed using a water activity meter (Aqualab Pawkit, Decagon Devices Pullman, Washington, USA). The colour of the flours was measured using a colorimeter (Chroma Meter CR-400, Konica-Minolta, Osaka, Japan) as L*, a* and b* values.

The bulk density of VitAto flours was assessed by measuring both the mass and the corresponding volume of the flour [9]. Approximately 50 g of flour was placed into a 100 ml graduated cylinder, and the bulk density was expressed as the ratio of the mass of the flour to the volume it occupied in the cylinder. For the measurement of tapped density, the graduated cylinder was repeatedly tapped onto a firm pad placed on the laboratory bench until the volume constant. The volume of the flour was used in mass/volume calculation.

2.3 Functional properties of VitAto flours

The functional properties of VitAto flours, including water absorption capacity (WAC), oil absorption capacity (OAC), wettability and swelling power, were determined according to standard methods. WAC and OAC were determined using the methods described by [10, 11]. For WAC, flour sample (1 g) was dispersed in 5 ml of water in a centrifugal tube. The slurry was vortexed, centrifuged (3000 rpm for 10 min) and the supernatant was discarded. WAC was expressed as the ratio of the weight of the sediment to the initial weight of the flour sample (g/g). For AOC, the flour was placed in a centrifuge tube and suspended in 5 ml of corn oil. Samples were vortexed (1 min), rested (30 min) and centrifuged (3000 rpm for 25 min). Then, the separated oil was removed, and the tube was inverted for 25 min to allow the oil to fully drain before reweighing. The OAC was expressed as weight of oil absorbed per weight of flour sample (g/g).

The wettability of VitAto flours was evaluated using the method described by [12]. One gram of each flour sample was transferred in a clean and dry 10 ml measuring cylinder. The open end of the cylinder was covered with a finger, and the cylinder was inverted and positioned 10 cm above the surface of a 600 ml beaker containing 500 ml of distilled water. The flour was gradually spread on the water surface from the cylinder at a moderate speed. The time required for the flour to become completely wetted was recorded as the wettability of the sample.

The swelling power of VitAto flours was evaluated following the method with slight modifications [10]. Sample (approximately 0.1 g) was placed into a pre-weighed 50 ml graduated centrifuge tube, and distilled water was added to reach a final volume of 10 ml. The mixture was vortexed at a low speed for 1 min at room temperature, followed by heating at 60°C for 30 min. The samples were centrifuged at 1600 rpm for 15 min after cooling to room temperature. The resulting sediment was then weighed and recorded.

2.4 Pasting properties of VitAto flours

The pasting properties of VitAto flours was evaluated with rapid visco analyser (RVA 4500, Perten, USA) based on method of [13]. The flour (3.5 g) was dispersed in 25 ml of distilled water subjected to a controlled heating-and-cooling cycle while maintaining constant shear. The temperature was initially maintained at 50°C for 1 min, followed by a linear temperature rise to 95°C in 5 min, where it was held for holding for 2 min and 30 sec. The system was then cooled to 50°C in 3 min. The data collected included peak viscosity (PV), breakdown viscosity (BV), setback viscosity (SV) and final viscosity (FV).

2.5 Statistical analysis

All measurements were performed in triplicate. To compare the mean values, one-way analysis of variance (ANOVA) followed by Duncan's multiple range test (at $p<0.05$) was conducted using the Statistical Package for the Social Sciences (SPSS) software, version 26.0 (SPSS, IBM, Illinois, USA).

3. Results and Discussion

3.1 Physical properties of VitAto flours

Table 1 presents the moisture content, water activity, colour and density of VitAto flours dried at different temperatures. Both moisture content and water activity of the flours decreased as the drying temperature increased due to greater water removal at higher temperatures. These parameters are critical indicators of flour stability and shelf life, as low moisture and water activity levels inhibit enzymatic activity and microbial growth [14]. In terms of colour, the L* and b* values of the flour increased, while the a* value decreased with increasing drying temperatures because of thermal degradation of β-carotene, resulting in a lighter and more yellowish colour of the flour (Figure 1). For bulk and tapped density, increasing drying temperatures led to lower densities. This phenomenon is attributed to the moisture level of flour, which significantly influences its flowability and compaction. Higher moisture content can promote particle agglomeration in the powder, resulting in clumpy powder with greater spaces between them, thereby reducing the tapped density [15].

Table 1. Physical properties of VitAto flours dried at different temperatures.

Physical properties	Dried at 50°C	Dried at 70°C	Dried at 90°C
Moisture content (%)	5.37±0.16^a	2.08±0.07^c	2.74±0.27^b
Water activity	0.48±0.01^a	0.31±0.01^b	0.29±0.02^b
Colour			
L* value	84.45±0.33^c	88.43±0.11^b	89.56±0.32^a
a* value	16.64±0.19^a	10.34±0.08^b	2.99±0.06^c
b* value	25.65±0.42^b	24.54±0.19^c	32.45±0.10^a
Bulk density (g/ml)	0.25±0.00^a	0.22±0.00^b	0.22±0.00^b
Tapped density (g/ml)	6.94±0.05^a	7.21±0.00^b	7.21±0.00^b

Values in the same row with different letters are significantly different ($p < 0.05$).

Figure 1. VitAto flours dried at different temperatures: a) 50°C, b) 70°C, and c) 90°C.

3.2 Functional and pasting properties of VitAto flours

Functional and pasting properties of VitAto flours dried under three different temperatures are listed in Table 2. The results revealed that the water absorption capacity of the flours slightly increased with increasing drying temperatures. Water absorption capacity, which enhances consistency of food products, is the ability of the flour to absorb water and swell [16]. Flour dried at 90°C recorded the highest water absorption capacity, potentially indicating reduced shelf life stability because of its tendency to absorb water from a humid atmosphere [17]. In contrast, oil absorption capacity, which involves the binding of fat by the proteins' non-polar side chains and the physical entrapment of oils, decreased with higher drying temperatures. A high oil absorption capacity is desirable for improving mouthfeel and flavour in food preparations [18].

Regarding the wettability, the time required to wet all flour particles decreased at higher drying temperatures. This variation in wettability can be attributed to factors such as polarity of the molecules, heterogeneity of the flour, contact surface and porosity [16]. Overall, the wettability of VitAto flours is comparable to that of sweet potato flour reported by other study [19]. The least swelling power was observed in the flour dried 50°C. The swelling capacity measures the ability of starch to absorb water molecules and swell, serving as an indicator of the strength of the associative forces within the starch granules [20].

Table 2. Functional and pasting properties of VitAto flours dried at different temperatures.

Measurements	Dried at 50°C	Dried at 70°C	Dried at 90°C
Functional properties			
Water absorption capacity (%)	1.91±0.02[b]	1.94±0.00[ab]	2.01±0.06[a]
Oil absorption capacity (%)	1.19±0.01[a]	1.05±0.00[b]	1.06±0.01[b]
Wettability (s)	36.33±1.53[c]	41.33±1.53[b]	49.00±2.00[a]
Swelling power (%)	5.24±0.15[b]	5.66±0.08[a]	5.82±0.10[a]
Pasting properties			
Peak viscosity (mPa.S)	393.67±10.97[c]	1919.00±0.01[b]	2436.33±85.44[a]
Breakdown viscosity (mPa.S)	219.33±8.14[c]	588.67±8.14[b]	1142.33±68.65[a]
Setback viscosity (mPa.S)	59.33±1.15[c]	573.33±8.33[a]	436.33±28.88[b]
Final viscosity (mPa.S)	233.67±4.16[c]	1903.67±23.63[a]	1730.33±45.39[b]

Values within the same row with different letters differ significantly ($p < 0.05$).

The results of pasting properties of VitAto flour show that lower drying temperature reduced the peak viscosity (PV), breakdown viscosity (BV), setback viscosity (SV) and final viscosity (FV). Another study reported similar effects [21]. The relatively low PV of VitAto flour dried at 50°C suggests its potential suitability for applications in products that requires low gel strength and elasticity [22]. The lower BV and FV suggest that the flour has a strong ability to form a viscous paste or gel upon cooking and cooling, while also demonstrating resistance to shear stress during stirring [22,23]. Furthermore, the higher SV observed in the flour dried at 90°C indicates a higher retrogradation level during cooling, leading to a firmer gel structure and thus higher staling in products made from this flour [10]. Consequently, VitAto flour produced at higher drying temperature, which results in a more viscose paste, may serve as a thickening and stabilizing agent in various food applications.

4. Conclusion

Based on the oven drying employed in this study, variation in drying temperatures significantly influenced the physical, functional and pasting properties of VitAto sweet potato flour, highlighting its potential applications in the food industry. Changes in drying temperature notably impacted parameters such as moisture content, water activity, colour, densities, water absorption capacity, oil absorption capacity, wettability, swelling and viscosities. These alterations can affect the suitability of the flour for various culinary and industrial applications by influencing the texture, stability and shelf life of food products.

Acknowledgements

This research was supported by Dana Penyelidikan Universiti 1.0 (Project No. R0357), Universiti Sultan Zainal Abidin, Terengganu, Malaysia.

References

[1] Chandrasekara, A., & Kumar, T. J. (2016). Roots and tuber crops as functional foods: a review on phytochemical constituents and their potential health benefits. International Journal of Food Science, 1, 1-15.

[2] Monteros-Altamirano, A., Paredes, D. & Buitrón-Bustamante, J. (2021). Genetic diversity of sweet potatoes [*Ipomoea batatas* (L) Lam.] in Ecuador. Genetic Resources and Crop Evolution, 68, 307–320.

[3] Storey, M. (2007). The harvested crop. In D. Vregdenhi, J. Bradshaw, C. Gebhardt, F. Govers, D. K.L. MacKerron, M.A. Taylor & H. A. Ross (Eds.) Potato biology and biotechnology (p. 441–470). Amsterdam: Elsevier.

[4] Mokhtar, M. N. (2022). Techno-economic evaluation of a process for the transformation of VitAto sweet potato into value-added products. Journal of Food Process Engineering, 45(8), e14043.

[5] Muhammad, R., Ikram, E., H., K., Sharif, M., S., Md. & Nor, N., Md. (2022). The physicochemical analysis and anthocyanin level of Malaysian purple sweet potato cracker. Current Research in Nutrition and Food Science, 10(3), 1030-1045.

[6] Zulkifli, N., A., Nor, Z., Omar, F., N. & Sulaiman, A. (2021). Proximate composition of Malaysian local sweet potatoes. Food Research 5(S1), 73-79.

[7] Lu, W., Maidannyk, A. & Lim, A.S.L. (2020). Carotenoids degradation and precautions during processing. In C.M. Galanakis (Ed.). Carotenoids: Properties, Processing and Applications (p. 223–258). New York: Academic Press.

[8] AOAC. (2005). The official method of analysis of the AOAC. 18th edition, Washington DC: Association of Officiating Analytical Chemists.

[9] Falade, K. O. & Christopher, A. S. (2015). Physical, functional, pasting and thermal properties of flours and starches of six Nigerian rice cultivars. Food Hydrocolloids, 44, 478-490.

[10] Tharise, N., Julianti, E. & Nurminah, M. (2014). Evaluation of physico-chemical and functional properties of composite flour from cassava, rice, potato, soybean and xanthan gum as alternative of wheat flour. International Food Research Journal, 21(4), 1641-1649.

[11] Oyewole, A. A., Abu, T. F. A. & Enujiugha, V. (2017). Emulsification and foaming properties of locust bean (*Parkiabiglobosa*) and pigeon pea (*Cajanuscajan*) seed flours and their protein isolates. Nutrition Food Science International Journal, 2(3), 1-9.

[12] Onwuka, G.I. (2005). Food analysis and instrumentation: theory and practice (p. 156-161). Lagos: Napthali.

[13] Chandanasree, D., Gul, K. & Riar, C. S. (2016). Effect of hydrocolloids and dry heat modification on physicochemical, thermal, pasting and morphological characteristics of cassava (*Manihot esculenta*) starch. Food Hydrocolloids. 52, 175–182.

[14] Carter, B., P., Galloway, M., T., Morris, C., F. & Weaver, G., L. (2015). The case for water activity as a specification for wheat tempering and flour production. Cereal Foods World, 60(4), 166-170.

[15] Hasmadi, M., Masri, H., Noraidah H., Zainol, K. & Jahurul, M., H. (2020). Functional properties of composite flour: a review. Food Research, 4(6), 1820-1831.

[16] Blessing, I.O. (2014). Chemical, functional & pasting properties of wheat (*triticumspp*) - walnut (*juglansregia*) flour. Food and Nutrition Sciences, 5, 1591-1604.

[17] Sittu, T. A., Sanni, L. O., Awonorin, S. O., Maziya-Dixon, B. & Dixon, A. (2007). Use of multivariate techniques in studying the flour making properties of some CMD resistant cassava clones. Food Chemistry, 101, 1606-1615.

[18] Berul, B.D. (2015). Effect of soaking & germination on proximate composition, mineral bioavailability & functional properties of chickpea flour. Food and Public Health, 5(4), 108-113.

[19] Ruke, S.R., Swant, A.A. & Kalse, S.B. (2022). Development and quality evaluation of biscuits from sweet potato and minor millets. International Journal of Science and Research, 11(4), 298-303.

[20] Twinomuhwezi, H., Awuchi, C.G. & Rachael, M. (2020) Comparative study of the proximate composition and functional properties of composite flours of amaranth, rice, millet and soybean. American Journal of Food Science and Nutrition, 6(1), 6-19.

[21] Desale, F. H. & Sasanatayart, R. (2017). Effect of drying temperature on functional and digestive properties of sweet potato flour. Proceedings of 63rd The IRES International Conference, 25-30.

[22] Abioye, V.F., Ade-Omowaye, B.I.O, Babarinde, G.O. & Adesigbin, M.K. (2011). Chemical, physico-chemical and sensory properties of soy-plantain flour. African Journal of Food Science 5(4), 176-180.

[23] Phattanakulkaewmorie, N., Paseephol, T. & Moongngarm, A. (2011). Chemical compsoitions and physico-chemical properties of malted sorghum flour and characteristics of gluten free bread. Word Academy of Science, Engineering and Technology, 57, 454-460.

IOP Conf. Series: Earth and Environmental Science **1460** (2025) 012028 doi:10.1088/1755-1315/1460/1/012028

Physical and Organoleptic Quality of Goat Meat Patties Using Red Onion Peel *(Allium cepa L.)* Powder

Aisna Rahma Tiarani, Herly Evanuarini* and Firman Jaya

Animal Product Technology, Faculty of Animal Science, Universitas Brawijaya, Malang, Veteran Street, Indonesia, 65145

*E-mail: herlyfpt@ub.ac.id

Abstract. Red onion peel powder is known as a naturally occurring antioxidant source and has high fiber content. This research was to develop goat patties using red onion peel powder as a source of natural antioxidants. The materials used were patties made from goat meat with addition red onion peel powder. The method used was a laboratory experiment using a Completely Randomized Design (CRD). The data were analyzed using ANOVA with 4 treatments and 5 replications. The treatments were addition onion peel powder 2% (T1), 4% (T2), 6% (T3) and without addition as a control (T0). The variables measured were pH value, water holding capacity (WHC), cooking loss, moisture content, antioxidant activity and organoleptics. The results showed that adding red onion peel powder can increase in water holding capacity (WHC) and antioxidant activity, while pH value, moisture content and cooking loss decreased, organoleptic quality including color, texture, taste and aroma were preffered by panelists. It can be inferred that goat patties with the addition of 6% red onion peel powder is considered the best based on its physical and organoleptic quality properties.

Keywords : goat meat, patties, red onion peel, resctructured meat

1. Introduction

Patties are a type of food made from minced meat and mixed with other ingredients such as spices and other seasonings, have a round shape, then processed by baking or frying [1]. The development of patties also provides an opportunity to increase the added value of meat products through the addition of functional ingredients such as dietary fiber, vegetable protein, and various vitamins and minerals. Quality assurance of a product can affect consumer appeal to consume. Mixing patties using starch or vegetable flour can improve the quality of low-fat meat products that function as binders or fillers. One important aspect in the development of patties is the use of high-quality ingredients and the right processing techniques to produce final products that meet food quality and safety standards. The weaknesses of patties products on the market are the lack of fiber, poor texture and easy to become rancid, so they have a short shelf life. One of the ingredients that can be used in making patties to improve the physicochemical quality and organoleptic quality of patties is red onion peel powder. Previous research related to red onion peel powder beef patties has been observed related to L a * b * color, lipid and protein oxidative stability during cold storage.

Red onion peel powder is generally considered as waste. The use of red onion peel powder as a source of fiber and antioxidants in patties not only increases the nutritional value of the product but can also reduce agricultural waste. Red onion peel powder can inhibit lipid oxidation, reduce oxidative stress, and has the potential to prevent chronic diseases like cancer and heart disease. [2]. The antioxidant content in red onion peel powder is quercetin flavonoid, which is found in free aglycone or in the form of mono and diglycosides. One of the main flavonoids that can be found in red onion peel powder is quertecin, which has anti-inflammatory and anticancer properties [3]. The proximate content of red onion peel powder includes moisture content (7-10%), protein content (1-3%), fat content (0.5-1%), ash content (4-6%) and crude fiber (50-60%) [4]. The advantages of red onion peel powder include being rich in fiber, containing high antioxidants, and having mineral and vitamin content that is beneficial for health. The purpose of this research was to assess the physical properties and organoleptic quality by utilizing red onion peel powder as a natural antioxidant in goat meat patties to improve the quality of goat meat patties.

2. Material and Methods

2.1 Material

The ingredients used to make goat meat patties are goat meat, tapioca flour, egg white, pepper, onion, garlic, salt, CMC bought from traditional market, Malang City, and ice cubes. Natural antioxidants added during the making of goat meat patties are red onion peel powder, processed with water washing and dried for 3 days using a natural sun drying method. After that, the red onion peel powder is ground using a dry mill to a size of 100 mesh and the natural antioxidant is ready to use.

pH analysis using pH meter [5]. WHC analysis using centrifugation method [5]. Cooking loss analysis using cooking and weighing [5]. Water content analysis using oven method [5]. Antioxidant activity analysis using DPPH method [5]. Organoleptic quality analysis using hedonic scale conducted by 5 semi-trained panelists [6].

2.2 Methods

In this research, the technique applied was a laboratory experiment using a Completely Randomized Design (CRD) and the data were analyzed using ANOVA with 4 treatments and 5 replications. If differences between treatments were obtained, continued with the Duncam Multiple Range Test (UJBD). The treatment of adding red onion peel powder to goat meat patties is as follows:

T0: without addition
T1: addition of red onion peel powder as much as 2% of the total ingredients
T2: addition of red onion peel powder as much as 4% of the total ingredients
T3: addition of red onion peel powder as much as 6% of the total ingredients

The process of making goat meat patties begins with grinding goat meat, salt and ice cubes using a meat grinder, then other ingredients such as tapioca flour, garlic, onion, egg white, pepper and CMC are added. Then the patties are formed into flat circles and baked at a temperature of 180±2°C, turned over until cooked.

3. Result and Discussion

3.1 pH Value

pH analysis is an indicator of the acidity or alkalinity of a product. pH analysis is one of the tests that can determine the level of freshness in meat products. The average results of the pH value of goat meat patties with the addition of red onion peel powder are presented in Figure 1.

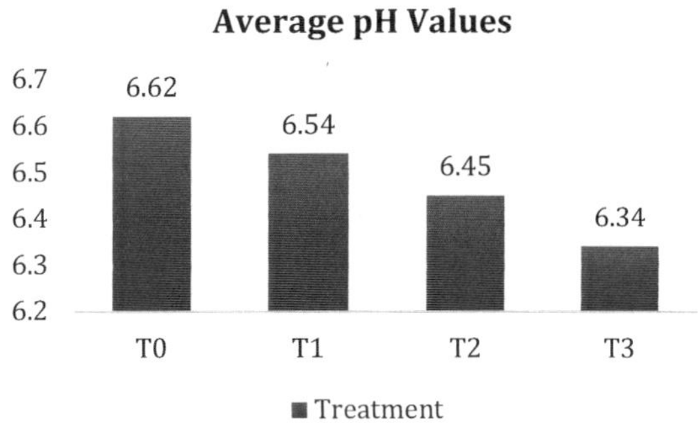

Figure 1. Average pH Values. Distinct letters within the figure indicate a statistically significant different (P<0.01)

According to Figure 1, it is recognized that the higher the addition of red onion peel powder to goat meat patties, the lower the pH value of the goat meat patties. This is because red onion peel has a pH value ranging from 6.8-6.3 [7]. Beef patties enhanced with additional ingredients of black cumin *(Nigella sativa)* as a synthetic antioxidant produce pH values ranging from 6.25-6.11, [8]. Another study stated that chicken patties with the inclusion of red beet peel flour *(Beta vulgaris L)* obtained a decrease in the pH of the patties, namely 6.08-5.76, this is influenced by the pH of the raw materials used, namely chicken meat and tapioca flour [9].

3.2 Water Holding Capacity (WHC)

WHC has a major impact on sensory acceptance such as juiciness, color and texture, thus it is a quality attribute of meat products. WHC levels can be affected by food additives such as salt because salt plays a role in inhibiting bacterial growth so that it can bind water components. The average results of water holding capacity (WHC) of goat meat patties with the addition of red onion peel powder are presented in Figure 2.

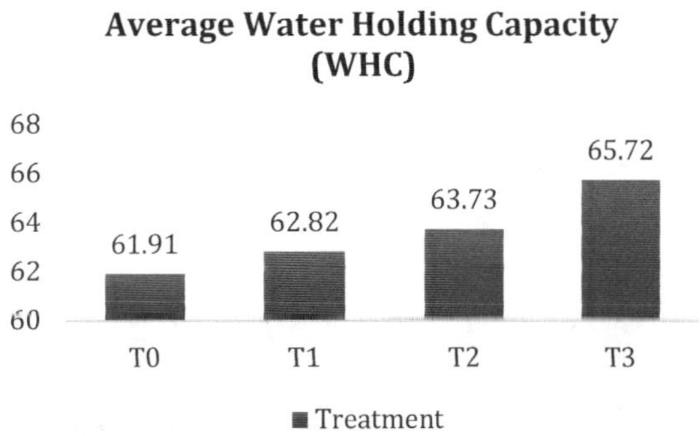

Figure 2. Average WHC. Distinct letters within the figure indicate a statistically significant difference (P<0.01)

According to Figure 2, it is known that the higher the addition of red onion peel powder to goat meat patties, the more it increases the water holding capacity (WHC). A high WHC value suggests that a food product has good water-holding capacity [10]. In a study on improving the gel characteristics pertaining to proteins within myofibrils derived from chicken meat patties combined incorporating dietary fiber from potatoes, it was found that the resulting water holding capacity (WHC) ranged from 52.1% - 62.19% [11]. Factors that can affect the WHC value are pH, fat content, protein content and the addition of dough ingredients to the meat. Another study stated that the addition of egg shell powder to beef patties obtained a water holding capacity (WHC) value of between 64.50% - 65.53%, this is because the calcium contained in egg shell powder can help increase the water binding capacity of beef patties [12].

3.3 Cooking Loss

Cooking loss occurs due to water evaporation, melted fat and loss of soluble protein which can be associated with changes in connective tissue during the heating process. Lower cooking temperatures and durations can minimize excessive cooking losses, by maintaining a more stable meat tissue structure so that it can maintain high water content. The average cooking loss results of goat meat patties enriched with the inclusion of red onion peel powder are presented in Figure 3.

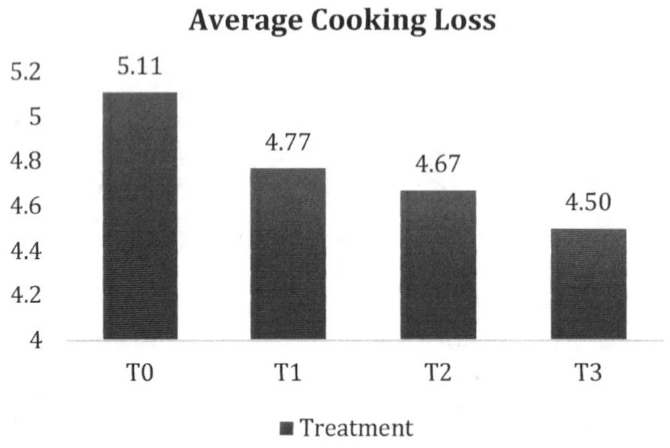

Figure 3. Average Cooking Loss. Distinct letters within the figure indicate a statistically significant difference (P<0.01)

According to Figure 3, it is recognized that the higher the addition of red onion peel powder to goat meat patties, the lower the cooking loss. The reduction in cooking loss is due to muscle protein denaturation resulting from water and fat loss [12]. In a study on adding eggshell powder to beef patties, cooking loss was found to range from 5.02% to 4.36%, attributed to the proteins present in beef and eggshell powder [12].

3.4 Moisture Content

Moisture content is an important factor determining product quality. Full-fat meat products will show higher water content or water activity. The average results of moisture content of goat meat patties enriched with the inclusion of red onion peel powder are presented in Figure 4.

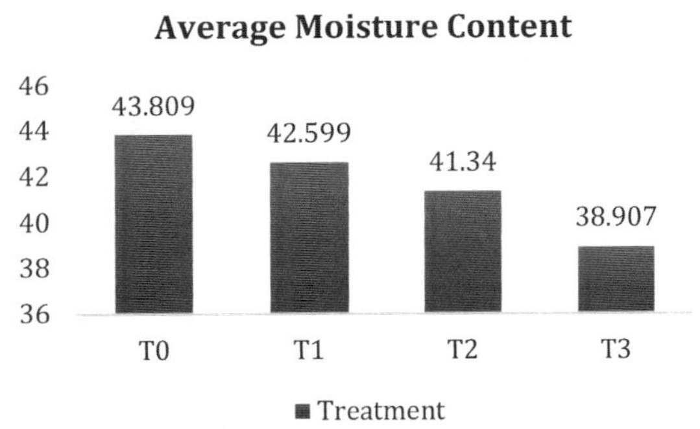

Figure 4. Average Moisture Content. Distinct letters within the figure indicate a statistically significant difference (P<0.01)

It's known that the higher the addition of red onion peel powder to goat meat patties, the lower the moisture content. In the study of the addition of egg shell powder to beef patties, the average moisture content ranged from 45.22% - 46.84%, the moisture content obtained decreased because the higher the percentage of egg shell powder addition, the lower the moisture content [12]. Research related to the impact of salt and polyphosphate on the oxidation of proteins in beef showed a moisture content of between 57.25% - 49.90%, this is due to the salt treatment during the grilling of patties which can retain more water [13].

3.5 Antioxidant Activity

The average results of antioxidant activity of goat meat patties enriched with the inclusion of red onion peel powder are presented in Figure 5.

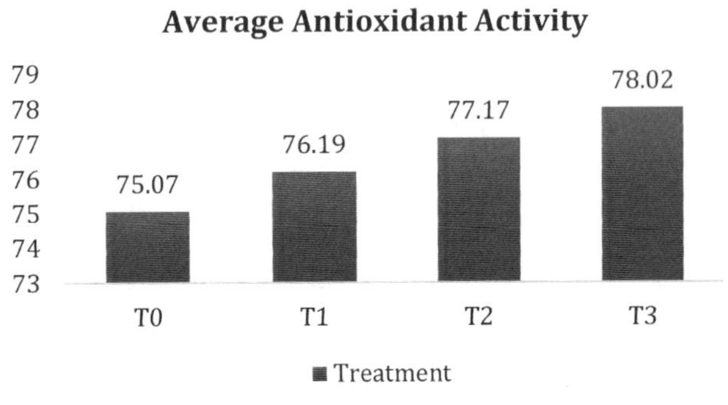

Figure 5. Average Antioxidant Activity. Distinct letters within the figure indicate a statistically significant difference (P<0.01)

According to Figure 5, it is recognized that the higher the addition of red onion peel powder to goat meat patties, the more it increases its antioxidant activity. This is because red onion peel powder has a high antioxidant activity of 78.88%, the high antioxidant activity is due to the presence of phenolic compounds [14]. In the study of chicken meat patties with the addition of rapeseed oil, the antioxidant activity was 75.46%, this can be attributed to the high concentration of vitamin E and unsaturated fatty acids [15].

3.6 Organoleptic Quality

Organoleptic evaluation is a method of assessment that involves the use of the five senses, including taste, sight, smell, hearing, and touch. Organoleptic quality assessment with a hedonic scale as follows: 1 (very dislike), 2 (dislike), 3 (enough), 4 (like), 5 (very like) [9]. The average results of the organoleptic quality of goat meat patties enriched with the inclusion of red onion peel powder are presented in Figure 6.

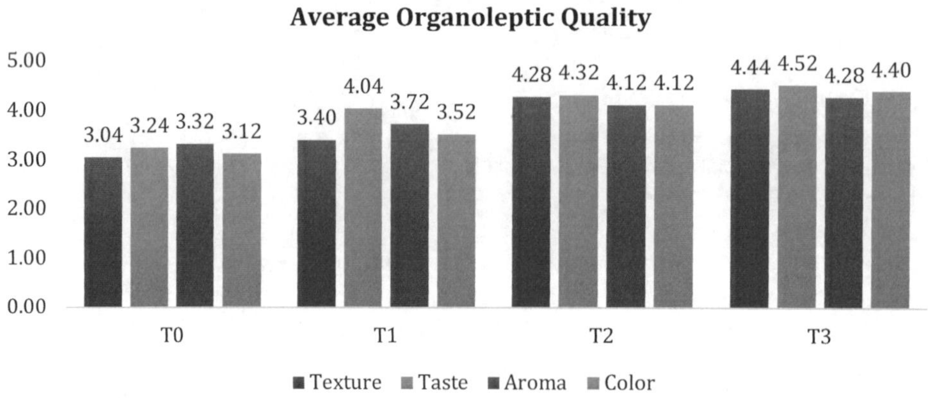

Figure 6. Average Organoleptic Quality

Goat meat patties with the addition of 6% red onion peel powder obtained the highest texture results. Red onion peel powder can help increase the compactness of patty products. This aligns with earlier research that stated that the addition of onion skin to beef patties provides a change in texture that is compact and juiciness [16]. Red onion peel powder can help cover or inhibit the development of fishy odors, so that it can increase the delicious taste of a product [17]. The appealing taste of goat meat patties enhanced with a 6% addition of red onion peel powder demonstrates this effect.

Goat meat patties enriched with the inclusion of 6% red onion peel powder have the highest acceptance score. This is caused red onion peel powder contains flavonoid compounds such as quercetin, in addition to functioning as an antioxidant, it also helps provide a delicious aroma [3]. Furthermore, the acceptance of the color score on patties with the addition of 6% red onion peel powder gets the highest results. Because when red onion peel powder is cooked and mixed with patty dough, it causes a natural pigment called anthocyanin to help provide red, purple or blue colors, so that patties become more attractive [18].

Conclusion

In this study, can be inferred that red onion peel powder serves like a natural antioxidant source for goat meat patties. Red onion peel powder has the best antioxidant activity content such as phenol. The best water holding capacity (WHC), pH value, cooking loss, moisture content and organoleptic quality are found in the addition of 6% red onion peel powder.

References

[1] Aviles, M. V., E. F. Naef, R. A. Abalos, L. H. Lound, and D. F. Olivera. 2024. Consumer Expectation and Perception of Ready-To-Eat Hybrid Chicken-Vegetable Patty. *International Journal of Gastronomy and Food Science*. 35 : 100847. DOI. https://doi.org/10.1016/j.ijgfs.2023.100847

[2] Benkeblia, N. 2020. Antioxidant activity and phenolic content of red onion peels: A potential bioactive source for functional food. *Journal of Food Biochemistry*. 44 (4) : e13119. DOI:10.1111/jfbc.13119

[3] Tyuryaeva, I. and O Lyublinskaya. 2023. Expected and Unexpected Effects Of Pharmacological Antioxidants. *International Journal of Molecular Sciences*. 24 (11) : 9303. DOI. https://doi.org/10.3390/ijms24119303

[4] Muzhahir, Z., Y. E. R. Unzilatirrizqi dan M. Fera. 2023. Analisa Proksimat Ekstrak Limbah Kulit Kedua Bawang Merah *(Allium Cepa L.)*. *Journal of Food and Agricultural Product*. 3 (2) : 114-123. DOI. https://doi.org/10.32585/jfap.v3i2.4608fey

[5] AOAC. (2005). Official Method of Analysis. 18th Edition. *Association of Official Analytical Chemistry International*. USA : AOAC Press. Maryland

[6] Agustina, R and R. Fadhil. 2021. Organoleptic test using the hedonic and descriptive methods to determine the quality of Pliek U. *IOP Conference Series: Earth and Environmental Science*. 644. 1 : (012006). IOP Publishing. DOI. 10.1088/1755-1315/644/1/012006

[7] Chadorshabi, S., S. Hallaj-Nezhadi, and Z. Ghasempour. 2022. Red onion skin active ingredients, extraction and biological properties for functional food applications. *Food Chemistry*. 386 : 132737. DOI. https://doi.org/10.1016/j.foodchem.2022.132737

[8] Rahman, M. H., M. S. Alam, M. M. Monir and K. Ahmed. 2021. Comprehensive effects of black cumin *(Nigella sativa)* and synthetic antioxidant on sensory and physicochemical quality of beef patties during refrigerant storage. *Journal of Agriculture and Food Research*. 4 : 100145. DOI. https://doi.org/10.1016/j.jafr.2021.100145

[9] Evanuarini, H., A. Susilo and D. Amertaningtyas. 2023. Improving the Physicochemical and Microstructural Qualities of Chicken Patties with the Addition of Red Beet Peel Flour *(Beta vulgaris L)* as a Filler. *Jurnal Ilmu-Ilmu Peternakan*. 33 (1). DOI. : 10.21776/ub.jiip.2023.033.01.07

[10] Gedikoğlu, A and A. D. Clarke. 2019. Quality attributes of citrus fiber added ground beef and consumer acceptance of citrus fiber added Turkish meatballs. *Food and health*. 5 (4) : 205-214. DOI. https://doi.org/10.3153/FH19022

[11] Feng, J., B. Kong, F. Sun and X. Xia. 2022. Effect of potato dietary fiber on the quality, microstructure, and thermal stability of chicken patty. *Foods*. 11 (24) : 3978. DOI. https://doi.org/10.3390/foods11243978

[12] Evanuarini, H., A. Susilo, U. Yanuhar and A. D. Kurniawati. 2023. Utilisation of chicken eggshell nanopowder on physicochemical quality, microstructure, and amino acid profile of beef patties. *Czech Journal of Food Sciences*. 41 (6). DOI. https://doi.org/10.17221/156/2023-CJFS

[13] Li, Y., C. Xue, W. Quan, F. Qin, Z. Wang, Z. He and J. Chen. 2021. Assessment the influence of salt and polyphosphate on protein oxidation and Nε-*(carboxymethyl)* lysine and Nε-*(carboxyethyl)* lysine formation in roasted beef patties. *Meat Science*. 177 : 108489. DOI. https://doi.org/10.1016/j.meatsci.2021.108489

[14] Shalaby, E. A., S. M. Shanab, S. A. Fayed, H. M. Abde Gawad, M. F Nasr and H. S. Gaballa. 2022. Chemical constituents and biological activity of successive extracts and silver nanoparticles from red onion peels. *Egyptian Journal of chemistry*. 65 (132) : 593-604. DOI. https://dx.doi.org/10.21608/ejchem.2022.136419.6010

[15] Rakasivi, K. G. J and K. B. Chin. 2022. Antioxidant activity of Cinnamomum cassia extract and quality of raw chicken patties added with C. cassia powder and Pleurotus sajor-caju powder as functional ingredients during storage. *Animal bioscience*. 35 (8) : 1279. DOI. https://doi.org/10.5851%2Fkosfa.2023.e71

[16] Wang, C., Y. Wang, Y. Song, M. Ren, Z. Gao and J. Ren. 2024. Effect of onion skin powder on color, lipid, and protein oxidative stability of premade beef patty during cold storage. *Scientific Reports*. 14 (1) : 20816. DOI. https://doi.org/10.1038/s41598-024-71265-x

[17] Bedrníček, J., J. Kadlec, I. Laknerová, J. Mráz, E. Samková, E. Petrášková and P. Smetana. 2020. Onion peel powder as an antioxidant-rich material for sausages prepared from mechanically separated fish meat. *Antioxidants*. 9 (10) : 974. DOI. https://doi.org/10.3390/antiox9100974

[18] Zhao, X., Z. Liang, Z. Wang, P. Zhang and Z. Fang. 2024. Effect of natural colorants on the quality attributes of pea protein-based meat patties. *Food Bioscience*. 59 : 103976. DOI. https://doi.org/10.1016/j.fbio.2024.103976

IOP Conf. Series: Earth and Environmental Science **1460** (2025) 012029 doi:10.1088/1755-1315/1460/1/012029

Enzymolyzed Enoki Mushroom Improves Textural and Functional Properties of Low-Grade Beef

F H Barido[1,2*]

[1]Faculty of Animal Science, Universitas Sebelas Maret, Jl. Ir Sutami 36A Kentingan, Surakarta, Jawa Tengah, Indonesia 57126
[2]Halal Research Center and Services (HRCS), Institute for Research and Community Service, Universitas Sebelas Maret, Surakarta, Jawa Tengah 57126, Indonesia

*E-mail : farouq.heidar@staff.uns.ac.id

Abstract. The quality and tenderness of low-grade beef are significantly influenced by its textural profiles, which are often compromised by factors such as inadequate nutrition, animal stress, and substandard handling. To address this, we investigated the impact of Enoki mushroom hydrolysate on the textural and functional properties of low-grade beef. Enoki mushrooms (*Flammulina filiformis*) were enzymatically hydrolyzed to produce a protein hydrolysate, which was then applied to beef samples through marination. The results showed that the hydrolyzed Enoki mushroom extract significantly reduced collagen content and shear force values, indicating improved tenderness. Additionally, the antioxidant capacity of the treated beef samples was enhanced, as evidenced by lower ABTS IC50 values. These findings suggest that enzymolysis of Enoki mushrooms can be a viable natural method for improving the quality and tenderness of low-grade beef, making it a promising alternative in the meat industry.

Keywords: beef, enoki mushroom, enzymatic hydrolysis, functional food, texture profiles

1. Introduction

Texture profiles significantly influence consumer decisions to repurchase meat and meat products. These profiles are the result of intricate factors that occur during various stages, including ante-mortem, rigour-mortis, and post-mortem animal handling [1]. Several factors may contribute to the development of tough meat textures, including inadequate nutrition, animal stress, the age at which they are slaughtered, and poor handling and processing practices. These elements can significantly affect the quality and texture of meat, leading to a less desirable culinary experience. Consequently, the meat industry employs various methods to improve the sensory attributes of meat, particularly texture profiles, through chemical and physical means [2]. However, both methods are often less favoured by consumers; physical methods lead to increased production costs and can negatively impact meat quality, while chemical methods may result in residues and metabolic byproducts that pose health risks. As a result, there is a growing interest in exploring natural alternatives to these methods [3].

The primary focus in enhancing meat texture profiles is the reduction of collagen content and actomyosin bonds, as well as the upregulation of endogenous proteolytic enzyme activity [2]. Several methods have been documented to achieve this improvement. Notably, the introduction of adenosine 5'-monophosphate (AMP) initiates a reaction that releases the actin-myosin bond, leading to the weakening of actomyosin cross-links and, as a result, an increase in meat tenderness [4]. Concurrently, the upregulation of proteolytic enzymes, specifically calpains, cathepsins, and caspases, results in the accelerated degradation of crucial proteins. Our previous research suggests that AMP significantly contributes to this upregulation, leading to a heightened myofibrillar fragmentation index and an elevated concentration of soluble proteins [5, 6].

Enoki mushrooms are gaining recognition in meat industries and general cuisine landscape for their role in enhancing flavor and texture of various culinary creations. These fungi are not merely a source of macronutrients like essential proteins and polysaccharides, which are vital for energy provision and tissue growth, but they also offer a rich spectrum of micronutrients, including vitamins, magnesium, phosphates, minerals, and bioactive compounds [7]. Such nutrients are crucial for bolstering overall health and elevating food quality. Recent research indicates that Enoki mushrooms can markedly contribute to the higher perception of quality and functionality of meat products. They are packed with bioactive compounds such as beta-glucans and glycoproteins, supported their positive effects on health, including antioxidant, anti-inflammatory, and immunomodulatory actions [8]. The mushrooms' capacity to improve texture is linked to their enzyme profile, particularly serine and metalloprotease enzymes, which aid in tenderizing. Moreover, the unique structure of Enoki mushrooms, characterized by their long, slender stems and small caps, contributes both a pleasant crunch and aesthetic enhancement to dishes, securing their place in a variety of culinary traditions.

The research gap in the adoption of protein hydrolysates for improving the textural properties and quality of low-grade beef is significant. This study was intended to elucidate the impact of protein hydrolysates derived from Enoki mushrooms on the quality and tenderness of low-grade beef. The findings could broaden the range of natural tenderizing options available to the meat industry.

2. Materials and Methods

2.1. Experimental preparation

The Enoki mushroom (*Flammulina filiformis*) was purchased from local market in Surakarta, Indonesia. The manufacturing of protein hydrolysate was referred to as protocol by [9] with minor alterations. Oven dried mushroom (60°C, 4 hours) was made into slurries using food blender. The slurries were brought to adjustment until pH value of 7.0 using 2.0 M NaOH. Following adjustment, flavourzyme at 1.5% from mushroom weight were introduced into slurries, wherein the enzyme was previously mixed with deionized water at 1:5 (w/v) ration.

Mixed mushrooms and enzymes were subsequently incubated at 40oC for 2 hours within a water bath system. Following hydrolyzation, slurries were cooled to room temperature, re-adjusted to neutral pH, and subjected to centrifugation with 4,000 rpm at 25oC for thirty minutes. The hydrolyzed Enoki mushrooms were then stored to -25oC prior to application to beef meat.

To understand its effect on beef quality and texture profiles, thirty-six uniformed-size beef samples (3 × 3 × 3 cm) were subjected to dipping marination into Enoki mushroom hydrolysate for one hour at 4 ± 2°C. The results were then compared to negative control (NC), receiving marination using saline buffer, and Enoki mushroom extracts (EMs). Each sample was prepared in three replicates.

2.2. pH value

The pH value serves as a fundamental quality parameter for assessing the quality of meat and its products. The current study evaluated the pH value post-marination at a specified time. The assessment was conducted by homogenizing a 5 g meat sample with 45 mL of distilled water using a PH91 homogenizer (SMT Co., Ltd., Chiba, Japan) at a speed of 10,000 rpm for one minute. Subsequently, the pH level was measured with a Seven Easy pH meter (Mettler-Toledo GmbH, Greifensee, Switzerland), which was previously calibrated using technical buffer solutions at pH 4.01, 7.00, and 9.00. Each measurement was performed in triplicate to ensure accuracy and reproducibility.

2.3. Antioxidant capacity

The 2,2′-azino-bis-(3-ethylbenzothiazoline-6-sulfonic acid) (ABTS) assay was chosen to evaluate the antioxidant capacity from beef sample. The protocol was referring previous study by [10] with minor modification. The evaluation of the ABTS assay was initiated by preparing a 0.1 mM ABTS solution in methanol. They were brought to homogenization at about 25 grams of beef in 225 mL of phosphate-buffered saline (PBS) to result a consistent beef extract. Following homogenization, centrifugation was conducted at 3000 rpm for 15 minutes to separate solids from the liquid extract. An aliquot extract at 1 mL was thoroughly homogenized with 1 mL of overnight ABTS solution in a test tube, let the mixture to stand at room temperature without illuminant for 30 min, and evaluated using UV-Vis spectrophotometer at 517 nm.

2.4. Shear force and collagen content

Collagen content was determined according to a method by [11], a 25-gram homogenized portion of the beef in 225 mL of phosphate-buffered saline (PBS) was centrifuged at 3000 rpm for 15 minutes to separate solids from the liquid extract. The total collagen in the raw muscle tissue is quantifiable using methods outlined in PN-ISO 3496:2000, where collagen content is expressed as a percentage by weight and in mg of collagen per g of meat, utilizing a coefficient of 7.25. To determine soluble collagen, heat the homogenate at 77 °C for 70 minutes, then centrifuge at 4000 × g for 10 minutes to isolate the soluble from the insoluble fraction.

Meanwhile for shear force evaluation, cooked beef samples with internal temperature of 70 °C was measured using a Warner-Bratzler shear press instrument, which involves cutting the cooked beef into cylinders and shearing perpendicular to the muscle fibers. Record the maximum force needed to cut through the beef in kgf/cm^2. This method is a standard in meat science for assessing tenderness, correlating to collagen content. Higher levels of total and insoluble collagen typically indicate tougher meat, as evidenced by increased shear force values.

2.5. Statistical analysis
Data for the specified variables, collected in a cold storage condition over a fifteen-day period, were subjected to a two-way multivariate analysis of variance (MANOVA) using R-studio version 3.6.1 (Vienna, Austria). The Duncan's multiple range test were subsequently employed to observe statistical differences, where P-values below 0.05 were indicative of statistically significant differences.

3. Results and Discussion

3.1. pH value
The beef samples' pH in this study were significantly influenced by the different treatment groups, as shown in Table 1. The pH values varied from 5.65 under NC treatment to 6.04 under EMs treated groups ($p<0.05$). Although significantly differed from NC, enzymolyzed mushroom did not show any significant difference when compared to that of EMs. This is in accordance with prior study, highlighting the importance of pH in determining meat quality through its impact on protein solubility, water-holding capacity, and flavor profile [12]. The variation in pH levels across the treatment groups is likely due to the enzymatic hydrolysis treatment, which can modify the biochemical milieu and enzymatic processes within the muscle tissue [13]. It is noteworthy that the elevated pH values exhibit a significant correlation with the cellular reactions occurring at the inner muscle environment [3, 5]. Additionally, measured level of pH from meat and meat products significantly affects multiple quality attributes such as grading characteristics, color, water-holding capacity, cooking loss, shrinkage, tenderness, as well as texture. Comprehending these relationships is important effort in order to elevate quality attributes and processing efficacy from meat and meat products. Our prior research indicated that the elevated pH levels in the treated samples may suggest an alkaline influence resulting from the mushroom extracts [6, 9].

Table 1. Beef samples' pH, collagen, and shear force value following treatment using various marinade solutions

Parameters	NC	EMs	EnEMs	SEM[2]	p-value
pH	5.65^{b}	6.04^{a}	6.01^{a}	0.02	<0.05
Collagen	2.64^{a}	1.89^{b}	1.21^{c}	0.12	<0.05
Shear force value (kgf/cm^2)	5.45^{a}	4.87^{b}	4.59^{Ba}	0.26	<0.05

[1]NC, negative control; EMs, marination with Enoki mushroom extract; EnEMs, enzymatically hydrolyzed Enoki mushroom extract
SEM, standard error of the mean
[a-b] superscripts at the row reflects statistical significances after treatment using various marinade solutions ($P<0.05$).

Collagen content was quantitatively assessed alongside shear force values to evaluate meat tenderness. The collagen levels in the treated samples exhibited significant variability, with specific values ranging from 1.21 to 2.64. Enzymolysis treatment showed to significantly lower collagen content, showing the lowest proportion among treatments ($p<0.05$). An increased collagen content correlates with greater meat toughness due to its fibrous structure, which can impede the tenderness of muscle fibers [14]. In the present study, treatments that led to reduced collagen levels were associated with enhanced tenderness, as evidenced by lower shear force measurements. These results substantiate the hypothesis that enzymolysis potentially diminishes collagen content, thereby improving the tenderness of meat products.

Shear force values were determined to assess the tenderness of beef samples across various treatment groups. A statistically significant differences was observed among the groups. Notably, the treatment group exhibiting the lowest shear force value of 4.59 kgf/cm^2 under EnEMs, and 4.87 kgf/cm^2, demonstrated superior tenderness in comparison to that of NC, which documented a shear force of 5.45 kgf/cm^2. These findings are consistent with existing literature that highlights a direct correlation between collagen content and shear force; specifically, as collagen levels decrease, meat tenderness tends to increase [15]. The mechanisms underlying these observations may involve the enzymatic breakdown of collagen during processing or cooking, which has been shown to enhance meat tenderness by converting resilient connective tissues into more soluble forms [16]. Furthermore, factors such as moisture retention and fat content are also likely to influence both collagen solubility and overall tenderness, contributing to the sensory attributes of the meat.

3.2. Antioxidant capacity

The antioxidant capacity of the hydrolyzed beef samples was measured using ABTS scavenging assay, performed 24 hours post-marination. This measurement was to evaluate the increase in antioxidant capacity resulting from the applied marination treatments. As depicted in Figure 1, it was revealed that samples treated with EnEMS exhibited the lowest IC50 value among all groups, reflecting significant enhancement in antioxidative potential. In contrast, the non-treatment group exhibited the highest ABTS IC50 value, recorded at 60.21 mg/mL. The IC50 values for EMs and EnEMS were 30.15 mg/mL and 25.41 mg/mL, respectively, demonstrating a statistically significant difference ($p<0.05$). The observed results indicate the efficacy of enzymolysis in augmenting and alleviating functional properties of low-grade beef.

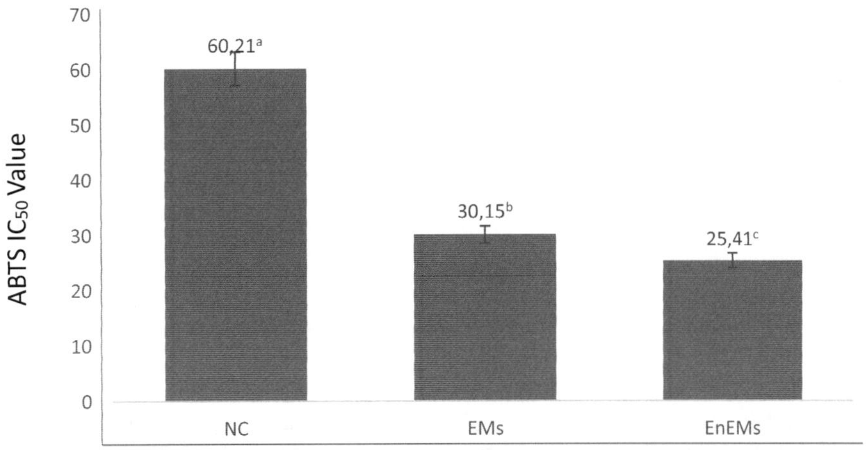

Figure 1. The ABTS IC_{50} value of the beef samples following treatment using various marinade solutions
[1]NC, negative control; EMs, marination with Enoki mushroom extract; EnEMs, enzymatically hydrolyzed Enoki mushroom extract

[a-b] superscripts at the displayed figure reflects statistical significances after treatment using various marinade solutions (P<0.05).

Antioxidant measurement using ABTS assay is a well-established method for assessing antioxidative power for broad range of substances, including meat extracts. It evaluates the capability of meat extract derivatives to neutralize unstable free radicals, thus reduce the risk of oxidative damage. The evaluation principle is based on the diminished activity of the ABTS radical cation (ABTS+), produced when ABTS reacts with an oxidizing agent, usually potassium persulfate [17]. The ABTS+ radical, known for its blue-green hue and specific absorption of illuminants at a specific wavelength, can be neutralized by antioxidants, which donate electrons, resulting in a reduced absorbance of the radical. The IC50 value in this assay indicates the concentration of an antioxidant needed to decrease the ABTS+ radical activity by 50%. A lower IC50 value suggests a higher antioxidant potential, indicating that less substance is required to exert the effect. The reduction in IC50 values after enzymolysis in this study is ascribed to both the increased release of polyphenolic and flavonoid counterparts, along with the deactivation of oxidative enzymes, located at the inner cell environment that possibly led to destruction, and degradation of antioxidative potentials in fresh mushrooms [18].

4. Conclusion

The findings from this study indicate that enzymolysis can effectively enhance essential beef attributes comparable to that of non-treated groups. Consequently, it is recommended that the immersion of beef in hydrolyzed Enoki mushroom combined serves as a viable alternative for improving the quality of beef. This approach not only augments the meat's antioxidant properties while maintains its textural properties, which is crucial for consumer acceptance.

Acknowledgements

The study of the enzymolyzed Enoki mushroom improves textural and functional properties of low-grade was fully funded by the Pusat Studi Halal Research Center and Services (HRCS) 2024, Universitas Sebelas Maret, Surakarta, under contract number of 194.2/UN27.22/PT.01.03/2024.

References

[1] Aalhus, J. L. (2020). In *Quality and Grading of Carcasses of Meat Animals* (pp. 51-81). CRC press.
[2] Barido, F. H., & Lee, S. K. (2021). *Animal Bioscience, 34*(11), 1859. https://doi.org/10.5713/ab.20.0831
[3] Cho, J., Barido, F. H., Kim, H. J., Kim, H. J., Kim, D., Shin, D. J., & Jang, A. (2023). *Food Science of Animal Resources, 43*(1), 25. https://doi.org/10.5851%2Fkosfa.2022.e53
[4] Chaosap, C., Sahatsanon, K., Sitthigripong, R., Sawanon, S., & Setakul, J. (2021). *Foods, 10*(10), 2319. https://doi.org/10.3390/foods10102319
[5] Barido, F. H., & Lee, S. K. (2021). *Poultry Science, 100*(5), 101056. https://doi.org/10.1016/j.psj.2021.101056
[6] Barido, F. H., Kang, S. M., & Lee, S. K. (2022). *Foods, 11*(3), 422. https://doi.org/10.3390/foods11030422
[7] Dhanapal, D., & Rajoo, B. (2023). *International Journal of Food Engineering, 19*(12), 573-591. https://doi.org/10.1515/ijfe-2023-0073
[8] You, S. W., Hoskin, R. T., Komarnytsky, S., & Moncada, M. (2022). *ACS Food Science & Technology, 2*(8), 1184-1195. https://doi.org/10.1021/acsfoodscitech.2c00107
[9] Ang, S. S., & Ismail-Fitry, M. R. (2019). *Food technology and biotechnology, 57*(4), 472. https://doi.org/10.17113%2Fftb.57.04.19.6294
[10] Kimatu, B. M., Zhao, L., Biao, Y., Ma, G., Yang, W., Pei, F., & Hu, Q. (2017). *Food chemistry, 230*, 58-67. https://doi.org/10.1016/j.foodchem.2017.03.030

[11] Kolar, K. (1990). *Journal of the Association of official Analytical Chemists*, *73*(1), 54-57. https://doi.org/10.1093/jaoac/73.1.54

[12] Cho, J., Barido, F. H., Kim, H. J., Kwon, J. S., Kim, H. J., Kim, D., Hur, S. J., & Jang, A. (2023). *Food Science of Animal Resources*, *43*(1), 139. https://doi.org/10.5851%2Fkosfa.2022.e67

[13] Dewi, M. M., Barido, F. H., Hertanto, B. S., Cahyadi, M., Kartikasari, L. R., Sujiwo, J., & Jang, A. (2024, June). In *IOP Conference Series: Earth and Environmental Science* (Vol. 1360, No. 1, p. 012032). IOP Publishing.

[14] Tang, C., Zhou, K., Xie, Y., Ma, Y., Zhu, Y., Zhang, W., Wang, Z., Zhou, H., & Xu, B. (2023). *Journal of Texture Studies*, *54*(2), 258-267. https://doi.org/10.1111/jtxs.12734

[15] Barido, F. H., & Lee, S. K. (2021). *Food Science of Animal Resources*, *41*(5), 869. https://doi.org/10.5851%2Fkosfa.2021.e43

[16] Roy, B. C., & Bruce, H. L. (2024). *Critical Reviews in Food Science and Nutrition*, *64*(25), 9280-9310. https://doi.org/10.1080/10408398.2023.2211671

[17] Echegaray, N., Pateiro, M., Munekata, P. E., Lorenzo, J. M., Chabani, Z., Farag, M. A., & Domínguez, R. (2021). *Molecules*, *26*(13), 3880. https://doi.org/10.3390/molecules26133880

[18] Bortnowska, G., Przybylska, S., & Iwański, R. (2021). *Journal of food science and technology*, *58*, 110-120. https://doi.org/10.1007/s13197-020-04519-x

IOP Conf. Series: Earth and Environmental Science **1460** (2025) 012030 doi:10.1088/1755-1315/1460/1/012030

Effect of Adding Sweet Corn Flour to Milk on Yoghurt Quality

Indra Maulani*, Anang Mohamad Legowo, Nurwantoro

Department of Animal Husbandry, Faculty of Animal Husbandry and Agriculture, Universitas Diponegoro, *Semarang, Jawa Tengah, 50275 – Indonesia*

*E-mail: lanifilandiyo1@gmail.com

Abstract. The purpose of research was to determine the effect of various different percentages of sweet corn flour on the total lactic acid bacteria (LAB), total soluble solids (TSS), total lactic acidity, pH, viscosity, syneresis, and hedonic quality of cow's milk yoghurt. This study used a completely randomized design with 4 treatments with addition of sweet corn flour: T0 (0%), T1 (2%), T2 (4%), and T3 (6%) and 5 replicates. The data were analyzed using Analysis of Variance (ANOVA) test followed by the Duncan Multiple Range Test (DMRT) and Kruskall Wallis test followed by Mann-Whitney test. The results showed that the addition of sweet corn flour had a significant effect ($p<0.05$) on TSS, total lactic acidity, pH, viscosity, and syneresis values, but had no significant effect ($p>0.05$) on the total LAB. The study concluded that adding sweet corn flour to cow's milk increased TSS, total lactic acidity, and viscosity of the yoghurt but lowers the pH value and the syneresis. The higher sweet corn flour added to cow's milk yoghurt up to 6% can improve the quality of the yoghurt, such as increasing the viscosity value and reducing syneresis value. In the hedonic quality tests showed an effect on color and texture, but no effect on aroma, taste, and overall preference.

Keywords : yoghurt, sweet corn flour, total lactic acid bacteria

1. Introduction

Cow's milk is a vital food that helps people meet their nutritional needs because it contains all the necessary nutrients (Resnawati, 2020). Some components contained in milk include fat, protein, lactose, ash, water, and various other substances such as enzymes, citrates, phospholipids, and vitamins like A, B, and C. The nutritional content of cow's milk consists of 3.8% fat, 3.2% protein, 87.25% water, 4.7% lactose, 0.855% ash, and 12.75% dry matter, forming its nutrient composition (Amen *et al.*, 2020). Milk with complete nutrient content provides many benefits for humans, such as increasing concentration, physical fitness, supporting growth, boosting immunity, and preventing osteoporosis (Suprapto *et al.*, 2014). However, milk is an excellent medium for bacterial growth, making it prone to spoilage (Susilaningrum *et al.*, 2022). Therefore, proper handling is required to enhance its functional value by processing it into various products. One easily applicable dairy product is yoghurt (Zulaikhah and Sidhi, 2021). Yoghurt is a dairy product resulting from fermentation by lactic acid bacteria (LAB) (Komalasari and Yoga, 2022). However, yoghurt production generally uses sweeteners like sucrose to provide sweetness. Sucrose is added in solid, powdered, and crystalline forms (Maharani and Ayuningtyas, 2018). Sweeteners such as sucrose can be replaced with sweet corn in the yoghurt-making process. Adding sweet corn flour in yoghurt production will yield yoghurt with different qualities. Flour is a thickening agent used in yoghurt to increase viscosity (Wijayanti *et al.,* 2015), and is expected to reduce syneresis (separation of solids and liquids) and enhance preference. Based on this, research is needed to study the effect of adding various amounts of sweet corn flour to yoghurt.

2. Materials and Methods

2.1 Materials

The materials used include cow's milk from Universitas Diponegoro dairy farm, homemade sweet corn flour, and a mixed bacterial starter (*S. thermophilus*, *L. bulgaricus*, and *L. acidophilus*) for yoghurt obtained from CV Citra Nasional. The main equipment used in this study includes an oven, grinder, pH meter, thermometer, analytic balance, centrifuge, stove, pipette and incubator.

2.2 Methods

The study involves the production of sweet corn flour, the production of yoghurt with added sweet corn flour, and the testing of parameters. The tested parameters are total LAB, TSS, total lactic acid, viscosity, syneresis, and hedonic quality. This research employs a completely randomized design with 4 treatments and 5 replications. The treatments for adding sweet corn flour are P0 (0%), P1 (2%), P2 (4%), and P3 (6%) based on the volume of milk (w/v).

2.3 Production of Sweet Corn Flour

The production of sweet corn flour follows the method of Irawan *et al.* (2017), where sweet corn is dried in an oven at 60°C for 48 hours or until the moisture content reaches ± 17%. After that, the dried sweet corn is shelled and dried again for 24 hours at 50°C in the oven or until the moisture content drops below 14%. The shelled sweet corn is then ground using a grinder and sieved to 80 mesh to produce finer sweet corn flour.

2.4 Production of Yoghurt with the Addition of Sweet Corn Flour

The yoghurt production follows the modified method of Alfaridhi *et al.* (2013) Fresh milk (10L) is divided into 20 cups, each with a volume of 500mL. Sweet corn flour is add full stop to the fresh milk in the treatments P0 (0%), P1 (2%), P2 (4%), and P3 (6%) of the milk volume (w/v). The

IOP Conf. Series: Earth and Environmental Science **1460** (2025) 012030 doi:10.1088/1755-1315/1460/1/012030

milk is pasteurized at 70°C for 15 seconds. The pasteurized milk is then cooled to 45°C. After that, a yoghurt bacterial starter is added to the pasteurized milk at 5% of the milk volume. The milk is incubated at 37°C for 6 hours. The finished yoghurt can be stored at 3-5°C. The total LAB (Huwae *et al.,* 2022), TSS (Wahyudi and Dewi, 2017), total lactic acid (Savitry *et al.,* 2017), pH value (Meilanie *et al.,* 2018), viscosity (Cahyani *et al.,* 2019), syneresis (Dewi *et al.,* 2019), and hedonic quality (Novidahlia *et al.,* 2019) were then measured.

2.5 Data Analysis

The data collected from the total LAB, TSS, total lactic acid, pH value, viscosity, and syneresis were analyzed using Analysis of Variance (ANOVA) followed by Duncan Multiple Range Test (DMRT). The hedonic quality (preference level) was analyzed using the Kruskal-Wallis test followed by the Mann-Whitney test.

3. Results and Discussion

3.1 Total Lactic Acid Bacteria (LAB)

The results of the total LAB in cow's milk yoghurt with the addition of sweet corn flour in different treatments are presented in Table 1. The analysis of variance revealed no significant effect ($p>0.05$) of adding sweet corn flour on the total LAB value. The total LAB in cow's milk yoghurt with the addition of sweet corn flour did not meet the standard of 10^5, as good yoghurt should have a minimum standard of 10^7 CFU/mL. This is supported by BSN (2009), which stated that the minimum total LAB in yoghurt is 10^7 CFU/mL.

The total LAB levels that are below the standard indicates that lactic acid bacteria cannot grow optimally. The growth of LAB is influenced by the availability of nutrients in the medium. LAB growth is suboptimal because LAB cannot efficiently utilize the sugars in sweet corn flour, making lactose the primary sugar source for LAB growth. This is in line with Umar *et al.* (2014), who stated that lactose plays the most crucial role during the LAB fermentation process, as it is used by LAB as a carbon and energy source. However, the total LAB decreases with increased sweet corn flour addition, possibly due to the declining pH and increasing acidity. This is supported by Rusdhi *et al.* (2023), who stated that a decrease in LAB count indicates bacterial activity, reflected in the pH reduction.

Table 1. Results of Total LAB, TSS, Total Lactic Acid, pH Value, Viscosity, and Syneresis

Parameter	P0	P1	P2	P3
LAB log(CFU/mL)[ns]	5.59 ± 0.075	5.51 ± 0.125	5.58 ± 0.227	5.38 ± 0.214
TSS (° Brix)	7.26 ± 0.195[a]	7.64 ± 0.182[b]	8.62 ± 0.303[c]	10.06 ± 0.279[d]
Total Lactic Acid (%)	0.73 ± 0.005[a]	0.88 ± 0.006[b]	1.01 ± 0.025[c]	1.25 ± 0.009[d]
pH Value	4.43 ± 0.027[d]	4.36 ± 0.017[c]	4.28 ± 0.017[b]	4.22 ± 0.020[a]
Viscosity (cp)	134.65 ± 7.047[a]	191.25 ± 4.778[b]	205.79 ± 10.625[c]	245.25 ± 8.387[d]
Syneresis (%)	55.61 ± 1.826[d]	47.28 ± 1.048[c]	36.59 ± 1.309[b]	18.79 ± 1.258[a]

Note: ns = non-significant (no significant difference at p<0.05).

Different superscripts in the same row indicate significant differences (p<0.05).

P0, P1, P2, and P3 = Concentrations of Sweet Corn Flour 0% (P0), 2% (P1), 4% (P2), 6% (P3) w/v.

3.2 Total Soluble Solid (TSS)

The results of total soluble solid in cow's milk yoghurt with the addition of sweet corn flour at different treatments are presented in Table 1. The analysis of variance revealed a significant effect ($p<0.05$) of adding sweet corn flour on the TSS value. The results indicate that the average TSS values with different treatments are above the standard. The minimum standard for total soluble solid is 8.2 °Brix (BSN, 2009).

The addition of sweet corn flour in making yoghurt causes an increase in the total soluble solid because, during fermentation, lactose and sucrose are broken down by LAB. This occurs because sweet corn contains sucrose and reducing sugars (glucose and fructose). Sugars are a major component of soluble solid, so adding them to yoghurt increases the TSS. This is supported by Fatmawati *et al.* (2017), who stated that sweet corn contains reducing sugars (glucose and fructose), sucrose, polysaccharides, and starch. This statement is further supported by Wiyono *et al.* (2023), who stated that high sugars from fruits or other sources added to yoghurt contribute to higher soluble solid components.

3.3 Total Lactic Acid and pH Value

The results of total lactic acid and pH value in cow's milk yoghurt with the addition of sweet corn flour in different treatments are presented in Table 1. The analysis of variance revealed a significant effect ($p<0.05$) of adding sweet corn flour on total lactic acid and pH values. As the total lactic acid value increases, the pH value decreases. The results show that the average total lactic acid values range between 0.73-1.25%. This indicates that the yoghurt quality requirements are met since the lactic acid standard, according to BSN (2009) is 0.5-2.0%.

Increased lactic acid is followed by a decrease in pH value. The results show that the average pH values range between 4.22-4.43. This means that the yoghurt quality requirements are met since the pH value standard according to BSN (2009) is 3.8-4.5. According to Oktavia *et al.* (2015), the acidity value is closely related to the pH value, where increased metabolic activity leads to higher lactic acid production and lower pH values.

3.4 Viscosity and Syneresis Values

The results of viscosity and syneresis values in cow's milk yoghurt with the addition of sweet corn flour in different treatments can be seen in Table 1. The analysis of variance revealed a significant effect ($p<0.05$) of adding sweet corn flour on viscosity and syneresis values. As the viscosity value increases, the syneresis value decreases. The results show that the average viscosity values range between 134.65-245.25 cp.

The increase in viscosity in the result is accompanied by a decrease in syneresis. The results of the study show that the average syneresis values range between 18.79-55.61%. Syneresis decreases with the increased addition of sweet corn flour. This is because sweet corn flour acts as a thickening agent in yoghurt. The greater the amount of sweet corn flour added, the greater the amount of water absorbed and bound, resulting in a stronger gel and reduced syneresis. According to Asasia and Sudarminto (2018), increasing the amount of thickening agent decreases syneresis because the thickening agent binds water by enhancing the hydrophilic properties of proteins. The hydrogen bonds between water molecules and proteins weaken, and the pores between casein molecules become looser, allowing free water to pass through, thereby reducing syneresis by adding a thickening agent to absorb water.

3.5 Hedonic Quality Test Results

The results of the hedonic quality test on cow's milk yoghurt with the addition of sweet corn flour in different treatments can be seen in Table 2. The analysis of variance revealed a significant effect ($p<0.05$) of adding sweet corn flour on the color and texture of cow's milk yoghurt. The color parameter shows significant differences ($p<0.05$) between P1 and P2, P1 and P3, P1 and P4, P2 and P3, P2 and P4. The texture parameter shows significant differences ($p<0.05$) between P1 and P2, P1 and P3, P1 and P4. However, the parameters of aroma, taste, and overall preference showed no significant effect ($p<0.05$) of adding sweet corn flour to cow's milk yoghurt.

Table 2. Hedonic Quality Test Results

Parameter	P0	P1	P2	P3
Aroma^{ns}	3.24 ± 0.899	3.61 ± 0.567	3.52 ± 0.740	3.72 ± 0.769
Color	4.21 ± 0.585^{c}	3.56 ± 0.548^{b}	3.04 ± 0.770^{a}	2.82 ± 0.822^{a}
Taste^{ns}	2.82 ± 0.860	2.58 ± 0.788	2.85 ± 0.900	2.74 ± 0.996
Texture	3.68 ± 0.794^{b}	3.16 ± 0.680^{a}	2.85 ± 0.855^{a}	2.80 ± 0.867^{a}
Overall^{ns}	3.20 ± 0.723	2.98 ± 0.843	3.11 ± 0.790	2.86 ± 0.801

Note: ns = non-significant (no significant difference at $p<0.05$).

Different superscripts in the same row indicate significant differences ($p<0.05$).

P0, P1, P2, and P3 = Concentrations of Sweet Corn Flour 0% (P0), 2% (P1), 4% (P2), 6% (P3) w/v.

3.5.1 Color

The results of the hedonic quality test for the color parameter revealed a significant effect ($p<0.05$) on the addition of sweet corn flour percentages. The study indicated that the color of cow's milk yoghurt with added sweet corn flour had an average score ranging from 2.82 to 4.21. The greater the amount of sweet corn flour added to the cow's milk yoghurt, the less the panelists liked the color. The white color of yoghurt without sweet corn flour was preferred by the panelists compared to the addition of sweet corn flour. This is because yoghurt with added sweet corn flour turned brown due to the caramelization process. This is supported by Wilberta *et al.* (2021), who stated that when sugar is heated, a caramelization reaction occurs, resulting in a brown color.

3.5.2 Texture

The results of the hedonic quality test for the texture parameter revealed a significant effect ($p<0.05$) on the addition of sweet corn flour percentages. The study indicated that the texture of cow's milk yoghurt with added sweet corn flour had an average score ranging from 2.8 to 3.68. The greater the amount of sweet corn flour added to the cow's milk yoghurt, the less the panelists liked the texture. This is likely because yoghurt with added sweet corn flour had a less appealing color for most panelists, which influenced their texture rating. Color is a quick response that creates a good impression of a product. If a product's color is less appealing, people may reconsider consuming it, even if the taste, texture, and aroma are good (Hanzen *et al.,* 2016).

3.5.3 Aroma

The results of the hedonic quality test for the aroma parameter revealed no significant effect ($p<0.05$) from the addition of sweet corn flour percentages. The study indicated that the aroma of cow's milk yoghurt with added sweet corn flour had an average score ranging from 3.24 to 3.72. The greater the amount of sweet corn flour added to the cow's milk yoghurt, the more the panelists liked the aroma. Aroma is the scent that arises from a product due to chemical stimuli, which can increase the deliciousness of the product's taste. This condition affects the level of panelists' acceptance of a particular food product (Luthfiyana, 2016). The preference score for the aroma of cow's milk yoghurt with added sweet corn flour was higher compared to yoghurt without sweet corn flour. This is because the addition of sweet corn flour resulted in a strong characteristic sweet corn aroma and a distinct sour aroma from yoghurt due to the breakdown of carbohydrates from milk and sweet corn flour into lactic acid, producing a uniform aroma. According to Salisa and Adi (2021), yoghurt with added sweet corn produces the characteristic yoghurt aroma with a slight sweet scent. The sweet aroma comes from the added sweet corn and is not lost during the yoghurt-making process.

3.5.4 Taste

The results of the hedonic quality test for the taste parameter revealed no significant effect ($p<0.05$) from the addition of sweet corn flour percentages. The study indicated that the taste of cow's milk yoghurt with added sweet corn flour had an average score ranging from 2.58 to 2.82. The addition of sweet corn flour did not affect the taste because the resulting flavor still exhibited the typical sour taste of yoghurt. According to Rahmawati *et al.* (2015), the sour taste in yoghurt is caused by the activity of LAB during the fermentation process.

3.5.5 Overall Preference

The results of the hedonic quality test for the overall preference parameter showed no significant effect ($p<0.05$) from the addition of sweet corn flour percentages. The study indicated that the overall preference for cow's milk yoghurt with added sweet corn flour had an average score ranging from 2.86 to 3.20. The lack of significant effect on the overall preference by panelists for cow's milk yoghurt with added sweet corn flour is due to the sour taste and aroma being relatively consistent, leading to a similar level of preference. According to Surahman and Wiwen (2021), the flavor profile of a food product consists of aroma, taste, and mouthfeel stimuli.

4. Conclusion

The addition of sweet corn flour affects TSS, total lactic acid, pH value, viscosity, and syneresis, but has no effect on total LAB. Adding sweet corn flour to milk increases TSS, lactic acid, and viscosity in the resulting yoghurt, but decreases pH and syneresis. Adding sweet corn flour to cow's milk yoghurt up to a level of 6% can improve the quality of yoghurt by increasing viscosity and reducing syneresis. The hedonic quality test showed that the addition of sweet corn flour only affected the color and texture, but had no effect on aroma, taste, and overall preference.

References

Alfaridhi, KK, AT, Lunggani & E, Kusdiyantini 2013, 'Penambahan filtrat tepung umbi dahlia (*Dahlia variabilis* Willd.) sebagai prebiotik dalam pembuatan yoghurt sinbiotik', *Bioma: Berkala Ilmiah Biologi*, vol. 15, no. 2, hlm. 64-72, DOI: 10.14710/bioma.15.2.64-72.

Amen, O, A, Jumiono & MA, Fulazzaky 2020, 'Penjaminan mutu dan kehalalan produk olahan susu', *Jurnal Ilmiah Pangan Halal,* vol. 2, no. 1, hlm. 42-48.

Asasia, PAA & Sudarminto, SY 2018, 'Pengaruh konsentrasi tepung maizena dan konsentrasi asam sitrat terhadap sifat fisik, kimia dan organoleptik selai mawar', *Jurnal Pangan dan Agroindustri*, vol. 6, no. 1, hlm. 64-74, DOI: 10.21776/ub.jpa.2018.006.01.8.

[BSN] Badan Standardisasi Nasional 2009, 'SNI 2981: Yoghurt', *Badan Standardisasi Nasional*, Jakarta.

Cahyani, SO, B, Dwiloka & H, Rizqiati 2019, 'Perubahan sifat fisikokimia dan mutu hedonik kefir air kelapa hijau (*Cocos nucifera* L.) dengan penambahan *High Fructose Syrup* (HFS)', *Jurnal Teknologi Pangan,* vol. 3, no. 1, hlm. 96-103, DOI: 10.14710/jtp.2019.22514.

Dewi, AP, T, Setyawardani & J, Sumarmono 2019, 'Pengaruh penambahan bunga telang (*Clitoria ternatea)* terhadap sineresis dan tingkat kesukaan yoghurt susu kambing', *Journal of animal Science and Technology,* vol. 1, no. 2, hlm. 145-151.

Fatmawati, Y, A, Purwantoro & P, Basunanda 2017, 'Keragaman morfologi dan molekuler empat kelompok kultivar jagung (*Zea mays* L.)', *Vegetalika,* vol. 6, no. 3, hlm. 50-64, DOI: 10.22146/veg.28017.

Hanzen, WE, US, Hastuti & B, Lukiati 2016, 'Kualitas yoghurt dari kulit buah naga berdasarkan variasi spesies dan macam gula ditinjau dari tekstur, aroma, rasa dan kadar asam laktat', *In Proceeding Biology Education Conference: Biology, Science, Enviromental, and Learning,* vol. 13, no. 1, hlm. 849-856.

Huwae, LMC, MA, Alfons, VC, Tanikwele, KJ, Linansera & PE, Linansera 2022, 'Pengaruh penambahan wine pala terhadap aktivitas antioksida dan antibakteri yoghurt sari buah sirsak (*Annona muricata L.*)', *Molucca Medica,* vol. 15, no. 2, hlm. 110-117, DOI: 10.30598/molmed.2022.v15.i2.110.

Irawan, F, MF, Sumual & J, Pontoh 2017, 'Pengaruh umur panen terhadap sifat fisik tepung jagung manis (*Zea mays saccharata* Sturt)', *Jurnal Teknologi Pertanian (Agricultural Technology Journal),* vol. 8, no. 1, hlm. 36-46, DOI: 10.35791/jteta.v8i1.16350.

Kiani, H, SMA, Mousavi & Z, Emam-Djomeh 2008, 'Rheological Properties of Iranian Yoghurt Drink, Doogh', *International J. of Dairy Sci,* vol. 3, no. 2, hlm. 71-78, DOI: 10.3923/ijds.2008.71.78.

Komalasari, H & WK, Yoga 2022, 'Potensi bakteri probiotik indigenous *Lactobacillus plantarum* Dad-13 sebagai starter pada pembuatan yoghurt fungsional: Kajian Pustaka', *Food Scientia: Journal of Food Science and Technolog,* vol. 2, no. 2, hlm. 199-217, DOI: 10.33830/fsj.v2i2.3694.2022.

Luthfiyana, N, NM, Nurilmala, E, Anwar & T, Hidayat 2016, 'Rasio bubur rumput laut *Eucheuma cottonii* dan *Sargassum* sp. sebagai formula krim tabir surya', *Jurnal Pengolahan Hasil Perikanan Indonesia,* vol. 19, no. 3, hlm. 183-195, DOI: 10.17844/jphpi.v19i3.14476.

Maharani, F & RD, Ayuningtyas 2018, 'Pelatihan pembuatan yoghurt di Kecamatan Pedurungan Kota Semarang', *Abdimas Unwahas,* vol. 3, no. 2, hlm. 5-9, DOI: 10.31942/abd.v3i2.2495.

Meilanie, RT, II, Arief & E, Taufik 2018, 'Karakteristik yoghurt probiotik dengan penambahan tepung bunga rosella (*Hibiscus sabdariffa* L) selama penyimpanan suhu dingin', *Jurnal Ilmu Produksi dan Teknologi Hasil Peternakan,* vol. 6, no. 1, hlm. 36-44.

Novidahlia, N, T, Rohmayanti & Y, Nurmilasari 2019, 'Karakteristik fisikokimia jelly drink daging semangka, albedo semangka, dan tomat dengan penambahan karagenan dan tepung porang (*Morphophallus muelleri* Blume)', *Jurnal Agroindustri Halal,* vol. 5, no. 1, hlm. 57-66, DOI: 10.30997/jah.v5i1.1694.

Oktavia, HM, N, Kusumawati & I, Kuswardhani 2015, 'Pengaruh lama penyimpanan selama distribusi dan pemasaran terhadap viabilitas bakteri asam laktat dan tingkat keasaman pada yoghurt murbei hitam (*Morus nigra L.*)', *Jurnal Teknologi Pangan dan Gizi (Journal of Food Technology and Nutrition),* vol. 14, no. 1, hlm. 22-30, DOI: 10.33508/jtpg.v14i1.1514.

Rahmawati, IS, E, Zubaidah & E, Saparianti 2015 'Evaluasi pertumbuhan isolat probiotik (L. *casei* dan L. *plantarum*) dalam medium fermentasi berbasis ubi jalar (*Ipomoea batatas* L.) selama proses fermentasi (kajian jenis isolat dan jenis tepung ubi jalar)', *Jurnal Aplikasi Teknologi Pangan,* vol. 4, no. 4, hlm. 133-141, DOI: 10.17728/jatp.v4i4.3.

Resnawati, H 2020 'Kualitas susu pada berbagai pengolahan dan penyimpanan', *Semiloka Nasional Prospek Industri Sapi Perah Menuju Perdagangan Bebas,* hlm. 497-502.

Rusdhi, A, P, Siswoyo & T, Adelia 2023, 'Uji kualitas kimia kefir dari imbangan susu kambing dan susu sapi dengan lama fermentasi yang berbeda', *Journal of Pharmaceutical and Sciences,* vol. 6, no. 4, hlm. 1626-1634, DOI: 10.36490/journal-jps.com.v6i4.284.

Salisa, W & AC, Adi 2021, 'Pengaruh substitusi sari kacang merah (*phaseolus vulgaris*) dan sari jagung manis (*Zea mays saccharata sturt.*) terhadap daya terima, nilai serat, dan total bakteri asam laktat yoghurt', *Jurnal Nutrisia,* vol. 23, no. 2, hlm. 68-75, DOI: 10.29238/jnutri.v23i2.87.

Savitry, NI, Nurwantoro & BE, Setiani 2017, 'Total bakteri asam laktat, total asam, nilai pH, viskositas, dan sifat organoleptik yoghurt dengan penambahan jus buah tomat', *Jurnal Aplikasi Teknologi Pangan,* vol. 6, no. 4, hlm. 184–187, DOI: 10.17728/jatp.272.

Suprapto, DA, R, Nurmalina & I, Fahmi 2014, 'Faktor-faktor yang memengaruhi keputusan konsumen dalam pembelian produk susu bubuk pertumbuhan', *Jurnal Ilmu Keluarga dan Konsumen,* vol. 7, no. 2, hlm. 113-122, DOI: 10.24156/jikk.2014.7.2.113.

Surahman, B & Wiwen, W 2021, 'Analisis pengaruh cita rasa terhadap kepuasan pelanggan hr coffe sp. empat bebesen', *Gajah Putih Journal of Economics Review,* vol. 3, hlm. 26-45, DOI: 10.55542/gpjer.v3i2.70.

Susilaningrum, DF, ASY, Wijaya, M, Zuliana, P, Ariani, AM, Firmansyah & T, Ujilestari 2022, 'Analisis pengaruh perbedaan teknik pemerahan susu sapi terhadap jumlah bakteri *Salmonella sp*', *Journal of Tropical Animal Research (JTAR),* vol. 3, no. 1, hlm. 1-9.

Umar, Razali & A, Novita 2014, 'Derajat keasaman dan angka reduktase susu sapi pasteurisasi dengan lama penyimpanan yang berbeda', *Jurnal Medika Veterinaria,* vol. 8, no. 1, hlm. 43-46.

Wahyudi, A & R, Dewi 2017, 'Upaya perbaikan kualitas dan produksi buah menggunakan teknologi budidaya sistem ToPAS pada 12 varietas semanga hibrida', *Jurnal Penelitian Pertanian,* vol. 17, no. 1, hlm. 17-25.

Wijayanti, IA, Purwadi & I, Thohari 2015, 'Pengaruh penambahan tepung sagu pada yoghurt terhadap viskositas, overrun, kecepatan meleleh dan total padatan es krim yoghurt', *Jurnal Ilmu dan Teknologi Hasil Ternak,* vol. 10, no. 2, hlm. 28-35 DOI: 10.21776/ub.jitek.2015.010.02.3.

Wilberta, N, NT, Sonya & SHR, Lydia 2021, 'Analisis kandungan gula reduksi pada gula semut dari nira aren yang dipengaruhi pH dan kadar air', *BIOEDUKASI: Jurnal Pendidikan Biologi,* vol. 12, no. 1, hlm. 101-108, DOI: 10.24127/bioedukasi.v12i1.3760.

Wiyono, AE, D, Rukmasari, E, Ruriani & Herlina 2023, 'Karakteristik mutu serbuk pewarna buah naga merah (*Hylocereus polyrhizus*) hasil foam mat drying dengan variasi rasio daging dan kulit buah', *Jurnal Agrointek,* vol. 17, no. 2, hlm. 412-422, DOI: 10.21107/agrointek.v17i2.14631.

Zulaikhah, SR & AH, Sidhi 2021, 'Pembuatan yoghurt susu sapi sebagai upaya meningkatkan nilai fungsional susu, gizi masyarakat dan pendapatan rumah tangga di masa pandemi', *Jurnal Pengabdian Magister Pendidikan IPA,* vol. 4, no. 3, hlm. 290-294, DOI: 10.29303/jpmpi.v4i3.924.

Optimization of goat milk cheese production using fig tree latex coagulant: A study on physicochemical characteristics and calcium content with RSM

Widitya Tri Nugraha[1], Tridjoko Wisnu Murti[1*], Yuny Erwanto[1], Nurliyani[1], Yustina Yuni Suranindyah[1], Muhlisin[1] and Dwi Larasatie Nur Fibri[2]

1 Faculty of Animal Science, Universitas Gadjah Mada, Indonesia
2 Faculty of Agricultural Technology, Universitas Gadjah Mada, Indonesia

*E-mail: tridjokomurti@mail.ugm.ac.id

Abstract. Plant-derived coagulants, such as fig tree latex extract (*Ficus carica L.*), offer a promising alternative in cheese production. Despite this potential, research on the properties of goat cheese made using fig tree latex remains limited. This study aims to optimize the goat milk cheese/goat cheese production using fig tree latex as a coagulant, employing Response Surface Methodology (RSM) and Central Composite Design (CCD). The variables considered include the concentration of fig tree latex extract (0.2 – 0.5 mL/100 mL of milk) and the incubation time (60-120 minutes). The optimization goal was to minimize the moisture content (%) and maximize the calcium content (%) of goat cheese. The optimal conditions were found to be a concentration of fig tree latex extract 0.5 mL/100 mL of milk and an incubation time of 120 minutes. Under these conditions, the predicted cheese moisture content was 55.864%, and the calcium content reached 803.930 mg/100 g, with a desirability value of 0.931. These optimized conditions were subsequently used to produce goat cheese, resulting in a physicochemical composition with a mean pH of 6.22 ± 0.07, moisture content of 56.46 ± 0.017%, ash content of 2.38 ± 0.039%, protein content of 17.369 ± 0.016%, and calcium content of 796.001 ± 10.061mg/100g. This study demonstrates the effectiveness of fig tree latex as a coagulant and underscores its potential in optimizing cheese characteristics.

Keywords: physicochemical composition, fig tree latex, goat cheese, RSM optimization

1. Introduction

Dairy products, especially cheese, are crucial for human nutrition because they are rich in essential nutrients, including fats, proteins, and minerals such as calcium. Furthermore, they contribute to various metabolic processes, ensuring the body operates efficiently. Regular consumption of dairy can help maintain a balanced diet, promoting overall development and health across different age groups [1].

Cheese, being a concentrated source of essential nutrients, offers numerous health benefits. It is an excellent source of calcium, a mineral essential for preserving bone and dental strength also supporting cardiovascular and muscular function. Additionally, the protein in cheese is key to muscle repair and development, making it an important part of a well-rounded diet. Consuming cheese and other dairy products in daily meals helps ensure an adequate intake of these essential nutrients, promoting overall body health and optimal function [2].

In particular, goat cheese has gained popularity in recent years, not only for its unique and appealing flavour but also for its nutritional advantages [3]. Goat cheese is often preferred for its high calcium and protein content, offering a nutritious alternative to traditional cow's milk cheese. These attributes make it a desirable option for health-conscious consumers seeking both taste and health benefits in their dietary choices [4].

Traditionally, animal-derived rennet has been the primary coagulant in cheesemaking. However, concerns over animal welfare, religious dietary restrictions, and the growing demand for vegetarian products have driven the search for plant-based coagulant alternatives [5].

Fig tree latex, a natural substance extracted from unripe figs, has been studied as a possible plant-derived coagulant for cheesemaking. Previous research has shown its coagulating properties and its ability to influence the physicochemical attributes of cheese [6]. However, the optimization of goat cheese production using fig latex has not been thoroughly explored.

Although research on plant-based coagulants is expanding, the specific application of fig latex in goat cheese production remains under researched. This research goals to fill this gap by cheese production process optimizing using Response Surface Methodology (RSM) to achieve minimum moisture and maximum calcium content in goat cheese. Additionally, it evaluates how fig latex as a coagulant impacts the physicochemical properties of the resulting cheese.

RSM is a widely used statistical instrument in food processing for optimizing complex processes involving multiple variables [7]. By applying RSM in this study, the aim is to systematically examine the effects of production process parameters such as fig latex concentration and incubation time on the quality of the final cheese. This approach will help identify optimal conditions for producing goat cheese using fig latex.

The findings from this study have the potential to contribute to more sustainable and innovative cheesemaking practices. By encouraging the use of plant-based coagulants, we can reduce reliance on animal-derived rennet, addressing consumer concerns regarding animal welfare and food safety. Additionally, the optimized production process could improve the quality and consistency of goat cheese.

2. Materials and methods

2.1 Materials

Raw Goat milk used for cheese production was sourced from a dairy goat farm in the Yogyakarta region of Indonesia. Prior to cheese making, the milk was filtered to remove any impurities or foreign particles.

2.2 Collection of fig tree latex.

The fig plants (*Ficus carica L.*) were sourced from a local plant store in the Yogyakarta region of Indonesia. Samples of fig latex were collected in the morning by cutting the stems and leaves of healthy plants until the latex flowed. Subsequently, the latex is transferred into a clean glass bottle, transported to the laboratory, and refrigerated until the commencement of the experiment [8].

2.3 Crude enzyme preparation.
Latex extracted from *Ficus carica L.* was homogenized with a phosphate buffer solution (pH 7) in a 1:9 ratio. The homogenized mixture was centrifuged at 3,200 rpm for 15 minutes at 4°C. The supernatant, containing the latex enzymes (referred to as "crude enzyme"), was used for further research [8].

2.4 Cheese production
For each experimental trial, approximately 100 mL of fresh goat's milk was heated to approximately ± 50°C for 5 minutes, adhering to the established protocols [9]; [10]. Fig tree latex extract was then added, following the specified treatment conditions. The mixture was incubated at ± 50°C for the durations outlined in the treatment plan. Finally, the mixture was filtered through cheesecloth to produce the final goat cheese product.

2.5 Experimental design
The goat cheese production process was optimize in this study utilized Response Surface Methodology (RSM) with Central Composite Design (CCD). RSM is a statistical technique for designing experiments, analyzing data, and understanding the relationship between independent and dependent variables. CCD is a popular experimental design within RSM for optimizing systems with multiple variables [11][12].

The experiment was conducted using a CCD design to determine the optimal combination of factors [13]. All variables used refer to [9,10] which were modified, namely the concentration of fig tree latex extract (2 - 5 mL of 1 mL of fig latex dissolved in 9 mL of phosphate buffer pH 7) and incubation time (60-120 minutes). The goal of optimizing goat cheese is to achieve a product with minimum moisture content and maximum calcium concentration.

Table 1. The coded and uncoded values used in the optimization of response

Point (code)	Code	Independent variable	
		Concentration (A) (mL/100mL milk)	incubation time (B) (minutes)
Minimum axial point	-α	0.1379	47.58
Minimal factorial point	-1	0.2	60
Center point	0	0.35	90
Maximum factorial point	1	0.5	120
Maximum axial point	α	0.5621	132.42

Experimental data was processed to calculate response values for each combination of factors. The most appropriate mathematical model was selected based on the response analysis. Finally, the optimal conditions were determined for maximizing the desired responses.

The Design-Expert 13.0.5.0 software was used to generate a Central Composite Design (CCD) model and analyze the experimental data using Response Surface Methodology (RSM). The CCD design included two independent variables with 13 experimental runs, comprising 5 center points, 4 factorial points, and 4 axial points. The independent variables were the concentration of fig tree latex extract and incubation time. The dependent variables measured were cheese moisture, and calcium content of cheese.

3. Result and discussion

3.1 CCD of RSM for optimization of goat cheese
The results of the characteristics of goat's milk cheese with different coagulant concentrations of fig latex extract and incubation time are presented in Table 2. The determination of the lower and

upper limits of the treatment is that the concentration ranges from 0.2 to 0.5 mL/100 mL of milk, and the incubation time ranges from 60 to 120 minutes. The aims of this research is to obtain the best concentration of fig latex and incubation time to produce minimum moisture content and maximum calcium content. The results indicate that the cheese has a moisture content ranging from 54.3% to 69.6%, and calcium content ranging from 665.73 mg/100g to 818.37 mg/100g.

Table 2. CCD of RSM for optimization of goat cheese

Run	Type	Variable Code		Variable Value		Response	
		A	B	Concentration (mL/100 mL)	Incubation Time (minute)	R1	R2
1	Center	0	0	0.35	90	64.7	714.59
2	Factorial	-1	-1	0.2	60	67.6	685.18
3	Center	0	0	0.35	90	62.4	737.13
4	Factorial	-1	1	0.2	120	65.0	711.64
5	Axial	0	1.414	0.35	132.42	56.6	795.34
6	Factorial	1	-1	0.5	60	64.2	719.88
7	Factorial	1	1	0.5	120	57.6	788.19
8	Center	0	0	0.35	90	62.5	736.30
9	Axial	0	-1.414	0.35	47.58	66.4	697.97
10	Center	0	0	0.35	90	61.6	745.37
11	Axial	-1.414	0	0.1379	90	69.6	665.73
12	Center	0	0	0.35	90	63.6	726.07
13	Axial	1.414	0	0.5621	90	54.3	818.37

[a] R1 = Moisture content response (%); R2 = Calcium content response (mg/100 g)

The lowest recorded moisture content in cheese, at 54.3%, was achieved using a treatment combination of 0.5621 mL/100mL milk concentration (A) and a 90-minute incubation time (B). Conversely, the highest moisture content observed in cheese, at 69.6%, resulted from a treatment with a lower concentration (A) of 0.1379 mL/100mL milk, also with a 90-minute incubation period. This data was instrumental in identifying optimal concentrations of fig tree latex (variable A) and the best incubation times (variable B) for moisture content. The 3D surface plot of moisture content, illustrating the correlation between the variables (A, B, and moisture content), is shown in Figure 1

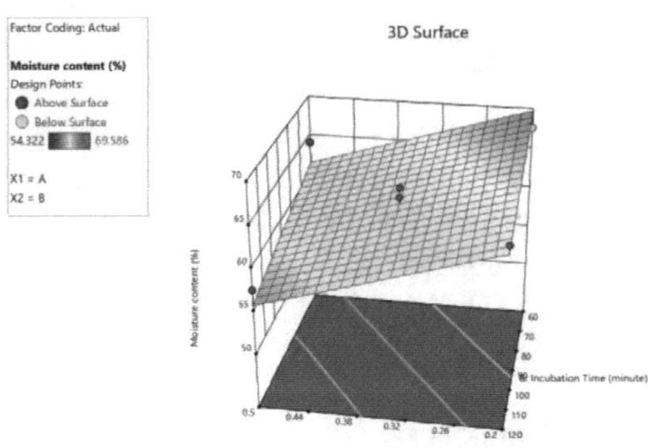

Figure 1. 3D surface plot of moisture and calcium content

In terms of calcium content, the lowest value measured in the cheese was 665.73 mg/100g, obtained with the treatment involving the lowest concentration (A) of 0.1379 mL/100mL milk and a 90-minute incubation time (B). The highest calcium content found in cheese was 818.371 mg/100g, which occurred with a higher concentration (A) of 0.5621 mL/100mL milk, also incubated for 90 minutes. The 3D surface plot of moisture content, illustrating the correlation between the variables (A, B, and Calcium content), is shown in Figure 2.

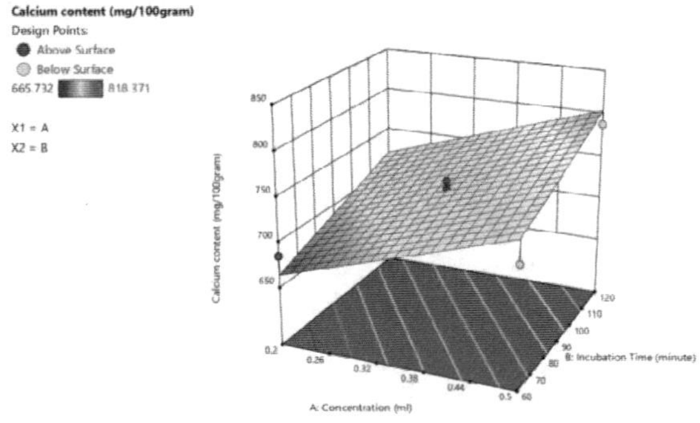

Figure 2. 3D surface plot of calcium content

Based on the research results above, it is known fig latex has the ability to coagulate milk. For centuries, fig leaves and fruit have been utilized as human food, and in certain countries, fig tree latex has been used in cheese production due to its proteolytic enzyme content, specifically ficin [14]. Fig tree latex extract has served as a key ingredient in traditional cheese-making, whether from cow's milk or the milk of small ruminants, particularly goat and sheep milk [15]. The protease from *Ficus carica L.*, commonly known as ficin, demonstrates broad specificity toward basic and neutral amino acids, including Gly, Val, Leu, Ala, Ser, Asn, Arg, and His. Ficin cleaves protein bonds at residues like Tyr, Phe, and Val, facilitating protein breakdown [6].

3.2 Model fitting and analysis of variance

Determining the research response model using the Design Expert application produces a fit summary of the research responses shown in Table 3. The recommended model for all responses is a linear model, as indicated by statistical testing.

Table 3. Fit Summary of research response

Response	*Sequential p-value*	*Lack of Fit p-value*	*Adjusted* R^2	*Predicted* R^2	*Suggested Model*
Moisture content	< 0.0001	0.1291	0.8229	0.7002	Linier
Calcium content	< 0.0001	0.1361	0.8305	0.7139	Linier

The analysis of variance (ANOVA) and model fit statistics for the response variables, including moisture content and calcium content in cheese, are thoroughly presented in Tables 4 and 5. This statistical evaluation reveals that the selected linear model is appropriately suited to analyze the response variables, as evidenced by meeting criteria related to model significance, residuals, and overall fit. This is particularly confirmed by examining the model's significance, lack of fit, and key fit statistics, which collectively indicate that the model structure aligns well with the observed data.

Table 4. Summary of ANOVA and fit statistics of research response models

Response	Source	Sum of Squares	df	Mean Square	F-value	p-value	Sig/not
Moisture content	Model	197.14	2	98.57	28.88	< 0.0001	significant
	Lack of Fit	28.51	6	4.75	3.38	0.1291	not significant
Calcium content	Model	20,131.05	2	10,065.53	30.39	< 0.0001	*significant*
	Lack of Fit	2,750.47	6	458.41	3.26	0.1361	*not significant*

Specifically, the p-values associated with factors A and B are both below the $\alpha = 0.05$ threshold, demonstrating their statistical significance in influencing the model. This implies that both factors significantly affect the response variables within the linear model. In contrast, the lack of fit statistic is greater than $\alpha = 0.05$, suggesting that any deviation from the model is not substantial enough to indicate inadequacy in fitting the data. This balance between significance and lack of fit suggests that the model's predictions are reliable within the specified parameters.

Table 5. Summary of fit statistics of research response models

Response	Adjusted R^2	Predicted R^2	Adeq Precision
Moisture content	0.8229	0.7002	15.5952
Calcium content	0.8305	0.7139	16.0012

Furthermore, all response variables meet the standards set for Adjusted R^2, with the difference between Predicted R^2 and Adjusted R^2 remaining under 0.2, indicating consistent predictive performance. Additionally, the Adequate Precision value surpasses the minimum threshold of 4 [16], underscoring the model's ability to differentiate between signal and noise effectively. Consequently, the linear model used for analysis is both statistically sound and reliable, with the final equation values for each response variable summarized in Table 6, providing a clear basis for interpreting the results.

Table 6. Actual response model equation

Response		*Actual Equation*
Moisture content	=	80,8425 – 27,0096*A – 0,09561*B
Calcium content	=	551.3946 + 272.5959**A* + 0.9686**B*

A= Concentration (ml/100ml milk) and B= Incubation time (minute)

The actual equations derived from the analysis serve as valuable tools for predicting responses at specific levels of each factor. It is crucial to ensure that factor levels are measured in their native units to ensure that the results are accurately interpreted. However, it is crucial to understand that these equations are not intended to assess the relative effect of each factor. This limitation occurs because of equations coefficients are modified to adjust varying measurement component, and the intercept may not be positioned at the center of the design.

In this study, moisture and calcium content evaluation in cheese was given the highest priority (scale 5), reflecting its critical importance to the research. This focus is a preliminary step towards the main investigation into the calcium bioavailability of goat's milk cheese. Consequently, optimizing the factors involved, such as fig latex concentration and incubation time, is essential to enhance both moisture content and calcium content. As outlined in Table 7, four potential solutions involving different combinations of these variables were proposed.

Table 7. Solutions for optimal response

Number	Concentration	Incubation Time	Moisture content	Calcium content	Desirability	
1	**0.500**	**120.000**	55.864	**803.930**	**0.931**	**Selected**
2	0.500	119.454	55.916	803.400	0.929	
3	0.500	117.976	56.058	801.969	0.922	
4	0.469	120.000	56.689	795.606	0.889	

A= Concentration (mL/100 mL milk) and B= Incubation time (minute)

Among the proposed solutions, the most favorable outcome was achieved with a fig latex concentration of 0.5 mL per 100 mL of milk combined with an incubation time of 120 minutes, which resulted in a high desirability value of 0.931. A desirability score near 1 suggests that this combination of factors is nearly aligned with the ideal conditions expected for the best outcomes. [17].

This optimal solution demonstrates a strong alignment with the expected outcomes, making it the preferred choice for further investigations. Pada beberapa penelitian ditemukan metode pembuatan keju segar dengan fig-plant latex dengan konsentrasi dan waktu yang berbeda-beda.

Therefore, this specific combination of variables is expected to produce cheese that meets the optimization goals. Projections suggest that the resulting cheese will have a moisture content of approximately 55.864% and a calcium content of 803.930 mg per 100 grams.

3.3 Validation of variable optimization results on response

The optimization results of the variables of fig latex concentration and incubation time influence the responses being studied. The validation of optimization response results is shown in Table 8.

Table 8. Validation of optimization response results

Solution 1 of 2 Response	95% PI low	Data Mean	95% PI high
Protein content (%)	52.1	56.5	59.6
Calcium content (mg/100g)	767.05	795.66	840.81

The validation results demonstrated that, after reapplying the recommended treatment combination, the mean data values for all responses fell within the 95% Prediction Interval (PI) Low and High ranges. This confirms that the results obtained with this combination are reliable. Overall, the test outcomes show that increasing both the concentration of fig latex and the incubation time will consistently produce cheese that aligns with the model predictions.

Until now, there has been no research that specifically explains the optimal concentration of fig tree latex and the incubation time required for producing goat milk cheese. However, several studies have mentioned the concentration and incubation time that can be used. In the production of fresh cheese, a concentration of 0.1 mL of fig tree latex per 10 mL of milk and an incubation time of 120 minutes have been used [9]. In another study on camel milk cheese, fig tree latex was used at a concentration of 1 mL per liter of milk with an incubation time of 24 hours [18,19].

It is important to recognize, however, that plant-based coagulants may sometimes give the cheese a bitter flavour. Therefore, ensuring that this level of optimization aligns with consumer preferences is vital. Achieving a balance between the cheese's physical, chemical, textural, and organoleptic properties is crucial to satisfy market demands and consumer expectations. [20].

3.4 Physiochemical goat cheese using fig latex coagulant

After selecting the combination of fig latex concentration and incubation time, cheese was produced according to the formulation. The results of this cheese production are presented in Table 9.

Table 9. Physiochemical goat cheese using fig latex coagulant

Parameter	Goat Milk
pH	6.22 ± 0.007
Mositure (%)	56.46 ± 0.017
Protein (%)	17.369 ± 0.016
Ash Content (%)	2.38 ± 0.039
Calcium (mg/100g)	796.001 ± 10.061

The characteristics of the cheese produced in this study closely resemble those observed in research on *Cacioricotta* cheese, which also used goat's milk and resulted in moisture content of 54.5 and protein content of 19.3% [21]. These variations underscore that numerous factors influence cheese characteristics, including the type and quality of milk used, the manufacturing process, the choice of coagulant, the addition of calcium compound, the presence of bacterial starters, and the ripening process. Each of these elements impacts not only the physiochemical properties of the cheese but also its texture and flavor [4,22–24].

Protein is a critical component in cheese, significantly affecting its texture, flavor, and consistency. The amount and type of protein can vary depending on the cheese variety, production methods, and milk source. Additionally, proteins in cheese provide essential amino acids and important nutrients, making cheese a valuable dietary component[2].

Ash content in cheese, which reflects its mineral or ash content, is influenced by the cheese type, manufacturing process, and milk composition. This content affects the cheese's taste, texture, and overall organoleptic characteristics, including flavor, aroma, and appearance. The ash content can vary widely, depending on the cheese type, and typically ranges from a few percent to several percent of the total cheese weight [25].

Calcium, a main mineral found in cheese, primarily originates from milk. It is essential for healthy bone and teeth also plays a role in various body biochemical processes. The calcium content in cheese varies by cheese type and milk source, with cheese being recognized as a major calcium source that significantly contributes to meeting daily calcium needs. Cheese is typically considered a rich source of calcium [4,23], and cheese consumption plays an important role in fulfilling human daily calcium needs [26].

4. CONCLUSION

In conclusion, the optimal combination for producing goat cheese was identified as a fig latex extract concentration of 0.5 mL/100 mL of milk and an incubation time of 120 minutes. Under these conditions, the predicted moisture content of the cheese was 55.864%, and the calcium content was 803.930 mg/100 g, with a desirability value of 0.931. Following this optimization, the cheese was produced, yielding a physicochemical composition with a mean pH of 6.22 ± 0.07, moisture content of 56.46 ± 0.017%, ash content of 2.38 ± 0.039%, protein content of 17.369 ± 0.016%, and calcium content of 796.001 ± 10.061mg/100g.

References

[1] Ünal G, El SN, Kiliç S. 2005. In vitro determination of calcium bioavailability of milk, dairy products and infant formulas. Int J Food Sci Nutr. 56:13–22. https://doi.org/10.1080/09637480500081423.

[2] Fox PF, Guinee TP, Cogann TM, McSweeney PLH. 2017. Fundamentals of Cheese Science. 2nd ed.

[3] Moises RMM, Salgueiro CC de M, Leitão MSP, Nunes JF. 2024. Exploring goat's milk cheese: A systematic review of production techniques and innovations (2013–2023). Braz J Food Technol. 27. https://doi.org/10.1590/1981-6723.15223.

[4] McSweeney PLH, Ottogalli G, Fox PF. 2004. Diversity of cheese varieties: An overview. p. 1–23. https://doi.org/10.1016/S1874-558X(04)80037-X.

[5] Nugraha WT, Murti TW, Novitasari IS, Kartika Sari T, Murcita G, Wijakangka GTR. 2015. Development of Halal Goat Cheese using Rennet Like from Vegetable Source to Replace Commercial Rennet Source. The 6th International Seminar on Tropical Animal Production Integrated Approach in Developing Sustainable Tropical Animal Production. p. 733–7.

[6] Aider M. 2021. Potential applications of ficin in the production of traditional cheeses and protein hydrolysates. JDS Commun. 2:233–7. https://doi.org/10.3168/jdsc.2020-0073.

[7] Sarabia LA, Ortiz MC. 2009. Response Surface Methodology. Comprehensive Chemometrics. Elsevier. p. 345–90. https://doi.org/10.1016/B978-044452701-1.00083-1.

[8] Mazri C, Soumia H el A, Siar H. 2018. Characterization and application of phytochemicals substances of the fig tree: Biological and sensory characterization of ficin and cheeses "fresh and soft." Asian J Appl Sci Eng. 7:71–8. https://doi.org/10.18034/ajase.v7i2.226.

[9] Abraha Z, Taddesse A, Yohannes M, Birhane E, Beyene G. 2018. Milk clotting characteristics of Solanum incanum, Ficus carica and Rhus natalensis for cheese making in Tigray, Northern Ethiopia. Livest Res Rural Dev. 30.

[10] Rana M, Hoque M, Rahman M, Habib R, Siddiki M. 2017. Papaya (Carica papaya) latex – an alternative to rennet for cottage cheese preparation. J Adv Vet Anim Res. 4:249. https://doi.org/10.5455/javar.2017.d218.

[11] Njoku CN, Otisi SK. 2022. Application of Central Composite Design with Design Expert v13 in Process Optimization.

[12] El-taweel RM, Mohamed N, Alrefaey KA, Husien S, Abdel-Aziz AB, Salim AI, Mostafa NG, Said LA, Fahim IS, Radwan AG. 2023. A review of coagulation explaining its definition, mechanism, coagulant types, and optimization models; RSM, and ANN. Curr Res Green Sustain Chem. 6:100358. https://doi.org/10.1016/j.crgsc.2023.100358.

[13] Montgomery DC. 2001. Design and Analysis of Experiments. 5th ed. New York: Wiley.

[14] Lazreg-Aref H, Gaaliche B, Ladhari A, Hammami M, Hammami SO. 2018. Co-evolution of enzyme activities and latex in fig (Ficus carica L.) during fruit maturity process. S Afr J Bot. 115:143–52. https://doi.org/10.1016/j.sajb.2018.01.022.

[15] Nouani A, Dako E, Morsli A, Belhamiche N, Belbraouet S, Bellal MM, Dadie A. 2009. Characterization of the purified coagulant extracts derived from artichoke flowers (Cynara scolymus) and from the fig tree latex (Ficus carica) in light of their use in the manufacture of traditional cheeses in Algeria. Medwell J Food Technol. 7.

[16] Variyana Y, Muchammad RSC, Mahfud M. 2019. Box-Behnken design for the optimization using solvent-free microwave gravity extraction of garlic oil from Allium sativum L. IOP Conf Ser Mater Sci Eng. 673:012005. https://doi.org/10.1088/1757-899X/673/1/012005.

[17] Gutema EM, Gopal M, Lemu HG. 2022. Temperature optimization by using Response Surface Methodology and Desirability Analysis of Aluminium 6061. Mater. 15:5892. https://doi.org/10.3390/ma15175892.

[18] Fguiri I, Sboui A, Ayeb N, Ziadi M, Guemri M, Arroum S, Dbara M, Khorchani T. 2020. Camel milk-clotting properties of latex protease from Ficus carica. J Anim Vet Adv. 19:99–106.

[19] Fguiri I, Sboui A, Ziadi M, Ayeb N, Arroum S, Dbara M, Hammadi M, Khorchani T. 2022. Optimization of camel milk cheese processing using protease from latex (Ficus carica). Res Aspects Agric Vet Sci. 5:27–36. https://doi.org/10.9734/bpi/raavs/v5/14538D.

[20] Nicosia FD, Puglisi I, Pino A, Caggia C, Randazzo CL. 2022. Plant milk-clotting enzymes for cheesemaking. Foods. 11:871. https://doi.org/10.3390/foods11060871.

[21] Faccia M, Apruzzese D, Passaro P. 2019. Making cheese with caprifig sap in Apulia, Italy: Possible rebirth of an ancient tradition. J Ethnic Foods. 6:6. https://doi.org/10.1186/s42779-019-0007-5.

[22] Hargrove RE, McDonough FE, Tittsler RP. 1967. Factors affecting characteristics, composition, and quality of skimmilk cheese. J Dairy Sci. 50:160–6. https://doi.org/10.3168/jds.S0022-0302(67)87382-X.

[23] McSweeney PLH, Ottogalli G, Fox PF. 2017. Diversity and classification of cheese varieties: An overview. Cheese. Elsevier. p. 781–808. https://doi.org/10.1016/B978-0-12-417012-4.00031-4.

[24] Uniacke-Lowe T, Fox PF. 2017. Chymosin, pepsins and other aspartyl proteinases: Structures, functions, catalytic mechanism and milk-clotting properties. Cheese. Elsevier. p. 69–113. https://doi.org/10.1016/B978-0-12-417012-4.00004-1.

[25] Vacca GM, Stocco G, Dettori ML, Bittante G, Pazzola M. 2020. Goat cheese yield and recovery of fat, protein, and total solids in curd are affected by milk coagulation properties. J Dairy Sci. 103:1352–65. https://doi.org/10.3168/jds.2019-16424.

[26] Pawlos M, Znamirowska A, Szajnar K. 2021. Effect of calcium compound type and dosage on the properties of acid rennet goat's milk gels. Molecules. 26:5563. https://doi.org/10.3390/molecules26185563.

IOP Conf. Series: Earth and Environmental Science **1460** (2025) 012032 doi:10.1088/1755-1315/1460/1/012032

Hedonic quality of beef sausage with the addition of pea flour (*Pisum sativum L.*) and ginger flour (*Zingiber officinale var Rubrum Rhizoma*)

Adi Magna Patriadi Nuhriawangsa*, Winny Swastike, Bayu Setya Hertanto, Farouq Haidar Barido and Lilik Retna Kertikasari

Research Group Food Technology of Animal Origin, Laboratory of Animal Product Industry and Processing, Faculty of Animal Science, Universitas Sebelas Maret, Surakarta, Indonesia

*E-mail: adimagna@staff.uns.ac.id

Abstract. The research aimed to analyze the hedonic quality of beef sausages using the addition of pea flour (PF) and ginger flour (GF) as a substitute for tapioca flour (TF). Hedonic quality analysis uses the 9-hedonic scale test method with 25 semi-trained panelists. Treatment used a combination of PF and GF at P0 = 0:0, P1 = 5:0.75, P2 = 10:1.5, P3 = 15:2.25 and P4 = 20:3. The results showed that the color, aroma, taste, taste, texture, overall and after taste had very significant differences ($P<0.01$), but the tenderness was not different in beef meatballs. Color, aroma, taste, taste, texture, overall and after taste began to experience a decrease in hedonic quality with the addition of 10% PF and 1.5% GF (P2) compared to the control (P0). The addition of PF and GF at levels 5 and 0.75% (P1) can be used to make beef sausages, because the hedonic quality can still be maintained with a neutral assessment by the panelists.

Keywords: hedonic quality, beef sausage, pea flour, ginger flour

1. Introduction

Research shows that there are beneficial physiological aspects of ginger utilization, especially due to its antioxidant, anti-inflammatory, antimicrobial, and anti-carcinogenic properties. Ginger contains several bioactive compounds that can improve health, so it is used in food products according to consumer demand. The use of ginger affects the oxidative stability, tenderness and sensory attributes of meat products [1]. Ginger facilitates the bioavailability of nutrients and provides aroma and flavor to food. Ginger is a natural preservative that can improve organoleptic properties and create visual appeal in food. Due to the potential aroma and flavor of ginger, ginger is consumed in the food and beverage industry. Ginger is used in food product development to improve organoleptic properties and to facilitate nutrient absorption [2]. The use of ginger in chicken soup shows a richer taste, but increases the bitter and astringent taste [3].

Peas (*Pisum sativum L.*) are one of the legumes that are widely known, popular and economically important. Peas are rich in macronutrients, including protein, starch, dietary fiber, and non-starch polysaccharides. Peas contain important bioactive substances such as polyphenols, especially flavonoids and phenolic acids. Peas and their products are developed as functional foods because they are rich in various bioactive nutrients that are beneficial for health, such as antioxidants, anti-inflammatory, antimicrobial, anti-renal fibrosis, and regulation of metabolic syndrome effects [4]. Peas have higher protein and fiber content than other legumes, and do not contain allergens [5]. The results of the study showed that the functional properties of pea protein isolate in making nuggets were acceptable with nutritional, texture, and sensory attributes, as a substitute for meat-based ingredients [6]. The addition of yellow pea flour to the burger formulation at a level of 10% can be well accepted by consumers [7].

Meat sausage is a product made from ground meat with or without the addition of other food ingredients and permitted food additives and inserted into sausage casings with or without a cooking process [8]. Consumer perceptions of the health of meat products have changed, so that meat products including sausages are made with healthier and more functional formulations to meet market and consumer demands. Sausage products that have health value can be produced by adding plant-based ingredients that contain functional nutrients. Plant-based ingredients are rich in polyphenols that can function as antibacterials and antioxidants [9].

The consumer acceptance factor is the highest factor, because consumer sensory acceptance of the product affects economic and business prosperity factors [5]. Thus, this study aimed to analyze consumer hedonic acceptance of sausage products with the addition of pea flour and ginger flour as sausage products as functional foods.

2. Materials and methods

2.1 Materials

The materials used for making sausages were beef on the commercial blade section purchased at a supermarket in Surakarta city. In addition, the sausage-making material formulation uses commercial ingredients with food grade quality such as tapioca flour, pea flour, ginger flour, ice cubes, egg white, spices, skim and beef stock purchased at a supermarket in Surakarta city. The sausage formulation follows as in [10] by using modifications to the ingredients and percentage of the formulation of the ingredients used. The formulation of the ingredients for making sausages is shown in Table 1.

Table 1. Beef sausage formulation with the addition of PF and GF as TF substitutes.

No	Sources of Sausages	Treatments (%)				
		P0	P1	P2	P3	P4
1	Beef meat (blade)	58.85	58.1	57.35	56.6	55.85
2	Tapioca flour	20	15	10	5	0
3	Pea flour	0	5	10	15	20
4	Ginger flour	0	0.75	1.5	2.25	3
5	Ice	10	10	10	10	10
6	Albumin egg	5	5	5	5	5
7	Spices	3.35	3.35	3.35	3.35	3.35
8	Skim milk	2	2	2	2	2
9	Beef broth	0.8	0.8	0.8	0.8	0.8
	Total	100	100	100	100	100

2.2 Method

Sausage making uses the method according to [10] which is modified by using a sausage making tool. Modifications to the time used in the sausage making stages. The sequence of sausage making with the following steps: cut the super kampung chicken meat into smaller pieces and grind it using a grinder machine with a size of 6 mm; ground meat with ice cubes, egg white, fish oil, fine spices are put in a mixer for 10 seconds then the mixer is turned off; tapioca flour and spirulina flour according to the formulation are put into the mixer and the mixer is turned on again until the 20th second, the mixer is turned off and the dough is leveled; the mixer is turned on again to homogenize the dough until the 35th second; the homogeneous dough is put into a casing machine with a casing size of 18 mm with each sausage weighing 50 g and the final stage is the sausage is steamed for 30 minutes at a temperature of 70-80°C.

Hedonic quality testing used the method according to [11] using 25 semi-trained panelists from students of the Faculty of Animal Science, Universitas Sebelas Maret, Surakarta. Panelists selection used a questionnaire with scoring. Panelists were taken with score results in the order of 1 to 25. Selected panelists filled out and signed the consent form, were given an explanation of the process of implementing the hedonic test method and carry out the hedonic test. Testing used a 9-point hedonic scale by giving a √ (tick) mark on the existing points according to the attributes at the level of panelist acceptance. Each sample used a 3-digit number when presented in the hedonic test. The attributes assessed were color, aroma, taste, flavor, tenderness, texture, overall and after taste. Hedonic assessment with a score of 1 = Dislike extremely, 2 = Dislike very much, 3 = Dislike moderately, 4 = Dislike somewhat, 5 = Neutral (neither like nor dislike), 6 = Slightly, 7 = Like moderately, 8 = Like very much and 9 = Dislike extremely [12].

2.3 Research design

The experimental design used a One-Way Pattern CRD and data analysis used analysis of variance (ANOVA). If there is a significant difference in the treatment, the study was continued with the Dunnet test [13] .

3. Results and discussion

The results of statistical analysis showed that there was a very significant difference ($P<0.01$) in color, aroma, taste, flavor, texture, overall and after taste, except for tenderness which showed no significant difference (**Table 2**; **Figure 1**). The attributes of color, aroma, taste, flavor, texture, overall and after taste showed a decrease in hedonic quality starting at P2 (combination of PF and GF = 10:1.5).

Aroma, taste and flavor decreased in quality with the addition of ginger flour at P2 to P4 (**Table 2**). This study had the same results as [14], aroma, taste and cohesiveness increased with the dominance of aroma, taste and cohesiveness of ginger being stronger in ground pork products compared to the control. Ginger extract is used in food processing to provide a different taste and aroma [2]. The pungent taste and aroma of ginger are caused by the presence of phenolic compounds and essential oils, while the spicy taste is caused by the presence of phenolics in the form of gingerol [15] . The combination of food ingredients with ginger products affects the taste and aroma and can reduce the product's preference value. The use of ginger has an adverse effect on organoleptic tests due to the strong and pungent odor and taste, so a certain concentration is required to be acceptable to consumers [16]. In this study, the smell, taste and aroma were acceptable to consumers at P1.

Table 2. Quality of beef sausage with the addition of PF and GF as a substitute for TG.

Variables	Treatment (*mean*±SD)					*p-value*
	P0	P1	P2	P3	P4	
Color	6.48±1.05[a]	5.48±1.12[a]	4.68±1.52[b]	3.68±1.28[c]	3.20±1.19[c]	0.001
Aroma	6.32±0.85[a]	5.88±1.01[a]	4.60±1.44[b]	3.88±1.24[c]	3.36±1.25[c]	0.001
Taste	6.00±1.23[a]	5.52±1.56[a]	4.36±1.47[b]	3.64±1.25[c]	2.72±1.31[c]	0.001
Flavor	6.04±1.51[a]	5.08±1.44[a]	4.28±1.34[b]	3.56±0.96[c]	2.88±1.36[c]	0.001
Tenderness	5.12±1.51	4.84±1.49	5.00±1.47	4.44±1.45	4.68±1.49	0.518
Texture	5.60±1.50[a]	5.32±1.22[a]	5.04±1.37[b]	4.68±1.46[b]	3.72±1.40[c]	0.001
Overall	5.92±1.47[a]	5.24±1.62[a]	4.36±1.35[b]	3.68±1.31[c]	3.00±1.26[c]	0.001
After taste	5.88±1.36[a]	5.04±1.43[a]	4.32±1.41[b]	3.56±1.19[c]	2.84±1.25[c]	0.001

[a-c]Different superscripts in the same row indicate highly significant differences ($P<0.01$).
P0 : Beef sausage with combination of PF and GF = 0.00 : 0.00.
P1 : Beef sausage with combination of PF and GF = 5.00 : 0.75.
P2 : Beef sausage with combination of PF and GF = 10.00 : 1.50.
P3 : Beef sausage with combination of PF and GF = 15.00 : 2.25.
P4 : Beef sausage with combination of PF and GF = 20.00 : 3.00.

The color is different with a decrease compared to the control with the addition of GF, this study is in accordance with [14] which states that there is a difference in color in minced pork with the addition of GF up to 2%. [17], the color decreases with the addition of GF to pork burgers and this decrease in color is due to the basic color of the flour in powder form according to the basic ingredients which is yellow in ginger. Flavor is the main component that determines the level of consumer preference [18]. Therefore, flavor is a factor that can influence the overall level of preference [19]. With this consideration, the overall and aftertastes of consumers are also influenced by flavor, so that the decrease in flavor attributes in this study will also be followed by a decrease in overall and after taste from consumers.

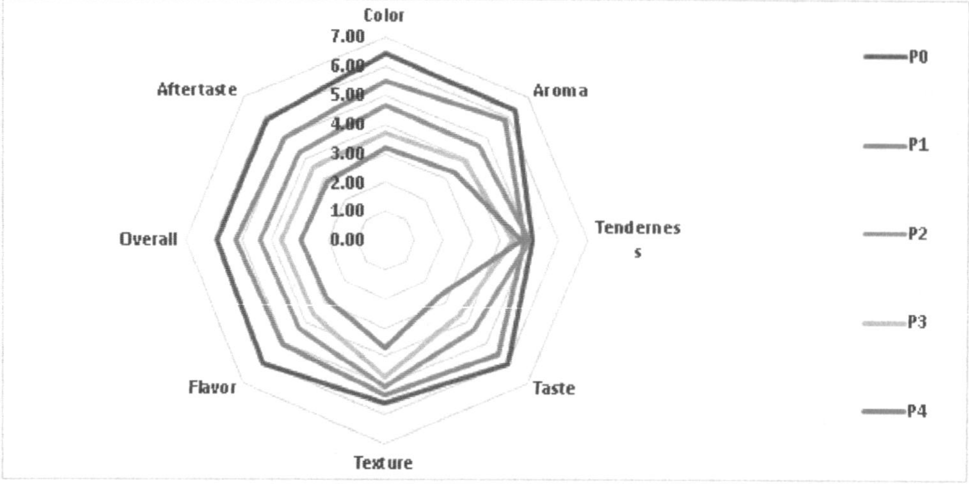

Figure 1. Spider graph of hedonic quality of beef sausage with the addition of PF and GF as a substitute for TF. P0 : Beef sausage with combination of PF and GF = 0.00 : 0.00; P1 : Beef sausage with combination of PF and GF = 5.00 : 0.75; P2 : Beef sausage with combination of PF and GF = 10.00 : 1.50; P3 : Beef sausage with combination of PF and GF = 15.00 : 2.25; P4 : Beef sausage with combination of PF and GF = 20.00 : 3.00.

Overall, the hedonic attributes decreased with the combination of PF and GF starting at P2. As described above, the GF factor is more dominant in reducing quality, so it can be covered by GF. This is due to the role of GF in covering the weaknesses of GF. Pea flour contains textured pea proteins which are good for improving the quality of texture, color and taste in meat products [20]. The use of pigeon peas up to 6% in sausages has good liking and preference values and is acceptable to consumers [21].

4. Conclusion

The addition of PF and GF at levels 5 and 0.75% (P1) can be used for making beef sausages, because the hedonic quality can still be maintained with a neutral assessment by the panelists.

5. Acknowledgement

The researcher would like to thank the UNS Rector for providing funding for this research through the LPPM UNS Non-APBN Fund Research Group Grant with contract number 194.2/UN27.22/PT.01.03/2024.

References

[1] Shaukat MN, A Nazir and B Fallico, 2023, Antioxidants, 12(11), 2015.
[2] Ersedo TL, TA Teka, SF Forsido, E Dessalegn, JA Adebo, M Tamiru and T Astatkie, 2023, International Journal of Food Properties, 26(01), 928-951.
[3] Duan W, L Liang, Y Huang, Y Zhang, B Sun and L Li, 2021, Foods, 10(7), 1456.
[4] Wu DT, WX Li, JJ Wan, YC Hu, RY Gan and L Zou, 2023, Foods, 12(13), 2527.
[5] Lara SW and A Tsiami, 2024, Foods, 13(14), 2290.
[6] Arshad M, S Anwar, I Pasha, F Ahmed and RM Aadil, 2022, International Journal of Food Science & Technology, 57(5), 3031-3037.
[7] Barker S and MB McSweeney, 2022, Journal of Food Science, 87(12), 5390-5401.
[8] BSN (Badan Standardisasi Nasional), 2015, SNI 3820:2015 tentang Sosis Daging, Badan Standardisasi Nasional, 1.
[9] Sirini N, PE Munekata, JM Lorenzo, MA Stegmayer, M Pateiro, JA Pérez-Álvarez, N Sepúlveda, ME Sosa-Morales, A Teixeira, J Fernández-López and L Frizzo, 2022, Foods, 11(8), 1128.
[10] Ratnaduhita A, AMP Nuhriawangsa and LR Kartikasari, 2021, Livestock and Animal Research, 19, 227-237.
[11] Kartikasari LR, 2013, Thesis, The University of Adelaide.
[12] Wichchukit S and M O'Mahony, 2015, Journal of the Science of Food and Agriculture, 95(11), 2167-2178.
[13] Steel RGD and JH Torrie, 1997, Principles and procedures of statistics: A biometrical approach, 3, McGraw-Hill, 666.
[14] Karpi´nska-Tymoszczyk M, A Draszanowska and M Danowska-Oziewicz, 2022, Foods, 11, 3563.
[15] AmoahRE, FD Wireko-Manu, I Oduro, FK Saalia, WO Ellis and E Owusu, 2022, Cogent Food & Agriculture, 8, 2123766.
[16] Kim TJ, KH Seo, JW Chon, HY Youn, D Jeong and KY Song, 2022, Journal of Dairy Science and Biotechnology, 40(3), 110-121.
[17] Mancini S, G Paci, F Fratini, B Torracca, R Nuvoloni, AD Bosco, V Roscini and G Preziuso, 2017, Meat Science, 129, 161-168.
[18] Ramalingam V, Z Song and I Hwang, 2019, Food Research International, 122, 174-182.
[19] Liu J, MP Ellies-Oury, S Chriki, I Legrand, G Pogorzelski, J Wierzbicki, L Farmer, D Troy, R Polkinghorne and JF Hocquette. 2020, Meat Science, 168, 108190.
[20] Maningat CC, T Jeradechachai and MR Buttshaw, 2021, Cereal Chemistry, 99(1), 37-66.
[21] Gomezulu AD, RJ Mongi and BM Kilima, 2021, Proceedings of the 2nd SUA Scientific Conference Held at SUA, 222-234.

Proteolytic Effects of Ragi Tempeh Starter and Soybean Derivative as Binders on The Chemical and Organoleptic Characteristics of Fermented Chicken Sausage

Mulia Winirsya Apriliyani[1,2*], Yuny Erwanto[1], Edi Suryanto[1], and Rusman[1]

[1]Departement of Animal Products Technology, Faculty of Animal Sciences, Universitas Gadjah Mada, Jl. Fauna No. 3, Bulaksumur, Yogyakarta 55281, Indonesia
[2]Departement of Animal Product Technology, Faculty of Animal Science, Universitas Brawijaya, Jl. Veteran, Malang, East Java 65145, Indonesia

*E-mail: muliawinirsyaapriliyani1088@mail.ugm.ac.id

Abstract. In Indonesia, ragi tempeh was used in tempeh making and is still rare in fermented chicken sausage whereas it has a proteolytic activity that degrades meat protein. A fermented sausage that consumers like has stable physical properties so it requires a binder. Tempeh flour, soy flour, and soy protein isolate as binders derived from vegetable ingredients are expected to have the ability to form a coherent and strong gel. This study aims to evaluate the proteolytic activity of ragi tempeh starter with different binders derived from soy products (tempeh flour, soy flour, and soy protein isolate) in the production of fermented chicken sausages, in terms of pH, a_w, moisture, protein, fat content, protease activity, dissolve protein, degree of hydrolysis, and organoleptic characteristics. The research method employed a completely randomized design with 6 replications. The experiment was set up using a completely randomized design (CRD). The data were analyzed using the Analysis of Variance (ANOVA) and the Least Significant Difference (LSD) test. The results of the statistical analysis of pH and a_w testing showed that they were significantly different ($P<0.05$), the water content, protein content, fat content, proteolytic activity, dissolved protein, degree of hydrolysis, hedonic and hedonic quality testing were not significantly different (P>0, 05). It can be concluded that using ragi tempeh with different binders from soybean derivatives affects fermented chicken sausage's chemical and organoleptic characteristics.

Keywords: fermentation, hydrolysis, proteolytic, ragi tempeh, sausage

1. Introduction

Sausage made through fermentation is a premium product. Sausage that has undergone fermentation was once considered a safe and healthful dish. Consuming fermented sausages has been linked in more recent times to health risks due to high levels of saturated fats and sodium chloride, the presence of nitrite and degradation products such as nitrosamines, and smoking, which can produce harmful substances including aromatic polycyclic hydrocarbons in the final goods [1]. Processing meat can also increase its functional value. Meat can be preserved via fermentation and ripening, which prevents the loss of important chemicals and enhances its functional value through biochemical changes (such as the production of bioactive peptides and the formation of bacterial strains that might be probiotics) [2]. Rhizopus molds improve the nutritional content and digestibility of tempeh while also giving the body essential living enzymes and advantageous microbes [3]. Using diversity starter microbes from *Rhizopus* sp. (*Rhizopus oligosporus* and *Rhizopus oryzae)* often used in tempeh fermentation for fermenting meat can be an interesting concept. *Rhizopus* fungi are known for their ability to ferment soybeans in tempe production, and they could potentially be adapted for use in meat fermentation to create functional products. Explore different strains of *Rhizopus* to determine which have the most suitable characteristics for meat fermentation. Consider factors like enzymatic activity, organoleptic quality, flavor profiles, and texture modification. In developing fermented sausages as a diversified functional food, incorporating a binder agent is essential to improve texture, stability, and overall quality. Binder agents help maintain the sausage's shape, enhance mouthfeel, and contribute to the product's nutritional profile. Because of their nutritional and functional quality, soy protein products, particularly SPI, have been employed as food components in various categories. In addition, soybeans are an inexpensive source of vegetable protein, which means that soy protein products provide benefits beyond the apparent cost advantages of vegetable proteins over animal proteins.

The conversion of soybeans into soy protein has produced goods with various culinary uses, including emulsification, binding, and texture [4]. Protein isolate's high protein content makes it a great binder for stabilizing emulsions, boosting meat products water-holding capacity, reducing the amount of shrinkage while cooking, enhancing the flavor and characteristics of product slices, raising a product's nutritional value, and lowering formulation expenses [5]. The main and supplementary ingredients used in the production of sausages are as follows: meat is the primary ingredient, binder and filler are the additives [6]. Due to their unique biochemical and structural characteristics, soybeans are particularly suitable for fermentation by *Rhizopus* sp. Each soybean derivative brings unique benefits to food formulations, making them valuable for different applications depending on nutritional goals, functional requirements, and flavor preferences. While soy protein isolate, soybean flour, and tempeh flour have benefits as binders, their impact on the quality of fermented chicken sausages should be evaluated in terms of pH, a_w, moisture, protein, fat content, protease activity, dissolved protein, degree of hydrolysis, and organoleptic characteristics.

2. Materials and Methods

2.1 The process of making fermented chicken sausage

The materials used for making fermented chicken sausages were chicken breast meat (80% w/w), fat (20% w/w), ice (20%), tapioca (10%), sugar (1.75%), ragi tempeh starter (1%), binder agents (soy protein isolate tempeh flour, soybean flour (each 1%), salt (1%), garlic (0.75%), Sodium Tripolyphosphate (0.6%), pepper (0.3%), and ginger (0.2%). The manufacturing process of fermented sausages involves preparing the meat, grinding, mixing with seasonings, binders, and starter ragi tempeh then stuffing it into casings, and fermenting at 35 °C for 24 h. After fermentation, the fermented chicken sausage was analyzed for pH, a_w, moisture, protein, fat content, protease activity, soluble protein, degree of hydrolysis, and sensory evaluation.

IOP Conf. Series: Earth and Environmental Science **1460** (2025) 012033 doi:10.1088/1755-1315/1460/1/012033

2.2 Determination of proteolytic activity

Proteolytic activity was assessed using a modified method [7]. A 1 mL portion of the substrate (1% casein dissolved in 0.01 M phosphate buffer, pH 7.6) was incubated for 20 minutes at 35°C. Prior to testing the protease activity, 1 gram of fermented chicken sausage extract was weighed and dissolved in 10 mL of phosphate buffer (pH 7.6) and vortexed. Then, 1 mL of the 1% casein substrate was combined with 0.5 mL of the sample and 0.5 mL of phosphate buffer (pH 7.6). The mixture was incubated for 20 minutes at 35°C. The reaction was terminated by adding 3 mL of 5% TCA, followed by vortexing. The mixture was then left to stand at room temperature for 30 minutes, after which it was centrifuged at 2500 rpm for 20 minutes to separate the filtrate. The filtrate was collected and its absorbance was measured using a UV-Vis spectrophotometer at 280 nm. For the blank solution, distilled water was used in place of the enzyme. The formula used is:

$$\text{Proteolitic activity} = \text{Abs of sample} - \text{Abs of blank}$$

2.3 Determination of dissolved protein

Sample measurements were carried out using reconstituted samples. A sample of 0.2 mL was added with 2 mL of Bradford reagent, then vortexed the solution until homogeneous, and it was allowed to stand for 15 min in a temperature room. Read the absorbance with a wavelength spectrophotometer at 595 nm [8].

2.4 Determination degree of hydrolysis

The measurement of the degree of hydrolysis (DH) is based on the method [9]. The samples were divided into two, namely with the addition of 6.25% TCA and without the addition of TCA. The sample added with TCA was incubated for 15 minutes and centrifuged at 8,000 g for 15 minutes. A total of 20 μL of filtrate was added to 150 μL of OPA solution, the mixture was homogenized using a vortex. The mixture was read on a spectrophotometer with a length of 340 nm. The DH value was calculated using the formula:

$$\text{DH (\%)} = \frac{\text{Absorbance of the sample dissolved in 6.25\% TCA}}{\text{Absorbance of the sample not dissolved in 6.25\% TCA}} \text{ x } 100\%$$

2.5 Determination of pH, a_w, moisture, protein, and fat content

The evaluation of pH value, a_w, moisture, protein, and fat content were determined based on AOAC [10].

2.6 Organoleptic characteristics

Organoleptic characteristics such as hedonic quality and hedonic tests were performed involving 20 semi-trained panelists. Hedonic quality is used to evaluate a specification of fermented chicken sausage using the modification method [1]. such as aroma (1= very not-aroma fermented sausage, 5= very aroma fermented sausage), color (1=white, 5= brownish), texture (1=very not-chewy, 5= very chewy) and taste (very not-taste fermented sausage, 5=very taste fermented sausage). Conversely, the hedonic test aims to discover acceptance and assess the quality of fermented chicken sausages, a grading scale from one to five (1= dislike extremely, 5=like extremely).

2.7 Data analysis

The data were analyzed using analysis of variance (ANOVA), followed by the least significant difference (LSD) test to determine the significant differences between the treatments.

3. Results and Discussion

3.1 pH and a_w

Table 1 shows that each fermented chicken sausage had a significantly different (P<0.05) in the pH and a_w. The range of pH was 4.53 to 5.02. The highest pH is fermented chicken sausage with tempeh flour (pH 5.02) because tempeh flour is dry and less active in terms of microbial fermentation. During the pre-fermentation of fermented chicken sausage, the protease of tempeh flour hydrolyzes proteins into peptides and amino acids. While soy protein isolate and soybean flour contribute to faster acidification, helping to break down carbohydrates and proteins by *Rhizopus* sp. (Ragi tempeh). Compared to Mediterranean nations, where a significant reduction in a_w is necessary, central and northern Europe benefit more from the fermentation fall pH value of 5.6-5.8 drops to 4.6-4.9 for the preservation of sausages [11].

The highest a_w value is with tempeh flour (0.93), followed by soybean flour (0.91), and lowest with soy protein isolate (0.89), this indicates that tempeh flour retains more moisture and has lower water-binding capacity compared to soy protein isolate, leading to higher free water and thus higher water activity in the final sausage product. Soybean flour has intermediate properties in moisture content and water-binding capacity. The protein soybeans can influence water interaction and water-holding capacity (WHC). WHC is a protein capacity matrix to absorb and hold against gravity water that is physically trapped, hydrodynamic, capillary, and bonded [12]. Fermented sausages prevent the growth of aerobic bacteria by establishing an anaerobic environment [13].

Table 1. Characteristics of fermented sausages chicken in pH, a_w, moisture content, protein content, fat content, proteolytic activity, dissolved protein, and degree of hydrolysis

Parameter	Soy protein isolate	Soybean flour	Tempeh flour
Ph	4.53±0.57^a	4.71±0.10^b	5.02±0.11^a
a_w	0.89±0.01^a	0.91±0.01^{ab}	0.93±0.01^b
Proteolytic activity (U/mL) ^{ns}	3.71±0.01	3.74±0.01	3.79±0.01
Dissolved protein (%)^{ns}	4.22±0.01	4.30±0.02	4.41±0.01
Degree of hydrolysis (%)^{ns}	19.04±0.05	19.28±0.02	19.97±0.01
Moisture content (%)^{ns}	66.54±0.01	67.19±0.01	66.75±0.01
Protein content (%)^{ns}	40.14±0.18	37.56±1.79	36.72±0.92
Fat content (%)^{ns}	14.12±1.68	16.36±0.39	14.63±0.53

$^{a\,b\,c}$: Values with different superscript letters in the line are significantly different
^{ns}: non-significant

3.2 Proteolytic activity, dissolved protein, and degree of hydrolysis

Table 1 shows that each fermented chicken sausage had not significantly different (P>0.05) in the proteolytic activity, dissolved protein, and degree of hydrolysis. The highest proteolytic activity in the fermented chicken sausage with tempeh flour would be the most effective, followed by soybean flour, and then soy protein isolate. Tempeh flour retains active proteolytic enzymes and a partially hydrolyzed protein matrix that can contribute to further proteolysis. As a starter, ragi tempeh which is the secretion of protease enzymes by the *Rhizopus* sp. Factor pH, temperature, and salt concentration impact the proteolytic activity of microbe [14]. The proteolytic process in fermented sausages using a ragi tempeh starter involves breaking down meat proteins into smaller peptides and free amino acids, contributing to the development of flavor, texture, and nutritional qualities of the final product. Proteolytic activity was used to evaluate the capability of hydrolyzing sarcoplasmic proteins [15]. Proteolysis is a critical aspect of fermentation, enhancing the sensory and nutritional attributes of fermented meats. Both endogenous and bacterial enzymes have an impact on the hydrolysis of sarcoplasmic proteins [16]. For example, high proteolytic activity was found in strains obtained from Doenjang, a traditional fermented cuisine from Korea, with a NaCl level of 7% or less [17]. Fermented chicken sausages with tempeh flour tend to produce the highest degree of hydrolysis and dissolved protein due to its pre-fermented nature and active enzymes, followed by soybean flour, with soy protein isolate. Generally, it has the highest degree of hydrolysis and dissolved protein because it contains active proteases and partially hydrolyzed proteins from the initial tempeh fermentation. A positive correlation was found between the degree of hydrolysis and neutral protease activity [18]. The percentage of peptide bonds that are cleaved is represented by the degree of hydrolysis, which is a measurement of the enzymatic breakdown of protein [9]. In general, a higher proteolytic activity is associated with an increased degree of hydrolysis. Research has demonstrated that as fermentation advances, there is a notable rise in both peptide concentration and free amino acid levels, suggesting efficient protein degradation [19, 20].

3.3 Moisture content, protein content, and fat content

Each fermented chicken sausage had not significantly different (P>0.05) in moisture, protein, and fat contents (Table 1). Fermented chicken sausages made with soybean flour have the highest moisture content because of their higher water-holding capacity than tempeh flour and soy protein isolate. Soy protein isolate had a lower moisture content because it binds less water and contains less moisture. Soy protein isolate in fermented sausages provides the highest protein content with minimal fat. The moisture content of the chicken nuggets was reduced by the addition of tempeh flour [21]. Tempe made from germinated soybeans had a higher lipid and carbohydrate content but a lower protein content [22]. Tempeh flour derived from germinated soybeans exhibited superior antioxidant activity compared to non-germinated ones [4]. Furthermore, the qualities of fermented sausages depend on the starter type used. Fermented products are simpler to digest and absorb because the protein content of the fermented sausage was hydrolyzed starter [23].

3.4. Organoleptic characteristics

Table 2 shows that each fermented chicken sausage had not significantly different (P>0.05) in the hedonic quality and hedonic test. This indicates that the ragi tempeh as a starter is dominant in using soy protein isolate, soybean flour, and tempeh flour so that the aroma produced by the sausage is slightly fermented. It can be attributed to the microbial and enzymatic activities of the ragi tempeh starter and the primary impact on texture and taste organoleptic properties. In fermented chicken sausages, the ragi tempeh as a starter is dominant in using soy protein isolate, soybean flour, and tempeh flour. It can be attributed to the microbial and enzymatic activities of the ragi tempeh starter and the primary impact on texture and taste organoleptic properties. In addition, similar fermentation effects, such as proteolytic and flavor development processes driven by the starter culture may overshadow differences contributed by the various ingredients. Consequently, the flavor and taste of soy protein isolate influenced significantly the final food products by the processes of soybean defatting and protein extraction [4]. There are advantages and disadvantages to using starting fungus for sausages. For instance, research has been done on the qualitative attributes of wild boar salami preserved with mushrooms that were prepared using various alternatives for the meat and fat tissue, as well as with and without commercial bacterial starting cultures [13]. Soy protein isolate is valuable for its high protein content and functional properties. Due to their nutrient profiles and impact on microbial activity, soybean flour, and tempeh flour play more direct roles in fermentation. Soy protein isolate is commonly used in the food industry because of its chemical composition, functional characteristics, and versatile applications [4]. In addition, soy protein isolate and soybean flour generally have a scent similar to soybeans [24]. The choice of protein in sausage formulations can influence the development of umami flavor. For example, using a mix of different meat sources or incorporating plant-based proteins such as tempeh can improve the umami taste, thanks to their distinct amino acid profiles and the formation of flavor-enhancing peptides during fermentation [25,26].

Table 2. Organoleptic characteristics of fermented chicken sausage with ragi tempeh starter and different binder agents

Parameter	Binder Agents	Hedonic quality score	Description	Hedonic score	Description
Aroma	Soy protein isolate	3.58±0.58	Slightly aroma of fermented sausage	3.15±1.08	Like slightly
	Soybean flour	3.64±0.88	Slightly aroma of fermented sausage	3.00±0.91	Like slightly
	Tempeh flour	3.82±0.63	Slightly aroma of fermented sausage	3.27±0.92	Like slightly
Color	Soy protein isolate	2.71±0.60	Pale yellow	3.52±0.81	Like slightly
	Soybean flour	2.61±0.21	Pale yellow	3.44±0.82	Like slightly
	Tempeh flour	2.91±0.66	Pale yellow	3.60±1.01	Like slightly
Texture	Soy protein isolate	3.51±0.89	Rather chewy	3.42±0.85	Like slightly
	Soybean flour	3.22±0.79	Rather chewy	3.32±1.03	Like slightly
	Tempeh flour	3.47±0.92	Rather chewy	3.24±1.27	Like slightly
Taste	Soy protein isolate	3.63±0.88	Slightly taste of fermented sausage	3.83±1.14	Like slightly
	Soybean flour	3.58±0.80	Slightly taste of fermented sausage	3.76±0.95	Like slightly
	Tempeh flour	3.11±0.68	Slightly taste of fermented sausage	4.12±0.86	Like

4. Conclusion

In conclusion, the differences in binder agents between soy protein isolate, soybean flour, and tempeh flour significantly impact the pH and activity of water. Generally, soy protein isolate results in a more stable product with a lower risk of microbial contamination and spoilage due to its lower pH and water activity. While soy protein isolate provides high protein content and affects texture, tempeh flour tends to have the highest proteolytic activity, degree of hydrolysis, and dissolved protein but also influences taste. During the fermentation of chicken sausages, the proteolytic enzymes from tempeh flour can remain active, and residual enzymes contribute to increasing the proteolytic activity, dissolved protein, and degree of hydrolysis. Tempeh flour has a significant impact on the taste of fermented chicken sausage.

Acknowledgment

This research was supported and funded by Beasiswa Pendidikan Indonesia (BPI), the Indonesia Endowment Funds for Education (LPDP), and the Education Financing Service Centre (Puslapdik).

References

[1] Lazic, I. B, Jovanic, J., Simunovic, S., Raseta, M., Trbovic, D., Baltic, I., and Ciric, J. (2019). Evaluation of sensory and chemical parameters of fermented sausages. *Meat Technology,* 60(2), 84–90. https://doi.org/10.18485/meattech.2019.60.2.2

[2] Zhu, C. Z., Zhang, W. G., Zhou, G. H., Xu, X. L., Kang, Z. L., and Yin, Y. (2013). Isolation and identification of antioxidant peptides from Jinhua Ham. *Journal of Agricultural and Food Chemistry,* 61, 1265–1271. https://doi.org/10.1021/jf3044764

[3] Pangastuti, A., Alfisah, R. K., Istiana, N. I., Sari, S. L. A., Setyaningsih, R., Susilowati, A., and Purwoko, T. (2019), *Biodiversitas Journal of Biological Diversity*, 20(4): 1106-1114. https://doi.org/10.13057/biodiv/d200423

[4] Astawan, M. and Prayudani, A. P. G. (2020). The overview of food technology to process soy protein isolate and its application toward food industry. *World.Nutrition Journal,* 4(S1):12-17. https://doi.org/10.25220/WNJ.V04.S1.0003

[5] Garba, U. and Kaur, S. (2014). Protein isolates: production, functional properties and application. *International Journal of Current Research and Review*, 6(3): 35-45.

[6] Canti, M., Murdiati, A., Naruki, S., and Supriyanto. (2021). Quality characteristics of chicken sausages using a combination of jack bean (*Canavalia ensiformis L.*) and soy protein isolate as a binder. *Food Research,* 5(3): 249-261. https://doi.org/10.26656/fr.2017.5(3).544

[7] Bergmeyer, H. U., Bergmeyer, J., and Graßl, M., (1983). *Methods of Enzymatic Analysis* (2nd ed.). Weinheim: Verlag Chemie.

[8] Utami, P., Lestari, S., and Lestari, S. D. (2016). Pengaruh metode pemasakan terhadap komposisi kimia dan asam amino ikan seluang (*Rasbora argyrotaenia*). *FishtecH – Jurnal Teknologi Hasil Perikanan*, 5(1): 73-84. http://ejournal.unsri.ac.id/index.php/fishtech

[9] Auwal, S. M., Zarei, M., Abdul-Hamid, A., and Saari, N. (2017). Optimization of bromelain-aided production of angiotensin I-converting enzyme inhibitory hydrolysates from stone fish using response surface methodology. *Marine Drugs.* 15 (4): 104 https://doi.org/10.3390/md15040104

[10] AOAC. (2005). *Official methods of analysis of the Association of Analytical Chemists* (19th ed.). Washington, D.C.: AOAC International.

[11] Mitrovic, R. R., Jankovic, V. V., Ciric, J. S., Djordjevic, V. Z., Juric, Z. Lj., Mitrovic-Stanivuk, M. R., and Baltic, B. M. *IOP Conf. Series: Earth and Environmental* Science, 854: 12021. http://doi.org/10.1088/1755-1315/854/1/012061

[12] Traynham, T. L., Myers, D. J., Carriquiry, A. L., and Johnson, L. A. (2007). Evaluation of water-holding capacity for wheat–soy flour blends. *Journal of the American Oil Chemists' Society,* 84(2):151-155. http://doi.org/10.1007/s11746-006-1018-0

[13] Laranjo, M., Elias, M., and Fraqueza, M. J. (2017). The use of starter cultures in traditional meat products. *Journal of Food Quality,* 1–18. http://doi.org/10.1155/2017/9546026

[14] Sun, F., Wang, H., Liu, Q., Kong, B., and Chen, Q. (2021). Effects of temperature and pH on the structure of a protease from *Lactobacillus brevis* R4 isolated from Harbin dry sausage and molecular docking of the protease to the meat proteins. *Food Bioscience*, 42: 101099. https://doi.org/10.1016/j.fbio.2021.101099

[15] Chen, X., Mi, R., Qi, B., Xiong, S., Li, J., Qu, C., Qiao, X., Chen, W., and Wang, S. (2021). Effect of proteolytic starter culture isolated from Chinese Dong fermented pork (Nanx Wudl) on microbiological, biochemical and organoleptic attributes in dry fermented sausages. *Food Science and Human Wellness,* 10 (1): 13-22. https://doi.org/10.1016/j.fshw.2020.05.012

[16] Shiling, L., Xianna, H., Yanbin, Y., Baokun, L., Chengjian, X., and Qingling, W, (2017). Proteolytic effect of starter culture during ripening of smoked horse sausage. *Food Science and Biotechnology,* 26(5): 1363–1369. https://doi.org/10.1007/s10068-017-0163-6

[17] Jeon, H. H., Jung, J. Y., Chun, B. H., Kim, M. D., Baek, S. Y., Moon, J. Y., Yeo, S. H., and Jeon, C. O. (2016). Screening and characterization of potential *Bacillus* starter cultures for fermenting low-salt soybean paste (Doenjang). *Journal of Microbiology and Biotechnology,* 26:666-674. https://doi.org/10.4014/jmb.1512.12014

[18] Zhao, C. J., Schieber, A., and Ganzle, M. G. (2016). Formation of taste-active amino acids, amino acid derivatives and peptides in food fermentations – A review. *Food Research International,* 89(1): 39-47. https://doi.org/10.1016/j.foodres.2016.08.042

[19] Wang, D., Cheng, F., Wang, Y., Han, J., Gao, F., Tian, J., Zhang, K., and Jin, Y. (2022). The Changes Occurring in Proteins during Processing and Storage of Fermented Meat Products and Their Regulation by Lactic Acid Bacteria. *Foods,* 11: 2427. https://doi.org/10.3390/foods11162427

[20] Li, H., Zheng, R., Zuo, F., Qian, C., Yao, Z., Dong, R., Zhao, D., and Li, C. (2022). Influence of proteolysis on the binding capacity of flavor compounds to myofibrillar proteins. *Foods,* 11: 891. https://doi.org/10.3390/foods11060891

[21] Murni, D., Asep, S., and Hidayat, T. (2014). Soy protein isolate sebagai bahan fungsional dalam produk pangan. *Berita Litbang Industri,* 3(2), 117-123.

[22] Kadar, A. D., Astawan, M., Putri, S. P., and Fukusaki. (2020). Metabolomics-based study of the effect of raw materials to the end product of tempe-an indonesian fermented soybean. *Metabolites,* 10(9): 367. https://doi.org/10.3390/metabo10090367

[23] Susilo, A., Widyastuti, E. S., Evanuarini, H., and Apriliyani, M. W. (2023). Comparison of the quality of fermented sausages with the use of yogurt starter and *Lactobacillus plantarum* (pH, aw, and proximate values). *Jurnal Penelitian Pendidikan IPA*, 9(5): 2319–2324. https://doi.org/10.29303/jppipa.v9i5.2463

[24] Herlina, Aji, S.B., and Purnomo. (2020). Physical, chemical, and sensory characteristics of chicken sausage with analog meat substitution. Industria: *Jurnal Teknologi dan Manajemen Agroindustri*, 10 (1): 67-77. https://doi.org/10.21776/ub.industria.2021.010.01.8

[25] Ju, M., Kim, J., Jang, H, Yeon, S., Hong, G., Park, W., Seo, H.G., and Lee, C. (2016). Changes of physicochemical and sensory properties of fermented sausage from sulfur-fed pork. *Korean Society for Food Science of Animal Resources,* 36(6): 729-736. https://doi.org/10.5851/kosfa.2016.36.6.729

[26] Susi, S., Al Hakim, H. M., and Gendrosari, S. (2023). Characteristics of seasoning powders from nagara bean tempeh and oyster mushroom for umami enhancer to reduce salt-use. *Asian Food Science Journal*, 22(6): 28-39. https://doi.org/10.9734/AFSJ/2023/v22i6640

Formulation of vegan chocolate with soybean and kidney bean: physical and sensory properties

Dimas Rahadian Aji Muhammad[1,2*], Diinah Salwa Kamiilah[1], Ardiba Rakhmi Sefrienda[3], Asri Nursiwi[1], Anastriyani Yulviatun[1,2],

[1] Dept. of Food Science and Technology, Universitas Sebelas Maret, Surakarta, Indonesia
[2] Research and Development Center for Food, Nutrition and Publich Health, Universitas Sebelas Maret, Surakarta, Indonesia
[3] Research Center for Food Technology and Processing, National Research and Innovation Agency (BRIN), Gunungkidul, Indonesia

*E-mail: dimasrahadian@staff.uns.ac.id

Abstract. The use of milk in chocolate formulation restricts vegan and lactose-intolerant consumer groups to consume the product. This study aims to determine the effect of using red bean and soybean powder on the sensory, physical, and chemical characteristics of vegan chocolate. A completely randomized design with one factor, namely the concentration of bean powder (0%, 4%, 8%, and 12%), was used in this study. Data analysis was conducted using SPSS with the Kruskal-Wallis test and One-way ANOVA, followed by the Duncan Multiple Range Test (DMRT). The results showed that the concentration of red bean and soybean powder had a significant in altering the textural and color properties of the chocolate formulated with soybean powder and decreasing hardness of the chocolate formulated with red kidney bean but no significant effect on the level of consumer acceptance. This research creates new opportunities for vegan chocolate formulated with soybean or red kidney bean.

Keywords : kidney bean, soybean, vegan chocolate

1. Introduction

The cocoa industry is one of the important industries for Indonesia because it acts as one of the country's foreign exchange earners as well as generating regional and agro-industrial development in Indonesia (Praseptiangga et al., 2020). Chocolate is one of the main cocoa-derived products that contains high phenolic compounds and is a source of flavonoid compounds that act as natural antioxidants (Muhammad et al., 2022a). The combination of potential health benefit and also indulgence effect of chocolate makes this product is so popular (Muhammad et al., 2023). Consequently, the market of this product is extremely competitive, and thus the cocoa and chocolate industries must innovate their products to win the market competition. There are several recent innovative cocoa-derived products that have been developed by the industry and the scientists, including chocolate enriched with herbs, spices, flower and other plant based functional ingredients (Muhammad et al., 2021a; Muhammad et al., 2022b; Lestari et al., 2021), probiotic (Muhammad et al., 2021b), nanocapsule (Muhammad et al., 2018), single origin chocolate (Muhammad et al., 2022c) and also vegan chocolate (Torregrosa et al., 2024).

There are three types of chocolate based on their composition: dark chocolate, milk chocolate, and white chocolate. The difference lays on the amount of cocoa butter, cocoa paste, sugar, milk, and other added ingredients used in its formulation (Praseptiangga et al., 2018). Milk chocolate consists of cocoa mass, cocoa butter, sugar, and milk powder, with or without the addition of other food ingredients. The milk content in milk chocolate is higher at ≥12% when compared to the milk content in semisweet or bittersweet dark chocolate (Ilmi et al., 2017; Katz et al., 2011; Brown et al., 2023). Milk plays an important role in producing the right texture and flavor in milk chocolate. However, there is a group of consumers who are intolerant of the lactose contained in cow's milk. Lactose intolerance is a condition where lactose is not hydrolyzed due to the lack of lactase enzyme in a person's body, so that when consuming milk or milk-derived products, several digestive symptoms arise, such as bloating, diarrhea, abdominal pain, vomiting, and flatulence (Vesa et al., 2000). In addition, there is currently a rapidly increasing trend toward plant-based (vegan) eating patterns. In America, the increase in the number of vegans from 2014–2017 increased by 500%. Moreover, the results of a global survey in 2019 reported that 40% of consumers are trying to reduce their consumption of animal protein, so sales of meat and dairy alternatives have increased. For instance, in America, sales of plant-based yogurt increased by 55%, plant-based cheese by 43%, and plant-based creamer by 131% (Alcorta et al., 2021). Therefore, a solution is needed so that these consumer groups can still enjoy milk chocolate safely, namely by making vegan milk-like chocolate products by replacing cow's milk content in chocolate.

Soybean powder (*Glycine max*) is a plant-based product potentially used for cow's milk substitution in chocolate formulation. It contains isoflavones having a role as antioxidants in inhibiting degenerative diseases and glycemic control and can reduce cholesterol levels in the blood (Arygunartha et al., 2022; Zarić et al., 2015). In addition, soybeans do not contain lactose or cholesterol, so they will not be too problematic if consumed by individuals who have dietary disorders or an intolerance to lactose (Osthoff et al., 2010). Red kidney bean powder (*Phaseolus vulgaris* L.) is another potential protein-rich plant-based product is vegan chocolate formulation. The protein content in 100 g of red kidney bean powder is 17.24 g. In addition, red bean powder contains 2.21 g of fat, 71.08 g of carbohydrates, 86.04 mg of phosphorus, and 39.15 mg of calcium (Perwita et al., 2021). Red kidney beans also contain anthocyanins, namely cyanidin and pelargonidin at a level of 1.2 mg/100 g and 2.4 mg/100 g, respectively (Sari et al. 2020).

Replacing cow's milk powder with soybean powder and red kidney bean powder in the process of making vegan chocolate will help lactose intolerance and vegan consumers for enjoying the chocolate. However, the differences in nutritional content and characteristics of soybean powder and red kidney bean powder potentially result in differences in the characteristics of the chocolate, including physical characteristics and sensory properties. Up till now, there is a limited study in the literature discussing the development of vegan chocolate. Therefore, this study aimed to study the effect of using soybean powder and red kidney bean powder as milk substitutes on the physical and characteristics of chocolate.

2. Materials and Methods

2.1 Sample preparation

Prior to vegan chocolate formulation, soybean powder and red kidney powder (ABT Store, Kendal, Indonesia) were roasted at 121°C for 2 h. The vegan chocolates were prepared based on the bean-to-bar ¬technique. The "grade A" dried fermented cocoa beans obtained from Gunungkidul (Indonesia), was roasted at 120oC for 40 minutes using a rotary drum roaster. Then the roasted beans were manually deshelled to produce chocolate nibs. These nibs were then pulverized using a grinding machine to obtain a thick, dark brown liquid known as cocoa paste. The next process was grinding using a melanger (Cocoatown, ECGC-12SL, Roswell, USA) to obtain cocoa liquor with smaller particle size. In this process, the cocoa liquor was mixed with other ingredients such as cocoa butter CAU Chocolates, PT Cau Coklat Internasional), soy lecithin (Naturindo, PT Naturindo Fresh), crushed sugar (Gulaku, PT Sweet Indolampung Sugar Group Companies) and soybean powder or red bean powder. The powder added into the chocolate formula was varied at a level of 0-12% from the weight of roasted cocoa beans. Meanwhile, the cocoa butter, sugar and soy lecithin were set at 14%, 19.35% and 0,65%, respectively, for each formula. The samples of chocolate containing 0% of red kidney beans or soybean can be considered as dark chocolate 80% used as the basic formula of the vegan chocolate. The grinding was continued up to 24 h. Afterward, the mixture was manually tempered by a professional chocolatier. The tempered chocolate was then poured into a mold and shaken so that the chocolate was evenly distributed avoiding bubbles. Next, the chocolate was stored in the refrigerator at 5°C for 30 minutes. After that, the solid chocolate was removed from the mold and packaged with aluminum foil.

2.2 Physical and sensory analysis

The color analysis was carried out using a Chromameter (Konica Minolta CR-400). The chroma value is obtained from the calculation using the L*, a*, and b* parameters, based on the formula of $C = \sqrt{a^2 + b^2}$; while the WI value is based on the formula of WI = $100 - \sqrt{(100 - L^*)^2 + a^2 + b^2}$; and the value of ΔE based on the formula of $\Delta E = \sqrt{\Delta L^2 + \Delta a^2 + \Delta b^2}$. Meanwhile, the texture analysis was carried out using a penetration test to evaluate the hardness based on the protocol of Muhammad et al. (2018). Universal Testing Machine Zwick/Roell BDO-FB0.5TS (Zwick GmbH & Co, Ulm, Germany) coupled with a probe (ϕ = 1 mm) was used. The moisture content was evaluated using thermogravimetry according to AOAC (2005).

The sensory test conducted in this study has fulfilled the ethical clearance aspect with the proof of ethical clearance letter issued by Moewardi Hospital, Surakarta, with letter number 101/I/HREC/2023. The sensory test was conducted using 50 untrained panelists with an adult age range of 18–25 years. The selection of the age range of panelists was based on the results of research by Kilian and Hamm (2021), which found that vegan consumers are dominated by adults with an age range under 40 years (18–39 years). The rating scale used in the sensory test in this study ranges from 1 to 7 (1: strongly dislike; 2: dislike; 3: somewhat dislike; 4: neutral; 5: somewhat like; 6: like; 7: strongly like).

2.3 Statistical analysis

The data was statistically analyzed using an SPSS software using the Kruskal-Wallis Method for the sensory test, and One-way ANOVA for the physical and chemical analysis with significance level at α = 5%. If there was a significance effect, the analysis was continued using Duncan Multiple Range Test (DMRT).

3. Results and Discussion

3.1 Color properties

Table 1 presents the results of the color test for vegan chocolate products formulated with red kidney bean and soybean. The lightness (L*), also known as brightness, of vegan chocolate samples formulated with red kidney bean varied significantly with the concentration of the bean. The L* value of chocolate increased with the concentration of red kidney bean which were 29.66, 30.15 and 31.04 for the red kidney bean of 4%, 8% and 12%, respectively. Meanwhile, vegan chocolates formulated with soybean showed no significant difference among samples as the concentration of soybean powder increased. The L* value of chocolate with 4%, 8%, and 12% of soybean were 30.57, 30.76, and 30.62, respectively.

The L* value of the control sample was 29.68. This result is consistent with the findings of Chire-Fajardo et al. (2017) regarding the brightness level of dark chocolate, which has a value of 29.72. The L* value of the control sample is lower compared to that of the vegan chocolates meaning that the bean can improve the lightness of dark chocolate. The control sample has a higher cocoa content compared to the other samples. As stated by Barišić et al. (2021), a higher cocoa content in chocolate results in a darker chocolate color.

L* value along with a* (redness) and b* (yellowness) values were then used to calculate the Chroma, Whiteness Index (WI) and the color difference between the formulated vegan chocolate and the control. Those three parameters were expectedly altered as the redness (a*) of the formulated vegan chocolates tent to decrease as the concentration of the beans increased. However, it was shown that there were no significant differences in the test results for the chroma (C*) value of the vegan chocolate samples formulated with red kidney bean as the concentration of red kidney bean powder increased, at the level of 10.37 to 10.43. Meanwhile, the chroma of vegan chocolate samples formulated with soybean showed significant differences that tended to decrease as the concentration of soybean increased. The C* value of vegan chocolates with 4%, 8%, and 12% of soybean were 12.36, 11.09, and 11.04, respectively.

Table 1. Color properties of vegan chocolates formulated with red kidney bean and soybean

Formula	Nilai					
	L*	a*	b*	C*	WI	ΔE
Control	29.68 ± 0.33^{a}	7.04 ± 0.05^{d}	8.48 ± 0.10^{ab}	11.02 ± 0.07^{b}	28.82 ± 0.32^{a}	-
Red kidney bean 4%	29.66 ± 0.35^{a}	6.03 ± 0.13^{a}	8.43 ± 0.10^{a}	10.37 ± 0.15^{a}	28.90 ± 0.36^{a}	1.06 ± 0.16
Red kidney bean 8%	30.15 ± 0.11^{b}	5.89 ± 0.02^{a}	8.60 ± 0.05^{b}	10.43 ± 0.04^{a}	29.38 ± 0.12^{b}	1.26 ± 0.11
Red kidney bean 12%	31.04 ± 0.11^{d}	5.90 ± 0.03^{a}	8.61 ± 0.09^{b}	10.43 ± 0.09^{a}	30.25 ± 0.12^{d}	1.78 ± 0.27
Soybean 4%	30.57 ± 0.26^{c}	7.63 ± 0.11^{e}	9.73 ± 0.04^{d}	12.36 ± 0.08^{c}	29.48 ± 0.27^{b}	1.64 ± 0.12
Soybean 8%	30.76 ± 0.11^{cd}	6.49 ± 0.14^{c}	8.99 ± 0.02^{c}	11.09 ± 0.09^{b}	30.00 ± 0.24^{cd}	1.43 ± 0.29
Soybean 12%	30.62 ± 0.09^{c}	6.33 ± 0.03^{b}	9.05 ± 0.04^{c}	11.04 ± 0.05^{b}	29.75 ± 0.09^{bc}	1.32 ± 0.31

Note: Different notations in each column indicate significant differences at the $\alpha = 0.05$ confidence level.

Table 1 also shows that the control sample has the lowest whiteness index (WI) value among all samples. This is because the WI value has a negative correlation with the cocoa content of chocolate. The higher the cocoa content in chocolate, the lower the WI value as stated by Fajardo et al. (2017). In this study, the cocoa content in the sample decreased as the concentration of the beans were increased in the formulation. The WI of vegan chocolate samples formulated with red kidney bean increased significantly with the concentration of red kidney bean. The WI of chocolate with 4%, 8%, and 12% red kidney bean was 28.90, 29.38, and 30.25, respectively.

Meanwhile, the whiteness index (WI) of vegan chocolate samples formulated with soybean did not show significant differences as the concentration of soybean powder increased. The WI value of vegan chocolates formulated with red kidney bean tent to be lower than that of samples formulated with soybean. This is because the a* and b* values of red bean flour are lower than those of soybean flour. According to Ratnawati et al. (2019), the WI value of red bean flour is lower than that of soybean flour.

Table 1 shows that the color difference (ΔE) between the control and the formulated vegan chocolates which was found at a range of 1.06 to 1.78. The ΔE value was categorized into five types based on the value range: <0.5 (changes are imperceptible); 0.5-1.5 (changes are slightly visible); 1.5-3.0 (changes are visible); 3.0-6.0 (changes are well visible); and >6.0 (changes are very visible). Thus, the study's findings on the ΔE of vegan chocolate samples formulated with either red kidney bean or soybean indicated visible, depending on the concentration of the beans incorporated in the formulation.

3.2 Texture

Figure 1a demonstrates the results of texture analysis for the formulated vegan chocolate products. As shown, there was a trend of significant decrease on the hardness of the formulated vegan chocolates as compared to the control. The hardness values for chocolate with red kidney bean concentrations of 4%, 8% and 12% were 22.54 N, 32.86 N and 34.8 N, respectively. Meanwhile, the hardness value of chocolate with 4% soybean powder concentration was 26.06, while that of 8% was 27.88, and that of 12% was 21.64. Afoakwa et al. (2008) stated that the texture of chocolate is influenced by various factors, including moisture content, fat content, particle size, and tempering level. An increase in fat content leads to a decrease in the hardness of chocolate (Do et al., 2011). In the formulated vegan chocolate, the fat content may be lower than the control because the beans were used to substitute the fat-rich cocoa liquor. The non-fat fraction may inhibit the crystallization of cocoa butter. Also, as explained by Hřivna et al. (2021), the other type of fat originated from the bean, may cause a eutectic effect that affects the tempering process and changes the cocoa butter crystals, which in turn affects the hardness of chocolate resulting in a softer texture compared to the control chocolate samples.

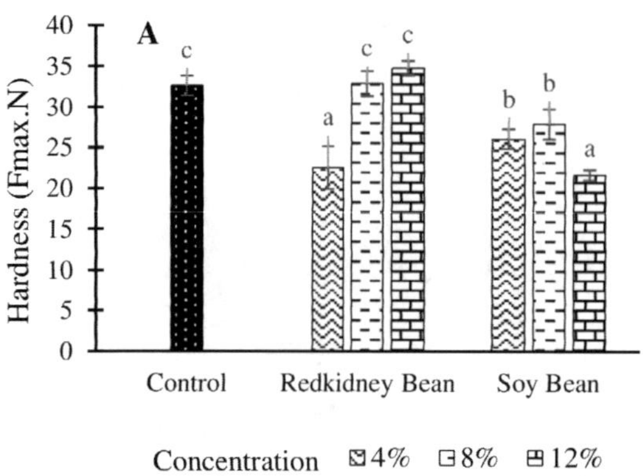

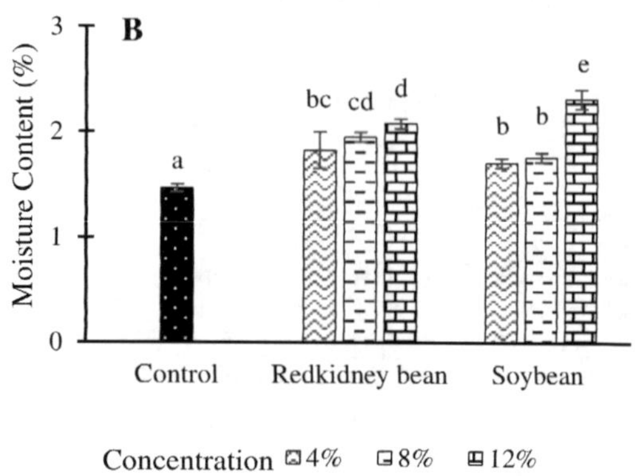

Figure 1. Textural properties (A) and moisture content (B) of vegan chocolates formulated with soybean and red kidney beans. Different notations indicate significant differences at the α = 0.05.

Figure 1b presents the moisture content of the formulated vegan chocolate products and the control sample. As shown, the moisture content of the formulated vegan chocolates increased along with the higher concentration of the beans. The moisture content of vegan chocolate samples formulated with red kidney beans at the level of 4%, 8% and 12% were 1.82, 1.94 and 2.08, respectively. The moisture content value of vegan chocolates with soybean also increased with the concentration of soybean powder added. Specifically, the moisture content of the sample increased to 1.70% for the chocolate with 4% of soybean, 1.76% for the chocolate with 8% of soybean and 2.32% for the chocolate with 12% of soybean. To investigate the reason of this rise, the moisture content of the red kidney beans and soybeans were analyzed. It was found that the moisture content of the beans was 10.34% and 6.51%, respectively, which is higher than chocolate control. Therefore, the increase of the moisture content of the vegan chocolate formulated with red kidney beans and soybean is reasonable as the moisture came from the beans

3.3 Consumer acceptance

Table 2 presents the results of the hedonic test of vegan chocolate formulated with soybean and red kidney bean powder. According to Table 2, the panelist acceptance for appearance texture showed no significant differences between the control and the sample formulated with red kidney bean powder. Similar result was obtained for the sample with soybean powder at the significance level of $\alpha = 0.05$. It was observed visually that each sample has a similar glossy level, and there is no fat blooming causing white spots on the surface. As a result, the liking test showed no significant difference in the appearance of all samples. Additionally, the appearance of the produced chocolate can be affected by both the manufacturing process and the fineness of the particles. The chocolate samples in this study underwent a thorough manufacturing process, ensuring that all ingredients, including the added powder, were perfectly mixed. As a result, the addition of red bean or soybean powder as an additive did not significantly impact the chocolate's appearance. Similar to the appearance, there was no significant difference between the control and the innovative vegan chocolates formulated with either red kidney beans or soybeans in terms of texture.

Table 2. Consumer acceptance level of vegan chocolate formulated with soybean red kidney bean

Formula	**Appearance**	**Texture**
Control	5.96 ± 0.88^{a}	5.20 ± 1.20^{a}
Red kidney bean 4%	6.20 ± 0.64^{a}	5.72 ± 0.97^{a}
Red kidney bean 8%	6.06 ± 0.62^{a}	5.54 ± 1.22^{a}
Red kidney bean 12%	6.16 ± 0.68^{a}	5.50 ± 1.33^{a}
Soybean 4%	5.94 ± 0.87^{a}	5.40 ± 1.13^{a}
Soybean 8%	6.02 ± 0.74^{a}	5.36 ± 1.19^{a}
Soybean 12%	5.98 ± 0.87^{a}	5.24 ± 1.17^{a}

Note: 1: strongly dislike; 2: dislike; 3: somewhat 4: neutral dislike; 5: somewhat like; 6: like; 7: strongly like. Different notations in each column indicate significant differences at the $\alpha = 0.05$ confidence level.

4. Conclusion

This study demonstrates that soybean and red kidney beans are potentially used in vegan chocolate formula. The use of soybean and red kidney bean resulted in alteration of physical characteristic of chocolate to some extent, but no had significant effect on the consumer acceptance in the attributes of appearance and texture. Further research investigating the chemical properties of formulted chocolate is important.

Acknowledgements

This research was funded by Universitas Sebelas Maret through the scheme of Penelitian Unggulan Terapan (PUT-UNS) 2023 (Contract No. 228/UN27.22/PT.01.03/2023).

References

Afoakwa, E. O., Paterson, A., Fowler, M., & Vieira, J. (2008). Particle size distribution and compositional effects on textural properties and appearance of dark chocolates. Journal of Food Engineering, 87(2), 181-190.

Alcorta, A., Porta, A., Tárrega, A., Alvarez, M. D., & Vaquero, M. P. (2021). Foods for plant-based diets: Challenges and innovations. Foods, 10(2), 293.

Arygunartha, G. Y., Setianingsih, N. L. P. P., & Sunarso, S. U. P. (2022). Pengaruh Proses Pengolahan terhadap Sifat Fisika dan Kimia Bubuk Kedelai: Literature Review. Jurnal Impresi Indonesia, 1(2), 89-94.

Barišić, V., Jozinović, A., Flanjak, I., Šubarić, D., Babić, J., Miličević, B., ... & Ačkar, Đ. (2021). Effect of addition of fibres and polyphenols on properties of chocolate–A review. Food reviews international, 37(3), 225-243.

Brown, A. L., Warren, E. R., Ingraham, B. W., Ziegler, G. R., & Hopfer, H. (2023). The effect of fat content on sensory perception and consumer acceptability of 70% cacao dark chocolate made from reconstituted cocoa liquor. Journal of Sensory Studies, 38(5), e12864.

Chire-Fajardo, G. C., Valdivia-Arrunategui, R. A., Orihuela-Rivera, C. A., & Ureña-Peralta, M. O. (2017). Assessment of physical and physicochemical quality of main chocolates traded in Peru. Acta Agronómica, 66(2), 164-171.

da Silva Medeiros, N., Koslowsky Marder, R., Farias Wohlenberg, M., Funchal, C., & Dani, C. (2015). Total phenolic content and antioxidant activity of different types of chocolate, milk, semisweet, dark, and soy, in cerebral cortex, hippocampus, and cerebellum of wistar rats. Biochemistry research international, 2015.

Do, T. A., Vieira, J., Hargreaves, J. M., Mitchell, J. R., & Wolf, B. (2011). Structural characteristics of cocoa particles and their effect on the viscosity of reduced fat chocolate. LWT-Food Science and Technology, 44(4), 1207-1211.

Hřivna, L., Machálková, L., Burešová, I., Nedomová, Š., & Gregor, T. (2021). Texture, color, and sensory changes occurring in chocolate bars with filling during storage. Food science & nutrition, 9(9), 4863-4873.

Ilmi, A., Praseptiangga, D., & Muhammad, D. R. A. (2017, April). Sensory attributes and preliminary characterization of milk chocolate bar enriched with cinnamon essential oil. In IOP conference series: Materials science and engineering (Vol. 193, No. 1, p. 012031). IOP Publishing.

Katz, D. L., Doughty, K., & Ali, A. (2011). Cocoa and chocolate in human health and disease. Antioxidants & redox signaling, 15(10), 2779-2811.

Lesschaeve, I., & Noble, A. C. (2005). Polyphenols: factors influencing their sensory properties and their effects on food and beverage preferences. The American journal of clinical nutrition, 81(1), 330S-335S.

Lestari, P. D., Kawiji, K., Yulviatun, A., Martien, R., & Muhammad, D. R. A. (2021). Physical and sensory characteristics of milk and white compound chocolate added with Asian pigeonwings flower (Clitoria ternatea). In E3S Web of Conferences (Vol. 332, p. 01001). EDP Sciences.

Muhammad, D. R. A., Anandito, R. B., Fauza, G., & Saputri, L. M. (2023, June). The effect of consumer knowledge and perceptions on the purchase decision of chocolate with sustainability label. In IOP Conference Series: Earth and Environmental Science (Vol. 1200, No. 1, p. 012054). IOP Publishing.

Muhammad, D. R. A., Fibri, D. L. N., & Prakash, S. (2022a). Improving the functionality of chocolate by incorporating vegetal extracts. Trends in sustainable chocolate production, 113-152.

Muhammad, D. R. A., Fibri, D. L. N., Saputro, A. D., & Sunarharum, W. B. (2022b). Investigating the elicited emotion of single-origin chocolate towards sustainable chocolate production in Indonesia. Open Agriculture, 7(1), 382-391.

Muhammad, D. R. A., Marettama, N. M., Fauza, G., & Affandi, D. R. (2022c). Can ingredients and information interventions affect the hedonic level and (emo-sensory) perceptions of the milk chocolate and cocoa drink's consumers?. Open Agriculture, 7(1), 847-856.

Muhammad, D. R. A., Rahayu, E. S., & Fibri, D. L. N. (2021b). Revisiting the development of probiotic-based functional chocolates. Reviews in Agricultural Science, 9, 233-248.

Muhammad, D. R. A., Saputro, A. D., Rottiers, H., Van de Walle, D., & Dewettinck, K. (2018). Physicochemical properties and antioxidant activities of chocolates enriched with engineered cinnamon nanoparticles. European Food Research and Technology, 244, 1185-1202.

Muhammad, D. R. A., Zulfa, F., Purnomo, D., Widiatmoko, C., & Fibri, D. L. N. (2021a). Consumer acceptance of chocolate formulated with functional ingredient. In IOP Conference Series: Earth and Environmental Science (Vol. 637, No. 1, p. 012081). IOP Publishing.

Osthoff, G., Hugo, A., Van Wyk, P., De Wit, M., & Meyer, S. (2010). Characterization of a spray-dried soymilk powder and changes observed during storage. Food science and technology international, 16(2), 169-178.

Perwita, E. S. (2021). Proporsi tepung kacang merah (Phaseolus vulgaris L.) dan bubuk daun kelor (Moringa oleifera L) terhadap sifat fisik organoleptic snack bar labu kuning. JTB, 10(2), 303-313.

Praseptiangga, D., Nabila, Y., & Muhammad, D. R. A. (2018). Kajian tingkat penerimaan panelis pada dark chocolate bar dengan penambahan bubuk kayu manis (Cinnamomum burmannii). Caraka Tani: Journal of Sustainable Agriculture, 33(1), 78-88.

Praseptiangga, D., Zambrano, J. M. G., Sanjaya, A. P., & Muhammad, D. R. A. (2020). Challenges in the development of the cocoa and chocolate industry in Indonesia: A case study in Madiun, East Java. AIMS Agriculture and Food, 5(4), 920-937.

Ratnawati, L., Ekafitri, R., & Desnilasari, D. (2019). Karakterisasi Tepung Komposit Berbasis Mocaf Dan Kacang-kacangan Sebagai Bahan Baku Biskuit MP-ASI (Characterization of Composite Flour Based on Mocaf and Beans Flour as Ingredient for Weaning Food). Biopropal Industri, 10(2), 65-81.

Sari, N. M. R. E., Wisaniyasa, N. W., & Wiadnyani, A. A. I. A. (2020). Studi kadar gizi, serat dan antosianin tepung kacang merah dan tepung kecambah kacang merah (Phaseolus vulgaris L.). J. Itepa, 9(3), 282-290.

Torregrossa, F., Cinquanta, L., Albanese, D., Cuomo, F., Librici, C., Lo Meo, P., & Conte, P. (2024). Vegan and sugar-substituted chocolates: assessing physicochemical characteristics by NMR relaxometry, rheology, and DSC. European Food Research and Technology, 1-10.

Vesa, T. H., Marteau, P., & Korpela, R. (2000). Lactose intolerance. Journal of the American College of Nutrition, 19(sup2), 165S-175S.

Wang, B., Zhang, Q., Zhang, N., Bak, K. H., Soladoye, O. P., Aluko, R. E., ... & Zhang, Y. (2021). Insights into formation, detection and removal of the beany flavor in soybean protein. Trends in Food Science & Technology, 112, 336-347.

Wiranata, I. G. A. G., Puspaningrum, D. H. D., & Kusumawati, I. G. A. W. (2017). Formulasi dan karakteristik nutrimat bar berbasis tepung kacang kedelai (glycine max. L) dan tepung kacang merah (phaseolus vulgaris. L) sebagai makanan pasien kemoterapi. Jurnal Gizi Indonesia (The Indonesian Journal of Nutrition), 5(2), 133-139.

Zarić, D. B., Pajin, B. S., Lončarević, I. S., Petrović, J. S., & Stamenković-Đoković, M. M. (2015). Effects of the amount of soy milk on thermorheological, thermal and textural properties of chocolate with soy milk. Acta Periodica Technologica, (46), 115-127.

Cocoa powder alkalisation in a small-scale pressurized vessel: Effect of sodium bicarbonate concentration and process duration

Dimas Rahadian Aji Muhammad[1,2*], Wisnu Pranatha[1], Setyaningrum Ariviani[1,2]

[1] Dept. of Food Science and Technology, Universitas Sebelas Maret, Surakarta, Indonesia
[2] Research and Development Center for Food, Nutrition and Publich Health, Universitas Sebelas Maret, Surakarta, Indonesia

*E-mail: dimasrahadian@staff.uns.ac.id

Abstract. Alkalisation is a vital process to strengthen the flavour or modify the colour of cocoa powder, prevent sedimentation, and reduce the acidity of cocoa powder. This study aims to determine the feasibility of a small-scale, environmentally friendly pressure vessel for cocoa powder for Alkalisation. The effect of sodium bicarbonate ($NaHCO_3$) concentration and Alkalisation time with pressurized vessels on cocoa powder's physical and antioxidant characteristics was evaluated. This study was compiled using a Completely Randomized Factorial Design with variations in NaHCO3 concentration and Alkalisation time. The data were analyzed using a Two-way Analysis of Variance (ANOVA) and Duncan's Multiple Range Test (DMRT). The result shows that the Alkalisation process significantly affects the physical and antioxidant properties of cocoa powder. The concentration of $NaHCO_3$ and the duration of the Alkalisation decrease the sedimentation index and the powder's lightness. However, these factors also decrease the cocoa powder's total phenolic and anti-radical scavenging activity. This research opens a new opportunity for small-scale industries to alkalize cocoa using a commercially pressured vessel.

Keywords: cocoa, alkalisation, pressurized vessel, sodium bicarbonate, process duratio

1. Introduction

The cocoa agro-industry shows economic significance in many countries, including in Indonesia (Praseptiangga et al., 2019). Cocoa powder is one of the main cocoa-derived products usually used for various purposes such as baking and making beverage (Aribah et al., 2020; Faiqoh et al., 2021; Fauza et al., 2021; Puchol-Miquel et al., 2021; Nurhayati et al., 2023). However, natural cocoa powder has some limitation on its characteristic due to its acidity, low dispersibility, suspension stability and solubility as well as unattractive color. Therefore, in the cocoa industry, alkalization, also known as "Dutching" process, is commonly carried out to improve the cocoa powder properties (Muhammad et al., 2019; Kongor & Muhammad, 2023). Beverage formulated with alkalized cocoa powder has been reported to have a better suspension stability, higher solubility, more attractive color and also better sensory characteristic (Muhammad et al., 2021; Juvinal et al.,

2023). Sensory properties are important quality attributes in cocoa drink significantly affecting the consumer acceptance (Muktiningrum et al., 2021; Muhammad et al., 2022). Up to date, alkalization is normally done in a large-scale cocoa industry. A limited literature was found related to the development of a small-scale alkalization process.

According to Valverde García et al. (2020)., several factors during alkalization significantly affect the final characteristic of alkalized cocoa powder, including duration, temperature, pressure, type of alkali and concentration of the alkali. Thus, finding a small-scale that is possibly able to control these factors is vital to innovate a small-scale alkalization. A commercial pressure vessel usually used in the household kitchen may be operational for alkalization process. Up till now, there is no report discussing the use of a commercial pressure vessel for alkalization. Recently, innovations in the alkalization process were conducted by combining the process with cold plasma and extrusion (Valverde et al., 2020; Valverde et al., 2021, Palabiyik et al., 2023).

Among several type of alkali, natrium bicarbonate (NaHCO3) is one of the potential materials used by a small-scale cocoa industry. Natrium bicarbonate which is also well-known as sodium bicarbonate is considerably cheap, easy to find and allowed to be used in food processing in the certain amount. However, it is also important to note that alkalization may affect the chemical properties of the cocoa powder particularly phenolic content and antioxidant activity (Miller et al., 2008; Li et al., 2012; Sioriki et al., 2022). Therefore, this research aimed to study the feasibility of a small-scale pressure vessel for cocoa powder alkalization. In particular, the effect of alkalization duration and the concentration of natrium bicarbonate on the physical and chemical properties of cocoa powder was investigated. This research may provide an important information for the small-scale cocoa industries regarding the applicability of a small-scale alkalization process using a small-scale pressure vessel.

2. Materials and Method

2.1 Sample preparation

Alkali solution was prepared by solubilizing NaHCO3 in distilled water at a level of 1%, 2% and 3% (w/v). Natural cocoa powder, obtained from UGM Cocoa Teaching Industry (Batang, Central Java), was mixed with the alkali solution at a ratio of 1:3 in a pressure vessel (Maxim Presto 4L). The alkalization process was conducted in an LPG stove for 5, 10 and 15 min, after the vessel made a sound. The alkalized cocoa was dried in cabinet dryer (60°C), continued by pulverization and sieving.

2.2 Physical and chemical analysis

Wettability, sedimentation index and color analysis followed the protocol explained by Muhammad et al. (2021). The L*, a*, and b* values obtained from colr analysis were then used to calculate the chroma, °Hue and color difference (ΔE) using Eq. 1., Eq., 2 and Eq. 3, respectively.

$$C = \sqrt{a^2 + b^2} \quad \text{(Eq. 1)}$$

$$°\text{Hue} = \text{arc tan } b/a \quad \text{(Eq. 2)}$$

$$\Delta E = \sqrt{\Delta L^2 + \Delta a^2 + \Delta b^2} \quad \text{(Eq. 3)}$$

To obtain a better conclusion regarding the physical characteristic of the alkalized cocoa powder, a microstructural analysis was conducted using a Scanning Electron Microscope JEOL JSM-T300.

In the determination of antioxidant properties, extraction of cocoa powder prior to the analysis was down by solubilizing 25 g of cocoa powder in 150 ml distilled water at 80°C. The analysis of total phenols (Folin-ciocalteu method), total flavonoids (Aluminum chloride method) and antioxidant activity (DPPH-radical scavenging activity method) of cocoa powder were conducted based on the protocol described by Muhammad et al. (2017) without any modification. Standard curves of gallic acid and quercetin were used to express the total phenolic content and total flavonoid content, respectively.

2.3 Experimental design

The experiment was conducted by Completely Randomized Factorial Designs (CRFD) with two factors namely alkali concentration (1-3%) and duration of alkalization (5-15 mins). Natural cocoa powder was used as control sample. The data was then statistically analyzed using the General Linear Model (Univariate Analysis of Variance) in an IBM SPSS 23 software. If there was a significant effect, Duncan's Multiple Range Test (DMRT) at a significance level of α= 0,05 was carried out

3. Results and Discussion

3.1. Color Properties

Table 1 shows the effect of alkalization using a small-scale pressure vessel on the color properties of cocoa powder. It is shown that the alkalization significantly decreased the lightness of the powder, from 54 to about 50, resulting in darker cocoa powder. Similar phenomenon was also shown in the parameter of chroma. The darkening effect of alkalization found in this study has a positive agreement with the study of Alasti et al. (2020), explaining that alkalization is able to change the browning index, causing darker cocoa powder in which the dark intensity may be modified by altering alkalization parameters. Li et al. (2014) stated that a high temperature and basic pH conditions during alkalization causes sugar degradation, Maillard reactions and anthocyanin polymerization forming dark components of brown polymers.

Table 1. Color properties of alkalized cocoa powder prepared using a small-scale pressure vessel

Alkalization Treatment		Lightness	°Hue	Chroma	ΔE
Control		54.00 ± 0.38^{Bc}	54.74 ± 0.11^{Bb}	18.66 ± 0.32^{Bb}	-
NaHCO3 (%)	1	50.27 ± 1.44^{A}	55.29 ± 0.80^{C}	14.32 ± 0.34^{A}	5.81 ± 1.02
	2	50.89 ± 0.94^{a}	54.22 ± 1.68^{A}	14.15 ± 0.60^{A}	5.54 ± 1.07
	3	50.80 ± 1.60^{A}	53.73 ± 0.95^{A}	13.93 ± 0.37^{A}	5.84 ± 1.20
Duration (Min)	5	50.99 ± 0.59^{b}	54.33 ± 1.89^{ab}	14.16 ± 0.37^{a}	5.47 ± 0.65
	10	49.24 ± 1.25^{a}	54.09 ± 1.19^{a}	13.92 ± 0.50^{a}	6.80 ± 0.85
	15	51.73 ± 0.46^{b}	54.82 ± 0.68^{b}	14.32 ± 0.47^{a}	4.92 ± 0.66

Note: Data of each NaHCO3 concentration was the mean of the data of three different duration at similar NaHCO3 concentration. Data of each duration was the mean of the data of three different NaHCO3 concentration at similar alkalization duration. Different notation shows significant different (α=0,05), with capital letter for the effect of NaHCO3 concentration and lower letter for the effect of alkalization duration.

In the parameter of °Hue, the value was about 54, meaning that the product was dominantly red. Alteration in the lightness as well as a* and b* resulted in a significant color difference (ΔE) between the alkalized cocoa powder and the control. According to Żyżelewicz et al. (2014), ΔE larger than 3 means that the difference was obvious. Interestingly, the concentration of NaHCO3 from 1% to 3% did not give rise to a significant effect on the colour properties of cocoa powder. Alkalization duration in the range of 5-15 minutes also did not result in a substantial influence on the color properties of the alkalized cocoa powder.

3.2. Wettability, Sedimentation Index and Microstructural Properties

Table 2 presents the effect of alkalization using a small-scale pressure vessel on the wettability and sedimentation index of cocoa powder. Wettability, also known as wetting time, is defined as the time required for all cocoa powder to become wet and disappear from the surface of the water. This study found that alkalization caused faster wetting time in which the more concentration alkali resulted in faster wetting time. The longer alkalization duration also instigated faster wetting time. This result is different with the previous study of Selamat et al. (1998). However, the reason behind this difference is still under-investigated.

Table 2. Wettability and sedimentation index of alkalized cocoa powder prepared using a small-scale pressure vessel

Alkalization Treatment		Wettability (s)	Sedimentation index
Control		88.54 ± 0.47^{Dc}	54.29 ± 0.96^{Cb}
$NaHCO_3$ (%)	1	81.37 ± 35.79^{C}	50.90 ± 3.17^{B}
	2	75.22 ± 35.46^{B}	49.98 ± 2.72^{AB}
	3	55.27 ± 16.61^{A}	47.35 ± 2.79^{A}
Duration (Min)	5	107.44 ± 23.79^{d}	51.00 ± 2.87^{ab}
	10	62.09 ± 12.21^{b}	49.15 ± 3.65^{a}
	15	42.33 ± 3.33^{a}	48.08 ± 2.52^{a}

Note: Data of each $NaHCO_3$ concentration was the mean of the data of three different duration at similar $NaHCO_3$ concentration. Data of each duration was the mean of the data of three different $NaHCO_3$ concentration at similar alkalization duration. Different notation shows significant different (α=0,05), with capital letter for the effect of $NaHCO_3$ concentration and lower letter for the effect of alkalization duration.

This study also found that alkalization using a small-scale pressure vessel resulted in a lower sedimentation index (SI). The SI was inversely proportional to the concentration of NaHCO3 and the duration of alkalization. The higher concentration of alkali and the longer alkalization duration successfully decrease the SI. This is notable because SI is the parameter of dispersion stability of cocoa powder in water. This finding agrees the statement of Muhammad et al. (2019) stating that cocoa beverage made of alkalized cocoa powder is more stable than that of natural cocoa powder. In addition to improving the dispersibility, alkalization may increase the solubility of cocoa powder. Thus, the non-soluble material in cocoa powder decreases and consequently the amount of sediment also decreases. To prove this hypothesis, a simple solubility test was conducted to control samples and the alkalized cocoa powder treated with 3% of NaHCO3 in 15 mins. It was found that the solubility of natural cocoa powder was 13.7%, while that of alkalized one was 22.6%.

To provide more insight related to physical properties of the alkalized cocoa powder, a microstructural properties analysis using SEM was conducted. It was found that the cocoa powder alkalized using a small-scale pressure vessel was structurally damaged. A clear difference of microstructural properties was shown before and after alkalization (Fig. 1). This may be the reason of increasing soluble matters in the alkalized cocoa powder and thus improving the dispersibility and dispersibility of the cocoa powder.

Fig. 1. Scanning Electron Microscopy image of cocoa powder with magnification of 1000x dan 2000x. (A) Natural cocoa powder and (B) alkalized cocoa powder (3% of NaHCO3, 15 min).

3.3. Antioxidant Properties

Table 2 demonstrates the effect of alkalization using a small-scale pressure vessel on the antioxidant properties of cocoa powder. It was clearly exposed that alkalization decrease significantly the total phenols and total flavonoids, and thus consequently decrease DPPH-radical scavenging activity of cocoa powder. It has been widely agreed that antioxidant activity of cocoa powder has a strong correlation with its phenolic and flavonoid content (Todorovic et al. 2017; Muhammad et al. (2017).

Table 3. Antioxidant properties of alkalized cocoa powder prepared using a small-scale pressure vessel

Alkalization Treatment		Total phenolic content (μg GAE/ml cocoa drink)	Total flavonoid content (μg QE/ml cocoa drink)	DPPH-radical scavenging activity (%)
Control		131,40 ± 5,14^{Dd}	11,96 ± 0,31^{Dd}	70,55 ± 1,04^{Dd}
$NaHCO_3$ (%)	1	81,14 ± 5,63^{C}	5,26 ± 0,83^{C}	64,65 ± 2,65^{C}
	2	73,32 ± 9,10^{B}	4,67 ± 1,19^{B}	58,49 ± 4,47^{B}
	3	69,01 ± 5,67^{A}	4,05 ± 0,87^{A}	54,60 ± 4,53^{A}
Duration (Min)	5	78,74 ± 4,86^{b}	5,75 ± 0,46^{c}	63,79 ± 3,31^{c}
	10	76,94 ± 8,74^{b}	4,49 ± 0,99^{b}	57,85 ± 5,28^{b}
	15	67,78 ± 7,35^{a}	3,73 ± 0,41^{a}	56,11 ± 5,47^{a}

Note: Data of each $NaHCO_3$ concentration was the mean of the data of three different duration at similar $NaHCO_3$ concentration. Data of each duration was the mean of the data of three different $NaHCO_3$ concentration at similar alkalization duration. Different notation shows significant different (α=0,05), with capital letter for the effect of $NaHCO_3$ concentration and lower letter for the effect of alkalization duration. GAE = Gallic Acid Equivalent. QE = Quercetin Equivalent.

The negative correlations between the concentration of NaHCO3 as well as the duration of alkalization and these parameters were shown. The higher concentration of alkali and the longer alkalization duration caused the more significant drop on the total phenols, total flavonoids and DPPH-radical scavenging activity. This finding is in a strong agreement with previous study by other research groups revealing that alkalization at higher pHs causes polyphenol oxidation and polymerization or the increased activity of polyphenol oxidase enzyme resulting in a lower phenolic content (Greño et al., 2022; Sioriki et al., 2021).

4. Conclusion

A small-scale pressured vessel is potentially applied for cocoa powder alkalization process as this technique can modify the physical properties of cocoa powder. However, development of the method is still required. The concentration of NaHCO3 and the duration of the alkalization significantly affect the physical and chemical properties of alkalized cocoa powder. In the antioxidant properties parameters, the degree of alkalization significantly impact on the degradation of phenols, flavonoids and radical scavenging activity of cocoa powder. This research opens a new opportunity of small-scale industry to alkalize cocoa by using a commercial pressure vessel.

Acknoledgement

This research was financially supported by Universitas Sebelas Maret through Hibah Kolaborasi Internasional 2023 (Grant no. 228/UN27.22/PT.01.03/2023).

References

Alasti, F. M., Asefi, N., Maleki, R., & SeiiedlouHeris, S. S. (2020). The influence of three different types and dosage of alkaline on the inherent properties in cocoa powder. Journal of Food Science and Technology, 57, 2561-2571.

Aribah, S. A., Sanjaya, A. P., Muhammad, D. R. A., & Praseptiangga, D. (2020, May). Sensorial and physical properties of chocolate beverage prepared using low fat cocoa powder. In AIP conference proceedings (Vol. 2219, No. 1). AIP Publishing.

Faiqoh, K. E. N., Muhammad, D. R. A., & Praseptiangga, D. (2021). Ginger-flavoured ready-to-drink cocoa beverage formulated with high and low-fat content powder: consumer preference, properties and stability. Food Res, 5(2), 7-17.

Fauza, G., Muhammad, D. R. A., Affandi, D. R., & Ariviani, S. (2021, July). Sensory profile analysis of steamed brownies using Quantitative Descriptive Analysis (QDA). In IOP Conference Series: Earth and Environmental Science (Vol. 828, No. 1, p. 012058). IOP Publishing.

Greño, M., Herrero, M., Cifuentes, A., Marina, M. L., & Castro-Puyana, M. (2022). Assessment of cocoa powder changes during the alkalization process using untargeted metabolomics. LWT, 172, 114207.

Juvinal, J. G., De Steur, H., Schouteten, J. J., Muhammad, D. R. A., De Leon, A. A., Dewettinck, K., & Gellynck, X. (2023). Physico-Chemical Property, Sensory Profile and Consumer Acceptability of Water Buffalo (Bubalus bubalis L.) Chocolate Milk Using Alkalized and Natural Cocoa Powder. Foods, 12(9), 1797.

Kongor, J. E., & Muhammad, D. R. A (2023). Processing of Cocoa and Development of Chocolate Beverages. In Natural Products in Beverages: Botany, Phytochemistry, Pharmacology and Processing (pp. 1-37). Cham: Springer International Publishing.

Li, Y., Feng, Y., Zhu, S., Luo, C., Ma, J., & Zhong, F. (2012). The effect of alkalization on the bioactive and flavor related components in commercial cocoa powder. Journal of Food Composition and Analysis, 25(1), 17-23.

Li, Y., Zhu, S., Feng, Y., Xu, F., Ma, J., & Zhong, F. (2014). Influence of alkalization treatment on the color quality and the total phenolic and anthocyanin contents in cocoa powder. Food Science and Biotechnology, 23, 59-63.

Miller, K. B., Hurst, W. J., Payne, M. J., Stuart, D. A., Apgar, J., Sweigart, D. S., & Ou, B. (2008). Impact of alkalization on the antioxidant and flavanol content of commercial cocoa powders. Journal of agricultural and food chemistry, 56(18), 8527-8533.

Muhammad, D. R. A., Gonzalez, C. G., Sedaghat Doost, A., Van de Walle, D., Van der Meeren, P., & Dewettinck, K. (2019). Improvement of antioxidant activity and physical stability of chocolate beverage using colloidal cinnamon nanoparticles. Food and Bioprocess Technology, 12, 976-989.

Muhammad, D. R. A., Kongor, J. E., & Dewettinck, K. (2021). Investigating the effect of different types of cocoa powder and stabilizers on suspension stability of cinnamon-cocoa drink. Journal of food science and technology, 58, 3933-3941.

Muhammad, D. R. A., Marettama, N. M., Fauza, G., & Affandi, D. R. (2022). Can ingredients and information interventions affect the hedonic level and (emo-sensory) perceptions of the milk chocolate and cocoa drink's consumers?. Open Agriculture, 7(1), 847-856.

Muhammad, D. R. A., Praseptiangga, D., Van de Walle, D., & Dewettinck, K. (2017). Interaction between natural antioxidants derived from cinnamon and cocoa in binary and complex mixtures. Food Chemistry, 231, 356-364.

Muktiningrum, T. A., Fauza, G., Ariviani, S., Muhammad, D. R., & Affandi, D. R. (2022). Sensory profile analysis of chocolate drinks using quantitative descriptive analysis (QDA). In E3S Web of Conferences (Vol. 344, p. 04005). EDP Sciences.

Nurhayati, R., Laila, U., Herawati, E. R. N., Prabawa, S., Muhammad, D. R. A., & Surwinanto, A. M. (2023). The developing of reduced-sugar ready-to-drink cocoa beverages: optimization of stabilizers and sugar replacers concentration. Food Science and Technology, 43, e107322.

Palabiyik, I., Kopuk, B., Konar, N., & Toker, O. S. (2023). Investigation of cold plasma technique as an alternative to conventional alkalization of cocoa powders. Innovative Food Science & Emerging Technologies, 88, 103440.

Praseptiangga, D., Zambrano, J. M. G., Sanjaya, A. P., & Muhammad, D. R. A. (2020). Challenges in the development of the cocoa and chocolate industry in Indonesia: A case study in Madiun, East Java. AIMS Agriculture and Food, 5(4), 920-937.

Puchol-Miquel, M., Palomares, C., Fernández-Segovia, I., Barat, J. M., & Perez-Esteve, É. (2021). Effect of the type and degree of alkalization of cocoa powder on the physico-chemical and sensory properties of sponge cakes. Lwt, 152, 112241.

Selamat, J., Hussin, N., Zain, A. M., & Man, Y. B. C. (1998). Effects of alkalized cocoa powder and soy lecithin on physical characteristics of chocolate beverage powders. Journal of food processing and preservation, 22(3), 241-254.

Sioriki, E., Lemarcq, V., Alhakim, F., Triharyogi, H., Tuenter, E., Cazin, C. S., Nolan, S. P., Pieters, L., Van de Walle, D., & Dewettinck, K. (2021). Impact of alkalization conditions on the phytochemical content of cocoa powder and the aroma of cocoa drinks. Lwt, 145, 111181.

Sioriki, E., Tuenter, E., Van de Walle, D., Lemarcq, V., Cazin, C. S., Nolan, S. P., ... & Dewettinck, K. (2022). The effect of cocoa alkalization on the non-volatile and volatile mood-enhancing compounds. Food Chemistry, 381, 132082.

Todorovic, V., Milenkovic, M., Vidovic, B., Todorovic, Z., & Sobajic, S. (2017). Correlation between antimicrobial, antioxidant activity, and polyphenols of alkalized/nonalkalized cocoa powders. Journal of Food Science, 82(4), 1020-1027.

Valverde García, D., Pérez Esteve, É., & Barat Baviera, J. M. (2020). Changes in cocoa properties induced by the alkalization process: A review. Comprehensive Reviews in Food Science and Food Safety, 19(4), 2200-2221.

Valverde, D., Behrends, B., Pérez-Esteve, É., Kuhnert, N., & Barat, J. M. (2020). Functional changes induced by extrusion during cocoa alkalization. Food Research International, 136, 109469.

Valverde, D., Sánchez-Jiménez, V., Barat, J. M., & Pérez-Esteve, É. (2021). The effect of extrusion on the physical and chemical properties of alkalized cocoa. Innovative Food Science & Emerging Technologies, 73, 102768.

Żyżelewicz, D., Krysiak, W., Nebesny, E., & Budryn, G. (2014). Application of various methods for determination of the color of cocoa beans roasted under variable process parameters. European Food Research and Technology, 238, 549-563.

Effects of Molar Ratio on Triacylglycerol Conversion in Glycerolysis Using Sodium Metasilicate and MgO Catalysts in a High Shear Compartment Reactor

Arief Rakhman Affandi[1,2], Chusnul Hidayat[2*]

[1] Food Technology Study Program, Universitas PGRI Semarang, Semarang, Indonesia
[2] Department of Food and Agricultural Product Technology, Universitas Gadjah Mada, Yogyakarta, Indonesia

*E-mail: chusnulhi@ugm.ac.id

Abstract. This research examines the influence of different molar ratios on the conversion of triacylglycerol (TAG) to monoglycerides (MAG) and diglycerides (DAG) in a solvent-free glycerolysis reaction conducted within a high shear compartment reactor. The reaction was catalyzed using a combination of sodium metasilicate and MgO at 120°C with a stirring speed of 1500 rpm. The highest TAG conversion was observed at a PKO to glycerol ratio of 1:3, yielding 56.96±3.39% MAG and DAG. However, increasing the glycerol content resulted in higher product viscosity, reaching up to 26.20±1.56 mPa.s on the MDAG product with a molar ratio of 1:5. Additionally, the lightness (L* value) of the liquid MDAG product increased with higher glycerol content, with a maximum L* value of 56.63 observed. An excessive glycerol ratio also increased the free glycerol in the MDAG product, with the highest level recorded at a 1:4 molar ratio, measuring 3.55±0.36%. These findings suggest that while a 1:3 PKO to glycerol ratio is optimal for maximizing TAG conversion, further increases in glycerol content may adversely affect product quality by increasing viscosity and free glycerol levels.

Keywords: glycerolysis, heterogeneous catalyst, HSCR, molar ratio, TAG conversion

1. Introduction

Glycerolysis is the process of converting triacylglycerol (TAG) and glycerol molecules into monoacylglycerol (MAG) and diacylglycerol (DAG). The immiscibility of the substrates results in a limited mass transfer rate in glycerolysis reaction. In general, the use of a considerable amount of glycerol in the glycerolysis reaction can increase the reaction rate. This is due to the fact that TAG molecules will interact more with glycerol, thereby producing more MAG and DAG (1). However, several problems can arise when using excess glycerol in the mixture, such as a low degree of substrate homogeneity (2). To overcome this problem, researchers applied a number of modification processes to optimize the interaction between TAG and glycerol. This approach involves adding organic solvents and surfactants, as well as using high stirring speeds which are key factors affecting the rate of glycerolysis reactions (3–5).

The combination of the mixing process with heterogeneous catalysts is a breakthrough in glycerolysis reaction. The advantages in terms of ease of product separation and reusability have encouraged researchers to use this type of catalyst in glycerolysis reactions (6). The implementation of a high-shear reactor enables the rapid production of a homogeneous mixture (7). This reactor effectively addresses the challenge of excess glycerol in the glycerolysis reaction, promoting a homogeneous substrate mixture. However, an excess of glycerol can result in the covering of the catalyst surface by the substrate, thereby preventing the interaction of other glycerol molecules with the catalyst surface (8). In addition, excess glycerol can also increase the viscosity of the reaction mixture, further inhibiting mass transfer (9).

This study aims to identify the effect of molar ratio on the rate of TAG conversion in solvent-free glycerolysis reactions using a combination of sodium metasilicate and MgO catalysts. The reaction will be carried out in a high-shear reactor equipped with a compartment serving as a heterogeneous catalyst container.

2. Materials and Methods

2.1 Materials

The RBDPKO oil was obtained from PT SMART, Tbk. The sodium metasilicate (SMS) was purchased from Sigma–Aldrich (Missouri, USA) and MgO catalyst was acquired form Merck KGaA (Darmstadt, Germany). Glycerol was purchased from P&G Chemicals (Kuala Lumpur, Malaysia). N-hexane, glacial acetic acid, comassie brilliant blue, and methanol were acquired from Merck (Germany). The primary equipment employed in this study was the JRJ300-1 high shear compartment reactor (TCH Instrument, China), featuring an impeller with a 2.7 cm diameter. The reactor design includes a compartment-shaped catalyst container with a volume of 75 cm^3, as depicted in Figure 1.

2.2 SMS-MgO Catalyst Preparation

MgO was synthesized using the wet impregnation method, followed by dry mixing the MgO powder with sodium metasilicate (SMS) at a ratio of 1:5 (w/w). Distilled water (5% v/w) was added to the catalyst mixture and mixed until a homogenous slurry was formed. The slurry was dried at 105°C for 6 h. Subsequently, the dried catalyst was crushed into granules with a mesh size of 15-30. This catalyst was identified as SMS-MgO (5:1).

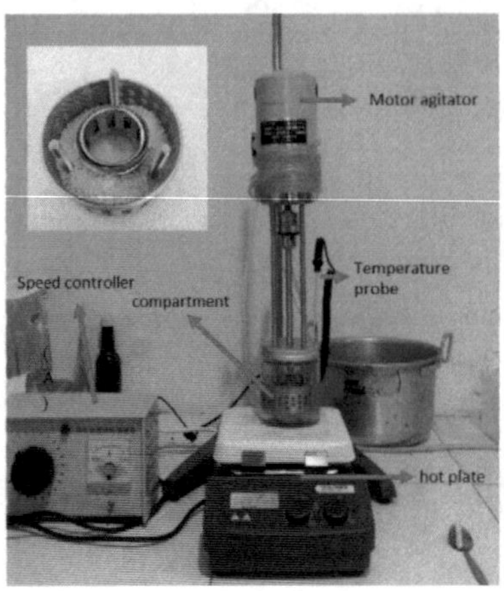

Figure 1. The design of the HSCR (the inserted picture is the catalyst-filled compartment).

2.3 Glycerolysis Reaction in HSCR

Palm kernel oil and glycerol were mixed in molar ratios of 1:2, 1:3, 1:4, and 1:5 (mol/mol). The mixture was heated to 120°C, and the compartment was filled with 20% (w/w by weight of the oil) catalyst granules. Once the substrate mixture reached 120°C, the compartment was placed in a chamber, and stirring was conducted at 1500 rpm. The glycerolysis reaction was conducted for a period of 360 minutes. The process was stopped by physically separating the compartment from the substrate. After the reaction was completed, the product was taken 100 mL into a beaker glass for viscosity analysis using an NDJ-8S viscometer.

2.4 Analysis of Acylglycerol Composition

The Thin Layer Chromatography (TLC) method was used to determine the concentration of acylglycerols in the product. Previously, the TLC plate was dried at 105°C for 1 hour. A total of 1µL of the sample was spotted on the TLC plate. Hexane, ethyl ether, and glacial acetic acid were added to the TLC chamber in the following ratio 80:20:2 (v/v/v) to serve as the developer solution for the elution procedure (10,11). The plate containing the sample spot was put into the chamber to facilitate the elution process. After the upper limit was reached, the plate was dried and then submerged in a solution containing 0.02 % (w/v) comassie brilliant blue. This solution was composed of a mixture of methanol, distillate water, and glacial acetic acid in the ratio of 30:60:10 (v/v/v). A Camag TLC Scanner and WinCATS Planar Chromatography software were used to assess the spots that developed on the TLC plate to calculate the amount of acylglycerol contained within the sample.

2.5 Analysis of Free Glycerol, Hardness, and Colour Measurement of MDAG Product

Analysis of free glycerol levels contained in MDAG products using the AOCS Ca 14-56 method (12). In the hardness analysis, 35 g of sample was melted at 80°C for 15 minutes before being transferred to a 50 mL bottle. The samples were allowed to harden at ambient temperature before being chilled for approximately 18 hours at 10°C. The samples were refrigerated and then allowed to thaw for about an hour at room temperature. Product hardness was assessed using a texture analyzer (Ametek Brookfield CTX050, USA) with a 4 mm diameter probe. The colour of the products was measured using a Konica Minolta Chromameter to find the lightness (L*).

2.6 Analysis of pH value of MDAG and Catalyst Basicity of Used Catalyst

The MDAG product is melted at 60-70°C then the electrode probe on the pH meter is inserted into the liquid. The pH value is recorded after the value is stable. The basicity of the solid catalyst was determined using the benzoic acid titration method (13).

3. Results and Discussion

3.1 Effect of Molar Ratio on Product Viscosity

Figure 2 illustrates the impact of varying the molar ratio of substrates from 1:2 to 1:5 (PKO:glycerol) upon the viscosity of the MDAG product. The statistical analysis showed that increasing the amount of glycerol in the mixture can significantly increase the viscosity of the product ($p<0.05$). The highest viscosity belongs to the MDAG product with a molar ratio of 1:5 (26.20 ± 1.56 mPa.s). The viscosity of MDAG products is influenced by several components, such as glycerol and acylglycerols concentration. Glycerol is a highly polar and hygroscopic molecule and can form hydrogen bonds with other molecules. These interactions create a strong network matrix within the liquid phase, inhibiting the free flow of molecules (14).

Based on the analysis, glycerol has a higher viscosity value than RBDPKO oil in the range of 8.5 ± 0.5 mPa.s, while RBDPKO has a viscosity of 5.33 ± 0.3 mPa.s (at 120°C). Similar findings were reported by Ferreira et al. indicated that glycerol has a viscosity of about 8.35 mPa.s at 120.85°C (15) . The difference in viscosity of oil and glycerol in the mixture affects the viscosity change of the mixture. Excess glycerol will result in high viscosity, thus limiting the diffusion and movement of substrate molecules on the active surface of the catalyst (16). Boas et al. also reported that in reactions with high concentrations of glycerol, increased viscosity can lead to poor mass transfer and decreased reaction rates (17). Slow molecular movement will decrease the collision frequency between glycerol and triglyceride molecules, thus slowing down the overall reaction kinetics (18).

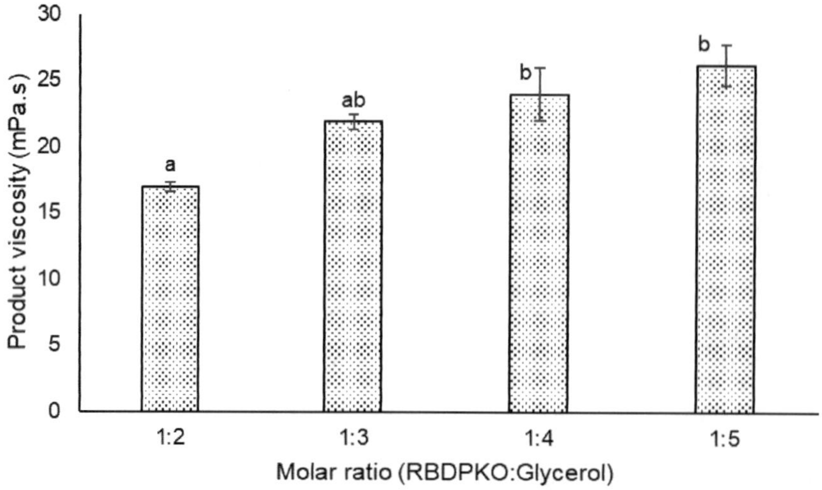

Figure 2. Effect of molar ratio on MDAG product viscosity (reaction conditions: temperature 120°C, catalyst concentration 20% (w/w based on the weight of oil), stirring speed 1500 rpm, time 360 min)

3.2 TAG Conversion Analysis

Figure 3 illustrates the impact of the molar ratio on the conversion rate of triacylglycerols (TAGs) to mono-diacylglycerols (MDAGs). The figure demonstrates that the highest TAG conversion value, 56.97 ± 3.39%, was attained in the glycerolysis reaction with a molar ratio of 1:3. In general, an increase in the molar ratio of substrate will result in a higher conversion rate of TAG to MDAG (10). This is attributed to the greater availability of glycerol as a substrate in the glycerolysis reaction, which enhances the interaction between glycerol and TAG molecules. Consequently, this condition accelerates the reaction rate, leading to the production of MAG and DAG.

Statistical analysis using SPSS version 25 indicated that variations in substrate molar ratio significantly influenced the conversion rate of TAG to MDAG ($p<0.05$). Increasing the amount of glycerol can alter the glycerolysis reaction rate. As shown in Figure 2, raising the glycerol content from a 1:2 to a 1:3 molar ratio enhances the glycerolysis reaction rate. However, excessive glycerol reduces the TAG conversion rate, as observed with molar ratios of 1:4 and 1:5, yielding TAG to MDAG conversion values of 51.99±2.40% and 48.30±0.54%, respectively. Similar findings were reported by Sun et al. where increased glycerol amounts produced more glycerolate anions, accelerating the reaction rate (19). In that study, the glycerolysis reaction reached optimal conditions at a mole ratio of 1:2 by producing MAG of 64.85 ± 2.13%, but at a molar ratio of 1:4 did not produce significant changes in MAG levels or TAG conversion rates.

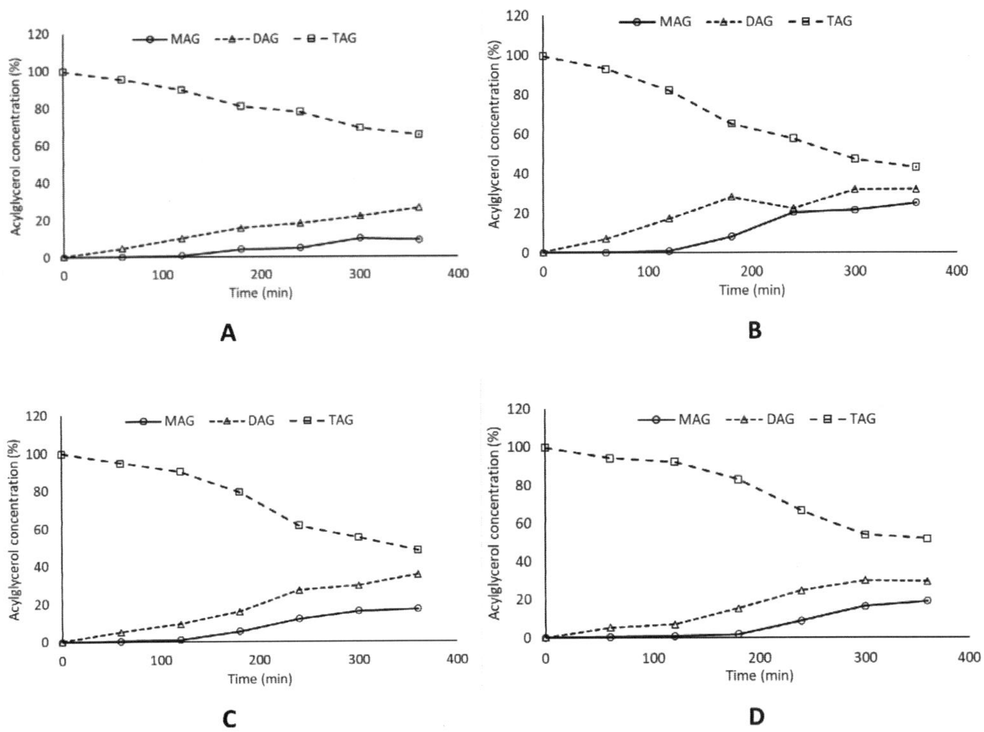

Figure 3. Changes in acylglycerol concentration in glycerolysis reactions with different molar ratios (a) 1:2, (b) 1:3, (c) 1:4, (d) 1:5 (reaction conditions: temperature 120°C, catalyst concentration 20% (w/w based on the weight of oil), stirring speed 1500 rpm, time 360 min)

This research also noted that using a molar ratio higher than 1:4 did not significantly affect TAG conversion or MAG synthesis and could even decrease the glycerolysis reaction rate due to increased substrate viscosity. Research by Boas et al. reported that MAG concentration dropped to around 20% with excess glycerol, likely due to impaired mass transfer to the active site and reduced substrate homogeneity, which lowered the reaction rate (17). The reduction in MDAG production may result from glycerol's hydrophilic nature, which can obstruct the active sites of the heterogeneous catalyst, thereby inhibiting its activity (20). An excess of glycerol can cover the catalyst surface, hindering the interaction between other substrate molecules and the catalyst (10,21).

3.3 Free Glycerol Content in MDAG Products

The free glycerol content in MDAG products with different mole ratio treatments is illustrated in Figure 4. Statistical analysis showed that different mole ratios significantly affected ($p < 0.05$) the free glycerol content in MDAG products. The analysis showed that the MDAG product synthesized with a molar ratio of 1:4 had the highest free glycerol content, which was 3.55 ± 0.36%. Generally, the glycerol content of raw MDAG products (before refining) is in the range of 4-12% (22,23). Glycerol molecules exhibit hydrophilic properties, enabling interaction with hydroxyl groups present in MAG and DAG molecules. The product synthesized with a molar ratio of 1:4 has a relatively high MAG and DAG content of 16.89 ± 1.12% and 35.10 ± 1.28%, respectively. In contrast, the product synthesized with a molar ratio of 1:2, which had the lowest MDAG content (34.90%), showed the lowest free glycerol content (1.60±0.02%) out of the samples.

According to the specifications outlined in Commission Regulation (EU) No. 231/2012 and JECFA, the maximum allowable free glycerol content in MDAG products intended as food additives is 7% (24). To achieve MAG purity levels of up to 90%, an effective glycerol separation process is necessary. Higher free glycerol content in MDAG products complicates the purification process, requiring more advanced techniques such as cream demulsification or molecular distillation (23,25).

3.4 Characteristics of Hardness and Colour in MDAG Products

The data presented in Table 1 and Figure 5 demonstrate the impact of varying molar ratios of RBDPKO and glycerol on the hardness and colour of the resulting product. As a consequence of an increase in the molar ratio of glycerol from 1:2 to 1:3, there was a notable rise in the product's hardness, from 1.33 N to 2.05 N. The increase in hardness was likely due to the significant rise in MAG content, which jumps from 8.93% to 24.99%, and the concurrent increase in DAG content

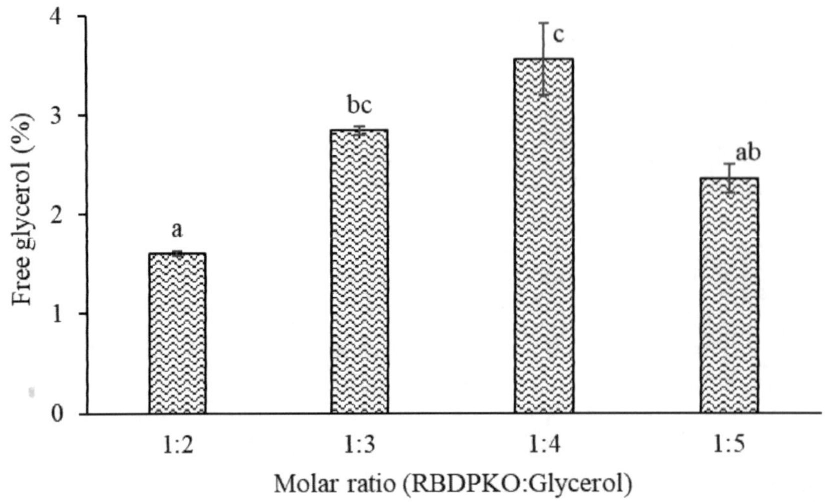

Figure 4. The effect of molar ratio on free glycerol content in MDAG products (reaction conditions: temperature 120°C, catalyst concentration 20% (w/w based on the weight of oil), stirring speed 1500 rpm, time 360 min)

from 25.97% to 31.98%. The higher concentration of MAG and DAG likely contributes to a more structured matrix within the product, leading to greater hardness (26).

Nevertheless, the hardness dropped to 1.82 N when the molar ratio was raised to 1:4 due to a drop in the MAG content to 16.89% (despite an increase in the DAG content to 35.10%). MAG plays a more critical role in determining the hardness of the product (27). At the highest molar ratio of 1:5, the hardness decreases further to 1.76 N. This decline correlates with a decrease in both MAG and DAG concentrations, which drop to 19.07% and 29.22%, respectively. The reduced levels of these glycerides likely weaken the structural integrity of the product, resulting in lower hardness (5,9).

Table 1. The effect of molar ratio on texture (hardness) and colour (lightness) in MDAG products.

Molar Ratio (RBDPKO: glycerol)	Hardness (N)	Lightness	
		solid	liquid
1:2	1.33 ± 0.11^{a}	37.20 ± 2.46^{a}	27.02 ± 0.42^{a}
1:3	2.05 ± 0.23^{b}	49.61 ± 1.25^{b}	33.06 ± 0.67^{b}
1:4	1.82 ± 0.10^{b}	53.05 ± 1.54^{bc}	38.91 ± 0.27^{c}
1:5	1.76 ± 0.10^{b}	56.63 ± 1.92^{c}	42.36 ± 0.25^{d}

Glycerolysis reaction: time 360 min; catalyst loading 20 wt% (mass of oil); temperature 120°C. Different letters indicated significantly different values for each column ($p < 0.05$)

A B

Figure 5. Photo of MDAG products produced using different substrate molar ratios. (a) solid form (b) liquid form

A statistical analysis indicated that the lightness value in MDAG products was considerably ($p < 0.05$) influenced by different mole ratios. The results showed that the use of excess glycerol in the glycerolysis reaction can increase the lightness of the solid and liquid MDAG products. As the molar ratio increased from 1:2 to 1:5, the lightness of the solid MDAG product increased significantly from 37.2 to 56.63. Similarly, the lightness value of the liquid MDAG product also increased from 27.02 to 42.36. The reaction between substrate and catalyst can change the color of the product. The change in colour is linked to the formation of the active catalyst, which is likely to be an intermediate glycerolate anion (28). For instance, during glycerolysis, glycerol can undergo side reactions such as dehydrogenation in the presence of a base catalyst, leading to the formation of coloured compounds like aldehydes or other compound, which can impart a darker colour to the product (29). The increase in lightness value can be attributed to the dilution effect of glycerol, which can reduce the concentration of coloured components in the mixture, leading to a lighter appearance.

3.5 The effect of molar ratio on pH value of the product and the used catalyst basicity

Figure 6 shows the effect of different molar ratios on the pH value of MDAG and the basicity of the used catalyst in the glycerolysis reaction. Statistical analysis showed that the molar ratio significantly influences the pH value of the product ($p < 0.05$), which ranges from 8.69 to 9.59. The highest pH value (9.59) is observed at a molar ratio of 1:3, suggesting a more alkaline environment compared to other ratios. During the reaction, sodium metasilicate can cause deprotonation of the glycerol molecule, producing glycerolate anions ($C_3H_5O_3^-$) (30). These glycerolate anions are highly nucleophilic and can readily attack the ester bonds of TAG. The formation of glycerolate anions contributes indirectly to maintaining or further increasing the pH of the reaction mixture.

The difference in the molar ratio of the substrate also has a significant effect on the change in the basicity value of the used catalyst ($p < 0.05$). At a molar ratio of 1:2, the highest basicity of 10.50 mmol.g^{-1} was observed. The glycerolysis process with the molar ratio of 1:2 shows that the catalyst has a low mass transfer rate between active compounds and substrate molecules. Thus, the active base compounds on the used catalyst are still present on the catalyst granules. However, as the molar ratio is further increased to 1:3 and 1:4, the basicity decreases to 8.83 and 8.60 mmol.g^{-1}, respectively. This reduction in basicity indicates a high mass transfer rate between the active moieties on the catalyst and the substrate molecules during glycerolysis reaction.

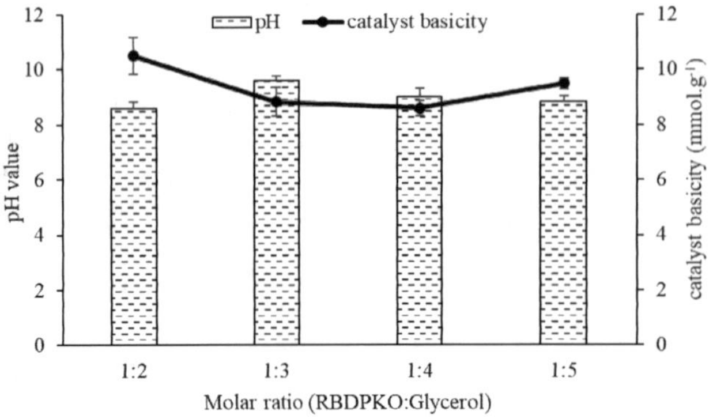

Figure 6. pH value and basicity of catalyst used in glycerolysis reaction with different molar ratios

4. Conclusion

The use of different substrate mole ratios resulted in significant changes in the conversion rate of TAG to MAG and DAG, product viscosity, free glycerol content, hardness, product colour, and used catalyst basicity. The use of 1:3 mole ratio (RBDPKO:glycerol) can produce the highest TAG conversion, 56.96%. The use of excess glycerol will result in higher product viscosity and free glycerol content. However, the use of too much glycerol will result in a lower pH of the liquid, which illustrates the low mass transfer between the substrate and the catalyst. This is because some of the glycerol molecules are adsorbed on the active surface of the catalyst, thus inhibiting the interaction of other substrate molecules in the reactor. The mole ratio of 1:3 is the preferred choice in the glycerolysis process carried out in the high shear compartment reactor because it can produce the highest MAG and DAG in the range of 24.99 and 31.98%, respectively.

Acknowledgments

The authors are grateful to the Directorate of General Education, Ministry of Education and Culture, Indonesia, who kindly provided all funding for this study through the doctor's dissertation research scheme (number of contract agreements: 122/E5/PG.02.00.PL/2023 and 3121/UN1/DITLIT/Dit-Lit/PT.01.03/2023).

References

1. Zhong N, Deng X, Huang J, Xu L, Hu K, Gao Y. Low-temperature chemical glycerolysis to produce diacylglycerols by heterogeneous base catalyst. European Journal of Lipid Science and Technology. 2014;116(4):470–6.
2. Solaesa ÁG, Sanz MT, Beltrán S, Melgosa R. Kinetic study and kinetic parameters of lipase-catalyzed glycerolysis of sardine oil in a homogeneous medium. Cuihua Xuebao/Chinese Journal of Catalysis. 2016 Apr 1;37(4):596–606.
3. Monte Blanco SFM, Santos JS, Feltes MMC, Dors G, Licodiedoff S, Lerin LA, et al. Optimization of diacylglycerol production by glycerolysis of fish oil catalyzed by Lipozyme TL IM with Tween 65. Bioprocess Biosyst Eng. 2015 Sep 21;38(12).
4. Chen W, Kou M, Lin S, Zhong N. Effects of solvents on the glycerolysis performance of the sba-15 supported lipases. J Oleo Sci. 2021;70(3):385–95.
5. Rohana NLRT, Millati R, Yanti R, Ardani AS, Hidayat C. Synthesis of Mono-and Diacylglycerol Using High Shear Reactor. Philipp J Sci [Internet]. 2023 Jul 18;152(4):1285–95. Available from: https://philjournalsci.dost.gov.ph/publication/regular-issues/next-issue/122-vol-152-no-4-august-2023/1908-synthesis-of-mono-and-diacylglycerol-using-high-shear-reactor
6. Reinoso DM, Damiani DE, Tonetto GM. Zinc glycerolate as a novel heterogeneous catalyst for the synthesis of fatty acid methyl esters. Appl Catal B [Internet]. 2014 Jan 1;144:308–16. Available from: http://dx.doi.org/10.1016/j.apcatb.2013.07.026
7. Sánchez-Cantú M, Morales Téllez M, Pérez-Díaz LM, Zeferino-Díaz R, Hilario-Martínez JC, Sandoval-Ramírez J. Biodiesel production under mild reaction conditions assisted by high shear mixing. Renew Energy. 2019;130:174–81.
8. Serafini MSA, Tonetto GM. Catalytic Synthesis of Monoglycerides by Glycerolysis of Triglycerides. International Journal of Chemical Reactor Engineering [Internet]. 2019 Nov 26;17(11). Available from: https://www.degruyter.com/document/doi/10.1515/ijcre-2019-0056/html
9. Subroto E, Supriyanto, Utami T, Hidayat C. Enzymatic glycerolysis–interesterification of palm stearin–olein blend for synthesis structured lipid containing high mono- and diacylglycerol. Food Sci Biotechnol. 2019 Apr 5;28(2):511–7.

10. Subroto E, Indiarto R, Wulandari E, Azimah HN. Oil to glycerol ratio in enzymatic and chemical glycerolysis for the production of mono- And diacylglycerol. International Journal of Engineering Trends and Technology. 2021 Aug 1;69(8):117–25.

11. Arum A, Hidayat C, Supriyanto. Synthesis of Emulsifier from Refined Bleached Deodorized Palm Stearin by Chemical Glycerolysis in Stirred Tank Reactor. KnE Life Sciences. 2019 Mar 10;4(11):130.

12. AOCS. AOCS Official Method Ca 14-56 Total, Free and Combined Glycerol Iodometric-Periodic Acid Method. 1997.

13. Guo F, Peng ZG, Dai JY, Xiu ZL. Calcined sodium silicate as solid base catalyst for biodiesel production. Fuel Processing Technology [Internet]. 2010 Mar;91(3):322–8. Available from: http://dx.doi.org/10.1016/j.fuproc.2009.11.003

14. Mirhosseini H, Tan CP, Taherian AR. Effect of glycerol and vegetable oil on physicochemical properties of Arabic gum-based beverage emulsion. European Food Research and Technology. 2008;228(1):19–28.

15. Ferreira AGM, Egas APV, Fonseca IMA, Costa AC, Abreu DC, Lobo LQ. The viscosity of glycerol. Journal of Chemical Thermodynamics [Internet]. 2017;113(June 2016):162–82. Available from: http://dx.doi.org/10.1016/j.jct.2017.05.042

16. Razak NNA, Pérès Y, Gew LT, Cognet P, Aroua MK. Effect of Reaction Medium Mixture on the Lipase Catalyzed Synthesis of Diacylglycerol. Ind Eng Chem Res. 2020 May 27;59(21):9869–81.

17. Bôas RNV, Lima R, Silva MVC, Freitas L, Aguiar LG, de Castro HF. Continuous production of monoacylglycerol via glycerolysis of babassu oil by immobilized Burkholderia cepacia lipase in a packed bed reactor. Bioprocess Biosyst Eng. 2021 Oct 1;44(10):2205–15.

18. Singh AK, Mukhopadhyay M. Olive oil glycerolysis with the immobilized lipase Candida antarctica in a solvent free system. Grasas y Aceites. 2012;63(2):202–8.

19. Sun S, Lv Y, Wang G, Chen X. Soybean oil-based monoacylglycerol synthesis using bio-compatible amino acid ionic liquid as a catalyst at low temperature. J Mol Liq. 2021 Oct 15;340.

20. Satriana, Arpi N, Lubis YM, Adisalamun, Supardan MD, Mustapha WAW. Diacylglycerol-enriched oil production using chemical glycerolysis. European Journal of Lipid Science and Technology. 2016 Dec 1;118(12):1880–90.

21. Zha B, Chen Z, Wang L, Wang R, Chen Z, Zheng L. Production of glycerol monolaurate-enriched monoacylglycerols by lipase-catalyzed glycerolysis from coconut oil. European Journal of Lipid Science and Technology. 2014;116(3):328–35.

22. Mujdalipah S, Sasmita AH, Amalia IK, Suryani A. Separation of glycerolysis product using hexane. In: IOP Conference Series: Materials Science and Engineering. Institute of Physics Publishing; 2016.

23. Mursalin M, Sahrial S, Yernisa Y. Purification engineering of mono-diacylglycerol using creaming demulsification technique in 20 kg-scale. Food Res. 2020 Aug 1;4(4):1011–7.

24. European Commission. Regulations II- laying down specifications for food additives listed in Annexes II and III to Regulation (EC) No 1333/2008 of the European Parliament and of the Council (Text with EEA relevance). Official Journal of the European Union [Internet]. 2012;83(231):1–295. Available from: https://eur-lex.europa.eu/legal-content/EN/TXT/PDF/?uri=CELEX:32012R0231

25. Solaesa ÁG, Sanz MT, Falkeborg M, Beltrán S, Guo Z. Production and concentration of monoacylglycerols rich in omega-3 polyunsaturated fatty acids by enzymatic glycerolysis and molecular distillation. Food Chem. 2016 Jun 25;190:960–7.

26. Wangi IP, Supriyanto, Sulistyo H, Hidayat C. Glycerolysis–interesterification in high-shear reactor using sodium silicate catalyst: effect of mixing rate on reaction kinetics. Reaction Kinetics, Mechanisms and Catalysis. 2023;136:881–99.

27. Fayaz G, Goli SAH, Kadivar M, Valoppi F, Barba L, Calligaris S, et al. Potential application of pomegranate seed oil oleogels based on monoglycerides, beeswax and propolis wax as partial substitutes of palm oil in functional chocolate spread. Lwt [Internet]. 2017;86:523–9. Available from: http://dx.doi.org/10.1016/j.lwt.2017.08.036

28. Brien RDO. Fats and Oils : Formulating and Processing for Applications. Florida: CRC Press; 2004.

29. Gonçalves D, Bozzi Barbeiro L, Tonon de Souza P, Caldas Batista EA, de Almeida Meirelles AJ. Thermal stability of glycerol containing compounds from the biodiesel production chain. J Therm Anal Calorim [Internet]. 2022;147(21):11857–74. Available from: https://doi.org/10.1007/s10973-022-11395-4

30. Luo J, Wang Z, Deng S, Zhang F, Bao G, Mao J, et al. Na2SiO3-Catalyzed Glycerolysis of Sacha Inchi (Plukenetia volubilis L.) Oil into Di- and Monoacylglycerols. JAOCS, Journal of the American Oil Chemists' Society. 2020 Jun 1;97(6):603–14.

IOP Conf. Series: Earth and Environmental Science **1460** (2025) 012037 doi:10.1088/1755-1315/1460/1/012037

Detection and characterization of praline chocolates with wet-based and fat-based fillings using X-ray CT images

Yoga Arif Firmansyah, Arifin Dwi Saputro and Bayu Nugraha*

Agricultural and Biosystems Engineering, Faculty of Agricultural Technology, Universitas Gadjah Mada, Jalan Flora 1, Bulaksumur, Depok, Sleman, Yogyakarta, Indonesia

*E-mail: bayu.nugraha@ugm.ac.id

Abstract. Praline chocolate is known to have poor stability and is susceptible to damage (deformation) due to interaction factors with the filling. X-ray Computed Tomography (CT) is a non-destructive instrument that can be used to identify the microstructure and the component of a product that affect the sensory of chocolate. X-ray CT scanning of praline chocolate is carried out at low resolution so that the interaction of the addition of filling on the final quality of chocolate can be known. The purpose of this study was to develop the method which can used to identify and visualize the internal component of praline chocolate due to the addition of fillings. The samples used consisted of three treatments, namely control (without filling), pineapple jam filling (wet-based) and peanut butter filling (fat-based). The grey value showed that there are three different regions including peanut butter region (lower), dark chocolate region, and pineapple jam region (higher). Filling regions were able to segmented by particular method. The better method for pineapple jam filling is MaxEntropy, while peanut butter is IJ_IsoData. This method is able to identify the component of fillings according to the grey value which will be essential to predict the deformation.

Keywords: chocolate, deformation, microstructure, praline, X-ray CT.

1. Introduction

Detection and characterization of food product morphology are necessary to evalute the implemented composition and processes. X-ray computed tomography (CT) is an imaging instrument allowing to depict physical structures of a material externally and internally without damaging the specimen object [1], [2]. The applications of X-ray CT on food products have been conducted for identification and visualization of macro- and microstructure of bread [3], ice cream [4], and cake [5], [6]. It is also utilized to identify the internal damage of onions [7], [8]. The implementation of using X-ray CT on chocolate has been done [9], but has not explored further regarding the identification of different types of chocolate and fillings.

Praline chocolate is one of the chocolate products that is very popular around the world. This product has unique characteristic, namely a layer of chocolate that covers a filling inside. The praline chocolate filling can vary such caramel, cream, fruit jam, to peanut butter others leading to a unique taste and texture [10]. However, this variations in praline chocolate fillings and processing compositions introduce new challenges, particularly concerning the stability and quality of the end product during storage periods. Praline chocolate is easily changed of structure and deformation, particularly when its production process uses the filling that has different physicochemical compositions to the chocolate layer [11]. The wet-based filling is expected to be reactive of chocolate layer due to water can increase the migration of humidity that can accelerate the damage of praline chocolate [11], [12]. Besides that, the moisture content within the wet-based filling also has the potentiality to grow the microorganism that can cause decay in praline chocolate [10]. This problems demand to emerge the method that capable to segmentate and identification the component which leads to damage phenomena of praline chocolate in microscale, therefore those problems will be able to identify the accurate solution.

Identification of praline chocolates using X-ray CT may help to evaluate the process and compositions used. However, as a first milestone of the whole X-ray CT based identification technique, detection of the chocolate filling matrix is highly required to calculate and visualize the precise area of the fillings, used for further analysis. To this end, three segmentation methods are implemented and evaluated in this research, providing an effective solution for identifying the microstructure of praline chocolate. The image processing method can explain the phenomenon of interaction between the filling and the layer of chocolate in praline chocolate. The result of X-ray CT scanning on microscale offers visualization that can identify phenomena of deformation, density change, and crack within the microstructure of praline chocolate. Furthermore, the result of X-ray CT is not only able to offer visualization but also enables quantitative analysis or 3D morphology studies, for instance, measurements of porosity volume, component distribution, and densities of filling and chocolate. This comprehensive approach allows for a deeper understanding of praline chocolate structure and composition.

This research aims to develop the method in order to segmentate particular component including pineapple jam (wet-based) and peanut butter (fat-based). This research is expected to provide the suitable method in order to predict the deformation of praline chocolate with different character of fillings.

2. Materials and Methods

2.1 Sample preparation

A dark chocolate, pineapple jam, and peanut butter were purchased at Intisari Bakers Mart in Yogyakarta, Indonesia. The dark chocolate, weighing 2 kg, was specifically selected for its compound formulation using natural fats, which provides enhanced stability in melting profile. For the fillings, 0.5 kg each of pineapple jam dan peanut butter were acquired. These fillings were chosen for their low moisture content to ensure stability and minimizing water interaction with praline chocolate shell. After purchased, all samples were stored at a cold temperature of 16 °C to maintain their stability and quality until processing.

2.2 Production of praline chocolate

A compound dark chocolate was prepared with temperature of 55 °C within the chocolate container in order to melt into the liquid phase. The liquid of chocolate was molded into the praline chocolate molds and vibrated in order to make the bubbles come out. The liquid compound chocolate within the molds was poured back into the liquid chocolate container, leaving a little chocolate sticking in the mold, and it was stored at cold temperature (16 °C), thus the chocolate shell had been molded. The fillings were prepared with 1.5 grams (15.79%) each weight and were placed in a chocolate shell. Then, the liquid of dark compound chocolate with approximately 9.5 grams was placed in praline chocolate molds in order to close the fillings and was vibrated. The praline chocolates were stored at cold temperature (16 °C). To maintain the quality of praline chocolate samples during transport to the X-ray CT Laboratory, praline chocolates were placed in a cooling box with ice gels surrounding the samples, preventing temperature-induced changes.

2.3 Scanning X-ray CT

The each of compound chocolate samples were arranged in the styrofoam holder due to has lower density and were placed in the X-ray CT instrument (Nanotom system at X-ray CT, Industrial Metrology Laboratorium, Cikarang, Indonesia). X-ray CT scanning works by passing X-ray beam through a rotating object and a detector captures the data, which is sent to a computer as a radiograph. Then, this radiograph processed and reconstructed into a greyscale image showing the internal structure **(Figure 1)**.

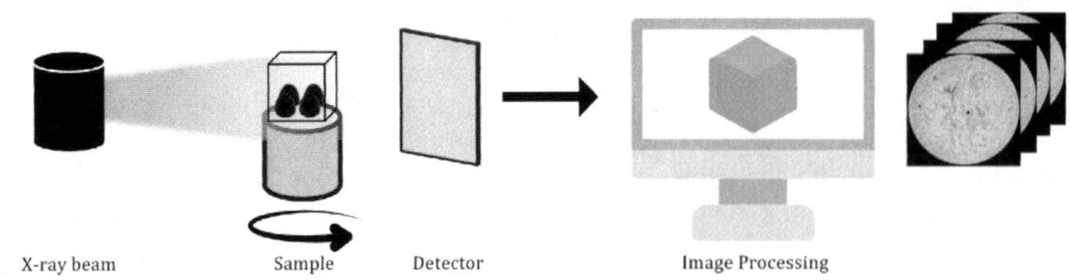

Figure 1. Schematic of X-ray computed tomography imaging

The scan was performed at a low resolution, namely 73.5 µm in order to obtain the detailed scanning. The system was operated including voltage and current. Those settings were determined depending by density of object which was food product, namely 140 kV and 200µA. The rotation angle was used at 0.5° in order to reduce the scanning times of X-ray CT. Those settings result short scanning times around 12 minutes. Scanning times are essential for food product for instance chocolate in order to reduce the potentiality of food change. This reduction is critical for praline chocolate's structure, particularly due to its wet-based filling and overall low melting point of approximately 34 °C. The shorter scan time minimizes heat exposure, helping to maintain the praline chocolate's structure and prevent undesired melting or texture changes. As the wet-based filling which has moisture content increases the praline's susceptibility to heat-induced [13], that could affect the accuracy of the CT scan results.

2.4 Image Processing

Several data which obtained from X-ray CT scanning were continued to image processing. Brightness and contrast were used to increase the visualization in order to show the better grey value. The bits of images were reduced from 16-bit into 8-bit to simplify processing and enhance compatibility with segmentation tools, and then a median filtered was applied to reduce noise and smooth the image, making the objects more clearly segmented. The components of the object resulting from CT image varied in the following order of density from low to high. The greyscale was expected to have several regions due to different components. Voxels with a low value are considered as an air fraction, while high value show the object (in this case; chocolate and fillings).

Otsu thresholding was used considering the threshold of the porous region, which is able to identify and visualize the microstructure, for instance pore size distribution [14]–[16].

Several methods were applied in order to segmentate the different component of fillings. Those formula's methods including Maximum Entropy (MaxEntropy) (1), IJ_IsoData (2), and Moments (3) for both of fillings (pineapple jam filling and peanut butter filling) were computed based on the following formula:

$$\varphi(t) = (-\sum_{i=0}^{t} \frac{P_i}{P_m} ln \frac{P_i}{P_m}) + (-\sum_{i=t+1}^{L-1} \frac{P_i}{1-P_m} ln \frac{P_i}{1-P_m}) \quad (1)$$

where $\varphi(t)$ is the optimal threshold, P_i represents the probability of a pixel having intensity value I, and P_m is the cumulative probability of pixels with intensity values less than or equal to the threshold t.

$$T_{n+1} = \frac{\mu 1(T_n)+\mu 2(T_n)}{2} \quad (2)$$

where T_{n+1} is the threshold at iteration n, $\mu 1(T_n)$ is mean of pixels with intensity values below T_n and $\mu 2(T_n)$ is the mean of pixels with intensity values equal to or above T_n.

$$\sum_{i=0}^{t} i^k . p_i = m_k . P_0 \quad (3)$$

where t represents the threshold to be determined, i is the pixel intensity, p_i is the probability of a pixel having intensity i, m_k is the k-th order moment of the entire image, P_0 is the cumulative probability up to threshold t.

3. Results and Discussion

3.1 Grey Value of dark chocolate with different treatments

This praline chocolate research with treatments including control, pineapple jam filling, and peanut butter filling resulted three regions with different grey value. Image processing technique plays an important role in resulting grey value that is able to divide three different regions based on density which namely low grey value (peanut butter), middle grey value (dark chocolate), and high grey value (pineapple jam). This process begins with image acquisition using X-ray CT, where the intensity of the X-ray penetrating the sample is translated into a 3D image based on absorption of the X-ray by material [17]. Each voxel in image is given a grey value based on how much absorption of X-ray within the voxel, which is a representation of the material density [2], [15], [17].

The result of X-ray scanning with three different treatments shows differences in image sections for each treatment. Control treatment showed an average grey value **(Figure 2A)**, while the filling treatments showed different visualization which caused by fillings. The pineapple jam filling **(Figure 2B)** has a lighter colour than the peanut butter filling **(Figure 2C)**, which attributes to the different density of material. Then, those treatments were processed to be a chart with range 0-255 grey value **(Figure 2D)** which can identify the region of each treatment. Peanut butter, which has a lower density due to its high fat content, produces a lower grey value, while pineapple jam, which has a higher water and sugar content, caused greater absorption, resulting in a higher grey value. The grey value of chocolate with a more varied solids content is in the middle grey value range.

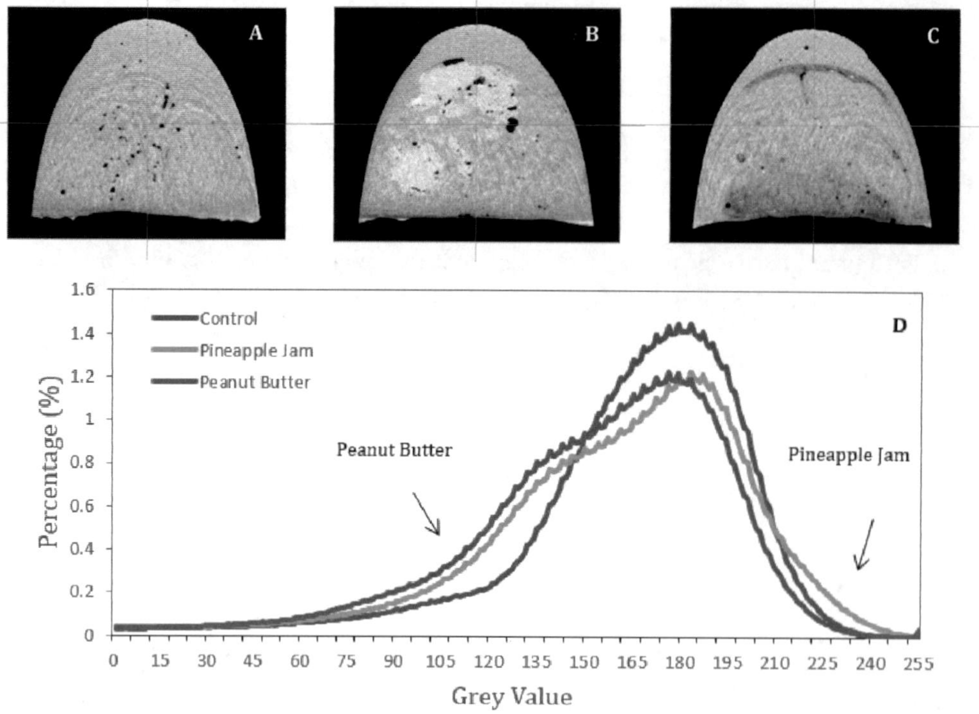

Figure 2. An X-ray CT image slice of praline chocolates without filling or control (A), treatment with a sample consist of consist of pineapple jam filling (B), and peanut butter filling (C), resulting in the different distributions greyscale values of chocolates (D).

3.2 Comparing methods of segmentation in filling treatments

Various segmentation algorithms were used to analyze the internal structure of praline chocolates, including a control (without filling) and samples with pineapple jam and peanut butter fillings. The objective was to differentiate materials based on their grey value intensities in X-ray CT images. Three distinct segmentation methods (MaxEntropy, IJ_IsoData, and Moments) were applied to each sample category. For the control sample, MaxEntropy **(Figure 3A)**, IJ_IsoData **(Figure 3D)**, and Moments **(Figure 3G)** algorithms were utilized. As anticipated for homogenous chocolate matrix without fillings, all three methods result irregular segmentation patterns, indicating to discernible region differentiation.

In the case of pineapple jam filling, MaxEntropy **(Figure 3B)**, IJ_IsoData **(Figure 3E)**, and Moments **(Figure 3H)** were applied. The MaxEntropy algorithm demonstrated the better segmentation for the pineapple jam region compared to IJ_IsoData and Moments. This result can be attributed to the MaxEntropy threshold's fundamental principle of maximizing the entropy between two threshold-separated sections [18], [19]. The method's proficiency in handling distinct bimodal histograms makes it particularly suitable for pineapple-jam filled pralines, where a high contrast exists between the dense, high grey value jam and the surrounding chocolate matrix.

On the other hand, peanut butter filling was also used three similar segmentation methods (MaxEntropy **(Figure 3C)**, IJ_IsoData **(Figure 3F)**, and Moments **(Figure 3I)**). The IJ_IsoData algorithm showed better segmentation compared to MaxEntropy and Moments for segmenting peanut butter regions. This result of IJ_IsoData can be explained by the lower grey value contrast between peanut butter and chocolate, resulting in a less pronounced bimodal distribution. The IJ_IsoData threshold is similar to the IsoData threshold, which allows it to minimize both the squared error function and intra-cluster variances [18]. IJ_IsoData's iterative approach allows it to converge on a suitable threshold even when intensity difference is not subtle. This makes it more suitable for segmenting fillings with grey values closer to those of the chocolate (in case of peanut butter filling).

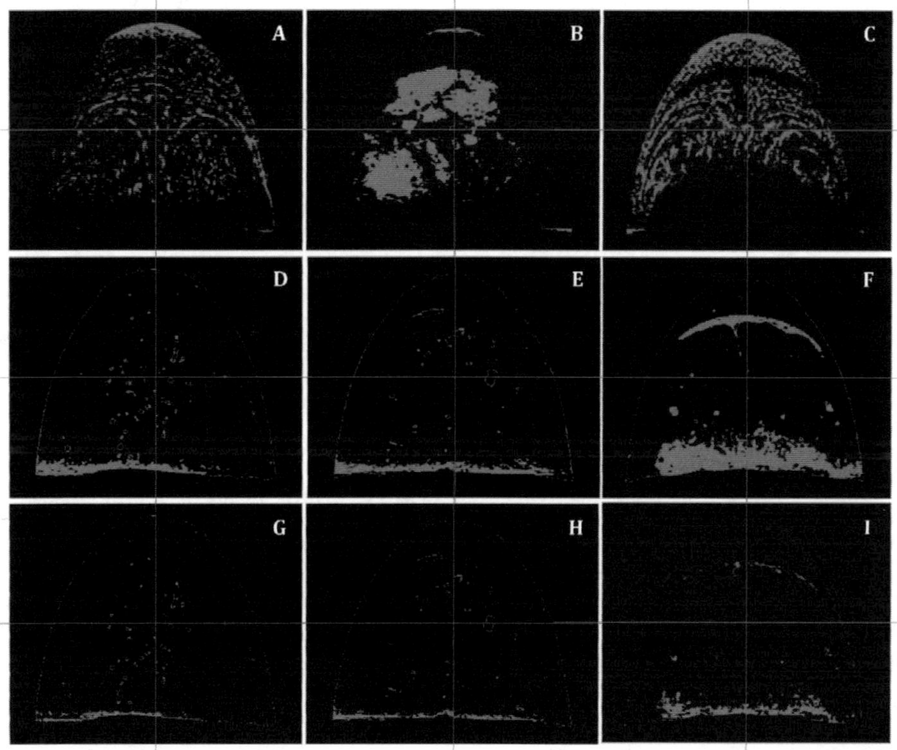

Figure 3. The different methods were applied to all praline chocolate treatments: control segmented with MaxEntropy (A), IJ_IsoData (D) and Moments (G); pineapple jam with MaxEntropy (B), IJ_IsoData (E), and Moments (H); and peanut butter with MaxEntropy (C), IJ_IsoData (F), and Moments (I). The cyan region represents the segmented area of chocolate, while the black region shows the unsegmented area, which varies depending on each method used.

3.3 Percentage result of filling segmentation

The measurement of different methods applied to each filling shows distinct performance results **(Table.1)**, which correlated with earlier figure **(Figure 2)**. For the pineapple jam filling, the MaxEntropy method demonstrates the highest effectiveness, yielding a filling percentage of

19.62% ± 6.90%. This result of segmentation can be attributed to MaxEntropy's ability to maximize the information or the pixel intensity distribution which performed to find the threshold value for segmentation [20]. Therefore, this method accurately segment regions with high greyscale values, corresponding to the high density of pineapple jam, as visually showed in **Figure 3B**. In contrast, for the peanut butter filling, the IJ_IsoData method proves most suitable, identifying 17.75% ± 1.96% as filling. The IJ_IsoData method is effective in identifying spectral clusters with minimal effort. IJ_IsoData, a variant of this algorithm, calculates the average pixel intensity above and below the threshold iteratively until convergence [18]. This method's effectiveness likely results from its capacity to detect subtle contrast images, enabling efficient thresholding detection which is the lower greyscale values of peanut butter as observed in **Figure 3F**. These results can be correlated with the actual composition of praline chocolate production which is closer to 17.64%.

Table 1. Percentage of filling and air space area segmented using different method.

	%Chocolate	%Bubbles	%Filling		
			MaxEntropy	IJ_IsoData	Moments
Control	100	0.28 ± 0.06	11.33 ± 11.51	9.80 ± 3.06	5.58 ± 2.97
Pineapple Jam	100	2.00 ± 0.75	19.62 ± 6.90	19.65 ± 10.14	6.45 ± 2.48
Peanut Butter	100	0.42 ± 0.16	8.64 ± 9.00	17.75 ± 1.96	6.93 ± 2.00

The Otsu thresholding method consistently provides reliable bubbles measurements across all samples, with notably low porosity in the control (0.28% ± 0.06%) and peanut butter (0.42% ± 0.16%) samples, while showing slightly higher porosity in the pineapple jam sample (2.00% ± 0.75%). These results can be attributes to the different type of components between pineapple jam as a wet-based and chocolate as a closer to fat-based. These porosity results complement the earlier segmentation findings, providing additional insights into the internal structure of the pralines.

4. Conclusion

This study demonstrates the effectiveness of X-ray CT imaging combined with three segmentation methods for analyzing the internal structure of praline chocolate with various fillings. Analysis of greyscale values revealed three distinct regions based on density: low grey value (peanut butter), middle grey value (dark chocolate), and high grey value (pineapple jam). The results highlight the importance of using particular segmentation algorithms to the specific characteristic of different fillings. MaxEntropy proved optimal for high density, high contrast fillings like pineapple jam, whie IJ_IsoData excelled with lower-contrast fillings such as peanut butter.

References

[1] C. J. Du and D. W. Sun, "Recent developments in the applications of image processing techniques for food quality evaluation," *Trends Food Sci. Technol.*, vol. 15, no. 5, pp. 230–249, 2004.
[2] B. Nugraha, P. Verboven, S. Janssen, Z. Wang, and B. M. Nicolaï, "Non-destructive porosity mapping of fruit and

vegetables using X-ray CT," *Postharvest Biol. Technol.*, vol. 150, no. November 2018, pp. 80–88, 2019.

[3] T. Van Dyck *et al.*, "Characterisation of structural patterns in bread as evaluated by X-ray computer tomography," *J. Food Eng.*, vol. 123, pp. 67–77, 2014.

[4] E. Guo *et al.*, "Revealing the microstructural stability of a three-phase soft solid (ice cream) by 4D synchrotron X-ray tomography," *J. Food Eng.*, vol. 237, no. May, pp. 204–214, 2018.

[5] B. Pareyt *et al.*, "The role of sugar and fat in sugar-snap cookies: Structural and textural properties," *J. Food Eng.*, vol. 90, no. 3, pp. 400–408, 2009.

[6] M. Assad-Bustillos, S. Guessasma, A. L. Réguerre, and G. Della Valle, "Impact of protein reinforcement on the deformation of soft cereal foods under chewing conditions studied by X-ray tomography and finite element modelling," *J. Food Eng.*, vol. 286, no. January, 2020.

[7] D. Priyanti, I. Ayuningsih, A. N. Fatkhur Rahmah, S. Rahayoe, and B. Nugraha, "X-ray CT applications to reveal physical defects of onion during storage," *BIO Web Conf.*, vol. 80, pp. 0–4, 2023.

[8] A. N. F. Rahmah, D. Priyanti, I. Ayuningsih, and B. Nugraha, "The Effect of Different Storage Temperatures and Internal Analysis of Onions with X-Ray CT on The Respiration Rate of Onions," *E3S Web Conf.*, vol. 425, 2023.

[9] S. K. Reinke *et al.*, "Synchrotron X-Ray microtomography reveals interior microstructure of multicomponent food materials such as chocolate," *J. Food Eng.*, vol. 174, pp. 37–46, 2016.

[10] C. L. Marvig, R. M. Kristiansen, M. G. Madsen, and D. S. Nielsen, "Identification and characterisation of organisms associated with chocolate pralines and sugar syrups used for their production," *Int. J. Food Microbiol.*, vol. 185, pp. 167–176, 2014.

[11] K. Franke, D. Middendorf, V. Heinz, and U. Bindrich, "Alcohol in praline fillings influences the water migration within the surrounding chocolate shell," *J. Food Eng.*, vol. 315, no. August 2021, p. 110805, 2022.

[12] H. Dilsad, V. Teresa, R. Foligni, and G. Zeppa, "Microstructural and rheological influence of different strategies to mitigate oil migration in chocolate pralines during storage in limiting conditions," *LWT*, vol. 208, no. August, p. 116672, 2024.

[13] A. D. Saputro, D. Van de Walle, S. Kadivar, M. D. Bin Sintang, P. Van der Meeren, and K. Dewettinck, "Investigating the rheological, microstructural and textural properties of chocolates sweetened with palm sap-based sugar by partial replacement," *Eur. Food Res. Technol.*, vol. 243, no. 10, pp. 1729–1738, 2017.

[14] M. A. Boone *et al.*, "3D mapping of water in oolithic limestone at atmospheric and vacuum saturation using X-ray micro-CT differential imaging," *Mater. Charact.*, vol. 97, pp. 150–160, 2014.

[15] E. Herremans *et al.*, "Microstructure-texture relationships of aerated sugar gels: Novel measurement techniques for analysis and control," *Innov. Food Sci. Emerg. Technol.*, vol. 18, pp. 202–211, Apr. 2013.

[16] B. Nugraha *et al.*, "Oxygen diffusivity mapping of fruit and vegetables based on X-ray CT," *J. Food Eng.*, vol. 306, no. April, 2021.

[17] V. Vicent, P. Verboven, F. T. Ndoye, G. Alvarez, and B. Nicolaï, "A new method developed to characterize the 3D microstructure of frozen apple using X-ray micro-CT," *J. Food Eng.*, vol. 212, pp. 154–164, Nov. 2017.

[18] Y. D. Shou, Z. Zhao, and X. P. Zhou, "Sensitivity analysis of segmentation techniques and voxel resolution on rock physical properties by X-ray imaging," *J. Struct. Geol.*, vol. 133, no. January, p. 103978, 2020.

[19] B. Wang, L. L. Chen, and J. Cheng, "New result on maximum entropy threshold image segmentation based on P system," *Optik (Stuttg).*, vol. 163, pp. 81–85, 2018.

[20] S. S. Khattak, G. Saman, I. Khan, and A. Salam, "Maximum Entropy Based Image Segmentation of Human Skin Lesion," *Int. J. Comput. Control. Quantum Inf. Eng.*, vol. 9, no. 5, pp. 667–671, 2015.

Sensory quality of beef sausage containing pea (*Pisum sativum*) flour and mackerel fish oil cooked to a well-done level

Lilik Retna Kartikasari[1,2*], Bayu Setya Hertanto[1,2], Farouq Heidar Barido[1,2], Winny Swastike[1,2], and Adi Magna Patriadi Nuhriawangsa[1,2]

[1] Faculty of Animal Science, Universitas Sebelas Maret, Surakarta, Indonesia
[2] Research group of Food Technology of Animal Origin, Department of Animal Science, Faculty of Animal Science, Universitas Sebelas Maret, Surakarta, Indonesia,

*E-mail: lilikretna@staff.uns.ac.id

Abstract. The objective of this research was to assess the impact of incorporating pea flour and mackerel fish oil on the sensory attributes of beef sausages when cooked to a well-done state. This study used a completely randomized design (CRD) in one direction using four treatments with panellist as replications. The research treatments were the substitution of tapioca flour with pea flour (PF) and the level of addition of fish oil (FO), including PF 0% and FO 0% (P0), PF 3% and FO 0.75% (P1); PF 6% and FO 1.5% (P2); and PF 9% and FO 2.25%. The sensory attributes observed included color, aroma, tenderness, taste, flavor, and aftertaste. The consumer preferences for various attributes of beef sausages were assessed using a 9-point hedonic scale by 21 semi-trained panelists. The sensory quality test data underwent Analysis of Variance to determine the significance of the treatments. Subsequently, the Tukey Test was employed for further analysis when significant effects were observed. The findings indicated that the incorporation of pea flour and mackerel fish oil into the beef sausages did not significantly influence the color, aroma, tenderness, texture, taste, flavor, aftertaste, or overall quality ($P<0.01$). The consumer preferences of beef sausages containing pea flour up to a level of 9% and mackerel fish oil at 2.25% were the same as the control group. The use of 9% pea flour and 2.25% mackerel fish oil can be applied and can maintain the sensory quality of the beef sausages.

Keywords: sensory assessment, sausage quality, fish oil, cooking methods

1. Introduction

The invention and reformulation of foods, even traditional ones, is driven by customer demand for healthier food items [1]. In a similar vein, there is an increasing need for delicious, nutritious, and reasonably priced meat products [2]. In an effort to combat health issues including obesity, colon cancer, and coronary heart disease, consumers are increasingly selecting healthier meat and poultry products [3]. One of the popular meat products is beef sausages. Manufacturers of sausage must select a range of ingredients in the right proportions to provide not only the intended level of safety and quality, but also cost effectiveness [4]. Therefore, the food sector faces a challenge in

IOP Conf. Series: Earth and Environmental Science **1460** (2025) 012038 doi:10.1088/1755-1315/1460/1/012038

creating healthier sausage that may help decrease health risks, such as replacing animal fats with vegetable fats and oils and acquiring foods low in saturated fats and cholesterol [5].

Be aware of this, many studies have been carried out in an effort to improve the quality and nutritional value of meat products by replacing or reducing ingredients that can pose a risk to human health [6]. In the processing of meat products, non-meat components are widely used with the aim of improving nutritional quality and certain qualitative characteristics, as well as cutting costs. Functional ingredients, including protein, starch, hydrocolloids, and fiber play a pivotal role in meat processing. They enhance the texture of meat, bind cuts, elevate cooking outcomes, and augment water/oil retention capabilities. Additionally, they are crucial in stabilizing fat and water elements within meat emulsions throughout food preparation and cooking phases [7]. Concerns about consumer safety have led to an increase in demand for reducing animal fat in meat products and replacing it with vegetable fats and oils [4], as well as marine oils [8]. Several studies have reported the use of marine oils to enhance processed meat products. Andrés et al. [9] created chicken sausages with low fat content using whey protein concentrate and pre-emulsified squid oil as a binder, resulting in good functional characteristics and sensory evaluation scores. Cáceres et al. [10] used pre-emulsified fish oil, water, and caseinates to replace beef tallow in cooked meat emulsions. The majority of the assessments of these strategies have revealed certain detrimental impacts on lipid oxidation and sensory characteristics because of the high sensitivity of ω-3 PUFA to oxidation as well as the rancid and fishy taste and odor [11].

Furthermore, the use of vegetable protein sources is worth investigating because it improves product quality. Protein helps to stabilize meat emulsions by retaining water and fat while also influencing organoleptic and textural characteristics [12,13]. Vegetable proteins are frequently utilized as extenders and binders in meat products, such as sausages. Petracci et al. [7] suggest that these proteins can improve fat emulsification in finely chopped meat and enhance the binding of meat particles and fat in coarsely ground meat products. This not only improves the meat's functionality but also its structural integrity. The incorporation of plant proteins into meat systems alters the physical and textural properties of the meat product, leading to the formation of a complex, heterogeneous structure [14]. Pea (*Pisum sativum* L.) is a protein-rich plant material that is highly sought after due to its agronomic and functional food benefits. The commercial application of pea protein products is currently limited due to their suboptimal sensory and functional properties [15]. Modifying proteins can be a viable approach to overcome these challenges and improve attributes such as emulsification, water/oil retention, solubility, and gelation [16]. Various studies have incorporated pea protein into processed meats. For instance, Broucke et al. [17] substituted 20% of pork in cooked sausages with pea protein isolate (PPI), low moisture extrudates (LME), and high moisture extrudates (HME). Additionally, Shen et al. [18] explored the use of pea protein as a functional extender in beef patties, enhancing the pea protein isolate with protein glutaminase and combining it with guar gum to improve both physical and sensory qualities.

However, product innovation in sausages must consider hedonic quality because not all consumers are willing to try new products. Product reformulation should always be consumer-centred, which means that sensory assessment of the reformulated product is critical to its success [6]. The goal of this research was to see how the combination of pea flour and fish oil in cooked sausages affected panellist preferences.

IOP Conf. Series: Earth and Environmental Science **1460** (2025) 012038 doi:10.1088/1755-1315/1460/1/012038

2. Materials and Methods

2.1 Research Materials

The materials used in the research included beef blades, commercial collagen casings measuring 23.55 mm with a length of 610 cm as sausage casings, tapioca flour, pea flour (*Pisum sativum*), ice cubes, beef stock, skim milk, salt, coriander powder, white pepper, garlic powder, mackerel fish oil (*Scomberomorus commerson*), egg white.

2.2 Research Design

The research design used in this study was a unidirectional completely randomized design (CRD). The sausage used comes from beef (blade) with 4 treatments with panelists as replications. The treatment is as follows:
P1: Beef sausage containing 0% pea flour and 0% fish oil
P2: Beef sausage containing 3% pea flour and 0.75% fish oil
P3: Beef sausage containing 6% pea flour and 1.5% fish oil
P4: Beef sausage containing 9% pea flour and 2.25% fish oil

2.3 Sausage Production

The formulation used in making beef sausages uses the substitution of pea flour (*Pisum sativum*) with the addition of mackerel fish oil (*Scomberomorus commerson*) can be seen in Table 1.

Table 1. Beef sausages formulation with the use of pea flour and mackerel fish oil

Ingredients	Treatments (%)			
	P1	P2	P3	P4
Beef meat	66.5	65.75	65	64.3
Tapioca flour	12	9	6	3
Pea flour	0	3	6	9
Mackarel fish oil	0	0.75	1.5	2.2
Ice cubes	10	10	10	10
White egg	5	5	5	5
Garlic	2	2	2	2
White pepper	0.5	0.5	0.5	0.5
Coriander	0.3	0.3	0.3	0.3
Salt	0.5	0.5	0.5	0.5
Skim milk	2	2	2	2
Beef broth	1.2	1.2	1.2	1.2
Total	100	100	100	100

The process of making beef sausage follows the procedure of Lenzum et al. [19] with several modifications regarding the use of tools and materials with the following stages:
1) Wash fresh beef with running water and weigh, then cut into smaller sizes and grind using a blender for 1 minute; 2) The ground meat is then grouped according to each treatment, then mixed well with all the spices and other additional ingredients using a chopper; 3) The sausage mixture, which has been thoroughly mixed, is placed in a 23.55 mm collagen casing; 4) The sausage is then steamed for approximately 30 minutes at a temperature of 70-80ºC until cooked well done; 5) The cooked sausages are cooled, then organoleptic quality testing is carried out.

2.4 Organoleptic Quality Testing

The evaluation was performed by involving a total of 25 semi-trained panelists. Panelists were recruited by administering a questionnaire to obtain information related to the consumption of beef sausage, including the possibility of allergies to beef and the ingredients used, level of preference and frequency of consuming beef sausage. The cooked sausages are put into a plastic container and then stored in the freezer. Before testing, the sausages were warmed for about 15 minutes. After heating, the sausage is transferred into a sealed container marked with three random digits. Prior to conducting the organoleptic assessment, the panelists received detailed instructions on the evaluation process of the sausage's organoleptic properties. They were then provided with consent forms, which they were required to read and sign before participating in the testing.

Panelists rated their level of liking for sausages using a 9-point hedonic scale. The levels or scores used in testing consumer preferences are extremely dislike (1), dislike very much (2), dislike (3), dislike slightly (4), neither like nor dislike (5), slightly (6), like (7), like very much (8), and extremely like (9). The attributes tested include color, aroma, tenderness, texture, juiciness, taste, taste and aftertaste. After evaluating one sample, panelists were asked to drink water and eat original crackers to neutralize the taste and panelists were given a break of 1 minute between each two samples [20].

3. Data Analysis

All data obtained in this study were analyzed using Analysis of Variance (ANOVA) with the Minitab 18 application to determine whether there was an effect of treatment using pea flour and mackerel on the organoleptic quality of beef sausages. Further tests were carried out if there was a real effect on the treatment using Tukey's test to determine the mean difference between treatments.

4. Results and Discussions

The results of statistical analysis of organoleptic quality are shown in Table 2. This present study focuses on investigating the sensory quality of beef sausages that were subjected to treatment using pea flour and mackerel fish oil. Utilizing 25 semi-trained panellists, with 9-point hedonic scale measurement on sensory evaluation, this study emphasis the evaluation on determined parameters including color, aroma, tenderness, texture, taste, flavor, and aftertaste. As seen on Table 2, the statistical analysis revealed that there were no significant differences in consumer preferences among the different sausage formulations. The color profiles were ranging from 5.76 in P2, to 6.04 in P1 ($p>0.05$). The perception for aroma was ranging from 5.84 in P3 to 6.64 under P1 treatment. With no significant differences, the overall acceptability of beef sausage in this study averaging 5.84 under P3 treatment, and 6.28 under P1 treatment ($p>0.05$). Interestingly, the beef sausages that were made using pea flour and fish oil resulted in similar sensory qualities when compared to that of control group. This suggests that these ingredients can potentially be added and incorporated to bring beneficial effects to beef sausages with no detrimental effects. Further, this study also revealed that incorporating up to 9% pea flour together with 2.25% mackerel fish oil (P4) did not lead to any significant changes in sensory attributes when compared to the control group ($P<0.01$), as illustrated in Table 2.

Table 2. Consumer preference of beef sausages containing pea flour and mackerel fish oil

Parameter	Treatments				*P-value*
	P1	P2	P3	P4	
Colour	6,040±1,098	5,760±1,300	5,840±1,248	5,920±1,382	0,879
Aroma	6,640±1,287	6,000±1,414	5,840±1,375	6,200±1,354	0,189
Tenderness	5,800±1,443	5,960±1,136	6,120±1,201	6,320±1,282	0,516
Texture	5,560±1,781	5,520±1,503	5,440±1,557	6,080±1,320	0,453
Taste	6,320±1,701	6,360±1,150	5,560±1,387	6,280±1,339	0,146
Flavour	6,600±1,384	6,120±1,130	5,880±1,236	6,040±1,485	0,252
Aftertaste	6,520±1,194	6,040±1,098	5,840±1,248	6,160±1,313	0,254
Overall	6,280±1,429	6,080±1,077	5,840±1,106	6,240±1,422	0,605

Notes:
P1: Beef sausages containing 0% pea flour and 0% mackerel fish oil
P2: Beef sausages containing 3% pea flour and 0.75% mackerel fish oil
P3: Beef sausages containing 6% pea flour and 1.5% mackerel fish oil
P4: Beef sausages containing 9% pea flour and 2.25% mackerel fish oil

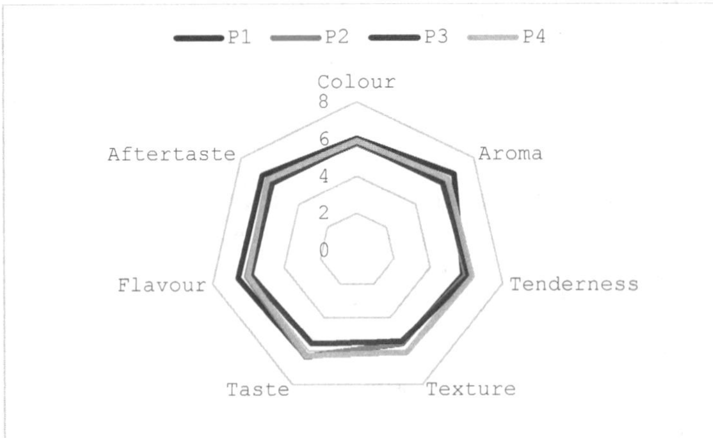

Figure 1. The spider charts of the sensory quality of beef sausages containing pea flour and mackerel fish oil. There were no significant effects on colour, aroma, tenderness, texture, taste, flavor, and aftertaste of beef sausages by including pea flour and mackerel fish oil up to a level of 9 and 2.25%, respectively.
P1: Beef sausages containing 0% pea flour and 0% mackerel fish oil; P2: Beef sausages containing 3% pea flour and 0.75% mackerel fish oil; P3: Beef sausages containing 6% pea flour and 1.5% mackerel fish oil; and P4: Beef sausages containing 9% pea flour and 2.25% mackerel fish oil

The observed results on consumer acceptances are consistent with prior studies focusing on the utilization of alternative ingredients in meat products. Argel et al. [21] demonstrated that adding plant-based fillers did not adversely affect the sensory properties of meat emulsions. Similarly, Marchetti et al. [8] found that meat emulsions with low lipid content and added fish oil maintained acceptable sensory qualities, which aligns with our observations regarding mackerel fish oil in beef sausages. In addition, this study has the same results as reported by Baune et al. [22], overall, the addition of vegetable ingredients including peas to artificial meat products has the same overall hedonic value. Likewise, according to Sogari et al. [23] there was no significant difference in preference between beef burgers and beef burgers with added peas.

The absence of significant differences in sensory attributes can be attributed to several factors. First, pea flour is rich in proteins and fibers, which may enhance the texture of meat products without compromising flavor [21]. Additionally, mackerel fish oil is high in omega-3 fatty acids, which not only provides health benefits but can also enhance flavor profiles when used appropriately [8]. The careful balance between fat content and moisture retention likely played a vital role in preserving tenderness and juiciness. Furthermore, both pea flour and fish oil possess emulsifying properties that improve the binding capacity of meat mixtures [14]. This leads to a cohesive texture that consumers generally find appealing.

This research is different from Mongi and Gomezulu [24], that the addition of peas had a significant effect compared to the control on beef sausage with the addition of pea extract. Some additional oil sources can improve the sensory quality of silver carp surimi sausages [16].

Some differences in the results of this study with other researchers can be caused by several factors. Basic ingredients and additional ingredients in making meat sausages can result in differences in consumer preferences [25]. Sensory quality or consumer preference for meat sausages is influenced by the amount of pea binding protein and content [24]. Differences in the source of fat used can also cause these differences in perception of the sensory attributes being evaluated. Some sources of fat in sausage ingredients can reduce the sensory quality of the product [26]. Overall, the study suggests that incorporating 9% pea flour and 2.25% mackerel fish oil into beef sausages is a viable option for maintaining sensory quality while enhancing nutritional value.

5. Conclusion

The consumer preferences of beef sausages containing pea flour up to a level of 9% and mackerel fish oil at 2.25% were the same as the control group. The use of 9% pea flour and 2.25% mackerel fish oil can be applied and can maintain the sensory quality of the beef sausages.

6. Acknowledgement

The researcher expresses gratitude to Universitas Sebelas Maret (UNS), Indonesia, for the financial support provided for this research under the LPPM UNS Non-APBN Fund (Research Group Grant) under number of contract 194.2/UN27.22/PT.01.03/2024.

References

[1] Olmedilla-Alonso B, Jim´enez-Colmenero F and S´anchez-Muniz F J 2013 *Meat Science* **95** 919–930 (https://doi.org/10.1016/j.meatsci.2013.03.030)
[2] Balestra F and Petracci M 2019 *Sustainable Meat Production and Processing* 45–68 (https://doi.org/10.1016/B978-0-12-814874-7.00003-1)
[3] Çelebi U and Erge A 2024 *International Journal of Gastronomy and Food Science* **36** 100907 (https://doi.org/10.1016/j.ijgfs.2024.100907)
[4] Savadkoohi S, Hoogenkamp H, Shamsi K and Farahnaky A 2014 *Meat Science* **97** 410–418 (http://doi.org/10.1016/j.meatsci.2014.03.017)
[5] Rinaldoni A N, Palatnik D R, Zaritzky N and Campderrós M E 2014 *LWT - Food Science and Technology* **55**(1) 139– 147. http://doi.org/10.1016/j.lwt.2013.09.003
[6] Rabadan A, Alvarez-Ortí M, Pardo-Giminez A, Zied D C and Pardo J E 2021 *LWT - Food Science and Technology* **136** 110307 (https://doi.org/10.1016/j.lwt.2020.110307)
[7] Petracci M, Bianchi M, Mudalal S, and Cavani C 2013 *Trends in Food Science & Technology* **33** 27–39 (https://doi.org/10.1016/J.TIFS.2013.06.004)
[8] Marchetti L, Andres S C and Califano A N 2013 *LWT – Food Science and Technology* **51** 514–523
[9] Andrés S C, García M E, Zaritzky N E and Califano A N 2006 *Journal of Food Engineering* **72** 311–319.

[10] Cáceres E, García M L, and Selgas M D 2008 *Meat Science* **80** 183–193 (https://doi.org/10.1016/j.meatsci.2007.11.018)
[11] Solomando J C, Antequera T, Perez-Palacios T 2020 *Meat Science* **162** 108031 (https://doi.org/10.1016/j.meatsci.2019.108031)
[12] Kamani M H, Meera M S, Bhaskar N, and Modi V K 2019 *Journal of Food Science & Technology* **56** 2660–2669 (https://doi.org/10.1007/s13197-019-03754-1)
[13] Kang K M, Lee S H and Kim H Y 2022 Effects of using soybean protein emulsion as a meat substitute for chicken breast on physicochemical properties of Vienna sausage *Food Science of Animal Resources* **42** 73 (https://doi.org/10.5851/kosfa.2021.e63)
[14] Kassama L S, Ngadi M O and Raghavan G S V 2003 *International Journal of Food Properties* **6** 519–529 (https://doi.org/10.1081/JFP-120021456)
[15] Zha F, Dong S, Rao J, and Chen B 2019 *Food Hydrocolloids* **92** 30–40 (https://doi.org/10.1016/J.FOODHYD.2019.01.046)
[16] Shen Z, Li S, Wu J, Wang F, Li X, Yu J, Liu Y, Ma X 2022a *Journal of Food Measurement and Characterization* 16 5007–5022 (https://doi.org/10.1007/s11694-022-01589-7)
[17] Broucke K, Van Poucke C, Duquenn B, De Witte B, Marie-Christin B, Lammers V b, Nino Terjung N, Ebert S, Gibis M, Weiss J and Van Royen G 2022 *Innovative Food Science and Emerging Technologies* **78** 102992 (https://doi.org/10.1016/j.ifset.2022.102992)
[18] Shen Y, Hong S, Du Z, Chao M, O'Quinn T and Li Y 2022b LWT - Food Science and Technology **154** 112774 (https://doi.org/10.1016/j.lwt.2021.112774)
[19] Lenzum, T., M. Sompie, dan S. E. Siswosubroto. 2021 *ZOOTEC* **41**(2) 340-347. doi:10.35792/zot.41.2.2021.34788
[20] Kartikasari L R, Geier M S, Hughes R J, Bastian S E P, Gibson R A 2024 *Journal of the Indonesian tropical animal agriculture* **49** 16-28 (https://doi.org/10.14710/jitaa.49.1.16-28)
[21] Argel N S, Ranalli N, Califano A N and Andrés S C 2020 *Journal of the Science of Food Agriculture* **100** 3932–3941
[22] Baune M C, Broucke K, Ebert S, Gibis M, Weiss J, Enneking U, Profeta A, Terjung N and Heinz V 2023 *Front. Nutr.* **10** 1101479 (https://doi: 10.3389/fnut.2023.1101479)
[23] Sogari G, Caputo V, Petterson A J, Mora C, and Boukid F 2023 *Food Research International* **169** 112813 (https://doi.org/10.1016/j.foodres.2023.112813)
[24] Mongi R J and Gomezulu A D 2022 *Heliyon.* **8** e10703 (https://doi.org/10.1016/j.heliyon.2022.e10703)
[25] Lazic I B, Jovanovic J, Simunovic S, Raseta M, Trbovic D, Baltic T dan Ciric J 2019 *Scientific Journal Meat Technology & quot.* **60** 84-90 (https://doi.org/10.18485/meattech.2019.60.2.2)
[26] de Carvalho F A L, Munekata P E, Pateiro M, Campagnol P C, Domínguez R, Trindade M A and Lorenzo J M, 2020 *LWT - Food Science and Technology* **122** 109052 (https://doi.org/10.1016/j.lwt.2020.109052)

Quality exploration of cocoa pod husk as an antioxidant source in powdered milk: effects on physicochemical properties

Premy Puspitawati Rahayu[1]*, Ria Dewi Andriani[1], Dwi Setiawan[2], and Citra Nurma Yunita[1]

[1] Animal Product Technology, Faculty of Animal Science, Universitas Brawijaya Malang, East Java, 65145, Indonesia
[2] Department of Biosystems Engineering, Faculty of Agricultural Technology, Universitas Brawijaya, Malang 65145, Indonesia

*E-mail: premypuspita@ub.ac.id

Abstract. This study investigated the potential of incorporating cocoa pod husk (CPH) extract into powdered milk to enhance its nutritional value and stability. The research aims to increase the added value of cocoa by-products, address the susceptibility of powdered milk to degradation during processing, and enhance the stability of powdered milk during storage. A completely randomized experimental design was employed with four different extract cocoa pod husk (CPH) concentrations (0% (control), 1.5%, 2%, and 2.5%) tested in triplicate. The evaluation included emulsion stability, emulsion activity, foaming capacity, sedimentation, water content, water activity (Aw), total polyphenol content, functional groups, and microstructure. The results revealed that powdered milk with a 2.5% CPH extract addition significantly improved casein stability, decreased Aw, and moisture. Additionally, the interaction between casein-catechin and other proteins in milk altered the resulting microstructure. These complex interactions impacted the microstructure of milk powder. The functional groups of bioactive compounds were detectable via FTIR analysis. These findings suggest that incorporating 2.5% CPH extract enhances powdered milk's stability and nutritional content, with casein effectively serving as a catechin carrier. This research underscores the potential of using cocoa by-products to improve the quality and functionality of powdered milk.

Keywords: nutraceutical, fortification, casein, whey protein, polyphenol

1. Introduction

Food products will always develop along with technological advances to produce quality products. Powdered milk is one of the processed milk products that experience a decrease in nutrition during the processing process so certain compounds are often added to increase its nutritional value, such as docosahexaenoic acid (DHA), arachidonic acid (AHA), vitamins, calcium and bioactive compounds that can maintain physicochemical properties, stability and have good benefits for health. Nutraceutical food is a product that provides more benefits and plays a role in providing nutrition and health [1].

Bioactive compounds can be obtained from plants including CPH, tea leaves, grapes, and gambir which are sources of polyphenols. Polyphenols are one of the secondary metabolite products derived from cocoa CPH by-products that have health benefits [2]. The free radical capacity contained in these compounds can reduce the risk of chronic diseases such as cancer, coronary heart disease, and stroke.

Cocoa production 2017 in East Java was 31,262 tons per year with an area of 34,734 Ha with the highest production in Trenggalek Regency, followed by Kediri and Blitar Regencies. CPH is a by-product that is used as animal feed. Cocoa fruit consists of 74% fruit skin, 2% placenta, and 24% seeds. The nutritional content of cocoa skin includes dry matter (DM) 88% crude protein (CP) 8%, crude fiber (CF) 40.1%, total digestible nutrients (TDN) 50.8%, and contains bioactive polyphenol compounds. Based on the data above, the total by-product of CPH was 23,133.88 tons in 2015 [3].

CPH can be extracted using the right method to produce polyphenols that can be used to make nutraceutical food products. Extraction can be done using the microwave-assisted extraction (MAE) method with microwaves [4]. Extraction is done by heating the solvent which causes the cell walls of the tissue to be damaged so that the bioactive compounds come out of the tissue and dissolve in the solvent.

The CPH contained polyphenols. Polyphenols are one of the bioactive compounds in plants. Polyphenols consist of several subclasses including flavonoids, isoflavones, flavanones, anthocyanidins, catechins, xanthones, and others. Polyphenols can interact with milk proteins and can increase milk stability due to the hydrophobic interaction of polyphenol components with hydrophobic proteins to form hydrogen bonds. Catechins easily interact with proline-rich proteins because proline has an open structure. The percentage of casein in cow's milk is 80% of the total milk protein which contains a lot of proline, is spread throughout the amino acid series, and has a relatively open structure so that cocoa skin catechins can interact with casein [5].

Milk is an ideal medium for delivering bioactive compounds to obtain new-generation milk products that have more benefits for human health, namely by adding polyphenol compounds from cocoa skin. Adding these compounds is expected to maintain powdered milk's stability, physicochemical, and organoleptic properties and maintain its nutrition and targets for health. Based on the explanation above, research is needed to obtain the right polyphenol extract to be applied in powdered milk.

2. Material and Methods

2.1 Materials

The materials used in this study were skim milk, and CPH (F1 Lindak hybrid) obtained from Blitar chocolate village. The chemicals needed include gallic acid, distilled water, Folin reagent, saturated sodium carbonate, KBr powder, Folin Ciocalteau reagent, Na_2CO_3.

The research equipment used was a Sharp Model R - 222Y (S) microwave with a frequency of 2450 MHz materials with a power setting button ranging from medium to high power levels, Mini spray dryer (Buchi), beakers, measuring cups, analytical balances, infrared spectrophotometers, UV Vis spectrophotometers (Hitachi 2900), automatic micropipettes, Whatman filter paper No. 4, bottles, water baths, refrigerators, centrifugators, rotary evaporator, infrared spectrophotometers and PSA (Particle Size Analyzer), Zetasizer Nano Serines Software Version 7.01, Malvern Instrument, SEM (Scanning Electron Microscopy).

IOP Conf. Series: Earth and Environmental Science **1460** (2025) 012039 doi:10.1088/1755-1315/1460/1/012039

2.2 Methods

2.2.1 Research Design

This study on powdered milk production employed an experimental approach using a completely randomized design, incorporating four treatments with varying concentrations of CPH extract additions (0% as a control, 1.5%, 2%, and 2.5%), each replicated three times. The results of data analysis that were significantly different or very significantly different were continued with the Duncan Multiple Range Test (DMRT). The evaluation included an assessment of emulsion stability, emulsion activity, foaming capacity, sedimentation, water content, water activity (Aw), total polyphenol content, functional groups, and microstructure.

2.2.2 Procedure CPH Extraction

The CPH extract was prepared using the MAE method [6] (Sharp Model R-222Y(S)) with slight modifications at a high power setting and a temperature of 70°C. A ratio of 3 g of CPH powder to 100 mL of pure ethanol was used, and the mixture was left to soak for 24 hours. Extraction was performed in a microwave oven for 10 minutes, with cycles of one minute on and two minutes off to prevent overheating. The resulting crude extract was filtered through Whatman filter paper no. 4. Ethanol was removed using a rotary evaporator set at a pressure of 55 mmHg and a temperature of 50°C. The extract was then combined with 2-propanol, and the mixture was filtered again using Whatman filter paper no. 4. Finally, the filtrate was concentrated with a rotary evaporator (IKA RV 10) to eliminate the solvent.

2.2.3 Procedure for making powdered milk

Casein was dissolved in 50 ml of pH 6.8 phosphate buffer at a temperature of 30-35°C for 10 minutes. Casein was added little by little while homogenized using a hand mixer for 1 minute. 2,5% CPH extract was added in the casein solution and homogenized using a stirrer for 5 minutes. Added the solution (casein and CPH extract) according to treatment 0; 1.5%; 2%; 2.5% into skim milk. Dried into powder form using a spray dryer set at an inlet temperature of 135°C and an outlet temperature of 75°C. Skim milk powder enriched with CPH extract was analyzed physicochemically.

2.2.4 Emulsion Stability and Activity

Emulsion stability and activity were measured according to [7] with modification. Soybean oil was added to the sample and then homogenized using a hand mixer for 1 min. Take 0.1 mL of sample and add 0.1% SDS-Page per 100 mL and stirred with a vortex for 1 min. Take 3 mL of sample and put it into a cuvette to be examined with a spectrophotometer at a wavelength of 500 nm. Record the absorbance value (A0). Wait for 10 minutes and perform a comparison test (A10). The results were measured using the formula:

$$\text{Emulsion Stability (\%)} = \frac{A10 \times 100}{A0}$$

Where, A= Absorbance, A0= A_{500} at time of 0 menit, A10= A_{500} at time of 10 menit

$$\text{Emulsion Activity } (m^2/g) = \frac{2 \times 2{,}303 \times A0 \times DF}{1 \times \phi \times C}$$

Where; A= Absorbance, A= A_{500}, DF= dilute factor (100), I= path length of cuvette (m), ϕ=oil volume fraction, C= sample concentration

2.2.5 Foaming Stability

The foaming power of the samples was determined according to the method described by Zhao et al. [8]. An aliquot of 20 mL (V) of the protein solution was homogenized using a homogenizer (T18BS25, IKA, Germany) at 10,000 rpm for 1 min to obtain foam. The foam volume was measured immediately (V0) and after 10 min (V10). The results were measured by the formula:

$$\text{Foaming Capacity (\%)} = \frac{V10 \times 100}{V0}$$

Where, V= volume at the time of 0 menit, V10= volume at time of 10 menit

2.2.6 Sedimentation

Sedimentation Index testing was carried out based on the method [9] with several modifications. A fiveteen ml sample was prepared and placed in a centrifuge tube. The suspension was centrifuged at a speed of 6000 rpm for 25 minutes using a centrifuge. The supernatant was separated from the sediment by slowly pouring the contents of the centrifuge tube and positioning the centrifuge tube upside down for 30 minutes until only the sediment was left attached to the walls of the centrifuge tube. The sedimentation index was calculated based on the weight of the sediment (A) divided by the weight of the initial solution (B), displayed in% (w/w).

$$\text{Sedimentation (\%)} = \frac{A}{B}$$

Where, A= weight of the sediment, B= weight of the initial solution

2.2.7 Water Content

The water content can be determined by looking at the difference in weight of the material before and after heating [10]. A porcelain cup is heated in an oven at a temperature of 100-150°C for 1 hour. Then the porcelain cup is removed and cooled in a desiccator, then weighed until the weight is constant (b). 2 g of powdered milk is put into a porcelain cup, then weighed (a). Then the sample and porcelain cup are put in an oven and heated at a temperature of 100-105°C. After 8 hours, it is removed and cooled in a desiccator. Then weighs until the weight is constant (c). Calculation with the formula :

$$\text{Water Content (\%)} = \frac{a - \{(b + c) - b\}}{a} \times 100$$

Where,

a = weight of the porcelain cup + sample before heating (g),
b = weight of the dry paper,
c = weight of the porcelain cup + sample after heating (g)

2.2.8 Water Activity (Aw)

Aw testing uses a calibrated Aw meter. The sample is inserted into the sample chamber, then the chamber is placed in the Aw meter and the automatic reading of the results will appear on the device display. The test results that appear on the device display are recorded [11].

IOP Conf. Series: Earth and Environmental Science **1460** (2025) 012039 doi:10.1088/1755-1315/1460/1/012039

2.2.9 Phenolic content

Prepared 2 mL of sample extract solution. After that, 5 mL of distilled water and 0.5 mL of 10% Folin Ciocalteu reagent were added. The blue solution was produced through Folin-Ciocalteau Reagent, which is useful for reducing or reducing hydroxyl groups in polyphenols. Then the solution was homogenized with a vortex and left for about 5 minutes. After that, 1 mL of 7% Na2CO3 was added and left for 60 minutes so that the solution was alkaline so that the Folin reagent and polyphenols could react optimally. Sample absorbance measurements were measured at 765 nm [12].

2.2.10 Functional Groups

Functional group observation was carried out using Fourier Transform InfraRed (FTIR) [13]. Infrared spectra were obtained using an FTIR spectrometer (Impact 420 model, Digilab, Mississauga, ON, Canada). One milligram of casein, with or without CPH extract treatment, was dried in a vacuum desiccator, finely ground, and thoroughly mixed with 200 mg of oven-dried KBr powder (Merck, DAC, USP). The mixture was pressed into a transparent disc using a mold. The spectra were examined over a range of 4000–600 cm^{-1} with a resolution of 2 cm^{-1} and 100 scans.

2.2.11 Microstructure

The surface microstructure of various powder samples produced was observed using (SEM-model SU8010, Hitachi High-Technologies Canada, Inc Toronto, ON, Canada) [14]. Electron microscopy was used to characterize the surface morphology of control casein and casein treated with catechin. The procedure. powder samples were placed on double-sided adhesive glass objects, which were mounted on aluminum discs to avoid water absorption. The samples were then coated with carbon for ± 30 seconds. The samples used a BAL-TEC SCD 005 sputter coater (Bal-Tec Co., Balzers, Liechtenstein). The samples were then observed using SEM with a voltage of 15 kV and a magnification of up to 5000 x.

3. Result and Discussion

This study explores various physicochemical parameters of powdered milk with fortified CPH as an antioxidant source. The Table 1. presented the results of measurements for emulsion stability, emulsion activity, foaming capacity, sedimentation, water content, and water activity (Aw). This data provides insight into the effects of CPH addition on the physicochemical quality of

Table 1. Result of Emulsion Stability, Emulsion Activity, Foaming Capacity, Sedimentation, Water Content, and Water Activity (Aw)

Parameters %	Treatment			
	T0	T1	T2	T3
Emulsion Stability	33.37±12.28^{a}	81.62±12.10^{b}	84.30±6.68^{b}	85.35±13.88^{b}
Emulsion Activity	4.34±0.25^{b}	1.49±0.09^{a}	1.26±0.05^{a}	1.16±0.15^{a}
Foaming Capacity	37.50±9.57^{a}	40±8.17^{b}	45±5.77^{c}	52.50±5.00^{d}
Sedimentation	1.40±0.12^{a}	1.69±0.51^{ab}	2.51±0.79^{ab}	2.77±0.32^{b}
Water Content	8.81±1.75^{b}	4.27±0.86^{a}	4.58±1.20^{a}	6.88±0.62^{a}
Water Activity (Aw)	0.44±0.08^{b}	0.35±0.01^{a}	0.33±0.04^{a}	0.37±0.01^{a}

^{a,b} Different uppercase letters in the same row indicated significant effect ($P < 0.05$).
T0: control (without CPH), T1: 1.5% CPH extract, T1: 2% CPH extract, T1: 2.5% CPH extract.

the product, serving as a reference to evaluate the potential of CPH as a functional additive in the powdered milk industry.

3.1 Emulsion Stability

The interaction between phenolic compounds and proteins significantly influenced emulsion stability. Among the treatments, adding 2.5% CPH extract to powdered milk resulted in the highest emulsion stability value of 85.35%. This improvement was attributed to the binding of milk proteins with polyphenols from the CPH extract, which enhanced emulsion stability [15]. The polyphenols in the CPH extract interacted with milk proteins, such as casein and whey, through hydrogen bonds, hydrophobic interactions, and covalent bonds. These protein-polyphenol complexes altered the structure and surface properties of proteins, improving their emulsifying capabilities. This was consistent with Wang et al. [16], who reported that proteins interacting with phenolic compounds formed stable emulsions due to non-covalent interactions, including hydrogen bonds, hydrophobic interactions, and Van der Waals forces.

3.2. Emulsion Activity

Based on Table 1, it can be seen that the magnitude of emulsion activity ranges from 1.16% to 4.34%. The addition of CPH extract in fairly high amounts (2.5%), can change the structure of casein protein in milk, which is an important component for emulsion activity. Changes in the structure of this protein can interfere with its ability to stabilize the emulsion so that the emulsion activity decreases. Polyphenols can interact with proteins through hydrogen bonds, hydrophobic interactions, and covalent bonds, which can result in changes in the secondary and tertiary structures of casein protein [17]. Changes in the structure of casein protein cause its ability to act on the emulsion to be disrupted or decreased.

3.3 Foaming Capacity

Based on Table 1, it can be seen that the foaming power ranges from 37.50% to 52.50%. The treatment of adding CPH extract to powdered skim milk by 2.5% has the highest foaming power value of 52.50% (T3) compared to other treatments. This is because the higher the percentage of polyphenols from CPH extract added to casein, the higher the foaming power, because polyphenols strengthen the interaction with casein protein, thus forming a stronger protective layer around the air bubbles. The foaming ability of casein added with polyphenols can encourage bubble formation so that the foaming power formed will increase [18]. Casein as a protein is amphiphilic which has the ability to reduce surface tension at the air-liquid interface, thus encouraging the formation of air bubbles. The addition of CPH extract polyphenols causes interactions between polyphenols and proteins which can increase the stability of the bubble interface. The control treatment without CPH extract (T0) produced lower foaming power compared to the treatment involving the addition of polyphenol, because the protein formed on the bubble surface was thinner and less stable, so the bubbles tended to break more easily and did not form a lot of foam. Based on research by [19], the foaming power of casein increased from 81.25% to 89.12% with the addition of 100 μ mol/L gallic acid. Gallic acid is a type of polyphenol compound that belongs to the phenolic acid group.

3.4 Sedimentation

Based on Table 1, the sedimentation rate ranges from 1.40% to 2.77%. The treatment involving the addition of 2.5% CPH extract to powdered skim milk resulted in the highest sedimentation value, at 2.77% (T3). Adding CPH extract to powdered milk increases the sedimentation value, as it can significantly impact the stability of the powdered milk emulsion system. A high sedimentation value indicates greater phase separation in the emulsion, which can occur due to altered interactions between casein proteins and other components within the emulsion system [20]. The added polyphenols may change the surface properties of particles in the emulsion, reducing the repulsion between particles, leading them to aggregate, which increases sedimentation. Higher polyphenol levels in the treatment can reduce emulsion stability and increase sedimentation. In the control treatment without CPH extract (T0), lower sedimentation values were observed compared to treatments with polyphenols. This is because the casein protein in powdered milk retained its function as an effective emulsion stabilizer. In the absence of polyphenols, there is no disruption to the structure or surface properties of casein, allowing casein to form a protective layer around fat particles. This layer maintains particle repulsion and even distribution within the emulsion, resulting in low sedimentation [21].

3.5 Water Content

Based on Table 1. it can be seen that the water content of 4.27% to 8.81%. The control treatment without the addition of CPH extract to the powdered milk has the highest water content, at 8.81% (T0). Adding CPH extract to the powdered milk decreases the water content because polyphenols exhibit hydrophobic properties, meaning they repel water. This hydrophobic nature of polyphenols limits water binding, reducing the amount of free water in the system. The interaction between polyphenols and proteins can form a denser network, which makes it more difficult for water to be trapped in the matrix. Unbound water or free water decreases, and finally, the total water content in powdered milk becomes lower [22]. The addition of CPH extract to powdered milk can reduce free water and increase the total solid content so that the total water content in powdered milk becomes lower. This is by research conducted that the drying process or production of powdered milk plays a role in reducing water content, but with the presence of polyphenols, the reduction in water content can be even more significant due to changes in the physical and chemical structure of the product [23].

3.6 Water Activity (Aw)

Based on Table 1. it can be seen that Aw ranges from 0.33% to 0.44%. The control treatment without the addition of CPH extract to powdered milk has the highest water content value of 8.81% (T0). The addition of CPH extract to powdered milk caused Aw to be lower when compared to the control treatment without the addition. This occured because the addition of polyphenols from CPH to powdered milk causes the water activity (Aw) to be lower compared to the control treatment without the addition of polyphenols. After all, polyphenols have hydrophobic properties (do not easily interact with water). Water activity (Aw) is a measure of the free water available to support the growth of microorganisms and the occurrence of chemical reactions in food products. The added polyphenols will interact with proteins and fats so that free water that is not bound in the product matrix is reduced. Polyphenols can change the physical and chemical structure of milk components, such as casein, and cause stronger bonds between milk components and water, resulting in less free water being available, thus reducing the Aw value [24].

IOP Conf. Series: Earth and Environmental Science **1460** (2025) 012039 doi:10.1088/1755-1315/1460/1/012039

3.7 Phenolic content

Table 2. Phenolic content

Treatment	Phenolic content (mg/kg)
T_0	96.378
T_1	672.181
T_2	1067.219
T_3	1262.058

T0: control (without CPH), T1: 1.5% CPH extract, T1: 2% CPH extract, T1: 2.5% CPH extract.

Based on Table 2 it can be seen that the phenolic content ranges from 96.378 – 1262.058 mg / kg. The control treatment without the addition of CPH extract to powdered milk had the lowest value of 96.378 mg / kg (T0). The addition of CPH extract to powdered milk caused the phenolic content to increase when compared to the control treatment. This occured because the CPH extract is absorbed into the milk protein, which causes an increase in total polyphenols. CPH extract had a rich content of polyphenols, especially compounds such as flavonoids, catechins, and phenolic acids. CPH extract added in the manufacture of powdered milk, these compounds are also absorbed into the milk matrix, causing a significant increase in total polyphenols [25]. A higher percentage of CPH extract addition is directly correlated with an increase in the total polyphenol content in the final product. The increase in total polyphenol content from the control treatment (without the addition of polyphenols) to the treatment with the addition of CPH extract by 2.5% resulted in a drastic increase. The treatment of adding CPH extract of 2.5% had the highest total phenol content value of 1262.058 mg/kg. The increase in phenolic content was related to the increase in the antioxidant potential of powdered milk. Polyphenols had antioxidant properties that can protect products from oxidative damage, so that in addition to increasing the total phenol content, it also extends the shelf life and stability of the product [26].

3.8 Functional Group

Analysis of functional groups of powdered milk added with polyphenols with various concentration treatments aimed to determine the functional groups that refer to the presence of compounds in powdered milk.

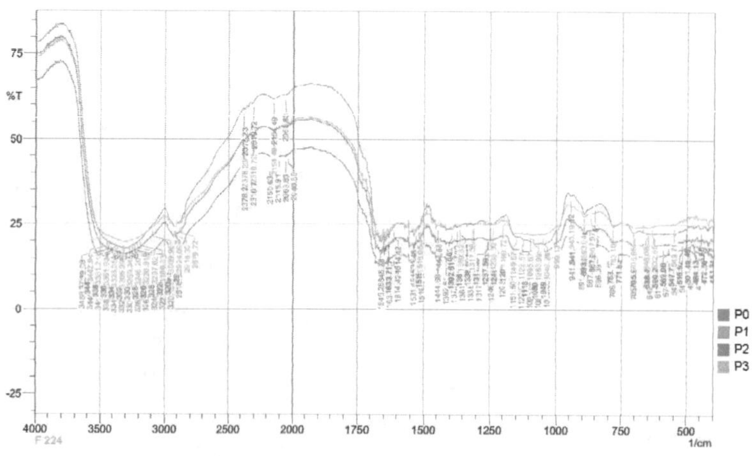

Figure 1. Functional Group Analyzed

T0: control (without CPH), T1: 1.5% CPH extract, T1: 2% CPH extract, T1: 2.5% CPH extract.

Based on the data obtained, it can be seen that there are several groups that begin to disappear during the processing of powdered milk. Treatments T0, T1 and T2 had groups that disappear, namely the C=C Stretch group and the O-H phenol number that disappears, but in treatment T3 with the addition of 2.5% CPH extract, the group is still visible. This can be caused by heat treatment in the processing process. The number group can change by time and temperature factors used during the process [27]. T3 is the best treatment in terms of observing functional groups. This is because the absorption area for important groups is still detected compared to other treatments.

Table 3. Results of observations of polyphenol functional groups in powdered milk using FTIR

CPH extract	T_0	T_1	T_2	T_3	Functional Group	Chemical Compound
	850.61	856.39	867.97	867.97	N-H	Amina
952.84	945.12	941.26	941.26	943.19	C-H	Alkana
				999.13	C-H	Alkana
1037.7	1045.42	1056.99	1049.28	1049.28	C-O stretch	Alkohol, ester, asam karboksilat
1228.88	1249.87	1246.02	1246.02	1238.30	C-N Strech	Aliphatic amines
	1311.59	1317.38	1311.59	1317.38	N-H	amida III
1456.26	1519.91	1519.91	1519.91	1516.05	C-C Strech (in ring)	Aromatik
1712.79		2150.63	2150.63	2154.49	C=O Strech	Karbonil
1766.8	2310.72	2310.72	2310.72	2310.72	C=C Strecth	Aromatik
2472.74	2378.23	2378.23	2380.16	2378.23	N-H	Amida A
					C-H	regangan
2883.58				2879.72	C-H Strech	Alkanes
	2918.30	2918.30	2918.30	2918.30	C-H Strech	Alkana
2970.38				2924.09	O-H. H-Bonded	Fenol, alkohol
	3207.62	3207.62	3209.55	3207.62	N-H	Amida
3344.57-3352.28	3439.08	3441.01	3442.94	3442.94	N-H	Amida

T0: control (without CPH), T1: 1.5% CPH extract, T1: 2% CPH extract, T1: 2.5% CPH extract.

3.9 Microstructure

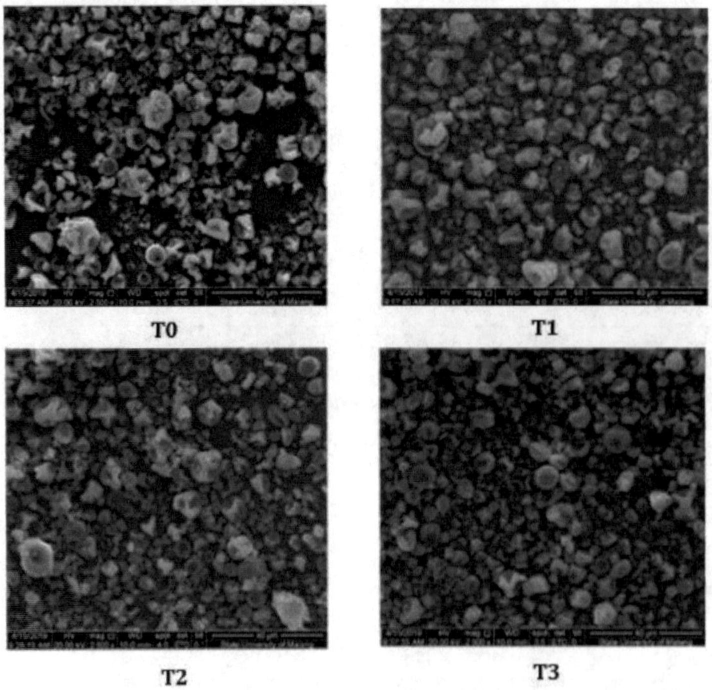

Figure 2. Microstructure at 2500x magnification
T0: control (without CPH), T1: 1.5% CPH extract, T1: 2% CPH extract, T1: 2.5% CPH extract.

Samples T0 to T3 had increasingly uniform microstructures as the amount of CPH extract added increased. T3 formed the most uniform granule size among the others. The resulting microstructure had an increasingly uniform size, the bonds formed was not intensive, because polyphenols bind to other proteins in milk, have a porous particle surface so that the solubility is high when the milk powder is dissolved in water. The interaction between casein-catechin with other proteins in milk can change the resulting microstructure. The complex interactions formed can affect the microstructure of milk powder. Milk protein contains a lot of proline which has an open structure so that it can form bonds with polyphenol components through hydrophobic interactions [28]. Casein had self-assembly properties so that when added to milk it can bind to other proteins causing its structure to become more complex. The complex formed nanoparticle sizes due to the role of phenolic OH in binding amino acids in casein [29].

4. Conclusion

It may be inferred that adding 2.5% CPH extract improves the stability and nutritional value of powdered skim milk, with casein acting as a phenolic carrier. The potential use of cocoa byproducts has been demonstrated to improve the quality and functionality of powdered skim milk.

References

[1] Olaiya, C. O., Soetan, K. O., and Esan, A. M. (2016). The role of nutraceuticals, functional foods and value add food products in the prevention and treatment of chronic diseases. *African Journal of food science*, 10(10): 185-193. https://doi.org/10.5897/AJFS2015.1402

[2] Pacyga, K., Pacyga, P., Topola, E., Viscardi, S., and Duda-Madej, A. (2024). Bioactive Compounds from Plant Origin as Natural Antimicrobial Agents for the Treatment of Wound Infections. *International Journal of Molecular Sciences*, 25(4), 2100. https://doi.org/10.3390/ijms25042100

[3] Haryono, D., Soetriono, S., Hartadi, R., and Aji, M. M. (2010). Analisis Daya Saing dan Dampak Kebijakan Pemerintah Terhadap Produksi Kakao di Jawa Timur. *JSEP (Journal of Social and Agricultural Economics)*, 5(2): 72-82. https://jurnal.unej.ac.id/index.php/JSEP/article/view/375/233

[4] Rafiee, Z., Jafari, S. M., Alami, and Khoimeri, M. (2011). Microwave asisted extraction of phenoliccompounds from olive leaves; a comparison with maceration. *Journal Animal and plant Sci*, 21(4): 738-745. https://thejaps.org.pk/docs/21-4/33.pdf

[5] Mulyatni, A. S., Budiani, A., dan Taniwiryono, D. (2012). Aktivitas antibakteri ekstraks kulit buah kakao terhadap *Escherchia coli, Bacillus subtilis*, dan *Staphylococcus aurens*. *Menara Perkebunan*. 80 (2): 77-84. https://doi.org/10.22302/iribb.jur.mp.v80i2.39

[6] Rahayu, P. P., Radiati, L. E., and Manab, A. (2015). Physico chemical properties of whey protein and gelatine biopolymer using tea leaf extract as crosslink materials. *Current Research in Nutrition and Food Science Journal*, 3(3): 224-236. https://dx.doi.org/10.12944/CRNFSJ.3.3.06

[7] Arioui F, Ait-Saada, D., and Cheriguene, A. (2017). Physicochemical and sensory quality of yogurt incorporated with pectin from peel of citrus sinensis. *Food Science & Nutrition.* 5(2):358-364. https://doi.org/10.1002/fsn3.400

[8] Zhao, X., Fan, X., Shao, X., Cheng, M., Wang, C., Jiang, H., and Yuan, C. (2022). Modifying the physicochemical properties, solubility and foaming capacity of milk proteins by ultrasound-assisted alkaline pH-shifting treatment. *Ultrasonics Sonochemistry*, 88: 106089. https://doi.org/10.1016/j.ultsonch.2022.106089

[9] Octaviani, P. A., Pramono, Y. B., and Pratama, Y. (2020). Karakteristik Fisik dan Sensoris Rice Milk Malt Beras Merah dengan Penambahan Bahan Penstabil dari Jenis yang Berbeda. *Jurnal Teknologi Pangan*, 4(1): 17-22. https://doi.org/10.14710/jtp.2020.23501

[10] Zuhra, Z., Sofyana, S., dan Erlina, C. (2012). Pengaruh kondisi operasi alat pengering semprot terhadap kualitas susu bubuk jagung. *Jurnal Rekayasa Kimia dan Lingkungan*, 9(1): 36-44. https://jurnal.usk.ac.id/RKL/article/view/252/238

[11] Rizqiati, H., Nurwantoro, N., Febrisiantosa, A., Shauma, C. A., dan Khasanah, R. (2020). Pengaruh isolat protein kedelai terhadap karakteristik fisik dan kimia kefir bubuk. *Jurnal Pangan dan Agroindustri*, 8(3): 111-121. https://doi.org/10.21776/ub.jpa.2020.008.03

[12] Shaidah, S. N., Afriliana, A., Wibowo, Y., Soekarno, S., dan Wiyono, A. E. (2024). Karakterisasi Roti Bolu Manis Berbasis Mocaf (*Modified Cassava Flour*) Terkomposit Serbuk Daun Kelor Sebagai Sumber Antioksidan Alami. *Jurnal Teknologi Pertanian*, 25(2): 175-194. https://doi.org/10.21776/ub.jtp.2024.025.02.7

[13] Zouari, A., Schuck, P., Gaucheron, F., Triki, M., Delaplace, G., Gauzelin-Gaiani, C., Lopezb, C., Attiaa, H., and Ayadi, M. A. (2020). Microstructure and chemical composition of camel and cow milk powders' surface. *Journal LWT*, 117 : 108693. https://doi.org/10.1016/j.lwt.2019.108693

[14] Čurlej, J., Zajác, P., Čapla, J., Golian, J., Benešová, L., Partika, A., Feher and Jakabová, S. (2022). The effect of heat treatment on cow's milk protein profiles. *J. Foods*, 11(7) : 1023. https://doi.org/10.3390/foods11071023

[15] Rahayu, P., Andriani, R. D., Manab, A., and Sawitri, M. E. (2022). Characteristics of whey protein-phylantin as abased for the development of immunomodulatory products. *Asian Food Science Journal*, 21(1): 12-20. https://doi.org/10.9734/afsj/2022/v21i130396

[16] Wang, X., and Zhao, Z. (2022). Improved encapsulation capacity of casein micelles with modified structure. *Journal of Food* Engineering, 333: 111138. https://doi.org/10.1016/j.jfoodeng.2022.111138

[17] Rodríguez, S. D., Von-Staszewski, M., and Pilosof, A. M. R. (2015). Green tea polyphenolswhey proteins nanoparticles: Bulk, interfacial and foaming behavior. *Food Hydrocolloids*, 50: 108–115. https://doi.org/10.1016/j.foodhyd.2015.04.015

[18] Chen, L., Chen, N., He, Q., Sun, Q., Gao, M. R., and Zeng, W. C. (2022). Effects of different phenolic compounds on the interfacial behaviour of casein and the action mechanism. *Food Research International*, 162: 112110. https://doi.org/10.1016/j.foodres.2022.112110

[19] Ye, A. (2021). Gastric colloidal behaviour of milk protein as a tool for manipulating nutrient digestion in dairy products and protein emulsions. *Food Hydrocolloids*, N 115: 106599. https://doi.org/10.1016/j.foodhyd.2021.106599

[20] Holkar, C. R., Jadhav, A. J., and Pinjari, D. V. (2019). A critical review on the possible remediation of sediment in cocoa/coffee flavored milk. *Trends in Food Science & Technology*, 86: 199-208. https://doi.org/10.1016/j.tifs.2019.02.035

[21] Valko, M., Leibfritz, D., Moncol, J., Cronin, M. T., Mazur, M., and Telser, J. (2007). Free radicals and antioxidants in normal physiological functions and human disease. *The international journal of biochemistry & cell biology*, 39(1): 44-84. https://doi.org/10.1016/j.biocel.2006.07.001

[22] George, S., Thomas, A., Kumar, M. V. P., Kamdod, A. S., Rajput, A., and Abdullah, S. (2023). Impact of processing parameters on the quality attributes of spray-dried powders: a review. *European Food Research and Technology,* 249(2): 241-257. https://doi.org/10.1007/s00217-022-04170-0

[23] Wijegunawardhana, D., Wijesekara, I., Liyanage, R., Truong, T., Silva, M., and Chandrapala, J. (2024). Process-Induced Molecular-Level Protein–Carbohydrate–Polyphenol Interactions in Milk–Tea Blends: *A Review.* Foods, 13(16): 2489. https://www.mdpi.com/2304-8158/13/16/2489

[24] Campos-Vega, R., Nieto-Figueroa, K. H., and Oomah, B. D. (2018). Cocoa (Theobroma cacao L.) pod husk: Renewable source of bioactive compounds. *Trends in Food Science & Technology*, 81: 172-184. https://doi.org/10.1016/j.tifs.2018.09.022

[25] Akca, S., and Akpinar, A. (2021). The effects of grape, pomegranate, sesame seed powder and their oils on probiotic ice cream: Total phenolic contents, antioxidant activity and probiotic viability. *Food Bioscience*, 42: 101203. https://doi.org/10.1016/j.fbio.2021.101203

[26] Dezhampanah, H., and Firouzi, R. (2017). An Investigation on intermolecular interaction between Bis (indolyl) methane and HSA and BSA using multi technique methods. *Journal of Biomolecular Structure and Dynamics*, 35(16). https://doi.org/10.1080/07391102.2016.1264890

[27] Mao, T., Wescombe, P., and Mohan, M. S. (2023). Predominance of non-covalent interactions of polyphenols with milk proteins and their health promoting properties. *Critical Reviews in Food Science and Nutrition*, 1-23. https://doi.org/10.1080/10408398.2023.2245037

[28] Li, Y. X., Du, X., and Zhao, Z. (2024). Structural Properties of β-Casein and Its Application as a Nano-Carrier of Bioactive Compounds: A Recent Update. *Food Reviews International*, 1-11. https://doi.org/10.1080/87559129.2024.2384633

[29] Yildirim-Elikoglu, S., and Erdem Y. K. (2017). Interactions between milk proteins and polyphenols: Binding mechanisms, related changes, and the future trends in the dairy industry. *Food Reviews International*. 34(7):665-697. https://doi.org/10.1080/87559129.2017.1377225

The Effects of Heat Treatment on the Stability of Oil-in-Water Nanoemulsions Stabilized by Nanocellulose

Ichlasia Ainul Fitri[1*], Chatarina Lilis Suryani[1], Amelia Murtisari[2]

[1] Department of Agricultural Product Technology, Faculty of Agroindustry, University of Mercu Buana Yogyakarta, Yogyakarta, Indonesia

[2] Department of Agribusiness, Faculty of Agriculture, Universitas Negeri Gorontalo, Gorontalo, Indonesia

*E-mail: ichlasia@mercubuana-yogya.ac.id

Abstract. Oil-in-water nanoemulsions are created by combining oil and water into nanosized mixtures. The stability of these emulsion systems can be significantly affected by processing and storage under uncontrolled temperatures. Investigating the effects of heat treatment and stabilizers on nanocellulose-stabilized oil-in-water nanoemulsions is essential. This study involves three different heating temperatures: 60°C, 70°C, and 80°C. Furthermore, the nanocellulose concentration was altered to 0%, 0.10%, and 0.20%. The findings indicated that elevated temperatures combined with reduced concentrations of nanocellulose diminish the lightness of emulsions. Nevertheless, the distinctions are invisible to human sight. At concentrations 0.20% and 70°C, particle size, and enhanced particle charge were reduced in oil-in-water nanoemulsions, indicating greater electrostatic repulsions. Viscosity escalates with elevated concentrations of the nanocellulose mixture and temperature. Consequently, the creaming index diminished, and the emulsion attained more stability. The amalgamation of NCC and NFC, particularly at 0.20%, serves as a natural stabilizer, enhancing the strength of o/w nanoemulsion to a temperature of 70°C.

Keywords: nanoemulsions, nanocrystalline cellulose, nano fibrillated cellulose, heat treatment

1. Introduction

An emulsion is a formulation of polar and non-polar solution, one of the phases is distributed finely and uniformly disseminated within the other, typically stabilized by an emulsifying agent. The oil-in-water phase is extensively utilized in the food sector. Oil-in-water emulsions include margarine, mayonnaise, milk, cream, and coconut milk [1]. Nonetheless, emulsion systems are typically unstable. Consequently, emulsifiers and stabilizers are required [2].

Appropriate and adequate emulsifiers and stabilizers are essential for stabilizing the emulsion. Mollet & Grubenmann [3] assert that emulsification requires chemical and physical energy. The emulsification process employs chemical energy by adding an emulsifier, whereas the physical process may utilize temperature or the pace of emulsification. Various emulsifiers exist, including CMC, Tween 20, and Tween 80. The purpose of using an emulsifier is to lower the surface tension from the dispersed and dispersing phases [4]. Numerous emulsifiers and stabilizers exist; nevertheless, utilizing natural components, such as nanocellulose, is advisable.

Nanocellulose is categorized as a natural fiber derived from cellulose. Nanocellulose serves as an emulsifier or stabilizer due to its lightweight, biodegradable nanofiber characteristics, including a low density of around 1.6 g/cm3 and exceptional tensile properties. Nanocellulose is categorized into three distinct types according to its structure. Initially, acid hydrolysis often isolates Nanocrystalline Cellulose (NCC) from cellulose fibrils. The amorphous component is hydrolyzed and eliminated by acid, whereas the crystalline component persists. NCC possesses a width from 3 to 20 nm [5]. Secondly, Nanofibrillated Cellulose (NFC) consists of elongated, flexible, and intertwined nanocellulose that can be generated from cellulose fibrils through mechanical processes. Typically, the diameter is under 100 nm [5]. Third, bacterial cellulose is derived from bacteria, exhibiting a width of around 2-20 nm[6].

Prior research utilized nanocellulose as a stabilizer in an o/w emulsion system, employing only a singular variety of nanocellulose. Research conducted by [7] indicates that nanocrystalline cellulose enhances the stability of the emulsion system. Similarly, using NFC nanocellulose, as shown in [8], stabilized emulsions can diminish the o/w interfacial tension, reducing emulsion droplet diameter and inhibiting coalescence. Limited research has been undertaken on emulsions that utilize heating temperature as a facilitator for the emulsification process.

Emulsions stabilized by nanocellulose exhibit structural stability, resulting in prolonged stability during storage. The emulsion exhibited commendable stability when nanocellulose without surface charge was employed as a stabilizer. Hopefully, it can be used to observe the optimal nanocellulose combination for enhancing the stability of o/w emulsions, given that most food items are subjected to heat treatment during processing.

2. Materials and methods

2.1 Materials

The equipment included Particle Size Analyzer (PSA) Model Microtac Nanotrac Wave II, a Colorimeter NH310 series, a Magnetic Stirrer, a Homogenizer IKA Model T25 Digital Ultra Turrax, a Viscometer DVE Brookfield series EB21BETA, a Glass Beaker, and a 20 ml Glass Bottle. This research utilized nanocrystalline cellulose (NCC) and nanofibrillated cellulose (NFC), both bought from Cellulose Lab Company, Canada. Tween 20 and sodium azide were acquired from the CV chemical store. Pratama Chem-Mix. Aquades. Palm oil was acquired at Indogrosir in Jogja.

2.2 Methods

The oil phase was effectively formulated by measuring 1 gram of soybean oil. Moreover, aqueous phase form from 1% (w/w) solution of Tween 20 and 0.01% (w/w) sodium azide to enhance its antimicrobial properties, and a ten mM sodium phosphate buffer solution was adjusted to a pH of 7. Then, it was efficiently carried out using a high-shear mixer for 2 minutes at ambient temperature (25 °C), ensuring a well-integrated mixture. To refine the emulsion, the coarse mixture underwent further processing with a shear mixer for 5 minutes at a power setting of 50%, employing a pulsed on/off cycle of 5 seconds throughout this time. Following this, the emulsion was allowed to equilibrate at room temperature (25 °C) for one hour.

2.2.1. Visual Creaming Stability

Fresh samples were promptly placed into transparent glass test tubes measuring 15 mm in diameter and 80 mm in height and sealed with plastic caps to ensure their integrity. To minimize external influences, the sample tubes were then maintained at a consistent 25 ◦C in a dark environment. Then, it systematically monitored the separation of the creaming layer at intervals of 0, 1, 3, and 7 days. This observation allowed us to calculate the creaming index, an important metric for assessing emulsion quality, using the following equation:

$$CI\ (\%) = \frac{Hs}{HT} \times 100 \quad (1)$$

Where HT is the total height of the emulsion in the tube and HS is the height of the creaming layer to be measured.

2.2.2. Color

The color measurement of the emulsion was carried out on day 0 using the L*, a*, and b* parameters with a 3NH NH310 colorimeter (Shenzhen Threenh Technology Co., Ltd, China). A 15 ml sample was transferred into an analysis vial, sealed, and then assessed using the colorimeter. In this context, L* represents brightness, while a* and b* denote color coordinates, with a* corresponding to red and b* to yellow. The following calculation quantifies the overall color variation:

$$\Delta E = \sqrt{(L^*_{ii} - L^*_{i})^2 + (a^*_{ii} - a^*_{i})^2 + (b^*_{ii} - b^*_{i})^2} \quad (2)$$

Where i is an emulsion sample without the addition of nanocellulose with varying heating temperatures, and ii is an emulsion sample with the addition of nanocellulose and varying heating temperatures at 60 °C, 70 °C, and 80 °C.

2.2.3. Particle Size

The particle size and distribution were analyzed by using the Mastersizer 2000 laser diffraction particle size analyzer (Malvern Instruments Ltd., Worcestershire, United Kingdom). Smples were thoughtfully diluted in a 10 mM phosphate buffer solution at pH 7.

2.2.4. ζ-Potential

The particle charges were analyzed by using a particle electrophoresis device (Zetasizer Nano ZS, Malvern Instruments Ltd., Worcestershire, United Kingdom). Furthermore, the sample dilution was conducted using 10 mM phosphate buffer at pH 7.

2.2.5. Apparent Viscosity Measurement

All samples' viscosity was analyzed using a Brookfield viscometer (PT Alfa Omega Indolab). First, a 50ml sample was added into a beaker glass, then analyzed using torch number 6.

2.2.6. Statistical Analysis

All findings from this study were presented as means and standard deviations. However, the significance differences continued by a one-way analysis of variance (ANOVA) followed by Duncan's multiple range test was employed, with significance set at $p<0.05$. The statistical analysis was performed using SPSS version 25 software.

3. Results and Discussions

3.1. Color

The emulsion sample () has a milky white hue and maintains a rather thin consistency due to the amalgamation of oil, water, Tween 20, sodium azide, and other nanocellulose components. Table 1 illustrates the impact of heating temperature on the concentration of nanocellulose with respect to color. The results indicated a substantial difference ($p<0.05$) in the L* values of samples at identical temperatures but varying concentrations, which diminishes as the concentration of additional nanocellulose increases. In addition to color features, the dimensions of the sample's oil particles might influence the emulsion's hue. L* is correlated with the particle size of the oil droplet. Emulsions comprise several oil droplet size ranges, resulting in differential scattering of light waves by each droplet size class, with larger droplets capable of absorbing lighter, thereby diminishing the L* value [9]. L* will diminish when NCC and NFC concentrations rise due to the increased viscosity of the emulsion at elevated NCC and NFC levels.

An increase in concentration is derived in a drop in the ɑ* value. At a temperature of 80° C, an increase in nanocellulose concentration correlates with a rise ɑ* value. For this ɑ* value, the optimal concentration is 0.20%, principally to ensure protection at a temperature of 70° C. As the concentration of nanocellulose increases, the b* value decreases. An increased concentration of nanocellulose in the sample resulted in a decrease in the b* color value, signifying less yellowness. The total color value ΔE exhibits an increase; when the concentration of nanocellulose rises, the total ΔE value also escalates. The nanocellulose concentration at 0.70° C has the highest value among all tested temperatures. An emulsion is considered stable if it has uniformly shaped droplets that retain their original color and resist degradation. Upon examining the values of the color parameters for concentration and temperature, a concentration of 0.20% at a temperature of 70° C is determined to be the most stable and effective for heating the emulsion.

Table 1. The effects of temperature and nanocellulose concentration on color (L*, ɑ*, b*) of oil-in-water emulsions.

Concentration	Temp (°C)	*L**	*ɑ**	*b**	*ΔE*
	60	56.83 ± 0.59[ef]	-0.85 ± 0.02[a]	1.62 ± 0.16[a]	-
0%	70	60.28 ± 0.41[g]	-0.74 ± 0.01[b]	1.92 ± 0.09[b]	-
	80	57. 11 ± 0.40[f]	-0.51 ± 0.01[fg]	1.74 ± 0.10[ab]	-
	60	55.08 ± 0.34[b]	-0.70 ± 0.09[bc]	1.83 ± 0.10[b]	1.82 ± 0.68[a]
0.10%	70	56.01 ± 0.24[cde]	-0.64 ± 0.02[cd]	1.87 ± 0.03[b]	4.27 ± 0.31[c]
	80	56.15 ± 0.10[de]	-0.47 ± 0.01[g]	2.22 ± 0.05[c]	1.38 ± 0.13[a]
	60	54.10 ± 0.24[a]	-0.56 ± 0.04[ef]	1.86 ± 0.14[b]	2.76 ± 0.37[b]
0.20%	70	55.13 ± 0.21[bcd]	-0.60 ± 0.05[de]	1.74 ± 0.03[ab]	5.16 ± 0.56[d]
	80	55.37 ± 0.56[bc]	-0.72 ± 0.07[bc]	1.78 ± 0.08[ab]	1.20 ± 0.42[ab]

Note: The test was carried out in three repetitions. The same column followed by different lowercase letters (a-g) indicates significant differences ($p<0.05$).

This is due to the structural characteristics of nanocellulose, which exhibits remarkable solubility in water or hydrophilic qualities, attributable to the abundance of hydroxyl groups capable of forming hydrogen bonds with water. However, this is inaccurate as nanocellulose is insoluble in water and also dissolves in various other solvents. The cause is the intricate and resilient structure of nanocellulose, along with the amorphous component, which is eliminated during the hydrolysis process by acid. This component contributes to the elevated crystallinity. The increased utilization of nanocellulose enhances the stability of the emulsion and facilitates the integration of nanocellulose and water, aided by the employed solvent [10] [11]. Additionally, the heating temperature element contributes to this, as elevated temperatures facilitate the complete dissolution of nanocellulose in the sample due to its excellent structural stability.

3.2. Particle size

Table 2 illustrates that the sample particle size at each temperature diminishes with increasing nanocellulose content. The incorporation of nanocellulose concentration may inhibit high-temperature flocculation, hence preventing the reduction of nanocellulose particle size within the emulsion. The diminutive particle size will yield a stable emulsion system, preventing any separation. Temperature affects the characteristics of the resultant water, the interfacial film, and the emulsifier's solubility in both oil and water. Various particle sizes were also achieved according to the formulation of nanocellulose concentration. At elevated concentrations, a reduction in particle size enhances the efficacy of nanocellulose in stabilizing the oil/water interface, attributable to its amphiphilic surface characteristics derived from the hydrophobic surfaces and hydrophilic edges of the cellulose chains. The diminutive particle size accelerates the occurrence of Brownian motion. The accelerated Brownian motion will inhibit sedimentation and yield a more transparent solution. The combination of high-temperature treatment and maximum nanocellulose concentration can diminish particle size, enhancing the oil-in-water emulsion's stability.

Table 2. Effect of temperature and nanocellulose concentration on particle size, particle charge, and viscosity in oil-in-water emulsions

Temp (°C)	Concentration (%)	*Particle Size* (nm)	Zeta Potential (mv)	Viscosity (cp)
60		657.66 ± 10.26^{f}	1.93 ± 1.51^{a}	0.85 ± 0.01^{a}
70	0	549.66 ± 8.50^{e}	1.86 ± 2.08^{a}	0.86 ± 0.01^{b}
80		199.23 ± 3.57^{d}	1.63 ± 1.02^{a}	0.87 ± 0.01^{d}
60		203.33 ± 3.34^{d}	3.06 ± 0.45^{a}	0.86 ± 0.01^{c}
70	10	171.40 ± 3.81^{c}	2.03 ± 0.80^{a}	0.87 ± 0.01^{f}
80		135.00 ± 3.11^{b}	1.66 ± 1.74^{a}	0.88 ± 0.01^{h}
60		175.90 ± 4.33^{c}	6.66 ± 0.20^{b}	0.87 ± 0.01^{e}
70	20	136.20 ± 4.33^{b}	2.93 ± 0.57^{a}	0.88 ± 0.01^{g}
80		102.30 ± 0.86^{a}	2.40 ± 0.45^{a}	0.88 ± 0.01^{i}

Note: The test was carried out in three repetitions. Different lowercase letters (a-f) on particle size, (a-b) on particle load and (a-i) on viscosity show significant differences ($p<0.05$).

3.3. Particle Charge

Table 2 illustrates the influence of heating temperature on the particle charge relative to nanocellulose content. The zeta potential is the potential difference between the layers on the surface that are closely associated with the electroneutrality in the formulation. This value helps ascertain the stability pattern and indicates the extent of attraction between neighboring particles with identical charges and scattered particles [12]. A zeta potential of ± 30 mV is considered adequate to guarantee the physical stability of the nanoemulsion. Zeta potential typically varies between +100 mV and -100 mV. Table 2 indicates a statistically significant difference in particle charge or zeta potential values ($p<0.05$) in samples with elevated nanocellulose concentration. The value of the particle charge will augment. The likelihood of flocculation or the aggregation of colloids from smaller to larger forms increases with a greater zeta potential value. Colloids exhibiting elevated zeta potential values are electrically stable, whereas those with diminished zeta potential are prone to coagulation or flocculation [13].

The emulsion's stability will enhance by incorporating suitable polymers in the dispersing phase and reducing the particle size of the dispersed phase. Cellulose fibers comprise crystalline and amorphous phases, with the crystalline phase being more prevalent than the amorphous phase. Applying heat can reactivate H^+ ions to interact with the amorphous phase of cellulose fibers; however, too high temperatures may lead to the carbonization of cellulose [14]. Using a moderate temperature combined with a brief heating duration and incorporating the greatest concentration of nanocellulose in this study effectively enhances the stability of oil-in-water emulsions since a high zeta potential inhibits flocculation.

3.4 Viscosity

Table 2 illustrates the impact of heating temperature on the viscosity of nanocellulose concentration. A high viscosity value may signify excellent emulsion stability. Emulsions undergoing flocculation exhibit increased viscosity due to the flocculation structure entraping a continuous phase (water) [15]. Table 2 indicates a substantial variation in viscosity values ($p<0.05$). Viscosity values rise with elevated temperature and increased nanocellulose concentration. Agi [1] also revealed that hydrophobic bonding may be responsible for this phenomenon at elevated temperatures, as the solution binding system enhances tensile strength and facilitates the formation of cohesive solid bonds between particles [16].

Eichie & Amalime [17] assert that an increase in mixing temperature enhances the energy available, facilitating the emulsifier's formation of a film layer over the dispersed droplets, resulting in reduced droplet size and increased viscosity. Additionally, An appropriate viscosity is characterized by a greater viscosity value, which complicates particle movement, hence enhancing sample stability. Stokes' law posits that a reduction in particle size prolongs the creaming process, thereby necessitating the minimization of particle dimensions. Smaller particle sizes result in increased surface free energy, hence rendering the emulsion more unstable. These two statements indicate that a specific particle size must be attained. Elevating the mixing temperature supplies energy for the droplet to mobilize and fragment into a smaller size, facilitating the formation of an emulsifier to envelop the droplet. Elevated mixing temperatures can diminish the surface tension at the phase interface, allowing the oil phase to be completely dispersed inside the dispersing phase (water) [18]. Furthermore, employing an excessively high temperature will disrupt the bonds between the droplets and the emulsifier; for instance, the hydrogen bonds between polyoxyethylene sorbitan fatty acid ester (polysorbate/tween) and water molecules will be severed, resulting in diminished emulsion stability [19].

The reduction in emulsion stability is also affected by unsuitable material composition, inadequate quantity and selection of emulsifiers, freezing, and mechanical stress or vibration during production. In addition to temperature, the presence of a stabilizer, specifically Tween 20, influences the stability of emulsion viscosity. With the suitable Tween 20, the emulsifier can inhibit droplets from coalescing from the dispersed phase into the continuous outer phase, hence preventing the formation of bigger droplets. The reduction of droplets (dispersed phase) to the continuous phase can decrease viscosity. An increased concentration of the inner phase (smaller droplet size) results in a higher phase volume ratio, hence elevating the viscosity of the emulsion. Consequently, with respect to the viscosity parameters, increased concentrations of NCC and NFC, along with elevated temperatures, result in higher viscosity values. Increased viscosity can inhibit the migration of oil droplets, signifying enhanced emulsion stability.

3.5 . Creaming Index

Figure 2 illustrates the impact of heating temperature on the concentration of nanocellulose on the creaming index. Creaming is a process of emulsion instability that initiates phase separation by the movement of droplets. The extent of phase separation in the emulsion system, represented by the creaming index, signifies a reduction in emulsion stability. As the concentration of the emulsifier increases, the strength of the resulting emulsion also increases until an ideal threshold is attained, beyond which a precipitate will form. Fat globules typically manifest on the emulsion's surface. This transpires because the density of fat is less than that of water. The emergence of fat on the emulsion's surface results in a clearly discernible cream

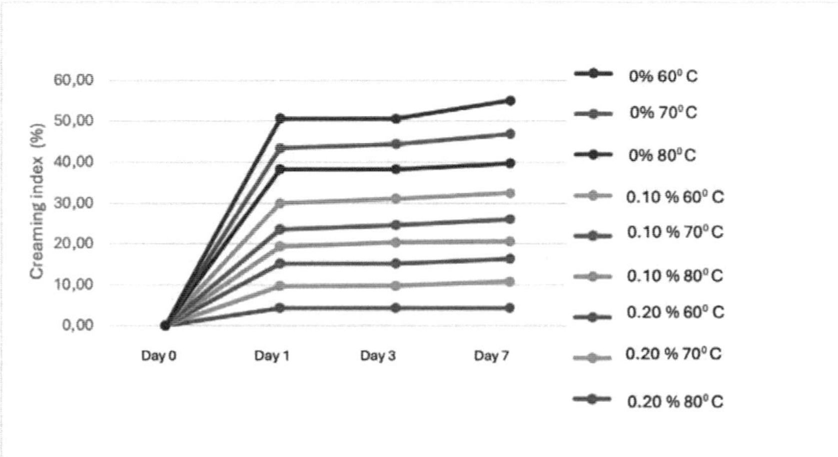

Figure 1. Effect of temperature and nanocellulose concentration on index creaming in oil-in water emulsions

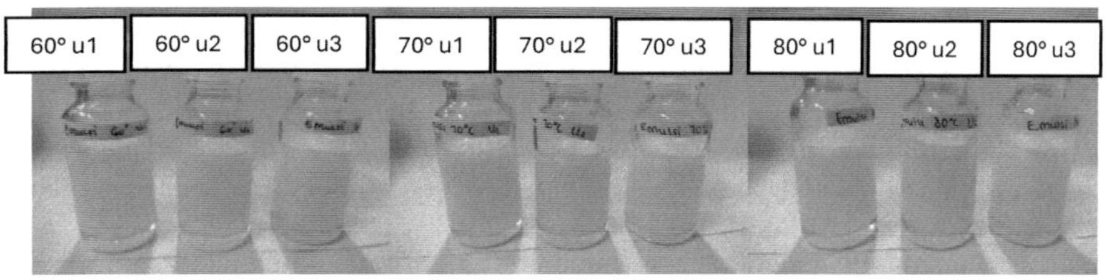

Figure 2. Creaming index at 60° C, 70° C, 80° C at concentration 0 % Day 0

On the first day of sample preparation, creaming had not commenced in samples with concentrations of 0%, 0.10%, and 0.20%. The creaming index in the samples was monitored until day 7, after which no additional creaming observations were conducted. The samples were stored for 30 days without any observations; they remained stable throughout this period. The stability of the emulsion phase is affected by the interaction of the emulsifier, Tween 20, in creating a firm and stable interfacial monolayer coating. The incorporation of nanocellulose into the emulsion network can enhance viscosity and reduce the rate of sedimentation and creaming caused by gravity forces. The relationship between nanocellulose and surfactants pertains to the attraction of non-ionic surfactants to the cellulose surface. This phenomenon can be ascribed to the hydrophobic contact between the surfactant's hydrophobic groups and the hydrophobic areas associated with the cellulose substance.

The graph of the creaming index rate indicates that at a concentration of 0% throughout all temperatures, the creaming index value is maximized. This may occur due to the sample concentration being 0% in the absence of nanocellulose addition. The creaming index value decreases with rising temperature and nanocellulose concentration. The incorporation of nanocellulose may lead to an increase in emulsion viscosity, hence impeding the ascent of oil droplets and diminishing the likelihood of creaming. The use of nanocellulose ensures prolonged shelf life and offers an aqueous dispersion with stability across a broad spectrum of pH and temperature conditions [28]. Consequently, augmenting the nanocellulose concentration in the emulsion and elevating the heating temperature can prolong its shelf life and preserve its stability.

4. Conclusions

Incorporating NCC and NFC at 0.20% and 70°C reduced particle size and enhanced particle charge in o/w nanoemulsions, resulting in greater electrostatic repulsions. The viscosity rises with increased concentrations of the nanocellulose mixture and temperature. Consequently, the creaming index diminished, resulting in enhanced emulsion stability. The amalgamation of NCC and NFC, specifically at 0.20%, serves as a natural stabilizer to enhance the stability of o/w nanoemulsion at a temperature of 70°C.

5. References

[1] Agi, A., Junin, R., Gbonhinbor, J., & Onyekonwu, M. (2018). Natural polymer flow behaviour in porous media for enhanced oil recovery applications: a review. Journal of Petroleum Exploration and Production Technology, 8(4), 1349–1362. https://doi.org/https://doi.org/10.1007/s1 3202-018-0434-7

[2] Al-Edresi S, Baie S. 2009. Formulation and stability of whitening VCO-in-water nano-cream. Int J of Phar. 373: 174-178.

[3] Mollet, H., Grubenmann, A., 2001, Formulation Technology: Emulsions, Suspensions, Solid Forms, 84, WILEY-VCH Verlag GmbH

[4] Tangsrianugul, N.; Winuprasith, T.; Suphantharika, M.; Wongkongkatep, J. Effect of hydrocolloids on physicochemical properties, stability, and digestibility of Pickering emulsions stabilized by nanofibrillated cellulose. *Food Funct.* **2022**, *13*, 990–999.

[5] Ioelovich, M. (2012). *Optimal Conditions for Isolation of Nanocrystalline Cellulose Particles*. *2*(2), 9–13. https://doi.org/10.5923/j.nn.20120202.03

[6] Missoum, K., Belgacem, M. N., & Bras, J. (2013). *Nanofibrillated Cellulose Surface Modification: A Review*. *Materials*, *6*(5), 1745–1766. https://doi.org/10.3390/ma6051745

[7] Fitri, i, a. , Mitbumrung, W., Akanitkul , P., Rungraung, N., Kemsawasd, V., Jain, S., Winuprasith , T. 2022. Encapsulation of β-Carotene in Oil-in-Water Emulsions Containing Nanocellulose: Impact on Emulsion Properties, In Vitro Digestion, and Bioaccessibility. 14: 1-18

[8] Kalashnikova I, Bizot H, Cathala B, Capron I (2012) Modultion sifat amphiphilic nanocrystals selulosa untuk menstabilkan antarmuka minyak / air. Biomakromolekul 13:267–275.https://doi.org/10.1021/bm201599j

[9] Bohren, C, F., Huffman, D, R. (1983). Adsorption and scattering of light by small particles, John Wiley and Sons, New York, NY. 10.1002/9783527618156

[10] Madaan, V, A. Chanana, M. K. Kataria and A. Bilandi. 2014. Emulsion technology and recent trends in emulsion applications. Int Res J. Pharm 6: 533-542.

[11] Kalashnikova I, Bizot H, Cathala B, Capron I (2012) Modultion sifat amphiphilic nanocrystals selulosa untuk menstabilkan antarmuka minyak / air. Biomakromolekul 13:267–275.https://doi.org/10.1021/bm201599j

[12] Dhankhar P. 2014. Homogenization fundamentals.IOSR J Engi 4(5): 01–08. DOI: h10.9790/3021- 04540108.

[13] Edén J, Dejmek P, Löfgren R, Paulsson M, Glantz M. 2016. Native milk fat globule size and its influence on whipping properties. Inte Dairy J 61: 176–181. DOI: 10.1016 j.idairyj.2016.06.004.

[14] Honary, S., Zahir, F. 2013. "Effect Of Zeta Potential On The Properties Of Nano-Drug Delivery Systems - A Review (Part 1)." Tropical Journal Of Pharmaceutical Research 12(2): 255–64.

[15] Singh, G., I.P.S. Kapoor, P. Singh, C.S. de Heluani, M.P. de Lampasona, C.A.N. Catalan. 2010. *Comparative study of chemical composition and antioxidant activity of fresh and dry rhizomes of turmeric (Curcuma longa Linn.). Food and Chemical Toxicology.* 48:1026-

1031.
[16] McClements, D.J. Emulsion design to improve the delivery of functional lipophilic components. *Annu. Rev. Food Sci. Technol.* **2010**, *1*, 241–269.
[17] Eichie, F. E., & Amalime, A. E. (2007). Evaluation of the binder effects of the gum mucilages of Cissus populnea and Acassia senegal on the mechanical properties of paracetamol tablets. African Journal of Biotechnology, 6(19), 2208– 2211. https://doi.org/10.5897/AJB2007.000- 2345.
[18] Claesson PM, Kjellin M, Rojas OJ, Stubenrauch C. 2006. Short-range interactions between non-ionic surfactant layers. Phys Chem Chem Phys. 8(47):5501-14.
[19] Schmitt, W. H. 1996. Skincare products. In: Williams, D.F. and W.H. Schmitt (Ed). 1996. Cosmetics and Toiletries Industry 2nd Ed. Blackie Academic and Profesional, London

Wheat flour partially substituted with heat-moisture-treated gembili (*Dioscorea esculenta* L.) flour affected physicochemical and organoleptic properties of pan bread prepared using straight-dough method

Ludwina Eugenia and Wilbur Donald Raymond Pokatong*

Food Technology Study Program, Faculty of Science and Technology, Universitas Pelita Harapan, Jl. M.H. Thamrin Boulevard 1100, Tangerang 15811, Banten, Indonesia

*E-mail: wilbur.pokatong@uph.edu

Abstract. To diminish reliance on wheat i.e. definitely imported from producing countries, gembili (*Dioscorea esculenta* L.) may be prepared into flour and partly replace wheat flour for preparing pan bread. However, modification is required because untreated gembili flour exhibits minus physicochemical characteristics. Research objectives were to investigate heat moisture treatment (HMT) temperature and time effect on gembili flour swelling power, solubility, lightness, then to choose the best; to investigate effect of ratios of wheat flour to modified gembili flour on straight-dough pan-bread physicochemical and organoleptic properties. Gembili flour produced was modified with HMT (78, 83, 88, 93 °C; 4, 6, 8, 10 h). Modified gembili flour (83 °C; 8 h) exhibited swelling power and solubility higher than control of 10.88±0.11 g/g and 26.10±0.35%, respectively; thus, it was selected. This selected HMT-modified gembili flour showed significantly higher moisture content (8.83±0.26%) than untreated one but lower protein, fat, carbohydrates, starch, amylopectin of 6.39±0.04%, 0.20±0.07%, 82.84±0.27%, 45.87±1.72%, 29.58±3.61%, respectively. Pan bread was prepared with ratios 90:10, 85:15, 80:20, 75:25, 70:30 where bread (90:10) was selected since it had similar volume, hardness, and overall acceptance of 864.17±16.55 cc, 174.66±33.94g, respectively, to control (100:0). The selected pan bread exhibited similar moisture, fat, ash, carbohydrates of 29.43±1.38%, 7.74±0.11%, 2.04±0.09%, 48.75±1.89%, respectively, to control (100:0), but significantly lower protein (12.03±0.40%). Thus, pan bread using 10% modified gembili flour may possibly be produced as an alternate to using only wheat in pan bread.

Keywords: gembili flour, heat-moisture treatment (HMT), pan bread, partial substitution, wheat flour

1. Introduction

Bread is one of the world's most ancient and vital primary foods. One of bread types is pan bread which is characterized as having no flavours and/or stuffings and baked in a bread pan to yield its typical form [1].

Wheat flour is a key ingredient in bread and as an imported commodity. From 2021 to 2022, it was reported that Indonesia imported 11.2 million tonnes of wheat [2]. Since wheat is imported, it is warranted to explore and utilize local food commodities as an alternative to substitute for wheat; thus, reducing dependence on this imported commodity.

Gembili (*Dioscorea esculenta* L.), or lesser yam, is a high-carbohydrate food commonly used as a rice substitute in parts of Indonesia [3]. However, its consumption and utilization are decreasing. To enhance its economic value, gembili can be processed into flour, which can serve as a gluten-free option and substitute for wheat flour in various food products, especially for those with gluten sensitivity [4]. However, due to its low gluten content, there are challenges for utilizing gembili flour for breadmaking. These include when used in bread-making, results in low volume and poor characteristics of its crust and crumb, which leads to lower overall acceptability of the bread [5] [6]. Therefore, flour modification is needed to improve pan bread volume and other attributes i.e. physicochemical and organoleptic.

Heat moisture treatment (HMT) is a method used to modify starch granules by heating them under low moisture conditions, which results in structural changes without damaging the granules. Various starch properties, such as solubility, swelling power, pasting temperature, peak viscosity, amylose leaching, and crystallinity are affected by this treatment [7] [8]. Research by Dudu and co-workers [9] showed that HMT-treated cassava flour improved pan bread by reducing amylopectin retrogradation and enhancing porosity of crumb and volume of bread. Applying HMT to gembili flour may similarly improve its physicochemical and sensory properties, making pan bread more acceptable to consumers. The objectives of this research were to investigate HMT temperature and time effect on gembili flour characteristics (swelling power, solubility, and lightness), and to examine different ratio effect of wheat flour to modified gembili flour on straight-dough pan-bread physicochemical and sensory characteristics.

2. Research methodology

2.1 Materials

In this research, materials used were gembili (*Dioscorea esculenta* L.) obtained from Talun in Cirebon, which was also sent to *Pusat Riset Biosistemika dan Evolusi, BRIN RI* for taxonomical verification. Other materials included rice grains, distilled water (Amidis), wheat flour (Cakra Kembar), heat-moisture modified gembili flour, skim milk powder (NZMP), shortening (Anchor), yeast (Fermipan), salt (Dolphin), sugar (Gulaku), bread improver (Baker's Bonus A), and drinking water, and required reagents/chemicals of technical and/or analytical grades for preparation and/or analysis.

In this research, equipment used were analytical balance (Sartorius), hot plate stirrer (Cimarec), centrifuge (Hettich Zentrifugen EBA 200), oven (Memmert), cabinet dryer, proofing box, baking oven, baking pan, rotary evaporator (Heidolph), CR-400 chromameter (Konica Minolta), TA-XT plus texture analyzer (Stable Micro Systems), spectrophotometer, plus required necessary lab apparatus/instruments for preparation and/or analytical purposes.

2.2 Research stages

2.2.1 Initial research

The gembili flour processing was adopted from existing methods [10] [4] with modifications. Gembili tubers (~11 kg) were cleaned, peeled and sliced into 2 mm thickness. The gembili slices were immersed in solution of 0.075% $Na_2S_2O_5$ for 1 h for browning prevention, then rinsed.

IOP Conf. Series: Earth and Environmental Science **1460** (2025) 012041 doi:10.1088/1755-1315/1460/1/012041

Afterwards, at 50 °C for 48 h, the slices were dried, ground, and sifted with a screen of 80 mesh to obtain gembili flour. The yield, proximate content, starch, amylose, amylopectin, swelling power, solubility, and lightness were determined.

2.2.2 Research stage I

In this research stage, gembili flour (~1 kg) was modified with Heat Moisture Treatment (HMT) based on existing methods [11] and [12] with modifications. The moisture content of gembili flour (~1 kg) was adjusted to 25% by adding water, based on the following formula (1):

$$(100\% - MC_1) \times W_1 = (100\% - MC_2) \times W_2 \quad ...(1)$$

Where:
MC_1 = initial moisture content of gembili flour
MC_2 = required moisture content of heat-moisture treatment (25%)
W_1 = initial weight of the flour
W_2 = final weight of gembili flour

When the moisture content adjustment was finished, the flour was chilled at 10 °C; 12 h, afterwhich modified with Heat Moisture Treatment (HMT) in an oven at 78, 83, 88 and 90 °C for 4, 6, 8, and 10 h. The HMT modifed flour was then dried (50 °C; 8 h), ground, and sifted with a screen of 80 mesh. Proximate compositions, and starch, amylose, amylopectin, swelling power, solubility, plus lightness were analysed.

2.2.3 Research stage II

In research stage II, HMT modified and selected gembili flour was used to partially substitute wheat flour for preparing pan bread employing the straight dough method from AACC international [13] with modifications [14]. Table 1 shows the ratios of substitution of wheat flour to modified gembili flour, and the ingredients in the formulation.

Table 1. Pan bread ingredients and formulation

Ratio* / Ingredients	Quantity (g)					
	100:0 (control)	90:10	85:15	80:20	75:25	70:30
Wheat flour	53.19	47.87	45.21	42.55	39.89	37.23
Modified gembili flour	0	5.32	7.98	10.64	13.30	15.96
Bread improver	0.27	0.27	0.27	0.27	0.27	0.27
Skim milk powder	3.19	3.19	3.19	3.19	3.19	3.19
Yeast	1.06	1.06	1.06	1.06	1.06	1.06
Sugar	4.26	4.26	4.26	4.26	4.26	4.26
Salt	0.80	0.80	0.80	0.80	0.80	0.80
Water	31.91	31.91	31.91	31.91	31.91	31.91
Shortening	5.32	5.32	5.32	5.32	5.32	5.32
Total	100	100	100	100	100	100

Source: AACC International [13] with modifications [14]
*: Ratio of wheat flour to HMT modified gembili flour

Each dry ingredient was weighed according to the formulation and then thoroughly mixed. Water was gradually poured while mixing at low and high speed for 2 min each. Then, shortening was added and mixed at high speed until homogenous. The dough was rounded into a ball shape, and proofed for 10 min. Afterwards, the dough was re-rounded then proofed (another 15 min). Then, the dough was rolled and formed in the baking pans. The panned dough was left for final proofing at 30-35 °C with RH of 80-85%, and baked i.e. 190 °C for 25 min. At room temperature, the baked bread was left for cooling and analysed for its physicochemical (volume, weight, texture, crust and crumb color, and proximate composition), plus organoleptic characteristics.

2.3 Procedures of analyses

2.3.1 Yield determination
Gembili flour yield of was measured according to AOAC [15] using the formula (2) below as follows:

$$\text{Yield (\%, db)} = \frac{\text{flour weight (g, db)}}{\text{raw material weight (g, db)}} \times 100\ \% \quad \text{...(2)}$$

2.3.2 Swelling power and solubility
Based on the existing procedure [4] with modification, swelling power and solubility were measured. Approximately 1 g of flour sample was mixed with 40 mL distilled water, and heated in a waterbath at 85 °C for 30 min. The sample was vortexed at minute 0, 10, 20, and 30 for 10 s each. After cooling at room temperature, the sample was centrifugated (2200 rpm; 15 min). The supernatant was separated, dried, and weighed for solubility, while the precipitate was weighed to obtain swelling power. Swelling power and solubility were determined by employing these formulas (3) and (4), respectively:

$$\text{Swelling power} = \frac{\text{Sediment weight (g)}}{\text{Sample weight (g)}} \quad \text{...(3)}$$

$$\text{Solubility} = \frac{\text{Soluble portion weight (g)}}{\text{Sample weight (g)}} \times 100 \quad \text{...(4)}$$

2.3.3 Lightness
Lightness of gembili flour, and crust and cumb of pan bread was dtermined using Chroma Meter CR-400. The lightness value is expressed as the L* value.

2.3.4 Total starch content
Starch content was determined using acid hydrolysis method based on AOAC [15]. A sample of approximately 0.5 g was washed using distilled water until 250 mL, and filtered. The obtained residue was washed using water (200 mL) and 25% HCl (20 mL). The solution was boiled for 2.5 h, cooled, and neutralized using 1N NaOH, afterwhich, diluted to 250 mL and filtered again. Starch content was calculated using a formula (5) below where X is the absorbance, DF is dilution factor, and 0.9 is conversion factor.

$$\text{Starch content (\%)} = \frac{\text{X x DF x 100 x 0.9}}{\text{weight of sample}} \quad \text{...(5)}$$

2.3.5 Amylose content

A sample (~100 mg) was put in a test tube then added 95% ethanol (1 mL) plus 1 N NaOH (9 mL). The suspension in the test tube was boiled in boiling water up until a gel was formed. The gel was then added 1 N acetic acid (1 mL) and iodine solution (2 mL), then topped up to 100 mL followed by resting for 20 min. The amylose content was calculated by measuring the colour intensity of the solution using a spectrophotometer at 625 nm wavelength [15].

2.3.6 Amylopectin content

Amylopectin was measured based on an AOAC method [15] by obtaining the value (%) of the difference between the starch and amylose content in the sample using the formula (6) below.

$$\text{Amylopectin (\%)} = \text{total starch (\%)} - \text{amylose (\%)} \;...(6)$$

2.3.7 Weight and volume

The weight of pan bread was measured using a table balance while the volume was measured using the seed displacement method. An empty container was filled with rice grains until it was full to obtain (V_1). The pan bread was placed in the empty container and the container was filled again with rice grain until it is full to obtain (V_2). The volume of pan bread was obtained by calculating the difference between V_1 and V_2 [16].

2.3.8 Texture (bread hardness)

Hardness of bread was measured by employing a TA-XT plus texture analyser. A uniform slice with a thickness of 20 mm was obtained from the middle of the loaf. The texture was evaluated by employing a 25 kg load cell equipped with a 36-mm probe of cylindrical aluminium. Texture analysis was performed with test speed (2 mm/s), post test speed (2 mm/s), trigger force (5 g), penetration distance (10 mm), with 5 s between compression cycle.

2.3.9 Proximate composition

Proximate analyses of moisture, protein, fat, and ash were carried out based on the AOAC methods [15].

2.3.10 Sensory analysis

Sensory analysis was conducted using a 6-point scale scoring test and 7-point scale hedonic test in 2 replications on 30 semi-trained panelists, both tests based on the existing method [17]. Scoring test was done to assess the perceived intensity of the color of crust and crumb, firmness of crumb, and aroma and taste of pan bread. Hedonic test was done to assess the acceptance of the pan bread. The parameters for the hedonic test were the color of crust and crumb, firmness of crumb, aroma and taste, plus overall acceptance of pan bread.

3. Results and discussion

3.1 Gembili taxonomical identity

Gembili used in this research was sent to *Pusat Riset Biosistemika dan Evolusi, BRIN RI* for taxonomical verification. According to the result, it was verified that the gembili used was indeed *Dioscorea Esculenta* L. from the Dioscoreaceae family.

3.2 Modified gembili flour swelling power
Table 2 shows the swelling power of modified gembili flour. The HMT temperature and time treatments did not show significant interaction (p>0.05) in influencing swelling power of the modified flour of gembili. HMT temperature individually did not affect significantly (p>0.05) swelling power either, with values from 9.01±0.61 to 10.74±0.75 g/g. However, HMT time did significantly affect (p≤0.05) swelling power of modified gembili flour.

Table 2. HMT time effect on swelling power of gembili flour

Time (h)	Swelling Power (g/g)
4	9.01±0.61[a]
6	9.13±0.33[a]
8	10.39±0.91[b]
10	10.74±0.75[b]

Note: Notation of different superscript on each value indicates significant difference (p≤0.05)

As shown in Table 2, swelling power is directly proportional to the time of HMT. Gembili flour HMT modified for 4 hours showed the smallest swelling power (9.01±0.61 g/g) while flour modified 10 hours had the highest swelling power of 10.74±0.75 g/g. The flour modified at 83 °C and 88 °C for 8 hours showed highest swelling power i.e. 10.88±0.11 g/g and 10.97±0.36 g/g, respectively. These values were higher than that of the untreated gembili flour of 8.48±0.58 g/g. Santosa and co-workers [18] observed a similar finding in HMT modified purple sweet potato flour.

Kumoro and co-workers [19] and Dudu and co-workers [9] reported that HMT can disrupt the starch-granule crystalline structure which contributes to improve starch swelling power. The internal reassociations of starch granules due to HMT may also lead an increased exposure of hydroxyl groups which promotes the elongation of amylopectin chains and improves the capacity of starch granules to absorb water and swell. Furthermore, Yan and co-workers [20] found that prolonged exposure to intense heat treatment can destroy the hydrogen bonds in the starch-granule amorphous and crystalline regions, leading to the destruction of starch double helix structure and decrease starch crystallinity which can further increase the swelling power of starch.

3.3 Modified gembili flour solubility
Figure 1 shows the solubility of modified flour of gembili. The HMT temperature and time had a significant interaction (p≤0.05) in influencing the modified flour solubility. Modified gembili flour (88°C; 8 h and 10 h) had the highest solubility of 26.10±0.35% and 27.94±0.77%, respectively. These were higher than the solubility of unmodified gembili flour which was 25.65±2.38%. In contrast, gembili flour modified at 93°C for 10 h had the lowest solubility of 16.84±0.48%.

Figure 1 shows that in general the the solubility of modified gembili flour initially increased, then decreased as HMT temperature and time increases. The initial increase in the solubility of modified gembili flour may be caused by the reduced crystallinity of the starch granules as it absorbed excess water and allowed amylose to dissociate and diffuse out of them. It was found that HMT done at higher temperatures can cause amylopectin chains to elongate at a faster rate and allow amylose to easily leach out during the swelling of starch granules [18]. Li and co-workers [21] also reported that the thermal degradation of amylopectin at higher HMT temperatures can improve the capacity of starch granules to swell, leading to an increase in amylose leaching.

The solubility of modified gembili flour decreased at higher HMT temperature was due to the reassociation of starch granules which increases in starch rigidity. These reassociations involves interactions i.e. more developed between amylose-amylopectin starch functional groups and amylopectin-amylopectin chains, more compact amylopectin clusters formation, complexes of amylose-lipid form, and double helices within starch granules which contribute to further restrictions of starch solubility [19].

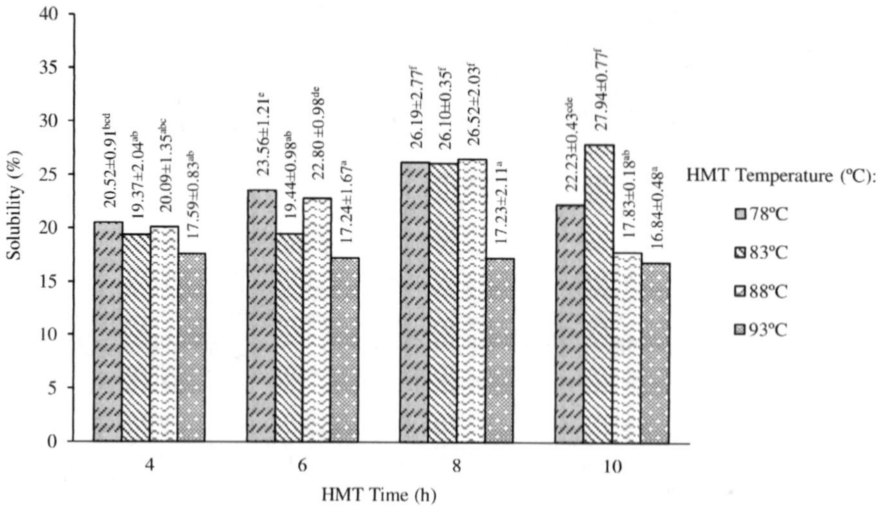

Figure 1. HMT temperature and time effect on the modified gembili flour solubility
Note: Notation of different superscript on each value indicates significant difference ($p \leq 0.05$)

3.4 Modified gembili flour lightness

Figure 2 displays the lightness of modified gembili flour. The HMT temperature and time had a significant interaction ($p \leq 0.05$) in influencing the modified gembili flour lightness.

Based on Figure 2, modified gembili flour (88°C; 8 h) had the highest lightness (67.44±1.47), while gembili flour modified at 88°C for 6 had the lowest lightness (63.33±1.14), which is lower than that of untreated gembili flour (67.34±0.78). The results also showed that lightness of modified gembili flour generally decreases with increases temperature for each time regime. Maillard reaction might be responsible for the darkening of flour as HMT temperature and time increases. This darkening process is due to the reaction between the reducing sugar carbonyl group and amino-acid amino group which leads to formation of brown pigments (melanoidins). As heating temperature and time increase, so will the rate of Maillard reaction [22].

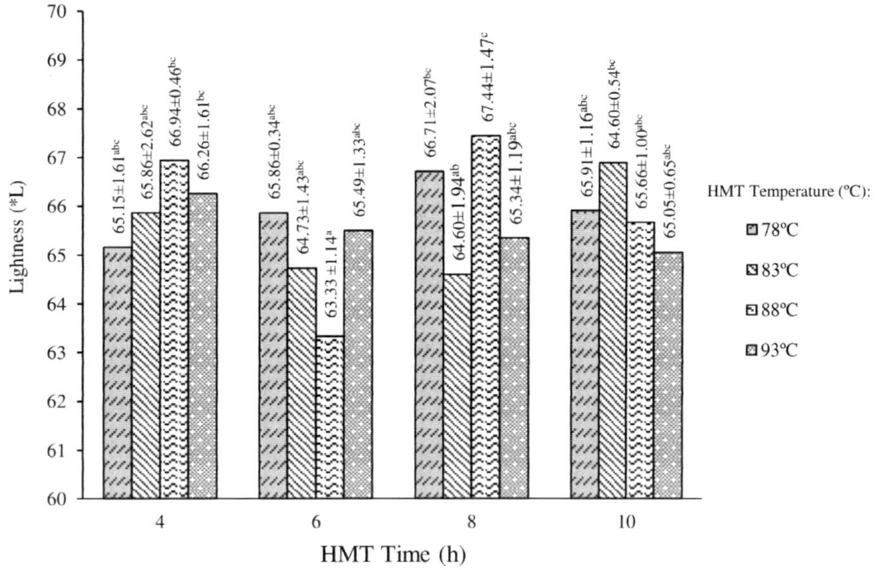

Note: Notation of different superscript on each value indicates significant difference ($p \leq 0.05$)

Figure 2. HMT temperature and time effect on the modified gembili flour lightness

3.5 Chosen modified flour of gembili on the basis of swelling power and solubility

Modified gembili flour of 83 °C; 8 h has been selected for partial substitution to wheat flour in the pan bread preparation due to its high swelling power and solubility of 10.88±0.11 g/g and 26.10±0.35%, respectively. Although modified gembili flour of 88 °C; 8 h was found to have a slightly higher swelling power and solubility (10.97±0.36 g/g) and (26.52±2.03%), respectively, statistical analysis showed that the swelling power and solubility were not significantly different to those of modified gembili flour of 83 °C; 8 h. Therefore, HMT modification at 83 °C for 8 h was selected as it requires less energy consumption for preparation. Higher swelling power and solubility indicate a higher water holding capacity which enables the starch granules of more absorption of water and more swelling. These swollen granules will raise the dough viscosity and contribute to a higher bread loaf volume [23] [24].

3.6 Chemical composition of native and chosen modified gembili flour

Table 3 presents the chemical composition of native and modified gembili flour (83 °C; 8 h). The moisture content of HMT modified gembili flour (8.83±0.26%) exhibited significantly higher ($p \leq 0.05$) than the native gembili flour moisture content (8.81±0.31%). Similar results were found in HMT modified sorghum flour[25]. This increase in moisture content of HMT modified gembili flour can be attributed to the moisture adjustment done before heating during the HMT modification.

The protein content of HMT modified gembili flour (6.39±0.04%) was significantly lower than that of native gembili flour (6.49±0.03%). According to Luga and Mironeasa [26], the heating in HMT modification can cause secondary structural changes in protein as well as protein

IOP Conf. Series: Earth and Environmental Science **1460** (2025) 012041 doi:10.1088/1755-1315/1460/1/012041

denaturation, degradation of amino acid, in turn loss of nitrogen in gaseous forms which lead to a lower protein content in HMT modified gembili flour.

The HMT modified gembili flour fat content (0.20±0.07%) exhibited significantly lower ($p \leq 0.05$) than the native gembili flour fat content(0.94±0.13%). Similar to protein, fat can be degraded during the heating in HMT modification process. Lipid-protein formation can also be formed during HMT, contributing to a reduction of fat in HMT modified flour [26].

Table 3. Chemical composition of native and HMT modified gembili flour

Component	Amount (%)	
	Unmodified gembili Flour	HMT Modified gembili Flour (83 °C; 8 h)
Moisture	6.81±0.31^{a}	8.83±0.26^{b}
Protein	6.49±0.03^{a}	6.39±0.04^{b}
Fat	0.94±0.13^{a}	0.20±0.07^{b}
Ash	1.68±0.13^{a}	1.74±0.04^{a}
Carbohydrate	84.07±0.28^{a}	82.84±0.27^{b}
Starch	77.22±2.48^{a}	45.87±1.72^{b}
Amylose	14.78±2.35^{a}	16.29±1.97^{a}
Amylopectin	62.44±0.18^{a}	29.58±3.61^{b}

Note: Notation of different superscript on each value indicates significant difference ($p \leq 0.05$)

The ash content of HMT modified gembili flour (1.74±0.04%) was observed to be higher than that of native gembili flour (1.68±0.13%). However, there was no significant difference ($p > 0.05$). HMT modification significantly decreases ($p \leq 0.05$) carbohydrate content from 84.07±0.28% in native gembili flour, to 82.84±0.27% in HMT modified gembili flour. This carbohydrate content reduction can be attributed to interaction increase between carbohydrates and other compounds such as lipids and protein during HMT modification, which reduces the amount of carbohydrate available in HMT modified gembili flour [26].

The starch content of HMT modified gembili flour (45.87±1.72%) exhibited significantly lower ($p \leq 0.05$) than the native gembili flour starch content (77.22±2.48%). Yan and co-workers [20] suggested that HMT destroys the double helix structure of starch granules which makes them harder to be quantified, resulting to the lower starch content in HMT modified gembili flour .

HMT modified gembili flour was found to have a nominally higher amylose content (16.29±1.97%) compared to native gembili flour (14.78±2.35%), but they were statistically not significantly different ($p > 0.05$). The insignificant amylose increase can be caused by amylose leaching from the starch granules during HMT modification [20].

HMT modification significantly decreases ($p \leq 0.05$) amylopectin content from 62.44±0.18% in native gembili flour to 29.58±3.61% in HMT modified gembili flour. This can be caused by the increase in amylopectin-amylopectin interactions, which makes amylopectin harder to be quantified [27].

3.7 Wheat flour to chosen modified gembili flour ratio effect on physicochemical characteristics of pan bread

3.7.1 Weight

Figure 3 shows effect of ratios of wheat flour : modified flour of gembili which exhibited no significant effect ($p>0.05$) on the pan bread weigh. Based on absolute values, the pan bread of a 100:0 flour ratio had the lowest weight (271.48±5.87g), whereas the highest weight was shown by the pan bread with a 70:30 flour ratio (279.90±3.93g).

According to Bakare and co-workers [28], bread dough lose weight during the proofing and baking process. This phenomenon is known as baking loss and is commonly caused by the evaporation of moisture and fermentation losses throughout the production process. In this case, the evaporation is exhibited similar patterns (amount) regardless of the ratios. Heavy bread weight is desirable for consumers due to its economic value. Since partial substitution of modified gembili flour had no effect in bread weight, the economic value of bread will not be reduced [29].

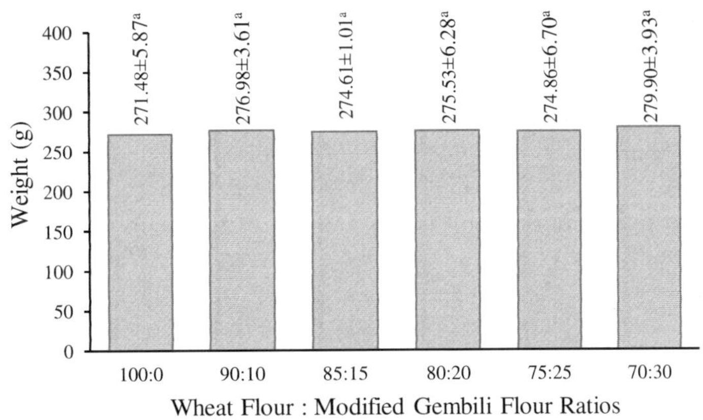

Note: Notation of different superscript on each value indicates significant difference ($p\leq0.05$)
Figure 3. Wheat flour : modified gembili flour ratio effect on pan bread weight

3.7.2 Hardness

Figure 4 shows the hardness of pan bread made with different wheat flour to modified gembili flour ratios. Wheat flour to modified flour of gembili ratios did affect significantly ($p\leq0.05$) the pan-bread hardness. The highest hardness was observed in the pan bread with 70:30 (632.06±32.04 g) and 75:25 (512.02±20.92 g) flour ratios. In contrast, the pan bread made using 100:0 and 90:10 flour ratio showed the lowest hardness of 156.68±2.81 g and 174.66±33.94 g, respectively.

Cauvain [30] found out that bread volume is directly related to the hardness of bread. Bread with smaller volumes are more compact, which negatively impacts the softness of bread crumb.

Increasing douh gas retention can improve bread dough expansion, which will lower the density of the centre crumb and increase its resistance to compression. High moisture also promotes soft bread texture.

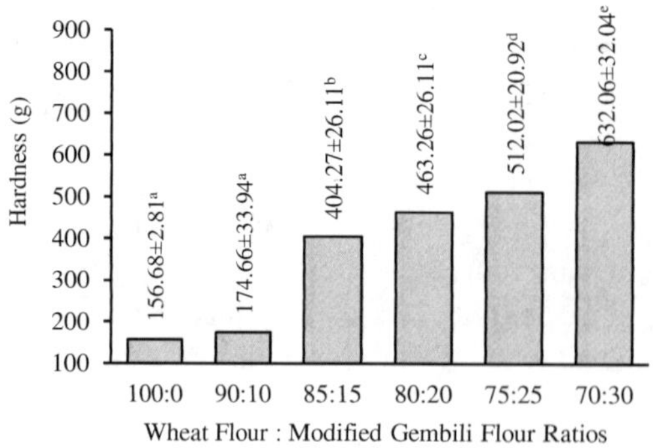

Note: Notation of different superscript on each value indicates significant difference (p≤0.05)
Figure 4. Wheat flour: modified gembili flour ratio effect on pan bread hardness

Gluten is responsible for the development of protein network, which traps gas and moisture. Glutenin in gluten helps develop dough cohesiveness and elasticity, contributing to a desirable bread volume and crumb hardness. Lower gluten content will result in a harder bread due to the dough's inability to retain sufficient amounts of moisture and gas [31].
This is reflected in the results, where the hardness of pan bread increases as the ratio of modified gembili flour increases. The lower amount of gluten in modified gembili flour contributes to poorer gas and moisture retention of bread dough, resulting in a pan bread with harder texture. Similar results were obtained in pan bread partial substitution with HMT modified flour of sweet potato [14].

3.7.3 Crust Lightness
Figure 5 shows the pan-bread crust lightness of different wheat flour to modified gembili flour ratios. Wheat flour to modified flour of gembili ratios did not affect significantly (p>0.05) the pan-bread crust lightness. Based on absolute values, the highest crust lightness was observed in the pan bread (70:30) of 55.90±1.44, while the lowest crust lightness was observed in the pan bread with 85:15 flour ratio (53.77±1.11) although all ratios were not significantly different.

Maillard reaction is responsible for the crust colour of pan bread. This reaction involves reducing sugars and amino compounds [30]. The higher crust lightness in the pan bread with higher modified gembili flour substitution may be caused by the lower protein content in gembili f lour, which means there are less amino groups available for Maillard reaction, resulting in a paler crust [32] [33].

3.7.4 Crumb Lightness
Figure 5 shows the pan-bread crumb lightness made using different wheat flour to modified gembili flour ratios. Wheat flour to modified flour of gembili ratios did not affect significantly (p>0.05) the pan-bread crumb lightness. The pan bread (70:30) had the highest crumb lightness

(70.34±0.43) while the pan bread (85:15) had the lowest crumb lightness (69.77±0.83) when seen in terms of absolute values although all ratios were not significantly different.

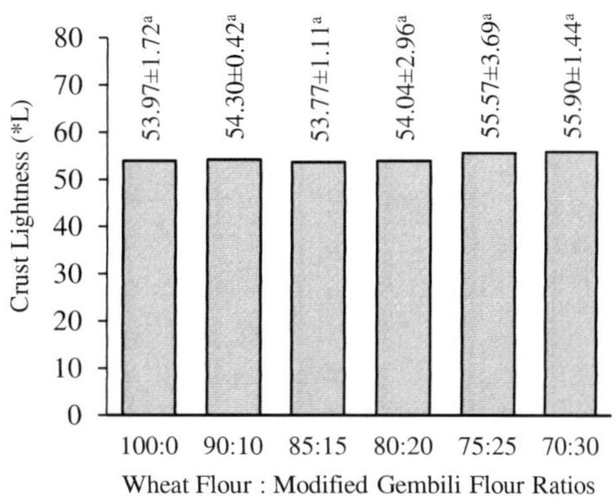

Note: Notation of different superscript on each value indicates significant difference (p≤0.05)
Figure 5. Wheat flour : modified gembili flour ratio effect on pan bread crust lightness

Pokatong and Julista [14] suggested that the lightness of flour used in making pan bread influences the resulting crumb lightness. Since wheat flour has a higher lightness value (74.94-76.38) compared to the modified gembili flour used in this study (64.60±1.94), it is expected that the pan bread made using higher substitution ratio of modified gembili flour will exhibit a darker crumb. However, this was not reflected in the results of this study. Maillard reaction also influences the lightness of bread crumb. Hence, the lower protein content in gembili flour compared to wheat flour means that there are less amino groups available to partake in the Maillard reaction, leading to a lighter colored crumb [30].

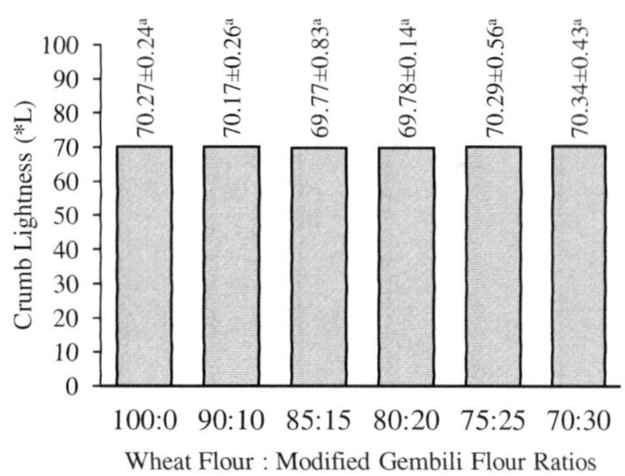

Note: Notation of different superscript on each value indicates significant difference (p≤0.05)
Figure 6. Wheat flour : modified gembili flour ratio effect on pan bread crumb lightness

3.7.5 Volume

The pan bread volume of different wheat flour to modified gembili flour ratios is presented in Figure 7. Wheat flour to modified flour of gembili ratios did affect significantly (p≤0.05) the pan-bread volume. The control pan bread with 100:0 (890.00±18.87 cc) and 90:10 (864.17±16.55 cc) flour ratio exhibited the highest volume. Additionally, the pan bread made using the 85:15 flour ratio had a volume of 839.17±15.07 cc, i.e. not significantly different with that of the 90:10 ratio pan bread. Pan bread with 75:25 (570.00±15.61 cc) and 70:30 flour ratios (542.50±13.23 cc) exhibited the lowest volume.

These results indicated that the volume of pan bread decreases as the amount of modified gembili flour increases. According to Cauvain [30], high loaf volume is desirable to consumers. Since bread loaf volume is determined by the carbon dioxide produced during fermentation of yeast and dough matrix ability in retaining the gas, the two main causes of smaller bread volume could be the lack of carbon dioxide production and poor gas retention. Although gembili flour has a high carbohydrate content that contributes to the production of carbon dioxide, it has a lower gluten content compared to wheat flour, which reduces its ability to form a gluten matrix to trap carbon dioxide and results in a lower bread volume [32] [34].

Bread loaf volume is also affected by the swelling power of starch. During baking, gluten denatures and loses its WHC (water holding capacity), remaining available water in the dough is absorbed by the starch, permitting granules to swell and gelatinize [36]. Starch granule ability to absorb water and raise dough viscosity influences the bread loaf volume where HMT structurally modifies starch granules, allowing them to take up more water during gelatinisation and swell further [23] [24]. This was reflected in the result, where the volume of pan bread 90:10 flour ratio was the closest (despite significantly different) to that of the control pan bread despite having a lower gluten content.

3.8 Wheat flour to chosen modified gembili flour ratio effect on pan-bread organoleptic characteristics

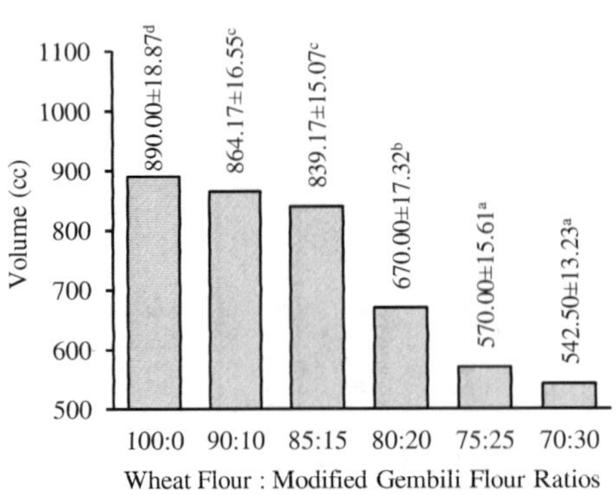

Note: Notation of different superscript on each value indicates significant difference (p≤0.05)

Figure 7. Wheat flour : modified gembili flour ratio effect on pan bread volume

3.8.1 Scoring results

To evaluate sensorically pan bread crust colour, crumb colour, firmness, aroma, and taste of pan bread subjected to partial substitution with modified gembili flour, scoring tests were carried out, and the results can be seen in Table 4. Wheat flour to modified flour of gembili ratios did affect significantly ($p \leq 0.05$) the crust color of pan bread. The 70:30 ration pan bread had the lowest score (2.92±0.07), indicating that it had the lightest crust colour. Meanwhile, the pan bread with 85:15 flour ratio had the highest score for crust colour (4.02±0.12), which suggested that it had the darkest crust colour. According to Cauvain (2017), caramelization and Maillard reaction are responsible for the darkening of crust colour. The degree of browning by caramelization depends on the amount of carbohydrates in the bread dough while Maillard reaction involves reducing sugars and amino compounds [36]. The partial substitution of modified gembili flour alters the carbohydrate and protein content of pan bread which contributes to the increase in crust lightness as the substitution ratio increases.

Similarly, statistical analysis showed crumb colour to be significantly affected ($p \leq 0.05$) by wheat to modified gembili flour ratios. The 70:30 ratio pan bread had a score of 4.05±0.16 which was the highest score, suggesting that the crumb colour is slightly yellow. In contrast, the control pan bread (100:0) and pan bread (90:10) had lowest scores of 3.17±0.09 and 3.22±0.02, respectively, which indicated that their crumb colour was slightly not yellow. Das and co-workers [37] suggested that the difference in crumb colour can be caused by the ingredients of the pan bread. Since modified gembili flour was slightly darker than wheat flour, the higher substitution of modified gembili flour would result in pan bread with more yellow crumb while the pan bread made with lesser amount of gembili flour would produce a paler crumb [38].

Firmness of pan bread was significantly affected ($p \leq 0.05$) by wheat to modified gembili flour ratios. Pan bread control (100:0) had the lowest score (2.80±0.07), indicating that it had a soft texture. Conversely, the the 70:30 ratio pan bread had a score of 4.12±0.11 which was the highest

Table 4. Scoring test results

Ratio (Wheat Flour : Modified Flour of Gembili)	Parameter				
	Crust Color	Crumb Color	Firmness	Aroma	Taste
100:0	3.72±0.02^{bcd}	3.17±0.09^{a}	2.80±0.07^{a}	1.85±0.02^{a}	1.70±0.05^{a}
90:10	3.42±0.12^{b}	3.22±0.02^{ab}	3.00±0.02^{ab}	1.87±0.05^{a}	1.98±0.01^{ab}
85:15	4.02±0.12^{d}	3.53±0.09^{bc}	3.33±0.07^{bc}	2.05±0.07^{a}	2.02±0.01^{b}
80:20	3.92±0.07^{cd}	3.62±0.07^{c}	3.55±0.11^{c}	1.95±0.16^{a}	2.22±0.01^{b}
75:25	3.58±0.26^{bc}	3.63±0.24^{c}	3.75±0.11^{cd}	2.50±0.05^{b}	2.63±0.12^{c}
70:30	2.92±0.07^{a}	4.05±0.16^{d}	4.12±0.11^{d}	2.45±0.07^{b}	2.67±0.07^{c}

Note: Notation of different superscript on each value indicates significant difference ($p \leq 0.05$)

value, suggesting that its texture was harder. Ukpabi [39] obtained similar results where along with the substitution of gembili flour increased, the bread firmness also increased. The increase in firmness may be because of lower content of gluten in modified flour of gembili. Since the protein matrix formed by gluten is responsible for trapping gas and moisture, pan bread with higher substitution of gembili flour would have a more compact structure and less moisture content, all of which contributes to a harder texture [31].

Both aroma and taste are significantly affected ($p \leq 0.05$) by wheat flour to modified gembili flour ratios. Pan bread control (100:0) had the lowest aroma and taste scores of 1.85±0.02 and 1.70±0.05, respectively. The pan bread with 90:10, 85:15, and 80:20 flour ratios showed no significance difference in aroma score with the control, indicating that all of the formulations did not exhibit any odd aroma. As for the taste scores, only pan bread with 90:10 flour ratio had no significant difference in taste score with the control. Pan bread with 75:25 and 70:30 flour ratio had the highest aroma score of 2.5±0.05 and 2.45±0.07, respectively. Correspondingly, the pan bread with 75:25 (2.63±0.12) and 70:30 (2.67±0.07) flour ratio had the highest taste score. The higher aroma and taste scores indicated that these bread formulations exhibit an aroma and taste that were odder than the control. Cho and Peterson [40] suggested that bread aroma is influenced by the ingredient and formulation, dough fermentation, as well as baking conditions. Thus, partial replacement of wheat flour using modified gembili flour would change its aroma and taste of pan bread produced that may be perceived as odd/off (i.e. not as bread aroma and taste).

3.8.2 Hedonic test results

Table 5 shows hedonic scores of pan bread. Wheat flour to modified gembili flour ratios affected significantly ($p \leq 0.05$) crust color acceptance. Pan bread control (100:0) had the highest score (5.63±0.07), which indicated high acceptance. The pan bread with 90:10 and 85:15 flour ratios with hedonic scores of 5.42±0.01 each, as compared to the control also exhibited insignificant difference in acceptance. Contrastingly, the 70:30 ratio pan bread with had the lowest hedonic score (4.82±0.06), indicating a lower acceptance. Therefore, it can be seen that the acceptance of crust colour decreases as the replacement of modified flour of gembili increases because of the lighter pan-bread crust colour.

Crumb colour acceptance was significantly affected ($p \leq 0.05$) by wheat to modified gembili flour ratios. Pan bread with 100:0 and 85:15 flour ratios had the highest hedonic scores of 5.77±0.09 and 5.57±0.06, respectively. Meanwhile, the 70:30 ratio pan bread had the lowest score (4.88±0.13). This shows that in general the acceptance of crumb colour decreases as the substitution ratio increases, which is caused by the increasingly yellow crumb colour.

Wheat to modified gembili flour ratios did significantly affect ($p \leq 0.05$) pan-bread firmness acceptance. Pan bread (100:0 and 90:10 flour ratios) had the highest hedonic score of 5.68±0.01 and 5.62±0.04, respectively, while the 70:30 ratio pan bread had the lowest hedonic score (4.22±0.15). This suggests that as the amount of modified gembili flour was increased in pan bread, the firmness acceptance of the resulting pan-bread decreased, this indicates that the panelists had higher preference for pan breads with softer texture.

Acceptance of pan bread aroma was significantly affected ($p \leq 0.05$) by wheat to modified gembili flour ratios. The highest hedonic score was observed in the pan bread control (100:0) of 5.80±0.02. Results also indicated the hedonic values of pan bread with 90:10 (5.62±0.11), 85:15

Table 5. Hedonic test results

Ratio (Wheat Flour : Modified Flour of Gembili)	Parameter					
	Crust Color	Crumb Color	Firmness	Aroma	Taste	Overall
100:0	5.63±0.07^{c}	5.77±0.09^{d}	5.68±0.01^{c}	5.80±0.02^{b}	5.80±0.07^{c}	5.95±0.16^{d}
90:10	5.42±0.01^{bc}	5.35±0.02^{bc}	5.62±0.04^{c}	5.62±0.11^{ab}	5.62±0.06^{bc}	5.73±0.05^{cd}
85:15	5.42±0.01^{bc}	5.57±0.06^{cd}	5.07±0.05^{b}	5.35±0.08^{ab}	5.50±0.05^{bc}	5.33±0.28^{bc}
80:20	5.33±0.02^{b}	5.32±0.01^{bc}	4.90±0.14^{b}	5.65±0.04^{ab}	5.60±0.02^{bc}	5.40±0.00^{b}
75:25	5.30±0.07^{b}	5.13±0.00^{ab}	4.77±0.05^{b}	5.48±0.11^{a}	5.33±0.09^{ab}	5.15±0.02^{b}
70:30	4.82±0.06^{a}	4.88±0.13^{a}	4.22±0.1^{a}	5.27±0.09^{a}	5.12±0.06^{a}	4.68±0.07^{a}

Note: Notation of different superscript on each value indicates significant difference ($p \leq 0.05$)

(5.35±0.08), and 80:20 (5.65±0.04) flour ratios exhibited similar values to that of the pan-bread control. Contrastingly, the lowest hedonic score was observed in the pan bread with 75:25 (5.48±0.11) and 70:30 ratio (5.27±0.09), indicating that the panelists prefer the pan-bread aroma with lower replacement of modified flour of gembili since those with higher substitution of modified gembili flour exhibit odd aroma.

Wheat to modified gembili flour ratios did significantly affect (p≤0.05) pan bread taste acceptance. The control pan bread (100:0) obtained the highest hedonic score of (5.80±0.07). This was not significantly different with the hedonic score obtained by the pan bread with 90:10 (5.62±0.06), 85:15 (5.50±0.05), 80:20 (55.60±0.02) flour ratio. In contrast, the pan bread with 75:25 and 70:30 had the lowest hedonic scores of 5.33±0.09 and 5.12±0.06, respectively. Similar to aroma, the lower pan-bread taste acceptance with higher replacement of modified gembili flour could be attributed to their odd taste (i.e. not a typical pan bread taste).

Overall acceptance was significantly affected (p≤0.05) by the wheat to modified gembili flour ratio. The highest hedonic score was obtained by the pan bread with 100:0 (5.95±0.16) and 90:10 (5.73±0.05) flour ratio, while the lowest score was observed in the 70:30 ratio pan bread (4.68±0.07). This shows overall acceptance is the lowest for pan bread with the highest replacement of modified flour of gembili (70:30) indicating a negative impact of higher substitutions of gembili in terms of overall reception by the panelists.

3.8.3 Chosen pan bread on the basis of physicochemical and organoleptic properties

Pan bread exhibiting most dominant better characteristics was chosen on the basis mainly of its volume, hardness, and entire acceptance. Pan bread produced from wheat to modified gembili flour ratio of 90:10 was selected since its volume, hardness, and overall hedonic acceptance since it showed these characteristics closely similar to those of the pan-bread control (100:0). Table 6 shows chemical composition of pan-bread control (100:0) and selected pan bread (90:10).

As an independent T-test revealed, the pan-bread control (100:0) and partially substituted pan bread (90:10) exhibited insignificant difference (p>0.05) for several compositions i.e. moisture, fat, ash, and carbohydrate contents. Conversely, the pan-bread control (100:0) and partially substituted pan bread (90:10) exhibited significant difference (p≤0.05) in terms protein content. The lower protein content in the partially-substituted pan bread (90:10) may be attributed due to lower protein seen in modified flour of gembili as compared to wheat flour as reported before [41].

Table 6. Chemical composition of pan bread (control and 90:10 ratio)

Wheat flour: Modified gembili flour	Amount (%, db)				
	Moisture	Protein	Fat	Ash	Carbohydrate (by difference)
100:0 (control)	30.12±0.84^{a}	12.66±0.16^{a}	7.46±0.88^{a}	2.02±0.13^{a}	47.74±2.20^{a}
90:10	29.43±1.38^{a}	12.03±0.40^{b}	7.74±0.11^{a}	2.04±0.09^{a}	48.75±1.89^{a}

Note: Notation of different superscript on each value indicates significant difference (p≤0.05)

4. Conclusion

Heat moisture treatment (HMT) modification of gembili flour and partial substitution of wheat flour with HMT modified gembili flour in making of straight-dough pan bread were successfully done. HMT modified gembili flour had improved swelling power and solubility despite reduced lightness. Modified gembili flour (83 °C; 8 h) was chosen as the best combination treatment and then used to partially substitute wheat flour for producing pan bread by employing the straight-dough method. The pan bread partly replaced with HMT modified flour of gembili (10%) exhibited similar results with the control in terms of volume, hardness, and overall acceptance score. For this reason, the pan bread with 90:10 wheat to HMT modified gembili flour ratio was chosen as the best formulation and analyzed for the chemical composition. The pan bread with 10% partial substitution of HMT-modified gembili flour exhibited a significantly lower protein content than the pan-bread control, however, showed insignificant difference in moisture, fat, ash, and carbohydrate contents. Taken all together, pan bread could be made by using 10% HMT-modified flour of gembili to partly replace wheat flour since it showed mostly comparable physicochemical and organoleptic properties to the pan-bread control.

References

[1] Ishida, Patricia Mello Garrido, and Caroline Joy Steel, 2014, Food Science and Technology 34, no. 4: 746-754.
[2] Meylinah, S., 2022, Indonesia: Grain and Feed Annual U.S. Department of Agriculture (USDA), Foreign Agricultural Service.
[3] Trimanto, Trimanto, and Lia Hapsari, 2015, AGRIVITA Journal of Agricultural Science 37(2): 97-107.
[4] Retnowati, Diah Susetyo, Ratnawati Ratnawati, and Andri Cahyo Kumoro, 2019, Reaktor 19(2):43-48.
[5] Bibiana, Igbabul, Num Grace, and Amove Julius, 2014, American Journal of Food Science and Technology 2, no. 4: 109-115.
[6] Trejo-Gonza´lez, A.S., Loyo-Gonza´lez, A.G., Mungui´a-Mazariegos, M.R., 2014, International Food Research Journal 21, no. 4:1683-1688.
[7] Ruiiz, Elvira, Khongsak Srikaeo, and Lucía Segovia de la Revilla, 2018, Food and Applied Bioscience Journal 6, no. 3: 140-153.
[8] Asranudin, Holilah, Andi Noor Kholidha Syarifin, Adi Setyo Purnomo, Ansharullah, and Ahmad Fudholi, 2021, Food Hydrocolloids 120.
[9] Dudu, Olayemi E., Ying Ma, Aminat Adelekan, Ajibola B. Oyedeji, Samson A. Oyeyinka, and Jessica W. Ogungbemi, 2020, LWT Food Science and Technology 130: 1-11.
[10] Jayakody, L., R. Hoover, Q. Liu, and E. Donner, 2007, *Carbohydrate Polymers* 69, No. 1: 148-163.
[11] Tanak, Yolivia, 2016, Jurnal Sains dan Teknologi Tadulako 5, No. 1: 39-49.
[12] Ayuningtyas, Laksmi Putri, Ashri Mukti Benita, and Desy Triastuti, 2022, Pangan Dari Agroindustri 10, no. 1: 1-8.
[13] AACC International, 2010, Approved Methods of Analysis 11th Edition. AACC International St. Paul, Minnesota.
[14] Pokatong, Wilbur Donald Raymond, and Evelind Pangestu Julista, 2022, Global Conference on Innovation in Science Technology Engineering and Mathematics 2021 1: 74-87.
[15] AOAC, 2005, Official Methods of Analysis of the Association Analytical Chemists (AOAC International, Washington DC., 2005)
[16] Mudau, Masala, Shonisani Ramashia, Mpho Mashau, and Henry Silungwe, 2021, Brazilian Journal of Food Technology 24.
[17 Heymann, H., and H.T. Lawless, 2013, Sensory Evaluation of Food: Principles and Practices: Springer US.
[18] Santosa, Herry, Noer Abyor Handayani, Hasbi Ashidqi Bastian, and Ilga Mega Kusuma, 2015, Metana, Vol. 11(01).
[19] Kumoro, Andri, D. Retnowati, Ratnawati Ratnawati, and M. Widiyanti, 2019, *Journal of Physics: Conference Series 1295.*
[20] Yan, Wenjia, Lijun Yin, Minghao Zhang, Meng Zhang, and Xin Jia, 2021, Gels 7 (4): 1-12.
[21] Li, R., Y. Liu, J. Shi, Y. Yu, H. Lu, L. Yu, Y. Liu, and F. Zhang, 2019, Biochim Biophys Acta Mol Cell Biol Lipids 1864, No. 8: 1124-1133.
[22] Lavanya, J. Perraulta, G. Gowthamraj, and Sangeetha Narayanasamy, 2021, *Journal of Food Processing and Preservation* 45 (09/29)
[23] Parwiyanti, Filli Pratama, Agus Wijaya, and Nura Malahayati, 2016, Jurnal Teknologi dan Industri Pangan 27: 185-192.
[24] Prameswari, I. K., G. J. Manuhara, B. S. Amanto, and W. Atmaka, 2018, Journal of Physics Conference Series, Vol. 1022.
[25] Ariyantoro, Achmad Ridwan, Bambang Sigit Amanto, and Kiswuri. 2020, IOP Conference Series: Earth and Environmental Science 518.
[26] Luga, Mădălina, and Silvia Mironeasa, 2021, Food Control 128: 108176.
[27] Fonseca, Laura Martins, Shanise Lisie Mello El Halal, Alvaro Renato Guerra Dias, and Elessandra da Rosa Zavareze, 2021, Carbohydrate Polymers 274: 1-15.
[28] Bakare, A. H., O. F. Osundahunsi, and J. O. Olusanya, 2016, Food Sci Nutr 4, no. 4 (Jul): 573-87.
[29] Cauvain, Stanley, 2015, Technology of Breadmaking. Third ed. New York: Springer International Publishing Switzerland.
[30] Cauvain, Stanley P, 2017, "Chapter 4 - Bread and Other Fermented Products." In *Baking Problems Solved (Second Edition)*, edited by Stanley P. Cauvain, 153-234: Woodhead Publishing.

[31] Demirkesen, Ilkem, Behic Mert, Gulum Sumnu, and Serpil Sahin, 2010, Journal of Food Engineering 96, no. 2 (2010/01/01/): 295-303.
[32] Prabowo, Aditya Yoga, Teti Estiasih, and Indria Purwantiningrum, 2014, Jurnal Pangan dan Agroindustri 2, No. 3: 129-135.
[33] Avramenko, N. A., R. T. Tyler, M. G. Scanlon, P. Hucl, and M. T. Nickerson. 2016, Food Reviews International 34 (11/21).
[34 Sareu, Pratiwi Listyana, Nurhaeni, Ahmad Ridhay, Moh. Mirzan, Syamsuddin, 2021, KOVALEN Jurnal Riset Kimia 7(1):51-58.
[35] Cauvain, S. P., 2016, "Bread: Breadmaking Processes." In Encyclopedia of Food and Health, edited by Benjamin Caballero, Paul M. Finglas, and Fidel Toldrá, 478-483. Oxford: Academic Press.
[36] Purlis, Emmanuel, 2010, Journal of Food Engineering 99: 239-249.
[37] Das, Lipi, Utpal Raychaudhuri, and Runu Chakraborty, 2012, Food Science and Biotechnology 21, no. 2 (2012/04/01): 425-433.
[38] Retnowati, Diah Susetyo, Andri Cahyo Kumoro, and Ratnawati Ratnawati, 2018, *Potravinarstvo Slovak Journal Of Food Sciences* 12, No. 1.
[39] Ukpabi, Joseph, 2010, Australian Journal of Crop Science 4(2): 68-73.
[40] Cho, In Hee, and Devin G. Peterson, 2010, *Food Science and Biotechnology* 19, no. 3 (2010/06/01): 575-582.
[41] Utami, Rohula, Esti Wahyu Widowati, and Annisa Dyah Ayu Retno Dewati, 2013, Jurnal Teknosains Pangan 2 (3).

Jelly candy with Javanese turmeric (*Curcuma xanthorrhiza* Roxb.) as a natural antioxidants source

Talita Alifa Nindiana[1], Dwiyati Pujimulyani [1*], Wisnu Adi Yulianto[1], Emi Windrayani[2]

[1]Departement of Food Technology, Faculty of Agroindustry, University of Mercu Buana Yogyakarta, Special Region of Yogyakarta, Indonesia

[2]Departement of Research and Development of CV Windra Mekar, Spesial Region of Yogyakarta, Indonesia

*E-mail: dwiyati@mercubuana-yogya.ac.id

Abstract. Javanese turmeric (JT) is a rhizome that has many health benefits. JT can be used as an additional ingredient in making jelly candy to increase consumer acceptance. This research aimed to determine the effects of adding JT and cooking temperature on JT jelly candy's antioxidant activity and organoleptic properties. The stages of jelly candy research are as follows: JT rhizomes are peeled, washed, given pressure blanching treatment for 5 minutes, dried with a cabinet dryer at 55 °C for 8 hours, milled, and extracted. Treatments varied in JT extract (10, 20, and 30 ml) and cooking temperature (60, 80, and 100 °C). The experimental design used was a Completely Randomized Design (CRD). All JT jelly candy products are tested for physical and liking levels. The data obtained was tested statistically using Analysis of Variance (ANOVA). If there were significant differences, it was further tested using the Duncan Multiple Range Test (DMRT) with an accuracy of 95%. Selected and controlled jelly candy products were chemically analyzed. Namely, water content and antioxidant activity were achieved using the 2,2-diphenyl-1-pickrylhydrazyl (DPPH) method, total phenolic content, flavonoids, and reducing sugar content. The results of this research show that the addition of 10 ml of JT extract and a cooking temperature of 60 °C was liked the panelists with a bright yellow color and had a water content of 14.5%, reducing sugar content of 1.92%, antioxidant activity of 34.14% RSA, total phenols content of 4.25 mg GAE/g db and flavonoids 0.55 mg QE/g db which meets SNI for jelly candy.

Keywords: antioxidant activity, cooking temperature, javanese turmeric, jelly candy, phenolic

1. Introduction

Javanese turmeric (Curcuma xanthorrhiza Roxb.) belongs to the Zingiberaceae family, with the part that is usually used as the rhizome. The JT is a native plant of Indonesia, found mainly in Java, Sumatra, Kalimantan, and Sulawesi (Prana, 2008). The JT rhizome contains antioxidants such as flavonoids, phenols, and curcumin (Jayaprakhasha, 2006). Research by Pujimulyani et al. (2022) shows that herbs such as white turmeric of the mango type also contain curcumin. The JT rhizome also contains starch, crude fiber, ash, protein, minerals, and essential oils consisting of d-camphor, cyclo isoprene, myrcene, tumoral, xanthorrhizol, zingiberene, gingerol (Wijayakusuma, 2007). Antioxidant activity in JT can increase due to changes in less active compounds into active compounds caused by the blanching process. Blanching is a process of heating at high temperatures for a short time with the aim of inactivating enzymes; blanching can also soften tissue and reduce contamination of harmful microorganisms so that good quality products are obtained that are dried, canned, and frozen. The blanching process aims to inactivate the polyphenolase enzyme to prevent color changes due to enzymatic browning reactions (Estiasih and Ahmadi, 2011). The blanching process can be done in two ways, namely direct heating with hot water (hot water blanching) and with steam (steam blanching) (Muchtadi, 1997). Blanching is one of the preliminary treatment stages that can increase JT's antioxidant activity. The blanching process is carried out on JT rhizomes before further processing.

The JT can be an additional ingredient in making jelly candy because it contains antioxidants. Candy is one of the snacks that is widely liked by the public, especially children because candy has a variety of flavors, colors, and attractive shapes. Candy made from fruits is one of the diversifications of food products widely known among the public. Candy circulating in the community consists of two types: hard candy and soft candy (Sahputra, 2018). The difference is based on the texture of the candy. Hard candy has a hard texture, while soft candy is characterized by its soft texture. Jelly candy is a soft candy made from fruit juice and gel-forming ingredients with a transparent appearance, specific texture, and elasticity (Jumri et al., 2015). Adding hydrocolloid components such as agar, gum, pectin, starch, carrageenan, and gelatin modify the texture to produce a chewy product. The gelatin has the properties and functions of forming gels, emulsifiers, thickeners, clarifiers, and water binders (Mahardika et al., 2014). Based on this explanation, jelly candy was studied with the addition of JT extract and variations in heating temperature.

2. Research Methodology

2.1 Tools

The tools used to make soft candy are digital scales, aluminum pans, pans, gas stoves, ovens, basins, spatulas, knives, impulse sealers, blancu cloths, thermometers, and measuring cups. The tools used for analysis are analytical scales, NH310 colorimeters, Whatman No.42 filter paper, 50 ml beakers, funnels, 50 ml measuring flasks, stirring rods, droppers, spatulas, weighing bottles, bottle clamps, Vortex 2, desiccators, Erlenmeyer flasks, UV-Vis spectrophotometers, electric stoves.

2.2 Materials

The main ingredient used in this study is JT rhizome. Other ingredients include gelatin, High Fructose Syrup, granulated sugar, and water. The chemicals used for the analysis were ethanol, BHT, 10% NaNO2, Folin-ciocalteu, 20% Na2CO3, Na2CO3, 10% NaOH, 0.1 mM DPPH solution, 10% AlCl26H2O, PB acetate, Nelson's reagent, arsenomolybdate reagent, and distilled water.

2.3 Jelly candy processing

Preparation of JT extract

JT powder begins with peeling the rhizome skin and then continues with washing to remove dirt. The rhizome is blanched under pressure at 120 °C for 5 minutes. The drying process of JT rhizome uses a cabinet dryer for 8 hours. The dried rhizome is ground using a blender to reduce the size. JT extract is made by mixing 25 g of JT powder and 500 ml of boiling water, then heated for 5 minutes. At a temperature of 60 °C. The last process is filtering the JT extract to obtain JT extract, which will be used to make jelly candy.

Jelly candy making

Jelly candy-making begins with a mixture of JT extract, high fructose syrup, gelatin, and granulated sugar. The heating is then heated for 10 minutes. The next stage is molding and cooling. The cooled jelly candy is cut into 2.5x2.5 cm^2 sizes.

2.4 Analysis

The analyses conducted in this study were physical, chemical, and preference levels (organoleptic test). The physical test conducted was a color test using coloumetry. The preference level test involved 20 panelists with color, aroma, taste, texture, and overall parameters. The assessment was conducted by scoring a preference level from 1 to 5 (1 = very dislike, 2 = dislike, 3 = somewhat like, 4 = like, 5 = very like). The chemical analysis conducted was: Water Content (AOAC, 1995), DPPH Antioxidant (Xu and Chang, 2007), Flavonoid (Dewanto et al., 2002), Total Phenol (Pujimulyani et al., 2010), Reducing sugar Nelson-Somogyi method.

2.5 Experimental design

The research method used a completely randomized design with two factors. The first factor is the addition of JT extract (variations of 10, 20, and 30 ml) and heating temperature (variations of 60, 80, and 100 °C). The experiment was conducted in 2 batches with three replications of sampling.

3. Results and Discussion

3.1 Color

The color test values of jelly candy are presented in Table 1.

Table 1. Lightness, Redness, and Yellowness values of JT jelly candy

Cooking Temperature (°C)	JT Extract (ml)	Color parameters		
		Lightness	*Redness*	*Yellowness*
60	0	49.72 ± 1.10^{e}	0.24 ± 0.31^{a}	3.84 ± 0.64^{a}
60	10	48.45 ± 0.18^{cd}	0.95 ± 0.34^{ab}	5.09 ± 0.07^{b}
80	10	47.75 ± 0.60^{bc}	1.21 ± 0.05^{bc}	4.75 ± 0.07^{b}
100	10	47.58 ± 1.63^{bc}	2.00 ± 0.58^{cd}	4.42 ± 0.28^{ab}
60	20	47.88 ± 0.46^{bcd}	1.49 ± 0.57^{bc}	5.10 ± 0.07^{b}
80	20	46.83 ± 0.79^{bc}	1.48 ± 0.14^{bc}	4.95 ± 0.14^{b}
100	20	46.36 ± 0.59^{ab}	2.06 ± 0.12^{cd}	4.80 ± 0.28^{b}
60	30	46.02 ± 0.28^{ab}	2.07 ± 0.08^{cd}	5.02 ± 0.07^{b}
80	30	45.88 ± 0.49^{ab}	2.39 ± 0.60^{d}	4.63 ± 0.71^{ab}
100	30	44.73 ± 0.93^{a}	2.43 ± 0.60^{d}	$4.48\pm0.64^{a b}$

Note: Numbers followed by different letter notations in the same column indicate significant differences at the 95% confidence level ($p<0.05$).

Lightness

The lightness color test showed that the treatment of adding JT extract and cooking temperature affected the lightness value of the jelly candy produced. The results of the lightness value measurement ranged from 44.73 to 48.45. The lowest lightness value was shown by jelly candy with a variation of 30 ml JT extract addition and a cooking temperature of 100 °C. In contrast, the highest lightness value was demonstrated by jelly candy with a variation of 10 ml JT extract addition and a cooking temperature of 60 °C.

The addition of JT extract significantly affected the lightness value of jelly candy. The addition of more JT extract decreased the lightness value of jelly candy. The cooking temperature also affected the lightness value of the jelly candy. Higher cooking temperatures can reduce the lightness value of jelly candy. This is thought to be caused by sugar in the jelly candy ingredients that undergo caramelization and non-enzymatic browning, namely the Maillard reaction. The Maillard reaction is formed between reducing sugars and amino acids and is stimulated by heat.

Redness

The redness color test ranged from 0.95 to 2.43. The lowest redness value was shown by jelly candy with 10 ml of JT extract and a cooking temperature of 60 °C, which was 0.95. The highest redness value was obtained by adding 30 ml of JT extract and a cooking temperature of 100 °C. The redness value of candy can be affected by the cooking temperature. Redness is affected by the heating process and the concentration of granulated sugar (Wicaksono and Elok, 2015).

Yellowness

The results of measuring the yellowness value of JT jelly candy ranged from 4.42 to 5.10. The yellowness of the jelly candy showed that the treatment of adding JT extract and heating temperature affected the yellowness value of the jelly candy produced. The lowest yellowness value was when adding 10 ml of JT extract and a heating temperature of 100°C. The highest yellowness value was when adding 20 ml of JT extract and a cooking temperature of 60 °C.

3.2 Preference level

The jelly candy preference level test values are presented in Table 2.

Table 2. Preference level test scores for JT jelly candy

Variations of JT Extract (ml)	Cooking Temperature (°C)	Parameter				
		Color	Flavor	Taste	Texture	Overall
0	60	3.05 ± 1.15^{b}	3.40 ± 1.05^{c}	3.55 ± 1.10^{c}	3.60 ± 0.75^{b}	3.45 ± 0.89^{cd}
10	60	4.05 ± 0.76^{e}	3.40 ± 0.91^{c}	3.55 ± 0.69^{c}	3.60 ± 1.14^{b}	3.75 ± 0.55^{d}
10	80	4.05 ± 0.67^{e}	3.35 ± 0.81^{c}	3.65 ± 0.67^{c}	3.30 ± 0.78^{b}	3.75 ± 0.72^{d}
10	100	3.20 ± 0.62^{b}	3.25 ± 0.64^{c}	3.25 ± 0.79^{bc}	2.25 ± 0.97^{a}	3.00 ± 0.72^{bc}
20	60	3.60 ± 0.82^{c}	3.20 ± 0.90^{c}	3.20 ± 0.89^{bc}	3.60 ± 1.09^{b}	3.45 ± 0.76^{cd}
20	80	3.90 ± 0.72^{d}	3.45 ± 0.76^{c}	3.05 ± 1.24^{bc}	3.45 ± 1.05^{b}	3.35 ± 1.09^{c}
20	100	3.05 ± 0.69^{b}	2.95 ± 0.60^{bc}	2.80 ± 0.83^{b}	2.05 ± 0.89^{a}	2.75 ± 0.64^{b}
30	60	3.50 ± 0.61^{b}	2.95 ± 0.76^{bc}	2.70 ± 0.92^{b}	3.50 ± 0.95^{b}	3.05 ± 0.69^{bc}
30	80	1.95 ± 0.83^{a}	2.45 ± 0.94^{ab}	2.70 ± 1.03^{b}	3.15 ± 0.87^{b}	2.70 ± 0.86^{b}
30	100	2.40 ± 0.68^{a}	2.30 ± 0.73^{a}	1.95 ± 0.94^{a}	1.85 ± 1.09^{a}	2.00 ± 0.72^{a}

Note: Numbers followed by different letter notations in the same column indicate significant differences at the 95% confidence level ($p<0.05$).

The addition of JT extract significantly affected the yellowness value of the jelly candy. The more JT extract was added, the lower the yellowness value because the jelly candy color became brown. The yellow color of the jelly candy is produced from the curcumin compound contained in the JT rhizome. According to research by Rosidi et al. (2010), the curcumin content in dried JT rhizome is 2.02%. The cooking temperature also affects the yellow color of the jelly candy; the higher the temperature, the more brown the color will be because of the sugar content in the candy ingredients that undergo caramelization.

Color

The jelly candy, with the addition of JT extract and different cooking temperatures, significantly affected the panelists' preference for the color of the jelly candy produced. The panelists' preference for color ranges from 1.95 to 4.05, which means that the panelists' assessment of the color parameters ranges from dislike to like. This means that the higher the value produced, the more it is liked by the panelists.

The jelly candy the panelists liked was made with 10 ml of JT extract and a cooking temperature of 60 °C. The jelly candy obtained was bright yellow due to the curcumin content in the JT rhizome, which gave the product color. The color of the jelly candy is more determined by the natural color of the raw material extract used and the browning results during the manufacturing process (Wardhani, 2022).

Flavor

The jelly candy, with the addition of JT extract and different cooking temperatures, has significantly affected the panelists' preference for the aroma parameters of the resulting jelly candy. Aroma parameters with panelists' preference levels range from 2.30 to 3.40, which means that the panelists' assessment of the aroma parameters is in the range of dislike to slightly like.

Based on the results of statistical tests conducted on the aroma parameters of jelly candy, it shows that the aroma preferred by the panelists is candy with the treatment of adding 10 ml of JT extract with a cooking temperature of 60 °C. Adding more JT extract to jelly candy products will affect the aroma of the resulting jelly candy. The JT contains curcuminoid compounds with a distinctive aroma and is non-toxic (Subagja, 2014). This compound is thought to strengthen the aroma of JT rhizome in the resulting JT jelly candy.

Taste

The most preferred taste by the panelists was jelly candy with 10 ml of JT extract at a cooking temperature of 60 °C. The addition of higher JT extract resulted in candy that was less preferred by the panelists because JT has a reasonably strong taste with a bitter aftertaste. The JT rhizome is divided into two main active ingredient fractions: curcuminoids and essential oils (Lukas et al., 2013). The essential oil content in JT is thought to cause the bitter aftertaste of JT jelly candy.

Texture

A product's texture significantly impacts consumer acceptance (Asropi et al., 2023). The panelists' preference level for the texture parameters of jelly candy ranged from 1.85 to 3.60, which means that the panelists' assessment of the texture parameters ranged from immensely dislike to quite like. The texture parameters showed the highest assessment preferred by the panelists was soft candy with the addition of 10 ml of JT extract and a cooking temperature of 60 °C. This is thought to be because the selected sample had a texture that was still chewy, resembling solid jelly. Meanwhile, the texture was quite complex and challenging for jelly candy with the addition of 30 ml of JT extract and a heating temperature of 100 °C.

Overall

The overall parameters were conducted to determine the panelist's response to jelly candy by adding JT extract and cooking temperature. The addition of JT extract and cooking temperature significantly affected the overall parameters of the jelly candy produced. The overall value of jelly candy ranged from 2.00 to 3.75, which means that the panelist's assessment of the overall parameters ranged from dislike to slightly like.

The results of statistical tests showed that the jelly candy preferred by the panelists for the overall parameters was soft candy with the treatment of adding 10 ml of JT extract and a cooking temperature of 60 °C. This treatment can produce jelly candy with a bright yellow color, a JT aroma that is not too intense, and a sweet taste with a slightly distinctive JT taste that gives a slightly bitter aftertaste. This chewy texture is less harsh and more attractive than other JT jelly candy treatments.

3.3 chemical Analysis

The chemical analysis values of jelly candy are presented in Table 3.

Table 3. Result of chemical analysis of JT jelly candy

Sample	Chemical Components	Condition	Analysis results
Selected V 10 T 60	Water content (% wb)	Max. 17.06 (SNI 01-3547-1994)	14.5
	Reducing sugar	Max. 25% (SNI 3547-2-2008)	1.92
	DPPH antioxidant activity (%RSA)	-	34.14
	Total phenol content (mg EAG/g db)	-	4.25
	Total flavonoid content (mg CE/g db)	-	0.55
control	Water content (% wb)	Max. 17.06 (SNI 01-3547-1994)	5.6
	Reducing sugar	Max 25% (SNI 3547-2-2008)	2.12
	DPPH antioxidant activity (%RSA)	-	5.00
	Total phenolic content (mg EAG/g db)	-	3.71
	Total flavonoid content (mg CE/g db)	-	0.45

Note: V= Variation of JT Extract (ml), T= Temperature

Water content

The results of the water content test carried out on jelly candy samples with 10 ml of JT extract and a cooking temperature of 60 °C were 14.5%. The water content of a product is thought to be influenced by the raw materials and additional materials used and the processing process carried out. Sucrose is one of the extra materials that can affect the water content of a product. Another raw material that significantly affects a product's water content is gelatin. This is thought to be because gelatin absorbs water. Gelatin is a hydrocolloid that can dissolve in water and can absorb water in relatively large amounts. Gelatin is a protein resulting from the hydrolysis of bone and skin collagen, and it is widely used for various industrial purposes. So far, various industries have used gelatine from animal skin and bones (Novitasari and Safriyanto, 2021).

Based on SNI 3547-2-2008, the maximum water content in jelly candy is 20%, so the water content in jelly candy with 10 ml JT extract and a cooking temperature of 60 °C still meets the requirements because it does not exceed the recommended quality standards.

Reducing sugar content

The results of the analysis of reducing sugar content produced in JT jelly candy with the addition of 10 ml of JT extract and a cooking temperature of 60 °C were 1.92%. Based on SNI 3547-2-2008, the value of reducing sugar content in jelly candy is a maximum of 25%, so the value of reducing sugar content in JT jelly candy with the addition of 10 ml of JT extract and a cooking temperature of 60 °C still meets the requirements because it does not exceed the recommended quality standards. According to Master (2017), the higher the cooking temperature, the greater the amount of inverted sugar. The reduced sugar content in candy is obtained from glucose and sucrose syrup, which convert into fructose and glucose. The reduced sugar content of candy is determined by the composition of the ratio of sucrose and glucose syrup (Mandei, 2014).

Antioxidant activity

The results of the analysis of antioxidant activity produced by the DPPH test on JT jelly candy with the addition of 10 ml of JT extract and a cooking temperature of 60 °C were 34.14% RSA. The antioxidant content in a product can be influenced by the type of material and processing process used in making a food product. JT is a rhizome that has antioxidant compounds in it. According to research (Susilowati & Ariviani, 2014), it states that in JT extract with a solvent proportion of 1: 1 (m/v), which is 47.23%. The JT also contains active substances in the form of essential oils of 3-12%. In addition, there are also curcuminoid compounds of 1-2% (Dermawaty, 2015). With these active substances, JT has anti-inflammatory properties and is a good antioxidant for health. Natural antioxidants are very sensitive to heat. Temperatures that are too high will damage antioxidants.

The antioxidant activity of JT is influenced by temperature; antioxidant activity is only detected at 100 °C. This is thought to be due to other compounds only extracted at 100 °C, which are responsible for increased antioxidant activity (Wardiastuti et al., 2019). So, the antioxidant activity figure is relatively low in the jelly candy sample with the addition of 10 ml of JT extract and a cooking temperature of 60 °C.

Total phenolic content

The results of the chemical analysis of JT jelly candy, the total phenol contained in the jelly candy with the addition of 10 ml JT extract and a cooking temperature of 60 °C was 4.24 mg EAG/g db. The phenol content in a product can be influenced by the use of ingredients and the processing process. According to research by Rosidi et al. (2021), the total phenol content in the JT extract was 261.9 mg EAG/g db. The total phenol content test results in JT jelly candy were low, allegedly due to the cooking process and heating in the jelly candy making process. The high temperature used during cooking will reduce the total phenol compound content because phenol compounds are susceptible to hot temperatures, resulting in phenol degradation. Phenol degradation occurs due to the oxidation of phenol compounds by oxygen, which is accelerated by temperature and heating duration (Dewata et al., 2017).

Total flavonoid content

The results of the chemical analysis of JT jelly candy, the flavonoid content in JT jelly candy with the addition of 10 ml of JT extract and a cooking temperature of 60 °C is 0.555 mg EK/g db. The results of flavonoid content in a product can be influenced by the ingredients and processing processes used. According to research by Khotimah (2023), the total flavonoid content in JT rhizome extract is 49.962 mg EK/g. The flavonoid content in JT jelly candy is low, possibly due to damage during the heating process. Flavonoids are one of the antioxidant compounds.

Antioxidant compounds are susceptible to heat and will be damaged when heated at high temperatures and for a long time.

4. Conclusions

The research concluded that JT jelly candy with 10 ml of JT extract and a cooking temperature of 60 °C has a bright yellow color, chewy texture, and a not-too-bitter solid taste of JT. The chemical composition of JT jelly candy is 14.5% water content, 1.92% reducing sugar, 34.14% RSA antioxidant activity, 4.25 mg EAG/g bk total phenol, and 0.55 mg EK/g bk flavonoids.

Acknowledgement

Thank you to CV Windra Mekar for providing research funds for the 2024 budget year so that the research runs smoothly.

References

Asropi, D., Ariani, R. P., & Masdarini, L. (2023). *Uji Organoleptik Modifikasi Kue Klemben Dengan Subtitusi Tepung Kelapa. 3.*

Dewata, I. P., Wipradnyadewi, P. A. S., & Widarta, I. W. R. (2017). *Pengaruh Suhu dan Lama Penyeduhan Terhadap Aktivitas Antioksidan dan Sifat Sensoris the Herbal Daun Alpukat (Persea americana Mill.). 6*(2).

Estiasih T. dan Ahmadi K. 2011. *Teknologi Pengolahan Pangan*. Jakarta: Bumi aksara.

Lukas, A., Purwanto, W., & Ridwam, A. Y. (2013). *Soft Candy dari Bahan Aktif Oleoresin Temulawak (Curcuma xanthorhiza Roxb.).* Jurnal Sains dan Teknologi Indonesia, 13(3). https://doi.org/10.29122/jsti.v13i3.889

Mandie, J. H. 2014. *Komposisi Beberapa Senyawa Gula Dalam Pembuatan Permen Keras Dari Buah Pala. 6*(1).

Mastur, M. (2017). *Respon Fisiologis Tanaman Tebu Terhadap Kekeringan, Physiological Responses of Sugarcane Plant to Drought. Buletin Tanaman Tembakau, Serat & Minyak Industri,* 8(2), 99. https://doi.org/10.21082/btsm.v8n2.2016.99-112

Muchtadi, T.R. 1997. *Teknologi Proses Pengolahan Pangan.* IPB-Press. Bogor.

Novitasari, M. dan Safriyanto. 2021. *Kadar Air dan Mutu Organoleptik Permen Jelly Rumput Laut Gelatin Sapi.*

Prana, M.S.2008. *The biology of temulawak (Curcuma xanthorrhiza Roxb). Proceeding of the first international symposium on temulawak.* Biopharma Research Center Bogor Agricultural University. p. 151-15.

Pujimulyani, D, Yulianto, W.A,. Setyowati, A,. Prastyo, P,. Windrayahya, S,. Maruf, A, 2022. White saffron (Curcuma mangga Val.) attenuates diabetes and improves pancreatic β-cell regeneration in streptozotocin-induced diabetic rats. Toxicology Reports Volume 9 Pages: 1213-1221

Pujimulyani, D., Raharjo, S., Marsono, Y., & Santoso, U. 2010. *Pengaruh Blanching Terhadap Aktivitas Antioksidan, Kadar Fenol, Flavonoid, dan Tanin Terkondensasi Kuni Putih (Curcuma mangga Val.). 30*(3).

Sahputra, M.B., Hamzah, F. dan Johan, V.S. 2018. *Rasio Buah Jambu Biji Merah (Psidium guajava L.) dan Buah Apel Hijau Manalagi (Mallus sylvestris Mill) terhadap Mutu Permen Jelly.* JOM UR, 5(2).

Subagja, H. P. 2014. *Waspada Kanker-Kanker Ganas Pembunuh Wanita. In Hira (Ed.), Waspada Kanker-Kanker Ganas Pembunuh Wanita (I).* Yogyakarta: FlashBooks.

Sudaryati. 2013. *Tinjauan Kualitas Permen Jelly Sirsak (Annona muricata Linn.) Terhadap Proporsi Jenis Gula dan Penambahan Gelatin.* Jurnal Rekapangan. 7 (2): 199-213.

Suripto, S., Maarif, M.S. dan Arkeman, Y., 2013. *Pengembangan Gula Cair Berbahan Baku Ubi Kayu Sebagai Alternatif Gula Kristal Dengan Pendekatan Sistem Inovasi*, Jurnal Teknik Industri, Volume 3 No 2, Juli 2013.

Susilowati, T., & Ariviani, S. 2014. *Kapasitas antioksidan dan kadar kurkuminoid ekstrak rimpang temulawak (Curcuma xanthorrhiza) menggunakan pelarut air dengan variasi proporsi pelarut dan metode pemanasan.*

Wardhani, A.P., (2022). *Karakteristik Sensori Permen Jelly dari Jeruk Siam Banjar (Citrus nobilis) dengan Variasi Konsentrasi Gelatin dan Agar.* Rawa Sains: Jurnal Sains STIPER Amuntai, 12(1), 16–22. https://doi.org/10.36589/rs.v12i1.202

Wijayakusuma, M. 2007. *Penyembuhan dengan Temulawak.* Sarana Pustaka Prima. Jakarta. 35 hal.

Properties of Turmeric (*Curcuma longa* L.) from Indonesia and Philippines

Dwiyati Pujimulyani[1*], Wisnu Adi Yulianto[1], Bayu Kanetro[1], Agus Setiyoko[1], Emi Windrayani[2], Marife R. Hilapad[3] and Edzil B. Lorenzo[3]

[1]Faculty of Agroindustry, University of Mercu Buana Yogyakarta, Indonesia
[2]Departement of Research and Development of CV Windra Mekar, Indonesia
[3]Capiz State University, Philippines

*E-mail: dwiyati@mercubuana-yogya.ac.id

Abstract. One of the ingredients that can be used for herbal medicines is the Curcuma longa L. rhizome. Citric acid as a blanching medium is reported to be able to chelate metals. This research aimed to determine the antioxidative properties of Curcuma longa L. from Indonesia and the Philippines with pressurized blanching treatment. These steps of the research of Curcuma longa: sorting, peeling, washing, pressurized blanching at a temperature of 120 °C with variations in blanching time of 0, 2.5, 5, 7 minutes, slicing, drying, grinding, sieving, and obtaining powder. The powder of Curcuma longa L. was analyzed for antioxidant activity using the DPPH (2,2-diphenyl-1-picrylhidrazil) method, total phenols, total flavonoids, and selected powders were tested for polyphenol profiles using HPLC. The selected turmeric was tested for phenolic profile by HPLC. This research concludes that turmeric was treated with pressurized blanching with 0.05% citric acid blanching media, and 7.5 minutes has the highest antioxidant activity. Turmeric with a slice thickness of 6 mm showed antioxidant activity using the DPPH method and total phenolic content significantly higher than those with a slice thickness of 2 mm. Antioxidant activity and total phenolic content in the main rhizome, the first branch of turmeric from Indonesia, were higher than those from the Philippines. The phenolic profile of selected turmeric (Indonesia) with HPLC method contains EGCG 46.55 mg/g, curcumin 146.21 mg/g, quercetin 5.84 mg/g, catechin 14.85 mg/g and rutin 0.31 mg/g.

Keywords: antioxidant activity, Curcuma longa L., pressurized blanching

1. Introduction

Indonesia, a country abundant in diverse flora and fauna, holds significant potential for the utilization of herbal medicines. Many of these resources remain untapped, offering vast opportunities as sources of medicinal compounds (Anonymous, 2000 in Sari, 2012). One of the ingredients that can be used for herbal medicines is the *Curcuma longa* L. rhizome obtained from the *Curcuma longa* L. plant (*Curcuma longa* L.). Along with the times, the efficacy of *Curcuma longa* L. as an alternative medicine is increasingly being recognized. It has been scientifically proven through various studies (Damayanti, 2008).

Curcuma longa L. is a medicinal plant belong to the rhizomes tribe (*Zingiberaceae*), and it is often used by the public both as a cooking spice and as a medicinal ingredient (Sudrajat, 2011). Medicinal plants, including rhizomes, roots, stems, leaves, flowers, and fruits, generally contain active compounds such as alkaloids, phenolics, terpenoids, essential oils, and glycosides. These compounds exhibit various properties, including antiviral, antibacterial, and immunomodulatory effects (Soedibyo, 1992). The phenolic compounds in *Curcuma longa* L. have antioxidant properties. Explains that the phenolic compounds in *Curcuma longa* L. are 95.76 mg GAE/g (Melannisa et al., 2011).

According to (Molyneux, 2004) high and low antioxidant activity is influenced by various factors, including its properties, which are easily damaged when exposed to oxygen, light, high temperature, and drying. *Curcuma longa* L. has been widely used in households and industries. The use of *Curcuma longa* L. in the industrial sector includes the food, beverage, medicine, textile, and cosmetic industries. Increasing the use of *Curcuma longa* L. in the pharmaceutical industry requires good processing techniques to improve its quality (Sembiring et al., 2006) (Xu & Chang, 2008).

According to Harris and Karmas (1989), Blanching is a process employed to deactivate enzymes that influence changes in ingredients' color, flavor, and nutritional content. Furthermore, this treatment helps eliminate oxygen from tissues, decreases the presence of fungi and bacteria, and accelerates the drying process. Blanching also softens cell walls, making them more permeable to water, facilitating rapid evaporation, and reducing drying time.

In addition, it turned out that the blanching treatment of frozen product vegetables was shown to have higher antioxidant activity than non-blanching (Barrett and Theerakulkait, 1995). It is hoped that the appropriate blanching time will produce quality processed products. The factors affecting heat penetration into the material to be blanched include maturity level, size, variety, temperature, and type of heating medium (Muljohardjo and Gardjito, 1973; Hui, 1992). According to Chandler and Schwartz (1998), sweet potato blanching treatment can cause the extraction of active components more efficiently than non-blanching materials because the cell walls become soft and more permeable.

Blanching can either reduce or enhance the antioxidant content in agricultural products. For instance, blanching red turi flowers lowers anthocyanin and vitamin C levels, as these antioxidant compounds dissolve into the blanching medium, and heat exposure is believed to cause oxidative damage (Wahyuningsih, 2008). Blanching red cabbage at a temperature of 94-96 °C for 3 minutes reduced the FRAP (Ferric Reducing Antioxidant Power) value to 42% and ORAC (Oxygen Radical Absorbance Capacity) by 51%, and the total phenol level decreased by 43% (Volden et al., 2008). Blanching lentil seeds with a pressure of 5 psi for 5 minutes decreased the inhibition of free radicals by 29.5%, ORAC by 11.1%, and total phenol levels by 67.9% (Xu & Chang, 2008). The decrease in antioxidant activity in the form of vitamin C is thought to be damaged during the heating process by blanching.

Meanwhile, research on several commodities showed that blanching could increase antioxidant activity.(Li et al., 2009) Puuponen-Pimia et al. (2003) reported that the antioxidant activity of cabbage was increased by 9% compared to that of cabbage without blanching. Heating the oven on the first day with a temperature of 100 °C wheat after harvesting showed an increase in the total phenol of wheat flour, such as ferulic acid, vanillic acid, and p-cumaric acid (Cheng et al., 2006). Likewise, blanching maize with an autoclave can increase total phenol (Randhir et al., 2008). The increase in antioxidant activity during the blanching process is determined by the type of antioxidant component, namely the phenolic component in the material.

The antioxidant activity of nuts, corn, and tomatoes, as measured by the DPPH method, increased after blanching (Kwan et al., 2007). Brussels sprouts (*Brassica oleracea* L.) with water blanching treatment at 100 °C for 2 and 3 minutes have higher antioxidant activity than non-blanched Brussels sprouts (Vina et al., 2007). The increase in antioxidant activity due to blanching is believed to result from the enhanced release of antioxidant components from cell tissues, improving extraction efficiency. For example, heated bilberry extract exhibits higher antioxidant activity than unheated extract, likely due to the conversion of glycosides into aglycones and sugars (Yue and Xu, 2008). This is by the opinion of Zhang et al. (2004) and Sadivola et al. (2006) that hydrolysis of anthocyanin glycosides under acidic conditions resulted in anthocyanidin aglycones. According to Li et al., (2009), the antioxidant activity of aglycones is higher than glycosides.

According to research by Khamidah et al., (2017), *Curcuma longa* L. shows antioxidant activity. Citric acid used as a blanching medium is reported to have the ability to chelate metals. Initial studies indicate that a 0.05% citric acid solution as a blanching medium yields the greatest increase in antioxidant activity.

2. Materials and Methods

2.1 Materials and tools

The primary material used in this study is 11-month-old *Curcuma longa* L. with a characteristic yellowness color, obtained from Sedayu, Bantul, Yogyakarta. The chemicals used are DPPH *(2,2-diphenyl-1-picrylhidrazil)*, gallic acid (GA = Gallic acid), quercetin, Folin-Ciocalteu reagent (Sigma Chemical Co., St. Louis), ethanol, methanol, HCl, buffer acetate, *2,4,6-tri- (2-pyridyl) -s-triazine* (TPTZ), $FeCl_3.6H_2O$, Na_2CO_3, $NaNO_2$, $AlCl_3.6H_2O$, NaOH, citric acid, acetone, acetic acid, vanillin, H_3PO_4, gas N_2, CH_3CN, ethyl acetate (E Merck), aquabidestilata (Ika Pharmindo) aquades, 96% ethanol, curcumin standard, and methanol.

The primary tools in this study is 11-month-old *Curcuma longa* L. with a characteristic yellowness color, obtained from Sedayu, Bantul, Yogyakarta. The chemicals used are DPPH *(2,2-diphenyl-1-picrylhidrazil)*, gallic acid (GA = Gallic acid), quercetin, Folin-Ciocalteu reagent (Sigma Chemical Co., St. Louis), ethanol, methanol, HCl, buffer acetate, *2,4,6-tri- (2-pyridyl) -s-triazine* (TPTZ), $FeCl_3.6H_2O$, Na_2CO_3, $NaNO_2$, $AlCl_3.6H_2O$, NaOH, citric acid, acetone, acetic acid, vanillin, H_3PO_4, gas N_2, CH_3CN, ethyl acetate (E Merck), aquabidestilata (Ika Pharmindo) aquades, 96% ethanol, curcumin standard, and methanol.

2.2 Research methods

This study examines the antioxidant activity of different parts of *Curcuma longa* L. using the DPPH method, focusing on the effects of blanching time and pressurized blanching media. The blanching durations tested include 0, 2.5, 5.0, and 7.5 minutes, with citric acid concentrations varying at 0%, 0.05%, and 0.10%. Additionally, the antioxidant activity and total phenolic content are analyzed using *Curcuma longa* L. slices with 6 mm, 4 mm, and 2 mm.

2.2.1 Antioxidant activity analysis of 2.2-diphenyl-1-picrylhydrazyl (DPPH) method (Rohman., 2020a)

A 0.2 ml sample was added to 3.8 ml of 0.1 mM DPPH solution, vortexed for a minute, and incubated at room temperature (27 °C) under conditions for 30 min. The absorbance was measured at λ 517 nm with a UV-Vis Spectrophotometer (Genesys). Blank using ethanol.

2.2.2 Total Phenolic Content analysis using the Folin-Ciocalteu method (Rohman., 2020a)

A 50 μl CX extract sample and 250 μl Folin-Ciocalteu solution were added and then allowed to stand for a minute. Then, 750 μl Na_2CO_3 20% was added, vortexed, and distilled water was added to a volume of 5 ml. After that, it was incubated for two hours at room temperature. The absorbance was measured at λ 760 nm.

2.2.3 Flavonoid analysis (Rohman., 2020b)

50 μl of CX extract was added with 4 ml distilled water and 0.3 ml of 10% $NaNO_2$ and let stand for 6 min. Add 0.3 ml $AlCl_3.6H_2O$ 10%, allow to stand for 5 min, then add 4 ml NaOH 10%. The mixture was added with distilled water up to 10 ml, vortexed for a minute, and left for 15 min. The absorbance was measured at λ 510 nm, and the blank was distilled water. Flavonoid content using quercetin standard is calculated as mg equivalent to Quercetin (QE)/g.

2.2.4 Determination of specific phenolic profile (EGCG, Curcumin, Quercetin, Catechin, Rutin) using the HPLC method (Violetta et al., 2012)

The analysis preparation involved weighing 1.0 g of the sample, dissolving it in methanol to a final volume of 10 mL, and sonicating the mixture in an ultrasonic bath. The solution was then filtered through a 0.4 μm Milex filter before injection into a Hitachi Lachrome Elite HPLC system. The column used was a C18 (4.6 mm diameter x 250 mm length) with a 5 μm particle size, equipped with a UV detector set at 254 nm. The eluent consisted of 0.1% acetic acid (A) and methanol (B) in a gradient system: at 0 minutes, 95% A, with increased polarity between 60–65 minutes at 5% A. The compounds were quantified by comparing the chromatogram peak areas of the sample with those of known standards for EGCG, curcumin, quercetin, catechin, and rutin, for which standard curves were prepared.

3. Results and Discussion

3.1 The treatment of citric acid concentration and blanching time

3.1.1 Antioxidant activity

Antioxidant activity of turmeric from Indonesia and Philipines showed in Table 1.

Table 1. Antioxidant activity of turmeric with DPPH method (Indonesia and Philippines)

Time (minutes)	Indonesia			Philippines		
	Citric acid concentration (%)			Citric acid concentration (%)		
	0	0.05	0.1	0	0.5	0.1
0	90.13^{a}	90.13^{a}	90.13^{a}	71.57^{a}	71.57^{a}	71.57^{a}
2.5	91.89^{b}	92.39^{bc}	90,65^{a}	79.59^{cd}	77.68^{bc}	81.89^{de}
5	92.92^{c}	92.56^{c}	91.92^{b}	77.81^{bc}	80.40^{cd}	80.59^{cd}
7.5	92.96^{c}	92.57^{c}	92.53^{c}	75.04^{b}	84.33^{e}	81.37^{d}

Note: Numbers accompanied by different letter notations within the same rows and columns denote significant differences at a 95% confidence level ($p<0.05$)

Table 1 shows that after pressure blanching at 120 °C, turmeric powder has a significantly greater free radical scavenging power than fresh turmeric. Indonesian and Philippine turmeric rhizomes blanched for 2.5, 5, and 7.5 minutes show higher antioxidant activity than fresh ones. The enhancement of antioxidant activity is believed to result from blanching treatment, which can break down glycosides into aglycones. This is by a study by Pujimulyani et al. (2020), which found that blanching white turmeric causes rutin glycosides to be degraded into quercetin aglycones. Furthermore, the rise in RSA value is believed to be attributed to the findings of a study. (El-Nafad et al., 2022) that there was an increase in RSA in organic and conventional green beans that were treated with hot water blanching or steam blanching methods. According to Gawlik-Dziki, (2008), blanching broccoli for 5 minutes can increase antioxidant activity compared to fresh. *Brassica oleracea* L. blanched at 100 °C for 2 min and 3 min had higher antioxidant activity than fresh (Vina et al. 2007).

The average antioxidant activity of blanched Indonesian turmeric ranged from 90.13% to 92.96% RSA. Analysis of variance revealed that blanching time, citric acid concentration, and their interaction significantly influenced antioxidant activity ($\alpha = 0.05$). As shown in Table 1, the lowest antioxidant activity (90.13% RSA) was observed with 0 minutes of blanching and citric acid concentrations of 0%, 0.05%, and 0.1%, while the highest activity (92.96% RSA) was recorded at 7.5 minutes of blanching with 0% citric acid. Antioxidants are compounds capable of slowing oxidation reactions that lead to rancidity and product degradation in fat-containing food materials (Putri & Susanto, 2015). Citric acid can be a chelating agent. It lowers pH to produce more H and regenerates antioxidant compounds by binding phenoxy radicals to form compounds again (Shinde et al., 2011).

The data above shows that the longer the blanching time, up to 7.5 minutes, can increase antioxidant activity and then decrease. According to research by Pujimulyani et al., (2020), total phenol and DPPH values will increase with increasing blanching time. The amount of citric acid added to the blanching medium, along with the blanching duration, can influence antioxidant activity. The data above shows that the aquadest blanching media with a 5 and 7.5 minute has high antioxidant activity.

Table 1 showed that the antioxidant activity of blanched Philippine turmeric ranged from 71.57% to 84.33% RSA. After blanching, the antioxidant activity was higher than that of fresh turmeric. The addition of citric acid helped preserve and enhance the %RSA of Philippine turmeric. These findings suggest that the antioxidant activity of turmeric in Indonesia is greater than Philippines, likely due to Indonesia's more favorable soil, climate, and weather conditions for turmeric cultivation.

3.1.2 Color

Table 2 presents the color values of turmeric with variations in blanching time and media. Lightness (L*) value indicates brightness (darkness of Color) (Winarno, 2004). The brightness value of blanched Indonesian turmeric powder ranges from 45.05–48.28. The analysis showed that the treatment of blanching time and citric acid concentration had a significant effect ($\alpha = 0.05$) on Color. Based on Table 2, the lowest lightness (L*) value was found in the treatment of 0 minutes blanching time and 0% citric acid concentration, which was 45.05, while the highest lightness (L*) value was found in 7.5 minutes blanching time with citric acid concentration of 0.1%, which was 48.28. The lightness value increased with the length of blanching time. The blanching treatment can increase the brightness of food ingredients (Ferdiansyah et al., 2015). This is suitable by the research by Fidela et al. (2021) that several types of ginger have increased brightness after the blanching process before further processing. The results of the variance analysis showed that the blanching time and the addition of citric acid had a very significant effect on the brightness of the turmeric. This is thought to be due to the blanching process, which can deactivate enzymes and prevent browning in turmeric.

Table 2 presented the color analysis results of Philippine turmeric, where the lightness values ranged from 20.58 to 55.78. Philippine turmeric powder, blanched with varying times and blanching media, exhibited significant differences. Blanched turmeric displayed a brighter color compared to fresh turmeric. The lowest lightness value of 20.58 was observed in turmeric powder with no blanching (fresh turmeric), while the highest value of 55.78 was found in turmeric treated with 5 minutes of blanching and 0.1% citric acid.

Table 2. Color value of pressurized blanching turmeric (Indonesia and Philippines)

		Indonesia			Philippines		
Treatment		Color			Color		
Citric Acid	*Blanching* (minutes)	L*	a*	b*	L*	a*	b*
0	0	45,05^{a}	13,93^{a}	27,65^{a}	20.58^{a}	23.40^{b}	24.99^{a}
0	2.5	45,82^{b}	16,42^{b}	30,12^{b}	49.58^{d}	25.97^{i}	53.16^{d}
0	5	48,66^{ef}	17,62^{c}	34,16^{f}	48.17^{b}	24.12^{e}	52.10^{c}
0	7,5	48,99^{f}	18,28^{d}	34,51^{g}	49.05^{c}	21.07^{a}	51.08^{b}
0.05	2.5	47,54^{c}	18,35^{d}	32,56^{de}	52.82^{h}	23.76^{c}	59.94^{h}
0,05	5	48,05^{d}	19,52^{e}	31,90^{c}	52.61^{g}	24.22^{f}	57.09^{e}
0,05	7,5	48,04^{d}	20,15^{f}	32,43^{de}	50.71^{f}	24.66^{g}	57.42^{f}
0,1	2.5	47,11^{c}	20,62^{fg}	32,32^{d}	50.68^{e}	26.55^{j}	58.36^{g}
0,1	5	47,51^{c}	21,28^{h}	32,71^{e}	55.78^{j}	24.00^{d}	63.69^{j}
0,1	7,5	48,28^{de}	20,93^{gh}	31,89^{c}	53.12^{i}	25.86^{h}	60.47^{i}

Note: Numbers accompanied by different letter notations within the same rows and columns denote significant differences at a 95% confidence level ($p<0.05$).

Based on these results, the lightness value of Indonesian turmeric is lower than that of the Philippines. The blanching treatment of Indonesian and Philippine turmeric powder showed a higher lightness value than fresh.

The (a*) value indicates the redness of turmeric powder. The (a*) value of blanched Indonesian turmeric ranges from 13.93 to 21.28. The analysis results show that the treatment of blanching time and citric acid concentration did not have a significant effect ($\alpha=0.05$). Based on Table 2, the lowest (a*) value was found in the 0-minute blanching time treatment with 0% citric acid concentration, which was 13.93, while the highest (a*) value was found in the 5-minute blanching time treatment. With a citric acid concentration of 0.1%, which was 21.28. As blanching time increases and citric acid concentration rises, the resulting (a*) value also increases. The curcumin compound remains stable in acidic conditions but is unstable in bright light and alkaline environments. Curcumin is stable at high temperatures and can be applied to heating products. According to Diannita et al. (2024), the redness of turmeric powder is due to the compound curcumin as a natural yellow-to-orange coloring agent.

Table 2 showed that the redness value of Philippine turmeric powder, ranging from 21.07 to 26.55. Philippine turmeric blanched powder with citric acid media showed a significant difference. The lowest (a*) value at 7.5 minutes blanching time with 0% citric acid media was 21.07, and the highest at 2.5 minutes blanching time with 0.1% citric acid media was 26.55.

Based on these results, it can be seen that the redness value of Indonesian turmeric is lower than that of Philippine turmeric. The blanching treatment of Indonesian and Philippines turmeric showed a higher redness value than fresh turmeric.

The yellowness value (b*) indicates the level of yellow Color of turmeric. The yellowness value of blanching turmeric ranges from 27.65-34.51. The analysis showed that the blanching time treatment and citric acid concentration had no significant effect ($\alpha=0.05$). Based on Table 2, the lowest (b*) value was found in the blanching time treatment of 0 minutes with 0% citric acid, namely 27.65, while the highest (b*) value was found in the blanching time treatment of 7.5 minutes with 0% citric acid concentration, namely 24.51. The yellow color in turmeric is due to its curcumin content. Curcumin in turmeric acts as a natural dye, imparting a yellow-orange hue (Abd El-Hack et al., 2021).

Table 2 showed the color test results for Philippine turmeric powder samples. The lightness value of turmeric blanched with citric acid ranged from 20.58 to 55.78, the redness value from 21.07 to 26.55, and the yellowness value from 24.99 to 63.69. These findings indicate that blanching with citric acid media influences the lightness and yellowness of turmeric. Blanched turmeric powder appears brighter and more yellow than fresh turmeric. Additionally, the data show that Philippine turmeric has a higher yellowness value compared to Indonesian turmeric.

3.2 Slice Thickness Treatment

The values of antioxidant activity and total phenolic content of the slice thickness variations are presented in Table 3 (Indonesia and Philippines).

Table 3. Results of antioxidant activity total phenolic content analysis using the DPPH method (Indonesia)

Thickness (mm)	Antioxidant activity (%RSA)	Total Phenol Content (mg EAG/g db)
Indonesia		
Sliced 6 mm	84,48^{b}±0.71	28.38^{b}±0.33
Sliced 4 mm	80,69^{ab}±0.57	25.99^{a}±0.66
Sliced 2 mm	75,84^{a}±0.28	24.08^{a}±0.92
Philippines		
Sliced 6 mm	76.96^{b}±1.70	19.67^{b}±0.82
Sliced 4 mm	74.64^{a}±1.43	18.49^{a}±0.68
Sliced 2 mm	74.06^{a}±2.30	18.01^{a}±0.21

Note: Numbers accompanied by different letter notations within the same rows and columns denote significant differences at a 95% confidence level ($p<0.05$).

Table 3 showed that different levels of turmeric rhizome slice thickness have different antioxidant activities; the smaller the thickness of the turmeric rhizome slice, the lower the antioxidant activity and total phenols. Thinly sliced turmeric has a larger surface area, so contact with air causes oxidation, which can damage the active antioxidant compounds. According to Radeep et al. (2016), a larger slice thickness can maintain its antioxidant activity better during drying. The highest antioxidant activity was found in samples with a slice thickness of 6 mm, namely 71.68% RSA, and the lowest was found in samples with a slice thickness of 2 mm, namely around 64.35% RSA.

Based on Table 3 it can be seen that the thickness of the turmeric rhizome slices has a difference in total phenol; the thinner the thickness of the turmeric rhizome slices, the smaller the total phenol content of the turmeric rhizome. The total phenol content of Indonesian turmeric ranges from 24.08-28.38 mg EAG/g. Turmeric with 2 mm slices has the lowest total phenol content of 24.08 mg EAG/g; the highest is in turmeric with 6 mm slices of 28.38 mg EAG/g. The phenol content of Philippine turmeric ranges from 19.67-18.01 mg EAG/g. Turmeric with 2 mm slices has the lowest total phenol content of 18.01 mg EAG/g; the highest is in turmeric with 6 mm slices of 19.67 mg EAG/g.

In contrast to Indonesian turmeric, the highest antioxidant activity results were found in the Philippines turmeric sample sliced 6 mm, namely 76.96% RSA, and the lowest antioxidant activity was found in the sample with 2 mm slices, 74.06% RSA. The highest total phenol content was found in the sample with 6 mm turmeric slices, namely 19.67 mg EAG/g, and the lowest in the sample with 2 mm slices, 18.01 mg EAG/g. The difference in results between Philippines turmeric and Indonesia turmeric is thought to be due to environmental factors.

3.3 Treatment of Rhizome Section

Table 4 (Indonesia) and Table 6 (Philippines) present the antioxidant activities of the first and second branches.

Table 4. Antioxidant activity of the main rhizome and first branch (Indonesia)

Sample	Antioxidant Activity (%RSA)
Main Rhizome (Filipina)	69.47^{a}±0.08
First branches (Filipina)	70.17^{a}±0.07
Main Rhizome (Indonesia)	87.43^{b}±1.63
First branches (Indonesia)	88.31^{b}±0.35

Note: Numbers accompanied by different letter notations within the same rows and columns denote significant differences at a 95% confidence level ($p<0.05$).

Table 4 shows that the highest antioxidant activity content is in the first branches and the lowest in the main rhizome. Antioxidant activity in the first branches is higher than that of the main rhizome due to the high concentration of active compounds such as curcumin and polyphenol compounds as antioxidants. Research conducted by Suprihatin et al. (2020) found that the compound with the highest content in turmeric powder is curcumin and its derivatives.

The level of antioxidant activity between the main rhizome and the first branches showed a significant difference. The highest antioxidant was found in the turmeric part, namely in the first rhizome of Indonesia, 88.31% RSA, and the lowest antioxidant was found in the main rhizome of Philipin, 69.47% RSA.

Table 5. Total phenolic content of the first rhizome and second branches (Philippines)

Sample	Total Phenolic Content (mg EAG/gdb)
Main Rhizome (Filipina)	18.56^{a}
First branches (Filipina)	17.25^{a}
Main Rhizome (Indonesia)	40.34^{c}
First branches (Indonesia)	37.5^{b}

Note: Numbers followed by different letter notations in the same column indicate significant differences at the 95% confidence level ($p<0.05$).

Table 5 shows that the total phenolic content in the main rhizome of Indonesia is 40.34 mg GAE/g db, and the lowest in the first branch of the Philippines is 17.25 mg EAG/g db. The specific phenolic content profile of turmeric blanched with 0.05 g citric acid media and 7.5 minutes using the HPLC method is presented in Table 6 (Indonesia).

Table 6. Specific phenolic content profile in *Curcuma longa* L. using the HPLC method (Indonesia)

Compound	Content
EGCG (mg/g)	56.55
Curcumin (mg/g)	146.21
Quercetin (mg/g)	5.84
Cathecin (mg/g)	14.85
Rutin (mg/g)	0.31

The EGCG (epigallocatechin gallate) analysis showed that the turmeric with pressure blanching treatment with 0.05% citric acid media and 7.5 minutes showed the highest EGCG value of around 46.55 mg/g. The hydrolysis of the phenol complex caused the increase in EGCG content in turmeric during the blanching process. The increase in antioxidant activity is thought to be because the blanching treatment can cause antioxidant components to be quickly released from cells, thereby increasing the extraction results. Pujimulyani et al., (2010) stated that blanching treatment in 0.05% citric acid media at a temperature of 100 °C can increase antioxidant activity, flavonoids, and total phenolic content levels.

Curcuminoid compounds are polyphenol compounds that are yellow in turmeric, temulawak, and several Zingiberaceae plants. The active compounds in curcuminoids include curcumin, demethoxycurcumin, and bidemethoxycurcumin (Christina et al., 2019). Curcumin is a phytochemical with various biological effects, such as antidyslipidemic, antioxidant, anti-inflammatory, antiviral, antifungal, and antibacterial properties, and it also offers liver protection (Vikri et al., 2022). The analysis showed that the highest curcumin content was found in the turmeric sample treated with pressure blanching with 0.05% citric acid media and a time of 7.5 minutes, which was 146.21 mg/g.

Quercetin is a group of flavonoid compounds that have phenolic compounds. The analysis showed that the value obtained in the turmeric sample treated with pressure blanching with 0.05% citric acid media and a time of 7.5 minutes was 5.84 mg/g. This is thought to be because flavonoids in the form of glycosides are hydrolyzed into aglycones. Previous research states that the hydrolysis of anthocyanin glycosides in an acidic environment produces anthocyanidin aglycones (Sadilova et al.,2006) . The results of Yue and Xu (2008) also showed that hydrolysis of glycoside compounds occurred in heated bilberry extract. Increasing flavonoid levels in blanched white turmeric can increase antioxidant activity. Flavonoids are antioxidants, antimutagenic, antineoplastic, and vasodilatory (Chopipah & Solihat, 2021).

Catechin is one of the polyphenol derivatives with high antioxidant properties. Likewise, turmeric also has antioxidant properties because it contains curcumin compounds (Setyowati and Suryani, 2013). The catechin content in the pressure blanching treatment with 0.05% citric acid media and 7.5 minutes was 14.85 mg/g.

Rutin is also known as rutiside, quercetin-3-rutinosidee, or sophorin. Generally, rutin can be isolated from plants by solvent extraction. Rutin can be easily oxidized by hydrolysis to its aglycone, quercetin. The analysis showed that the rutin content in the turmeric sample blanching treatment with 0.05% citric acid and 7.5 minutes blanching time was 0.31 mg/g. Based on the data, it can be seen that the main rhizome with the HPLC method of turmeric has the most significant antioxidant activity compared to the main rhizome of turmeric. The first branch has an antioxidant activity of 70.17% RSA and the largest total phenol, namely in the main rhizome of turmeric at 18.56 mg EAG/g bk. The first branch has the least curcumin content compared to the main rhizome, 52.40 mg/100 g.

4. Conclusion

This study concludes that turmeric treated with pressurized blanching in a 0.05% citric acid blanching medium for 7.5 minutes exhibited the highest antioxidant activity. Turmeric slices with a thickness of 6 mm demonstrated significantly higher antioxidant activity and total phenolic content, as measured by the DPPH method, compared to those with a 2 mm thickness. The antioxidant activity and total phenolic content of turmeric from the main rhizome and the first branch of Indonesia were greater than those from the Philippines. The phenolic profile of the selected turmeric from Indonesia, analyzed using the HPLC method, included EGCG (46.55 mg/g), curcumin (146.21 mg/g), quercetin (5.84 mg/g), catechin (14.85 mg/g), and rutin (0.31 mg/g).

Acknowledgements
Thank you to Mercu Buana Yogyakarta University (UMB Yogyakarta) and Capiz State University (CAPSU) Philippines for providing research funds for the 2021 fiscal year so that the research runs smoothly.

References

Abd El-Hack, M. E., El-Saadony, M. T., Swelum, A. A., Arif, M., Abo Ghanima, M. M., Shukry, M., Noreldin, A., Taha, A. E., & El-Tarabily, K. A. (2021). Curcumin, the active substance of turmeric: Its effects on health and ways to improve its bioavailability. *Journal of the Science of Food and Agriculture*, *101*(14), 5747–5762.

Barrett DM, Theerakulkait, C., 1995. Indikator kualitas dalam sayuran yang direbus, dibekukan, dan disimpan. *Teknologi Pangan.* Vol 49 hal 62-65

Chandler LA, Schwartz, SJ, 1998. Isomerisasi dan kehilangan trans-p-karoten dalam ubi jalar yang dipengaruhi oleh perlakuan pemrosesan . *J Agric Food Chem*, 36: 129-133.

Chopipah, S., & Solihat, S. S. (2021). Aktivitas Antioksidan Senyawa Flavonoid pada Daun Benalu, Katuk, Johar, dan Kajajahi: Review. *Tropical Bioscience: Journal Of Biological Science*, *1*(2), Article 2.

Christina, I. A. M., Kencana, I. N., & Permana, I. D. G. M. (2019). Pengaruh Metode Pengeringan dan Jenis Pelarut terhadap Rendemen dan Kadar Kurkumin Ekstrak Kunyit (Curcuma domestica Val). *Jurnal Ilmiah Teknologi Pertanian Agrotechno*, *3*(2), 319. https://doi.org/10.24843/JITPA.2018.v03.i02.p02

Damayanti, R. (2008). *Uji Efek Sediaan Serbuk Instan Rimpang Temulawak (Curcuma xanthorrhiza Roxb) Sebagai Tonikum Terhadap Mencit Jantan Galur Swiss Webster* [PhD Thesis, Universitas Muhammadiyah Surakarta].

Diannita, E. M., Pujimulyani, D., & Murti, S. T. C. (2024). Characteristics of Physicochemistry and Sensory of Arrowroot-wheat Flour Katte Tong with White Turmeric Powder and Carboxymethyl Cellulose Addition: Karakteristik Fisik, Kimia dan Kesukaan Katte Tong Pati Garut-terigu dengan Penambahan Bubuk Kunir Putih (Curcuma mangga Val.) dan Carboxymethyl Cellulose. *JITIPARI (Jurnal Ilmiah Teknologi dan Industri Pangan UNISRI)*, *9*(1), 42–53.

El-Nafad, R., El-Bastawesy, A., Ghoniem, G., & El-Refai, A. (2022). Effect of Blanching Process and Frozen Storage on Bioactive Compounds and Some Quality Parameters of Organic and Conventional Green Bean. *Journal of Food and Dairy Sciences*, *13*(8), 125–131. https://doi.org/10.21608/jfds.2022.150884.1069

Ferdiansyah, M. K., Retnowati, E. I., Muflihati, I., & Affandi, A. R. (2015). Peningkatan derajat putih tepung umbi suweg (Amorpophalus oncophilus) dengan kombinasi proses blanching dan bleaching menggunakan larutan sodium metabisulfit. *Jurnal Pangan Dan Gizi*, *5*(1).

Fidela, Z., Muflihati, I., Purwati Nurlaili, E., & Rakhman Affandi, A. (2021). Efek Jenis Jahe dan Pre Treatment Blansing Terhadap Karakteristik Fisikokimia Minuman Jahe Instan. *Journal of Food and Culinary*, *4*(2), 99–110.

Gawlik-Dziki, U. (2008). Effect of hydrothermal treatment on the antioxidant properties of broccoli (Brassica oleracea var. Botrytis italica) florets. *Food Chemistry*, *109*(2), 393–401.

Harris, Robert S., dan Endel Karmas. 1989. Evaluasi Gizi Pada Pengolahan Bahan Pangan. Penerbit Institut Teknologi Bandung (ITB). Bandung

Khamidah, A., Antarlina, S. S., & Sudaryono, T. (2017). Ragam produk olahan temulawak untuk mendukung keanekaragaman pangan. *Jurnal Litbang Pertanian*, *36*(1), 1–12.

Kwan, Y., I., Apostolidis, E., Shetty, K., 2007. Diet tradisional orang Amerika untuk penatalaksanaan diabetes dan hipertensi . *J. Makanan Obat*, 10: 266-275.

Li, C., Du, H., Wang, L., Shu, Q., Zheng, Y., Xu, Y., Zhang, J., Zhang, J., Yang, R., & Ge, Y. (2009). Flavonoid Composition and Antioxidant Activity of Tree Peony (Paeonia Section *Moutan*) Yellow Flowers. *Journal of Agricultural and Food Chemistry*, *57*(18), 8496–8503. https://doi.org/10.1021/jf902103b

Melannisa, R., Da'i, M., & Rahmi, R. T. (2011). *Uji aktivitas penangkap radikal bebas dan penetapan kadar fenolik total ekstrak etanol tiga rimpang genus curcuma dan rimpang temu kunci (Boesenbergia pandurata).*

Molyneux, P. (2004). Penggunaan Radikal Bebas Difenilpikril Hidrazil (DPPH) Yang Stabil Untuk Pendugaan Aktivitas Antioksidan. *Songklanakarin J. Sci Teknologi, 26*(2), 211–219.

Muljohadjo, M.Gardjito, M., 1973. *Penyimpanan waktu blanching yang optimal untuk bermacam-macam sayuran dan buah-buahan* . Universitas Gadjah Mada: Yogyakarta.

Pradeep, P. and P. Pal, 2016. A review on potential application of antioxidants. *Int. J. Indigenous Herbs Drugs*., 1: 29-39

Pujimulyani, D., Raharjo, S., Marsono, Y., & Santoso, U. (2010). Pengaruh blanching terhadap aktivitas antioksidan, kadar fenol, flavonoid, dan tanin terkondensasi kunir putih (Curcuma mangga Val.). *Agritech, 30*(3). https://journal.ugm.ac.id/agritech/article/view/9665

Pujimulyani, D., Santoso, U., Luwihana, S., & Maruf, A. (2020). Orally administered pressure-blanched white saffron (Curcuma mangga Val.) improves antioxidative properties and lipid profiles in vivo. *Heliyon, 6*(6). https://www.cell.com/heliyon/fulltext/S2405-8440(20)31063-X

Puuponen-Pimia, R., Hakkinen, ST, Aarni, M., Suorlti, T., Lampi, AM, Eurola, M. 2003. Blanching Dan Pembekuan Jangka Panjang Mempengaruhi Berbagai Senyawa Bioaktif Sayuran Dengan Cara Yang Berbeda . *J. Sci Food Agric,* 62 (3):259-265.

Putri, A. C. K., & Susanto, W. H. (2015). PENGARUH PENAMBAHAN EKSTRAK DAUN TEH SEGAR (Camellia sinensis) TERHADAP KARAKTERISTIK KIMIA PANGAN SERTA ORGANOLEPTIK KACANG PRES GORENG SELAMA PENYIMPANAN [IN PRESS APRIL 2015]. *Jurnal Pangan Dan Agroindustri, 3*(2), Article 2.

Randhir, R., Kwan, YI, Shetty, K. 2008. Pengaruh Pemrosesan Termal Pada Fenolat, Aktivitas Antioksidan Dan Fungsi Yang Relevan Dengan Kesehatan Dari Kecambah Dan Bibit Biji-Bijian Pilihan. *Sains Makanan Inovatif dan Teknologi Baru*, 9: 355-364.

Rohman A, Bisa AT, Irnawati, Rafi M, Lukitaningsih E, Fadzilah NA. 2020a. Analisis Komponen Utama Aktivitas Antioksidan, Kandungan Total Fenolik, dan Kandungan Total Flavonoid Kunyit (Curcuma Longa L.). Int J Pharm Res 12: 2966-2972.

Rohman A, Widodo H, Lukitaningsih E, Windarsih A, Rafi M, Nurrulhidayah AF. 2020b. Review Aktivitas Antioksidan In vitro Spesies Temulawak yang Biasa Digunakan Sebagai Komponen Herbal di Indonesia. Resolusi Makanan 4(2): 286–93.

Sadilova E, Stintzing, EC, Carle R., 2006. Degradasi Termal Anthocyanin Terasilasi Dan Nonasilasi. *J. Ilmu Pangan* , 71: C504-512.

Sari, L.O.R.K., 2012. Pemanfaatan Obat Tradisional Dengan Pertimbangan Manfaat dan Keamanananannya.Pharm.Sci.Res.PSR 3.

Sembiring, BS, Ma'mun, Ginting, EI, 2006. Pengaruh Kehalusan Bahan Dan Lama Ekstraksi Terhadap Mutu Ekstrak Temulawak (*Curcuma xanthorrhiza* Roxb). *Buletin Littro* , 17 (2): 53-58

Setyowati A., Suryani CL 2013. Peningkatan Kadar Kurkuminoid Dan Aktivitas Antioksidan Minuman Instan Temulawak Dan Kunyit. *Agritech,* 33 (4).

Shinde, A. D., Chikhali, U. S., Patil, S. B., Naikwade, N. S. 2011. Antitumor, Antioxidant And Cytotoxic Activity Of Saraca Indica Linn. Leaves Extract. International Journal of Drug Formulation and Research Vol.2(3) hal 312-324

Sudrajat, H., Azar, FA, 2011. Uji Aktivitas Antifungi Minyak Atsiri Rimpang Temulawak (*Curcuma xanthorrhiza* Roxb.) Secara In Vitro Terhadap Kandida Albikan. Balai Besar Penelitian dan Pengembangan Tanaman Obat dan Obat Tradisional Badan Penelitian dan Pengembang Kesehatan Kementerian Kesehatan RI.

Suprihatin, T., Rahayu, S., Rifa'i, M., & Widyarti, S. (2020). Senyawa pada Serbuk Rimpang Kunyit (Curcuma longa L.) yang Berpotensi sebagai Antioksidan. *Buletin Anatomi dan Fisiologi, 5*(1), 35–42.

Vikri, M., Sholih, M. G., & Gatera, V. A. (2022). *Identifikasi Kadar Kurkumin pada Minuman Serbuk Berbahan Temulawak dengan Metode Spektrofotometri Uv-Vis. 3*(2).

Vina, Sonia Z., Daniela F., Olivera, Claudio M., Marani, Ricardo, M. Ferreyra, Alicia Mugridge, Alicia R. Chaves, dan Rodolfo H. Mascheroni. 2007. Quality of Brussels sprouts (Brassica oleracea L. gemmifera DC) as affected by blanching method. *Journal of Food Engineering* 80 (1) hal 218-225

Volden, J, GIA Borge, GB Bengtsson, M. Hansen, IE Thygesen, T. Wicklund, 2008. Pengaruh Perlakuan Panas Pada Glukosinolat Dan Terkait Antioksidan .

Wahyuningsih, D., 2008. Pengaruh Cara Dan Waktu Blanching Terhadap Kadar Antiosianin Dan Vitamun C Bunga Turi Merah (*Sesbania grandiflora* L.(Pers)). *Skripsi*. Fakultas Agroindustri, Universitas Mercu Buana, Yogyakarta.

Winarno, F.G. 2004. Kimia Pangan dan Gizi. Jakarta: Gramedia Pustaka Utama, 246 hlm

Xu, B., & Chang, S. K. (2008). Effect of soaking, boiling, and steaming on total phenolic contentand antioxidant activities of cool season food legumes. *Food Chemistry*, *110*(1), 1–13.

Yue X., Xu, Z., 2008. Perubahan Antosianin, Antosianidin, dan Aktivitas Antioksidan Pada Ekstrak Bilberry Selama Pemanasan Kering. *J. Makanan Sci,* Vol. 73.

Zhang, Z. Kou X., Fugal K., Mclaughlin J., 2004. Perbandingan Metode HPLC Untuk Penentuan Anthocyanin Dan Anthocyanidins Dalam Ekstrak Bilberry. *J. Makanan Pertanian Chem* 52: 688-91.

Physical and Sensory Properties of Analog Rice Made from Modified Cassava Flour and Substituted by Mung Bean Flour

Nur Annisa Metya Novikasari and Iffah Muflihati

Department of Food Technology, Faculty of Engineering and Informatics, University of PGRI Semarang, Semarang, 50232, Indonesia

*E-mail: annisametya@gmail.com

Abstract. Analog rice is artificial rice made from non-rice raw materials, such as from cassava flour and mung bean flour. Modification of starch by annealing, heat moisture treatment, and autoclaving-cooling was conducted to improve the characteristics of analog rice. This study purpose was to compare the effectiveness of physical modification of cassava flour and the ratio of modified cassava flour and mung bean flour on physical and organoleptic properties of analog rice. Analog rice was subjected to physical analyses of water absorption, cooking loss, cooking time, and expansion volume, while sensory properties was evaluated by descriptive method. The results showed that the type of modification and the ratio of modified cassava flour and mung bean flour improved improved water absorption capacity (121.44 to 161.19%), cooking loss (11.12 to 20.80%), and swelling volume (114.05 to 133.51%). Might lead to shorter cooking time (11.27 – 14.38 minutes), as well as affected sensory descriptive properties of analog rice.

Keywords: analog rice, cassava flour, mung bean flour, annealing, HMT, autoclaving-cooling

1. Introduction

Rice position as the main staple food in most part of Indonesia is almost irreplaceable. Problem arises when domestic supply cannot meet demand due to various reasons which leads to increasing import. Researches have been conducted to improve food diversification, particularly to create food that similar to rice for sustainable consumption, such as analog rice. Analog rice is artificial rice which can be made from local non-rice food ingredients [1]. Among potential ingredients, such as cassava and mung bean.

Agusman et al. [2] reported that analog rice made from modified cassava and seaweed powder has cylindrical form not similar to rice with low protein content, such as protein of 0.86%. The disadvantages of previous research must be improved to produce nutritious analog rice with preferable physical appearance. Physical modification on starch to create analog rice with similar form to the original through various methods is preferable as it chemical-free without any unwanted residue [3]. Some of the methods are annealing, heat moisture treatment (HMT), and autoclaving-cooling.

Annealing was reportedly able to reduce swelling power and starch solubility while hamper gelatinization [4]. In another research, heat moisture treatment (HMT) was able to reduce solubility and swelling power of sorghum powder while increase its viscosity, gelatinization degree, and crystallinity [5]. Autoclaving-cooling which consist of high temperature heating followed by cooling was able to change characteristics of starch gelatinization. The problem of low nutrients can be substituting mung bean flour. The protein content of mung bean flour is very high at 28.38% [6]. According to Kanetro et al. [7], the addition of green beans in the manufacture of analog rice from oyek flour can affect the hardness of rice to approach rice (control) which is 1061.50 g. This research aimed to measure the effect of several cassava modifications as well as the ratio of cassava flour and mung bean flour on physical properties and sensory characteristics of analog rice.

2. Material And Methods

2.1 Materials
The main material was Adira 4 variety white cassava harvested after 6 months cultivation from Gunungpati, Semarang District, and Vima 1 variety mung bean harvested after 60-80 days cultivation from Karangtengah, Demak. Several instruments used were cabinet dryer, analytical balance (OHAUS), autoclave (All American 25X cap. 13L), refrigerator (Samsung), oven (MEMMERT BF-201), muffle furnace (Thermolyne), hot plate stirrer (IKA C- MAC HS 7 IKAMAG), twin screw extruder (Berto BEX-DS-2256), chromameter (Konica Minolta CR-400), texture analyzer (Brookfield CT3), and water bath (Wise Bath).

2.2 Analog Rice Preparation
After cassava flour preparation [8], the flour was then modified using several methods of annealing [9], heat moisture treatment (HMT) [5], and autoclaving-cooling [10], followed by mung bean flour preparation [11], and analog rice production [12].

2.3 Experimental Design
Factorial design was used on two factors of modification methods and the ratio of modified cassava flour and mung bean flour. Annealing (ANN), Heat moisture treatment (HMT) and Autoclaving-cooling (ACC) and three ratios of modified cassava and mung bean flour (1:1, 2:1 and 3:1) were applied. Peeled rice was used as control.

2.4 Analysis of Analog Rice
Produced analog rice was subjected to analyses of water absorption capacity, cooking loss [13], cooking time [14], swelling volume [15] and descriptive sensory analysis by 15 trained panelists.

2.5 Data Analysis
Data was statistically analyzed by analysis of variance (ANOVA) using SPSS 21. When significant difference was found among means, analysis was followed by Duncan Multiple Range Test (DMRT) at 5% significance level.

3. Results And Discussion

3.1 Water Absorption Capacity
The levels of Water absorption capacity of analog rice in the range of 121.44 to 161.19%. Analog rice had higher water absorption capacity than control (Table 1). This was caused by higher protein content which able to absorb more water due to its hydrophilic property. Water absorption capacity can also be affected by carbohydrate, crude fiber, and other components [16]. Owolabi et al. [17] reported that modification using HMT and ACC at high temperature (105°C and 121°C) increased water absorption capacity. This was due to the increasing hydrophilic property of starch from slightly swelling of amorphous structure that break the crystalline bond and increase bonding to hydrogen from water. In ANN modification, hydrated

granules with more than 60% moisture content and increasing annealing temperature enable increasing compactness of amorphous structure and amylopectin double-helix structure.

Table 1. Water absorption capacity (%) of analog rice.

Ratio of Modified Cassava Flour: Mung Bean Flour	Cassava Flour Modification Method		
	ANN	HMT	ACC
1:1	138.95±1.92[d]	157.41±2.32[a]	121.44±2.84[f]
2:1	148.04±1.81[c]	138.33±2.91[d]	127.49±2.34[e]
3:1	161.19±2.51[a]	122.06±2.59[f]	153.00±2.42[b]
Control		107.58±2.24[g]	

ANN: Annealing; HMT: Heat moisture treatment; ACC (Autoclaving-cooling), control (peeled rice). Different notation indicates significant difference at significance level (α) of 5% using Duncan Multiple Range Test.

The increasing of starch's initial temperature of gelatinization from annealing was caused by amorphous amylose transformation into helix, interaction enhancement among amylose chain, and interaction modification between crystal and amorphous matrix during annealing [17] that increase water absorption capacity of ANN modified rice. Increasing water absorption capacity in analog rice with mung bean addition was caused by high amylose content which able to bind water due to its hydrophilic property. Water absorption capacity can be affected by several factors, including carbohydrate, protein, crude protein content and other components [16].

3.2 Cooking Loss

Significant difference in cooking loss was found among different modification methods and flour proportions (Table 2). ACC (3:1) had the highest cooking loss of 20.80% while ANN (3:1) was the lowest at 11.12%. Cooking loss of analog rice made using ANN, HMT and ACC modified cassava flour was caused by the modification methods. Different with other methods, higher modified cassava flour proportion reduced cooking loss. ANN increased compactness of starch's amorphous structure and amylopectin double-helix structure while reduced swelling ability that decreased solubility of starch component [18].

In ACC modified analog rice, starch structure was possibly damaged due high temperature heating that became more tenuous; thus, its particle became more soluble in water and increased cooking loss. Cassava modification using HMT at 110°C was able to increased solubility. Hydrogen bond of amylose in starch granule break down into water soluble short-chain amylose fractions [19]. Different portions of mung bean affected cooking loss of analog rice as it contains starch and fiber that able to absorb water unabsorbed by protein [20].

Table 2. Cooking loss (%) of analog rice.

Ratio of Modified Cassava Flour: Mung Bean Flour	Cassava Flour Modification Method		
	ANN	HMT	ACC
1:1	14.91±0.20[e]	11.92±0.12[g]	16.03±0.31[d]
2:1	12.12±0.27[g]	13.16±0.29[f]	20.16±0.52[b]
3:1	11.12±0.16[h]	18.97±0.47[c]	20.80±0.29[a]
Control		2.37±0.34[i]	

ANN: Annealing; HMT: Heat moisture treatment; ACC (Autoclaving-cooling), control (peeled rice). Different notation indicates significant difference at significance level (α) of 5% using Duncan Multiple Range Test.

3.3 Cooking Time

There was significant difference among cooking time due to cassava modification (Table 3). Analog rice in this research had faster cooking time than control rice. This can be affected by water absorption capacity and protein content. According to [21], high water absorption might lead to shorter cooking time.

Table 3. Cooking time (minute) analog rice.

Ratio of Modified Cassava Flour: Mung Bean Flour	Cassava Flour Modification Method		
	ANN	HMT	ACC
1:1	11.39±0.07^{e}	14.21±0.02^{c}	13.34±0.04^{d}
2:1	11.33±0.04^{f}	14.24±0.01^{bc}	13.33±0.01^{d}
3:1	11.27±0.04^{f}	14.38±0.01^{b}	13.32±0.02^{d}
Control		18.33±0.02^{a}	

ANN: Annealing; HMT: Heat moisture treatment; ACC (Autoclaving-cooling), control (peeled rice). Different notation indicates significant difference at significance level (α) of 5% using Duncan Multiple Range Test.

In ANN modified cassava analog rice, shorter cooking time was produced by higher cassava and lower mung bean ratio, with the fastest of 11.27 – 11.39 minutes. This was due to high water absorption capacity (138.95- 161.19%), as also reported by previous research [21]. Adebowale et al. [22] also reported that ANN modified flour has higher water absorption capacity that increase hydrophilic property of starch granule.

In the other hand, HMT modified analog rice, higher modified cassava ratio and lower mung bean flour had longer cooking time. Increasing cooking time caused longer and lower water absorption, made analog rice became more brittle in longer cooking time [23]. Mondragon et al. [24] mentioned that longer cooking time for analog rice might lead to hollow structure that facilitate heat penetration into analog rice and accelerate gelatinization. Consequently, longer gelatinization causes longer cooking time. Mung bean addition might affect cooking time related to nutrition composition of its cotyledon and peel [25].

3.4 Swelling Volume

Swelling volume of analog rice in this research had ranges between 114.05 to 133.51% (Table 4). Higher ANN modified cassava and lower mung bean flour proportion increased swelling volume. Whereas, for HMT and ACC at 1:1 and 2:1 ratio, swelling volume decreased and increased again at ratio of 3:1. Swelling in analog rice was caused by water absorption into starch granule. Thus, swelling volume is proportional to water absorption capacity. Moreover, swelling volume can be affected by amylose content, as higher amylose leads to higher swelling volume, due to water binding capacity from its hydrophilic property [26].

Table 4. Swelling volume (%) of analog rice.

Ratio of Modified Cassava Flour : Mung Bean Flour	Cassava Flour Modification Method		
	ANN	HMT	ACC
1:1	114.39±0.40^{f}	133.51±0.31^{a}	117.44±0.27^{e}
2:1	118.65±0.29^{d}	132.16±0.67^{b}	114.05±0.41^{f}
3:1	122.34±0.27^{c}	132.79±0.27^{b}	122.69±0.59^{c}
Control		117.49±0.34e	

ANN: Annealing; HMT: Heat moisture treatment; ACC (Autoclaving-cooling), control (peeled rice). Different notation indicates significant difference at significance level (α) of 5% using Duncan Multiple Range Test.

3.5 Descriptive Sensory Properties of Analog rice

Based on brown color parameter, panelists rated ANN modified rice (3:1) as the highest brown scale while HMT (3:1) had the lowest (Table 5). It was indicated that ANN (3:1) analog rice had the darkest brown color among samples, caused by high proportion of modified cassava flour. Modified cassava flour had brown color with diverse brightness intensity that might affect the color of analog rice. Mung bean flour has brighter color of yellowish white with L* 83.37 [26], that lower mung bean flour proportion might affect the color and brightness of analog rice. The color can be possibly affected by maillard reaction occurs to modified cassava flour during modification using heat process [27].

Table 5. Descriptive sensory properties of analog rice.

Treatment	Brown Color	Mung Bean Aroma	Cassava Aroma	Brightness
ANN (1:1)	3.15±1.14[ab]	2.08±0.76[a]	2.15±0.55[ab]	2.08±0.76[c]
ANN (2:1)	3.23±1.09[ab]	1.92±1.04[a]	2.08±0.76[ab]	2.46±0.97[abc]
ANN (3:1)	3.54±0.88[a]	1.46±0.52[a]	2.08±0.86[ab]	2.38±0.87[bc]
HMT (1:1)	3.15±1.14[ab]	1.92±0.86[a]	2.00±0.71[ab]	3.00±0.58[ab]
HMT (2:1)	2.62±0.96[bc]	1.69±0.63[a]	1.85±0.80[ab]	2.92±0.76[ab]
HMT (3:1)	2.00±0.82[c]	1.38±0.51[a]	2.00±0.82[ab]	3.15±0.69[a]
ACC (1:1)	3.08±1.04[ab]	1.92±0.86[a]	2.23±0.83[ab]	2.85±0.80[ab]
ACC (2:1)	2.62±0.77[bc]	1.85±0.69[a]	2.54±1.13[a]	2.85±0.80[ab]
ACC (3:1)	3.38±0.96[ab]	2.00±0.91[a]	1.69±0.75[b]	2.62±0.96[abc]

ANN: Annealing; HMT: Heat moisture treatment; ACC (Autoclaving-cooling), control (peeled rice). Scale: 1 (very pale brown) – 5 (very dark brown), Mung bean Aroma:1 (very mild smell) – 5 (strong smell), cassava aroma: 1 (very mild smell) – 5 (strong smell), brightness: 1 (dark) – 5 (very bright). Different notation indicates significant difference at significance level (α) of 5% using Duncan Multiple Range Test.

In aroma, ANN (1:1) and HMT (1:1) analog rice had milder smell of mung bean aroma than those with lower mung bean concentration in each treatment, whereas ACC modified rice samples had the otherwise (Table 5). Panelists rated cassava aroma intensity between 1.69 to 2.54. The highest was obtained by ACC (2:1) at 2.54, indicated mild to slightly smelled cassava aroma. The lowest cassava aroma was obtained by ACC (3:1) analog rice at 1.69, indicated very mild to mild smell. Modified cassava flour has stronger aroma than mung bean flour. Therefore, increasing modified cassava proportion increased typical cassava aroma. Brightness score was between 2.08 to 3.15 which indicated that panelists rated analog rice as not bright to slightly bright. Higher modified cassava proportion and lower mung bean flour might affect analog rice brightness with no particular pattern.

4. Conclusion

Physical modification of cassava flour using annealing (ANN), Heat moisture treatment (HMT), and autoclaving-cooling (ACC) significantly affected physical and sensory properties of analog rice. Analog rice made by several ratio flour (modified cassava flour and mung bean flour) improved water absorption capacity (121.44 to 161.19%), cooking loss (11,12 to 20,80%), and swelling volume (114.05 to 133.51%). Might lead to shorter cooking time (11.27 – 14.38 minutes), as well as affected sensory descriptive properties of analog rice.

Acknowledgements
Authors thank PT. Indofood CBP Sukses Makmur Tbk., for providing research fund through Indofood Riset Nugraha 2020-2021. We also thank Maju Jaya Women Farmer Group, Purwodadi District, for their support during research.

References

[1] S. Sumardiono, I. Pudjihastuti, A. R. Poerwoprajitno, and M. S. Suswadi, Physichocemical properties of analog rice from composite flour: cassava, green bean and hanjeli, World App. Sci. J. 32, 1140–1146 (2014).

[2] Agusman, A., Apriyani, S. N. K. and Murdinah, M. (2014) 'The Use of Eucheuma cottonii Flour in the Processing of Rice Analogues from Modified Cassava Flour (Mocaf)', *Jurnal Pascapanen dan Bioteknologi Kelautan dan Perikanan*, 9(1), pp. 1–10. doi: 10.15578/jpbkp.v9i1.94.

[3] C. Pukkahuta, B. Suwannawat, S. Shobsngob, and S. Varavinit, Comparative study of pasting and thermal transition characteristics of osmotic pressure and heat-moisture treated corn starch, J. Carb. Poly. 72, 527–536 (2008).

[4] T. A. Siswoyo and N. Morita, Influence of annealing on gelatinization properties, retrogradation and susceptibility of breadfruit starch (*Artocarpus communis*), Int. J. Food Prop. 13, 553–561 (2010).

[5] Q. Sun, Z. Han, L. Wang, and L. Xiong, Physicochemical differences between sorghum starch and sorghum flour modified by heat-moisture treatment, Food Chem. 145, 756–764 (2014).

[6] Blessing, I. A. and Gregory, I. O. (2010) 'Effect of Processing on the Proximate Momposition of the Dehulled and Undehulled Mungbean (Vigna radiata (L.(Wilczek) Flour', Pakistan Journal of Nutrition, 9(10), pp. 1006–1016.

[7] Kanetro, B. et al. (2017) 'Characteristics of Low Glycemic Index Artifisial Rice from Oyek with Addition of Various Legumes', Agritech, 37(3), pp. 256–262. doi: doi.org/10.22146/agritech.31538.

[8] E. Eriksson, "Flour from three local varieties of cassava (*Manihot esculenta* Crantz): Physico-chemical properties, bread making quality and sensory evaluation," Ph.D. thesis, Swedish University of Agricultural Sciences, 2013.

[9] Y. Nakazawa and Y. J. Wang, Acid hydrolysis of native and annealed starches and branch-structure of their Naegeli dextrins, J. Carb. Res. 338, 2871–2882 (2003).

[10] U. Lehmann, C. Rössler, D. Schmiedl, and G. Jacobasch, Production and physicochemical characterization of resistant starch type III derived from pea starch, Nahrung/Food 47, 60–63 (2003).

[11] R. Yenrina, R., W. Surya, and N. N. Putri, Mocaf bread enriched with mung bean (*Vigna radiata* L.) as a source of protein, Asia Pacific J. Sustain. Agri. Food and Energy 1, 10–13 (2013).

[12] Pudjihastuti, I. *et al.* (2018) 'Quality analog rice composite flour: Modified starch, Colocasia esculenta, Canna edulis Ker high protein', *AIP Conference Proceedings*, 1977(1), pp. 030017-1–030017-4. doi: 10.1063/1.5042937.

[13] H. Z. Tan, Z. G. Li, and B. Tan, Starch noodles: History, classification, materials, processing, structure, nutrition, quality evaluating and improving, Food Res. Int. 42, 551–576 (2009).

[14] N. H. Oh *et al.* (1985) 'Noodles. II. The Surface Firmness of Cooked Noodles from Soft and Hard Wheat Flours', *American Association of Cereal Chemists*, 62(6), pp. 431–436.

[15] Crosbie, G. B. *et al.* (1992) 'Further evaluation of the flour swelling volume test for identifying wheats potentially suitable for Japanese noodles', *Journal of Cereal Science*. Academic Press Limited, 15(3), pp. 271–280. doi: 10.1016/S0733-5210(09)80125-2.

[16] Awuchi, C. G., Echeta, C. K. and Victory, I. S. (2019) 'the Functional Properties of Foods and Flours', *International Journal of Advanced Academic Research | Sciences, Technology and Engineering*, 5(11), pp. 139–160

[17] Olu-Owolabi, B. I., Afolabi, T. A. and Adebowale, K. O. (2011) 'Pasting, thermal, hydration, and functional properties of annealed and heat-moisture treated starch of sword bean (Canavalia gladiata)', *International Journal of Food Properties*, 14(1), pp. 157–174. doi: 10.1080/10942910903160331.

[18] Jayakody, L. and Hoover, R. (2008) 'Effect of annealing on the molecular structure and physicochemical properties of starches from different botanical origins - A review', *Carbohydrate Polymers*, 74(3), pp. 691–703. doi: 10.1016/j.carbpol.2008.04.032.

[19] I. Muflihati, D. W. Marseno, and Y. Pranoto, Oxidation of oven-dried cassava starch using hydrogen peroxide and UV-C irradiation to improve frying expansion, J. Indonesian Food and Nutr. Prog. 16, 1–14 (2019).

[20] Liu, Y. *et al.* (2018) 'The compositional, physicochemical and functional properties of germinated mung bean flour and its addition on quality of wheat flour noodle', *Journal of Food Science and Technology*. Springer India, 55(12), pp. 5142–5152. doi: 10.1007/s13197-018-3460-z.

[21] Sozer, N. and Kaya, A. (2008) 'The effect of cooking water composition on textural and cooking properties of spaghetti', *International Journal of Food Properties*, 11(2), pp. 351–362. doi: 10.1080/10942910701409260.

[22] K. O. Adebowale, B. I. Olu-owolabi, O. O. Olayinka, and S. Olayide, Effect of heat moisture treatment and annealing on physicochemical properties of red sorghum starch, African J. Biotech. 4, 928– 933 (2016).
[23] S. Sukamto and D. G. Patria, The utilization of flour made of the non-milled rice as analog rice ingredients, Food Res. 4, 1427–1434 (2020).
[24] Mondragon, M. *et al.* (2004) 'Effect of Cooking Time, Steeping and Lime Concentration on Starch Gelatinization of Corn During Nixtamalization', *Starch*, 56(6), pp. 248–253. doi: 10.1002/star.200200190.
[25] P. K. Dahiya, A. R. Linnemann, M. A. J. S. van Boekel, N. Khetarpaul, R. B. Grewal, and M. J. R. Nout, Mung Bean: technological and nutritional potential, Critical Reviews in Food Sci. and Nutr. 55, 670–688 (2015).
[26] Ratnawati, L. *et al.* (2019) 'Evaluation of Physicochemical, Functional and Pasting Properties of Soybean, Mung Bean and Red Kidney Bean Flour as Ingredient in Biscuit', *IOP Conference Series: Earth and Environmental Science*, 251(1). doi: 10.1088/1755-1315/251/1/012026.
[27] Ariyantoro, A. R., Amanto, B. S. and Kiswuri (2020) 'Effect of Heat Moisture Treatment (HMT) on Physicochemical Characteristics of Sorghum Flour (Sorghum Bicolor L. moench)', IOP Conference Series: Earth and Environmental Science, 518(1), pp. 1–7. doi: 10.1088/1755-1315/518/1/012067.

The Effect of Rabbit Skin Gelatin Concentration on Frozen Yoghurt Quality

Sri Mulyani*, Amaliya Putri, and Yeni Ferianto.

Food Technology, Diponegoro University, Semarang, Indonesia

* E-mail: srimulyani@lecturer.undip.ac.id

Abstract. Frozen yogurt during storage can experience recrystallization, which results in the texture not being soft, so a stabilizer needs to be added, one of which is rabbit skin gelatin. The research aims to determine the effect of adding rabbit skin gelatin concentration to total solids, overrun, melting time, lactic acid bacteria profile, and sensory frozen yogurt. Frozen yogurt is made by varying the concentration of adding rabbit skin gelatin, namely 0%, 0.3%, 0.6%, and 0.9%, for five repetitions. The manufacturing method for frozen yogurt includes pasteurizing ingredients, mixing dough, aging, mixing and freezing 3 times, hardening, and then storing for 1 week. The research results showed the total solids value was 34.16 – 35.44%, overrun 16.01 – 20.24%, melting time 11.59 – 19.10 minutes, total BAL 7.9x107 - 2.6x108 CFU/ml, white color score 2.24 – 3.16, sour taste 2.56 – 2.96 , soft texture 2.40 – 3.56, and overall 3.04 – 3.48. The research concludes that adding 0.6% rabbit skin gelatin to frozen yogurt can optimally prevent recrystallization, thus producing a very soft texture, inhibiting melting at room temperature, and maintaining the number of lactic acid bacteria and the panelists liked.

Keywords: lactic acid bacteria, stabilizer, soft texture, recrystallization

1. Introduction

Frozen yogurt can undergo recrystallization during storage due to temperature fluctuations in the freezer [1]. When the temperature increases, tiny ice crystals will melt, increasing the amount of unfrozen water. When the temperature drops again, the water will freeze, but it cannot restore the shape of the ice crystals as before and will form large and coarse ice crystals. Recrystallization causes the texture of frozen yogurt to become less soft during storage. Therefore, a stabilizer is needed, one of which is gelatin, because it has water-binding solid properties, so the amount of bound water will be more significant. Recrystallization can be prevented [2]. Gelatin stabilizer in ice cream is better than other stabilizers, such as gum Arabic, in terms of melting time, total solids, and sensory values [3].

Rabbit skin holds significant potential as a raw material for gelatin due to its high protein content, approximately 28.04%, most of which is collagen, a primary source of gelatin. However, its application in food products still needs to be improved, encouraging further research to explore its effectiveness in improving food quality, such as in frozen yogurt. Previous studies on Rex rabbit skin gelatin have shown promising results, including 70.3% protein content, 11.17 cP viscosity, 187 g Bloom gel strength, and 8.26% yield [4]. Despite this potential, rabbit skin waste still needs to be utilized.

In Indonesia, gelatin research has predominantly focused on halal sources like fish, such as African catfish (*Clarias gariepinus*) [5], while rabbit skin is less explored. Utilizing rabbit skin as a sustainable source of gelatin addresses waste reduction while offering competitive alternatives for applications in frozen yogurt, ice cream, and gelato. This study aims to evaluate the impact of rabbit skin gelatin concentrations on total solids, overrun, melting time, lactic acid bacteria profiles, and sensory attributes in frozen yogurt, contributing to innovative and sustainable uses of underutilized resources.

2. Materials and Methods

2.1 Material

The materials used were rabbit skin, water, distilled water, 2% lime water, 0.25M NaOH (Merck KGaA, Germany), 8% butanol (Merck KGaA, Germany), 0.3M citric acid (Merck KGaA, Germany), fresh cow's milk (FPP UNDIP Farm, Indonesia), freeze-dried yogurt starter (Yogourmet, Canada), buffer solutions 4 and 7, skim milk (Indoprima, Indonesia), granulated sugar (Gulaku, Indonesia), whipping cream (Pondan, Indonesia) and de Man Rogosa Sharpe Agar (MRSA) medium (Merck KGaA, Germany).

2.2 Method

The study included making rabbit skin gelatin and frozen yogurt, parameter testing, and data analysis. The experimental design used a Completely Randomized Design (CRD) with four treatments and five replications. The treatment variations added 0%, 0.3%, 0.6% and 0.9% rabbit skin gelatin.

2.2.1 Making Rabbit Skin Gelatin

The manufacture of rabbit skin gelatin follows a previously established method with modifications. The rabbit skin is washed and cleaned of residual meat and fat, then soaked in a 2% (w/v) lime solution for 24 hours [6]. The rabbit skin is cleaned from residual hair, washed, and soaked in 0.25M NaOH (1:3; w/v) for 2 hours, then rinsed with distilled water to a pH of 5–6. Rabbit skin is soaked in 8% butanol for 18 hours, then washed and cut into pieces of ±1 cm2. The rabbit skin is soaked in 0.3M citric acid (demineralization) for 4 hours with a ratio of 1:4 (w/v), then washed with distilled water to a pH of 5–6. Rabbit skin is extracted in a water bath (Memmert, Germany) with 1:3 (w/v) distilled water in stages 3 times, namely at temperatures of 65°C, 68°C, and 70°C, each for 5 hours. The extraction results were filtered to obtain the filtrate, then dried in an oven at 55°C for ± 72 hours until dry and ground with a blender (Phillips, Indonesia) to obtain gelatin powder. The characteristics of the gelatin powder were tested, including pH, gel strength, and viscosity [7].

2.2.2 Making Frozen Yogurt

The making of frozen yogurt was carried out following a previously established method with modifications [8]. The ingredients needed include 105 g yogurt, 32 g skim milk, 75 g whipping cream, 30 g granulated sugar, 58 g water, and rabbit skin gelatin with variations of additions, namely 0% (0 g), 0.3% (0.9 g), 0.6% (1.8 g), and 0.9% (2.7 g). Skim milk and granulated sugar were dissolved in water and pasteurized at 72°C for 15 seconds. The pasteurized mixture was mixed with other ingredients, such as whipping cream, yogurt, and gelatin, dissolved in water at 50–60°C with a magnetic stirrer. The mixture was stirred with a mixer (Phillips, Indonesia) for 10 minutes, then cooled (aging) at 4°C for 4 hours. The dough is stirred with a mixer for 5 minutes and frozen in the freezer for 1 hour. The mixing process until freezing is repeated up to 3 times, and then the frozen yogurt is packaged and hardened for 24 hours in the freezer. The following process is continued by storing in the freezer for 1 week at a temperature range of -18°C before being tested.

2.2.3 Frozen Yoghurt Quality Analysis

Frozen Yogurt quality testing includes total solids [9], melting time [10], overrun [11], total lactic acid bacteria [12], and sensory [13] with modifications involving 25 panelists, where the sensory

IOP Conf. Series: Earth and Environmental Science **1460** (2025) 012045 doi:10.1088/1755-1315/1460/1/012045

assessments carried out included taste, color, texture and overall. The scale is 1 to 4, which states the product's characteristics and overall preference.

2.3 Statistical Analysis

Data were analyzed using the SPSS 26.0 application at 5% significance. Total solids, overrun, and melting time data were analyzed using the Analysis of Variance (ANOVA) test; if significantly different, the Duncan Multiple Range Test (DMRT) was used. Sensory data were analyzed using the Kruskal test Wallis; if significantly different, the Mann-Whitney test was used.

3. Results and Discussion

3.1 Quality Characteristics of Rabbit Skin Gelatin

The pH value of rabbit skin gelatin (Table 1) meets the standards of Gelatin Manufacturers of America (GMIA) (2012), which is 3.8–6.0 for type A gelatin (acid) [14]. The pH value of the rabbit skin gelatin produced is relatively low, which is 5.10, because, during the demineralization process, an acid solution is used, namely 0.3 M citric acid. This aligns with the finding that the acid-soaking process tends to produce gelatin with a low pH. The low pH is also due to citric acid remaining trapped in the collagen tissue during washing, making it difficult to neutralize and causing it to persist in the extraction process [15]. Rabbit skin gelatin (Table 1) has a reasonably high viscosity, which is thought to be because the amino acid chain in the gelatin is still quite long because the solution used during soaking is citric acid, which is classified as a weak acid and the temperature used for extraction is not too high, thus making the gelatin viscosity high. This is supported by the previous research of Saputra et al. (2015), which found that higher extraction temperatures lead to further hydrolysis of collagen into gelatin, breaking the bonds between amino acids and resulting in lower gelatin viscosity [16]. The gel strength value of rabbit skin gelatin (Table 1) has met the GMIA (2012) standard, which ranges from 50–300 bloom [14]. The high gel strength of rabbit skin gelatin is thought to be because the extraction temperature and acid concentration used are correct, so the breakdown of amino acids at the right size can make the gel strength high. This is in accordance with the opinion of Gerungan et al. (2019) that the optimum temperature in the gelatin extraction process can break down the amino acid polymer chain at the right size so that the gel strength will improve [17].

Table 1. Characteristics of rabbit skin gelatin quality

Variable	Mark
pH value	5.10 ± 0.18
Viscosity (cP)	10.76 ± 1.57
Gel Strength (*bloom*)	161.18 ± 0.51

Note: Data are shown as mean ± standard deviation (n=4)

3.2 Total Solids

The addition of rabbit skin gelatin concentration gave a significant difference ($p < 0.05$) to the total solids value of frozen yogurt, where the more significant the addition of rabbit skin gelatin, the greater the percentage of total solids of frozen yogurt. This is in accordance with the opinion of Sembiring et al. (2019) that the total solids of ice cream increase with the increasing amount of stabilizer used because it can bind water so that the amount of water contained in ice cream is reduced [18]. The total solids value of frozen yogurt (Table 2) has met SNI 01-3713-1995 regarding the total solids quality of ice cream, which is set at a minimum of 34% [41]. The lowest total solids were found in the addition of 0% gelatin. In comparison, the highest total solids were found in the addition of 0.9% gelatin because gelatin has the property of binding to free water. This is in accordance with the opinion of Ariska & Utomo (2020) that the higher the stabilizer added, the higher the total solids value will be because the stabilizer can bind free water [19].

According to Ayudiarti et al. (2020), ingredients such as whole cream milk, sugar, and skim milk can affect the total solids of ice cream because they have high total solids [20].

Table 2. Total solids of *frozen yoghurt* with added concentration of rabbit skin gelatin

Concentration of rabbit skin gelatin addition	Total solid (%)
0%	34.16 ± 0.46 [a]
0.3%	34.45 ± 0.25 [ab]
0.6%	34.94 ± 0.46 [bc]
0.9%	35.44 ± 0.32 [c]

Note: Data are shown as mean ± standard deviation (n=5).
[ad] *Different superscripts in the same column indicate significant differences (p<0.05).*

3.3 Overrun

The addition of rabbit skin gelatin concentration gave a significant difference ($p<0.05$) to the overrun of frozen yogurt, where the more significant the addition of rabbit skin gelatin, the smaller the percentage of frozen yogurt overrun. This is in accordance with the opinion (Goff & Hartel, 2013) that the higher the concentration of gelatin used in ice cream dough, the thicker the dough will be, and the surface tension will be higher so that it is difficult to expand. The ice cream overrun value will be smaller [11]. Overrun Frozen yogurt with the addition of rabbit skin gelatin (Table 3) does not meet the household scale overrun standards of 35-50% and industry of 70-80% [21]. The low overrun value is likely due to the raw material used, namely yogurt, which has a gel-like shape characteristic, increasing the viscosity of the frozen yogurt dough and making the overrun small. This is in accordance with the opinion of Yulistiani et al. (2017) that yogurt has a semi-solid texture which is caused by an increase in acidity due to the accumulation of lactic acid, which causes casein to become unstable and coagulate to form yogurt gel [22].

The most considerable frozen yogurt overrun value is found in the addition of 0% gelatin. In comparison, the lowest overrun is found in the addition of 0.9% gelatin, which is caused by gelatin being able to bind water so that the viscosity of the dough becomes high, making frozen yogurt challenging to expand. According to Saputri et al. (2015), ice cream dough should not be too thick because it will inhibit the development of ice cream, where the air that enters the ice cream dough is getting less because of the narrow space between particles so the resulting overrun is getting lower [23].

Table 3. *Overrun frozen yoghurt* with added concentration of rabbit skin gelatin

Concentration of rabbit skin gelatin addition	Overrun (%)
0%	20.24 ± 1.92 [b]
0.3%	18.04 ± 1.94 [a]
0.6%	17.20 ± 0.93 [a]
0.9%	16.01 ± 1.21 [a]

Note: Data are shown as mean ± standard deviation (n=5).
[ad] *Different superscripts in the same column indicate significant differences (p<0.05).*

3.4 Melting Time

The addition of rabbit skin gelatin concentration gave a significant difference ($p<0.05$) to the melting time of frozen yogurt, where the more significant the addition of rabbit skin gelatin, the longer the melting time of frozen yogurt. This is in accordance with the opinion of Nofrida et al. (2018) that a higher concentration of stabilizer used in ice cream will result in a longer melting time of the ice cream [24]. The melting time of frozen yogurt (Table 4) is in accordance with the Failisnur (2013) standard, which is between 10-20 minutes at room temperature [25]. The fastest melting time was found with the addition of 0% gelatin, while the longest melting time was found with the addition of 0.9% gelatin because gelatin has the property of binding water so that it can slow down the melting time of frozen yogurt. This is in accordance with the opinion of Zahro & Nisa (2015) that gelatin can affect the speed of ice cream melting time by binding water, causing water molecules to be trapped in the gel structure so that the melting time of the ice cream is longer [26]. According to Mulyani et al. (2017), ingredients such as skim milk, full cream milk, and the type of stabilizer can affect the melting speed of ice cream [27].

Table 4. *Frozen yoghurt melting time* with added rabbit skin gelatin concentration

Concentration of rabbit skin gelatin addition	Frozen melting time (min)
0%	11.59 ± 0.52 [a]
0.3%	14.40 ± 0.60 [b]
0.6%	16.48 ± 0.17 [c]
0.9%	19.10 ± 0.49 [d]

Note: Data are shown as mean ± standard deviation (n=5).
[ad] *Different superscripts in the same column indicate significant differences (p<0.05).*

3.5 Lactic Acid Bacteria (LAB) Profile

The number of lactic acid bacteria in the addition of 0% gelatin was less than the concentration of 0.3% gelatin (Table 5); this is thought to be because, in the addition of 0.3% gelatin, the number of large ice crystals is small. After all, gelatin can bind free water so that bacterial damage by ice crystals is less. This is supported by Jayus et al. (2022), that the death of microbes in synbiotic ice cream is caused by the presence of large ice crystals during the frozen storage process, where the ice crystals can damage the microbial cell walls due to the difference in pressure between the fluid outside and inside the microbial cells [28]. According to Bezerra et al. (2015), the presence of cryoprotectants found in frozen yogurt mixtures, such as casein, sucrose, and lactose, is thought to be able to maintain probiotic cells during frozen storage for 90 days [29].

Based on the analysis results (Table 5), the number of lactic acid bacteria has met the SNI 2981-2009 standard regarding fermented products, namely a minimum of 1.0 x 107 CFU/ml of lactic acid bacteria [42]. The addition of rabbit skin gelatin to frozen yogurt shows that rabbit skin gelatin can maintain the number of lactic acid bacteria at a certain level. It will decrease again if the percentage of rabbit skin gelatin in frozen yogurt is continued. This is thought to be because gelatin can protect lactic acid bacteria so that they do not come into direct contact with ice crystals during frozen storage. According to Bilang et al. (2018), hydrocolloids added to the manufacture of symbiotic ice cream function to protect microbes from direct contact with ice crystals formed at freezing temperatures [30].

Table 5. Lactic acid bacteria profile of *frozen yoghurt* with added rabbit skin gelatin concentration

Concentration of rabbit skin gelatin addition	LAB (CFU/ml)
0%	1.8×10^{8}
0.3%	2.6×10^{8}
0.6%	1.3×10^{8}
0.9%	7.9×10^{7}

Note: The test was only performed once.

3.6 White Color Sensory

The addition of rabbit skin gelatin concentration (Table 6) gave a significant difference (p<0.05) to the white color of frozen yogurt. The white color of frozen yogurt is influenced by the raw materials used in its manufacturing, such as yogurt, skim milk, granulated sugar, and whipping cream. The addition of 0% and 0.3% gelatin has a whiter color than the addition of 0.6% and 0.9% gelatin (Table 6); this shows that the more rabbit skin gelatin is added, the whiter the frozen yogurt color. This is thought to be due to the presence of yellowish-brown gelatin and sugar that undergoes a Maillard reaction during the pasteurization process of the ingredients, which can affect the color of frozen yogurt. This is in accordance with Putra's opinion (2016) that the caramelization reaction (browning) can occur due to the interaction between sugar and high temperatures [31]. This is also supported by Eletra et al. (2013) that the higher the concentration of gelatin added to a product, the more yellow-brown the product will be [32].

Table 6. *Frozen yoghurt* sensory test results

Gelatin addition concentration	White	Sour taste ^{ns}	Soft texture	*Overall* ^{ns}
0%	3.16 ± 0.69 ^{a}	2.56 ± 0.77	2.40 ± 0.58 ^{a}	3.04 ± 0.61
0.3%	2.96 ± 0.79 ^{a}	2.68 ± 0.75	3.08 ± 0.57 ^{b}	3.28 ± 0.61
0.6%	2.48 ± 0.77 ^{b}	2.88 ± 0.78	3.56 ± 0.51 ^{c}	3.48 ± 0.50
0.9%	2.24 ± 0.72 ^{b}	2.96 ± 0.89	3.24 ± 0.60 ^{bc}	3.16 ± 0.47

Description: Data are shown as mean ± standard deviation (n=25).
^{ad} *Different superscripts in the same column indicate significant differences (p<0.05).*
Color: 1(not white), 2(slightly white), 3(white), 4(very white).
Flavor: 1(not sour), 2(slightly sour), 3(sour), 4(very sour).
Texture: 1(not soft), 2(slightly soft), 3(soft), 4(very soft).
Overall: 1(dislike), 2(slightly like), 3(like), 4(very much like).

3.7 Sour Taste Sensory

The difference in the concentration of rabbit skin gelatin addition (Table 6) did not provide a significant difference (p>0.05) in the taste of frozen yogurt. Sensory testing of all treatments of rabbit skin gelatin addition did not change; the average sensory value of the sour taste of frozen yogurt had the criteria of a sour taste. This is in accordance with the opinion of Febriana et al. (2021) that gelatin has the characteristics of being odorless, colorless, or slightly yellow in appearance, and tasteless so that it does not affect the taste of a product. The taste of frozen yogurt produced is influenced by the raw materials used in the manufacturing process [33]. According to Agustina et al. (2015), the typical sour taste of yogurt is due to the presence of lactic acid content produced from the activity of lactic acid bacteria in breaking down lactose into glucose and galactose with the help of the lactase enzyme. The glucose is hydrolyzed into lactic acid [35].

3.8 Soft Texture Sensory

The addition of rabbit skin gelatin concentration (Table 6) gave a significant difference (p<0.05) to the soft texture of frozen yogurt. According to Karami et al. (2018), stabilizers can form a smooth texture in ice cream because they have a high water binding capacity [35]. The treatment of adding 0% rabbit skin gelatin was significantly different from the addition of 0.3%, 0.6%, and 0.9% rabbit skin gelatin. According to Putri et al. (2016), factors that affect the softness of ice cream include the volume of the ice phase, overrun, fat instability, and ice crystal size [36]. The use of cream also affects the texture of the frozen yogurt produced because fat can reduce the formation of ice crystals during freezing and trap air, resulting in a smooth texture in frozen yogurt.

The highest soft texture score was found in the treatment of adding 0.6% rabbit skin gelatin, while the lowest soft texture score was found in the treatment of adding 0% gelatin. The low soft texture score in the addition of 0% gelatin is thought to be due to the absence of gelatin addition, where gelatin can inhibit the recrystallization process due to temperature fluctuations so that the ice crystals formed are larger and the texture becomes rougher. According to Darma et al. (2013), stabilizers can bind fat and water by forming micro-sized membranes and stabilizing air molecules in the dough so that frozen water can be inhibited [37]. The soft texture score in the addition of 0.9% rabbit skin gelatin decreased from the treatment of adding 0.6% rabbit skin gelatin, which was caused by the addition of too much gelatin, which could bind the water content in the dough and during the freezing process larger ice crystals would form so that the texture of frozen yogurt became rougher [38].

3.9 Sensory Overall
The difference in the concentration of rabbit skin gelatin addition (Table 6) did not provide a significant difference ($p>0.05$) in the overall frozen yogurt. According to Nuryati et al. (2020) the level of product preference is influenced by the personal taste of each panelist with its sensory properties [38]. Panelist preferences can be influenced by the raw materials used in making the product and the feelings of the panelists after swallowing the product. Other sensors, such as taste, color, and texture also determine frozen yogurt preferences. This is in accordance with the opinion of Siswati et al. (2019) that ice cream preferences are influenced by the taste, color, and texture of the ice cream, where consumers prefer ice cream with delicious taste characteristics, attractive colors, and smooth and firm textures [39].

4. Conclusion
The higher the addition of rabbit skin gelatin concentration, the more the total solids and melting time can increase, and the overrun value and white color intensity can be reduced. The addition of gelatin can improve the soft texture and produce frozen yogurt with a total BAL that is still high according to standards. The addition of 0.6% rabbit skin gelatin to frozen yogurt can optimally prevent the recrystallization process so that it can produce a very soft texture, inhibit melting at room temperature, and maintain the number of living lactic acid bacteria that panelists like.

References
[1] Satria, R., E. Rossi, & N. Harun. 2017. Study of packaging types and storage time on the quality of soyghurt ice cream. Online Journal of Animal Husbandry Faculty Students 4(2):1–15.
[2] Wijayanti, IA, Purwadi, & Imam Thohari. 2015. Effect of adding sago flour to yoghurt on viscosity, overrun, melting speed and total solids of yoghurt ice cream. Journal of Animal Science and Technology 10(2):28–35. DOI: 10.21776/ub.jitek.2015.010.02.3.
[3] Mulyani, S., N. Rosyidah, S. Susanti, & YB Pramono. 2020. The physical sensory characteristics of ice cream enriched corn oil using different stabilizers. The International Journal Of Science & Technology 8(5):90–95. DOI: 10.24940/theijst/2020/v8/i5/ST2005-023.
[4] Putro, PA., M. Wachid, & N. Harini. 2019. Ekstraksi gelatin dari kulit kelinci lokal jawa (*Lepus negricollis*) dengan variasi jenis pelarut dalam suhu ekstraksi serta aplikasinya pada bakso kelinci. Journal Food Technology and Halal Science 2(2): 265-281. DOI: 10.22219/fths.v2i2.12991
[5] Hidayah, UN., DR. Affandi, & AM. Sari. 2017. Microstructure, physical and sensory characteristic of ice cream with dumbo catfish bone gelatin (*Clarias gariepinus* sp) as stabilizer. Jurnal Teknologi Hasil Pertanian 2(2): 89-98. DOI: 10.20961/jthp.v10i2.29070
[6] Mulyani, S., FMCS Setyabudi, Y. Pranoto, & U. Santoso. 2017. The effect of pretreatment using hydrochloric acid on the characteristics of buffalo hide gelatin. Journal of Indonesian Tropical Animal Agriculture 42(1):14–22. DOI: 10.14710/jitaa.42.1.14-22.
[7] Wardhani, DH, E. Rahmawati, GT Arifin, & H. Cahyono. 2017. Characteristics of demineralized gelatin from lizardfish (Saurida spp.) scales using NaOH-NaCl solution. Journal of Renewable Natural Materials 6(2):132–142. DOI: 10.15294/jbat.v6i2.9621.
[8] Haryanti, N., & A. Zueni. 2015. Identification of physical, chemical and organoleptic quality of mangosteen skin (Garcinia mangostana L.) ice cream with variations of milk cream. AGRITEPA 1(2):143–156. DOI: 10.37676/agritepa.v2i1.103.

[9] Purdi, TS, YB Pramono, & VP Bintoro. 2020. Total solids, hedonic quality test of color and aroma of soursop fruit velva (Annona muricata Linn) with the use of different types of stabilizers. Journal of Food Technology 4(2):144-148. DOI: 10.14710/jtp.v4i2.26577.
[10] Umela, S. 2016. Analysis of the quality of green bean ice cream (Phaseolus radiatus L) and fresh cow's milk. Technopreneur Journal 4(2):131–137. DOI: 10.30869/jtech.v4i2.68.
[11] Goff, HD, & RW Hartel. 2013. Ice Cream. Edition 7. Springer, New York.
[12] Septiani, AH, Kusrahayu, & AM Legowo. 2013. Effect of adding skim milk in the process of making frozen yogurt based on whey on total acid, pH and number of lactic acid bacteria. Animal Agriculture Journal 2(1):225–231.
[13] Rahardjo, M., M. Sihombing, & MK Anggraeni. 2021. Physical and sensory characteristics of ice cream with the addition of honey caramel. Journal of Food Science and Technology 6(2):3785–3798.
[14] [GMIA] Gelatin Manufacturers Institute of America. 2012. Gelatin Handbook. Pages: 1–25.
[15] Suptijah, P., SH Suseno, & C. Anwar. 2013. Analysis of gel strength of jelly candy products from shark skin gelatin with the addition of carrageenan and seaweed. Indonesian Journal of Fisheries Product Processing 16(2):183-191 . DOI: 10.17844/jphpi.v16i2.8053.
[16] Saputra, RH, I. Widiastuti, & A. Supriadi. 2015. Physical and chemical characteristics of catfish (Pangasius pangasius) skin gelatin with a combination of various acids and temperatures. Journal of Fisheries Product Technology 4(1):29–36. DOI: 10.36706/fishtech.v4i1.3496.
[17] Gerungan, D., M. Sompie, JM Soputan, & A.Dp. Mirah. 2019. Effect of different extraction temperatures on gel strength, viscosity, yield and pH of pig skin gelatin. Zootec Journal 39(1):93–100. DOI: 10.35792/zot.39.1.2019.23761.
[18] Sembiring, CI, AM Legowo, & A. Hintono. 2019. Effect of adding porang tuber flour (Amorphophallus oncophyllus) as a stabilizer on the physical, chemical, and organoleptic properties of jackfruit ice cream. Journal of Food Technology 3(2):241–246. DOI: 10.14710/jtp.v3i2.23870.
[19] Ariska, SD, & D. Utomo. 2020. Quality of instant lemongrass (Cymbopogon citratus) powdered drink with foam mat drying method. Media Information and Scientific Communication of Agricultural Technology 11(1):42–51. DOI: 10.35891/tp.v11i1.1903.
[20] Ayudiarti, DL, Suryanti, & DA Oktavia. 2020. The effect of different types and gelatin concentrations on ice cream quality. E3S Web of Conferences 147(03026):1–8. DOI: 10.1051/e3sconf/202014703026.
[21] Devitasari, ISA, I. Kentjonowaty, & OR Puspitarini. 2021. The effect of adding baking powder on overrun and melting power of ice cream. Journal of Animal Engineering Dynamics 4(3):346–349.
[22] Yulistiani, R., T. Mulyani, & P. Minja. 2017. The role of starter and cryoprotectant on the quality of frozen pepino fruit yogurt. Journal of Food Technology 6(2):1–9.
[23] Saputri, E., E. Rossi, & U. Pato. 2015. Making functional ice cream from soyghurt and low-fat milk. Journal of Agricultural Product Technology 3(1):1–13.
[24] Nofrida, R., Y. Sulastri, YR Widyasari, MA Zaini, & A. Nasrullah. 2018. The effect of adding natural stabilizer based on local tubers to improve the physical and chemical properties of red dragon fruit ice cream (Hylocereus polyrhizus Sp.). Scientific Journal of Technology Agrotechno Agriculture 3(1):298-306. DOI: 10.24843/JITPA.2018.v03.i01.p06.
[25] Failisnur. 2013. Characteristics of jicama ice cream using several types of milk. Journal of Industrial Research and Development 3(1):11–20. DOI: 10.24960/jli.v3i1.623.11-20.
[26] Zahro, C., & FC Nisa. 2015. The effect of adding grape juice (Vitis vinifera L.) and stabilizer on the physical, chemical and organoleptic characteristics of ice cream. Journal of Food and Agroindustry 3(4):1481–1491.
[27] Mulyani, DR, EN Dewi, & RA Kurniasih. 2017. Characteristics of ice cream with the addition of alginate as a stabilizer. Journal of Fishery Products Processing and Biotechnology 6(3):36–42.
[28] Jayus, A. Nafi, & MN Prawintasari. 2022. Characterization of low-fat synbiotic ice cream with different concentrations of banana and edamame extracts. Journal of Research and Community Service 1(9):488–500. DOI: 10.36418/comserva.v1i9.107.
[29] Bezerra, M., A. Araujo, K. Santos, & R. Correia. 2015. Caprine frozen yoghurt produced with fresh and spray dried jambolan fruit pulp (Eugenia jambolana Lam) and Bifidobacterium animalis subsp. lactis BI-07. Food Science and Technology 62:1099–1104. DOI: 10.1016/j.lwt.2015.01.049.
[30] Bilang, M., M. Tahir, & D. Haedar. 2018. Studying the viability of probiotic cell encapsulation (Lactobacillus plantarum and Streptococcus thermophilus) in ice cream. Food Technology, Nutritions, and Culinary Journal 1(1):41–52. DOI : 10.20956/canrea.v1i1.21.
[31] Putra, INK 2016. Efforts to improve the color of palm sugar by adding Na-Metabisulfite. Journal of Food Technology Applications 5(1):1–5. DOI: 10.17728/jatp.v5i1.2.
[32] Eletra, Y., Susilawati, & S. Astuti. 2013. Effect of gelatin concentration on organoleptic properties of goat milk jelly candy. Journal of Industrial Technology and Agricultural Products 18(2):185–195.
[33] Febriana, LG, NAS Sannia, AN Fitriani, & NA Putriana. 2021. Potential of gelatin from fish bones as an alternative to halal capsule shells: characteristics and pre-formulation. Pharmaceutical Journal 6(3):223–233. DOI: 10.24198/mfarmasetika.v6i3.33183.

[34] Agustina, Y., Kartika, R., and Panggabean, A. S. 2015. The Effect of Fermentation Time Variations on Lactose, Fat, pH, and Acidity Levels in Cow's Milk Fermented into Yogurt. Mulawarman Chemistry Journal. Vol 12, No 2

[35] Karami, A., T. Rahayuni, & S. Priyono. 2018. Effect of carrageenan and sago starch formulation on physical, chemical, and organoleptic properties of purple sweet potato ice cream. Journal of Food Technology 1(1):42–49. DOI: 10.26418/jft.v1i1.31396.

[36] Putri, D., Y. Wulandari, & N. Suhartati. 2016. Physicochemical and sensory characteristics of red bean ice cream (Phaseolus vulgaris L.) with variations in the addition of rosella flower petal powder. Journal of Food Technology and Industry 1(1):47–53.

[37] Darma, G.S., Puspitasari, D., Noerhartati, E, 2013. Media Teknologi dan Manajemen Agroindustri, 1, 46-53.

[38] Prastiti, R., S. Mulyani, & Masykuri. 2013. Microbiological characteristics and texture of probiotic ice cream with fresh milk as raw material combined with purple sweet potato (Ipomoea batatas). Animal Agriculture Journal 2(4):127–136.

[39] Nuryati, C., AM Legowo, & Nurwantoro. 2020. Physical and sensory characteristics of red bean ice cream (Phaseolus vulgaris L.) with the addition of gembili tuber flour (Dioscorea esculenta L.) as a stabilizer. Journal of Agrotechnology 14(2):199–207. DOI: 10.19184/j-agt.v14i02.17615.

[40] Siswati, OD, VP Bintoro, & Nuewantoro. 2019. Characteristics of purple sweet potato (Ipomoea batatas var Ayamurasaki) ice cream with the addition of gembili tuber flour (Dioscorea esculenta L.) as a stabilizer. Journal of Food Technology 3(1):121–126. DOI: 10.14710/jtp.v3i1.23254.

[41] [BSN] National Standardization Agency. 1995. SNI 01-3713-1995. Ice Cream Quality Requirements. National Standardization Board, Jakarta.

[42] [BSN] National Standardization Agency. 2009. SNI 2981-2009. Yogurt Quality Requirements. National Standardization Board, Jakarta.

Characteristics of lemon-infused water added by turmeric and red ginger with different immersion storage temperature

Titri Siratantri Mastuti*, Vania Felicia Lisandi

Food Technology Study Program, Pelita Harapan University, Jl. M.H. Thamrin Boulevard 1100, Kelapa Dua, Tangerang 15811, Banten, Indonesia

* E-mail : titri.mastuti@uph.edu

Abstract. Infused water is drinking water containing pieces of fruit or spices immersed for several hours before consumption. Lemon is commonly used in the production of infused water. In this study, turmeric and red ginger were added to increase the functionality of lemon-infused water. This research aimed to determine the effect of the ratio of lemon-turmeric, immersion time, and immersion storage temperature to produce infused water made by a mix of lemon, turmeric, and red ginger, with the best characteristics. In the first stage, infused water was prepared with lemon-turmeric ratios (1:2, 1:1, 2:1), with 5 hours of immersion in storage temperatures of 4ºC and 25ºC. The amount of red ginger remained the same. In the second stage, infused water was prepared with different immersion times of 1 hour, 3 hours, 5 hours, 7 hours, and 9 hours. The result was that infused water with a 1:1 ratio of lemon and turmeric, stored at 4ºC for 3 hours, had the best characteristics. The selected infused water has antioxidant activity with 73.24% RSA, total phenolics of 0.13 mg GAE/ml, total flavonoids of 0.13 mg QE/ml, 1.74 mg / 100 ml vitamin C, a yellow color, and pH 4.05. It could accepted by consumers based on the hedonic test.

Keywords : infused water, lemon, storage temperature, turmeric

1. Introduction

The role of drinking water is very important for the human body, including in regulating metabolism, digestion, body temperature, and blood flow that carries nutrients to body cells. The average recommended water consumption for adults is 2 -2.5 liters or the equivalent of 8 glasses per day [1]. Drinking water is generally colorless and tasteless, which can affect people's preferences and influence consumption levels.

Infused water can be an alternative to encourage people to drink more water [2]. Infused water is known as water that is infused with pieces of fruit or spices to provide an enjoyable drinking water experience and health benefits [3]. Infused pieces of fruit or spices impart certain characteristics to drinking water such as aroma, flavor, and freshness sensation [4]. Active compound components in fruits or spices can increase the health benefits of drinking water such as the content of vitamins, minerals, and phenolic compounds [5].

Infused water generally uses lemon pieces as the basic ingredient [6]. Infused water with lemon as the basic ingredient has higher antioxidant activity and vitamin C when compared to other types of citrus with the antioxidant activity value of infused water lemon and black grapes amounting to 68.42% RSA DPPH and vitamin C infused water lemon and kiwi fruit amounting to

246.40 mg / 100 g [7]. According to [8], lemons have vitamin C (ascorbic acid) content of about 40-50 mg/100 g and 3.7% of citric acid. Citric acid in lemons causes the juice to have a low pH of 2.74, so it is very acidic when consumed directly [9]. The process of making infused water with fruit that is only cut and immersed in mineral water without any extraction or squeezing of lemon makes infused water still less popular, especially in terms of taste.

Red ginger has an essential oil content of 1.5-3.8% [10]. Essential oils contain active compounds such as zingeron, shogaol, gingerol, which are known to have a role as natural antioxidants [11]. Based on [12], red ginger juice has a total flavonoid of 73.43 ± 2.40 mgQE/L. Red ginger-infused water has an antioxidant activity of 88.26% RSA DPPH which is greater than the use of white ginger and large yellow ginger [13]. Although ginger has a spicy taste sensation, its essential oil component produces a distinctive aroma of ginger.

Turmeric is a bright yellow rhizome plant that contains curcuminoid compounds that are widely used as natural dyes [14). Turmeric also has essential oils, phenolic compounds, and flavonoids [15]. Using turmeric as an ingredient of infused water can increase antioxidant activity and sensory, especially the color of infused water.

The production of infused water by immersing the ingredients can extract active compounds in the ingredients [16]. The final characteristics of infused water can be influenced by several factors, such as immersion temperature, storage temperature, immersion time, and the amount of material immersed [6,17]. Infused water is generally made from lemon as the main ingredient, so the addition of red ginger and turmeric is expected to increase antioxidant activity, vitamin C, aroma, color, and taste of infused water. The purpose of this study was to determine the effect of the ratio of lemon-turmeric, immersion time, and immersion storage temperature to produce infused water made by lemon, turmeric, and red ginger, with the best characteristics.

2. Materials and Methods

2.1 Materials

The main materials used in this study were red ginger (*Zingiber officinale* Roscoe) with a length of 6-7 cm obtained from the BSD modern market, turmeric (*Curcuma longa* L.) with a length of 7-8 cm obtained from Grand Wisata modern market, local lemon (*Citrus limon* (L.) Osbeck) with a length of 9-10 cm and a diameter of 6-7 cm obtained from Market City, drinking water "Amidis".

The supporting materials were DPPH, ascorbic acid standard, indophenol, gallic acid, Folin-Ciocalteu reagent, and quercetin. All the materials mentioned previously are obtained from Merck, Germany. Other materials were ethanol p.a. (Smart Lab, Indonesia) and distilled water. The types of equipment used in this research were a hot plate stirrer, thermometer, analytical balance "Ohaus", micropipette, chromameter "Minolta CR400", pH meter "Ohaus Starter 3100", and spectrophotometer "DLAB SP-V1000".

2.2 Methods

This research consisted of two stages. The first stage aims to determine the best lemon: turmeric ratio and immersing storage temperature in making infused water. Infused water was made by immersing lemon, turmeric, and red ginger slices in 90°C water, referring to previous procedures with modifications [7,13]. The formulation of infused water can be seen in Table 1 with variations in the ratio of the amount of lemon-turmeric and a fixed amount of red ginger.

Table 1. Formulation of infused water

Ingredient	Amount
Lemon:turmeric	40 g (1:2, 1:1, 2:1)
Red ginger	20 g
Water	200 ml

The mixture of lemon, turmeric, and red ginger slices was immersed for 5 hours and stored at room temperature (25°C) and refrigerator temperature (4°C). After that, the mixture was filtered to obtain infused water. The flowchart of the first stage of making infused water can be seen in Figure 1.

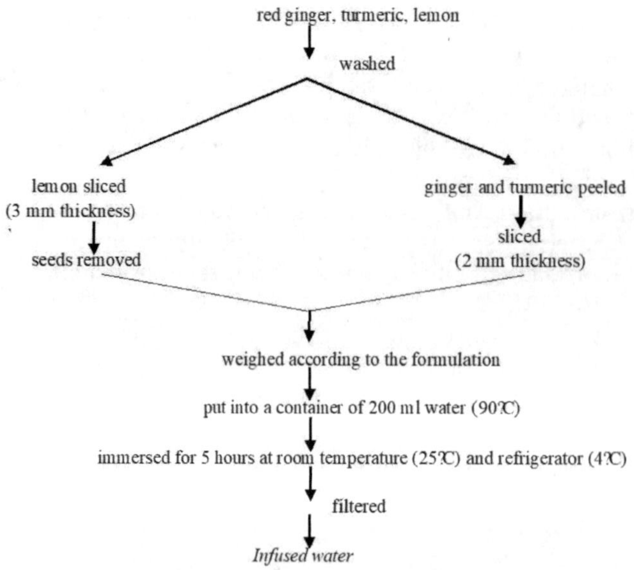

Figure 1. Infused water procedure flowchart of first stage

The second stage aims to determine the best immersing time for infused water containing a mix of lemon, turmeric, red ginger. Modification of the infused water making procedure was carried out at the lemon mixing stage. Lemon was added when the temperature of the water used to immerse the turmeric and ginger mixture had dropped to 25°C. Immersing of the three mixtures was carried out for 1 hour, 3 hours, 5 hours, 7 hours, and 9 hours at the selected storage temperature of stage I. The flowchart of the second stage of making infused water can be seen in Figure 2.

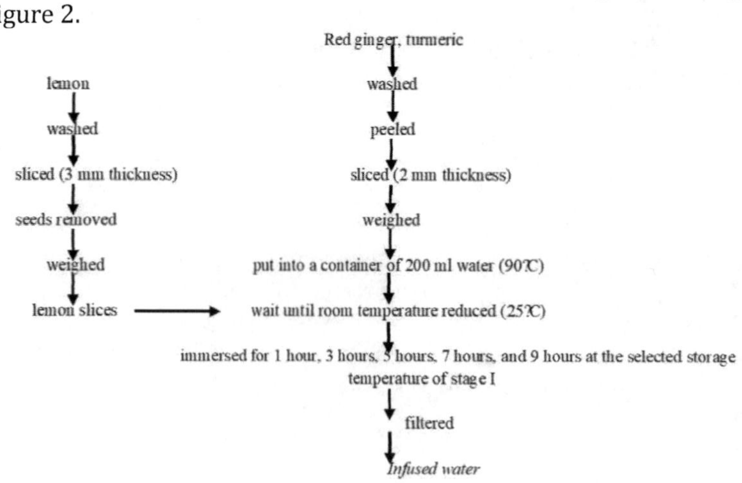

Figure 2. Infused water procedure flowchart of second stage

2.3 Parameter analysis
The analysis conducted on infused water was antioxidant activity [18], total of vitamin C, color [19], total of phenolics [20], total of flavonoid [21], pH [22], and organoleptics test [23,24]. The organoleptic test in this study includes a scoring test (scale 1-6) and a hedonic test (scale 1-5) with color, turmeric aroma, lemon aroma, sour taste, spicy sensation, and overall acceptance parameters of 30 semi-trained panelists.

2.4 Data analysis
This research used a Completely Randomized Design (CRD) with two factors and 3 replications in the first stage. The analysis used two-way ANOVA with factors of lemon: turmeric ratio (1:2, 1:1, and 2:1) and immersing storage temperature (25°C and 4°C). In the second stage, the research used a Completely Randomized Design (CRD) with one factor and 4 replications. The analysis used one-way ANOVA with factors of immersing time, namely 1 hour, 3 hours, 5 hours, 7 hours, and 9 hours. Analysis of variance ANOVA was performed using the SPSS 25.0 software. Significant differences were detected using Duncan's test at the p≤ 0.05 level.

3. Result and Discussion

3.1 Effect of lemon-turmeric ratio and immersing storage temperature to infused water characteristics
Infused water in this research is made with a mix of lemon, turmeric, and red ginger. Each material has different properties that could affect the characteristics of infused water. Some physical, chemical, and sensory analyses that could describe the properties of infused water include pH, color, vitamin C, phenolic and flavonoid content, and organoleptic test. In this study, the amount of red ginger is constant. Therefore, the infused water characteristics in the first stage of research focused on the effect of the different lemon and turmeric ratios.

3.1.1 Acidity (pH) of infused water
Normally drinking water has a pH neutral around 6,5 – 8,5 [25]. Based on Figure 3, the use of lemon and turmeric slices reduces the pH of water with a range between 3.6 ± 0.02 – 4.34 ±0.01 categorized as acid pH. Interaction between the ratio of lemon and turmeric with immersed storage temperature had a significant effect (p≤0.05) on the pH value of infused water. The lowest pH value was found in the (lemon: turmeric) 2:1 ratio treatment with an immerse storage temperature of 4ºC, meanwhile the highest was found in the (lemon: turmeric) 1:2 ratio treatment with a storage temperature of 25ºC.

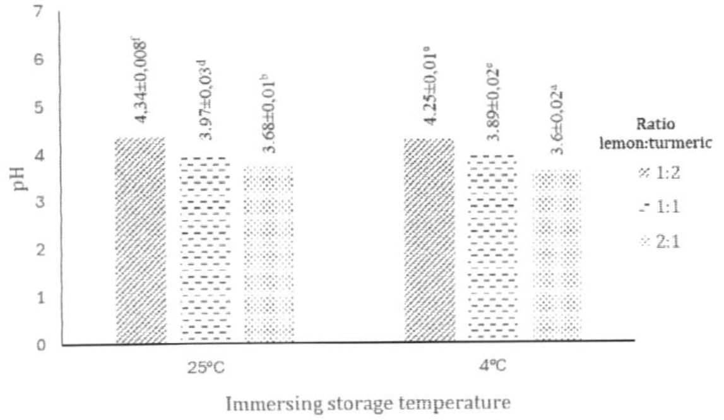

Note : Different superscript letters indicate a significant difference (p ≤0,05)
Figure 3. Effect of lemon-turmeric ratio and immersing storage temperature to infused water pH

IOP Conf. Series: Earth and Environmental Science **1460** (2025) 012046 doi:10.1088/1755-1315/1460/1/012046

The citric acid in lemons causes lemons to have a low pH of 2.74 [9], therefore the more the amount of lemon will reduce the pH of infused water. It happens because of the protons released by the carboxyl group from citric acid so the higher the concentration of citric acid, the lower the acidity of the infused water [26]. Turmeric solution has a higher pH of around 7.5 [27], therefore the presence of turmeric can increase the pH of infused water. Besides it, temperature is a critical factor for some bioactive compounds. The high temperature can accelerate the oxidation process of citric acid which can affect the infused water pH [28]. Therefore the interaction between the 4ºC immersed storage temperature and the increasing lemon-turmeric ratio will produce a lower pH of infused water.

3.1.2 Color

The color of a product can be expressed by degree or Hue value. Based on the results analysis for all infused water treatments, the ºHue value ranges from 87.49-102.06º which indicates the color of the infused water is yellow [29]. The yellow color is produced from the contribution of curcumin content from turmeric [14]. Curcumin tends to be yellow in solutions with acidic to neutral (pH 3-7) and become reddish at higher pH (base) due to the formation of phenoxide anion from the reaction between phenolic groups with hydroxyl groups in basic solutions [30].

The ratio between lemon and turmeric gives a significant effect in the infused water lightness ($p \leq 0.05$) with a range of 47.54 - 50.12. The lightness level of infused water will decrease with more lemon in the mixture of lemon and turmeric. Carotenoids in lemon peel can contribute to lowering the lightness level [31].

3.1.3 Vitamin C

Infused water is favored because of the nutrients derived from the raw materials. Vitamin C is one of the components of lemon as a source of antioxidants. Based on the results of ANOVA analysis, lemon: turmeric ratio and immersing storage temperature had a significant effect ($p \leq 0.05$) on total vitamin C.

In Figure 4, the 2:1 lemon: turmeric ratio has the highest vitamin C value of 2.24 mg/100 ml. The more the number of lemons, the higher the vitamin C content in the infused water. Infused water at 4ºC refrigerator storage temperature had a higher vitamin C content (2.13 mg/100 ml) than storage at 25ºC room temperature (1.72 mg/100 ml). Vitamin C or ascorbic acid is very easily oxidized mainly due to the influence of temperature and light. Therefore, storage at room temperature causes a greater decrease in vitamin C compared to low-temperature storage [17].

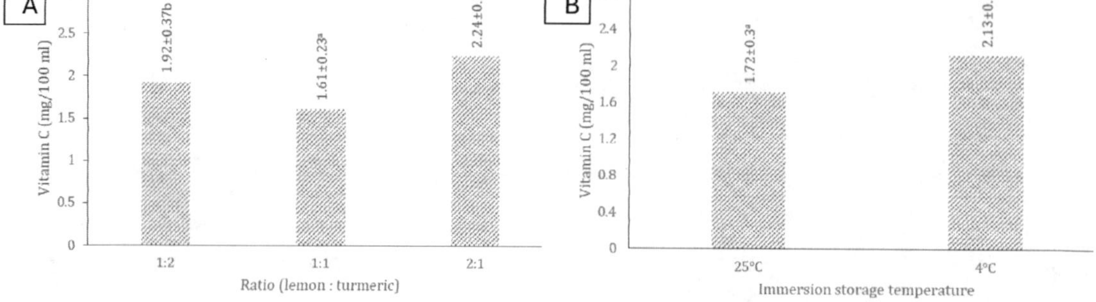

Note : Different superscript letters indicate a significant difference (p ≤0,05)

Figure 4. Effect of lemon-turmeric ratio (A) and immersing storage temperature (B) to infused water Vitamin C

3.1.4 Total of fenolic and total of flavonoid

Lemon, turmeric, and red ginger are fruits and spices famous for having phenolic compounds that can enhance antioxidant activity. Phenolic compounds have various types that can be grouped based on differences in the structure and function of the phenol ring. They are phenolic acids, flavonoids, stilbene, and lignans. Among the four, flavonoids have more variations [32]. Infused water is the simplest beverage product preparation process because it only needs immersing fruit slices or spices in water. This could affect how much phenolic components can be extracted from it.

Based on Figure 5, it can be seen that infused water made by immersing lemon, turmeric, and red ginger for 5 hours can extract phenolic compounds. The interaction between the lemon-turmeric ratio and the storage temperature during the immersion of the ingredients affects the total phenolic content. The ratio of lemon and turmeric (1:1) produced the highest total phenolic (0.20 mg GAE/ml) at 25ºC storage temperature. The effect of this ratio is due to differences in the content of each ingredient. Lemon has a fairly large total phenolic content in the fruit, namely 1.1025 mg GAE/ml [33], and in the peel namely 9.7 mg GAE/g [34]. Based on [35], turmeric has a total phenol of 0.52 mg GAE/g. Therefore, the ratio of lemon: turmeric 1:2 has the lowest total phenolic content.

However, the 2:1 lemon: turmeric ratio did not produce the highest total phenolics. This can be due to differences in the composition or type of phenolic compounds in the ingredients. The largest phenolic component in lemon is the Hesperidin compound (79 ppm in peel) which is a flavonoid group [32]. Whereas in turmeric besides Curcumin as the main phenolic component, there are also Quercetin and caffeic acid [36].

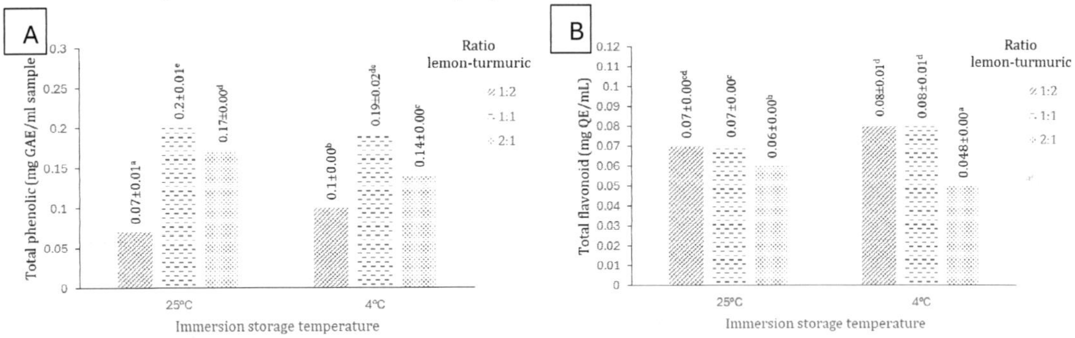

Note : Different superscript letters indicate a significant difference (p ≤0,05)

Figure 5. Effect of lemon-turmeric ratio and immersion storage temperature to infused water total phenolic (A) and total flavonoid (B)

The presence of quercetin in turmeric affects the total flavonoids of infused water. Despite having lower total phenolics, based on Figure 5, it can be seen that the total flavonoids in the lemon:turmeric ratio (1:2) is not significantly different (p>0.05) from the 1:1 ratio, both at 25°C and 4°C storage. However, the immersed storage temperature at 4°C produced the highest total flavonoids (0.08 mg QE/ml). According to [37], flavonoid content can be better maintained at low temperatures. The flavonoid content of infused water in this study is higher than goji berry-infused water of 0.04 mg QE/ml [38].

3.1.5 Organoleptic

The organoleptic test was conducted using scoring and hedonic test on 30 panelists. Based on the Two Way Anova statistic analysis, only the lemon-turmeric ratio factor has an effect (p≤0.05) on the scoring assessment of color, lemon aroma, turmeric aroma, and sour taste of infused water, which can be seen in Table 2. The immersion storage temperature affect ($p \leq 0.05$) only on the color and spicy sensation scoring of infused water, as can be seen in Table 3.

Table 2. Effect of lemon-turmeric ratio to infused water scoring test

Ratio lemon-turmeric	Color	Aroma of lemon	Aroma of turmeric	Sour taste
1 : 2	5.18±0,73^{b}	2.72±1,06^{a}	4.77±1,02^{b}	3.7±1,14^{a}
1 : 1	5.15±0,82^{b}	2.92±1,11^{ab}	4.53±1,18^{b}	3.95±1,14^{a}
2 : 1	4.18±0,91^{a}	3.25±1,31^{b}	4.13±1,07^{a}	4.87±1,17^{b}

Note: Note : Different superscript letters indicate a significant difference ($p \leq 0{,}05$) on each column/parameter
Color : score 1 – 6 (very not yellow - very yellow)
Aroma of lemon : score 1 – 6 (very not lemon aroma - very lemon aroma)
Aroma of turmeric : score 1 – 6 (very not turmeric aroma - very turmeric aroma)
Sour taste : score 1 – 6 (very not sour - very sour)

Based on Table 2, it can be seen that the ratio of ingredients produces different infused water characteristics. The increase of turmeric makes the infused water yellow and the reduced amount of turmeric will reduce the intensity of the color so that the panelists detected infused water with a lemon:turmeric ratio of 2:1 with a slightly yellow assessment. According to [14], turmeric is a type of plant that gives a bright yellow color to drinks or food due to the presence of curcumin as an active component of turmeric. The presence of carotenoids in turmeric and lemon, especially β-Carotene further increases the intensity of the yellow color of infused water [36,39].

The aroma of turmeric is also more dominant in the infused water from this study, resulting in a score ranging from 4.13 to 4.77, which means it has a slight turmeric aroma to turmeric aroma. According to [40], the more turmeric is added, the stronger the turmeric aroma will be due to the presence of essential oils and curcumin, which consist of several compounds such as ketones, turmerone, zingiberene, cineole, and several other compounds. Lemon contains the compound limonene, which gives it a distinctive lemon scent [41]. However, the stronger aroma of turmeric makes the lemon scent less detectable, resulting in a rating of 2.72 - 3.25, which indicates a somewhat lemon-less aroma. However, the use of lemon in infused water tends to sour taste. The intensity of the sour taste decreases with the increasing amount of turmeric used. This result is linear with the analysis of infused water pH, which shows a higher pH as the amount of turmeric used increases.

Based on Table 3, it can be seen that the infused water made by immersing lemon, turmeric, and ginger at a temperature of 4 °C produces a yellower color and a stronger spicy sensation when consumed. Both curcumin and carotenoids are more susceptible to degradation at room temperature storage, which can reduce the intensity of the resulting color [42].

Table 3. Effect of immersion storage temperature to infused water scoring test

Immersion storage temperature	Color	Spicy sensation
25 °C	4.61±1,01^{a}	3.04±1,46^{a}
4 °C	5.07±0,80^{b}	3.68±1,44^{b}

Note: Different superscripts indicate there is a significant difference ($p \leq 0.05$) on each column/parameter
Color : score 1 – 6 (very not yellow - very yellow)
Spicy sensation : score 1 – 6 (very no spicy sensation – very a spicy sensation)

Turmeric rhizome not only provides the distinctive aroma of turmeric but also adds a spicy flavor to beverages [30]. In addition, red ginger also has a distinctive taste and spicy sensation, and it can provide warmth due to the chemical compounds in ginger such as oleoresin, gingerol, zingerone, and shogaol [43]. In this study, the spicy sensation in infused water is not influenced by the lemon-turmeric ratio, so the spiciness from ginger may contributes to the overall spicy sensation. Nevertheless, the assessment of the spicy sensation score in the range of 3.04 – 3.68

indicates that the infused water only has a slight spicy sensation, which may be due to the presence of lemon contributing to a more dominant sour taste in the infused water. In addition, cold temperatures can affect the taste sensors of food products, such as sweetness and spiciness, so at a storage temperature of 4°C, infused water has a more pronounced spicy sensation compared to infused water storage at room temperature (25°C) [44].

The hedonic test carried out using a 1-5 rating scale (very dislike – very like) indicates that the ratio of lemon-turmeric and immersion storage temperature does not affect the panelists' preference for infused water ($p>0.05$). A preference for the aroma of lemon, the aroma of turmeric, sour taste, and spicy sensation has a neutral preference. Meanwhile, the assessment of the hedonic color of infused water received a higher score in the range of 3.63-4.07, which means liking infused water made by lemon, turmeric, and red ginger. The hedonic test for overall acceptance scored between 3.17-3.67, which means neutral to like. This indicates that the infused water from the research is liked and accepted by consumers.

3.2 Effect of immersion time to infused water characteristics

According to the analysis result of the first stage of research, it was determined that infused water with a lemon-turmeric ratio of 1:1 and an immersion storage temperature of 4°C is the best treatment based on total phenolic, flavonoids, and vitamin C produced. Infused water with the 1:1 ratio has a lower acidity level based on scoring tests and pH.

Additionally, the scoring and preference for the color of the infused water are also higher at 4°C. However, the total phenolic, flavonoid, and vitamin C content of the selected infused water is relatively low, with an antioxidant activity (IC50) is 1939 ppm based on total dissolved solids. Therefore, in the second stage of the research, a modification was made where lemon was added when the immersion water of the turmeric and ginger mixture had cooled down (from 90°C to a temperature of 25°C). After all the ingredients were immersed, they were stored in a refrigerator at 4°C with varying immersing durations of 1, 3, 5, 7, and 9 hours.

3.2.1 Chemical analysis of infused water

Based on the One Way Anova statistical analysis, the immersion time in the preparation of infused water has a significant effect ($p \leq 0.05$) on the total phenolic content, total flavonoids, vitamin C, and antioxidant activity. According to Figure 6, it can be seen that the total phenolic, total flavonoid, vitamin C, and antioxidant activity in infused water increased with the duration of immersing time. However, in immersing for more than 5 hours there is a decrease. The highest total phenolic content of infused water (0.13 mg GAE/ml) was found in the immersing time of 3 hours. The highest flavonoid content of infused water (0.13 mg QE/ml) is found in the immersing time of the material for 3 hours. The highest vitamin C content (1.77 mg/100 ml) was found in the 5-hour immersing time which was not significantly different from the 3-hour immersing time (1.74 mg/100 ml). The content of active compounds can affect the antioxidant activity of a product [45]. Therefore, the antioxidant activity of infused water expressed as a percent inhibition of RSA has the highest value when immersing the ingredients for 3 hours, which is 73.24%.

Immersing lemon, turmeric, and red ginger in water for making infused water is a simple extraction method. There should be an increase in the amount of chemical compounds extracted as the extraction time increases. However, prolonged immersing time also affects the reactions that may occur with the chemical compounds within it. Several phenolic compounds that have been extracted can undergo damage or degradation if immersed for a long time, such as curcumin and carotenoids [30,46]. This result is similar to infused water made by a mix of butterfly pea flower and andaliman, which experiences a decrease in flavonoids throughout the immersing time [5]. The amount of flavonoid compounds in infused water is linier with the total phenolics in the infused water.

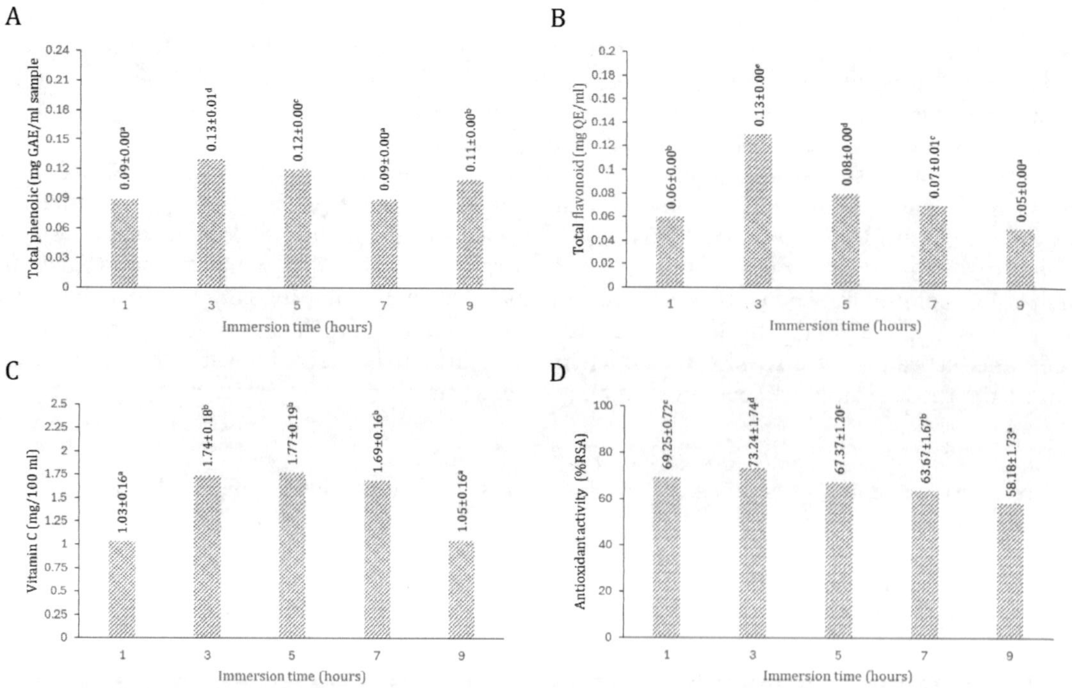

Note : Different superscript letters indicate a significant difference (p ≤0,05)

Figure 6. Effect immersion time to infused water chemical analysis, (A) total phenolic, (B) total flavonoid, (C) vitamin C, (D) antioxidant activity

Compared to the first stage result, the total flavonoids and vitamin C in the second stage have higher values, although the total phenolics are lower. This shows that the method of making infused water, such as the selection of ingredients and the immersing process, can affect the characteristics of the infused water. Lemon is added to the immersing water containing turmeric and red ginger at a temperature of 25°C to inhibit the degradation of vitamin C, particularly ascorbic acid in the lemon. This method also optimizes the extraction of flavonoid compounds from turmeric and red ginger. However, room temperature can also result in less effective extraction of phenolic compounds outside the flavonoid group, such as gallic acid [32,39]. The antioxidant activity of infused water in this study has a higher inhibition value (%RSA) compared to lemon-black grape fruit infused water (68.42%) and lemon infused water (39.78%) immersed for 4 hours [7]. The type of ingredients and method of making infused water affect the antioxidant activity of the final product [6].

3.2.2 Color of infused water

Color is one of the important parameters because it can visually affect the appearance of a product. Attractive colors can also influence the interest and preference of consumers towards a product. Based on the scoring test results in Figure 7A, the infused water produced shows a yellow color. Immersing time affects the results of the color-scoring test ($p \leq 0.05$). The highest score for the assessment of yellow-infused water is at an immersing time of 3 hours (5.5 ± 0.51), which is not significantly different from the results of a 5-hour. Longer immersing times reduce the intensity of the assessment of the yellow color. This is linear with the visual appearance of the color-infused water that can be seen in Figure 7B. The difference in assessment of yellow color intensity may be due to differences in color brightness. Infused water after immersion for 3 hours appears darker and looks lighter after immersing in longer.

The yellow color in the infused water in this study primarily comes from the compound curcumin found in turmeric and carotenoids from both turmeric and lemon. Curcumin and carotenoids are compounds that are easily degraded, influenced by the duration of storage or soaking, which affects the yellow color, causing it to fade. [39,42].

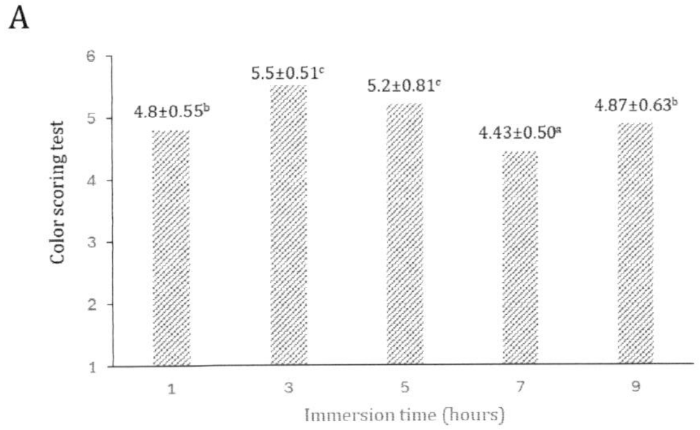

Note : Different superscript letters indicate a significant difference (p ≤0,05)
Color : score 1 – 6 (very not yellow - very yellow)
Figure 7. Color-infused water color, (A) Scoring test, (B) Visual color (left to right: 1-9 hours)

3.2.3 Overall preference
Overall acceptance in the hedonic test describes the overall preference for the sensory attributes of the infused water. The results of this study showed that immersing time did not affect the overall assessment score ($p>0.05$). The value of panelists' level of liking for overall acceptance ranged from 3.27-3.83 which indicates neutral to like. The treatment of immersing time affects the chemical compounds extracted, but visually and sensory such as color, taste, and aroma in infused water immersed for 1 to 9 hours is not much different from the first stage infused water so it can be said that infused water with various treatments can be accepted by consumers.

3.2.4 Characteristics of the best sample infused water
The research results indicate that infused water made with slices of lemon, turmeric, and red ginger, using a lemon-turmeric ratio of 1:1, an immersing time of 3 hours, and storage of the infusion in a refrigerator at 4°C, provides the best characteristics of infused water compared to other treatments. This treatment was chosen because it produced infused water with the highest levels of phenolic, flavonoid, antioxidant activity, and vitamin C. The selected infused water has a pH value of 4.05, which means it is less acidic compared to the pH of red ginger-infused water combined with lemon and mint leaves [47].

The antioxidant activity of selected infused water has a %RSA value of 73.24% and an IC50 of 1743.75 ppm (based on total dissolved solids). This infused water's antioxidant activity is derived from a vitamin C content of 1.74 mg/100 ml, total phenolics of 0.13 mg GAE/ml, and total flavonoids of 0.13 mg QE/ml. The color of the infused water is visually yellow, with a hue value of 79.80 and a lightness of 47.70. Infused water has a turmeric aroma, a slightly sour taste, and a slightly spicy sensation. Based on the overall acceptance, the preference score of the panelists is 3.67 indicating that the panelists' assessment tends to like and can accept the infused water.

4. Conclusions
Characteristics of infused water made with a mix of lemon, turmeric, and red ginger are affected by the ratio of lemon-turmeric, immersion storage temperature, and time of immersion. The color

of the infused water will be more yellow as the number of turmeric increases. Meanwhile the infused water will become sour as the lemon increase. The immersion storage temperature and immersion time of all materials can increase the total flavonoid and vitamin C content.

The ratio of lemon-turmeric (1:1), refrigerator temperature (4°C) for storage immersion, and 3 hours of immersion time provide the best characteristics of infused water. The infused water has a yellow colour, a turmeric aroma, a slightly sour taste, and a slightly spicy sensation. It has a total phenolic content of 0.13 mg GAE/ml, a total flavonoid content of 0.13 mg QE/ml, a vitamin C content of 1.74 mg/100m, a pH of 4.05 and an antioxidant activity of 73.24% RSA. Infused water is favoured by the panellists, so it can be an alternative beverage to increase healthy daily water consumption.

References

[1] Setyawan FB and Sofyan I. 2024. Pentingnya Minum Air Putih bagi Kesehatan Siswa Sekolah Dasar. *J Basicedu.* 8(1):299–305. doi:10.31004/basicedu.v8i1.7081.

[2] Soraya N. 2014. Infused Water. Penerbit Penebar Swadaya, Jakarta.

[3] Surati S and Qomariah N. 2017. Tingkat Keamanan Minuman Infused Water dengan Diversifikasi Penyimpanan yang Berbeda. *J Ris Kesehat.* 6(1):13–19.

[4] Haitami H, Ulfa A, Muntaha A. 2017. Kadar Vitamin C Jeruk Sunkist Peras dan Infused Water. *Med Lab Technol J.* 3(1):22. doi:10.31964/mltj.v3i1.149.

[5] Sakdiah H, Nurminah M, Karo-Karo T. 2022. Characteristic infused water of butterfly pea (Clitoria ternatea Linn) with the addition of traditional spice andaliman (Zanthoxylum acanthopodium DC) based on sustainable local resources. *IOP Conf Ser Earth Environ Sci.* 977(1). doi:10.1088/1755-1315/977/1/012082

[6] Trisnawati I, Hersoelistyorini W, Nurhidajah N. 2019. Tingkat Kekeruhan Kadar Vitamin C dan Aktivitas Antioksidan Infused Water Lemon Dengan Variasi Suhu Dan Lama Perendaman. *J Pangan dan Gizi.* 9(1):27. doi:10.26714/jpg.9.1.2019.27-38

[7] Harifah I, Mustofa A, Suhartatik N. 2016. Aktivitas Antioksidan *Infused Water* dengan Variasi Jenis Jeruk (Nipis, Lemon, dan *Baby*) dan Buah Tambahan(Stroberi, Anggur Hitam, dan Kiwi). *Jurnal Teknologi dan Industri Pangan*, 1(1) : 54-58.

[8] Kristanto F. 2013. Kekerasan Permukaan Enamel Gigi Manusia Setelah Kontak dengan Air Perasan Citrus Limon. Skripsi. Universitas Airlangga, Surabaya.

[9] Manners, H. I., Buker, R. S., Smith, V. E., Ward, D., Elevitch, C. R. 2006. *Citrus (citrus) and Fortunela (kumquat). Species Profiles for Pacific Island Agroforestry*, 2.1 : 1-35

[10] Febriani Y, Riasari H, Winingsih W, Aulifa DL, Permatasari A. 2018. The Potential Use of Red Ginger (*Zingiber officinale* Roscoe) Dregs as Analgesic. *Indonesian Journal of Pharmacceutical Science and Technology*, 1(1):57-64.

[11] Aryanta IWR. 2019. Manfaat Jahe untuk Kesehatan. *E-Jurnal Widya Kesehatan*, 1(2):39-43.

[12] Mastuti TS, Setiawanto AF. 2022. Characteristic of Red Ginger Jelly Stick with Variation Type of Gelling Agent. *Proc 6th Int Conf Food, Agric Nat Resour (IC-FANRES 2021)*. 16:199–207. doi:10.2991/absr.k.220101.026.

[13] Wahyuningsih N, Suhartatik N, Mustofa A. 2018. Aktivitas Antioksidan *Herbs Infused Water* dengan Variasi Lama Perendaman dan Varietas Jahe (*Zingiberoffcinale*) Emprit, Gajjah, dan Merah. *Jurnal Agrisaintifika*, 160-166.

[14] Shan CY and Iskandar Y. 2018. Studi Kandungan Kimia dan Aktivitas FarmakologiTanaman Kunyit (*Curcuma longa L.*). *Jurnal Farmaka*, 16(2):547-555

[15] Simanjuntak P. 2012. Studi Kimia dan Farmakologi Tanaman Kunyit (*Curcuma longa* L.) sebagai Tumbuhan Obat Serbaguna. *Agrium*, 17(2):103-107.

[16] Kiswandono AA. 2011. Skrining Senyawa Kimia dan Pengaruh Metode Maserasi dan Refluks pada Biji Kelor (*Moringa oleifera*, Lamk) terhadap Rendemen Ekstrak yang Dihasilkan. *Jurnal Sains Natural Universitas Nusa Bangsa*, 1(2):126-134.

[17] Kartikawati E, and Yudi YHC. 2020. Pengaruh Waktu dan Suhu Penyimpanan terhadap Kadar Vitamin C *Infused Water* Buah Lemon (*Citrus lemon* (L.). *Jurnal Sabdariffarma*, 8(1):28-34.

[18] Pratiwi P, Suzery M, Cahyono B. 2010. Total Fenolat dan Flavonoid dari Ekstrak dan Fraksi Daun Kumis Kucing (*Orthoshipon stamineus* B.) Jawa Tengah serta Aktivitas Antioksidannya. *Jurnal Sains dan Matematika*, 18(4):140-148.

[19] Nielsen SS. 2010. Food Analysis Laboratory Manual. 2nd ed. Springer Science+Business *Media*, LLC, New York.

[20] Andriani D, and Murtisiwi L. 2018. Penetapan kadar fenolik total ekstrak etanol bunga telang (Clitoria Ternatea L.) dengan spektrofotometri UV VIS, *Cendekia Journal of Pharmacy*, 2(1) : 32- 38.

[21] Nugraheni M, Santoso U, Windarwati. 2018. Phytochemical compounds and antioxidant activity of Coleus tuberosus flesh and peel on different solvent, *Food Research*, 2(5) : 460-467.

[22] AOAC. 2005. Official Methods of Analysis. Maryland: Assocoation of Official Analytical Chemist.

[23] Tarwendah IP. 2017. Studi Komparasi Atribut Sensoris dan Kesadaran Merek Produk Pangan. *Jurnal Pangan dan Argoindustri*, 5(2):66-73.

[24] Martiyanti MAA and Vita VV. 2018. Sifat Organoleptik Mi Instan Tepung Ubi Jalar Putih Penambahan Tepung Daun Kelor. *Jurnal Teknologi Pangan*, 1(1):1-13.

[26] Permenkes. 2010. Peraturan Menteri Kesehatan Republik Indonesia Nomor 492/Menkes/per/IV/2010 tentang Persyaratan Kualitas Air Minum. Depkes RI, Jakarta.

[27] Kristianingrum S, Siswani ED, Fillaeli. 2011. Pengaruh Jenis Asam pada Sintesis Silika Gel dari Abu Bagasse dan Uji Sifat Adsorptifnya Terhadap Ion Logam Tembaga. Prosiding Seminar Nasional Kimia, Yogyakarta.

[28] Amalia M, Raharjo D, Priyono S. 2019. Pengaruh konsentrasi ekstrak kunyit dan lama perendaman terhadap daya simpan kerupuk basah. *J Teknol Pangan*. 3(2):273–280.

[29] Reda SY. 2011. Evaluation of antioxidants stability by thermal analysis and its protective effect in heated edible vegetable oil. Ciênc. Tecnol. Aliment., Campinas, 31(2): 475-480

[30] Hutchings JB. 1999. Food Color and Appearance. 2nd ed. Springer, United State.

[31] El-Saadony MT, Yang T, Korma SA, Sitohy M, Abd El-Mageed TA, Selim S, Al Jaouni SK, Salem HM, Mahmmod Y, Soliman SM, et al. 2023. Impacts of turmeric and its principal bioactive curcumin on human health: Pharmaceutical, medicinal, and food applications: A comprehensive review. Front Nutr. 9. doi:10.3389/fnut.2022.1040259.

[32] Nianti EE, Dwiloka B, Setiani BE. 2017. Pengaruh Derajat Kecerahan, Kekenyalan, Vitamin C, dan Sifat Organoleptik pada Permen Jelly Kulit Jeruk Lemon (Citrus medica var Lemon). *Jurnal Teknologi Pangan*, 2(1):64-69.

[33] Irkin R, Dogan S, Degirmenioglu N, Diken ME, Guldas M. 2015. Phenolic content, antioxidant activities and stimulatory roles of citrus fruits on some lactic acid bacteria. *Arch Biol Sci.* 67(4):1313–1321. doi:10.2298/ABS140909108I

[34] Permata AN, Kurniawati A, Lukiati B. 2018. Screeing Fitokimia, Aktivitas Antioksidan dan Antimikroba pada Buah Jeruk Lemon (Citrus limon) dan Jeruk Nipis (Citrus aurantiifolia). *Jurnal Ilmiah Ibnu Sina*, 3(1):64-76.

[35] Agarwal M, Gupta R, Upadhyaya, S. 2012. Extraction of Polyphenol, Flavonoid from Emblica officinalis, Citrus limon, Curcumis sativus and Evaluation of Their Antioxidant Activity. *Oriental Journal of Chemistry*, 28(2):993-998

[36] Wahyuningtyas SEP, Permana IDGM, Wiadnyani AAIS. 2017. Pengaruh Jenis Pelarut Terhadap Kandungan Senyawa Kurkumin dan Aktivitas Antioksidan Ekstrak Kunyit (Curcuma domestica Val.). *Jurnal ITEPA*, 6(2):61-70

[37] Suprihatin T, Rahayu S, Rifa'i M, Widyarti S. 2020. Senyawa pada Serbuk Rimpang Kunyit (Curcuma longa L.) yang Berpotensi sebagai Antioksidan. *Bul Anat dan Fisiol*. 5(1):35–42. doi:10.14710/baf.5.1.2020.35-42.

[38] Yahya. 2015. Perbedaan Tingkat Laju Osmosis antara Umbi Solonum Tuberosum dan Doucus Carota. *Jurnal Biology Education*, 4(1):196-206.

[39] Putri EB, Putri FK, Sulaiha, S. 2020. Perbandingan Kadar Flavonoid dan Vitamin C pada Infused Water Goji Berry (Lycium barbarum) dan Air Nabeez Kurma (Pheonix dactylifera L.). *Medical Technology and Public Health Journal*, 4(1):32-37. https://doi.org/10.33086/mtphj.v4i1.1458

[40] Uçan F, Ağçam E, Akyildiz A. 2016. Bioactive compounds and quality parameters of natural cloudy lemon juices. *J Food Sci Technol*. 53(3) : 1465–1474. doi:10.1007/s13197-015-2155-y

[41] Mulyani S, Harsojuwono BA, Puspawati, G. 2014. Potensi minuman Kunyit Asam sebagai Minuman Kaya Antioksidan. *Jurnal Agritech*, 34(1):65-71.

[42] Ekaputri F. 2018. Pengaruh Perbandingan Kulit dan Sari Lemon dan Konsentrasi Kayu Manis terhadap Karakteristik Selai Lemon (Citrus limon burm f.) secara Organolpetik. Skripsi. Universitas Pasudan, Bandung.

[43] Tensiska, Nurhadi B, Isfron AF. 2012. Kestabilan Warna Kurkumin Terenkapsulasi dari Kunyit (Curcuma domestica Val.) dalam Minuman Ringan dan Jelly pada Berbagai Kondisi Penyimpanan. *Bionatura Jurnal Ilmu-Ilmu hayati dan Fisik*, 14(3):198-206.

[44] Srinivasan K. 2017. Ginger Rhizomes (Zingiber officinale): A Spice with Multiple Health Beneficial Potencials. *Pharma Nutrition*, 5(1):18-28.

[45] Kapaun CL, and Dando R. 2017. Deconvoluting physical and chemical heat: Temperature and spiciness influence flavor differently. *Physiol Behav*. 170:54–61. doi:10.1016/j.physbeh.2016.12.015

[46] Dhurhania E, and Novianto A. 2018. Uji Kandungan Fenolik Total dan Pengaruhnya terhadap Aktivitas Antioksidan dari Berbagai bentuk Sediaan Sarang Semut (Myrmecodia pendens). *Jurnal Farmasi dan Ilmu Kefarmasian*, 5(2):62-63.

[47] Jahangiri Y, Ghahremani H, Torghabeh JA, Salehi EA. 2011. Effect of Temperature and Solvent on The Total Phenolic Compounds Extraction from Leaves of Ficus carica. *Journal of Chemical and Pharmaceutical Research*, 3(5):253-259

[48] Sepriani R, and Deswandi. Biochemical analysis of ginger-infused water with a combination of lemon and mint leaves. 2021. *Proceeding E3S Web Conf ICFTNSA.* 332, 06002:1-8. https://doi.org/10.1051/e3sconf/202133206002

Thermal stability of crude pectin methyl esterase from potato widely cultivated in Indonesia

Setya Budi Muhammad Abduh*, Arif Bestari, Ghina Husna Ulaya Nasution, Novi Sya'bania, Tri Zahra Desvina Wandani, Margaretha Ruri Kristiani, Joey Beatrix Bernadette, Saraswati, Fitra Inna Istifara, Keisha Fayza Putri Artya, Dzikri Nurazizi, Fadhila Tri Julia Putri, Farah Diba Chairani, Maria Widya Tasika Putri, Ahmad Ni'matullah Al-Baarri, Rafli Zulfa Kamil, Yoga Pratama, Swastika Dewi

Department of Food Technology, Faculty of Animal and Agricultural Sciences, Universitas Diponegoro, Semarang 50275, Indonesia

*E-mail: setya.abduh@live.undip.ac.id

Abstract. The behaviour of pectin in controlling the texture of potato tuber is influenced by pectin methyl esterase (PME). Since thermal processing is necessary for potato prior consumption, understanding the thermal stability of PME is important. This study investigates the thermal stability of PME from Granola potato, a cultivar that is widely cultivated in Indonesia. Crude PME was prepared from potato puree passing through centrifugation and precipitation using tris(hydroxymethyl)aminomethane buffer, NaCl, and ammonium sulphate. Half mL of PME was transferred into capillary tubes and heat-treated for 5 min at temperatures of 25, 50, 55, 60, and 72 °C in a water bath. The PME activity was determined by mean of 0.01 M NaOH titration on mixtures of crude PME and pectin solution (0.35% (w/v) at pH at 7. The PME activity ranges from 0.0108 to 0.0258 unit/mL depending on the temperature of the heat-treatment. The results showed that PME activity increased as the temperature raise to 50°C but decreased when the temperature exceeded 50°C. This pattern suggests that PME is initially activated by heat but loses activity at temperature higher than 50°C indicating thermal activation–inactivation behaviour. This finding is potentially used as a basis to control the textural development of potatoes during cooking.

Keywords: granola potato, heat-treatment, thermal-activation–inactivation, texture

1. Introduction

Potato (*Solanum tuberosum* L.) is a tuber widely consumed as one of staple foods. Indonesia cultivates several potato varieties, with the granola variety being one of the most widely grown due to its high productivity, yielding 30-35 tons per hectare [2]. Granola potatoes are often prepared as cooked potatoes because of their dense pore structure that retains moisture, resulting in a water content exceeding 80% [3]. Texture is an important product characteristic of cooked potato that is developed during cooking. One of important factor in texture development is pectin system. While fresh potato tends to have a firm texture that is attributed to their low-methoxyl pectin structure (approximately 1.55% methoxyl per 100 grams) [1], cooked potato is less firm as affected by starch gelatinisation and cellular disintegration. However, overly soft, cooked potatoes are often undesirable, since it may lead to a higher glycaemic index of starch and are often less preferred by consumers.

The cellular disintegration of potato during cooking is controlled by cell wall pectic system. Apart from thermal degradation, cell wall pectin disintegration is influenced by pectinase enzymes, such as polygalacturonase (PG) and pectin methylesterase (PME), which modify texture during storage and processing [4].

Thermal processing increases the activity of PME leading to firmer texture but temperatures above 70°C decreases cell wall hardness by degrading pectin structures [5]. The firmer structure is suggested being associated with the release of carboxyl groups, which in turn bind calcium ions to form a stronger pectin complex and lead to a firmer textural property as occurred in Japanese Radish [6]. The PME inactivation occurs in tomato juice treated with high-pressure carbon dioxide at 55°C after 90 minutes at 20 MPa [7] and in citrus fruit heat-treated at above 70°C [8].

Whilst Granola potato is an important commodity for local economy [9], very limited study investigated the thermal stability of its PME. Therefore, this study investigates the temperature range of PME stability of local Granola potato to provide scientific evidence for better processing and utilization.

2. Materials and Methods

Potato (*Solanum tuberosum* L.) of Granola cultivar were obtained from Agro Lestari Merbabu Co. Ltd. in Magelang, Central Java. Upon receival, the potato was immediately washed and cleaned to prevent biological damage, then peeled.

2.1 Pectin methylesterase extraction

Pectin methylesterase (PME) was extracted from the potato following the method from [10]. Briefly, peeled potato was shredded and homogenized into a puree, mixed with 0.05% cold sodium bisulphite (0,6L/kg puree) to prevent enzymatic browning and subsequently centrifuged at 10,000g for 30 min at 4°C (FC5706 Frontier Multi Centrifuge, OHAUS, China). After the supernatant being discarded to obtained pellet, it was washed with cold sodium bisulphite. Then, the PME was extracted using tris(hydroxymethyl)aminomethane buffer (pH 8.0) with 1M NaCl. After overnight end-over-end shaking at 4°C, the mixture was centrifuged again at 18,000g for 30 min. Thirty per cent ammonium sulphate was then added on the obtained supernatant to precipitate the large protein and being centrifuged 18,000g for 30 min. Subsequently, a lower concentration (25%) of ammonium sulphate was added on the supernatant to precipitate PME. The obtained pellet was dissolved in Tris-buffer (pH 7.0) is considered as crude PME and stored at -40°C until further analysis.

2.2 Thermal Treatments

Frozen crude PME was thawed and transferred into capillary tubes (2 mm external diameter, 100 mm long, Ae Lab, UK). Several tubes containing 0,5 mL of crude PME were immersed in water for 5 min at temperatures of 25, 50, 55, 60, and 72°C using a water bath (DWB 18L-2R-6H, B-One, China). Immediate cooling in ice bath was conducted after each heat treatment to stop the heating.

2.3 Pectin methylesterase activity assay

The activity of crude PME was determined on its reaction with 0.35% (m/v) of apple pectin (65-70% DE, PCT0411, Himedia) at pH 7. After the reaction, the mixture was titrated using 0.01 N NaOH solution to maintain the pH within period no longer than 20 min. The activity of crude PME was calculated based on the volume of NaOH using the following equation.

$$PME\ activity\ \left(\frac{unit}{mL}\right) = \frac{(mL\ NaOH)(NaOH\ molarity)(1000)}{(titration\ time)(mL\ sample)}$$

3. Results and discussion

Figure 1 shows the change in the activity of crude PME as affected by heating temperature. The activity of crude PME consistently increased from 0,011 U/mL to 0,028 U/mL with increasing temperature of heating from 25 to 50 °C. However, the activity of crude PME decreased after being heated at 70 °C.

It is suggested that heating for 5 min at temperature range from 25 to 50 °C is associated with thermal induced activation of the enzyme [11]. Oppositely, heating for the same period at 60 °C is associated with irreversible denaturation of the enzyme, where the protein structure is disrupted, leading to loss of function [12, 13, 14, 15, 16, 17]. The specific temperature at which inactivation occurs can vary depending on factors such as pH, water activity, and the presence of other compounds [18]. Thermal-induced activation and inactivation of PME as presented in Figure 1 have significant implications for the textural properties of potatoes during cooking. The initial activation of the enzyme can contribute to the softening of potato tissue by catalysing the de-esterification of pectin process [19].

Previous studies report that the temperature range of PME activation is up to 65 °C [20, 21] and even 70 °C [10, 21], higher temperatures than the current study. However, it is important to note that the heat treatment conducted in the current study is limited to 5 min period hence requires further investigation for different period to represent the actual condition of local Granola processing. Therefore, a kinetic study investigating both time-dependent and temperature-dependent under realistic processing condition is necessary to obtain a solid data useful for the basis of process optimisation of local Granola potato. The optimum processing condition is relevant not only for textural properties of potato but also for nutritional properties and sustainability [9].

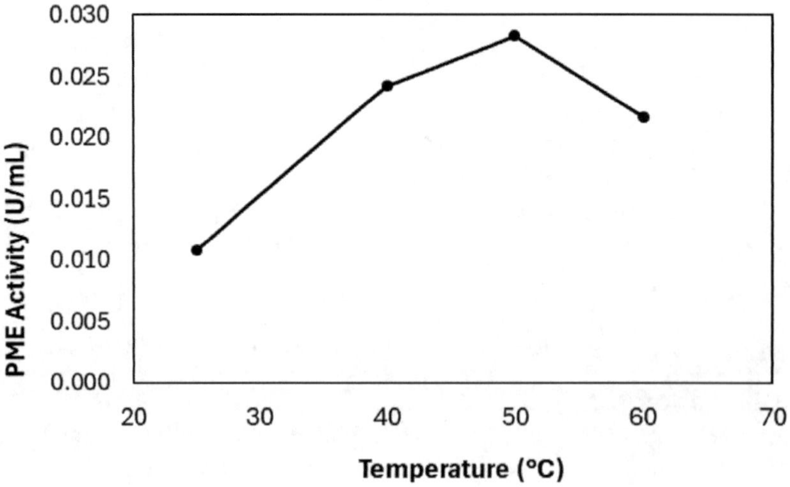

Figure 1. Activity of crude PME from potato granola being heated at different temperatures for 5 min

4. Conclusion

Being heat-treated for 5 min at temperature from 25 to 60 °C, the activity of crude PME from Granola potato locally cultivated in Magelang Regency, Central Java, Indonesia is found to increase with increasing temperature up to 50 °C and found to decrease after being treated at 60 °C, indicating the temperature range of activation of up to 50 °C. Furthermore, a kinetic study is essential to investigate the thermal sensitivity of the PME activation-inactivation cycle under condition realistic to daily processing practices to provide solid data for process optimisation of potato.

Acknowledgement

This research was financially funded by Universitas Diponegoro.

Reference

[1] Mandei JH, AM Nuryadi. Pengaruh Cara Perendaman dan Jenis Kentang Terhadap Mutu Keripik Kentang. J. Penelitian Teknologi Industri. 2017;9(2):123-136.

[2] Sahara D, Wulanjari ME. Cara Pemupukan yang Menguntungkan Usaha Tani Kentang di Kabupaten Banjarnegara, Jawa Tengah. J. Ilmu Pertanian Indonesia. 2022;27(4): 473-480.

[3] Parhusip A, Neysha A, Halim L, Iwantoro FO. Aplikasi *Pre-heating* dan *Edible Coating* untuk Peningkatan Kualitas Keripik Kentang. J. Mutu Pangan: Indonesian Journal of Food Quality. 2017;8(1):53-62.

[4] Lyu J, Yu X, Wang F, Xie J, Bi J. Innovative Food Science and Emerging Technologies Cell Wall Modifications During Storage: Changes in The Texture Behavior of The Canned Yellow Peach. 2023

[5] Liu J, C Wen, M Wang, S Wang, N Dong, Z Lei, B Zhu. Enhancing the hardness of potato slices after boiling by combined treatment with lactic acid and calcium chloride: Mechanism and optimization. J. Food chemistry. 2020;308:124832.

[6] Ando, Y, S Hagiwara, H Nabetani. Thermal Inactivation Kinetics of Pectin Methylesterase and The Impact of Thermal Treatment on the Texture, Electrical Impedance Characteristics and Cell Wall Structure of Japanese Radish (*Raphanus sativus* L.). Journal of Food Engineering. 2017;199: 9-18.

[7] Illera, AE, Sanz, MT, Trigueros, E, Beltran, S, Melgosa, R. Effect of High Pressure Carbon Dioxide on Tomato Juice: Inactivation Kinetics of Pectin Methylesterase and Polygalacturonase and Determination of Other Quality Parameters. 2018.

[8] Aghajanzadeh, S, Mohammad Ziaiifar, A, Kashaninejad, M, Maghsoudlou, Y, Esmailzadeh, E. Thermal Inactivation Kinetic of Pectin Methylesterase and Cloud Stability in Sour Orange Juice. 2016

[9] Abduh SBM, Nurwantoro N, Mulyani S, Kamil RZ, Al-Baarri ANM. Commercial grading and domestic processing of Indonesian MZ potato: implications for nutritional and textural properties. Acta Sci Pol Technol Aliment. 2024;23(3):301-14.

[10] Moens, LG, E De Laet, JC Van Ceunebroeck, AM Van-Loey, ME Hendrickx. Thermal Inactivation of Pectin Methylesterase from Different Potato Cultivars (*Solanum tuberosum L.*). Lwt. 2021;138: 110600.

[11] González-Martínez G, Ahrné L, Gekas V, Sjöholm I. Analysis of temperature distribution in potato tissue during blanching and its effect on the absolute residual pectin methylesterase activity. J Food Eng. 2004;65(3):433-41.

[12] Ünal MÜ, Şener A. Extraction and characterization of pectin methylesterase from Alyanak apricot (Prunus armeniaca L). J Food Sci Technol. 2015;52(2):1194-9. doi:10.1007/s13197-013-1099-3.

[13] Dirix C, Duvetter T, Loey AV, Hendrickx M, Heremans K. The in-situ observation of the temperature and pressure stability of recombinant *Aspergillus aculeatus* pectin methylesterase with Fourier transform IR spectroscopy reveals an unusual pressure stability of beta-helices. Biochem J. 2005;392(Pt 3):565-71. doi:10.1042/BJ20050721.

[14] Daniel RM, Dines M, Petach HH. The denaturation and degradation of stable enzymes at high temperatures. Biochem J. 1996;317(Pt 1):1-11. doi:10.1042/bj3170001.

[15] Drozłowska E, Weronis M, Bartkowiak A. The influence of thermal hydrolysis process on emulsifying properties of potato protein isolate. J Food Sci Technol. 2020;57(3):1131-7. doi:10.1007/s13197-019-04148-z.

[16] Andlinger DJ, Röscheisen P, Hengst C, Kulozik U. Influence of pH, temperature and protease inhibitors on kinetics and mechanism of thermally induced aggregation of potato proteins. Foods. 2021;10(4):796. doi:10.3390/foods10040796.

[17] Wu HC, Bulgakov VP, Jinn TL. Pectin Methylesterases: Cell Wall Remodeling Proteins Are Required for Plant Response to Heat Stress. Front Plant Sci. 2018 Nov 6;9:1612. doi: 10.3389/fpls.2018.01612. PMID: 30459794; PMCID: PMC6232315.

[18] Soni, A, Bremer, P, Brightwell, G. A comprehensive review of variability in the thermal resistance (D-values) of food-borne pathogens—a challenge for thermal validation trials. Foods, 2022;11(24), 4117.

[19] Kumar, R, Meghwanshi, G K, Marciano, D, Ullah, S F, Bulone, V, Toffolatti, S L, Srivastava, V. Sequence, Structure and Functionality of Pectin Methylesterases and Their Use in Sustainable Carbohydrate Bioproducts: A Review. International Journal of Biological Macromolecules. 2023;244, 125385.

[20] Abu-Ghannam N, Crowley H. The effect of low temperature blanching on the texture of whole processed new potatoes. J Food Eng. 2006;74(3):335-44.

[21] Anthon GE, Barrett DM. Kinetic parameters for the thermal inactivation of quality-related enzymes in carrots and potatoes. J Agric Food Chem. 2002;50(14):4119-25.

Antioxidant Activity and Preferences Level of Kenikir Leaves (*Cosmos caudatus*) Cendol

Yuliana Jakiyah and Siti Tamaroh*

Department of Agricultural Product Technology, University of Mercu Buana Yogyakarta, Karanglo 55753, Indonesia.

*E-mail: tamaroh@mercubuana-yogya.ac.id

Abstract. Kenikir is a vegetable that has a high antioxidant content. So far, kenikir has only been used as a vegetable for fresh vegetables. There is not much processing of kenikir as functional food. Kenikir has a distinctive and pungent aroma so it is not liked. Making cendol with the addition of kenikir leaves is expected to produce cendol with antioxidant functional properties and prferences by panelists.In this research, cendol was made by adding kenikir leaf powder. Kenikir leaf powder is made by blanching and without blanching. Kenikir powder is added to the cendol mixture at 1%, 3%, 5%, and 7% w/w. The cendol produced was tested for color, water content, total phenols, flavonoids, antioxidant activity and preferent level. The data obtained were subjected to statistical tests using a Randomized Group Design and ANOVA test. If there is a fundamental difference, continue with Duncan's Multiple Range Test at a confidence level of 95%. The results of the research showed that the blanching treatment and the amount of kenikir powder had a significant effect on the color, chemical components and level of preference of kenikir cendol. Kenikir cendol with blanching treatment and the addition of 5% kenikir powder had an antioxidant activity of 19.38% (RSA); total phenol content was 0.0478 mg GAE/g db; flavonoid content was 0.0296 mg EQ/g db ; and color L = 41.34, a = 1.11, b = 7.51 was the most preferred.

Keyword: antioxidant activity, cendol, kenikir leaves

1. Introduction

One type of vegetable that contains antioxidants is kenikir leaves (*Cosmos caudatus*). Young kenikir leaves and shoots can be used as vegetables or eaten raw. Kenikir leaves contain active phenolic compounds, flavonoids, flavones and flavanones, polyphenols, saponins, tannins, alkaloids and essential oils. The flavonoids in kenikir leaves include myricetin, quercetin, kaempferol, luteolin and apigenin.

Currently, the majority of kenikir leaves are used as vegetables. Important components contained in kenikir leaves include flavonoids, which are natural antioxidants that are very effective in warding off free radicals. The research results explain the total content of flavonoids and quercetin in fresh kenikir leaves and kenikir leaf powder [1]. The total flavonoids and quercetin in fresh kenikir leaves were 143.00 mg/100 g and 51.30 mg/100 g, while the total flavonoids and quercetin in kenikir leaf powder were 1089.79 mg/100 g and 390.95 mg/100 g.

Cendol is an ingredient for making ice that is in great demand by the public. The raw material for making cendol is rice flour or other types of flour as a source of carbohydrates. The color of cendol is usually green which comes from the natural ingredients of pandan leaves and suji leaves. There are many sources of natural green dye apart from pandan leaves and suji leaves, namely kenikir leaves.

Blanching is a short heat treatment that is usually applied to vegetables or grains before further processing to improve quality [2]. Blanching has benefits such as killing existing microbes, improving color, texture and maintaining the quality of vegetable or grain products. Blanching is the use of hot media which has long been used as an initial treatment and is a very important stage in drying vegetable and fruit products. The use of hot water as a blanching medium for kenikir leaves aims to deactivate enzymes and remove air from the cell tissue of the material and to avoid changes in the color and aroma of the material. Blanching can also increase mass transfer within the tissue and affect the drying process [3].

One alternative to increase consumption of antioxidant source materials is to enrich antioxidant components in foods or drinks that many people like, for example cendol drinks. However, it is necessary to consider the appropriate proportion of additions so that it does not physical properties of the product which could reduce its level of favorability. This research aims to determine the effect of preparation (blanching) and the amount of kenikir leaf powder on the acceptability of cendol and producing functional food as a source of natural antioxidants.

2. Methods

2.1 Material

The material used in this cendol research was fresh young kenikir leaves aged 30-45 days. The parts used are the first 4 leaves, which were purchased at Kranggan Market, Yogyakarta, hun we (merk Jeruk), rice flour (Rose Brand), salt (merk Revina), citric acid, ice, and vanilla. The chemicals used are the free radical 2,2-diphenyl-1-picrylhydrazyl (DPPH), Folin–ciocalteu reagent (Merck, Germany), gallic acid (GA) and quercethin (Sigma Chemical Co., St Louis, United States), ethanol, methanol (E. Merck), distilled water, $NaCO_3$, $NaNO_2$, $AlCl_3.6H_2O$, NaOH.

The equipment used in this research was analytical scales (Ohaus), 1000 ml measuring glass (Pyrex), 100 ml beaker glass (Pyrex), 100 ml Erlenmeyer (Pyrex), 100 ml measuring flask (Pyrex), blender (Philips), sieve size 60 mesh, oven (Memmert), cabinet dryer, UV-Vis spectrophotometer (Shimadzu UV-752N), and colorimeter (High-Quality Colorimeter NH310).

2.2 Research stage

2.2.1 Production of kenikir powder

Making kenikir leaf powder is done in 2 ways: by blanching and without. For kenikir leaves that are blanched as follows, young kenikir leaves are selected, then washed and blanched. The blanching process was carried out using citric acid with a concentration of 0.05%, then dried in a cabinet dryer at a temperature of 50 ^{o}C for 16 hours, crushed, and sieved with a 60 mesh sieve. For kenikir leaves that are not treated with blanching, the young kenikir leaves are washed then dried in a cabinet dryer at a temperature of 50 ^{o}C for 16 hours, crushed then sieved with a 60 mesh sieve.

2.2.2 Production of kenikir cendol

The ingredients for making cendol are as follows: 100 g of hun kwe flour, 10 g of rice flour, 5 g of salt and 1.25 g of vanilla and 200 ml of water, stir until evenly mixed, then heated at 100 ^{o}C for 10 minutes, and a gel is formed. Next, printing is carried out using a coarse filter. The dough formed is soaked in ice water for about 5 minutes and drained. The treatment in this study was the

addition of kenikir leaf powder which was treated with blanching and without blanching, with treatment variations being 1%, 3%, 5% and 7%. The resulting cendol was tested for color, preference and the panelists liked the cendol the most.
The most preferred cendol from the treatment with the addition of blanched and unblanched kenikir leaf powder was tested for total phenolic content, flavonoids and antioxidant activity.

2.2.3 Determination of Total Phenolic Content
Total phenolic levels were determined by the Folin-Ciocalteu method [4] using gallic acid as the standard.

2.2.4 Determination of flavonoid
The total flavonoid content was determined using visible spectrophotometry accordingly with the absorbance on wavelength 510 nm [5]. The total flavonoid content was expressed as mg quercetin equivalent/g of subfraction (mg QE/g).

2.2.5 DPPH Method of Free Radical Scavenging Test
The antioxidant activity test was carried out by calculating the capacity of DPPH free radical scavenging [6]. Around 0.2 ml of the sample plus 3.8 ml of 0.1 mM DPPH solution, was vortex for 3 min and observed the absorbance at 30 minutes with a spectrophotometer at a wavelength of 517 nm [15]. The calculation of scavenging capacity of free radicals was calculated and expressed in percent (%) RSA =% Radical Scavenging Activity, which was the of DPPH bleaching.
% RSA=1-(absorbance of sample)/(absorbance of control) X 100%

2.3.Experimental Design
The research was conducted using a factorial Completely Randomized Design with 2 factors, namely the influence of the method of preparing the ingredients (blanching and without blanching) and the proportion of added kenikir powder (1%, 5% and 7%) in making kenikir cendol. The method for determining the influence of preparation and the amount of powder added between treatments used the univariate test with SPSS, then the significant difference between samples were determined using Duncan's Multiples Range Test (DMRT).

3. Results and Discussion

3.1. Cendol color
The cendol color in this study was interpreted into three color parameters, namely L, a and b. The lightness (L) value in this study was used on a scale of 0 (dark or black) to 100 (light or white), the greater the lightness value, the brighter the product. The color of cendol from the research results can be seen in Table 1.

Table 1 shows that the blanching and non-blanching treatments and the number of additions produce significantly different L colors. Adding more kenikir powder produces cendol with a smaller L color value, which means it is darker. In this study, the kenikir cendol that had the darkest color value was cendol kenikir with treatment without blanching and the proportion of kenikir powder added was 7%. This is following research, the higher the concentration of green grass jelly leaf extract added to cendol, the lower the lightness value because there is more color pigment, namely chlorophyll, in cendol so the color is darker [7].

Table 1. Color profile of Kenikir cendol

Treatment	Concentration of added kenikir powder (% b/b)	L	a	b
Blanching	1	46.84^{f}	0.75^{bc}	7.53^{cd}
	3	42.58^{d}	0.88^{cd}	7.69^{d}
	5	41.34^{c}	1.11^{ef}	7.51^{cd}
	7	39.20^{b}	1.92^{g}	6.89^{b}
Non Blanching	1	46.11^{e}	0.58^{ab}	7.68^{d}
	3	42.20^{d}	1.02^{de}	7.51^{cd}
	5	41.47^{c}	1.25^{f}	7.42^{c}
	7	38.26^{a}	2.04^{g}	7.35^{c}

Note: Numbers followed by different notation letters in the same column indicate significant differences at the 95% confidence level.

The color value a is an identification of red and green. If a value is more than 1, the resulting color becomes greener, and if it is close to 1, the resulting color becomes redder [8]. In cendol, the results of this research show the color value a, which is that the greater the addition of kenikir powder, the greater the value a. Blanching and non-blanching treatments have no effect on color value a. The greater the color value a indicates the color green. The green color comes from the chlorophyll found in kenikir leaves.

The color value b, shows yellow to blue. If the value of a is more than 10 values, the color becomes yellower, if less than 10 values are produced, the color becomes bluer [8]. Table 1 shows that the color b values are not significantly different when adding 1%, 3% and 5% kenikir leaf powder. The color value a was lower when 7% kenikir leaf powder was added, which showed that the color was getting bluer or darker, due to the presence of a greater chlorophyll component.

3.2. Kenikir cendol preference test results

The preferences test was carried out using the Hedonic method which the panelists carried out. In this study, 25 panelists who gave assessments on a scale of 1 to 5 were used. The value parameters were strongly dislike (1), dislike (2), somewhat like (3), like (4) and like very much (5). The quality attributes or parameters assessed in kenikir cendol products in the research conducted included color, aroma, texture, taste and overall quality. The results of the panelists' preference test for cendol kenikir can be seen in Table 2.

Table 2. Kenikir cendol preference test results

Treatment	Concentration of added kenikir powder (%)	color	Aroma	Texture	taste	Overall
Blanching	1	3.32^{ab}	3.76^{b}	3.72^{b}	3.56^{d}	3.80^{cde}
	3	3.68^{bc}	3.84^{bc}	3.72^{b}	3.40^{bc}	3.64^{bcd}
	5	4.20^{d}	4.32^{c}	3.80^{b}	3.96^{d}	4.12^{e}
	7	3.48^{abc}	3.80^{b}	3.08^{a}	2.84^{a}	3.08^{a}
Non Blanching	1	3.12^{a}	3.88^{bc}	3.56^{b}	3.40^{bc}	3.56^{bc}
	3	3.48^{abc}	3.84^{bc}	3.04^{a}	3.08^{abc}	3.20^{ab}
	5	3.56^{abc}	3.28^{a}	3.08^{a}	2.64^{a}	2.92^{a}
	7	3.44^{abc}	3.04^{a}	3.04^{a}	2.96^{ab}	2.96^{a}

Note: Numbers followed by different notation letters in the same column indicate significant differences at the 95% confidence level.

IOP Conf. Series: Earth and Environmental Science **1460** (2025) 012048 doi:10.1088/1755-1315/1460/1/012048

Based on Table 2. In the assessment of color, the panelists preferred the sample with blanching treatment and the addition of 5% kenikir leaf powder, namely the value = 4.02, which means they liked it. Meanwhile, other treatments were rated lower. Blanching treatment can affect the color of food ingredients. Blanching the kenikir leaves can maintain the original color of the kenikir leaves. Kenikir leaves that were treated with steam water blanching with the addition of 0.05% citric acid for 5 minutes did not change color.

The addition of 0.05% citric acid in the blanching process can prevent and reduce the occurrence of enzymatic browning reactions in kenikir leaves. The purpose of adding citric acid to food is as an antioxidant and to inactivate oxidizing enzymes as well as regulate pH and prevent browning reactions in kenikir leaves. Citric acid can also inhibit browning by lowering the pH like acetic acid the PPO enzyme becomes inactive [9].

Meanwhile, the panelists' assessment of the aroma showed that the treatment of kenikir leaves without blanching with the addition of 5% and 7% was significantly different from other treatments with lower values (3.28 and 3.04), it is possible that there was a pleasant flavor produced by the kenikir leaves. not blanched. The effect of preparing kenikir leaves by blanching gives a cendol aroma that is not too typical of kenikir leaves. Meanwhile, kenikir leaves without blanching give a pungent cendol aroma typical of kenikir.

The results of sensory tests on cendol texture did not differ between treatments. The texture of cendol is formed due to the gelatinization process of the starch of the raw material for making cendol, which depends on the proportion of amylose and amylopectin. In this study, it is estimated that the proportions of amylose and amylopectin are the same, so the texture tends to be rated the same, namely in the range of 3.04 – 3.80.

The results of the taste preference test showed that the addition of kenikir powder with blanching treatment was more preferred (average preference value = 3.44), compared to without blanching (average preference value = 3.02). This is possible because the taste of cendol with kenikir powder without blanching still has the typical kenikir taste.

The results of the liking test on overall liking parameters showed that there was a real difference between the treatment of making cendol with the addition of powder with blanching treatment and without blanching treatment. Cendol with the addition of blanched kenikir powder was more preferred than (average preference test value = 3.66), cendol added with powder without blanching treatment (average preference test value = 3.11). This is because cendol with the addition of kenikir powder without blanching, produces cendol with a distinctive kenikir aroma.

3.3. Total levels of phenols, flavonoids and antioxidant activity of selected cendol

From the results of the preference test, cendol was determined as the most preferred, and compared between the addition of blanched and unblanched kenikir powder. The treatment is cendol with the addition of 5% kenikir powder. Next, the selected cendol was tested for total phenols, flavonoid content test and antioxidant activity test (% RSA). The total phenol test results for cendol can be seen in Table 3.

Table 3. Total phenol content of kenikir cendol

Treatment	Concentration of added kenikir powder (%)	Total phenol (mg GAE/g db)
Blanching	5	0.0478[a]
Non Blanching	5	0.0272[b]

The results of this study showed that the total phenol content in cendol with the addition of kenikir powder treated with blanching was greater than without blanching treatment. The results of this research are in accordance with the research of making purple yam flour which was treated with blanching, the total phenol content was 1.6 times greater without blanching treatment [10]. This result is in line with some previous research, including the research that stated that the blanching treatment will increase the total phenolics in sweet potatoes [11]. Blanching on some green vegetables showed that the total phenolic will increase due to blanching treatment [12]. This research conducted with the blanching treatment on white saffron also showed the increasing content of phenolic compounds [13]. The increase in total phenolic is considered due to the damaged tannins into phenol. In addition, phenolic compounds do not experience enzymatic oxidation due to blanching treatment.

The results of the flavonoid content test can be seen in Table 4. Table 4 shows that the flavonoid content of cendol treated with kenikir blanching powder is greater than that added with kenikir powder without blanching.

Table 4. Flavonoid content of kenikir cendol

Treatment	Concentration of added kenikir powder (%)	Flavonoid content (mg QE/g db)
Blanching	5	0.0296[a]
Non Blanching	5	0.0179[b]

The test results for flavonoid levels in cendol with the addition of winged bean powder treated with blanching showed that it was higher than without blanching treatment. This research is in accordance with research on white turmeric after blanching, the flavonoid content is greater than that of fresh white turmeric [13]. In addition, the blanching process under acidic conditions is thought to result in flavonoid compounds not undergoing enzymatic oxidation. The results of the cendol antioxidant activity test in this study can be seen in Table 5.

Table 5. Antioxidant activity of cendol kenikir

Treatment	Concentration of added kenikir powder (%)	% (RSA)
Blanching	5	19.38[a]
Non Blanching	5	11.40[b]

The antioxidant activity test results of cendol added with powder with blanching treatment showed that the % RSA value was higher than cendol added with kenikir powder without blanching. Several researchers say that antioxidant activity is determined by the phenolic compounds that make up it. This study showed that cendol added with kenikir powder with blanching treatment had greater levels of total phenols and flavonoid levels than cendol added with kenikir powder without blanching. Several studies that support this include research, which states a positive correlation exists between total phenol content and antioxidant activity in making purple yam flour [10]. Research conducted by [14] stated an antioxidant activity in raspberries, blackberries, and red currants determined by total phenol content.

4.Conclusion

The results showed that adding 5% blanched kenikir powder was preferred by the panelists and had the potential as a source of antioxidants. The criteria for cendol from this research are as follows: antioxidants of 19.38% RSA; total phenol content was 0.0478 mg GAE/g db; flavonoid 0.0296 mg QE/g db; color L = 41.34, a = 1.11 and b = 7.51.

References

[1] Sahid, A., P., N. & Mubawani,E . 2016 . Pengaruh Bubuk Kenikir (*Cosmos caudatus*) Terhadap Kadar Glukosa Darah Tikus Diabetes Diinduksi Streptozotocin . Journal of Nutrition College. 5 (2) : 51 - 57.

[2] Saidi, I. A., & Wulandari, F. E. 2019. Pengeringan Sayuran dan Buah-buahan. Umsida Press, 1–35.

[3] Pratiwi P, A. D., Nurdjanah, S., & Utomo, T. P. 2020. Pengaruh Suhu dan Lama Pemanasan Saat Proses Blansing Terhadap Sifat Kimia, Fisikokimia dan Fisik Tepung Ubi Kayu. Jurnal Penelitian Pascapanen Pertanian, 17(2), 117. https://doi.org/10.21082/jpasca.v17n2.2020.117-125

[4] Roy, M.K., Juneja, L.R., Isobe, S. & Tsushida, T. 2009. Steam Processed Broccoli (*Brassica Oleracea*) has High¬er Antioxidant Activity in Chemical and Cellular Assay Systems. Food Chemistry.114 (1) : 263-9.

[5] Zou, Y., Lu, Y. & Wei,D. 2004. Antioxidant activity of Flavonoid-rich extract of *Hypericum perforatum* L in vitro. Journal of Agriculture and Food Chemistry. 52(16):5032-9. doi: 10.1021/jf049571r.

[6] Xu, B.J. & Chang, S.K.C. 2007. A comparative study on Phenolic Profiles and Antioxidant Activities of Legumes Affected by extraction. Journal of Food Science. 72: SI 59-66. DOI: 10.1111/j.1750- 3841.2006.00260.x.

[7] Lestari, B. P. 2019. Karakteristik Fisik dan Sensoris Cendol Instan dengan Penambahan Cincau Hijau (*Cyclea barbata* L.). Orbital: Jurnal Pendidikan Kimia, 3(1), 65-80. https://doi.org/10.19109/ojpk.v3i1.3369

[8] Nabila, H. 2019. Pengaruh Jenis Teh dan Penambahan Sari Nangka Terhadap Sifat Fisik, Kimia, dan Tingkat Kesukaan Teh Kombucha. Skripsi Fakultas Agroindustri. Universitas Mercu Buana. Yogyakarta

[9] Moon, K.M., Kwon, E.B., Lee, B & Kim, C.Y. 2020. Recent Trends in Controlling the Enzymatic Browning of Fruit and Vegetable Products. Molecules , 25, 2754 p. 2-15. doi:10.3390.

[10] Tamaroh, S., Raharjo, S. , Murdiati, A. & Anggrahini, S. 2017. The Effect of Purple Yam (*Dioscorea alata*, L.) Blanching Time on Anthocyanins Content and Antioxidant Activity. The International Journal of Science & Technoledge : 5 : 8 : 83- 88.

[11] Huang, Y.C., Chang, Y.H. & Shao, Y.Y. 2006. Effects of Genotype and Treatment on The Antioxidant Activity of Sweet Potato in Taiwan . Food Chemistry. 98: 529–538.

[12] Oboh, G. 2005. Effect of Blanching on The Antioxidant Properties of Some Tropical Green Leafy Vegetables. Biochemistry Department, Federal University of Technology, LWT.38:13–517.

[13] Pujimulyani, D., Raharjo. S., Marsono, Y. & Santosa, U. 2010. Pengaruh Blanching Terhadap Aktivitas Antioksidan, Kadar Fenol, Flavonoid dan Tanin Terkondensasi Kunir Putih (*Curcuma mangga* Val.). Agritech. 30 (3) :141-147.

[14] Pantelidis, G.E., Vasilakakis, M., Manganaris, G.A., & Diamantidis, Gr. 2007. Antioxidant Capacity, Phenol, Anthocyanin and Ascorbic Acid Contents in Raspberries, Blackberries, Red Currants, Gooseberries and Cornelian Cherries.Food Chemistry.102: 777–783.

The effect of honey pineapple concentrate proportion on hedonic value, color, moisture, and total dissolved solid in Panna Cotta dessert

Hega Bintang Pratama Putra, Valentinus Priyo Bintoro, Damar Jati, Ulil Afidah, and Siti Susanti*

Food Technology Program Study, Department of Agriculture, Faculty of Animal and Agricultural Sciences, Diponegoro University, Indonesia

*E-mail: sitisusanti@live.undip.ac.id

Abstract. Panna Cotta is a widely popular dessert characterized by its high sugar content, derived from dairy products; however, the range of available flavors remains limited. This study aimed to investigate the effects of incorporating pineapple concentrate on the hedonic value, color, moisture content, and total dissolved solids of Panna Cotta. The materials utilized include pineapple concentrate, UHT milk, and cream, with varying formulations for each treatment. A Completely Randomized Design (CRD) was used, comprising five treatments with four replications, involving pineapple concentrate proportions of 0%, 5%, 10%, 15%, and 20% in the milk. Data analysis was conducted using Analysis of Variance (ANOVA) at a 5% significance level. The results indicated that the addition of pineapple concentrate significantly influenced ($p < 0.05$) the hedonic value, color, moisture content, and total dissolved solids of the Panna Cotta. Increasing the concentration of pineapple concentrate resulted in a decrease in moisture content, lightness, and redness, while simultaneously increasing yellowness and total dissolved solids. Moreover, the incorporation of pineapple concentrate in the formulation markedly affected the organoleptic properties of Panna Cotta, particularly its flavor and texture.

Keywords: hedonic, panna cotta, pineapple, water content, milk cream

1. Introduction

Honey pineapple (*Ananas comosus (L) Merr*) represents a notable variety of pineapple that flourishes in tropical regions, particularly reaching its peak production during the summer season. Over the past five years, the production of honey pineapple in Indonesia has experienced a significant increase, rising from 1.8 tons to 3 tons [1]. From a nutritional perspective, honey pineapple is rich in vitamin C, vitamin A, calcium, phosphorus, magnesium, iron, sodium, potassium, and sucrose [2] .

As an agricultural commodity characterized by high yield and nutritional quality, the price of honey pineapple from farmers remains relatively low, ranging from IDR 1,000.00 to IDR 1,500.00 per fruit [3]. Furthermore, the processing of honey pineapple into various food products with higher economic value has not been extensively explored. Its utilization in the food industry is largely confined to relatively simple processing methods, which fail to cater to the preferences of modern consumers who seek convenience and variety.

A prominent example of the modern lifestyle in urban areas is the social activity centered around coffee shops and restaurants, which typically offer a range of contemporary dairy-based menu items. Among the popular desserts is Panna Cotta, a pudding made from cream cooked with gelatin and infused with various flavors. Its unique taste has contributed to its popularity in modern society, with prices ranging from IDR 25,000.00 to IDR 50,000.00 per serving.

To date, there has been limited exploration of flavor variations for Panna Cotta among dessert enthusiasts. Panna cotta is dairy dessert with high in fat content and calorie also mostly consumed as an indulgent choice [4]. Generally, Panna Cotta produced by some food industries developed by creative entrepreneurs is primarily available in a plain variant. The incorporation of certain fruits, particularly pineapple juice, has been largely restricted to condiments such as sauces and toppings at the point of service [5]. Hien at al. stated that fermented beverage from unqualified pineapple for canned, frozen or fresh consumption Pineapple processing into the specific use of honey pineapple juice has yet to be investigated as an additive in the batter for product extension of Panna Cotta [6].

This study examined the impact of incorporating honey pineapple concentrate (HPC) into the batter on the hedonic value, color, moisture content, and total dissolved solids of Panna Cotta dessert. The findings were expected to provide valuable information regarding the application of horticultural processing technology to enhance the value-added potential of honey pineapple within the food industry.

2. Research Methods

2.1 Production of Pineapple Panna Cotta (PPC)

The production of honey pineapple concentrate (HPC) was conducted according to Ashoush et al. [7] with some modifications and the ratio of ingredients was referred to Handani et al. [8]. 1) HPC was produced by washing and cutting pineapple fruit, then crushing and filtering to obtain pineapple honey juice. 2) The juice was heated in a pan and stirred until get 15°brix. 3) Milk and cream (1:1), gelatin (1.5%) and sugar (7.5%) were mixed and heated to a temperature of 70ºC (mixture I). Some of milk was set aside to hydrate the gelatin with a ratio of 1:15 (gelatin:milk) and left for 15 minutes (mixture II). Mixture II was mixed with mixture I and heated to a temperature of 70ºC for 1 minute. Pineapple concentrate was added with various concentrations (0, 5, 10, 15, and 20%). The pineapple Panna Cotta mixture was poured into plastic cup then stored in refrigerator ±5ºC for 24 hours.

2.2 Data Analysis

The moisture content, total dissolved solids, and color of the Panna Cotta were analyzed using Analysis of Variance (ANOVA). Duncan's new multiple range tests was used to know the differences among treatments. Hedonic evaluation was analyzed using Kruskal-Wallis and Mann-Whitney was used to know the differences among treatments.

3. Result and Discussion

3.1 Hedonic Value of Pineapple Panna Cotta

The hedonic test was conducted to assess consumer preferences for a product. The results indicated that the proportion of honey pineapple concentrate (HPC) in the cream significantly affected all hedonic attributes of the resulting Panna Cotta, including flavor, texture, and color (Table 1). The proportion of HPC 5-10% in the cream produced a distinctive sweet and sour flavor profile favored by the panelists. However, the panelists' preference significantly decreased when the HPC proportion exceeded 10%, as the Panna Cotta began to exhibit a more pronounced sour taste. This dominant sourness, which intensified with increasing HPC proportion, can be attributed to the high levels of citric acid and malic acid present in honey pineapple, measured at approximately 14.24 and 9.41 g/L, respectively [9].

Regarding texture, the proportion of HPC 5-15% resulted in a Panna Cotta that was sufficiently jiggly yet soft and melting in the mouth, which was well-received by the panelists. In contrast, proportions exceeding 15% led to a less jiggly and somewhat mushy texture, resulting in a significant decline in panelist preference. The high moisture content of honey pineapple (84%) contributed to a more liquid Panna Cotta as the HPC proportion increased to replace the cream in the batter. The total solids content of the cream, approximately 12.11-16.80%, comprises fats and proteins, both of which can function as emulsifiers, physically imparting a thickening effect to the batter [10].

Furthermore, with increasing proportions of HPC in the cream, the Panna Cotta exhibited a more pronounced yellow color compared to the control (without HPC), thereby enhancing panelist preference ($p < 0.05$). The yellow hue of the product is attributed to the carotenoid and xanthophyll pigments in HPC, which are recognized as natural colorants that enhance the visual appeal of food products [11].

Table 1. The hedonic value of Panna Cotta dessert with varying proportions of honey pineapple concentrate (HPC)

Attribute	Honey pineapple concentrate (%)				
	0	5	10	15	20
Taste	3,20±0,69^{a}	3,28±0,92^{a}	3,4±0,69^{a}	2,88±0,77^{b}	2,48±0,90^{c}
Texture	2,92±0,84^{a}	3,24±0,86^{a}	3,08±0,69^{a}	3,4±0,49^{a}	2,4±0,75^{b}
Color	2,88±0,86^{a}	2,88±0,71^{a}	3,12±0,77^{ab}	3,16±67^{b}	3,24±0,65^{c}

Notes: The data are presented as mean values ± standard deviation. The percentage indicates the proportion of pineapple concentrate relative to the total volume of milk in the mixture. Different superscripts from values in the same row indicate significant differences ($p < 0.05$).

3.2 Color of Pineapple Panna Cotta

The carotenoid and xanthophyll pigments present in pineapple, along with the chemical reactions occurring during the production of Panna Cotta, influence the resulting color intensity. The study indicated that the proportion of honey pineapple concentrate (HPC) in the cream significantly affected the color intensity of the produced Panna Cotta, specifically in terms of lightness (L*), redness (a*), and yellowness (b*) (Table 2). The proportion of HPC 10-20% in the cream resulted in a decreasing intensity of L* in the Panna Cotta ($p < 0.05$). The carotenoid and xanthophyll pigments in pineapple impart a range of colors from yellow to red-orange. When a higher amount of HPC is added, these pigments can become dominant in the mixture, leading to a reduction in L* intensity. Additionally, caramelization may significantly influence the L* value in the product. Caramelization is a complex chemical reaction that occurs when sugar is subjected to high heat.

As the degree of caramelization increases, the product tends to darken, which subsequently lowers the L* value [12].

Redness (a*) indicates a chromatic mixture of red-green, with values of +a* ranging from 0 to +80 for red and -a* from 0 to -80 for green. The proportion of HPC 20% in the cream led to an increase in a* intensity ($p < 0.05$). Research by [13] measured the beta-carotene content in Queen variety pineapples using spectrophotometry, finding it to be 11.72 μg/g. When elevated levels of honey pineapple concentrate (HPC) were added, the pigment content within the product correspondingly increased, potentially enhancing a* intensity. Yellowness (b*) represents a chromatic mixture of blue-yellow, with +b* values ranging from 0 to +70 for yellow and -b* from 0 to -70 for blue. The proportion of HPC 5-20% in the cream resulted in an increasing intensity of b* ($p < 0.05$), attributed to the yellow pigments found in honey pineapple. The yellow color in Panna Cotta arose from the carotenoid and xanthophyll pigments present in honey pineapple. The addition or increase of carotenoid content in the product can elevate the b* value [14].

Carotenoid and xanthophyll pigments found in pineapple and chemical reactions in the panna cotta making process affect the color. The results showed that the proportion of HPC in milk cream had a significant effect on the intensity of the resulting panna cotta color including lightness (L*), redness (a*), and yellowness (b*) (Table 2). The proportion of HPC 10-20% in milk cream resulted in a decreasing L* intensity in panna cotta products ($p<0.05$). Carotenoid and xanthophyll pigments in pineapple provide a yellow to orange-red color. When HPC is added in large amounts, these pigments can become dominant in the mixture, thereby reducing the intensity of the L* color. In addition, the caramelization process can be a significant factor in changing the L* value of food products. Caramelization is a complex chemical reaction that occurs when sugar is exposed to high heat. As the level of caramelization increases, the product tends to become darker, which will reduce the L* value [12]. Redness (a*) indicates a red-green mixed chromatic color with a +a* value of 0 to +80 for red and a -a* value of 0 to -80 for green. The proportion of 20% HPC in milk cream produces an increasing a* intensity ($p<0.05$). Research by [13]tested the beta-carotene content of queen pineapple using a spectrophotometry method of 11.72μg/g. When high HPC is added, the pigment content in the product also increases, which can increase the a* intensity. Yellowness (b*) indicates a blue-yellow mixed chromatic color with a +b* value of 0 to +70 for yellow and a -b* value of 0 to -70 for blue. The proportion of 5-20% HPC in milk cream produces an increasing b* color intensity ($p<0.05$). This is due to the yellow pigment found in honey pineapple. The yellow color of panna cotta comes from carotenoid and xanthophyll pigments found in honey pineapple. Addition or increase in carotenoid content in the product can increase the b* value of the product [14].

Table 2. The color value of Panna Cotta desserts with varying proportions of honey pineapple concentrate (HPC)

Color value	Honey pineapple concentrate (%)				
	0	5	10	15	20
L*	91,62±0,75^{a}	89,4±0,75^{b}	87,39±0,14^{c}	86,49±0,59^{c}	82,69±0,22^{d}
a*	2,7±0,44^{a}	2,96±0,79^{a}	2,44±0,07^{a}	2,28±0,66^{a}	4,33±0,04^{b}
b*	5,61±0,56^{a}	9,35±0,21^{b}	10,13±0,54^{c}	15,05±0,32^{d}	17,77±0,17^{e}

Notes: The data are presented as mean values ± standard deviation. The percentage indicates the proportion of pineapple concentrate relative to the total volume of milk in the mixture. Different superscripts from values in the same row indicate significant differences ($p < 0.05$).

3.3 Moisture of Pineapple Panna Cotta

Moisture content is a critical factor influencing food products, significantly impacting texture, flavor, shelf life of the product. The study indicated that the proportion of honey pineapple concentrate (HPC) in the cream of Panna Cotta has a significant effect on the moisture content of the Panna Cotta (Figure 1). The proportion of HPC 15-20% in the cream resulted in a decreasing moisture content ($p < 0.05$). Honey pineapple concentrate is a product that has undergone a process to significantly reduce its water content. This process aims to create a denser concentrate, which possesses a considerably lower moisture content than fresh pineapple. Consequently, the addition of HPC to the mixture increases the solids without significantly augmenting the moisture content [15]. Furthermore, the concentration of sucrose can bind moisture [16].

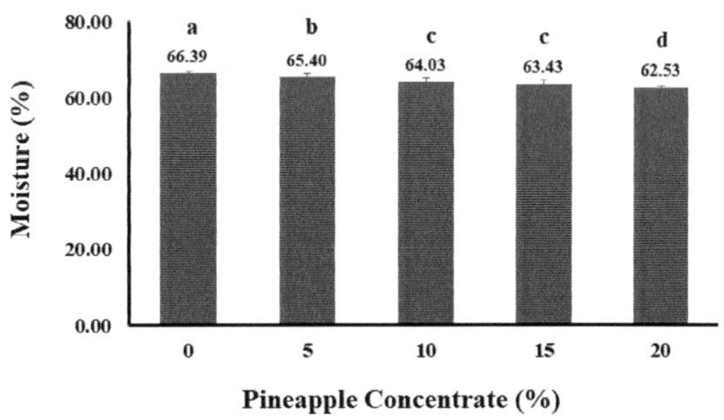

Figure 1. The water content of Panna Cotta desserts with varying proportions of honey pineapple concentrate (HPC). The data are presented as mean values ± standard deviation. The percentage indicates the proportion of pineapple concentrate relative to the total volume of milk in the mixture. Different superscripts from values in the bar graph indicate significant differences ($p < 0.05$).

3.4 Total Soluble Solid of Pineapple Panna Cotta

The analysis of total dissolved solids (TDS) aims to measure the total amount of solids dissolved in food or beverages. In food products, TDS refers to the quantity of soluble compounds present, such as salts, sugars, amino acids, minerals, vitamins, and other chemicals in dissolved form [17]. The study indicated that the proportion of honey pineapple concentrate (HPC) in the cream significantly affected the total dissolved solids content in the Panna Cotta (Figure 2). The proportion of HPC 10-20% in the cream increased the total dissolved solids of the Panna Cotta. The addition of HPC enhanced the TDS due to the various compounds contained in the mixture. Additionally, HPC contains natural sugars, particularly fructose, and glucose, which have a high solubility in water. This is consistent with the statement by [18], honey pineapple contains 1.79 grams of glucose, 4.59 grams of sucrose, and 1.94 grams of fructose per 100 grams of fruit. When HPC is added to the solution, these sugars dissolve and contribute to the increase in TDS [19].

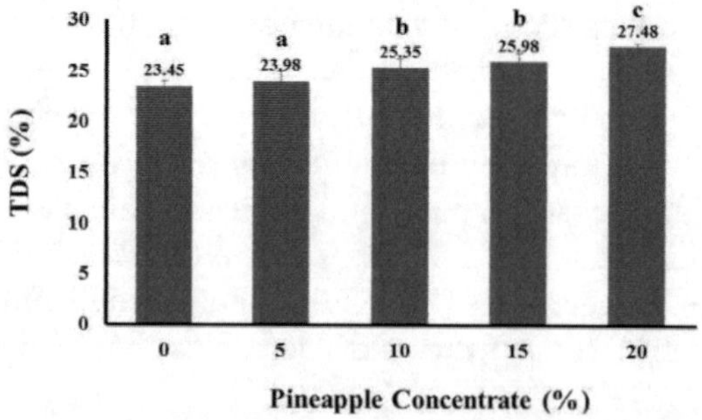

Figure 2. Total dissolved solid (TDS) of Panna Cotta desserts with varying proportions of honey pineapple concentrate (HPC). The data are presented as mean values ± standard deviation. The percentage indicates the proportion of pineapple concentrate relative to the total volume of milk in the mixture. Different superscripts from values in the bar graph indicate significant differences ($p < 0.05$).

4. Conclusion

The proportion of honey pineapple concentrate (HPC) combined with milk influenced the hedonic value, color, moisture content, and total dissolved solids of Panna Cotta. An increase in the concentration of honey pineapple concentrate resulted in a decrease in moisture content, lightness, and redness of the Panna Cotta, while yellowness and total dissolved solids increased. The addition of honey pineapple concentrate to the mixture significantly affected the organoleptic properties of Panna Cotta, including flavor and texture. The recommended treatment for the preparation of Panna Cotta is a concentration of 10% honey pineapple concentrate (HPC).

References

[1] Badan Pusat Statistik, *Produksi Tanaman Buah-Buahan 2022*. Jakarta, 2022.

[2] R. Febriani, K. R. Kuswanto, and L. Kurniawati, "KARAKTERISTIK SELAI FUNGSIONAL YANG DIBUAT DARI RASIO BUAH NAGA MERAH (Hylocereus polyhizus)-JAMBU BIJI MERAH (Psidium guajava)-NANAS MADU (Ananas comosus) DENGAN VARIASI PENAMBAHAN GULA," *JITIPARI (Jurnal Ilmiah Teknologi dan Industri Pangan UNISRI)*, vol. 2, no. 1, pp. 46–52, 2017, [Online]. Available: https://ejurnal.unisri.ac.id/index.php/jtpr/article/view/1535

[3] L. M. Tyas, B. Santoso, and Irawati, "Urgensi Pendaftaran Pada Nanas Madu Asal Kabupaten Pemalang Sebagai Potensi Komoditas Indikasi Geografis," *Notarius*, vol. 14, no. 1, pp. 302–317, 2021, doi: 10.14710/nts.v14i1.38918.

[4] J. Borreani, I. Hernando, and A. Quiles, "Cream replacement by hydrocolloid-stabilized emulsions to reduce fat digestion in panna cottas," *Lwt*, vol. 119, no. November 2019, 2020, doi: 10.1016/j.lwt.2019.108896.

[5] T. P. Andrea and D. S. Pramesti, "Inovasi Rempah Pada Makanan Penutup.," *Journal of Tourismpreneurship, Culinary, Hospitality, Convention and Event Management*, vol. 3, no. 2, pp. 33–50, 2020, doi: 10.31857/s0320930x20040088.

[6] T. T. Hien, V. C. Khang, N. Van Muoi, and T. T. Truc, "Production of a fermented beverage from pineapple (Ananas comosus) byproduct crumbs," *Mater Today Proc*, vol. 60, pp. 2034–2042, 2022, doi: 10.1016/j.matpr.2022.01.270.

[7] I. S. Ashoush and M. Gadallah, "Effects of Different Heating Methods on the Quality Characteristics of Pomegranate Juice Concentrates," no. January, 2012.

[8] Y. Handani, A. M. Sutedja, and Y. Trisnawati, "Pengaruh Konsentrasi Gelatin dan Gula terhadap Sifat Fisikokimia dan Organoleptik Panna Cotta," *Jurnal Teknologi Pangan dan Gizi* , vol. 15, no. 2, pp. 72–78, 2016.

[9] C. Prakitchaiwattana, K. Boonin, and P. Kaewklin, "De-acidification of fresh whole pineapple juice wine by secondary malolactic fermentation with lactic acid bacteria," *Int Food Res J*, vol. 24, no. 1, pp. 223–231, 2017.

[10] L. A, A. F, A. K, and F. T, "A Comparative Study on the Physicochemical Parameters of Milk of Camel, Cow and Goat in Somali Regional State, Ethiopia," *Chem Sci J*, vol. 08, no. 04, 2017, doi: 10.4172/2150-3494.1000171.

[11] R. Efendi, Yusmarini, and Zulkifli, "Pembuatan permen jelly dari wortel dan bonggol nanas," *Prosiding Seminar Nasional Fakultas Pertanian*, pp. 404–417, 2018.

[12] F. Indriaty, "Pengaruh Penambahan Sari Buah Nenas Pada Permen Keras," *Jurnal Penelitian Teknologi Industri*, vol. 8, no. 2, p. 159, 2018, doi: 10.33749/jpti.v8i2.2223.

[13] U. M. Putri, R. S. Ningrum, and W. Lindasari, "Analisis Beta Karoten Pada Nanas (Ananas comosus (L.) Merr) Varietas Queen dan Cayenne Menggunakan Spektrofotometri," in *Seminar Nasional Sains , Teknologi dan Analisis Ke-1*, 2018, pp. 212–218.

[14] M. R. Siregar, N. Harun, and Yusmarini, "Pemanfaatan buah belimbing manis (Averrhoa carambola L.) dan buah nanas (Ananas comosus L.) dalam pembuatan permen jelly," *Jurnal Online Mahasiswa Fakultas Pertanian Universitas Riau*, vol. 3, no. 1, pp. 1–7, 2016, [Online]. Available: http://www.tjyybjb.ac.cn/CN/article/downloadArticleFile.do?attachType=PDF&id=9987

[15] W. Marsigit, T. Tutuarima, and R. Hutapea, "EFFECT OF ADDITION SUGAR AND CARRAGENAN ON PHYSICS, CHEMICAL AND ORGANOLEPTIC CHARACTERISTICS OF SOFT CANDY CYTRUS CALAMANSI (Citrofortunella microcarpa)," *Jurnal Agroindustri*, vol. 8, no. 2, pp. 113–123, 2019, doi: 10.31186/j.agroind.8.2.113-123.

[16] D. L. Engka, J. Kandou, and T. Koapaha, "Effect Concentration Sucrose and Glucose Syrup Of Chemical And Sensory Properties Hard Candy Starfruit (Averrhoa bilimbi. L)," *Cocos* , vol. 1, no. 1, pp. 1–9, 2016.

[17] N. Benny, R. Shams, K. K. Dash, V. K. Pandey, and O. Bashir, "Recent trends in utilization of citrus fruits in production of eco-enzyme," *J Agric Food Res*, vol. 13, no. January, 2023, doi: 10.1016/j.jafr.2023.100657.

[18] Y. Herlambang, D. M. Kurniawati, and M. A. Ali, "Pengaruh Jus Nanas Madu Terhadap Denyut Nadi Dan Tekanan Darah Pada Siswa Sekolah Sepak Bola Pasca Lari Jarak Jauh 10 Km," *Journal of Nutrition College*, vol. 11, no. 3, pp. 182–187, 2022, doi: 10.14710/jnc.v11i3.33136.

[19] Daniel, Z. Lubis, and E. Yusraini, "Pengaruh Persentase Carboxy Methyl Cellulose Dan Persentase Gula Terhadap Mutu Selai Jagung," *Jurnal Rekayasa Pangan dan Pertanian*, vol. 5, no. 1, pp. 66–72, 2017.

Valorizing Waste: The Effects of Ultrasonic Extraction Time on Jackfruit Rind Pectin Properties

Swastika Dewi[1], Wiwid Rizqi Maryati[1], Oktavia Winda Dewi[1], Agung Dewantoro Mulyono[2], Yoga Pratama[1*]

[1] Food Technology Study Program, Faculty of Animal and Agricultural Sciences, Universitas Diponegoro, Semarang
[2] Department of Chemical Engineering, Faculty of Engineering, Universitas Diponegoro, Semarang

*E-mail: yogapratama@live.undip.ac.id

Abstract. The waste from jackfruit (Artocarpus heterophyllus) processing is estimated between 65-80% from the total weight, in which, about 30% comes from the rind. Pectin, a natural biopolymer and widely used gelling agent, has been the focus of several jackfruit rind valorization studies. The current work aims to evaluate if ultrasonic treatment is effective in improving the jackfruit rind pectin extraction, both in yield and its physical-chemical properties. Ultrasonic-assisted extractions (40 kHz) were carried out with time variables of 20 to 35 minutes. The yield increases significantly with the increasing extraction time, with the range of 7.84 - 11.02%; moisture content 10.04-10.1%; equivalent weight 333.55-656.02; methoxyl content 1.34-5.65; and viscosity 5.08-12.76 cP. Extraction time treatments showed significant differences ($p \leq 0.05$) to yield, equivalent weight, methoxyl content, and viscosity but not to moisture content ($p \geq 0.05$). According to SEM data, ultrasound caused observable alterations in the pectin's microstructure. The best pectin was obtained by 20 minutes extraction time according to the best yield and less impurities.

Keywords: extraction time, jackfruit rind, pectin, ultrasonic, waste

1. Introduction

Jackfruit (Artocarpus heterophyllus) is a tropical fruit that mostly growing in Indonesia. Processing of jackfruit into chips obtained waste about 65-80% of jackfruit's total weight. Jackfruit waste involve 39% rind, 27% straw (dami), 14% core, and 20% seed (1). So far, utilization of the waste is only as animal feeds. Pectin is a complex heteropolysaccharides that mostly found in plants. The central lamella and cell wall of plant tissues contain the most pectin (2). Most of protopectin in jackfruit is contained in rind (3); (4).

Pectin is a common natural food ingredient that is used in the pharmaceutical and food industries as a thickening, gelling agent, stabilizer, and texturizer (5). The presence of pectin in jackfruit rind is characterized by several factors, including its anhydrouronic acid content (AUA), methoxyl content, degree of esterification, and acetyl values (6). That several factors could affect its utilization as a commercial pectin. Along with the rapid development of industry nowadays, the need for pectin in industry is increasing as well. According to the data from (7), total pectin that was imported by Indonesia are 455.035 kg in 2021; 534.124 kg in 2022 and 467.152 kg in 2023. Pectin can be obtained by extracting it from its source. To acquire pectin suitable for specific uses, the appropriate extraction method must be applied. To recover pectin from fruits and biowaste, a number of sophisticated methods have been developed, including physical field-based extraction, microwave-assisted extraction, and ultrasound-assisted extraction (8). Compared to conventional methods, these unconventional methods significantly shorten the extraction time and lower the temperature, while still providing relatively competitive extraction efficiency.

Factors that can affect the characteristics of pectin are including raw materials, type of used solvents, extraction time, extraction temperature and extraction methods (8). Several methods of pectin extraction have been reported in the literature. Extraction method using conventional techniques require huge amount of energy and potentially can reduce the quality of pectin (9). One of the modern extraction techniques is extraction method using ultrasonic waves. The Ultrasound-Assisted Extraction (UAE) method can greatly shorten the extraction time, leading to improved extraction efficiency. Ultrasound-assisted extraction (UAE) employs ultrasonic energy combined with solvents to extract specific compounds from various plant materials (10). Ultrasound refers to mechanical waves with frequencies above 20 kHz, surpassing the audible range of human hearing, which spans from 20 Hz to 20 kHz. Ultrasonic-assisted extraction method has been used in previous studies including pectin extraction of banana peels (11); tomato waste (12) and apple peel (13), but there is no study yet about ultrasonic-assisted extraction method of pectin from rind of jackfruit especially in Indonesia. This study aims to evaluate the effect of extraction time using the ultrasound-assisted method on the characteristics of pectin derived from jackfruit rind.

2. Materials and Methods

2.1 Materials

Jackfruits (Artocarpus heterophyllus) rind were obtained from Jati Banyumanik traditional market, citric acid (Indrasari), the chemicals (ethanol, NaCl, NaOH, HCl) from Indrasari, Semarang.

2.2 Methods

2.2.1 Sample preparation

After washing, the rinds was separated from other parts (fruit flesh and dami), then cut into 1x1 cm and dried by cabinet dryer at 55°C (Maksindo, MKS-FDH16, 1500 watt) for 24 hours. Dried jackfruits rind were reduced in size using blender (Philips Blender series 5000 HR2223/60) and sieved (Mesh No. 80) to get uniform size. Jackfruit rind powder is stored in airtight jar at -18°C for further analysis.

2.2.2 Extraction of Pectin

Pectin extraction was performed based on the method described by (14), with modifications. Jackfruit rind powder and acid solution (pH 1.5, adjusted with citric acid) were combined with solid-liquid ratio of 1:20, w/v. The mixture was then placed in a sonicatorbath (Branson2510) with a 40 kHz frequency, temperature of 60°C, with various extraction time (20 minutes, 25 minutes, 30 minutes, and 35 minutes). The extract was then filtered using filter paper, and the filtrate was precipitated with 96% ethanol at a ratio of 1:3 for 14 hours at room temperature. The precipitate was filtered and washed sequentially with 75% ethanol and 95% ethanol to eliminate acids and soluble impurities. Wet pectin is dried at 55° C for 6 hours and mashed using a blender (Philips Blender series 5000 HR2223/60).

2.2.3 Pectin Yield

The pectin yield was calculated using the method described by (15), equation 1:

$$\text{YPectin (\%)} = \frac{\text{Weight (g)of dried pectin}}{\text{Weight (g)of dry raw material}} \text{ x } 100\% \qquad (1)$$

2.2.4 Moisture Content

The moisture content of pectin was determined using the gravimetric method, as outlined by (16) method. Two grams of pectin are dried at temperature of 130°C to constant material weight that is for 4 hours. Moisture content is calculated by the formula, equation 2:

$$\text{Moisture content (\%)} = \frac{\text{B-(C-A)}}{\text{B}} \text{ x } 100\% \qquad (2)$$

A: weight of porcelain (g)
B: weight of sample (g)
C: constant weight of sample after drying (g)

2.2.5 Equivalent Weight

The equivalent weight was determined according to (15). In brief, 0.5 g of the pectin sample was weighed and placed in a 250 ml conical flask, to which 5 ml of ethanol was added. Subsequently, 1 g of sodium chloride (NaCl) and 100 ml of distilled water were added to the mixture, followed by 5 drops of phenolphthalein indicator. The solution was then titrated with 0.1 N sodium hydroxide (NaOH), with the endpoint indicated by a purple color. The equivalent weight was calculated using the following equation (3):

$$\text{Ea.W(mg/mol)} = \frac{\text{ml NaOH x N NaOH x 31}}{\text{Weight of sample (mg)}} \qquad (3)$$

2.2.6 Methoxyl content

For methoxyl content (MeO) content, the neutral solution obtained from the equivalent weight determination was collected, and 25 ml of 0.25 N sodium hydroxide (NaOH) was added. The mixture was left at room temperature for 30 minutes. Afterward, 25 ml of 0.25 N hydrochloric acid (HCl) was added and titrated with 0.1 N NaOH. The endpoint was the same as that in the equivalent weight titration, indicated by the same color change (17). The methoxyl content was calculated using the following equation (4):

$$\text{Methoxyl content (\%)} = \frac{\text{ml NaOH x N NaOH x 31}}{\text{Weight od sample (mg)}} \text{ x } 100\% \qquad (4)$$

Where the number 31 denotes the molecular weight of methoxyl.

2.2.7 Viscosity
To determined viscosity, the pectin solutions (1%) was added into the ostwald pipe. The rubber bulb was used to boost the solutions until the upper marks of ostwald pipe. Then, the flow time of the solution to reach the lower mark was calculated and the sample was compared with the aquadest. For measurement of density, picnometer is used, (18). The sample viscosity is calculated by the formula, equation 5:

$$\frac{\eta^0}{\eta} = \frac{\rho^0 x t^0}{\rho t} \text{ x } 100\% \quad (5)$$

η^0 = viscosity of the comparative solution (cP)
η = viscosity of sample solution (cP)
ρ^0 = density of comparative solution (g/ml)
ρ = density of sample solution (g/ml)
t^0= flow time of comparator solution (sec)
t = flow time of sample solution (sec)

2.2.8 Microstructure morphological analysis
The graphical performance test was conducted using the method described by (19) with a Scanning Electron Microscope (SEM). The dry powdered pectin samples were analyzed using SEM (JEOL JSM-6510LA, Japan) with magnification used in this SEM analysis of 1000×, 20 kV, Working distance (WD) 15mm, and spot size (SS) 30.

2.2.9 Data analysis
The test data were analyzed using Analysis of Variance (ANOVA) at a 5% significance level. If a significant difference was found, a Duncan test was then performed. All data were analyzed with SPSS 23 application except Morphological analysis performed.

3. Results and Discussion

3.1 Pectin Yield
Table 1 demonstrates that the yield of jackfruit pectin produced in this study ranged from 7.84% to 11.02%. A significant effect ($P \leq 0.05$) of extraction time on pectin yield was observed. There is a positive correlation between extraction time and pectin yield. This indicates that the longer contact between the solvent with the ingredients, more pectin is dissolved. This statement is in accordance with (20) which states that The duration of extraction is associated with an elevation in the kinetic energy of the solution, thereby facilitating enhanced solute diffusion into the tissue cells. It causes more pectin being detached from tissue cells.

Optimal pectin yield of 11.02% was realized through extraction at a duration of thirty-five minutes. The yield of this study was higher than research has been done by (21) where the pectin extract from the jackfruit waste using a sulfuric acid solvent pH 2.5 (1:40 g / ml) at 80°C with time extraction of 120 minutes has a yield about 8.94%. The extraction using ultrasound-assisted method produce higher yield than the maceration extraction method using shaker (22).

The pectin of this study is lower than the research of (3) where extract pectin of jackfruit rind by Microwave Assisted Extraction method (MAE) has a yield about 17.28%. The MAE method usesfrom 300 MHz to 300 GHz waves are higher than Ultrasonic Assisted Extraction (UAE) 40 khz (8). The use of ultrasonic techniques has a basic principle whereby ultrasonic waves are propagated through the medium. As the wave is propagated through the medium, the medium will be vibrate. Vibration will provide an intensive agitation of the extraction process, thereby enhancing the osmosis process (8).

Table 1. Yield of the pectin with various variations of extraction time

Time (min)	Yield (%)
20	7.84±0.131[c]
25	8.04±0.935[c]
30	9.23±0.335[b]
35	11.02±0.616[a]

Note: Values within the same column that were followed by different superscript letters indicated significant differences at a 95% confidence level.

3.2 Moisture Content

The moisture content of fruit jackfruit pectin exhibited a narrow range, spanning from 10.04 to 10.10%. Statistical analysis revealed that extraction duration had no statistically significant impact ($P \geq 0.05$) on the moisture content of the obtained pectin. The results are not significantly different due to the temperature and duration of drying and the same storage conditions. It has the same result as (23) research where the moisture content of citrus pectin extract with the same temperature and drying time did not show significantly different results. The moisture content of this study has met the commercial pectin standard based on the International Pectin Producers Association (IPPA) quality standard which requires pectin to have a maximum moisture content of 12% (24).

Table 2. Moisture content of the pectin with various variations of extraction time

Time (min)	Moisture content (%)
20	10.10±0.916
25	10.04±0.346
30	10.10±0.907
35	10.07±0.676

3.3 Equivalent Weight

Pectin content of jackfruit rind equivalent weights varied between 333.55 and 656.02 units (Table 3). According to the results from SPSS 23, there was a significant effect ($P < 0.05$) of extraction time on the equivalent weight of jackfruit rind pectin. The equivalent weight reflects the total free galacturonic acid (un-esterified) in the pectin chain (24). The pectin equivalent weight after 20 minutes of extraction was 656.05, which meets the International Pectin Producers Association (IPPA) quality standard, which specifies that commercial pectin should have an equivalent weight between 600 and 800.

Pectin with a 25, 30, and 35 minute extraction time has a lower equivalent weight than IPPA quality standard. The low equivalent weight occurs because the extraction process is carried out too long so that the reaction of depolymerization of pectin to pectat acid. In accordance with (25) the low equivalent weight can be obtained due to the long duration of extraction resulting in de-polymerization and de-esterification reactions. In addition, low solvent pH also increases depolymerization.

Table 3. Equivalent weight of the pectin with various variations of extraction time

Time (min)	Equivalent weight
20	656.02±36.49[a]
25	541.67±19.02[b]
30	413.08±36.68[c]
35	333.55±9.45[d]

Note: Values within the same column that were followed by different superscript letters indicated significant differences at a 95% confidence level.

The equivalent weight of pectin is influenced by the quality of the raw material. Pectin extracted from material with excess maturity conditions will have a low equivalent weight. In accordance with the study of (15) in which the extracted pectin of the immature, mature and over ripen lemons extracted had equivalent weight of 1,175; 1.632; and 368. The quality of pectin can be assessed based on its equivalent weight. This aligns with the findings of (26), which state that a lower equivalent weight is a result of significant partial degradation of pectin. It makes the quality of pectin down so that it also affects the strength of gel formation is lower.

3.4 Methoxyl Content

The methoxyl produced in this stucontent of extracted pectin ranged from 1.34 to 5.65%. The results of the test using SPSS 23 showed a significant effect ($P < 0.05$) of extraction time on the methoxyl content of jackfruit rind pectin. Methoxyl level indicates the amount of methanol presents in pectin. Increased levels of methoxyl in pectin as more carboxyl groups are esterified with methanol (24).

Table 4. Methoxyl of pectin with various variations of extraction time

Time (min)	Methoxyl (%)
20	1.34±0.14[c]
25	1.88±0.13[c]
30	4.46±1.02[b]
35	5.65±0.31[a]

Note: Values within the same column that were followed by different superscript letters indicated significant differences at a 95% confidence level.

The methoxyl content is essential in controlling the setting time of pectin, its ability to bind with metal ions, its beneficial properties for forming low-solids gels, and its capacity to disperse in water and gels. Based on the methoxyl content values in this study, jackfruit rind pectin is classified as low-methoxyl, with a methoxyl value of less than 7.12%. Low-methoxyl pectin has the ability to form gels in the presence of polyvalent cations, such as calcium. The low methoxyl content may be influenced by the quality of the raw materials used. The results showed a significant decrease in methoxyl content with an increase in extraction time. As fruits ripen, their sugar content increases, leading to a decrease in methoxyl content (27). In line with the study by (15), pectin extracted from immature lemons has a high methoxyl content, whereas pectin from mature and overripe lemons has a low methoxyl content.

3.5 Viscosity

Viscosity value is a representation of gel strength formed from pectin. The relative viscosity of jackfruit pectin produced ranged from 5.08 - 12.76 cP. The results of the test using SPSS 23 showed a significant effect ($P < 0.05$) of extraction time on the viscosity value of pectin. The result showed that extraction time with the viscosity value of jackfruit rind pectin has an inverse relation. The viscosity value decreases with increasing time of extraction due to depolymerized pectin thus minimizing the viscosity value of the pectin solution. According to (28) the decrease of viscosity in pectin extraction is due to the depolymerization of pectin into shorter chains.

Table 5. Viscosity of pectin with various variations of extraction time

Time (min)	Viscosity (cP)
20	12.76 ± 0.450^{a}
25	11.63 ± 0.707^{b}
30	6.30 ± 0.870^{c}
35	5.08 ± 0.332^{d}

Note: Values within the same column that were followed by different superscript letters indicated significant differences at a 95% confidence level.

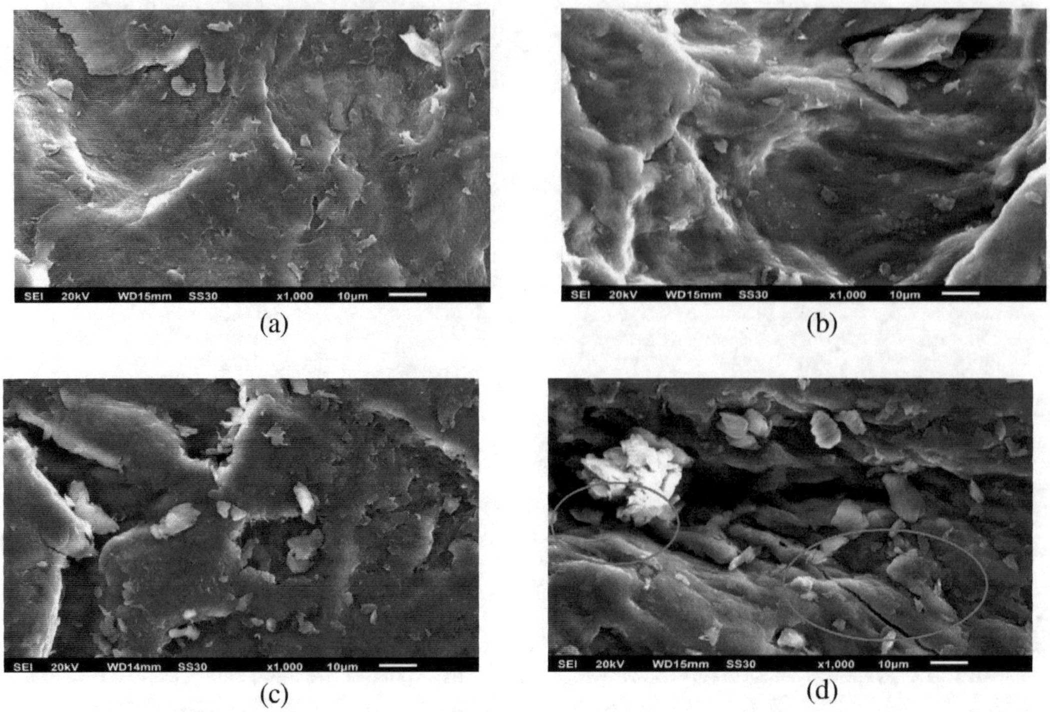

Figure 1. Micrograph of jackfruit rind pectin, (a) Extraction time of 20 mins ; (b) 25 mins ; (c) 30 mins ; (d) 35 mins

The viscosity value of the pectin solution is affected by the purity of the pectin. Low viscosity may be due to extracted impurities such as water soluble proteins and water-soluble non-pectin polysaccharides (16). Increased duration of extraction time will affect the hydrolysis of conversion from protopectin to pectin, but also followed by the increase of ash content along with pectin. Based on research conducted by (24) indicated that the increased ash content of pectin extract produced will be followed by a decrease in the relative viscosity value of pectin. Ash content increases with overcooked raw material conditions (15). Pectin viscosity is also influenced by external factors such as temperature, solution concentration, pH, stirring time, and the presence of salt.

3.6 Microstructure morphological analysis
After extraction using ultrasound with difference time, the morphology of pectin was observed. Fig.1, 2, 3, and 4 shows the SEM images of samples that have been done with different time extraction (20, 25, 30, 35 min). Figures 1 and 2 show that the surface of the pectin remained compact and sturdy after extraction. Extraction time for 30 min represented in Fig.3 has a structure flaky in the shape and little cracked. However, the pectin has been done extracted for 35 min (Fig. 4) showed that The surface morphology underwent a noticeable alteration. The tissue became much looser and cracked. The damage caused by too longtime extraction using ultrasound. The results indicate that prolonged extraction time may result in the rupture of the parenchymal cells in the plant material. In line with the study by (29), extraction causes the rupture of parenchymal cells, allowing the target compounds within the cells to be quickly released into the surrounding extraction solvents.

4. Conclusion

Based on the results of the research, it can be concluded that there is a significant effect ($p \leq 0.05$) between different treatment time of different extraction to yield, equivalent, methoxyl, and pectin viscosity. However, there was no significant difference of moisture content parameter ($p \geq 0{,}05$). The SEM results showed that ultrasound heating caused noticeable changes in the microstructure of the pectin. Overall treatment with a 20 minute extraction time is the optimal duration of time in pulp jackfruit extract with 40 kHz ultrasonic as source of heat.

Bibliography

[1] Nilta Shabrinal Itsnani SBW. OPTIMASI PROSES EKSTRAKSI PEKTIN DARI KULIT DAN JERAMI NANGKA(Artocarpus heterophyllus) MENGGUNAKAN KURVA RESPON PERMUKAAN. 2017;5(1):14–6.

[2] Thi N, Tran K, Bao V, Tran T Van, Thi T, Nguyen T. Microwave-assisted extraction of pectin from jackfruit rags : Optimization , physicochemical properties and antibacterial activities. Food Chem [Internet]. 2023;418(March):135807. Available from: https://doi.org/10.1016/j.foodchem.2023.135807

[3] Windiarsih C, Nugroho WA, Argo BD. OPTIMASI PEKTIN DARI KULIT BUAH NANGKA (Artocarpus heterophyllus) EKSTRAKSI DAN KONSENTRASI PELARUT) Optimization of Microwave Assisted Extraction (MAE) of Pectin from Jackfruit Peels (Artocarpus heterophyllus) (S. J Bioproses Komod Trop. 2015;3(1):39–49.

[4] Leong CM, Noranizan MA, Kharidah M, Choo WS. Physicochemical properties of pectin extracted from jackfruit and chempedak fruit rinds using various acids. Int Food Res J. 2016;23(3):973–8.

[5] Kumar M, Potkule J, Tomar M, Punia S, Singh S, Patil S, et al. Jackfruit seed slimy sheath, a novel source of pectin: Studies on antioxidant activity, functional group, and structural morphology. Carbohydr Polym Technol Appl [Internet]. 2021 Dec;2:100054. Available from: https://linkinghub.elsevier.com/retrieve/pii/S2666893921000220

[6] Maskey B, Dhakal D, Pradhananga M, Shrestha NK. Extraction and process optimization of bael fruit pectin. Food Sci Nutr [Internet]. 2018 Oct 23;6(7):1927–32. Available from: https://onlinelibrary.wiley.com/doi/10.1002/fsn3.761

[7] Ummah MS. Buletin Statistik Perdagangan Luar Negeri [Internet]. Vol. 11, Sustainability (Switzerland). 2019. 1–14 p. Available from: http://scioteca.caf.com/bitstream/handle/123456789/1091/RED2017-Eng-8ene.pdf?sequence=12&isAllowed=y%0Ahttp://dx.doi.org/10.1016/j.regsciurbeco.2008.06.005%0Ahttps://www.researchgate.net/publication/305320484_SISTEM_PEMBETUNGAN_TERPUSAT_STRATEGI_MELESTARI

[8] Adetunji LR, Adekunle A, Orsat V, Raghavan V. Advances in the pectin production process using novel extraction techniques: A review. Food Hydrocoll. 2017 Jan;62:239–50.

[9] Megawati M, Machsunah EL. EKSTRAKSI PEKTIN DARI KULIT PISANG KEPOK (Musa paradisiaca) MENGGUNAKAN PELARUT HCl SEBAGAI EDIBLE FILM. J Bahan Alam Terbarukan [Internet]. 2016 Jun 20;5(1):14–21. Available from: https://journal.unnes.ac.id/nju/index.php/jbat/article/view/4177

[10] Freitas de Oliveira C, Giordani D, Lutckemier R, Gurak PD, Cladera-Olivera F, Ferreira Marczak LD. Extraction of pectin from passion fruit peel assisted by ultrasound. LWT - Food Sci Technol [Internet]. 2016 Sep;71:110–5. Available from: https://linkinghub.elsevier.com/retrieve/pii/S0023643816301633

[11] Adhiksana A. PERBANDINGAN METODE KONVENSIONAL EKSTRAKSI PEKTIN DARI KULIT BUAH PISANG DENGAN METODE ULTRASONIK. J Res Technol. 2017 Dec 30;3(2):80–7.

[12] Grassino AN, Brnc˘ic´ M, Vikic´-Topic´ D, Roca S, Dent M, Brnc˘ic´ SR. Ultrasound assisted extraction and characterization of pectin from tomato waste. Food Chem [Internet]. 2016 May;198:93–100. Available from: https://linkinghub.elsevier.com/retrieve/pii/S0308814615302405

[13] Villamil-Galindo E, Piagentini AM. Sequential ultrasound-assisted extraction of pectin and phenolic compounds for the valorisation of 'Granny Smith' apple peel. Food Biosci [Internet]. 2022 Oct;49:101958. Availablefrom: https://linkinghub.elsevier.com/retrieve/pii/S2212429222004187

[14] Moorthy IG, Maran JP, Ilakya S, Anitha SL, Sabarima SP, Priya B. Ultrasound assisted extraction of pectin from waste Artocarpus heterophyllus fruit peel. Ultrason Sonochem [Internet]. 2017 Jan;34:525–30. Available from: https://linkinghub.elsevier.com/retrieve/pii/S1350417716302140

[15] Azad AKM. Isolation and Characterization of Pectin Extracted from Lemon Pomace during Ripening. J Food Nutr Sci [Internet]. 2014;2(2):30. Available from: http://www.sciencepublishinggroup.com/journal/paperinfo.aspx?journalid=154&doi=10.11648/j.jfns.20140202.12

[16] Nurhayati N, Maryanto M, Tafrikhah R. Ekstraksi Pektin dari Kulit dan Tandan Pisang dengan Variasi Suhu dan Metode (Pectin Extraction from Banana Peels and Bunch with Various Temperatures and Methods). J Agritech [Internet]. 2016 Dec 21;36(03):327. Available from: https://journal.ugm.ac.id/agritech/article/view/16605

[17] Roy MC, Alam M, Saeid A, Das BC, Mia MB, Rahman MA, et al. Extraction and characterization of pectin from pomelo peel and its impact on nutritional properties of carrot jam during storage. J Food Process Preserv [Internet]. 2018 Jan;42(1):e13411. Available from: https://onlinelibrary.wiley.com/doi/10.1111/jfpp.13411

[18] Joshi C, Sarasan G. A Study on Density and Viscosity of Binary Solutions of n- Propyl Acetate and IsoPropyl Alcohol at Various Temperatures. Iosr. 2017;10(12):36–40.

[19] Liew SQ, Chin NL, Yusof YA. Extraction and Characterization of Pectin from Passion Fruit Peels. Ital Oral Surg [Internet]. 2014;2:231–6. Available from: http://dx.doi.org/10.1016/j.aaspro.2014.11.033

[20] Tuhuloula A, Budiyarti L, Fitriana EN. KARAKTERISASI PEKTIN DENGAN MEMANFAATKAN LIMBAH KULIT PISANG MENGGUNAKAN METODE EKSTRAKSI. Konversi [Internet]. 2013 Apr 1;2(1):21. Availablefrom: https://ppjp.ulm.ac.id/journal/index.php/konversi/article/view/123

[21] Begum R, Aziz MG, Yusof YA, Burhan M. Extraction and characterization of pectin from jackfruit (Artocarpus heterophyllus Lam) waste. 2017;12(6):42–9.

[22] dos Santos EKR, Azoubel PM, Gouveia ER. Better Pectin Yield From Passion Fruit Peel (Passiflora edulis f. flavicarpa): From Shaker or Ultrasound? A Comparison. Waste and Biomass Valorization [Internet]. 2017 Apr 21;8(3):905–10. Available from: http://link.springer.com/10.1007/s12649-016-9611-4

[23] Ayora-Talavera T, Ramos-Chan C, Covarrubias-Ca´rdenas A, Sa´nchez-Contreras A, Garcí´a- Cruz U, Pacheco L. N. Evaluation of Pectin Extraction Conditions and Polyphenol Profile from Citrus x lantifolia Waste: Potential Application as Functional Ingredients. Agriculture. 2017 Mar 15;7(3):28.

[24] Roikah S, Rengga WDP, Latifah L, Kusumastuti E. EKSTRAKSI DAN KARAKTERISASI PEKTIN DARI BELIMBING WULUH (Averrhoa bilimbi,L). J Bahan Alam Terbarukan. 2016;5(1):29– 36.

[25] Nazaruddin Ramli. Effect of ammonium oxalate and acetic acid at several extraction time and pH on some physicochemical properties of pectin from cocoa husks (Theobroma cacao). African J Food Sci [Internet]. 2011 Dec 15;5(5). Available from: http://www.academicjournals.org/ajfs/abstracts/abstracts/abstract2011/15 Dec/Ramli and Asmawati.htm

[26] Alfa P, Rose E, Abilasha D. Extraction and Characterization of Pectin from Lemon Peel. Int J Adv Sci Res [Internet]. 2016;1(12):12–5. Available from: www.allsciencejournal.com

[27] Sirisakulwat S, Nagel A, Sruamsiri P, Carle R, Neidhart S. Yield and Quality of Pectins Extractable from the Peels of Thai Mango Cultivars Depending on Fruit Ripeness. J Agric Food Chem [Internet]. 2008 Nov 26;56(22):10727–38. Available from: https://pubs.acs.org/doi/10.1021/jf802173c

[28] Erika C. Ekstraksi Pektin dari Kulit Kakao (Theobroma cacao L.) Menggunakan Amonium Oksalat. J Teknol dan Ind Pertan Indones [Internet]. 2013 Jun 1;5(2). Available from: https://jurnal.usk.ac.id/TIPI/article/view/1001

[29] Rahmati S, Kang O L, Momeny E, Ramli N. Morphological studies on pectin extracted from dragon fruit peel using microwaveand conventional extraction methods. Int Conf Halal Gums (IcoHaG 2012) [Internet]. 2012;(December 2012):175–80. Available from: https://www.researchgate.net/publication/315810541

Comparing Colour, Proximate, and Hedonic Quality of Floss Made from Retired Layings Hens Meat with Various Sugar Types

Siti Susanti*, Kartini Anggraeni Silalahi, Heni Rizqiati, and Ahmad Ni'matullah Al-Baarri, Yasmin Aulia Rachma

Food Technology, Department of Agriculture, Faculty of Animal and Agricultural Sciences, Diponegoro University, Semarang, Indonesia

*E-mail: sitisusanti@live.undip.ac.id

Abstract. Chicken floss is one of the diversifications of processed chicken meat into dry food products with long shelf-life. In addition to broiler meat, retired laying hens' meat is now an alternative material for floss production because of its abundant availability and tough texture. In chicken floss production, sugar is one of the additional ingredients to enhance flavour, improve appearance, and preserve product. Some types of sugar that are common on the market are granulated sugar, coconut sugar, and palm sugar. This study aims to compare the quality (colour, proximate, and hedonic) of floss made from retired laying hens' meat (RLHM-floss) with 3 various sugar types (granulated sugar, coconut sugar, palm sugar). The results showed that type of sugar significantly affected the quality in terms of colour, proximate, and hedonic. The use of granulated sugar for RLHM-floss production resulted a product with the brightest yellow in appearance (the lowest a* value with the highest L*, b*, and °hue values) compared to coconut sugar and palm sugar. Palm sugar resulted the darkest brown product (lowest L*, b*, and °hue values with higher a* values), while coconut sugar produced a product with a yellowish brown (golden brown). From the proximate aspect, the use of coconut sugar resulted RLHM-floss with the highest moisture, ash, and protein content, while the fat content is lower than granulated sugar but higher than palm sugar ($p<0.05$). From the hedonic aspect, compared to granulated sugar, coconut sugar and palm sugar produce RLHM-floss that is more preferred by panellists ($p<0.05$), in terms of colour (golden brown), flavour (fragrant, savoury, and sweet) and texture (smooth, slightly moist, and soft). Compared to commercial chicken floss, RLHM-floss is more golden brown in colour and higher in moisture, ash, and protein, so it was preferred by panellists. Thus, coconut sugar is the most proper sugar among other sugar to produce the quality RLHM-floss in terms of colour, proximate, and hedonic value.

Keywords: colour, floss, hedonic, RLHM, sugars, proximate

1. Introduction

Chicken Floss (CF) is one of the diversifications of processed chicken meat into dry food products with long shelf life. People from many ages are likes CF because it tastes good. CF is made by boiling or steaming and frying with an addition of spices [1]. The CF general characteristics are dark brown in colour, stringy texture, and special aroma. The material commonly used for making CF is broiler chicken meat.

Retired laying hens' meat (RLHM) is now an alternative ingredient for floss production besides broiler meat. RLHM is meat obtained from laying hens that have reached their retirement period. RLHM has abundant availability. Data from Badan Pusat Statistik recorded that RLHM production reached 170,575.94 tonnes in 2023. RLHM is suitable for use in making CF because of its tough texture. The texture of RLHM tends to be tough due to the formation of fibre bonds as the chicken ages [2]. RLHM Also have a sticky texture due to the fat content causes low meat quality , coarse muscle fiber, and low binding percentage [3]. This kind of texture supports the floss quality as a product with a fibrous texture.

Production of CF involves sugar as an additional ingredient. The utilization of sugar can help to enhance flavour, improve appearance, and preserve the product. Some types of sugar that common on the market are granulated sugar, coconut sugar, and palm sugar. There is no reference that regulates the use of sugar types in CF. Sugar can come from different raw materials, so that can cause differences in the quality of CF. The aim of this study is to compare the quality (colour, proximate, and hedonic) of floss made from retired laying hens' meat (RLHM-floss) with three various types of sugar (granulated sugar, coconut sugar, and palm sugar). This study is expected to provide information on the quality of RLHM-floss with the best sugar type variation compared to commercial CF.

2. Materials and Methods

2.1 Production of RLHM-Floss with Various Sugar Types

Production of RLHM-floss started with preparation of the meat, seasoning, mixing formula, and frying. RLHM that already cleaned are cooked in the pressure cooker for 40 minutes. The RLHM was then separated from its skin and bones and shredded. Floss seasoning that includes shallots, garlic, ginger, galangal, turmeric, and coriander were mashed. Seasoning such as bay leaves and lime leaves are separated from its stems and lemongrass is crushed. Shredded RLHM, spices, chicken broth, salt, flavouring, and sugar are mixed. The formulation is mixed into a batter then cooked until the water shrinks. The batter was then fried for 3 minutes, removed, and drained from the oil. Frying processed is continued in an air fryer for 2 hours at 120°C while it turned every 30 minutes.

2.2 Parameter Testing

Parameter tested in this study includes colour (L*, a*, b*), proximate values (water, ash, fat, protein), and hedonic quality (colour, aroma, flavour, texture, overall). Color was tested with a colour reader. The colour reader is calibrated first, the L*a*b* scale measurement menu is selected, and the start button is pressed. Data collection was carried out at three different points in each replication. Calculation of chromatic colour areas (°hue) is calculated by the following equation.

$$°hue = \tan^{-1}\left(b^*/a^*\right)$$

Water content was tested with an oven (gravimetry method). The porcelain cup was dried for 1 hour at 105 °C, cooled, and weighed (A). The CF sample was weighed as much as 2 g, put into a porcelain cup, and weighed (B). The porcelain cup with the CF sample was dried again for 4 hours at 105 °C, cooled in a desiccator, and weighed. The porcelain cup was dried again for 1 hour until a constant weight was achieved (weight difference <±0.02 g). The cup is weighed again until a constant weight (C) is obtained. Water content is calculated by the following equation.

$$Water\ content = \left(B - C/_{B - A}\right) \times 100\%$$

Ash content was tested with samples from water content testing. The porcelain cup containing the sample is inserted into the furnace. Sample is ashed for 6 hours until the ashing process is complete at a temperature of 600°C. Then, removed the porcelain cup from furnace, cooled in a desiccator, and weighed (D). Ash content is calculated by the following equation.

$$Ash\ content = \left(D - A/_{B - A}\right) \times 100\%$$

Fat content test is carried out with the Soxhlet method. The fine filter paper is cut and then placed in the oven for 1 hour at 105 °C. The sample was macerated, weighed 1 g and wrapped (E). The filter paper containing the CF sample dried in an oven for 4 hours at 105 °C. The sample was dried again for 1 hour at

105 °C, weighed, and the results were recorded when a constant weight was reached (F). Samples were inserted into a Soxhlet extraction apparatus. An ether solution of 2.5–3.0 times the extraction volume was added to the extraction flask containing the sample. The condenser is installed and then connected to the heater. Extraction is carried out for 5–6 cycles or ±6 hours. After extraction is complete, the heater is turned off, and the sample is removed and aired. Next, dried the samples in an oven for 1 hour at 105 °C. The sample is cooled in a desiccator and then weighed (G). Fat content is calculated by following equation.

$$Fat\ content = \left(F - G/_{E}\right) \times 100\%$$

Protein content tested with the Kjeldahl method. The sample was weighed, ground, and placed in a Kjeldahl flask. Mixture of selenium and H_2SO_4 (concentrated) was added to the flask. The sample was destroyed in a fume hood until a clear green colour was formed. The Kjeldahl flask is then cooled for ±10–20 minutes. After cooling, distilled water and NaOH 45% (w/w) were carefully added. Distillation was carried out with the distillate collected in an Erlenmeyer flask containing 4% H_3BO_3 mixed with MR+MB indicator. Distillation was carried out until 40 ml of distillate was obtained. The distillate was then titrated with 0.1 M HCl until a light purple colour was formed. A blank solution is prepared by replacing the sample with distilled water, destruction, distillation, and titration are performed as for the sample. The calculation of protein content is formulated in the following equation. (*conversion factor = 6.25*)

$$Protein\ content = \left(\left(\frac{(ml_{HCl\ sample} - ml_{HCl\ blanko}) \times M_{HCl} \times 14{,}01}{m_{sample}}\right) \times 100\%\right) \times conversion\ factor$$

The hedonic test was carried out with reference to previous research [4]. This test was carried out using a scoring method on 30 semi-trained panellists. Each sample is given a three-digit number as a sample code. The panellists' liking level was assessed for the quality attributes of colour, aroma, texture, taste and overall quality. Each panellist was given RLHM-floss with various types of sugar and commercial CF. Panellists were asked to rate the samples according to their level of liking and then write them down on the form sheet provided. The panellists' liking scale consists of a scale of 1–4, namely Dislike (1), Like Slightly (2), Like Moderately (3), and Like Extremely (4). Mineral water is provided to neutralize the panellists' taste buds when testing product liking levels.

2.3 *Comparing RLHM-Floss with the Best Sugar Types and Commercial CF*

RLHM-floss with various sugar types was compared for each parameter test. RLHM-floss was determined to be the best treatment due to the panellists' preference level. RLHM-floss has the best variety of sugar types compared to commercial CF. Comparisons were made to see the effect of RLHM substance on the raw material for broiler chicken meat in making commercial CF and the effect of sugar type on the quality of RLHM-floss. Comparison of RLHM-floss with the best sugar type and commercial CF was carried out on colour parameters (L*, a*, b*, °hue), proximate values (water, ash, fat, protein), and hedonic quality (colour, aroma, taste, texture, overalls).

2.4 *Data Analysis*

Data in this study were analysed in two stages. The first stage of data analysis was carried out on RLHM-floss samples with various sugar types. This stage uses data analysis by SPSS Statistics 26.0 at a significance level of 5%. Parametric data that include data of colour and proximate were analysed using ANOVA (Analysis of Variance) test, if there was a significant influence, it was tested further using the DMRT (Duncan Multiple Range Test). Non-parametric data for hedonic quality were tested using the Kruskall Wallis method, followed by the Mann-Whitney test if a significant influence was found ($p<0.05$). The second stage of study consisted of a comparison of RLHM-floss with the best sugar types and commercial CF using descriptive data analysis.

IOP Conf. Series: Earth and Environmental Science **1460** (2025) 012051 doi:10.1088/1755-1315/1460/1/012051

3. Result and Discussions

3.1 Comparing Colour of RLHM-floss with Various Sugar Types

The ANOVA test showed that there is a real influence ($p<0.05$) of variations in the type of sugar on the degree of lightness (L*) of RLHM-floss. The L* value is related to pigmentation which can reach maximum value of 100, the higher the L* value, the brighter the product [5]. The RLHM-floss with the highest brightness is the RLHM-floss with granulated sugar which has a L* value of 56.18. The difference in the L* value of RLHM-floss is influenced by the physical characteristics of the sugar used. Granulated sugar or table sugar has a white colour with a crystalline shape [6]. The addition of granulated sugar in food products, included RLHM-floss, caused the colour lighter. RLHM-floss with coconut sugar have lower L* values compare to granulated sugar treatment, but higher than the palm sugar treatment. Coconut sugar has a light brown colour with sweet taste and unique coconut aroma [7]. RLHM-floss with palm sugar be the least bright treatment. Palm sugar have darker colour than other types of sugar affect the colour of the resulting RLHM-floss. Palm sugar became darker caused by caramelisation and Maillard reaction that occur during production process [8].

The ANOVA test results on the red-green value (a*) showed a significant effect ($p<0.05$) of sugar type variations on RLHM-floss. The a* values are in order from highest to lowest, RLHM-floss with coconut sugar (6.96), palm sugar (6.05), and granulated sugar (4.52). The a* value represents the redness level of the product with a range between -100 to +100 with negative values referring to green while positive values refer to red [9]. The non-enzymatic browning reaction in the form of the Maillard reaction of the product tends to increase the a* value in the positive direction. The Maillard reaction occurs due to the condensation of reducing sugars in the form of fructose and glucose which contain free amine groups [10]. The fructose and glucose content of coconut sugar tends to be higher than palm sugar and granulated sugar, making Maillard reactions occur more frequently. This triggers a change in product colour to become browner and the a* value becomes greater.

The ANOVA test results on the yellow-blue value (b*) showed a significant effect ($p<0.05$) of sugar type variations on RLHM-floss. The b* values of each treatment from the highest to the lowest are granulated sugar, coconut sugar, and palm sugar. The b* values of each treatment are 34.96, 27.25, and 22.30. The difference in the b* value of the product can be affected by type and quality of raw material, storage condition, temperature, and chemical process [11]. The b* value of RLHM-floss with granulated sugar is the highest because the use of granulated sugar does not trigger the formation of brown pigments (melanoidins). Caramelization in palm sugar and brown sugar makes the resulting product colour least yellow (brown). RLHM-floss with palm sugar has the lowest b* value compared to other sugar treatments. Brown colour in palm sugar arises more intensively due to high temperature and long heat caused higher hydroxymethylfurfural formation and sucrose inversion [12].

The ANOVA test results on the basic colour (°hue) show that there is a significant effect ($p<0.05$) of sugar type variations on RLHM-floss. The °hue value of RLHM-floss with granulated sugar is 82.62, with coconut sugar 75.66, and with palm sugar 75.82. The °hue value indicates the angle over the captured scene interpreted as the important colour [13]. The °hue value expressed in the form of an angle of 0–360° with a value between 60–120 being yellow. The °hue value <60 will be closer to red, while the °hue value >120 will be closer to green. The yellow colour of RLHM-floss comes from the use of turmeric that contain curcumin. Curcumin is a yellow polyphenols pigment that used in century as food colouring agents [14]. Coconut sugar and palm sugar have colours ranging from brown to blackish brown. The addition of brownish ingredients causes the °hue value to decrease so that the product is yellow-reddish.

Table 1. Colour of RLHM-floss with Various Sugar Types

Colour	Sugar Types		
	Granulated Sugar	Coconut Sugar	Palm Sugar
L*	56.18 ± 0.606[c]	52.48 ± 0.361[b]	44.27 ± 0.783[a]
a*	4.52 ± 0.208[a]	6.96 ± 0.465[c]	6.05 ± 0.362[b]
b*	34.96 ± 0.963[c]	27.25 ± 0.883[b]	22.30 ± 0.735[a]
°hue	82.62 ± 0.482[b]	75.66 ± 1.074[a]	74.82 ± 0.901[a]

[a-c] Different superscript in the same line indicate significant differences (p<0.05) among samples.

3.2 Comparing Proximate of RLHM-floss with Various Sugar Types

Table 2. Proximate of RLHM-floss with Various Sugar Types

Proximate	Sugar Types		
	Granulated Sugar	Coconut Sugar	Palm Sugar
Water Content	4.95 ± 0.053[a]	6.79 ± 0.138[c]	5.80 ± 0.097[b]
Ash Content	0.71 ± 0.076[a]	1.15 ± 0.307[b]	0.51 ± 0.129[a]
Fat Content	11.35 ± 0.300[c]	10.54 ± 0.125[b]	7.23 ± 0.335[a]
Protein Content	31.97 ± 0.555[b]	36.36 ± 0.721[c]	26.81 ± 0.559[a]

[a-c] Different superscript in the same line indicate significant differences (p<0.05) among samples.

The ANOVA test results showed a significant effect of water content ($p<0.05$) of variations in sugar types on the water content of RLHM-floss. RLHM-floss with coconut sugar has the highest water content reached 6.79%. The water content of RLHM-floss with palm sugar and granulated sugar is 5.80% and 4.95%. Sugar is a material that can affect water content because of its hygroscopic properties. The hygroscopic properties of sugar is even greater than carbohydrates and proteins [15]. RLHM-floss with coconut sugar has the highest water content because the sugar composition is in the form of reducing sugars, glucose and fructose. Reducing sugars, especially fructose, have hygroscopic properties so that they can bind more water [16]. Granulated sugar and palm sugar are composed of the main sugar in the form of sucrose. Sucrose as a disaccharide has a more complex structure so that it is less able to interact with water than monosaccharides such as fructose. The sucrose content in palm sugar is lower than in the granulated sugar. High sucrose content makes water content of the product decreased by osmotic dehydration [17].

Based on the ANOVA test for ash content there is a significant effect ($p<0.05$) of sugar type variations on RLHM-floss. The RLHM-floss ash content from the largest to the smallest was 1.15%, 0.71%, and 0.51%. RLHM-floss with coconut sugar became treatment that highest in ash content. The ash content in floss influenced by mineral content due to type and quality of material, process of making, and nutritional value [18]. The ash content can also be increased by the addition of preservatives. The preservative commonly used in coconut sugar is $CaCO_2$ which functions to increase the pH of the sugar so that yeast growth is inhibited. The ash content of coconut sugar also comes from the raw material of sap and the coconut sugar processing process related to minerals that can increase the ash content.

Based on the ANOVA test for fat content there is a significant effect ($p <0.05$) of sugar type variations on RLHM-floss. Granulated sugar is the RLHM-floss treatment with the highest fat content, reached 11.35%. The coconut sugar and palm sugar treatments have fat content of 10.54% and 7.23%. Fat content is an indication of the vulnerability of shredded meat to a decrease in quality due to the oxidation process [19]. Fat content have high correlation with moisture content. Low water content indicates high fat content [20]. Compared to coconut sugar and palm sugar, granulated sugar has low water content. Frying floss in oil causes the monosaccharide components of the sugar to form bonds with fat [21]. Oil fills the empty space in the material along with the water content so that the fat content increases. RLHM-floss with coconut sugar has more fat content compare to RLHM-floss with palm sugar due to its hygroscopic properties. These properties allow the absorption of oil as a source of fat in RLHM-floss.

Based on the ANOVA test on protein content, it showed that sugar type variations have a significant effect ($p<0.05$) on RLHM-floss. The protein content of RLHM-floss with granulated sugar, coconut sugar, and palm sugar treatments respectively 31.97%, 36.36%, and 26.81%. RLHM-floss with with coconut sugar was the treatment with highest protein content. Protein in RLHM-floss mostly comes from its main ingredients, RLHM. The RLHM-floss was made with adjustment to the sweetness level of each type of sugar. Coconut sugar has the highest sweetness level compared to granulated sugar and palm sugar. A high proportion of meat formula in a floss gives a higher protein content [22]. The addition of granulated sugar relative does not affect to RLHM-floss protein. RLHM-floss with palm sugar has the lowest protein content among the other treatments. Palm sugar is less sweet than granulated sugar, means that it is added in large quantities for RLHM-floss. This reduces the proportion of meat in this treatment. Palm sugar is composed of water, sugar, carbohydrates, protein, fat, ash, energy, calcium, phosphorus, and iron [23]. Minerals and bioactive components of palm sugar affect the stability of floss protein. Palm sugar as a reducing sugar that undergoes Maillard reaction triggers a decrease in the amount of protein. Maillard reaction reduces protein through the destruction of lysine and cystine due to reactions with dicarbonyl and aldehyde, reduces the availability of amino acids, prevents enzyme penetration into protein substrates [24].

3.3 Comparing Hedonic Quality of RLHM-floss with Various Sugar Types

Table 3. Hedonic Quality of RLHM-floss with Various Sugar Types

Hedonic Values	Sugar Types		
	Granulated Sugar	Coconut Sugar	Palm Sugar
Colour	1.90 ± 0.759^{a}	3.40 ± 0.932^{b}	1.83 ± 0.950^{a}
Aroma	2.40 ± 0.894^{a}	3.30 ± 0.915^{b}	2.40 ± 1.070^{a}
Taste	2.63 ± 0.718	3.07 ± 1.048	3.00 ± 0.947
Texture	2.27 ± 0.828	2.83 ± 0.950	2.70 ± 1.149
Overall	2.37 ± 0.718^{a}	3.10 ± 0.995^{b}	2.77 ± 0.898^{ab}

$^{a-c}$ Different superscript in the same line indicate significant differences ($p<0.05$) among samples.

Based on the hedonic colour analysis of RLHM-floss, there is a significant difference ($p<0.05$) in panellists' preferences between the treatment. The further test with Mann-Whitney method showed that the colour of the coconut sugar treatment sample was significantly different from the samples with granulated sugar and palm sugar. Colour is an important aspect that influences product appearance that often implies the sensory properties [25]. Panellists prefer RLHM-floss with coconut sugar because it has a golden-brown colour. The preference value for RLHM-floss with coconut sugar reached 3.40, namely Like Moderately – Like Extremely. RLHM-floss with granulated sugar and palm sugar had preference values of 1.90 and 1.83, namely Dislike – Like Slightly. Panellists' preference for RLHM-floss was too bright or dark. The colour that is too bright or dark gave unusual atmosphere which affected to quality perceptions [26].

Based on the hedonic aroma analysis, there is a significant difference ($p<0.05$) in RLHM-floss with granulated sugar and palm sugar compared to RLHM-floss with coconut sugar. The hedonic aroma value of

RLHM-floss with granulated sugar and palm sugar was 2.40, while RLHM-floss with coconut sugar was 3.30. The RLHM-floss with coconut sugar has a special aroma that panellists liked. Cooking of RLHM-floss triggers the emergence of caramel aroma in coconut sugar. The aroma can be associated with the protein content of floss. The more intensive the aroma produced indicates the higher the protein content [27].

Taste is a hedonic assessment attribute taken from the response of tongue as the sense of taste. Taste response can be influenced by some factors such as chemical compounds, concentration, interaction with other taste components, and temperature. Based on the analysis by Kruskal Wallis method, it can be seen that the variations of sugar types on RLHM-floss have no significant effect ($p \geq 0.05$) on the hedonic attribute of taste. Taste can be influenced by various additional factors in meat floss, such as spices. Spices are used to influence the taste, enhance the flavor so that it can be accepted by consumers [28]. The variation of sugar types in RLHM-floss also has no significant effect ($p \geq 0.05$) on the hedonic quality attribute texture. Texture is a hedonic assessment attribute of the pressure sensation response either by taste or touch. Texture preference can be affected by individual difference, cultural norms, and sensory experiences [29]. The texture preference scores for RLHM-floss with granulated sugar, coconut sugar, and palm sugar were 2.27, 2.83, and 2.70 which included in the criteria of Like Slightly – Like Moderately.

Overall is the attribute of the overall value of the panellists' preference for the RLHM-floss sample. The data analysis on hedonic quality attribute overall showed a significant difference ($p<0.05$) between treatments. The overall score of RLHM-floss with granulated sugar was 2.37 and was significantly different from the RLHM-floss sample with coconut sugar with an overall score of 3.10. RLHM-floss with palm sugar had an overall score of 2.77, not significantly different compared two other treatments. The overall value is subjective so it can differ depending on the likes and preferences of each individual [30].

The best sugar variation in RLHM-floss was determined through an assessment of the hedonic quality test results. Hedonic quality test is carried out to measure the food preferences in the form of informative assessment by panellists. This test is an important parameter in considering the diversification of food products [31]. The results of the hedonic quality test analysis showed that RLHM-floss with coconut sugar more preferable compared to other two treatments. This caused by the combination of colour (golden brown), aroma (caramel), and overall. RLHM-floss with coconut sugar was high in protein. Protein is a major parameter for CF. These results indicate that processing RLHM into a floss can contribute to the provision of protein as an important nutrient for humans.

3.4 Comparing RLHM-floss and Commercial CF

Table 4. Comparison of RLHM-floss processed with coconut sugar and commercial CF

Parameter	RLHM-floss with Coconut Sugar	Commercial CF
L*	52.48	58.26
a*	6.96	3.97
b*	27.25	26.56
°hue	75.66	81.45
Water Content	6.79	2.08
Ash Content	1.15	0.60
Fat Content	10.54	10.56
Protein Content	36.36	31.29
Color	3.40	3.30
Aroma	3.30	3.10
Taste	3.07	2.10
Texture	2.83	2.90
Overall	3.10	2.43

Based on the comparison of the two types of the CF, it can be concluded that RLHM-floss with coconut sugar has higher in colour value (a* and b*), proximate (water, ash, and protein), and hedonic quality (colour, aroma, taste, and overall). The L* value of RLHM-floss with coconut sugar is lower than the L* value of commercial CF. This is due to the use of a mixture of granulated sugar and coconut sugar in commercial CF. The characteristics of granulated sugar which is bright in colour increases the brightness of the product when added to the product. The a* value of RLHM-floss with coconut sugar is higher than the a* value of commercial CF. The increase in the a* value is proportional to the decrease in L* value and the increase in water content. The °hue values of both samples are in the yellow base colour area. The decrease in °hue is due to the addition of yellow-orange coloured ingredients and Maillard reaction [32]. The colour of RLHM-floss with coconut sugar looks more golden-brown compared to commercial CF.

The water content of RLHM-floss with coconut sugar is higher than commercial CF with a difference of 4.71%. The higher water content in RLHM-floss with coconut sugar indicates a shorter shelf life than commercial CF. However, the higher water content in RLHM-floss makes its texture smoother, slightly moist, and soft. RLHM-floss with coconut sugar has a higher ash content than commercial CF. The use of coconut sugar in RLHM-floss causes high ash content in the RLHM-floss. The fat content of the two samples tends to be the same with a difference of only 0.02%. The protein content of RLHM-floss with coconut sugar reaches 36.36%, while the protein content of commercial CF is only 31.29%. The higher protein content of RLHM-floss with coconut sugar encourages faster fulfilment of protein needs for body.

The hedonic quality value of the colour, aroma, taste, and overall attributes of RLHM-floss with coconut sugar were preferred by panellists. The range of panellists' preference for RLHM-floss was between Like Moderately – Like Extremely. The texture quality attribute of RLHM-floss with coconut sugar showed a smaller value than commercial CF. Panellists preferred commercial CF because its texture was not too lumpy.

The lumpy texture of the floss was thought to be due to the fact that it was not properly broken-down during frying. The overall quality attribute of RLHM-floss with coconut sugar was greater than commercial CF on the Like Moderately – Like Extremely scale.

4. Conclusions

The variation of sugar types affected colour, proximate, and hedonic quality RLHM-floss. Coconut sugar is the most appropriate sugar to produce floss from RLHM among granulated sugar and palm sugar based on considerations of panellist preference. Compared to commercial CF, RLHM-floss with coconut sugar is more golden-brown colour, higher in moisture, ash, and protein, and preferred by panellists.

References

[1] R. Harianti and F. S. Tanberika, "Pemberdayaan wanita tani melalui produksi abon ikan lele," *Jurnal Pendidikan dan Pemberdayaan Masyarakat*, vol. 5, no. 2, pp. 167–180, 2018, doi: 10.21831/jppm.v5i2.21071.

[2] M. Masyitah, Z. M. Lesti, S. Sukinem, and I. Yani, "Pelatihan pemanfaatan daging ayam bertelur afkir menjadi produk abon dengan penambahan keluwih (Artocarpus communis) di Gampong Pasie Lamgarot Kecamatan Ingin Jaya Kabupaten Aceh Besar," *Jurnal Ilmiah Peternakan Abdi Masyarakat*, vol. 2, no. 1, pp. 1–5, 2022, doi: 10.24815/petamas.v2i1.25773.

[3] B. V. Raksasiri, S. Khianngam, M. Nopparatmaitee, W. Injana, and P. N. Chiangmai, "The Effect of the aging period on meat quality and consumption properties of spent laying hens," *Journal for the Agriculture, Biotechnology, and Education*, vol. 1, no. 1, pp. 1–6, 2021.

[4] M. Sigit, M. Akbar, and L. Fianti, "Kualitas organoleptik abon ayam yang diberi perlakuan substitusi kacang tanah (Arachis hypogaea L.)," *Jurnal Fillia Cendekia*, vol. 2, no. 1, pp. 1–8, 2017.

[5] B. C. K. Ly, E. B. Dyer, J. L. Feig, A. L. Chien, and S. Del Bino, "Research Techniques Made Simple: Cutaneous Colorimetry: A Reliable Technique for Objective Skin Color Measurement," *Journal of Investigative Dermatology*, vol. 140, no. 1, pp. 3-12.e1, 2020, doi: 10.1016/j.jid.2019.11.003.

[6] H. Patil and M. Peshave, "Study on the Types of Sugar Used in Various Culinary Products," *Shodh Sarita*, vol. 8, no. 9, pp. 314–318, 2021.

[7] S. Riyanti, C. Cahyana, and A. Kandriasari, "Perbandingan Kualitas Bolu Kukus Mekar Gula Aren dan Gula Kelapa," in *Garina*, 2024, pp. 58–66. doi: 10.69697/garina.v16i2.123.

[8] A. D. Saputro, D. Van de Walle, and K. Dewettinck, "Physicochemical properties of coarse palm sap sugars as natural alternative sweetener," *Food Bioscience*, vol. 38, no. 1, pp. 1–8, 2020, doi: 10.1016/j.fbio.2020.100780.

[9] F. A. Sitanggang, M. M. Machfoedz, and M. A. F. Falah, "Comparison of Color Quality Measurement Using Chromameter and Image Processing for Dehydrated Strawberry Products," in *Proceedings of the 2nd International Conference for Smart Agriculture, Food, and Environment (ICSAFE 2021)*, Atlantis Press International BV, 2023, pp. 4–17. doi: 10.2991/978-94-6463-090-9.

[10] V. Andragogi, P. V. Bintoro, and S. Susanti, "Pengaruh Berbagai Jenis Gula terhadap Sifat Sensori dan Nilai Gizi Roti Manis," *Jurnal Teknologi Pangan*, vol. 2, no. 2, pp. 163–167, 2018.

[11] Y. Wen *et al.*, "Relationship between wheat flour's quality characteristics and color of fresh wet noodles," *International Journal of Food Properties*, vol. 26, no. 1, pp. 290–300, 2023, doi: 10.1080/10942912.2022.2161565.

[12] T. Kurniawan, Jayanudin, I. Kustiningsih, and M. Adha Firdaus, "Palm Sap Sources, Characteristics, and Utilization in Indonesia," *Journal of Food and Nutrition Research*, vol. 6, no. 9, pp. 590–596, 2018, doi: 10.12691/jfnr-6-9-8.

[13] L. L. P. Nguyen, L. Baranyai, D. Nagy, P. V. Mahajan, V. Zsom-Muha, and T. Zsom, "Color analysis of horticultural produces using hue spectra fingerprinting," *MethodsX*, vol. 8, no. 101594, pp. 1–8, 2021, doi: 10.1016/j.mex.2021.101594.

[14] J. Sharifi-Rad *et al.*, "Turmeric and Its Major Compound Curcumin on Health: Bioactive Effects and Safety Profiles for Food, Pharmaceutical, Biotechnological and Medicinal Applications," *Frontiers in Pharmacology*, vol. 11, no. 1, pp. 1–23, 2020, doi: 10.3389/fphar.2020.01021.

[15] K. Khalisa, Y. M. Lubis, and R. Agustina, "Uji Organoleptik Minuman Sari Buah Belimbing Wuluh (Averrhoa bilimbi.L)," *Jurnal Ilmiah Mahasiswa Pertanian*, vol. 6, no. 4, pp. 594–601, 2021, doi: 10.17969/jimfp.v6i4.18689.

[16] Q. Ma, S. Wang, and G. Dong, "Macroscale liquid superlubricity achieved with mixtures of fructose and diols," *Wear*, vol. 484–485, no. 204037, pp. 1–8, 2021, doi: 10.1016/j.wear.2021.204037.

[17] W. S. Chen, D. C. Liu, and M. T. Chen, "Effects of high level of sucrose on the moisture content, water activity, protein denaturation and sensory properties in Chinese-style pork jerky," *Asian-Australasian Journal of Animal Sciences*, vol. 15, no. 4, pp. 585–590, 2002, doi: 10.5713/ajas.2002.585.

[18] T. Rihayat *et al.*, "Effect of Determination Temperature on Nutrition and Organoleptic Tuna Fish Floss," in *1st South Aceh International Conference on Engineering and Technolgy*, 2019, pp. 1–9. doi: 10.1088/1757-899X/506/1/012055.

[19] A. Bujang, S. S. Jamil, N. F. Basar, and N. M. Nor, "Nutrient content, Sensory and Storage Stability of Meat floss Incorporated with Cassava leaves," *IOSR Journal of Environmental Science, Toxicology and Food Technology*, vol. 10, no. 09, pp. 30–34, 2016, doi: 10.9790/2402-1009013034.

[20] J. An, Y. Li, C. Zhang, and D. Zhang, "Rapid Nondestructive Prediction of Multiple Quality Attributes for Different Commercial Meat Cut Types Using Optical System," *Food Science of Animal Resources*, vol. 42, no. 4, pp. 655–671, 2022, doi: 10.5851/kosfa.2022.e28.

[21] S. H. Qonitah, R. A. Dian, and Basito, "Kajian Penggunaan Hight Fructose Syrup (HFS) Sebagai Pengganti Gula Sukrosa Terhadap Karakteristik Fisik dan Kimia Biskuit Berbasis Tepung Jagung (Zea mays) dan Tepung Kacang Merah (Phaseolus vulgaris L.)," *Jurnal Teknologi Hasil Pertanian*, vol. IX, no. 2, pp. 9–21, 2016.

[22] A. B. Omojola, O. R. Kassim, O. O. Olusola, P. O. Adeniji, and J. O. Aremo, "Development and Quality Evaluation of Danbunama (Meat Floss) - a Nigerian Shredded Meat Product," *British Journal of Applied Science & Technology*, vol. 4, no. 26, pp. 3862–3873, 2014, doi: 10.9734/bjast/2014/10055.

[23] S. Sudiarto, R. Saleh, S. Sawab, and I. Widowati, "Nutritional Content of Functional Herbal Drinks Made from Palm Sugar Powder and Spice Powder," *AGRITEKNO: Jurnal Teknologi Pertanian*, vol. 11, no. 2, pp. 61–71, 2022, doi: 10.30598/jagritekno.2022.11.2.61.

[24] N. Suryani, Rosita, and U. Hasanah, "Perbedaan Kadar Protein dan Kadar Lemak Ikan Patin (Pangasius hypophtalmus) yang Diolah secara Digoreng, Dipanggang dan Direbus," *Jurnal Kesehatan Indonesia*, vol. 6, no. 1, pp. 39–45, 2015.

[25] H. Blackmore, C. Hidrio, and M. R. Yeomans, "How sensory and hedonic expectations shape perceived properties of regular and non-alcoholic beer," *Food Quality and Preference*, vol. 99, no. 104562, pp. 1–11, 2022, doi: 10.1016/j.foodqual.2022.104562.

[26] E. Özkul, B. Bilgili, and E. Koç, "The Influence of the color of light on the customers' perception of service quality and satisfaction in the restaurant," *Color Research and Application*, vol. 45, no. 6, pp. 1217–1240, 2020, doi: 10.1002/col.22560.

[27] M. A. Ridhani, I. P. Vidyaningrum, N. N. Akmala, R. Fatihatunisa, S. Azzahro, and N. Aini, "Potensi Penambahan Berbagai Jenis Gula Terhadap Sifat Sensori Dan Fisikokimia Roti Manis: Review," *Pasundan Food Technology Journal*, vol. 8, no. 3, pp. 61–68, 2021, doi: 10.23969/pftj.v8i3.4106.

[28] A. Dougkas, M. Vannereux, and A. Giboreau, "The impact of herbs and spices on increasing the appreciation and intake of low-salt legume-based meals," *Nutrients*, vol. 11, no. 2901, pp. 1–20, 2019, doi: 10.3390/nu11122901.

[29] M. Baingana, "Food Texture Perception and Its Influence on Consumer Preferences," *Journal of Food Sciences*, vol. 5, no. 2, pp. 43–55, 2024, doi: 10.47941/jfs.1847.

[30] Y. F. Daeli, Y. B. Pramono, and S. Mulyani, "Perbedaan karakteristik fisik dan hedonik daging sapi dengan metode perebusan yang berbeda," *Jurnal Ilmu dan Teknologi Peternakan Tropis)*, vol. 9, no. 1, pp. 300–305, 2022, doi: 10.33772/jitro.v9i1.18428.

[31] I. Ismawati and R. D. Putri, "Uji Kesukaan Dan Kelayakan Usaha Produk Nugget Ikan Lele," *Jurnal Pertanian Cemara*, vol. 15, no. 2, pp. 1–5, 2018, doi: 10.24929/fp.v15i2.651.

[32] M. V. P. Rani and W. H. Susanto, "Pengaruh Lama Pengukusan Serta Proporsi Tepung Mocaf Dan Pasta Labu Kuning Terhadap Sifat Fisik Kimia Organoleptik Kerupuk Cekeremes Effect Of Steaming Time And Mocaf Flour And Pumpkin Yellow Pasta ' S Proportion To Physical Chemistry And Organoleptic Cek," *Pangan dan Agroindustri Vol.*, vol. 3, no. 3, pp. 1062–1070, 2015.

IOP Conf. Series: Earth and Environmental Science **1460** (2025) 012052 doi:10.1088/1755-1315/1460/1/012052

Modulating Satiety Through Dietary Intervention: Exploring the Synergistic Effects of Fibre and Chili Consumption

Hanis Gani[1*], Scarlet Lee[2], Caroline Orfila[3] and Peter Ho[3]

[1] School of Food Industry, Faculty of Bioresources and Food Industry, Universiti Sultan Zainal Abidin, 22200 Besut Terengganu Malaysia
[2] Singapore Institute of Technology, 10 Dover Drive, Singapore 138683
[3] School of Food Science and Nutrition, Faculty of Environment, University of Leeds, Woodhouse Lane, Leeds, LS2 9JT, United Kingdom

*E-mail: hanissyazwani@unisza.edu.my

Abstract. Chronic overconsumption, defined as consistently exceeding dietary requirements, presents a critical global public health issue by disrupting energy homeostasis and contributing to obesity. Developing satiety-enhancing functional foods offers a promising strategy for individuals with low intuitive eating behaviours to manage their intake. Dietary components like dietary fibre and capsaicin have shown potential in extending satiety duration. This study examined the effects of dietary fiber (BG28) and chilli compound capsaicin enrichment on satiety responses in healthy individuals consuming ad libitum noodle meals. Using a single-blinded, randomized, crossover design, twenty-three participants consumed four distinct ad libitum test meals on four separate testing sessions. The intervention meals included a control, BG28-enriched noodles, capsaicin-enriched noodles (CAP), and a combination of BG28 and capsaicin-enriched noodles (BG28CAP). Compared to the control, BG28 fibre and capsaicin (CAP) supplementation in the noodle formulation significantly reduced ($p < 0.05$) both self-reported Mental Hunger scores and ad libitum intake. Additionally, the combined BG28CAP meal further reduced intake, indicating a synergistic effect between fibre and capsaicin in enhancing satiety. Understanding these synergistic interactions between macronutrients and specific food compounds could advance the development of dietary interventions and satiety-promoting functional foods to address weight management challenges.

Keywords: Capsaicin, dietary fibre, intuitive eating, satiety.

1. Introduction

Noodles have gained global popularity as a staple in numerous dishes, making them effective carriers for essential nutrients [1]. Nutrient fortification, including dietary fibre (DF), can be incorporated directly into noodle formulations to enhance their nutritional profile. However, such additions may affect the physicochemical properties and sensory acceptance of noodles and pasta in varying ways.

DF has attracted considerable interest for its effectiveness to promote various health benefits. Studies have associated DF intake with a decreased risk of coronary heart disease,

IOP Conf. Series: Earth and Environmental Science **1460** (2025) 012052 doi:10.1088/1755-1315/1460/1/012052

colorectal cancer, other cancers, and obesity [2-4] by reducing plasma cholesterol and glycaemic index, improving lipid metabolism and promoting satiety. The extensive research documenting the satiating properties of DF leaves little room for doubt regarding its effectiveness in this area.

Herbs and spices are widely incorporated into food products to enhance their sensory attributes (e.g., flavour, aroma) and improve consumer acceptability [5-6]. Additionally, these ingredients possess preservative and medicinal properties. Among the most prevalent spices globally is chili pepper (*Capsicum spp.*), a foundation in culinary traditions that span both Eastern and Western cuisines [7-11]. The pungent characteristic of chili peppers is attributed to capsaicin (CAP), an active component. Previous studies suggested that CAP markedly enhances energy expenditure and promotes lipid oxidation [12-15]. This mechanism promotes early fat oxidation, prioritising it over protein and carbohydrate oxidation before the next meal [16], potentially leading to suppressed energy intake, body weight management, appetite regulation, and enhanced satiety [14,16,17-19]. While evidence suggests that both DF and CAP independently influence satiety, a critical gap exists in the current literature: very limited research, to our knowledge, has explored the potential synergistic interaction between the combination of these two ingredients on satiety in healthy adults

2. Materials and methods

2.1 Test meals and fixed preloads

The control noodles (Control) were made with 100% high-protein wheat flour. In the fibre-enriched noodles, 25% (w/w) of BG28 fibre (Naturex, Sweden) were used to substitute wheat flour. Fresh fixed preloads for each satiety trial were sourced from local retailers in Leeds, UK, on the morning of each session. A 10-ml capsaicin (CAP) stock solution was infused into vegetable-flavoured broth for both CAP and BG28CAP noodles.

2.2 Satiety trials

The trial employed a single-blind, randomized crossover design, with participants assigned to consume ad libitum noodle test meals across four separate test days. A total of twenty-three participants were recruited for the satiety trial. Each meal was initially served as a single portion, with participants able to request additional servings until they felt comfortably satiated. Participants completed a series of Mental Hunger items [20] at five intervals, beginning pre-meal and continuing up to three hours post-meal.

2.2 Rasch analysis

Data from the satiety trial were analysed using Rasch repeated measures analysis with the Winsteps Version 4.3.3 software package [21]. In this analysis, item measures also referred to as “item difficulties” were held fixed, while the person measures or “person abilities” were allowed to vary. Step count was evaluated according to the procedures outlined by [22].

3. Results and Discussion

Enrichment with DF and CAP had a significant effect ($p < 0.001$) on ad libitum intake at lunch ($F_{(3,19)} = 10.64$), with higher consumption observed for the Control noodles (Figure 1). CAP enrichment alone did not significantly reduce energy intake compared to the Control, while BG28 and BG28CAP (combined DF and CAP) both led to reductions in intake. Notably, the BG28CAP noodles resulted in an obvious drop in intake as opposed to both the Control and the single-enrichment groups, suggesting a potential synergistic effect between DF and CAP.

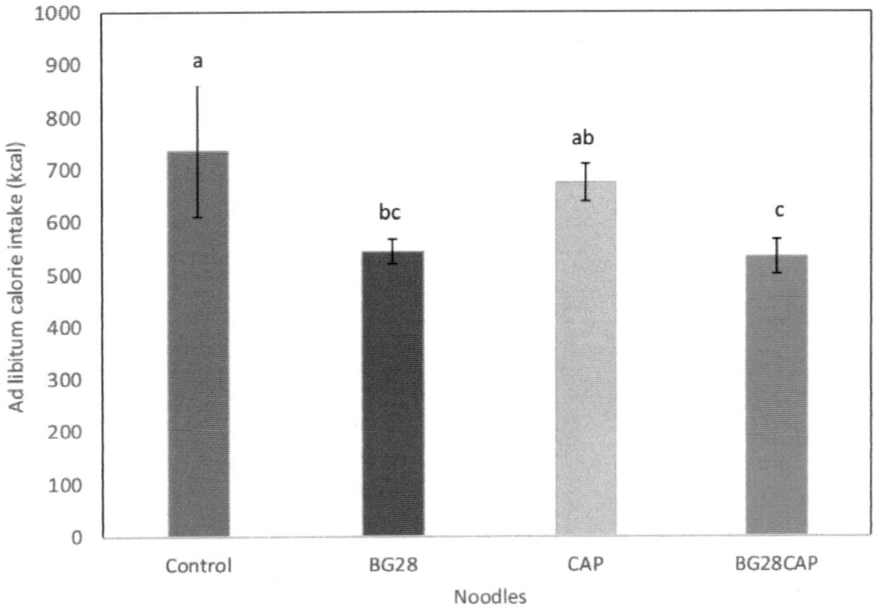

Figure 1. Ad libitum calorie intake (kcal) for the four noodle formulations shown as mean values (N=23), with standard deviations depicted by vertical bars. Bars labelled with the same lowercase letters indicate no significant difference at $p < 0.05$.

Satiation and satiety, as measured by Mental Hunger (MH), varied significantly across the four noodle formulations (Figure 2). MH measures decreased pre-ingestively and increased post-ingestively ($p < 0.05$), with the lowest values observed immediately after lunch. One-hour post-meal, MH scores rose significantly, though BG28CAP noodles sustained a lower MH score (28.82 ± 8.02) compared to the Control. Across five time points, BG28CAP significantly reduced MH more than other formulations, indicating greater satiety effects. These results align with earlier findings that combining chili and fibre increases fullness and reduces hunger more effectively than either component alone [23].

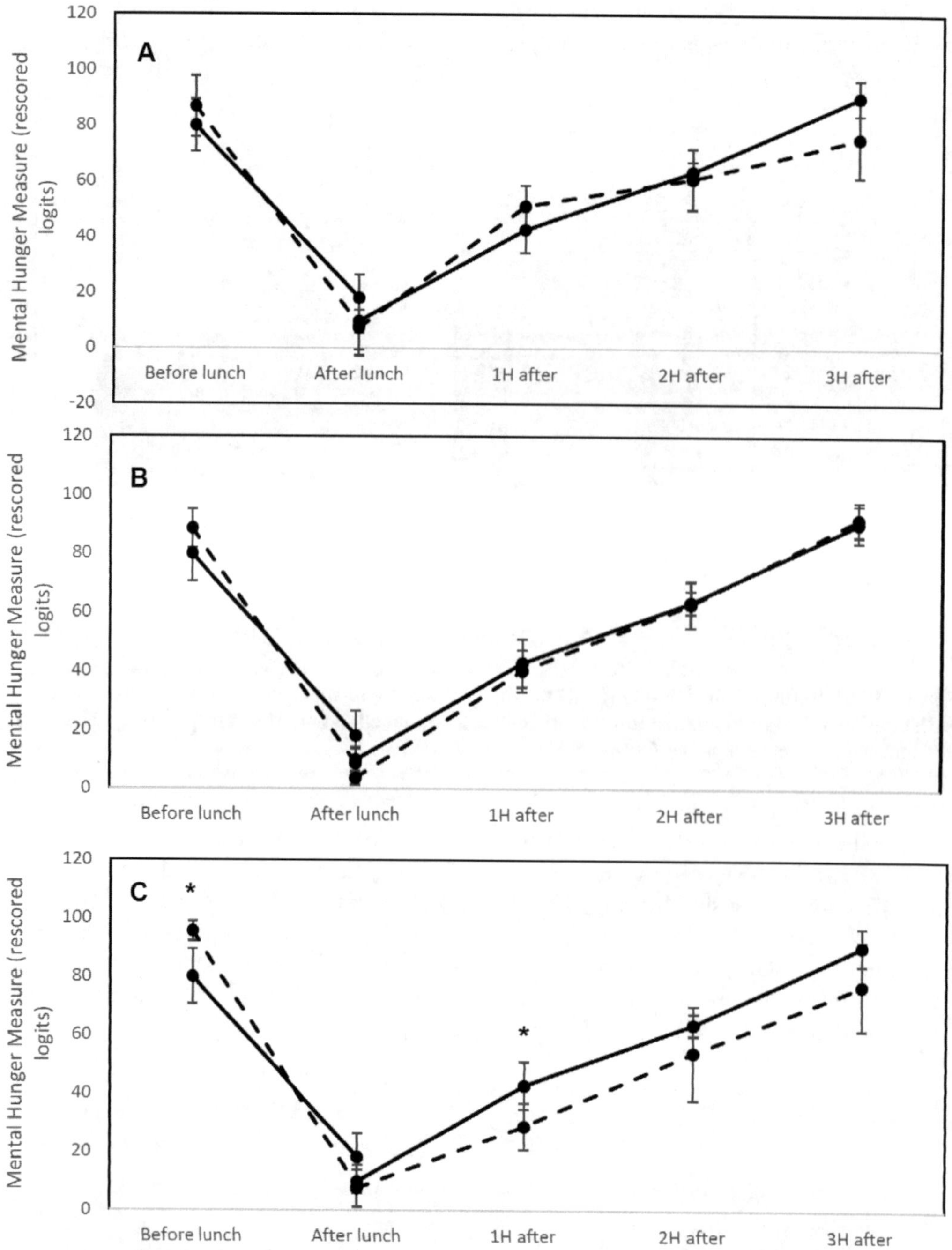

Figure 2. Plots of the mean ± SD of Mental Hunger (MH) measures are shown for both the pre-ingestive period (from before to after lunch) and the post-ingestive period (from after lunch to three hours later) for the Control noodle meal (solid black line) and treatment noodles (dashed lines: A = BG28, B = CAP, C = BG28CAP). MH measures, expressed in logits, were rescaled to a 0-100 scale. An asterisk (*) indicates a significant difference in MH measures at the specified time point ($p < 0.05$).

Previous studies have extensively examined the effects of dietary fibre (DF) and capsaicin (CAP), both individually and in combination with other ingredients, on objective and subjective satiety assessments and also biochemical responses. The observed decreases in Mental Hunger (MH) measures and energy consumption for the BG28CAP suggest a potential synergistic effect between DF and CAP in promoting satiety.BG28 fibre, derived from oat grain, is high in β-glucan, a fibre with both soluble and insoluble properties. These properties, together with the dietary fibres in BG28, help modulate satiety through physicochemical characteristics such as water-holding capacity, food structures, viscosity and fermentability [24]. The differing satiating effects of macronutrients—including protein, carbohydrates, fats, and fibres—clarify the reason of caloric content influences satiety to varying extents, underscoring the role of these factors in developing satiety-enhancing foods [25]. Additionally, CAP may influence satiety by promoting thermogenesis, enhancing lipid oxidation, and reducing adipose tissue [18, 26]. The effects of CAP on diet management are dose-dependent, with studies testing various CAP concentrations to assess its impact on appetite [14,16,19]. Further research is warranted to identify optimal ingredient levels for significant outcomes. However, the synergistic mechanism between DF and CAP remains uncertain, requiring more evidence for a thorough understanding.

4. Conclusion

In conclusion, fibre and chilli seem to interact synergistically to enhance satiety, especially at higher fibre levels. Future research will aim to identify the optimal combination of fibre and capsaicin for maximising this effect and to explore their impact over extended post-meal periods.

Acknowledgements

This research was supported by Dana Penyelidikan Universiti 1.0 (Project No. R0355), Universiti Sultan Zainal Abidin, Terengganu, Malaysia. The authors wish to thank Swedish Oat Fiber (Naturex), Sweden, for providing BG28 fibre.

References

[1] Zhang, N., & Ma, G. (2016). Noodles, traditionally and today. *Journal of Ethnic Foods*, 3(3), 209–212. https://doi.org/10.1016/j.jef.2016.08.003

[2] Izydorczyk, M. S., & Dexter, J. E. (2008). Barley β-glucans and arabinoxylans: Molecular structure, physicochemical properties, and uses in food products-a Review. *Food Research International*, 41(9), 850–868. https://doi.org/10.1016/j.foodres.2008.04.001

[3] Rebello, C. J., O'Neil, C. E., & Greenway, F. L. (2016). Dietary fiber and satiety: The effects of oats on satiety. *Nutrition Reviews*, 74(2),131-47. https://doi.org/10.1093/nutrit/nuv063

[4] Slavin, J. L. (2013). Health aspects of dietary fibre. In J. A. Delcour & K. Poutanen (Eds.), *Fibre-Rich and Wholegrain Foods: Improving Quality*. https://doi.org/10.1533/9780857095787.1.61

[5] El-Sayed, S. M., & Youssef, A. M. (2019). Potential application of herbs and spices and their effects in functional dairy products. *Heliyon*, 5(6), e01989. https://doi.org/10.1016/j.heliyon.2019.e01989

[6] Tapsell L.C., Hemphill, I., Cobiac, L., Patch, C.S., Sullivan, D.R., Fenech, M., Roodenrys, S., Keogh, J.B., Clifton, P.M., Williams, P.G., Fazio, V.A., & Inge, K.E. (2006). Health benefits of herbs and spices: the past, the present, the future. *The Medical Journal of Australia*., 185(S4) S1-S24. https://doi.org/10.5694/j.1326-5377.2006.tb00548.x

[7] Lv, J., Qi, L., Yu, C., Yang, L., Guo, Y., Chen, Y., ... Li, L. (2015). Consumption of spicy foods and total and cause specific mortality: Population based cohort study. *BMJ (Online)*, 351, 1–10. https://doi.org/10.1136/bmj.h3942

[8] Mongkolporn, O., & Taylor, P. W. J. (2011). Capsicum. In C. Kole (Ed.), *Wild Crop Relatives: Genomic and Breeding Resources* (pp. 43–57). Heidelberg, Berlin: Springer. https://doi.org/10.1007/978-3-642-20450-0_4

[9] Mózsik, G. (2016). It remains a mystery why people living in hot climates consume spicier food. *Temperature*, 3(1), 50–51. https://doi.org/10.1080/23328940.2015.1131033

[10] Sativa, M., Harianto, H., & Suryana, A. (2017). Impact of Red Chilli Reference Price Policy in Indonesia. *International Journal of Agriculture System*, 5(2), 120. https://doi.org/10.20956/ijas.v5i2.1201

[11] Shi, Z., Riley, M., Brown, A., & Page, A. (2018). Chilli intake is inversely associated with hypertension among adults. *Clinical Nutrition ESPEN*, 23, 67–72. https://doi.org/10.1016/j.clnesp.2017.12.007

[12] Balaji, M., Ganjayi, M. S., Hanuma Kumar, G. E. N., Parim, B. N., Mopuri, R., & Dasari, S. (2016). A review on possible therapeutic targets to contain obesity: The role of phytochemicals. *Obesity Research and Clinical Practice*, 10(4), 363–380. https://doi.org/10.1016/j.orcp.2015.12.004

[13] Lejeune, M. P. G. M., Kovacs, E. M. R., & Westerterp-Plantenga, M. S. (2003). Effect of capsaicin on substrate oxidation and weight maintenance after modest body-weight loss in human subjects. *British Journal of Nutrition*, 90(3), 651–659. https://doi.org/10.1079/BJN2003938

[14] Ludy, M. J., & Mattes, R. D. (2011). The effects of hedonically acceptable red pepper doses on thermogenesis and appetite. *Physiology and Behavior*, 102(3–4), 251–258. https://doi.org/10.1016/j.physbeh.2010.11.018

[15] Matsumoto, T., Miyawaki, C., Ue, H., Yuasa, T., Miyatsuji, A., & Moritani, T. (2000). Effects of capsaicin-containing yellow curry sauce on sympathetic nervous system activity and diet-induced thermogenesis in lean and obese young women. *Journal of Nutritional Science and Vitaminology*, 46(6), 309–315. https://doi.org/10.3177/jnsv.46.309

[16] Westerterp-Plantenga, M. S., Smeets, A., & Lejeune, M. P. G. (2005). Sensory and gastrointestinal satiety effects of capsaicin on food intake. *International Journal of Obesity*, 29(6), 682–688. https://doi.org/10.1038/sj.ijo.0802862

[17] Golzarand, M., Toolabi, K., & Aghasi, M. (2018). Effect of green tea, caffeine and capsaicin supplements on the anthropometric indices: A meta-analysis of randomized clinical trials. *Journal of Functional Foods*, 46(May), 320–328. https://doi.org/10.1016/j.jff.2018.04.002

[18] Whiting, S., Derbyshire, E., & Tiwari, B. K. (2012). Capsaicinoids and capsinoids. A potential role for weight management? A systematic review of the evidence. *Appetite*, 59(2), 341–348. https://doi.org/10.1016/j.appet.2012.05.015

[19] Yoshioka, M., Imanaga, M., Ueyama, H., Yamane, M., Kubo, Y., Boivin, A., St-Amand, J., Tanaka, H. & Kiyonaga, A. (2004). Maximum tolerable dose of red pepper decreases fat intake independently of spicy sensation in the mouth. *British Journal of Nutrition*, 91(6), 991–995. https://doi.org/10.1079/bjn20041148

[20] Karalus, M., & Vickers, Z. (2016). Satiation and satiety sensations produced by eating oatmeal vs. oranges. a comparison of different scales. *Appetite*, 99, 168–176. https://doi.org/10.1016/j.appet.2016.01.012

[21] Linacre, John M. (2019). Winsteps® Rasch measurement computer program User's Guide. Beaverton, Oregon: www.winsteps.com.

[22] Ho, P. (2019). A new approach to measuring Overall Liking with the Many-Facet Rasch Model. *Food Quality and Preference*, 74, 100–111. https://doi.org/10.1016/j.foodqual.2019.01.015

[23] Gani, H., Fearnley, J., Ho, P. & Orfila, C. (2018). Oat fibre andchilli promote satiety synergistically. *Proceedings of the Nutrition Society*, 77(OCE4), E135. https://doi.org/10.1017/S0029665118001416

[24] Clark, M. J., & Slavin, J. L. (2013). The effect of fiber on satiety and food intake: a systematic review. *Journal of the American College of Nutrition*, 32(3), 200–211. https://doi.org/10.1080/07315724.2013.791194

[25] Dericioglu, D., Oldham, S., Methven, L., Shafat, A., & Clegg, M.E. (2023). Macronutrients effects on satiety and food intake in older and younger adults: A randomised controlled trial. *Appetite*, 189, 106982. https://doi.org/10.1016/j.appet.2023.106982

[26] Ludy, M.J., Moore, G.E. & Mattes, R.D. (2012). The Effects of Capsaicin and Capsiate on Energy Balance: Critical Review and Meta-analyses of Studies in Humans, *Chemical Senses*, 37(2), 103–121. https://doi.org/10.1093/chemse/bjr100

IOP Conf. Series: Earth and Environmental Science **1460** (2025) 012053 doi:10.1088/1755-1315/1460/1/012053

Profile of Fecal Microflora in Relation to Food Habit of Male Adult from Pekanbaru, Indonesia

Usman Pato[1*], Yusmarini[1], Emma Riftyan[1], Wafiatul Akbar[1], Yiohana Enzelina Marsoit[1], Agrina[2], Aqila Sakura Usman[3], and Suyanto[3]

[1] Faculty of Agriculture, Universitas Riau, Pekanbaru 28293, Indonesia
[2] Faculty of Nursing, Universitas Riau, Pekanbaru 28131, Indonesia
[3] Faculty of Medicine, Universitas Riau, Pekanbaru 28131, Indonesia

*E-mail: usmanpato@yahoo.com

Abstract. The composition of fecal microflora indicates its digestive system and eating habits. The precise composition can differ significantly based on a person's diet. This study examined the relationship between 30 adult men's dietary habits and their fecal microbiota composition. Secondary data were obtained through questionnaires filled out by the subjects. Primary data, including total plate count (TPC), lactic acid bacteria (LAB), Bifidobacterium spp., *E. coli*, Salmonella spp., and Shigella spp., were analyzed using specific media. The number of microorganisms was analyzed using the ANOVA procedure of SPSS Version 26. The least significant difference procedure was used to determine whether a statistically significant difference existed between the means. Multivariate analysis was conducted to assess the effect of eating habits on the number of fecal microbes. The quantity of microflora, including TPC, LAB, Bifidobacterium spp., *E. coli*, Salmonella spp., and Shigella spp., varied significantly among participants, according to ANOVA. There were 10.10–12.86 log CFU/g of TPC, 7.72–10.16 log CFU/g of LAB, 3.98–7.42 log CFU/g of Bifidobacterium spp., 7.73–10.16 log CFU/g of *E. coli*, and 5.29–10.80 log CFU/g of Salmonella spp. and Shigella spp. The study's findings also revealed that, despite suspicions, all of the bacteria may impact the quantity of TPC, *E. coli*, Salmonella spp., and Shigella spp., no particular food intake category had an effect. Regarding the individuals' diet of fruits, vegetables, and nuts, there was a correlation between the variation in LAB and Bifidobacterium counts.

Keywords: adult, food habit, microflora profile, male, Pekanbaru

1. Introduction

Millions of microorganisms are present in a healthy human body. The microbiome is the collective term for these creatures. The microbiome comprises bacteria, dead viruses, fungi, archaea, and protozoa. Most microorganisms in the human body are bacteria. Approximately 75–200 trillion unique organisms comprise the human body's microbiome, primarily bacteria [1].

In contrast, the somatic cell count in the human body is believed to be between 50 and 100 trillion [2]. Thus, the human body is essentially made up of a mixture of genes and cells from microbes and humans, resulting in a mix of traits from both species. The gut microbiome, the largest and most complex bacterial population, is found in the intestinal digestive system [1]. The connection between diet and fecal composition is a well-documented area of research, offering insight into how food intake influences gastrointestinal health, nutrient absorption, and microbial activity. Dietary choices directly affect the gut microbiota composition, impacting fecal characteristics such as consistency, color, and composition [3]. Dietary components, including macronutrients like fiber, fats, and proteins, influence digestion's digestive process and end products. For instance, high-fiber diets have been shown to increase stool bulk and promote regular bowel movements. Fibers resist digestion in the small intestine and are fermented by gut bacteria in the colon, contributing to the formation of short-chain fatty acids (SCFAs) that affect stool consistency and health outcomes [4]. In contrast, diets high in animal fats and low in fiber are associated with firmer, less frequent stools and a different microbial composition [5].

The role of the gut microbiota in determining fecal composition cannot be overstated. Gut bacteria metabolize undigested dietary components and produce metabolic by-products excreted in the feces. Differences in diet, particularly the intake of plant-based foods versus animal-based foods, result in varying fecal microbiota compositions. Studies have indicated that diets rich in plant-based foods, such as fruits and vegetables, are associated with a higher diversity of gut bacteria linked to healthier fecal profiles [6]. Conversely, diets high in processed and fatty foods can lead to dysbiosis or an imbalance in gut bacteria, resulting in altered fecal characteristics and an increased risk of gastrointestinal disorders [7].

The relationship between diet and fecal composition is crucial for digestive health and offers diagnostic potential. Variations in stool characteristics, such as undigested food particles, changes in color, or abnormal consistency, can indicate malabsorption or digestive disorders, often traced back to dietary habits. Thus, understanding the dietary impact on fecal composition is critical for health maintenance and disease prevention [3], [8]. Younger participants had more significant bacterial populations than older counterparts. Clostridium was the predominant bacterial group in these patients, followed by Bacteroides, Atopobium, Bifidobacterium, and Prevotella. The two tribes' gut microbiotas were comparable in this study. However, because of their food and way of life, Yogyakarta subjects had about ten times less *Clostridium perfringens* than Balinese subjects [9]. This study investigated how adult males' dietary practices affect their fecal microbiota composition.

2. Materials and Method

2.1 Subjects

This study was the initial stage of intervention research on oral administration of UHT milk and UHT milk containing probiotics to subjects on fecal microflora and SCFA, blood profile of total cholesterol, sugar and uric acid levels, and immune modulation. Therefore, the main requirement for selecting subjects was not suffering from lactose intolerance. For secondary data, subjects were asked to complete a questionnaire regarding their eating habits, frequency of eating, and types of food they consumed, including sources of carbohydrate, protein, oil, vegetables, and fruits, as well as functional foods containing LAB and Bifidobacterium spp. and probiotic consumption in the month before stool sampling. On day 0, before the intervention, stool samples were taken from 30 subjects. Stool samples were collected and stored in sterile bottles. Immediately after collection, stool samples were stored in a refrigerator at a temperature of 4-6°C. Subjects of male adults aged 18-21 who participated in this study were voluntary and had made and signed a written consent letter before the study was carried out. The research procedure was approved by the Faculty of Nursing Ethics Committee, Universitas Riau, Pekanbaru, Indonesia, with Ethical Approval number 1179/UN19.5.1.8/KEPK.FPp/2024.

2.2 Chemicals and Media
The chemicals used are Plate Count Agar (PCA) medium, Merck, Germany to calculate total plate count (TPC), deMan Rogosa Sharp Agar (MRSA), Merck, Germany to calculate total lactic acid bacteria (LAB), deMan Rogosa Sharp L-cysteine hydrochloride Agar (MRS-LCA), Merck, Germany to estimate the number of Bifidobacterium spp., and Eosin Methylene Blue Agar (EMBA), Himedia, India to enumerate the number of *E. coli*, Salmonella spp., and Shigella spp. The chemical used is sodium chloride.

2.3 Observation
The parameters of the fecal microflora profile tested include total plate count and the number of LAB, Bifidobacterium spp., *E. coli*, Salmonella spp., and Shigella spp.

2.4 Sterilization of equipment
Equipment such as test tubes, Petri dishes, Erlenmeyer, droppers, measuring cups, and beakers are washed, rinsed with detergent, and dried to avoid dust or dirt. The equipment is sterilized in an autoclave at a temperature of 121°C for 15 minutes with a pressure of 1 atm. The ose needle is sterilized by burning it over a Bunsen lamp until it glows.

2.5 Preparation of medium
The preparation of PCA media, MRSA, MRS-LCA, and EMBA refers to the instructions for making media according to the procedures specified by the manufacturer of the press concerned (Merck Indonesia). Furthermore, the media is put into a flask of glass and covered with aluminum foil. The medium is sterilized using an autoclave for 15 minutes at a temperature of 121°C with a pressure of 1 atm. The medium was poured into Petri dishes of approximately 25 ml each and allowed to solidify before being used to enumerate fecal microflora.

2.6 Analysis of the number of fecal microflora
The number of fecal microflora calculations refers to Sutrisna et al. [10] using the spread surface plate method. The number of microbes was analyzed after the medium was incubated for 24 hours at 37°C. The number of microflora was calculated by taking 0.1 ml of fecal sample and then putting it in 1 ml of 0.85% physiological saline for a dilution of 10^{-1} and continued until a dilution of 10^{-9}. Furthermore, 0.1 ml was taken starting from a dilution of 10^{-5} to 10^{-9} depending on the type of microflora counted and inoculated on PCA, MRS agar, MRS-LC agar, or EMB agar media by dripping the sample on a plate containing the agar media and then spreading it evenly over the entire surface of the medium with a sterile hockey stick. The inoculation process was carried out using laminar airflow. The inoculated Petri dishes were then incubated in an incubator for 24 hours to count *E. coli*, Salmonella spp., and Shigella spp., and 48 hours for TPC, LAB, and Bifidobacterium spp, at 37°C in an inverted state. The growing microbial colonies were counted using a colony counter. The number of each microflora was expressed in log CFU/g of feces and calculated using the formula:

$$\text{Number of microflora per g} = \text{Number of colonies} \times 1/\text{Dilution factor} \times 10 \quad (1)$$

2.7 Data analysis
Data on the number of microflora, including TPC, BAL, Bifidobacterium spp., *E. coli*, Salmonella spp., and Shigella spp., were analyzed statistically using the SPSS version 25 application with the Analysis of Variance (ANOVA) test. If Fcount ≥ Ftable, further testing was carried out using Duncan's New Multiple Range Test (DNMRT) at a level of 5%. The existence or absence of a relationship between the number of microflora and the subjects' eating habits was analyzed using bivariate analysis using the SPSS version 26.

3. Result and Discussion

3.1. Total plate count

Total plate microflora of fecal microflora refers to the enumeration of viable bacteria in a fecal sample. It is a microbiological method used to quantify the total number of live microorganisms in feces, providing insight into the overall bacterial load in the gut. The variance analysis results showed significant differences in TPC among subjects, as presented in Table 1.

Table 1. Fecal total plate count of adult male subjects

Fecal samples of subjects	Number of TPC (log CFU/g) ±SD
S1	12.58 ±0.08 ^{klmn}
S2	11.08 ± 0.13 ^{fg}
S3	11.50± 0.04 ^{h}
S4	11.35 ± 0.19 ^{gh}
S5	11.60± 0.14^{h}
S6	12.89 ± 0.17^{o}
S7	12.70 ± 0.06 ^{lmno}
S8	10.10 ± 0.01^{b}
S9	10.21 ± 0.11 ^{bc}
S10	12.14 ± 0.01^{ij}
S11	10.85 ± 0.11^{ef}
S12	12.11± 0.18^{i}
S13	11.34 ± 0.03^{gh}
S14	12.51 ± 0.02^{klm}
S15	12.54 ± 0.11^{klm}
S16	12.18 ± 0.17^{ij}
S17	9.63 ± 0.18^{a}
S18	12.79 ± 0.06^{mno}
S19	12.80 ± 0.04^{mno}
S20	12.86± 0.04^{no}
S21	12.42 ±0.05^{jkl}
S22	10.49 ± 0.23^{cd}
S23	10.21±0.17^{bc}
S24	12.78± 0.14^{mno}
S25	10.73 ± 0.09^{de}
S26	12.35 ±0.16^{ijk}
S27	12.05 ± 0.18^{i}
S28	12.17 ± 0.21^{ij}
S29	12.12 ± 0.19^{ij}
S30	11.62± 0.08^{h}

^{zA,b,c,d,e,f,g,h,i,j,k,m,n,o}Means in the same column followed by different superscript letters indicate significant ($p<0.05$)

The data in Table 1 shows that the amount of TPC between one subject and another varies. The TPC ranged from 9.36 log CFU/g in S17 to 12.89 log CFU/g in S6. The amount of TPC is most likely due to the type and variety of food consumed by the subjects. However, the multivariate test did not specifically indicate the type of food and frequency of food consumption by the subjects that affect the amount of TPC in each subject based on the secondary data obtained from the questionnaire. This fact illustrates that all food sources, especially carbohydrates, protein, fat, fruit and vegetables, and fermented milk products, are thought to contribute to the amount of TPC in the subject's feces. Fecal total plate count refers to the number of viable bacteria present in a sample, often used as an indicator of gut microbial load. Different types of diets can significantly influence the microbial composition and abundance in the gastrointestinal tract, which is reflected in the fecal total plate count [11]. The results of this study indicate that subjects with a higher number of LAB and Bifidobacterium spp. produce a higher number of TPC. High-fiber diets generally result in a higher fecal TPC due to increased bacterial growth. Fiber acts as a prebiotic, stimulating microbial fermentation and producing short-chain fatty acids (SCFA), further supporting gut bacteria proliferation. Diets high in animal protein and fat can shift the gut microbiota composition towards species more adept at metabolizing proteins and fats, e.g., Bacteroides [12].

3.2. Number of lactic acid bacteria and Bifidobacterium

LAB and Bifidobacterium spp. count in fecal microflora refers to the enumeration of live lactic acid-producing bacteria in a fecal sample. LAB and Bifidobacterium spp. are a group of beneficial bacteria that play a crucial role in maintaining gut health by fermenting carbohydrates into lactic acid, which helps maintain a slightly acidic environment in the gut and inhibits the growth of harmful pathogens [13][14]. The variance analysis results showed significant differences in the number of LAB and Bifidobacterium spp. among subjects, as presented in Table 2. The smallest number of LAB was found in S12 at 7.73 log CFU/g and the largest in S3 at 10.16 log CFU/g. At the same time, the smallest number of Bifidobacterium was obtained in S6 at 3.98 log CFU/g and the largest in S11 at 7.33 log CFU/g. The relationship between diet type and fecal LAB and Bifidobacterium spp. count is significant because these two genera are crucial in maintaining gut health, fermenting carbohydrates, and producing beneficial metabolites such as lactic acid and other organic acids. The type of diet can significantly influence the abundance of these bacteria in the gastrointestinal tract, which is reflected in their count in fecal samples. Multivariate test results indicated that the quantity of LAB and Bifidobacterium spp. was influenced by respondents' frequent consumption of high-fiber meals, such as fruits, vegetables, nuts, and fermented milk. LAB and Bifidobacterium spp. were detected in more significant quantities in these subjects than in those who ate these items infrequently.

Diets rich in dietary fibers (from fruits, vegetables, legumes, and whole grains) promote the growth of lactic acid bacteria. These bacteria ferment the non-digestible carbohydrates, mainly prebiotic, to produce lactic acid and other SCFA. A high-fiber diet tends to increase the fecal LAB and Bifidobacterium spp. count because fiber, especially prebiotics, selectively encourages the growth of LAB and Bifidobacterium spp. [15]. Specifically, Lactobacilli and Bifidobacterium spp. thrive in such environments, leading to higher fecal counts of these beneficial bacteria. Diets rich in fermented foods (e.g., yoghurt, kefir, sauerkraut, kimchi) introduce live LAB directly into the gut. These foods contain high Lactobacillus spp. and Bifidobacterium strains, which can colonize the gut or pass through transiently. Bifidobacterium spp. are a group of beneficial, Gram-positive, anaerobic bacteria that significantly maintain gut health. They are among the first colonizers of the human gut, particularly in breastfed infants. They are essential throughout life to support digestion, enhance immune function, and maintain a healthy gut microbiome [4].

Table 2. Fecal lactic acid bacteria and Bifidobacterium spp. of adult male subjects

Fecal samples of subjects	Number of LAB (log CFU/g) ± SD	Number of Bifidobacterium spp. (Log CFU/g) ± SD
S1	8.27 ± 0.04^{d}	7.11 ± 0.23^{c}
S2	8.14 ± 0.09^{bcd}	6.83 ± 0.19^{c}
S3	10.16 ± 0.08^{l}	4.94 ± 0.42^{ab}
S4	7.96 ± 0.16^{abc}	6.96 ± 0.28^{c}
S5	8.85 ± 0.04^{gh}	6.47 ± 0.33^{bc}
S6	8.63 ± 0.09^{efg}	3.98 ± 4.14^{a}
S7	9.79 ± 0.12^{k}	6.68 ± 0.05^{bc}
S8	8.62 ± 0.09^{efg}	6.72 ± 0.35^{bc}
S9	9.80 ± 0.04^{k}	7.42 ± 0.01^{c}
S10	8.62 ± 0.06^{efg}	5.67 ± 0.03^{bc}
S11	8.62 ± 0.06^{efg}	7.33 ± 0.01^{c}
S12	7.73 ± 0.12^{a}	6.88 ± 0.03^{c}
S13	8.86 ± 0.04^{gh}	7.21 ± 0.06^{c}
S14	9.11 ± 0.09^{hi}	6.79 ± 0.08^{c}
S15	9.54 ± 0.05^{jk}	5.55 ± 0.08^{abc}
S16	9.11 ± 0.08^{hi}	6.99 ± 0.01^{c}
S17	9.36 ± 0.04^{ij}	7.18 ± 0.02^{c}
S18	9.73 ± 0.16^{k}	6.97 ± 0.08^{c}
S19	7.86 ± 0.04^{ab}	7.17 ± 0.05^{c}
S20	8.35 ± 0.02^{de}	7.18 ± 0.01^{c}
S21	9.64 ± 0.06^{jk}	7.03 ± 0.13^{c}
S22	9.68 ± 0.01^{k}	6.81 ± 0.08^{c}
S23	7.89 ± 0.02^{ab}	6.96 ± 0.06^{c}
S24	8.68 ± 0.12^{fg}	6.73 ± 0.08^{bc}
S25	7.87 ± 0.47^{ab}	6.88 ± 0.01^{c}
S26	7.99 ± 0.11^{abc}	7.06 ± 0.08^{c}
S27	8.88 ± 0.06^{gh}	6.85 ± 0.08^{c}
S28	8.66 ± 0.02^{efg}	7.11 ± 0.04^{c}
S29	9.08 ± 0.06^{hi}	7.26 ± 0.05^{c}
S30	8.39 ± 0.42^{def}	5.82 ± 0.12^{bc}

^{a,b,c,d,e,f,g,h,i,j,k}Means in the same column followed by different superscript letters indicate significant ($p < 0.05$)

3.3. Number of *E. coli*

E. coli counts in fecal microflora refer to quantifying *E. coli* bacteria in a fecal sample. *E. coli* is a Gram-negative, rod-shaped bacterium commonly found in the intestines of humans and animals [16]. While most *E. coli* strains are harmless and are part of the normal gut flora, some strains can be pathogenic, causing gastrointestinal diseases. As shown in Table 3, the analysis of variance results indicates significant variations in the quantity of *E. coli* among the participants. The smallest number of *E. coli* was found in S5 at 7.73 log CFU/g, and the largest number in S3 at 9.67 log CFU/g. The relationship between diet type and fecal *E. coli* count is complex because *E. coli* exists as both commensal (beneficial or neutral) and pathogenic strains in the gut. While some strains of *E. coli* are essential for healthy digestion, others can lead to health issues if they overgrow or become pathogenic [8]. The multivariate test findings demonstrated that the quantity of *E. coli* was not influenced by a single food source type. The data indicates that the amount of *E. coli* in the subject's feces was influenced by different food sources. Diet plays a significant role in determining the abundance of *E. coli* in the gut, affecting the overall microbial environment and nutrient availability. A high-fiber diet rich in fruits, vegetables, legumes, and whole grains promotes the growth of beneficial bacteria, like Lactobacillus spp. and Bifidobacterium spp., while potentially suppressing pathogenic or opportunistic bacteria like *E. coli*. Fiber, particularly soluble fiber and prebiotics, undergo fermentation by beneficial gut bacteria, producing SCFA that can inhibit the growth of some harmful bacterial species [17]. Diets rich in animal proteins and low in fiber, such as ketogenic or carnivore diets, can alter the gut microbiota by promoting bacteria that thrive on proteins and fats, including certain strains of *E. coli*.

Table 3. Fecal *E. coli* of adult male subjects

Fecal samples of subjects	Number of *E. coli* (log CFU/g) ± SD
S1	9.23 ± 0.02^{fgh}
S2	8.31 ± 0.01^{b}
S3	9.67 ± 0.03^{ij}
S4	9.40 ± 0.01^{fghij}
S5	7.28 ± 0.19^{a}
S6	9.33 ± 0.04^{fghi}
S7	8.69 ± 0.12^{cd}
S8	9.52 ± 0.28^{ghij}
S9	9.21 ± 0.06^{fgh}
S10	9.45 ± 0.07^{fghij}
S11	9.51 ± 0.17^{ghij}
S12	8.75 ± 0.13^{cd}
S13	9.72 ± 0.04^{j}
S14	9.65 ± 0.01^{ij}
S15	7.39 ± 0.02^{a}
S16	7.46 ± 0.07^{a}
S17	9.20 ± 0.01^{fg}
S18	9.13 ± 0.02^{ef}
S19	8.70 ± 0.27^{cd}
S20	9.39 ± 0.12^{fghij}
S21	8.87 ± 0.04^{de}
S22	9.56 ± 0.13^{hij}
S23	9.25 ± 0.35^{fgh}
S24	9.26 ± 0.01^{fgh}
S25	8.54 ± 0.44^{bcd}
S26	8.67 ± 0.03^{cd}
S27	8.47 ± 0.01^{bc}
S28	8.86 ± 0.01^{de}
S29	9.29 ± 0.17^{fgh}
S30	8.85 ± 0.15^{de}

^{a,b,c,d,e,f,g,h,i,j}Means in the same column followed by different superscript letters indicate sigsnificant ($p<0.05$)

Protein fermentation in the gut produces by-products like ammonia, hydrogen sulfide, and amines, which can foster an environment favorable for the growth of *E. coli*. Diets high in simple sugars, commonly found in processed foods and sugary drinks, can fuel the growth of opportunistic bacteria, including some strains of *E. coli*. Sugars are readily fermentable and, in excess, can create conditions that promote bacterial overgrowth, including *E. coli*. Sugar-rich diets can lead to an increase in fecal *E. coli* counts. Excess sugars can disrupt gut bacteria's balance, reducing beneficial microbes' populations and creating an environment where *E. coli* can grow well [11].

3.4. Number of Salmonella and Shigella

Salmonella spp. and Shigella spp. count in fecal microflora refers to quantifying Salmonella and Shigella species in a fecal sample.

Table 4. Fecal Salmonella sp. and Shigella spp. of adult male subjects

Fecal samples of subjects	Number Salmonella and Shigella spp. (Log CFU/g) ± SD
S1	10.66 ± 0.07^{i}
S2	9.44 ± 0.042^{g}
S3	9.27 ± 0.13^{fg}
S4	10.21 ± 0.05^{h}
S5	7.43 ± 0.18^{c}
S6	10.67 ± 0.05^{i}
S7	9.18 ± 0.01^{efg}
S8	5.33 ± 0.04^{a}
S9	6.28 ± 0.15^{b}
S10	9.58 ± 0.27^{g}
S11	8.57 ± 0.01^{d}
S12	8.86 ± 0.09^{def}
S13	8.79 ± 0.15^{de}
S14	10.16 ± 0.13^{h}
S15	8.64 ± 0.01^{d}
S16	7.17 ± 0.56^{c}
S17	6.19 ± 0.11^{b}
S18	10.79 ± 0.04^{i}
S19	9.19 ± 0.11^{efg}
S20	10.67 ± 0.09^{i}
S21	8.40 ± 0.14^{d}
S22	9.58 ± 0.04^{g}
S23	5.29 ± 0.21^{a}
S24	10.23 ± 0.02^{h}
S25	10.77 ± 0.01^{i}
S26	8.77 ± 0.05^{de}
S27	9.54 ± 0.52^{g}
S28	7.09 ± 0.03^{c}
S29	7.17 ± 0.56^{c}
S30	8.59 ± 0.07^{d}

^{a,b,c,d,e,f,g,h,i}Means in the same column followed by different superscript letters indicate significant ($p<0.05$)

Both are pathogenic, Gram-negative bacteria that can cause severe gastrointestinal infections. Unlike commensal bacteria such as *E. coli*, Salmonella spp. and Shigella spp. are not part of the normal gut flora, and their presence in feces typically indicates an active infection. The analysis of variance suggests significant differences in the amount of Shigella spp. and Salmonella spp. present in each participant, as shown in Table 4. The number of Salmonella and Shigella ranged from 5.29 log CFU/g in S23 to 10.79 log CFU/g in S18. The relationship between diet type and fecal Salmonella spp. and Shigella spp. count is critical in understanding how dietary patterns affect the presence of these harmful, pathogenic bacteria in the gastrointestinal tract [8][5]. The multivariate test results showed that no single type of food source affected the number of Salmonella spp. and Shigella spp. The data indicates that all types of food play a role in Salmonella spp. and Shigella spp. growth in the digestive tract of subjects was detected in their feces. Salmonella spp. and Shigella spp. are well-known causes of foodborne illness, leading to infections like salmonellosis and shigellosis [18]. Their presence in fecal matter indicates active or past infection, and certain dietary factors can influence their colonization, growth, and survival in the gut. High-protein, low-fiber diets may increase the risk of Salmonella spp. and Shigella spp. infections, leading to elevated fecal counts of these pathogens [19][20]. When not completely digested, proteins can undergo putrefaction in the colon, producing by-products like ammonia and hydrogen sulfide, compromising gut barrier function and increasing the risk of pathogen overgrowth. Additionally, animal-based diets may increase exposure to contaminated food sources, raising the likelihood of infection. Diets high in simple sugars can disrupt the balance of the gut microbiota by promoting the growth of opportunistic pathogens, including Salmonella spp. and Shigella spp. Sugars provide an easy energy source for some harmful bacteria, which may encourage their proliferation [8][3].

4. Conclusion

These findings suggest significant individual differences in the quantity of TPC, LAB, Bifidobacterium spp., Salmonella spp., Shigella spp., and *E. coli* microflora. This investigation also showed that the amount of TPC, *E. coli*, Salmonella spp., and Shigella spp. was not correlated with any particular food category; however, the total amount of food was likely a contributing factor to the bacterial load. However, the subjects' diet, which included fruits, vegetables, and nuts, impacted the quantity of LAB and Bifidobacterium spp.

References

[1] A. F. Ahmad, X. G. Dwivedi, F. O. Gara, J. Caparros-martin, and X. N. C. Ward, "The gut microbiome and cardiovascular disease : current knowledge and clinical potential ATHEROSCLEROSIS," 2024, doi: 10.1152/ajpheart.00376.2019.

[2] R. Sender, S. Fuchs, and R. Milo, "Revised Estimates for the Number of Human and Bacteria Cells in the Body," pp. 1–14, 2016, doi: 10.1371/journal.pbio.1002533.

[3] A. M. Dinsmoor, M. Aguilar-lopez, N. A. Khan, and S. M. Donovan, "REVIEW A Systematic Review of Dietary Influences on Fecal Microbiota Composition and Function among Healthy Humans 1 – 20 Years of Age," pp. 1734–1750, 2021.

[4] J. Slavin, "Fiber and Prebiotics: Mechanisms and Health Benefits," pp. 1417–1435, 2013, doi: 10.3390/nu5041417.

[5] R. K. Singh *et al.*, "Influence of diet on the gut microbiome and implications for human health," *J Transl Med*, pp. 1–17, 2017, doi: 10.1186/s12967-017-1175-y.

[6] L. A. David *et al.*, "HHS Public Access," vol. 505, no. 7484, pp. 559–563, 2014, doi: 10.1038/nature12820.Diet.

[7] Z. Huang, K. Liu, W. Ma, D. Li, T. Mo, and Q. Liu, "The gut microbiome in human health and disease — Where are we and where are we going? A bibliometric analysis," no. December, pp. 1–15, 2022, doi: 10.3389/fmicb.2022.1018594.

[8] L. Wang, M. Chen, and C. Care, "A review of the interaction between diet composition and gut microbiota and its impact on associated disease," *Journal of Future Foods*, vol. 4, no. 3, pp. 221–232, 2024, doi: 10.1016/j.jfutfo.2023.07.004.

[9] E. S. Rahayu *et al.*, "Gut microbiota profile in healthy Indonesians," vol. 25, no. 12, pp. 1478–1491, 2019, doi: 10.3748/wjg.v25.i12.1478.

[10] R. Sutrisna, C. N. Ekowati, S. Farisi, and H. V. Setyawan, "Uji Viabilitas Bakteri Asam Laktat dari Usus yang Dipreparasi dalam Ransum Unggas," *Jurnal Ilmiah Peternakan Terpadu*, vol. 5, no. 3, pp. 53–57, 2017, doi: 10.23960/jipt.v5i3.1735.

[11] P. Zhang, "Influence of Foods and Nutrition on the Gut Microbiome and Implications for Intestinal Health," 2022.

[12] E. Rinninella *et al.*, "What is the Healthy Gut Microbiota Composition ? A Changing Ecosystem across Age , Environment , Diet , and Diseases," 2019, doi: 10.3390/microorganisms7010014.

[13] S. Salminen, A. Von Wright, and A. Ouwehand, *Lactic Acid Bacteria: Microbiology And Functional Aspects, 2nd edn.* New York: Marcel Dekker, Inc, 2004. doi: 10.1046/j.1365-2621.1998.33201914.x.

[14] A. Sarkar and S. Mandal, "Review article: Bifidobacteria- Insight into clinical outcomes and mechanisms of its probiotic action," *Microbiol Res*, 2016, doi: 10.1016/j.micres.2016.07.001.

[15] Y. Zheng, C. Qin, M. Wen, and L. Zhang, "The Effects of Food Nutrients and Bioactive Compounds on the Gut Microbiota : A Comprehensive Review," pp. 1–21, 2024.

[16] M. Ingerson and A. Reid, "E. coli: Good, bad, & deadly," *American Academy of Microbiology*, pp. 1–14, 2011.

[17] Y. Nie, F. Luo, and Q. Lin, "Dietary nutrition and gut microflora: A promising target for treating diseases," *Trends Food Sci Technol*, 2018, doi: 10.1016/j.tifs.2018.03.002.

[18] B. Lamichhane *et al.*, "Salmonellosis : An Overview of Epidemiology , Pathogenesis , and Innovative Approaches to Mitigate the Antimicrobial Resistant Infections," 2024.

[19] B. Senghor, C. Sokhna, R. Ruimy, and J. Lagier, "Gut microbiota diversity according to dietary habits and geographical provenance," *Hum Microb J*, 2018, doi: 10.1016/j.humic.2018.01.001.

[20] N. Tecce, R. Pivonello, D. Menafra, M. Progan, and A. Colao, "Human Nutrition & Metabolism Diet and gut microbiome : Impact of each factor and mutual interactions on prevention and treatment of type 1 , type 2 , and gestational diabetes mellitus," vol. 38, no. July, 2024, doi: 10.1016/j.hnm.2024.200286.

Household food security of women farming group members in city and village during the Covid-19 pandemic in Lampung Province

Yaktiworo Indriani*, Wuryaningsih Dwi Sayekti, Maya Riantini, Maharani Anggia Rizkyta, Rendi Setiawan

Agribussiness Department, Faculty of Agriculture, Lampung University, Bandar Lampung, Indonesia

*E-mail: yaktiworo.indriani@fp.unila.ac.id

Abstract. This research aims to analyse the food security level and the influencing factors on household (HH) food security level of the Women Farming Group (WFG) members in the City and Village in Lampung Province. The village of Sumber Agung, Kemiling District, Bandar Lampung City and The Village of Giri Tunggal, North Pagelaran District, Pringsewu Regency were chosen as the research location. There were 60 HHs of WFG member involved as research samples. Data collection was carried out in 2021 when the Covid-19 pandemic was still occurring. The FSL in this study was analysed by cross-classifying between the level of energy adequacy and the percentage of food expenditure. Factors affecting HHs' food security level were analysed using the logit loglinear analysis. The results showed that both in the city and in the village the majority of HHs (>50%) were categorized as food secure. Some HHs were less food secure (30%), vulnerable (10%) and food insecure (7%) in the city, while in the village there were HHs that are food vulnerable (37%). Factors that affect FSL were the quantity of HHs' members, the tempeh price, age of wife and location.

Keywords: Covid-19, food security level, HHs, pandemic, WFG

1. Introduction

As a basic need, food must be met by every human being for their survival. Food is a basic component in the realization of quality human resources, and even becomes the main pillar in national development to maintain economic, political, and social stability, so it can be said that food security is an important point for national development to form quality humans [1]. However, in reality, not everyone is able to meet their food needs because of some reasons. The pandemic of Coronavirus disease 2019 (Covid-19) that broke out in early 2020 not only caused public health problems, but also had an impact on all sub-sectors in Indonesia, including the agricultural sector. Policies aimed at preventing the Covid-19 pandemic from spreading further, such as Large-Scale Social Restrictions (LSSR), have also had a broad impact, including the disruption of food circulation, obstruction of the access to food of physical and economic, and others which ultimately disrupt individual, family, regional to national food security. Therefore, Therefore, the Agency of Food Security of the Agricultural Ministry [2] at that time prepared solution, one of which was to implement the Sustainable Food Yard (SFY) program. Activities of

SFY are carried out to support programs implemented by the government to address areas that are prioritized to receive stunting intervention and improvements to escape food insecurity or become food secure. The activities are carried out through the utilization of yards, sleeping land, or non-productive empty land, as food makers to meet household (HH) nutrition and food needs and can be sold in the market to increase HH income. The program of SFY was attended by woman farming group (WFG) members, in both cities and rural areas. Based on data from the Covid-19 Task Force, one City of the areas that had red zone status or high Covid-19 risk is Bandar Lampung. The status as a red zone required the city government to tighten health protocols. This had an impact on FSL in Bandar Lampung. One of WFG in Bandar Lampung City that had the opportunity to participate in this program was WFG of Mekar Agung which is located in Sumber Agung Village, Kemiling District. Whereas, WFG of Mekar Jaya in Pekon Giri Tunggal, Pagelaran District of Pringsewu Regency was also selected to receive the benefits of the SFY program in 2020. In addition to the SFY program, there must be other influencing factors on the achievement of HH food security level. Based on above description, this study aims to analyse HH food security level and the factors that influence HH food security level of WFG members in cities and villages in Lampung Province during the Covid-19 pandemic.

2. Methods

This study was conducted by survey method. The research location was chosen purposively in WFG of Mekar Agung, Sumber Agung Village, District of Kemiling, Bandar Lampung and in WFG of Mekar Jaya located in Pekon Giri Tunggal, Pagelaran District, Pringsewu Regency. Both are WFGs selected receiving SFY benefits in cities and villages in Lampung Province. Research data was collected from March to July 2021 when Covid-19 pandemic was still ongoing. The research samples were 60 households (HHs) consisting of all 30 HHs of Mekar Agung members in Bandar Lampung City and all 30 HHs of Mekar Jaya members in Pringsewu Regency.

The first objective is analysed by performing a classified cross-calculation between the percentage of food expenditure and the level of energy adequacy [3]. The percentage of food expenditure is the comparison between HH's food expenditure to its total expenditure. Household food consumption data was obtained by conducting a two days 24-hour food recall on non-consecutive days, calculating the energy content, then averaging it in kilocalories per day. The amount of energy content of each food ingredient can be determined by calculating the nutritional content of the food ingredient using the nutritional content calculation template based on the Indonesian Food Composition Table [4]. Furthermore, the energy adequacy level is calculated quantitatively, namely the percentage of real energy intake to the recommended energy adequacy figured in percent units. Measurement of the level of food security is carried out using cross-tabulation between the portion of food expenditure and the level of energy sufficiency as presented in Table 1.

The second objective is statistically analysed by logit regression to determine the effect of the number of HH members (X1), husband's education level (X2), housewife's education level (X3), cooking oil price (X4), eggs price (X5), tempeh price (X6), HH income (X7), mother's age (X8) and location (Dummy) on the food security level of WFG members in the city and in the village. Food security level is used as a dummy variable with D1 = 1 for food security in the category of food secure and food shortage, D1 = 0 for food security in the category of food vulnerable and food insecure. Location was the second dummy variable with D2 = 0 for city and D2 = 1 for village locations.

Table 1. Cross-classification of food security levels [3]

Energy adequacy level (EAL)	Share of food expenditure	
	Low (< 60% total expenditure)	High (≥ 60% total expenditure)
Enough (> 80% of EAL)	Food Secure	Food Vulnerable
Lack (≤ 80% of EAL)	Food shortages	Food insecurity

3. Result and Discussion

3.1 Respondent Characteristic

The respondents of this study were wives aged between 30-60 years (city) and 27-63 years (village), with the majority being in the 41-50 year age range (Figure 1).

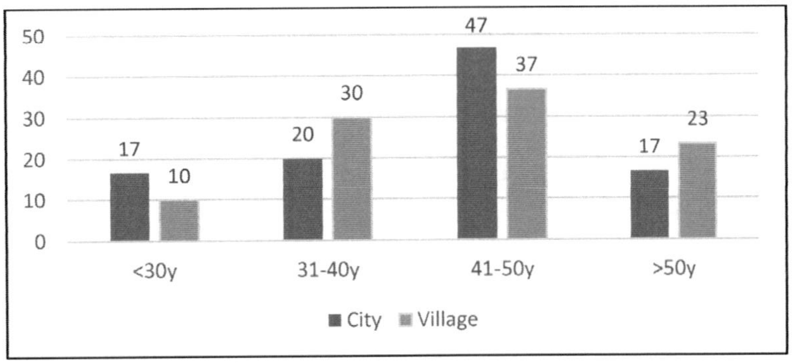

Figure 1. Wife's age distribution (%)

Most of the last education level of the husband and the wife were only up to Elementary School (SD) followed by Junior High School (SMP) in the village and Senior High School (SMA) in the city (Figure 2). This shows the lack of awareness of respondents and their husbands about the importance of education, due to various factors that influence moving up to a higher level, one of which is the low economic capacity of the HH.

Most HHs in both cities and villages, have 4-5 members. The majority of heads of HHs work as farmers and the majority of housewives do not work. The HH income per month in the city is in the range of IDR 1,500,000 - IDR 6,000,000 with an average of IDR 3,130,000.00; while in the village between IDR 1,620,000 - IDR 2,866,000 with an average of IDR 2,285,533.33. Based on the 2020 Central Statistics Agency's [5] prosperous family indicator, based on their consumption or expenditure, HHs are divided into high welfare levels (> IDR 5,000,000), medium welfare levels (IDR 1,000,000 - IDR 5,000,000), and low welfare (<IDR 1,000,000). Based on these criteria, all HHs in cities and villages are included in the medium welfare category. There is only one HH in the city that is classified as high welfare.

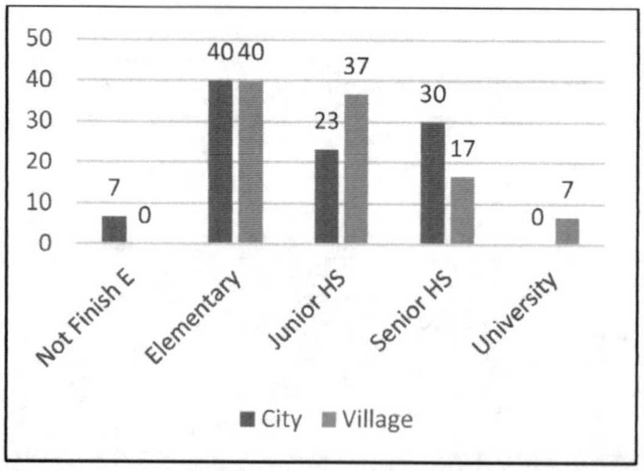

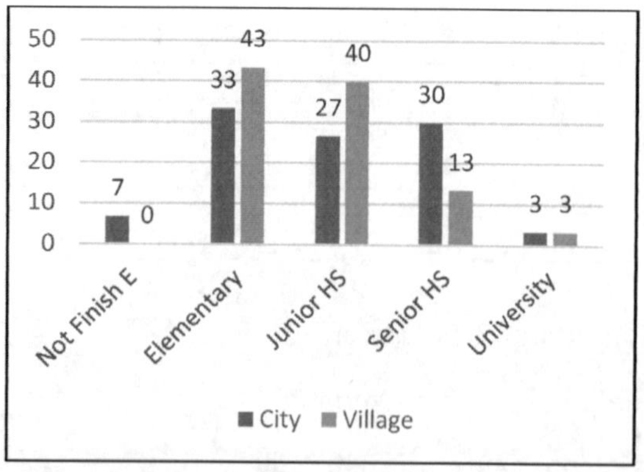

Figure 2. Distribution of husband's education (left) and wife's education (right) in %

3.2 Analysis of HH Food Security level

Food security level in this study is the result of calculating the percentage of food expenditure that was crossed-classification with energy sufficiency that was first introduced in 1991 [3]. The percentage of food expenditure is the comparison between food expenditure and total expenditure. Food expenditure consists of expenditure on rice, instant noodles, tubers, fish, chicken/beef, eggs, tofu, tempeh, vegetables, fruits, spices, beverage ingredients, processed food/drinks, cigarettes, wheat flour, cooking oil, and refilled water. The average HH's expenditure of food and non-food is presented in Table 2.

The largest food expenditure in HHs in both cities and villages is expenditure to buy rice. The average food expenditure for rice in WFG members in cities is 19.25 percent of total food expenditure, while in WFG members in villages it is 17.80 percent. The large expenditure to buy rice is because all HHs still rely on rice as their staple food, while carbohydrate sources other than rice such as tubers are only used as snacks so that rice consumption remains high. Other large HH food expenditures in urban in order are expenditures for animal food, cigarettes, vegetables and beverages. Meanwhile, in rural in order are cigarettes, vegetables, beverages and animal food. The average HH food expenditure for cigarettes in cities is 13.80 percent (third highest) of total food expenditure, while in villages is 12.94 percent (second highest). The high expenditure for cigarettes shows that HHs have not been able to allocate their expenditures for food better. Research conducted at corn farmers during Covid-19 in village of Blumbungan, District of Larangan, Pamekasan Regency also found almost the same result, that the largest food expenditure was for rice (34,28%), followed by animal food (18,07%) and cigarettes (16,57%) [6].

The average percentage of food expenditure for vegetable protein sources in urban is higher than in rural areas, but for vegetables it is lower. This shows that vegetable consumption in rural areas is higher than in urban areas, so that even though they have planted vegetables, they still have to buy vegetables, the percentage of which even exceeds that for animal food. On the other hand, those living in urban areas have higher percentage of expenditure on animal food compared to vegetables. Here, it implies that animal food consumption in urban areas is higher, while vegetable consumption is lower compared to those in rural areas. In addition to food expenditure, there are also non-food expenditures. Non-food expenditures consist of LPG gas, health/beauty hygiene, education, electricity and water, fuel, credit/quota, savings/arisan, taxes/insurance and cleaning materials.

Table 2. Average household food expenditure by location (Rp/month)

No	Types	City		Village	
		Rp/month	%	Rp/month	%
	A. Foods				
1	Rice	200,086.21	11.34	170,233.33	10.05
2	Tubers	9,862.07	0.56	45,900.00	2.71
3	Wheat Flour	12,724.14	0.72	14,600.00	0.86
4	Cooking Oil	70,241.38	3.98	66,000.00	3.90
5	Animal Foods	183,736.45	10.41	85,000.00	5.02
6	Tempeh	61,655.17	3.49	43,400.00	2.56
7	Tofu	59,206.90	3.35	19,866.67	1.17
8	Vegetables	91,724.14	5.20	118,466.67	6.99
9	Fruits	16,344.83	0.93	16,333.33	0.96
10	Spices	66,034.48	3.74	76,766.67	4.53
11	Beverages	86,598.52	4.91	89,400.00	5.28
12	fast foods/drinks	37,948.28	2.15	86,300.00	5.09
13	Cigarettes	143,482.76	8.13	123,666.67	7.30
	Total Foods	1,039,645.33	58.91	955,933.34	56.42
	B. Non-foods				
1	Health/cosmetics	22,844.83	1.29	48,700.00	2.87
2	Education	27,689.66	1.57	111,500.00	6.58
3	Electricity & water	92,620.69	5.25	64,700.00	3.82
4	Fuels	273,206.90	15.48	158,233.33	9.34
5	Quota	108,689.66	6.16	113,566.67	6.70
6	Saving	77,206.90	4.37	149,666.67	8.83
7	Tax/assurance	9,672.41	0.55	24,600.00	1.45
8	Cleanliness	113,275.86	6.42	67,433.33	3.98
	Total Non-food	725,206.91	41.09	738,400.00	43.58
	Total Expenditure	1,764,852.24	100	1,694,333	100.00

The largest non-food expenditure in both urban and rural areas is fuel expenditure, which consists of petroleum for transportation and gas for cooking. The average fuel expenditure of urban HHs is 37.6 percent of total non-food expenditures, while in rural areas it is 21.4 percent. The large expenditure on fuel, especially petroleum, is due to the fact that most HHs have more than one motorized vehicle. These motorized vehicles are usually used to go to work, pick up and drop off school children, and other activities. Expenditure on fuel is greater than others because most HHs work as farmers who have more than one vehicle, so fuel expenditure is relatively high. These results are similar to research in South Lampung and Tanggamus Regencies that expenditure on fuel was the second largest expenditure after education at the family of corn and coffee farmers [7][8]. Other large non-food expenditures in sequence are expenditures on clothing and hygiene, credit and quota, electricity and water, and savings and donations for those in the city, while those in the village are savings and donations, quota, education, and clothing and hygiene. The average HH expenditure on credit and quota in the city is 14.99 percent of the non-food expenditure total, while the average expenditure on credit and quota for HHs in the village is 15.3 percent. The large expenditure on credit and quota is influenced by the Covid-19 pandemic which requires online learning so that HHs need more quota/credit for their children to go to school online. After spending on credit and quota, the next large expenditure of non-food expenditure is electricity expenditure. The average electricity expenditure for HHs in the city is 12.7 percent of non-food expenditure total, while in the village it is 8.7 percent. Based on the food and non-food expenditure, the total expenditure of member HHs in the city and in the village can be seen in Table 2.

Households in the city, 58.9 percent of the total expenditure is spent on food, while the remaining 41.09 percent is for non-food. Whereas, HHs in the village, 56.42 percent of the total expenditure is spent on food, while the remaining 43.58 percent of the total expenditure is for non-food. The average percentage of food expenditure of HHs in the city is greater than that of in the village.

Tabel 3. Distribution of percentage food expenditure based on location

Percentage food expenditure	Category	City		Village	
		n (HH)	%	n (HH)	%
< 60%	Low	21	70,00	19	63,33
≥ 60%	High	9	30,00	11	36,67
Total		30	100,00	30	100,00

Overall, 66.67 percent of HHs fall into the low percentage of food expenditure criteria, while the remaining 33.33 percent fall into the high percentage of food expenditure criteria. The amount of food and non-food expenditure is influenced by the HH income itself. The low percentage of food expenditure indicates that HH income is relatively capable of meeting all needs. There are slightly more HHs in the city that fall into the low percentage food expenditure criteria compared to those in the village. This difference is because respondents in the city who are 50 years old and over with few family members are more numerous than those in the village. Older age and fewer members result in less food expenditure. Therefore, the low percentage of food expenditure of HH in the city is relatively greater than those in the village.

The energy sufficiency level is categorized into sufficient and insufficient. The energy sufficiency level is categorised to be enough if the value reaches more than 80 percent, while it is categorized to be insufficient if the value is 80 percent or less than 80 percent. Data in Table 4 shows that energy adequacy level of the HH in the village is all classified as sufficient, but in the city, there are 16.67 percent who are lacking. Overall, categorization of the HH energy adequacy level is showed in Table 4.

Table 4. Household energy adequacy levels based on location

Energy Adequacy Level (EAL)	Category	City		Village	
		n (HH)	%	n (HH)	%
> 80%	adequate	19	83,33	30	100,00
≤ 80%	Less	11	16,67	0	0,00
Total		30	100,00	30	100,00

Food security level of the HH during Covid-19 pandemic is presented in Table 5. Overall, the most HHs in cities and villages are categorized as food secure. The data in Table 5 shows that there are more HHs in villages that are categorized as food secure compared to those in cities. In the food insecure category, there are HHs in cities, namely 6.67 percent, while in villages there are none that are food insecure. These results show that urban HHs are more affected by the pandemic of Covid-19, as indicated by greater number of HHs that are food insecure. In overall, the HH food security level in villages is better than that of in cities. Based on these conditions, it can be said that HHs in cities feel the negative impact of the Covid-19 pandemic on HH food security level more than those in villages. One of the reasons is that as a farmer, the head of a HH in rural areas continues to work on his farm as usual and still gets agricultural products to support his family's food security. As for those in urban areas, almost all of their food must be purchased, but transportation access to markets or stalls is limited during the Covid-19 pandemic.

Table 5. Household food security level based on location

Category	City		Village	
	n (HH)	%	n (HH)	%
Food Insecure	2	6,67	0	0,00
Food Vulnerable	3	10,00	11	36,67
Food shortages	9	30,00	0	0,00
Food Secure	16	53,33	19	63,33
Total	30	100,00	30	100,00

In a study also conducted in the recipient area of the sustainable food yard program in Sleman Regency, DIY showed difference result. There was found that the majority of HHs were in the category of food shortages (75%) of the total sample, followed by food security (20.45%) and the rest were food insecure (4.55%). The majority of HHs in the category of food shortages, namely a low percentage of food expenditure and a low energy consumption level, meaning that HHs are able to use income for a smaller proportion of food expenditure, but HH food consumption has not implemented food selection according to the type and amount needed by the body so that the level of energy consumption is still lacking. This can be caused by a lack of knowledge about food and nutrition, especially the selection of food consumed according to the type and amount so that the food consumed is not yet able to meet the need for sufficient energy so that energy sufficiency is lacking [9].

3.3 Influencing Factors of HH Food Security Level

Influencing factors of HH food security level in this research are analysed by statistical tests using logit regression with factors of number of HH (X1), education level of husband (X2), education level of wife (X3), price of cooking oil (X4), price of eggs (X5), price of tempeh (X6), HH income (X7), age of the mother (X8) and location of the HH (Dummy) of Mekar Agung members in the city and Mekar Jaya in the village. The results of the logit regression for influencing factors of HH food security level is shown in Table 6.

Table 6. Influencing factors of household food security level

Variables	Unstandardized Coefficients		Standardized Coefficients		
	B	Std. Error	Beta	t	Sig.
(Constant)	2.244	1.155		1.943	0.058
HH members (X1)**	0.163	0.064	0.340	2.549	0.014
Husband Education (X2)	0.016	0.035	0.103	0.472	0.639
Wife Education (X3)	-0.041	0.035	-0.238	-1.169	0.248
Cooking Oil Price (X4)	-3.991E-5	0.000	-0.097	-0.802	0.427
Egg's Price (X5)	-4.549E-5	0.000	-0.199	-1.070	0.290
Tempeh's Price (X6)*	-5.304E-5	0.000	-0.612	-1.897	0.064
Household Income (X7)	9.270E-8	0.000	0.169	1.087	0.282
Wife's Age (X8)*	-0.016	0.008	-0.304	-2.107	0.040
Location (Dummy)*	0.594	0.314	0.616	1.890	0.065

Overall, the logit regression analysis shows that there are four variables that have a significant effect on the HH food security level, namely the number of HH members and wife's age which are significant at $P < 0.05$; whereas, the price of tempeh, and the location which are significant at $P < 0.10$. The number of HH members (X1) and the location have significant positive effects on HH food security level; while the price of tempeh and the wife's age have negative effects. That the number of HH members (X1) has a positive effect on the level of HH food security can be interpreted that the more HH members, the greater the opportunity for the HH to achieve secure of food security. This is similar to the research that states the more family members, the greater the percentage of food expenditure because the needs of food will be more diverse. This is because of each HH member has different tastes [10]. However, other researches [11][12] stated that the number of HH members had negative effect on the food security level of farmer HHs. The variable of the age of wife (X8) has negative effect on the HH food security level, it means that the older wife, the lower food security level. This is not the same with the finding result study that the age of wife has positive effect on HH food security level [13]. The variable of the price of tempeh (X6) also has negative effect on the HH food security level, it means that the higher price of tempeh, the smaller opportunity for HHs to achieve secure food security. Tempeh is vegetable protein source that is preferred by HHs compared to tofu. The HH expenditure on tempeh is higher than tofu. The increase price of tempeh will lower HH food security level. Meanwhile, location has positive effect on the HH food security level. Those who live in villages are more food secure. The variables of the level of husband education (X2) and wife education (X3) do not effect on the level of HH food security level because the level of confidence is below 90 percent. The level of husband education mostly only up to elementary school (SD). Usually, HH decision-maker on food expenditure and determining the menu are entirely regulated by housewife. The knowledge of husband and wife in managing income to achieve food security does not only come from formal education.

The variables of oil price (X4) and chicken egg price (X5) do not affect HH food security level. This means that changes in the price of oil and eggs do not cause HHs to change their spending on these two commodities. This is because most of the food menus consumed by HHs at the research location use a lot of cooking oil and the most easily obtained animal side dishes at affordable prices are chicken eggs. Other researches also find that food security level of the HH was not affected by housewife education level [9, 14], age of wife [15] and oil price [14].

4. Conclusion

The majority of household of WFG members (>50%) are categorized as food secure. There are households that are less food secure (30%), vulnerable (10%) and food insecure (7%) in the city, while in the village there are households at the food vulnerable level (37%). Factors that significantly affected food security level of household are the number of household members, age of wife, price of tempeh and location of WFG (City/Village).

References

[1] Indriani, Y. 2019. *Gizi dan Pangan*. Aura Publishing. Bandar Lampung.

[2] BKP [Badan Ketahanan Pangan]. 2020. *Petunjuk Teknis Bantuan Pemerintah Kegiatan Pekarangan Pangan Lestari (P2L) Tahun 2020.* Badan Ketahanan Pangan Kementrian Pertanian. Jakarta.

[3] Jonsson, U., and D. Toole. 1991. *Household food security and nutrition: A conceptual analysis.* New York: United Nations Children's Fund.

[4] Direktorat Jenderal Kesehatan Masyarakat Direktorat Gizi Masyarakat. 2018. *Tabel Komposisi Pangan Indonesia.* Kementerian Kesehatan RI. 2018. Jakarta.

[5] Badan Pusat Statistik. 2020. *Indikator Kesejahteraan Rakyat 2020*. Badan Pusat Statistik. Jakarta

[6] Ningsih, K., Sustiyana, S. 2022. Analisis Ketahanan Pangan Rumah Tangga Petani Pada Masa Pandemi Covid-19 Melalui Berusahatani Dari Rumah (Farm From Home). *SEPA: Jurnal Sosial Ekonomi Pertanian dan Agribisnis,* Vol 19(1):114-26. https://jurnal.uns.ac.id/ sepa/article/view/55830

[7] Adelina, F.E., Indriani, Y., Adawiyah, R. 2019. Pendapatan dan ketahanan pangan rumah tangga petani lahan sawah tadah hujan pola tanam padi dan jagung di kecamatan Kalianda Kabupaten Lampung Selatan. *Jurnal Ilmu-Ilmu Agribisnis*, Vol 7 (4): 575-585. *https://jurnal.fp.unila.ac.id/index.php/JIA/article/view/3875/2843*

[8] Anggitasari, E.D., Indriani, Y., Prasmatiwi, F.E. 2021. Ketahanan pangan rumahtangga petani kopi di Kecamatan Ulu Belu Kabupaten Tanggamus. *Jurnal Ilmu Agribisnis: Journal of Agribusiness Science*, Vol 9 (3): 531-538. *https://jurnal.fp.unila.ac.id/index.php/JIA/ article/view/5349/pdf*

[9] Senjawati ND, Azizah AF. 2024. Analisis Ketahanan Pangan Rumah Tangga pada Program Pekarangan Pangan Lestari. *Jurnal Sosial Ekonomi Pertanian*, Vol. 20 (1): 93-102. https://doi.org/10.20956/jsep.v20i1.30975

[10] Rahmi RD, Suratiyah K, dan Mulyo JH. 2013. Ketahanan Rumah Tangga Petani di Kecamatan Ponjong Kabupaten Gunungkidul. *Agro Ekonomi.* Vol 24 (2): 190-201. *http://etd.repository.ugm.ac.id/index.php?mod=penelitian_detail&sub=PenelitianDetail&act=view&typ=html&buku_id=65155*

[11] Hernanda, E.N.P., Indriani, Y., Kalsum, U. 2017. Income and Food Security of Rice Farmer Households in Food Insecurity Village. *Jurnal Ilmu Agribisnis (JIIA),* Vol 5 (3): *283-291.* https://jurnal.fp.unila.ac.id/index.php/JIA/article/view/1641/1467

[12] Sari, D. Indriani, Y. and Hasanudin, T. 2022. Ketahanan Pangan Rumah Tangga Petani Padi, Petani Jagung dan Nonpetani Di Kota Metro Provinsi Lampung. *Jurnal Ilmu Agribisnis: Journal of Agribusiness Science*, 10 (1). pp. 164-171. https://jurnal.fp.unila.ac.id/index.php/JIA/article/view/5692/pdf

[13] Nurdiani, Ulfah, & Widjojoko, T. 2016. F aktor-faktor yang Mempengaruhi Ketahanan Pangan Rumah Tangga Miskin di Wilayah Perkotaan Kabupaten Banyumas. *Jurnal Agrin,* 20 (2), 169-178. https://jurnalagrin.net/index.php/agrin/article/view/324/248

[14] Saputro, W.A., Fidayani, Y. 2020. Faktor-Faktor Yang Mempengaruhi Ketahanan Pangan Rumah Tangga Petani Di Kabupaten Klaten. *JURNAL AGRICA* , Vol. 13 (2):115-123. https://ojs.uma.ac.id/index.php/agrica/article/view/4078

[15] Sumekar, W. and Legowo, A.M. 2023. Analysis of factors affecting food security of fisherman's household in Tayu district, Pati Regency. In *AIP Conference Proceedings* (Vol. 2586, No. 1). AIP Publishing.

IOP Conf. Series: Earth and Environmental Science **1460** (2025) 012055 doi:10.1088/1755-1315/1460/1/012055

Critical safety points in handling fresh fruits and vegetables throughout the supply chain

Dhany Hermansyah[1*], Machfud[2], Muhammad Romli[2] and Muslich[2]

[1] Postgraduate Programme Agro-Industrial Engineering, IPB University, Bogor, Indonesia
[2] Department of Agro-Industrial Technology, IPB University, Bogor, Indonesia

*E-mail: bangdhanyhermansyah@apps.ipb.ac.id

Abstract. Fruits and vegetables, categorized fresh plant-derived food products, were highly susceptible to physical, chemical, and biological deterioration and contamination throughout their supply chain. While previous research has highlighted risks at various stages of the supply chain, there needs to be more in-depth analysis of how these risks interconnect and which should be prioritized for remediation based on a comprehensive criticality assessment. This study aims to identify critical safety points in the supply chain of fresh fruit and vegetable products using the Failure Mode Effect and Criticality Analysis (FMECA) method. FMECA was proposed as a comprehensive risk assessment tool to systematically analyze potential failure modes and their causes, impacts, and criticality levels in food-handling processes. The study results indicated that critical food safety points for fruits and vegetables occurred during postharvest handling during the cultivation stage at the farmer level. Noncompliance with hygienic sanitation requirement during postharvest handling is a primary risk factor for the safety of fresh fruits and vegetables. This noncompliance significantly elevates the risk of foodborne illnesses, undermining consumer trust and public health. Additionally, the improper use of pesticides is identified as the second most severe critical point and is categorized as medium-high risk. Strict supervision, implementation of good sanitation practices, and proper environmental control are essential measures that must be taken to prevent contamination and maintain product quality throughout the supply chain.

1. Introduction

Ensuring food safety for consumers has always been a significant concern in the food industry, particularly among policymakers and researchers worldwide. Food safety risks to human health can arise from chemical, biological, or physical hazards [1]. Fruits and vegetables are perishable agricultural products prone to physical, chemical, and biological contamination. At every stage of the supply chain, fresh fruits and vegetables were at risk of contamination, rendering them unsafe for consumption.

Fresh fruits and vegetables were contaminated at different points along the production chain. Generally, sources of contamination are categorized into pre-harvest and postharvest sources [2]. Pre-harvest contamination can result from soil, irrigation water, improper use of fertilizers and pesticides, and pests or nuisance animals. Meanwhile, postharvest contamination can come from

IOP Conf. Series: Earth and Environmental Science **1460** (2025) 012055 doi:10.1088/1755-1315/1460/1/012055

harvesting equipment, handling personnel, insect or animal pests, water used for washing, transportation, and packaging, and processing equipment [3].

Vegetables were the most frequently involved fresh food products in food safety incidents, with outbreak frequencies of 34.1% in Europe and 47.4% in North America [4]. Fresh food products like vegetables and fruits are the primary causes of foodborne illness outbreaks due to contamination by pathogenic microbes such as *Salmonella* sp., *L. monocytogenes*, and *E. coli* [5].

Aspects such as food ingredients origin, quality, and safety are critical parameters considered by business actors and the government as the supervisory authority. Therefore, various policies and standards regarding quality management and food safety through traceability systems have been implemented in several countries worldwide [6]. Traceability refers to obtaining recorded information on food ingredients and additives throughout the production chain. A traceability system enables monitoring of the supply chain to ensure the safety of the produced products [7].

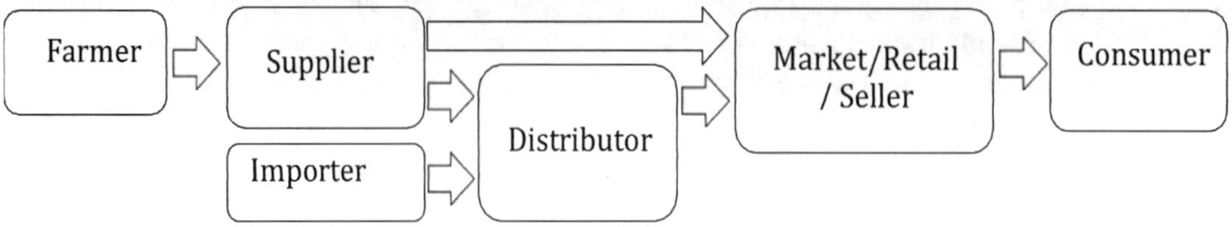

Figure 1. The Supply Chain of Fruits and Vegetables Products.

The supply chain is not limited to factories and suppliers; it also involves examining the conditions of distributors, warehouses, retailers, and consumer needs [8]. At each stage of the supply chain, fresh food products are susceptible to contamination, potentially rendering them unsafe for consumption [9]. Generally, the handling process for fruits and vegetables must meet hygienic sanitation requirements, covering the facilities, equipment used, and personnel handling the products. Providing safe and quality food to the public requires a joint commitment between the government and business actors [10].

As an agricultural country in the tropics, Indonesia has great potential for producing fruits and vegetables and serves as a strategic market for these commodities [11]. Despite the recognized importance of food safety, research on the effectiveness of traceability systems in Indonesia's context, particularly for fresh fruits and vegetables, still needs to be completed. Current food safety management practices often fail to dynamically assess and adapt to emerging risks throughout the supply chain, particularly in Indonesia's diverse agricultural practices and environments.

The results of monitoring fresh food in circulation carried out by competent food safety authorities at the central and regional levels throughout Indonesia showed that microbiology contamination was the contamination parameter that most often does not meet the requirements.

Table 1. Result of post-market food safety inspection in Indonesia*.

Parameter	Number of samples	Complies	Not Complies
Pesticide Residue	10388	9380 (90.3%)	1008 (9.7%)
Metal	4936	4528 (91.7%)	408 (8.3%)
Microbiology	903	687 (76.1%)	216 (23.9%)
Mycotoxin	3232	2713 (83.9%)	519 (16.1%)
Total	19459	17308 (88.9%)	2151 (11.1%)

*Source: National Food Agency, Republic of Indonesia in 2023.

While effective to a degree, existing traceability systems often need more granularity to effectively identify and mitigate risks at each stage of the supply chain. Given the identified gaps in understanding the dynamic nature of food safety risks within Indonesia's fruits and vegetables supply chain, this study aims to leverage the Failure Mode Effect and Critical Analysis (FMECA) tool not only to identify critical points of contamination but also to provide a framework for ongoing risk assessment and adaptation. Unlike traditional risk assessment methodologies that often provide a static snapshot of potential hazards, FMECA offers a dynamic framework for ongoing assessment, allowing for identifying emerging risks and adapting control measures [12].

2. Methodology

Food safety risks can emerge at various stages within the supply chain, making it essential to identify critical points where safety measures should be reinforced. This study utilizes Failure Modes, Effect, and Criticality Analysis (FMECA), an extension of Failure Mode and Effect Analysis (FMEA), to systematically assess the risks involved in the handling of fresh fruit and vegetable products [13]. FMECA is a method that combines two separate analyses, namely FMEA and Critical Analysis (CA), using a criticality matrix. FMECA is used to identify factors that cause damage/failure in a production process and identify preventive measures to avoid such damage/failure [14].

2.1 Scope of Study

This study focuses on fresh fruit and vegetable supply chains, explicitly examining imported apples and locally sourced lettuce leaves. Apples are fruits typically consumed raw without undergoing processing, thus requiring effective post-harvest handling to minimize contamination by pathogenic microbes [15]. Additionally, according to Indonesia's Agricultural Statistics 2022, apples are among the fruits imported by Indonesia in substantial volumes compared to other fruits, totalling 190,754 tons. Apples are a popular fruit commodity in several countries and have been associated with outbreaks of microbial contamination, such as the cases reported in the United States [16]. Lettuce, another fresh produce item, has also been implicated in various foodborne illness outbreaks. Contamination of lettuce leaves can occur at various points, including production, harvesting, distribution, storage, packaging, and household handling [17]. In Indonesia, lettuce is typically consumed raw, either as a side dish (*lalapan*) or in salads.

The selection of imported apples and locally sourced lettuce in this study was based on their differing supply chain characteristics, enabling a comparative risk analysis of food safety in these two contexts. Apples, as an imported product, involve a long and complex supply chain, including storage, transportation, and repackaging processes, which may increase the potential for both

IOP Conf. Series: Earth and Environmental Science **1460** (2025) 012055 doi:10.1088/1755-1315/1460/1/012055

chemical and biological contamination. In contrast, locally sourced lettuce represents a highly perishable product that is susceptible to microbiological contamination due to improper cultivation and post-harvest handling practices.

Table 2. Fruits and vegetables handling process

Business Actor	Handling Process	Contamination
Farmers	Cultivation Harvesting Washing Sortation/grading Storage	Chemical, microbiology, physical
Supplier/importer	Receiving and sortation Packaging Storage Delivery	Chemical, microbiology, physical
Distributor	Receiving and sortation Storage Delivery	Chemical, microbiology, physical
Retail/seller	Receiving and sortation Sell/display Storage	Microbiology, physical

The research was conducted across key locations in Jakarta, West Java, and Banten, Indonesia, encompassing farms, importers, and both traditional and modern markets. For apples, the study follows the supply chain from importers to retailers, while for lettuce, the chain is mapped from cultivation through postharvest handling to market distribution and retailers.

Each stage in these supply chains presents potential safety hazards identified through observations and structured interviews with key stakeholders. Based on initial observations at the research site and interviews with several stakeholders, the general stages of fruit and vegetable product handling as well as their potential contaminants were identified, as presented in Table 2. The maximum allowable contamination levels comply with applicable regulations, such as Minister of Agriculture Regulation No. 53 of 2018, SNI 7338:2009, and Codex Alimentarius standards.

2.2 FMECA

FMECA was applied to each stage of the fruit and vegetable handling process to identify potential failure modes that could systematically jeopardize product safety. Failure modes in this context include biological, chemical, or physical contamination and quality degradation resulting from improper handling. The FMECA process follows three main steps:

2.2.1 Identification of Failure Modes

Potential failure modes were identified at each stage of the supply chain based on observational data, stakeholder interviews, and historical records. Failure modes were defined as any process deviation that could lead to safety hazards, including contamination or spoilage.

2.2.2 *Assessment of Severity, Occurrence, and Detectability*

For each identified failure mode, the severity (S), occurrence (O), and detectability (D) were evaluated using predefined criteria. Severity measures the potential impact on food safety and public health, while occurrence evaluates the likelihood of the failure happening based on frequency data. Detectability refers to the likelihood of detecting the failure before it impacts the end consumer. Each parameter was rated on a scale of 1 to 10, as shown in Table 3.

2.2.3 *Calculation of Risk Priority Number (RPN)*

The RPN was calculated for each failure mode by multiplying the ratings for severity, occurrence, and detectability ($RPN = S \times O \times D$). This metric allows for prioritizing failure modes based on their potential risk. Higher RPN values indicate critical points where immediate corrective actions are necessary. Data for the FMECA analysis were collected through qualitative and quantitative methods, including field observations and interviews with food safety officers and supply chain actors. A total of 50 respondents participated in this study, consisting of stakeholders from the lettuce supply chain, including representatives from two lettuce farmer groups (2 individuals), distributors or collectors (2 individuals), regional food safety inspectors (10 individuals), representatives from two modern retail companies (2 individuals), and traditional market vendors (10 individuals). Meanwhile, the respondents for the apple supply chain included representatives from two apple importing companies (4 individuals), central food safety inspectors (6 individuals), a distributor company (2 individuals), representatives from two modern retail companies (2 individuals), and traditional market vendors (10 individuals).

Respondents were selected based on their experience in the fruit and vegetable supply chains and their direct involvement in product management at various stages. The data collected included insights into potential food safety risks, operational procedures, and compliance levels with good food safety handling practices. The assessment had been complemented by audits and conformity checks against established food safety standards.

Table 3. Rating scale of severity, occurrence, and detectability for calculating RPN

Rated aspect	Score	Description
Severity (S) (Regulation of MoA No.15/2021)	1,2	No failures/deviations in the process
	3,4	Failures/deviations that, if not corrected or allowed to continue to occur, will have the potential to affect food safety
	5,6	Failures/deviations will potentially affect food safety.
	7,8	Failures/deviations can directly compromise food safety.
	9,10	Failures/deviations will disrupt food safety and human health
Occurrence (O) [18]	1	Never been failure
	2,3	Small chance of failure; very few occur (once a year)
	4,5,6	The probability of failure is relatively high (once a month or once in three months)
	7,8	There is a high probability of failure, a high number of failures (once a week or once every two weeks)
	9,10	It often happens, the number of failures is very high (it happens every day)
Detectability (D) [18]	1	Easy to detect
	2	High chance of being detected
	3	Low chance of detection
	4	Difficult to detect

5	Very difficult to detect

The table provides detailed criteria for scoring failure modes, enabling a consistent approach to evaluating risk across different supply chain stages. These scores guide the assignment of priorities for risk mitigation measures.

3. Results and Discussion

Failure modes in handling fresh fruit and vegetable products were identified through direct field surveys, observations, and interviews with industry stakeholders. In this study, the primary failure modes identified were risks to food safety and product damage. Theoretically, potential food safety risks and product damage can occur at any stage of the supply chain [19]. Observational results revealed several failure modes with the potential to become critical control points in food handling processes across different supply chain actors. The supply chain flow for lettuce and imported apples, along with their respective handling processes, is illustrated in Figure 2.

The identification results showed 30 failure modes from 6 stages of handling fruits and vegetables. The calculation of the RPN (Risk Priority Number) for each failure mode across product handling stages, derived from the multiplication of severity, occurrence, and detectability scores. The S, O, and D values were assessed by business actors, particularly those in quality control roles, and verified by food safety inspectors. The results of the failure mode identification and its RPN value are presented in Table 4.

Figure 2. Supply chain flow of lettuce and imported apple.

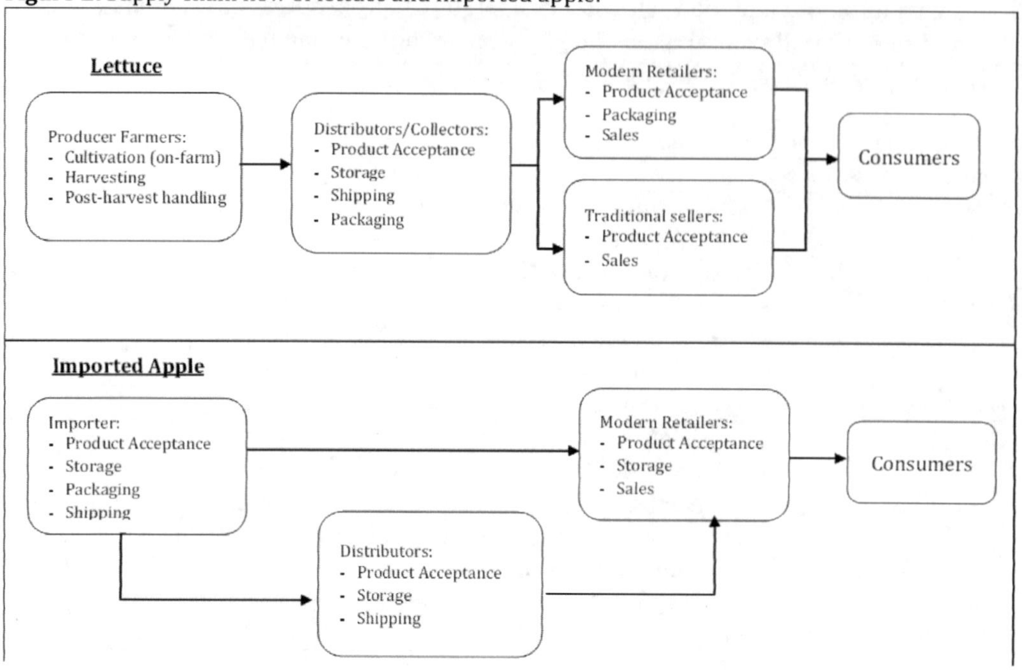

Table 4. Failure modes in the fruits and vegetables handling process

Process stages	ID Number	Failure modes	S	O	D	RPN
Cultivation on farm	1.1	Planting media contains dangerous and toxic ingredients	5	3	3	45
	1.2	Water irrigation contains hazardous and toxic waste	7	5	2	70
	1.3	Using fertilizer does not comply with the recommendation	7	6	3	108
	1.4	Using pesticides does not comply with the recommendation	9	8	2	144
	1.5	Postharvest handling does not comply with the hygiene sanitation principle	9	9	2	162
Product Acceptance	2.1	Defect products/ rotten/ do not meet requirements	9	4	2	72
	2.2	The product falls to the floor	7	4	2	56
	2.3	No recording of products received	5	5	3	75
	2.4	Cross-contamination occurs	7	4	4	112
	2.5	Dropping process takes too long time	4	5	2	40
Storage	3.1	The storage room floor does not use pallets	8	5	2	80
	3.2	Unsuitable storage temperature	8	3	3	72
	3.3	Dirty storage area	6	5	2	60
	3.4	Limit product stack	4	3	4	48
	3.5	Not implement the fi-fo mechanism (first in - first out)	4	5	4	80
	3.6	Overcapacity storage	7	6	2	84
Packaging	4.1	Product fall scattered	6	3	2	36
	4.2	Unhygienic packaging room and equipment	8	5	2	80
	4.3	Broken packaging	7	4	2	56
	4.4	Uncomplying label	7	4	2	56
	4.5	Unhygienic personnel	9	6	2	108
Delivery/ shipping	5.1	Unsuitable temperature in the refrigerated car	8	4	2	64
	5.2	Contamination occurs during shipping	8	5	3	120
	5.3	Too long-time delivery	7	5	2	70
	5.4	Overcapacity	6	5	3	90
	5.5	Unneat product pile in container	7	5	3	105
	5.6	Dirty/unhygienic container	8	5	2	80
Product sales	6.1	Unhygienic sales places	8	6	2	96
	6.2	Unsuitable product placement that potentially contaminated	7	5	2	70
	6.3	Unsuitable temperature at the product place sales	8	5	3	120

3.1 RPN analysis

Table 4 revealed critical failure modes with the highest Risk Priority Numbers (RPN) occurring during the cultivation stage. Specifically, "Postharvest handling does not comply with the hygiene sanitation principle" (RPN: 162) and "Using pesticides does not comply with the recommendation" (RPN: 144) are identified as the most significant risks. These failure modes underscore the challenges faced by local farmers in adhering to proper sanitation and pesticide application practices. This finding is consistent with research which emphasizes that inadequate sanitation during postharvest handling is a primary source of microbiological contamination, including *E. coli* and *Salmonella* [20]. Locally sourced lettuce is particularly vulnerable due to the lack of infrastructure and training among farmers. Improper cleaning of harvesting equipment, unclean washing water, and poor worker hygiene are the dominant contributors to this failure mode.

Moreover, in the product sales stage, differences in temperature control are apparent. The RPN for unsuitable temperature at product display for both commodities is high (RPN: 120), but imported apples generally experience better temperature management due to the use of refrigerated storage along the supply chain. Conversely, lettuce, often handled in open and uncontrolled environments, shows a higher tendency for microbial growth and physical deterioration, as corroborated by previous studies [21].

These RPN values were then sorted from highest to lowest using a Pareto chart, a tool designed to identify and visualize the most critical factors in a given problem, process, or system. This chart provides a clear visual distribution of issues, aiding data-driven decision-making and strategy development for improvements [22].

After sorting with the Pareto chart in Figure 3, it was found that the failure mode with the highest RPN (162) was associated with non-compliance with hygiene and sanitation principles during harvesting and postharvest handling. The progressively lower values appear to the right, with the lowest value being 36, representing the failure mode in which products fall and scatter during packaging.

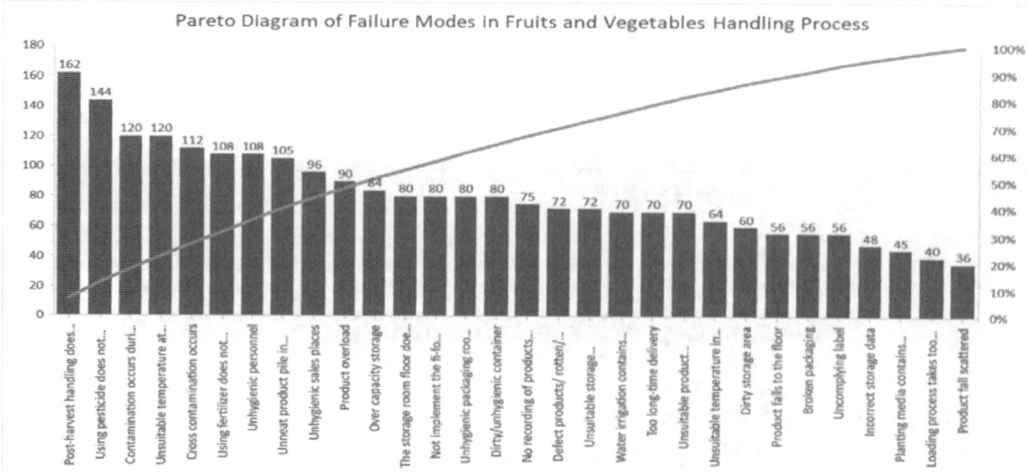

Figure 3. Pareto diagram of failure modes.

Table 5 presented the ten failure modes with the highest priority based on the Pareto diagram. These ten priority failure modes are distributed across five stages of fruit and vegetable handling: on-farm cultivation and postharvest handling, shipping, sales, product reception, and packaging.

Table 5. Prioritize failure modes with the highest risk value

Stage	ID Number	Process	RPN
Cultivation on farm	1.5	Harvesting and postharvest handling do not comply with the hygiene sanitation principle	162
Cultivation on farm	1.4	Using pesticides does not comply with the recommendation	144
Delivery/shipping	5.2	Contamination occurs during shipping	120
Product Sales	6.2	Unsuitable temperature at the product place sales	120
Product Acceptance	2.4	Cross-contamination occurs	112
Cultivation on farm	1.3	Using fertilizer does not comply with the recommendation	108
Packaging	4.5	Unhygienic personnel	108
Delivery/shipping	5.5	Unneat product pile in container	105
Product Sales	6.1	Unhygienic sales places	96
Delivery/shipping	5.4	Product overload	90

3.2 Criticality analysis

Criticality Analysis is a fundamental risk evaluation technique that prioritizes failure modes within a system based on their potential impact (severity) and the likelihood of occurrence. This technique is commonly applied in risk management to identify, analyze, and prioritize potential failures based on their impact (severity) and likelihood of occurrence [23]. By combining these two dimensions, this analysis helps identify failure modes that pose the greatest threat to consumer health, product safety, and overall quality in the food supply chain. The outcome of this analysis is crucial for developing targeted mitigation strategies to address the most critical risks [24].

In this study, severity (S) and occurrence (O) values were assigned to each failure mode, with scores ranging from 1 (low impact/likelihood) to 10 (high impact/likelihood). The criticality level was then determined by plotting these values on a 10x10 matrix, resulting in a risk categorization that guides the necessary interventions. Table 6 presents the five risk categories derived from this criticality analysis.

Table 6. Risk categorization

Category	Value Range	Color	Action
Low Risk	1 – 20		Minimal monitoring
Low-Medium Risk	21 – 40		Routine monitoring
Medium Risk	41 – 60		Further analysis
High-Medium Risk	61 – 80		Mitigation action required
High Risk	81 – 100		Direct intervention required

Next, a criticality graph is created to categorize the ten priority failure modes into risk categories, as shown in Figure 4. According to the graph, the primary critical point in handling fruits and vegetables is failure mode 1.5, "Harvesting and postharvest handling does not comply with hygiene sanitation principles," occurring at the on-farm cultivation stage with a high-risk category. The second critical point is failure mode 1.4, "Using pesticides does not comply with recommendations," which also falls under the same stage but is classified as high-medium risk.

Additionally, four failure modes are classified under the medium-risk category and four under the low-medium-risk category. None of the failure modes are classified as low risk.

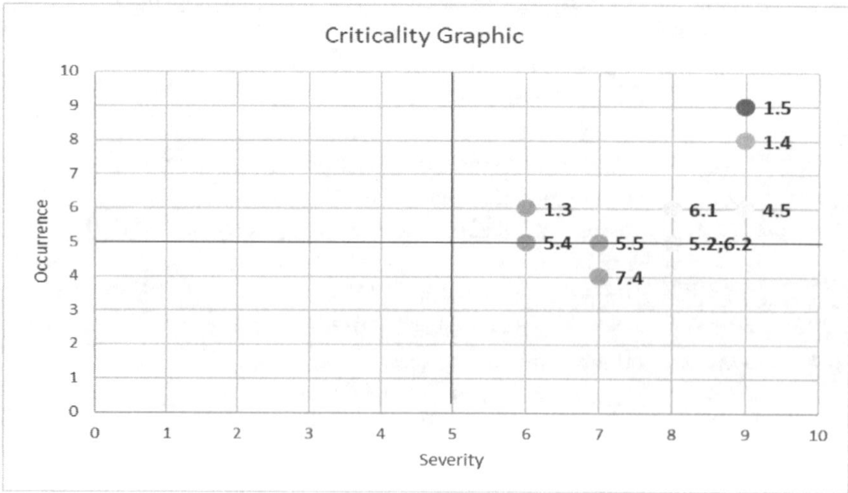

Figure 4. Criticality graphic

The criticality analysis identified that the cultivation on-farm and delivery/shipping stages are the most vulnerable points in the supply chain. Failure modes from these stages consistently exhibited high criticality scores, underscoring the importance of targeted interventions, specifically:

a. Cultivation on-farm

At this stage, three critical failure modes were identified, including pesticide misapplication and noncompliance with postharvest sanitation protocols. These issues significantly elevate the risk of contamination, requiring comprehensive training programs and more rigorous inspection regimes.

Noncompliance with hygiene and sanitation principles during postharvest handling was a significant risk frequently observed among local farmers. This issue arises from several factors, including a lack of knowledge about proper sanitation practices (ignorance), limited access to infrastructure and resources (inability), and resistance to adopting new methods due to cultural habits (unwillingness).

Previous studies have shown that low adoption of Good Agricultural Practices (GAP) is often driven by economic barriers and inadequate support systems for smallholder farmers [11]. Addressing these factors requires tailored interventions, such as providing accessible training programs, offering subsidies for sanitation infrastructure, and implementing incentive-based policies to encourage compliance [20].

b. Delivery/shipping

This stage also contained three high-risk failure modes, highlighting vulnerabilities in product handling during transportation. Inadequate load management, overcapacity, and insufficient container hygiene present substantial risks, indicating a need for enhanced logistics management and transportation protocols.

Shipping processes for local farmers' products and imported products follow distinctly different Standard Operating Procedures (SOPs), which significantly impact contamination risks. Local farmers' products are often transported using open vehicles with minimal temperature control and basic packaging, exposing them to higher risks of microbial and physical contamination. In contrast, imported products typically follow stricter international SOPs, including refrigerated transport and standardized packaging, which minimize contamination risks. Field observations revealed that contamination during local shipping is more frequent due to inadequate hygiene in vehicles and prolonged delivery times. This aligns with findings by Osaili *et al.* and Shewfelt & Prussia, which highlight the critical role of cold chain management in preventing contamination during shipping [25][21].

3.2 Food safety analysis

Although the primary aim of this study is to identify critical food safety risks in each supply chain, the collected data provide insights into the tendencies of food safety in local farmers' and imported products. Observations and interviews revealed notable differences in food safety risks between the two supply chains. Locally sourced lettuce, for instance, exhibited higher risks of contamination, particularly during cultivation and postharvest handling stages. Conversely, imported apples demonstrated lower risks at these stages due to stricter adherence to international SOPs, though moderate risks were identified in shipping and storage.

Food safety control measures for fresh food products must be implemented at all points along the supply chain. A comprehensive understanding of the sources of contamination and their causal factors is crucial to ensuring food safety is maintained [26]. The supply chain analysis for fresh produce reveals four key actors: farmers/importers, producers, distributors/exporters, and retailers. Among these, the critical points for food safety are mainly associated with producers, distributors/exporters, and retailers [27].

In this supply chain, these business actors operate at medium and large scales and are equipped with adequate warehousing and shipping facilities. The critical points in product handling for these actors are cultivating on-farm, receiving goods, packaging, storage, and shipping. In these processes, handling must consider product characteristics [28]. Proper treatment of the products and maintaining hygienic sanitation for the facilities, equipment, and handling personnel are essential to ensuring the safety and quality of fresh fruits and vegetables [29].

3.2.1 Cultivating on-farm and postharvest handling

During cultivation, fruits and vegetables can become contaminated by polluted water, pathogen-laden soil, or unsafe use of pesticides and fertilizers. Therefore, following Good Agricultural Practices (GAP) is crucial, including choosing safe water sources, monitoring pesticide residues, and controlling the types of fertilizers used [30]. Microbial contamination, such as *E. coli*, *Salmonella*, and *Listeria*, is a frequent concern at this stage. The misuse of pesticides (e.g., exceeding maximum residue limits or using banned pesticides) represents a high RPN failure mode. Strict regulations, training on pesticide use, and implementing better monitoring programs are necessary to ensure compliance with food safety standards.

The inaccuracy of farmers in cultivation and the absence of Good Agricultural Practices (GAP) are influenced by a combination of factors, including ignorance, inability, and a low adoption culture of GAP. Field observations and interviews revealed that many farmers lack sufficient knowledge about GAP and its importance in ensuring food safety, which aligns with findings by Sukayat (2023) indicating that smallholder farmers often have limited access to formal agricultural training [31]. Studies by Macieira (2021) highlight that economic limitations hinder farmers from adopting improved practices, even when they are aware of their benefits [5].

Postharvest handling involves various activities such as harvesting, sorting, washing, storage, and transportation, all of which significantly impact the quality and safety of the products [32]. During postharvest processes, products are vulnerable to microbiological contamination from wash water, unclean equipment, and improper handling by workers. Physical contamination may also occur due to using non-standard tools [33]. Ineffective postharvest handling can increase the risk of pathogen contamination, such as *Salmonella* and *Listeria*, which are linked to serious food safety and public health issues [34].

Noncompliance with hygienic sanitation requirements during postharvest handling is one of the significant risk factors for the safety of fresh fruits and vegetables. This is in line with research by Fulano, *et al.* (2021) which showed that failure to adhere to good sanitation practices can lead to microbiological, chemical, and physical contamination, ultimately threatening consumer health [35]. Addressing this risk requires the implementation of Good Handling Practices (GHP), including proper worker training, effective monitoring systems, and using facilities and equipment that meet sanitation standards.

3.2.2 Product Acceptance

In the product reception process, it is important to ensure that incoming products meet the specified quantity and quality standards. The receiving officer must verify that the quantity on the order note or delivery document matches the goods received [36]. To check the safety and quality of the products, the receiving officer must review the laboratory test result documents from the previous supply chain actor and randomly take product samples to verify that the quality meets the company's requirements.

3.2.3 Packaging

Packaging is a critical step in preventing contamination during distribution. Non-sterile packaging materials or poor hygiene practices by workers handling the products can lead to contamination. The packaging or repackaging of fresh fruits and vegetables must consider hygienic sanitation aspects related to packaging machinery and equipment, packaging personnel, and packaging areas [37].

Additionally, packaging materials must be food-grade and comply with food packaging and labeling requirements. Preventive measures include using safe and standardized packaging materials, implementing Good Manufacturing Practices (GMP), and providing sanitation training to workers.

3.2.4 Storage

The storage conditions for products from both local farmers and imports revealed varying levels of compliance with established standards. Observations and interviews indicated that storage practices among local farmers often do not meet the recommended standards, such as the use of pallets, proper ventilation, and temperature control. For instance, improper storage temperature and the absence of pallets to prevent contact with the ground were common issues identified in local supply chains. These findings align with the standards outlined by Salehi (2020), which emphasize the critical role of maintaining temperatures between 0–2°C for most fresh produce and ensuring adequate separation to prevent cross-contamination [38].

Imported products typically showed better adherence to storage standards, benefiting from cold chain systems that ensure consistent temperature control and hygienic storage conditions. However, certain risks were still identified, such as overcapacity in storage facilities and contamination from improperly cleaned containers. These issues suggest that while imported products generally meet the guidelines, lapses in handling at distribution centre can compromise product quality and safety.

Proper storage practices include organizing product placement, controlling temperature and humidity, labelling products with reception times and expiration dates, and ensuring the storage area is clean and separated between fresh products and other materials to prevent cross-contamination [39]. Most fruits and vegetables require storage temperatures ranging from -2°C to 2°C, while cold-sensitive crops need temperatures between 5–15°C [40].

Table 7. Temperature for fruits and vegetables storage [40]

Commodity	Temperature	Humidity
Fruit and Vegetables	0 – 2° C	90 – 98%
Subtropical fruit	7 – 10° C	85 – 95%
Tubers, pumpkins, melons, tropical fruit	13 – 18° C	85 – 95%

3.2.5 Delivery/shipping

The delivery process connects the actors within the supply chain and is a crucial stage in food safety due to the risk of contamination. During delivery, choosing appropriate transportation modes and routes is essential to ensure that products reach their destination promptly [41]. Products may become contaminated during shipping if temperatures are not controlled or if they come into contact with hazardous or contaminated materials.

Dirty shipping equipment can also be a source of contamination. The transport mode and its facilities must consider the characteristics of the products being shipped. Use clean, well-maintained vehicles with controlled temperatures. Using refrigerated transport vehicles is an effort to maintain the safety and quality of products during shipping [15]. Ensure no cross-contamination during transportation by separating fresh products from other potentially hazardous materials.

3.2.6 Product sales

When fresh fruits and vegetables are displayed for sale, they are at high risk of contamination from the surrounding environment, direct contact with consumers, or temperature changes. The sanitation quality of the sales area is crucial for food safety [25]. Ensure product displays are kept at appropriate temperatures to slow down microbial growth [21]. Regularly clean the sales area and avoid direct exposure from unhygienic consumers. Additionally, Good Hygiene Practices (GHP) standards can be developed in markets, such as providing handwashing facilities, sanitation tools, and hygiene guidelines for traders and consumers.

Competent food safety authorities should also conduct routine inspections to ensure traders/retailers comply with food safety guidelines. Monitoring cleanliness and storage conditions is essential to prevent contamination. Traders who meet food safety standards should be given incentives or certifications to build consumer trust and boost businesses that comply with regulations [42].

4. Conclusion

This study identified critical food safety points in the handling of fresh fruits and vegetables, specifically within the supply chains of locally sourced lettuce and imported apples, using the Failure Mode Effect and Criticality Analysis (FMECA) method. This approach facilitates a systematic analysis of potential failure modes, their impacts, and the prioritization of risks based on Risk Priority Numbers (RPN).

The FMECA analysis revealed that the most vulnerable risks in the local farmers' supply chain occur during the cultivation and postharvest handling stages. Noncompliance with hygiene and sanitation principles (RPN: 162) and improper pesticide usage (RPN: 144) were identified as the primary risk factors. Meanwhile, risks in the imported apple supply chain were more concentrated during transportation and storage stages, including contamination during shipping (RPN: 120) and overcapacity in storage facilities (RPN: 84). These findings indicate that local supply chains exhibit higher risk levels compared to imported products, which benefit from adherence to international standards.

The FMECA approach not only identified the key failure modes but also provided a clear prioritization for interventions in the most critical areas. This method has proven effective in quantitatively assessing risks and offers a framework for data-driven decision-making.

5. Recommendations

Based on the findings of this study, the following targeted recommendations are proposed to address the identified food safety risks and improve overall product safety in both local and imported supply chains:

- For local farmers: Provide regular training programs tailored to smallholder farmers, focusing on hygiene, proper pesticide use, and postharvest handling. These programs should be facilitated by agricultural extension officers under the supervision of local agricultural departments. Offer financial assistance or subsidies for low-cost cold storage units and proper harvesting tools. This initiative should be coordinated by the Ministry of Agriculture and supported by local cooperatives or farmer associations.
- For the government: Enforce compliance with GAP and Good Handling Practices (GHP) by developing monitoring systems at regional and national levels. The government should also collaborate with food safety authorities to conduct periodic inspections of farming and handling practices.
- For importers and distributors: Require regular audits and adherence to international food safety standards, such as Codex Alimentarius. Importers should integrate digital tools, such as blockchain or IoT-based systems, to monitor temperature and hygiene during shipping and storage.
- For retailers: Retailers should ensure that fresh produce is displayed in sanitary environments with proper temperature control. These standards should be part of retailer agreements with suppliers, monitored by food safety inspectors.
- For researchers: Developing more comprehensive and adaptive risk analysis methods, applying new technologies for monitoring and risk mitigation, and gaining a better understanding of market behaviour and dynamics.

These recommendations aim to operationalize interventions and ensure effective implementation. These measures are expected to reduce contamination risks and improve food safety across both local and international supply chains.

References

[1] Mu W, van Asselt E D and van der Fels-Klerx H J 2021 Towards a resilient food supply chain in the context of food safety *Food Control* **125**

[2] Ehuwa O, Jaiswal A K and Jaiswal S 2021 Salmonella, food safety and food handling practices *Foods* **10** 1–16

[3] Gil M I, Selma M V., Suslow T, Jacxsens L, Uyttendaele M and Allende A 2015 Pre- and Postharvest Preventive Measures and Intervention Strategies to Control Microbial Food Safety Hazards of Fresh Leafy Vegetables *Crit. Rev. Food Sci. Nutr.* **55** 453–68

[4] Aiyedun S O, Onarinde B A, Swainson M and Dixon R A 2021 Foodborne outbreaks of microbial infection from fresh produce in Europe and North America: a systematic review of data from this millennium *Int. J. Food Sci. Technol.* **56** 2215–23

[5] Macieira A, Barbosa J and Teixeira P 2021 Food safety in local farming of fruits and vegetables *Int. J. Environ. Res. Public Health* **18**

[6] Ma B, Han Y, Cui S, Geng Z, Li H and Chu C 2020 Risk early warning and control of food safety based on an improved analytic hierarchy process integrating quality control analysis method *Food Control* **108**

[7] Gallo A, Accorsi R, Goh A, Hsiao H and Manzini R 2021 A traceability-support system to control safety and sustainability indicators in food distribution *Food Control* **124**

[8] Chopra S and Peter M 2014 *Supply chain management: strategy, planning, and operation - third edition*

[9] Sadati A K, Nayedar M, Zartash L and Falakodin Z 2021 Challenges for food security and safety: a qualitative study in an agriculture supply chain company in Iran *Agric. Food Secur.* **10** 1–7

[10] Ramirez-Hernandez A, Galagarza O A, Álvarez Rodriguez M V., Pachari Vera E, Valdez Ortiz M del C, Deering A J and Oliver H F 2020 Food safety in Peru: A review of fresh produce production and challenges in the public health system *Compr. Rev. Food Sci. Food Saf.* **19** 3323–42

[11] Hendriadi A, Sulistiyorini and Devilana M R 2021 Pesticides residues in fresh food of plant origin: Case study in indonesia *Agrivita* **43** 285–99

[12] Lee J, Daraba A, Voidarou C, Rozos G, Enshasy H A El and Varzakas T 2021 Pareto). The Case Study of Listeria monocytogenes and Correlation with Microbiological Criteria *Foods* **10** 2–24

[13] Magableh G M, Mumani A A, Obaidat S F and Mistarihi M Z 2024 Fuzzy-FMEA Theory Approach for Prioritizing Supply Chain Nervousness Factors *Appl. Sci.* **14**

[14] Ciani L, Guidi G and Patrizi G 2019 A Critical Comparison of Alternative Risk Priority Numbers in Failure Modes, Effects, and Criticality Analysis *IEEE Access* **7** 92398–409

[15] Frankish E J, Bozkurt H, Ross T, Phan-Thien K-Y, Luning P A, Bell T L and McConchie R 2021 An observational assessment of Australian apple production practices for microbial control *Food Control* **123**

[16] Pietrysiak E, Smith S and Ganjyal G M 2019 Food Safety Interventions to Control Listeria monocytogenes in the Fresh Apple Packing Industry: A Review *Compr. Rev. Food Sci. Food Saf.* **18** 1705–26

[17] Jensen D A, Friedrich L M, Harris L J, Danyluk M D and Schaffner D W 2015 Cross contamination of Escherichia coli O157:H7 between lettuce and wash water during home-scale washing *Food Microbiol.* **46** 428–33

[18] Aleksic B, Djekic I, Miocinovic J, Miloradovic Z, Memisi N and Smigic N 2022 The application of Failure Mode Effects Analysis in the long supply chain – A case study of ultra filtrated milk cheese *Food Control* **138** 109057

[19] Zhang H, Yamamoto E, Murphy J and Locas A 2020 Microbiological safety of ready-to-eat fresh-cut fruits and vegetables sold on the Canadian retail market *Int. J. Food Microbiol.* **335** 108855

[20] Murray K, Wu F, Shi J, Jun Xue S and Warriner K 2017 Challenges in the microbiological food safety of fresh produce: Limitations of post-harvest washing and the need for alternative interventions *Food Qual. Saf.* **1** 289–301

[21] Shewfelt R L and Prussia S E 2021 Challenges in handling fresh fruits and vegetables *Postharvest Handl. A Syst. Approach* 167–86

[22] Xhafka E, Avrami E, Canton G and Mazreku A 2024 Applying the Ishikawa Diagram and Pareto Chart for Defect Reduction in the Manufacturing Industry. A Case Study from a Textile-Producing Company BT - Proceedings of the Joint International Conference: 10th Textile Conference and 4th Conference on Enginee ed G Guxho, T Kosova Spahiu, V Prifti, A Gjeta, E Xhafka and A Sulejmani (Cham: Springer Nature Switzerland) pp 45–55

[23] Giardina M, Tomarchio E, Buffa P, Palagonia M, Veronese I and Cantone M C 2022 FMECA Application in Tomotherapy: Comparison between Classic and Fuzzy Methodologies *Environ. - MDPI* **9**

[24] La Fata C M, Giallanza A, Micale R and La Scalia G 2022 Improved FMECA for effective risk management decision making by failure modes classification under uncertainty *Eng. Fail. Anal.* **135** 106163

[25] Osaili T M, Hasan F, Al-Nabulsi A A, Olaimat A N, Ayyash M, Obaid R S and Holley R 2023 A worldwide review of illness outbreaks involving mixed salads/dressings and factors influencing product safety and shelf life *Food Microbiol.* **112** 104238

[26] Bhilwadikar T, Pounraj S, Manivannan S, Rastogi N K and Negi P S 2019 Decontamination of Microorganisms and Pesticides from Fresh Fruits and Vegetables: A Comprehensive Review from Common Household Processes to Modern Techniques *Compr. Rev. Food Sci. Food Saf.* **18**

[27] Aworh O C 2021 Food safety issues in fresh produce supply chain with particular reference to sub-Saharan

Africa *Food Control* **123**

[28] Zupaniec M A, Schafft H A, Pieper R, Lindemann A K and Mader A 2022 A Conceptual Framework for the Identification of Food Safety Risks in Global Commodity Flows Exemplified by Agricultural Bulk Commodities *Oper. Supply Chain Manag.* **15** 79–92

[29] Pace B and Cefola M 2021 Innovative preservation technology for the fresh fruit and vegetables *Foods* **10** 8–11

[30] Prapamontol T, Hongsibsong S, Naksen W, Kerdnoi T, Kawichai S, Polyiem W, Pakvilai N, Phansawan B and Santasup C 2021 Multiple Pesticide Residues Found in Vegetables and Fruits from Rural and Urban Markets in Upper Northern Thailand *Chiang Mai Univ. J. Nat. Sci.* **20** 1–25

[31] Sukayat Y, Setiawan I, Suharfaputra U and Kurnia G 2023 Determining Factors for Farmers to Engage in Sustainable Agricultural Practices: A Case from Indonesia *Sustain.* **15** 1–14

[32] Opara U L 2009 *Quality Management: An Industrial Approach to Produce Handling*

[33] Abdelfattah A, Whitehead S R, Macarisin D, Liu J, Burchard E, Freilich S, Dardick C, Droby S and Wisniewski M 2020 Effect of washing, waxing and low-temperature storage on the postharvest microbiome of apple *Microorganisms* **8** 1–21

[34] Bolívar A, Saiz-Abajo M J, García-Gimeno R M, Petri-Ortega E, Díez-Leturia M, González D, Vitas A I and Pérez-Rodríguez F 2023 Cross contamination of Escherichia coli O157:H7 in fresh-cut leafy vegetables: Derivation of a food safety objective and other risk management metrics *Food Control* **147**

[35] Fulano A M, Lengai G M W and Muthomi J W 2021 Phytosanitary and technical quality challenges in export fresh vegetables and strategies to compliance with market requirements: Case of smallholder snap beans in Kenya *Sustain.* **13** 1–21

[36] Wahyuni H C, Rosid M A, Azara R and Voak A 2024 Blockchain technology design based on food safety and halal risk analysis in the beef supply chain with FMEA-FTA *J. Eng. Res.*

[37] Palumbo M, Attolico G, Capozzi V, Cozzolino R, Corvino A, de Chiara M L V, Pace B, Pelosi S, Ricci I, Romaniello R and Cefola M 2022 Emerging Postharvest Technologies to Enhance the Shelf-Life of Fruit and Vegetables: An Overview *Foods* **11** 1–29

[38] Salehi F 2020 Recent Advances in the Modeling and Predicting Quality Parameters of Fruits and Vegetables during Postharvest Storage: A Review *Int. J. Fruit Sci.* **20** 506–20

[39] Ziv C and Fallik E 2021 Postharvest storage techniques and quality evaluation of fruits and vegetables for reducing food loss *Agronomy* **11** 1–5

[40] Makule E, Dimoso N and Tassou S A 2022 Precooling and Cold Storage Methods for Fruits and Vegetables in Sub-Saharan Africa—A Review *Horticulturae* **8** 1–15

[41] Panghal A, Yadav D N, Khatkar B S, Sharma H, Kumar V and Chhikara N 2018 Post-harvest malpractices in fresh fruits and vegetables: food safety and health issues in India *Nutr. Food Sci.* **48** 561–78

[42] L'Abbé M, Schermel A, Minaker L, Kelly B, Lee A, Vandevijvere S, Twohig P, Barquera S, Friel S, Hawkes C, Kumanyika S, Lobstein T, Ma J, Macmullan J, Mohan S, Monteiro C, Neal B, Rayner M, Sacks G, Sanders D, Snowdon W, Swinburn B and Walker C 2013 Monitoring foods and beverages provided and sold in public sector settings *Obes. Rev.* **14** 96–107

Correlation Between Quality Index of Fresh Mackerel Tuna (*Euthynnus affinis*) and Consumer Acceptance of Fried Mackerel Tuna

Muhammad Hizbullah Al Anshary[1], Andi Irawan[2], Kuncoro Harto Widodo[2], Makbul Hajad[3], Andriati Ningrum[1], Diah Puspitasari[4], Lintar Sekar Pembayun[2], Manikharda[1*]

[1]Universitas Gadjah Mada, Faculty of Agricultural Technology, Department of Food and Agricultural Product Technology, Jl. Flora No. 1, Bulaksumur, Yogyakarta 55281, Indonesia
[2]Universitas Gadjah Mada, Faculty of Agricultural Technology, Department of Agroindustrial Technology, Jl. Flora No. 1, Bulaksumur, Yogyakarta 55281, Indonesia
[3]Universitas Gadjah Mada, Faculty of Agricultural Technology, Department of Agricultural and Biosystems Engineering, Jl. Flora No.1 Bulaksumur, Yogyakarta 55281, Indonesia
[4]Universitas 'Aisyiyah Yogyakarta, Faculty of Health Sciences, Indonesia, Nutrition Study Program, Jl. Ringroad Barat 63 Mlangi, Nogotirto, Gamping, Sleman, Yogyakarta 55292, Indonesia

*E-mail: manikharda@ugm.ac.id

Abstract. Mackerel tuna (*Euthynnus affinis*) has broad appeal in the fisheries market. Maintaining the quality of fish from the Scombridae family is essential for the industry, consumer health, and satisfaction, especially during handling, storage and processing. The Quality Index Method (QIM) is a suitable tool for measuring freshness, its impact on sensory properties, and fish safety. This study aims to verify the relationship between the Quality Index (QI) scores of fresh mackerel tuna and consumer acceptance of the fish after frying, a popular household cooking method. Results showed that QI scores did not change significantly from day 2 to day 10 during ice storage but increased sharply between days 10 and 16. Nevertheless, histamine levels remained within safe limits throughout storage, with concentrations below the 50 mg/kg threshold. Sensory assessments of fried mackerel tuna included evaluations of aroma, texture, and flavor, which showed minimal declines over the 16-day period. However, overall acceptability showed no significant change during the first 8 days, with significant declines observed starting on day 8 as freshness diminished. Despite this decrease, the product remained acceptable to panelists through the end of the 16-day storage period. The study highlights a strong positive correlation (R^2=0.7998) between QI scores and consumer acceptance of fried mackerel tuna, supporting the usefulness of QIM in assessing mackerel tuna quality. The QIM method developed in this study provides a practical model for evaluating mackerel tuna quality.

1. Introduction

Euthynnus affinis, commonly known as mackerel tuna, is an important species in commercial fisheries due to its high nutritional value and versatility in cooking. Properly chilling mackerel tuna is essential to protect health, extend shelf life, and maintain desirable sensory properties for potential buyers [1]. In Indonesia, from 2005 to 2015, histamine was primarily responsible for 1,176 food poisoning incidents, leading to foodborne outbreaks. Fish species such as tuna, mackerel tuna (tongkol), and skipjack (cakalang) are often linked to histamine fish poisoning (HFP) [2]. Consequently, the fishing industry faces pressure to maintain high-quality standards for fish, given its perishable nature.

Fish begins to decay immediately after death, making proper handling and storage essential. Accurate evaluation of seafood freshness is therefore critical in determining the appropiate handling processes. The fisheries sector primarily relies on sensory evaluation for this purpose, using it to establish quality standards and guide product development and optimization [3,4]. Quality Index Method (QIM) was developed to provided a structured and objective method for evaluating fish quality based on sensory parameters [5]. This method assigns scores based on various attributes such as skin appearance, odor, and texture, with lower scores indicating fresher fish. A study by Bernardi et al [6] demonstrated that QIM is an effective tool for assessing the shelf life of fish species. The research emphasized that QIM not only provides an objective measure of freshness but also predicts remaining shelf life accurately. By offering a species-specific approach, QIM establishes detailed criteria that allow for each species, reliable assessment compared to other methods.

Throughout storage, fish undergo biochemical and microbiological changes that can impact the sensory characteristics of both fresh and processed products. The role of QIM in monitoring these changes has been validated across several fish species, including Turbot and Goldlined seabream, where QIM scores have been shown to correlate strongly with quality degradation indicators like microbial load and lipid oxidation [7,8]. These studies highlight QIM's robustness as a quality assessment tool for the seafood industry.

This study investigates the relationship between panelists' acceptance and the correlation between the quality index (QI) of fresh mackerel tuna and consumer acceptance of fried mackerel tuna. Understanding this correlation can provide valuable insights for the seafood industry, guiding best practices in harvesting, processing, and marketing. Key quality indices, including freshness, texture, odor, and color, play a pivotal role in determining overall seafood acceptability [9,10,11]. These indices, assessed through both objective measurements and sensory evaluations, offer a comprehensive understanding of product quality. As consumers become more discerning, their acceptance of seafood increasingly depends on these quality attributes, influencing purchasing and consumption choices.

2. Materials and method

2.1 Materials

The sample in this study is mackerel tuna (*Euthynnus affinis*) obtained from a fisherman at Jetis Beach (Cilacap, Central Java, Indonesia). The material used in this study includes cooking oil (Tropical, PT Bina Karya Prima, East Java, Indonesia), citral, benzyl acetate, thymol, and vanillin (PT Firmenich Indonesia, Jakarta, Indonesia). Other ingredients include sucrose (granulated sugar), sodium chloride (table salt), caffeine (black coffee), citric acid, tannic acid, and ferrous gluconate from (Sangobion, Central Java, Indonesia).

2.2 Preparation sample

Freshly caught mackerel tuna was stored in ice. Before frying, the freshness of the mackerel tuna was assessed using the Quality Index Method (QIM)—furthermore, sensory evaluation involved removing the internal organs and head from the fresh mackerel tuna. The fish was filleted and cut into pieces approximately 2 cm in thickness. The fish pieces were then cooked without seasoning at 160°C for 8 minutes.

2.3 Histamine content analysis

Histamine analysis was performed by ISO 19343:2017 in the PT. Saraswanti Indo Genetech (Semarang, Central Java, Indonesia). Fresh fish flesh was ground using a mixer. Transfer the test portion of 5 g ± 0.1 g of fish and add 10 ml perchloric acid and 100 µl of 1,7-diaminoheptane homogenate into a centrifuge tube. After completing the centrifuge, transfer 100 µl of supernatant into a tube and add 300 µl of sodium carbonate solution and 400 µl of Dansyl chloride solution. The sample was homogenized and incubated for 5 minutes at 60°C without light. The tubes should be cooled using cold tap water and 100 µl of L-proline solution. Homogenized and positioned the tube in a dark room for 15 min.

2.4 Sensory evaluation

Sensory evaluation was conducted by 14 trained panelists, selected and trained in according to ISO 8586:2012 guidelines for sensory assessors. Panelists were recruited based on specific criteria: a minimum age of 18 years, absence of allergies to fish, preference for consuming fish, good general health, willingness to attend testing sessions at the Sensory Testing Laboratory of FTP UGM, and commitment to complete the entire sensory evaluation process. Panelist selection involved a series of tests to assess sensory capabilities, including a matching test, an identification test for basic tastes and odors, and color-blindness test to ensure accurate perception of sample characteristic.

Following selection, selected panelists underwent training to familiarize them with the sensory attributes of the samples and to practice using the 9-point hedonic scale. Training also included a duo-trio test to help panelists develop the ability to differentiate between samples, focusing on attributes like aroma, texture, flavor, and overall acceptability. This training ensured that panelists could consistently identify and rate sensory qualities with accuracy.

For sensory evaluation, panelists assessed the samples using a 9-point hedonic scale, covering parameters of aroma, texture, flavor, and overall acceptability. Testing was conducted once every 2 days for 16 days period to monitor sensory changes in the samples over time.

2.5 Statistical analysis

The one-way Analysis of Variance (ANOVA) test was used to analyze the hedonic test at a significance level of 95%. Using one-way ANOVA, panelists were asked to rate the differences in eight varieties of fried tool fish samples' scent, flavor, texture, and overall attribute values. The panelists' evaluation of the fish's acceptance and liking was the independent variable, and the fish's shelf life was the control. When the effect was significant, the Duncan test was also performed to determine the significance of the differences. IBM SPSS Statistics 25 (IBM Corp., Chicago, IL, USA) was used to analyze the hedonic test statistically.

2.6 Ethical statement

This study, "Correlation between Quality Index of Fresh Mackerel Tuna (*Euthynnus affinis*) and Consumer Acceptance of Fried Mackerel Tuna," was conducted strictly to ethical guidelines and principles for research involving human participants. The research protocol received approval from the Ethical Committee of Medical and Health Research Ethics Committee, Faculty of Medicine, Public Health and Nursing, Universitas Gadjah Mada, Dr. Sardjito General Hospital, with approval number KE/FK/0344/EC/2023.

3. Result and discussion

3.1 Evaluation of quality index method for raw material

The attributes used in the QIM evaluation were appearance, eyes, belly, gills, and flesh. Quality Index (QI) values were tested within 16-days storage period (Table 1), with 10 trained panelists conducting the QIM testing. The QI value is calculated by summing the scores for each attribute, including overall appearance, eyes, abdomen, gills, and flesh. Figure 1 illustrates the variation in fresh mackerel tuna based on the total Quality Index (QI) value.

Table 1. Alteration process of mackerel tuna during process

Storage Time (day)	Parameter				
	Appearance	Eye	Stomach	Gill	Flesh
0			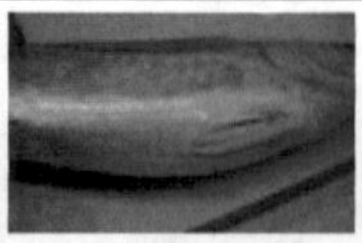	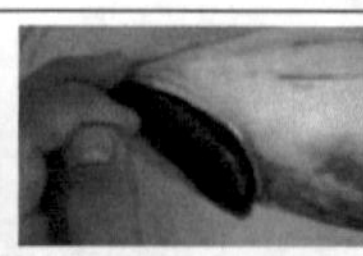	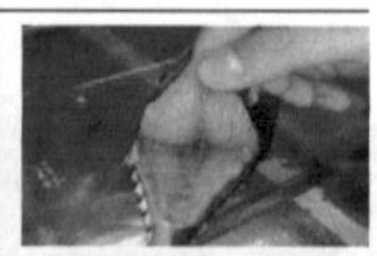
2	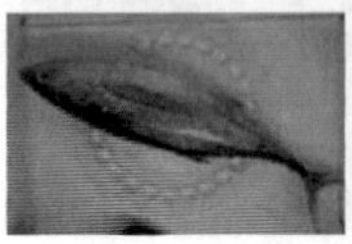			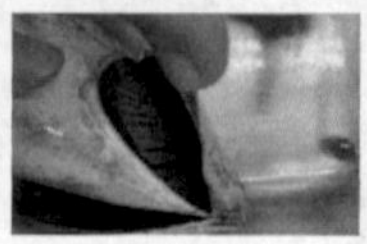	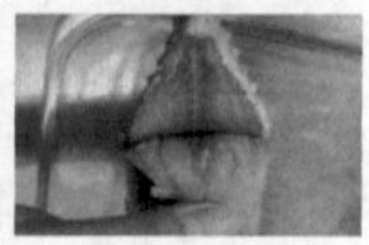
4	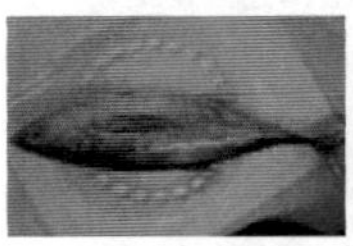	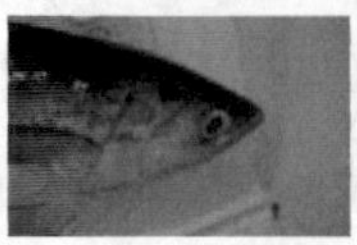	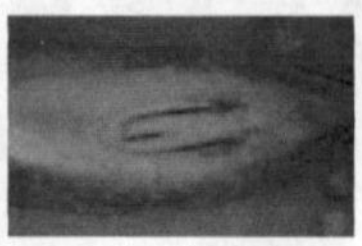		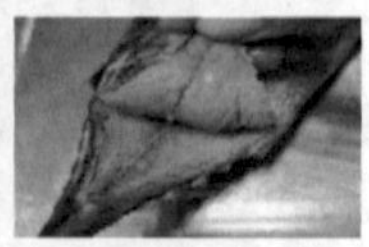
6	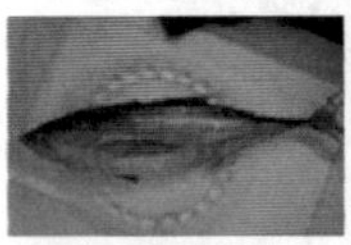	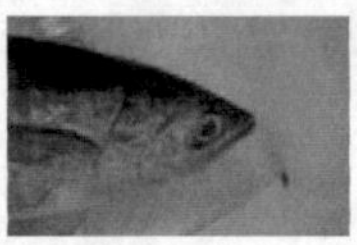	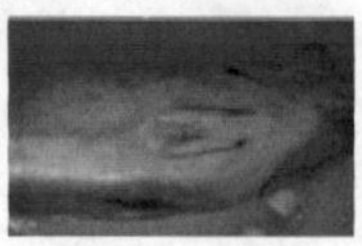	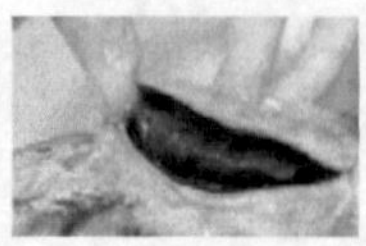	
8	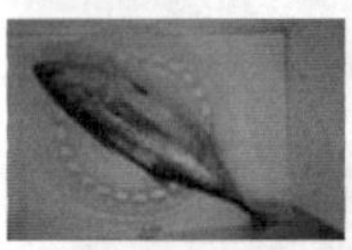		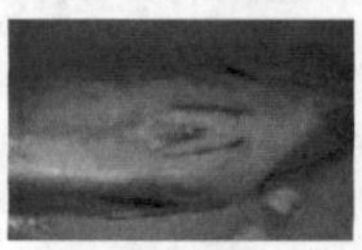	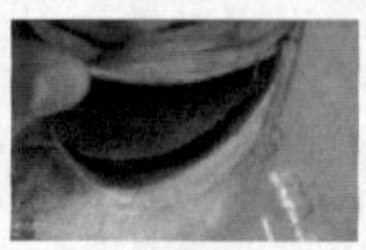	
10	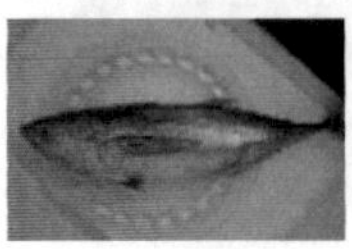		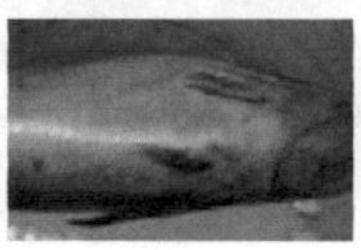	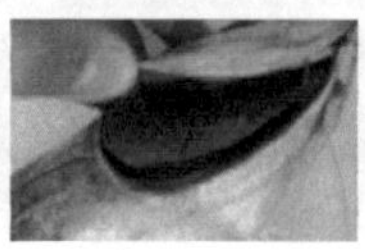	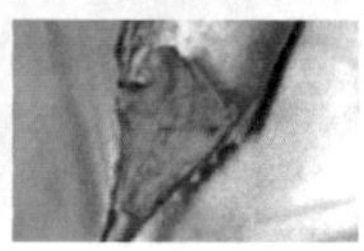

12	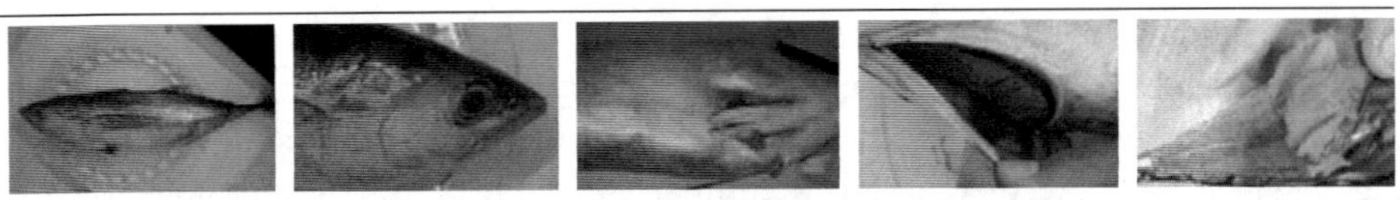
14	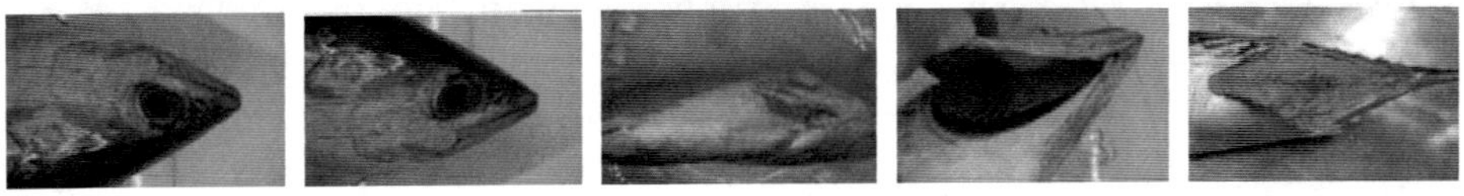
16	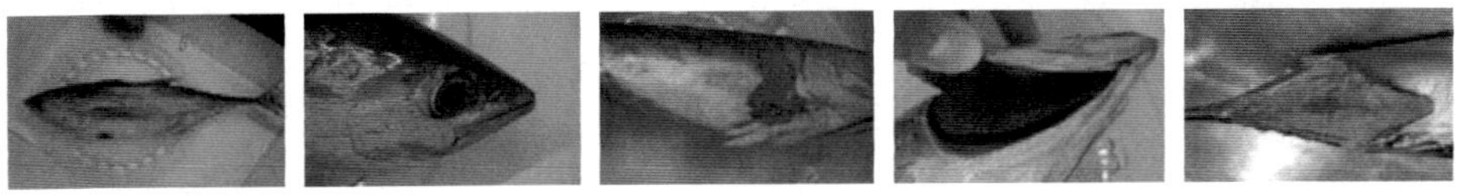

On day 2, the QI score of 2 indicated freshness with a fresh odor and firmness in the rigor mortis phase. By day 8 (QI score of 7), a fishy odor and post-rigor phase began. On day 16, a QI score of 21 showed a rancid odor, flattened eyes, and faded flesh color.

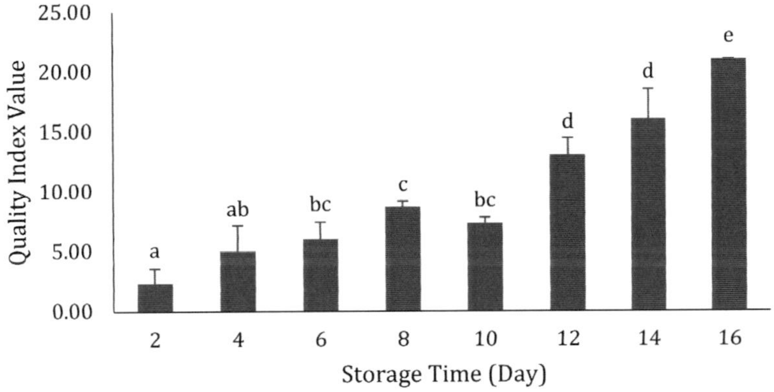

Figure 1. Average Quality Index (QI) values of fresh mackerel tuna during storage (16 days) Data was expressed as average±standard deviation (n = 10). Different letters in the bars showed statistically significant differences based on one-way ANOVA and Duncan post hoc test (*p*<0.05).

The study results show that the quality index increase with extended storage duration. The longer the storage, the higher the fish's quality index value, indicating a decline in quality. An increase in QI value suggests deteriorating fish quality. A deviation was observed on the 10th day of storage, where the QI value was lower than on previous day. This anomaly can be attributed to the inconsistency in the composition of panelists for each test, resulting in fluctuations in the overall QIM score due to disparities in individual evaluations. In a previous study, on the 10th day, the chilling temperature storage little tuna fillet (*Euthynnus affinis*) was spoiled [12]. Despite this deviation, the results remain satisfactory, and this methodology may serve as a reliable reference for assessing the quality of fried mackerel tuna.

3.2 Histamine content of fresh mackerel tuna

Consumers are not at risk of histamine contamination and subsequent food illness when consuming freshly caught tuna. Table 2 shows a notable increase in histamine levels of mackerel tuna during ice storage. According to the findings from days 2 to 8, fresh mackerel tuna showed no presence of histamine. However, starting from day 10, histamine was detected, and its levels continued to rise until day 16. Elevated histamine levels result from bacterial growth that produces histamine and the activity of histidine decarboxylase. As storage duration increases, the histamine levels also increase [13]. The variation in histamine levels across the samples can be ascribed to several factors, including the fish's treatment, preservation, and processing conditions and the time elapsed since it was caught. In comparison, studies on other pelagic species, like sardine, have observed that histamine content remained relatively stable, ranging from 2.19 to 5.02 ppm during 6 days of ice storage [14]. Similarly, another study found that fresh tuna stored at 4°C showed undetectable levels of histamine within the first 6 days, reinforcing the effectiveness of low-temperature storage in delaying histamin formation [15].

Table 2. Histamine content of fresh mackerel tuna

Storage time (Day)	Histamine (mg/kg)	Limit detection (mg/kg)
2	n.d	0.33
4	n.d	0.33
6	n.d	0.33
8	n.d	0.33
10	1.69 ± 0.01^{a}	0.33
12	2.45 ± 0.03^{b}	0.33
14	2.91 ± 0.01^{c}	0.33
16	3.70 ± 0.01^{d}	0.33

Note: The number shows average ± standard deviation. Different superscript letters on the same row indicate statistically significant differences based on the one-way ANOVA followed by the Duncan post hoc Test ($p<0.05$).

Nevertheless, the histamine level until day 16 of storage (Table 2) remained below the FDA's maximum allowable for histamine content in fish, which is set at 50 ppm or 50 mg/kg (FDA). This indicates that the samples in this study were safe and free from contamination bacteria that degrade histidine. High levels of histidine in food do not automatically result in histamine formation but it depends on the presence of specific bacteria capable of decarboxylating histidine [15]. The consistent finding of low histamine levels under controlled ice storage also aligns with previous research showing effective histamine control in fish stored at low temperatures, highlighting the importance of cold storage in preventing histaming accumulation.

3.3 Sensory characteristics of fried mackerel tuna

The acceptance test on fried mackerel tuna was conducted to evaluate changes in sensory attributes throughout ice storage. The sensory evaluation focused on four parameters: aroma, texture, flavor and overall acceptability. Table 3 presents the acceptance value based on the sensory evaluation from 14 trained panelists. The acceptability value remained within the acceptable range until day 16, as it consistently fell within the 5 out of 9 score range. The QIM results for fresh mackerel tuna suggest that the quality remains adequate during cold storage over day 16.

3.3.1 Aroma

A crucial factor in assessing the sensory qualities of food is its aroma. The acceptance value of the aroma of fried mackerel tuna decreased as the storage duration increased, although this decrease was not statistically significant. On day 6, there was a notable decline, characterized by a shift in scent from fresh to mildly fishy, accompanied by the emergence of an earthy aroma. By day 8, the characteristic scents of fish, marine, shellfish, and fresh were still noticeable. Subsequently, between 10 and 16 days of storage, the fish's characteristic smell diminished, the level of freshness declined, and rancidity became noticeable. During days 14 and 16, the olfactory analysis revealed the presence of scents reminiscent of charred paper, fermented yogurt, and medicinal substances. Despite these changes, the panelists found the overall aroma of the fried fish stored on ice for 16 days to be satisfactory.

The primary volatile components for the fresh fish aroma are aldehydes, ketones, and alcohols groups that might be derived from the breakdown of unsaturated fatty acids. The volatile components might be responsible for the fish aroma consist of 1-octen-3-ol, 1-octen-3-one, 1-cis-5-octadien-3-ol, and 1-cis-5-octadien-3-one [16]. These chemicals experience oxidation while being stored, forming unpleasant-smelling molecules with earthy, nutty, and fishy odors. More precisely, the presence of volatile molecules such as 1,5-octadien-3-ol is responsible for the sense of an earthy scent [17]. Similarly, the component (E)-oct-2-enal is responsible for the nutty aroma [18]. Trimethylamine (TMA), derived from the degradation of trimethylamine oxide (TMAO), is the causative agent of the fishy odor [16].

3.3.2 Texture

Generally, the acceptance for fried mackerel tuna's texture decreased as the storage duration increased, although this decline was not statistically significant. From day 2 to day 6, a fibrous texture was observed, consistent with the findings of Xiao et al. [19], who reported that cooked fish flesh becomes firm, succulent, and resilient during the rigor phase. Between day 8 to day 12 of storage, the fibrous texture gradually decreased and eventually disappeared by the final day of testing. The panelists did not detect a firm texture on the first four days, but it became noticeable starting on day 8. The fish's texture exhibited increased toughness from day 10 onwards and maintained a dry and hard consistency throughout the remainder of the test. This change is attributed to the denaturation of proteins during storage. Nakazawa & Okazaki [20] stated that the water retention capacity of myofibrillar proteins diminishes upon denaturation. Disruption of cell structure can lead dehydration of muscle fibers, resulting in a dry and tough texture that is difficult to consume [21]. Overall, the texture of mackerel tuna decreased over time during storage, with a slight rise observed on the final day of testing. However, this increase was not statistically significant. The fish's texture remained acceptable to the panelists, with ratings ranging from "like" to "somewhat like" even after 16 days of ice storage.

3.3.3 Flavor

The acceptance of fried mackerel tuna's flavor decreased as the storage duration increased, although this decrease was not statistically significant (Table 3). The fish's savory umami and mildly sweet flavor remained detectable after 10 days. Following 24 hours of storage, the fish's savory flavor underwent a noticeable change, becoming more dull. Between day 10 and day 16, panelists reported negative flavors, specifically metallic, earthy, raw egg-like, and uncooked rice-like flavors. Nevertheless, the panelists still recognized the overall taste of mackerel tuna stored in ice for 16 days, with ratings ranging from "very like" to "somewhat like". Additionally, longer storage periods led to increased microbial growth, which in turn altered the composition of flavor compounds, resulting in a less distinct taste, similar to that of brown rice.

The changes in taste perception are influenced by chemical compounds. Substances such as 1-octen-3-ol, 1-octen-3-one, 1-cis-5-octadien-3-ol, and 1-cis-5-octadien-3-one are responsible for the metallic flavor. Furthermore, microbial proliferation induces autolysis, which alters the composition of flavor components [22].

3.3.4 Overall acceptability

The study's results indicate that during the 16-day ice storage period, there was a decline in the overall acceptance for fried mackerel tuna (Table 3). There was no significant change from days 2 to 8, although there was a significant decrease in the overall acceptability of fried mackerel tuna from days 10 to 14 ($p<0.05$). The sensory panel used a structured 9-point hedonic scale, with a score of 4 considered unacceptable. Despite the decline, the overall acceptability score remained at 5 throughout the 16 days, indicating that the fried mackerel tuna was still generally acceptable to the panelists. These findings suggest that while there is a noticeable decline in sensory qualities during the first week, the product remains consumable for more than two weeks under proper ice storage conditions.

Table 3. Sensory evaluation of fried mackerel tuna

Storage Time (day)	Parameter			
	Aroma	Texture	Flavor	Overall
2	7.44 ± 0.31^{a}	7.19 ± 0.55^{a}	7.44 ± 0.57^{a}	7.74 ± 0.38^{a}
4	7.89 ± 0.60^{a}	7.28 ± 0.49^{a}	7.00 ± 0.64^{ab}	7.33 ± 0.43^{ab}
6	6.90 ± 0.52^{bc}	7.17 ± 0.60^{a}	6.97 ± 0.35^{ab}	7.20 ± 0.40^{b}
8	7.29 ± 0.42^{ab}	7.00 ± 0.47^{a}	6.88 ± 0.58^{ab}	6.92 ± 0.14^{b}
10	6.63 ± 0.51^{bc}	6.67 ± 0.55^{ab}	6.44 ± 0.47^{bc}	6.48 ± 0.47^{c}
12	6.18 ± 0.46^{cd}	6.06 ± 0.57^{ab}	5.94 ± 0.58^{cd}	5.88 ± 0.41^{d}
14	6.05 ± 0.65^{cd}	5.43 ± 0.68^{c}	5.29 ± 0.45^{e}	5.19 ± 0.50^{e}
16	5.71 ± 0.31^{d}	5.54 ± 0.60^{c}	5.42 ± 0.55^{de}	5.54 ± 0.37^{de}

Note: The number shows average ± standard deviation. Different superscript letters on the same row indicate statistically significant differences based on the Duncan post hoc Test ($p< 0.05$).

During the frying process, sensory quality can be affected by instability and inconsistency. The following is due to changes in the composition of compounds found in fish, which are influenced by various factors, including the type of oil, frying temperature and frying time. According to Wang et al. [23], water evaporation, the Maillard reaction, caramelization, and lipid oxidation play significant roles in determining the sensory attributes of fried food.

Research indicates that consumers often develop specific sensory expectations based on previous experiences with a product. When the sensory profile, such as aroma, texture, or flavor shifts due to storage, this change can lead to a decrease in perceived quality because it no longer aligns with the consumers' sensory memory expectations [24]. This phenomenon, often referred to as sensory disconfirmation, can diminish acceptance as consumers may psychologically associate unexpected flavors or textures with spoilage or lower quality [25].

Cultural factors can also shape sensory preferences, particularly when it comes to the perceived freshness of seafood. In Asian culture, freshness in seafood is highly prized, and even subtle shifts in aroma or flavor can trigger aversion due to cultural norms that equate such changes with spoilage or impurity [26]. As a result, consumers from these backgrounds may be more sensitive to changes in sensory attributes and thus rate stored seafood products lower on acceptability scales compared to consumers from another cultural backgrounds with different expectations.

The psychological context of food consumption, including emotions tied to certain sensory experiences, play a role in consumer acceptance. Studies show that consumers associates specific sensory characteristics with positive or negative emotional responses, which can influence acceptability. For instance, a fresh, oceanic aroma might evoke pleasant memories or feelings of satisfaction, while an earthy, stale smell can lead to a perception of unpleasantness, impacting overall acceptability [24]. This emotional responses is reinforced over repeated exposures, leading to a gradual decrease in acceptance when sensory attributes deviate from the ideal.

3.4 Correlation of quality index value to acceptance value

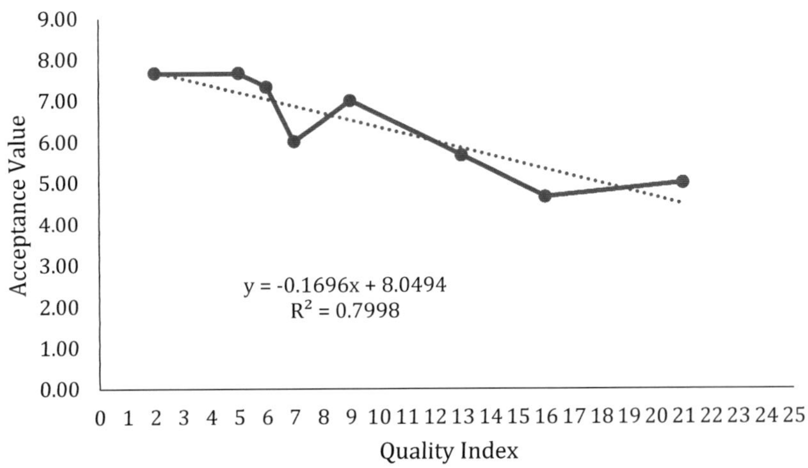

Figure 2. Correlation of Quality Index (QI) value to the acceptance value of fried mackerel tuna

The degree of correlation between two variables is measured by the Pearson correlation coefficient (R), which ranges from -1 to +1. A value close to +1 indicates a strong positive correlation, meaning that as one variable rises, the other variable also increases. The range of a strong correlation coefficient is +0.5 to +1. An R^2 value of 0.7998 was obtained from the linearity study between the Quality Index (QI) value graph and the acceptance value of mackerel tuna during ice storage. This suggests a significant positive relationship between the QI and acceptance values, where lower acceptance values correlated with higher quality index values of mackerel tuna. According to the panelists, fried mackerel tuna with a QI value of 21 can be stored on ice for up to 16 days. Extrapolating the correlation graph using a linear equation is essential to determining the QIM value at which panelists would reject fried mackerel tuna. The QI value obtained from the extrapolation was 23.88, suggesting that panelists are unlikely to find fried mackerel tuna acceptable if the QIM value exceeds 23.88.

4. Conclusion

The acceptability of ice-stored tuna was found to alter in terms of aroma, texture, flavor and overall acceptability over time. In general, the aroma, texture and flavor of fried tuna decreased insignificantly, while the overall acceptability decreased significantly. However, the decrease in acceptability was still acceptable to the panelists. Furthermore, the quality index value of tuna was found to have a positive correlation with the acceptability value of the panelists.

Acknowledgments

We express our gratitude to the Faculty of Agricultural Technology, Universitas Gadjah Mada, for funding this research and Sarah Fathia from PT Firmenich for providing the standard fragrance compounds.

Funding Information

This study was funded by Hibah Rekognisi Tugas Akhir Universitas Gadjah Mada Tahun Anggaran 2024 program, No. 5286/UN1.P1/PT.01.03/2024.

References

[1] Chávez-mejía, A. C., Magaña-lópez, R., Durán-álvarez, J. C., & Jiménez-cisneros, B. E. (2019). The Quality of Fresh Mackerel Tuna (*Euthynnus affinis*) Preserved with Different Icing Methods. 1878(November), 54–60. https://doi.org/10.22161/ijeab

[2] Rachmawati, N., & Triwibowo, R. (2022). Histamine Fish Poisoning (HFP) in Indonesia: Current status and challenges. E3S Web of Conferences, 344, 1–14. https://doi.org/10.1051/e3sconf/202234405001

[3] Getu, A., & Misganaw, K. (2015). Post-harvesting and Major Related Problems of Fish Production. Fisheries and Aquaculture Journal, 06(04). https://doi.org/10.4172/2150-3508.1000154

[4] Siddiqui, S. A., Singh, S., Bahmid, N. A., & Sasidharan, A. (2024). Applying innovative technological interventions in the preservation and packaging of fresh seafood products to minimize spoilage - A systematic review and meta-analysis. Heliyon, 10(8), e29066. https://doi.org/10.1016/j.heliyon.2024.e29066

[5] Freitas, J., Vaz-Pires, P., & Câmara, J. S. (2021). Quality Index Method for fish quality control: Understanding the applications, the appointed limits and the upcoming trends. Trends in Food Science and Technology, 111(March), 333–345. https://doi.org/10.1016/j.tifs.2021.03.011

[6] Bernardi, D. C., Mársico, E. T., & de Freitas, M. Q. (2013). Quality index method (QIM) to assess the freshness and shelf life of fish. Brazilian Archives of Biology and Technology, 56(4), 587–598. https://doi.org/10.1590/S1516-89132013000400009

[7] Ahmadi Shalhe, M., Khodanazary, A., & Hosseini, S. M. (2018). Development of a quality index method (QIM) scheme for whole Goldlined seabream Rhabdosargus sarba stored in ice. International Journal of Food Properties, 21(1), 2539–2549. https://doi.org/10.1080/10942912.2018.1536147

[8] Li X., Chen Y., Cai L., Xu Y., Yi S., Zhu W., Mi H., Li J., Lin H. (2016). Freshness assessment of turbot (Scophthalmus maximus) by Quality Index Method (QIM), biochemical, and proteomic methods. LWT-Food Science and Technology, 78(2017), 172-180. http://dx.doi.org/10.1016/j.lwt.2016.12.037

[9] Abbas, K. A., Mohamed, A., Jamilah, B., & Ebrahimian, M. (2008). A review on correlations between fish freshness and pH during cold storage. American Journal of Biochemistry and Biotechnology, 4(4), 416–421. https://doi.org/10.3844/ajbbsp.2008.416.421

[10] Barrett, D. M., Beaulieu, J. C., & Shewfelt, R. (2010). Color, flavor, texture, and nutritional quality of fresh-cut fruits and vegetables: Desirable levels, instrumental and sensory measurement, and the effects of processing. Critical Reviews in Food Science and Nutrition, 50(5), 369–389. https://doi.org/10.1080/10408391003626322

[11] Basto, A., Marques, A., Silva, A., Sá, T., Sousa, V., Oliveira, M. B. P. P., Aires, T., & Valente, L. M. P. (2023). Nutritional, organoleptic and sensory quality of market-sized European sea bass (Dicentrarchus labrax) fed defatted Tenebrio molitor larvae meal as main protein source. Aquaculture, 566(July 2022), 739210. https://doi.org/10.1016/j.aquaculture.2022.739210

[12] Hizbullah, H. H., Sari, N. K., Nurhayati, T., & Nurilmala, M. (2019). Quality changes of little tuna fillet (*Euthynnus affinis*) during chilling temperature storage. IOP Conference Series: Earth and Environmental Science, 404(1). https://doi.org/10.1088/1755-1315/404/1/012015

[13] Noor, T. A., Sudaryati, Ni Luh, G., & Widyantari, A. A. A. S. S. (2019). Evaluation of the histamine levels of tuna (*Euthynnus affinis*) based on the length of storage and its implications for health. Indonesian Journal of Global Health Research, 2(4), 1975–1984. https://doi.org/10.37287/ijghr.v2i4.250

[14] Khatouf R., Dahani S., Hariri O.E., Bouchriti N. (2024). The Relationship of Histamine Content in European Pilchard (Sardina pilchardus) with Freshness, Temperature, and Storage Duration. World Vet J, 14(1):66-73. https://dx.doi.org/10.54203/scil.2024.wvj9

[15] Altafini, A., Roncada, P., Guerrini, A., Sonfack, G. M., Accurso, D., & Caprai, E. (2022). Development of Histamine in Fresh and Canned Tuna Steaks Stored under Different Experimental Temperature Conditions. Foods, 11(24), 1–19. https://doi.org/10.3390/foods11244034

[16] Li, X., Tu, Z., Sha, X., Li, Z., Li, J., & Huang, M. (2022). Effect of coating on flavor metabolism of fish under different storage temperatures. Food Chemistry: X, 13 buk(February), 100256. https://doi.org/10.1016/j.fochx.2022.100256

[17] Morita, K., Kubota, K., & Aishima, T. (2003). Comparison of aroma characteristics of 16 fish species by sensory evaluation and gas chromatographic analysis. Journal of the Science of Food and Agriculture, 83(4), 289–297. https://doi.org/10.1002/jsfa.1311

[18] Jones, B. C., Rocker, M. M., Keast, R. S. J., Callahan, D. L., Redmond, H. J., Smullen, R. P., & Francis, D. S. (2022). Systematic review of the odorous volatile compounds that contribute to flavour profiles of aquatic animals. Reviews in Aquaculture, 14(3), 1418–1477. https://doi.org/10.1111/raq.12657

[19] Xiao, H., Yu, J., Song, L., Hu, M., Guo, H., Xue, Y., & Xue, C. (2022). Characterization of flesh firmness and ease of separation in the fermentation of sea bass in terms of protein structure, texture, and muscle tissue structural changes. Food Research International, 162(PA), 111965. https://doi.org/10.1016/j.foodres.2022.111965

[20] Nakazawa, N., & Okazaki, E. (2020). Recent research on factors influencing the quality of frozen seafood. Fisheries Science, 86(2), 231–244. https://doi.org/10.1007/s12562-020-01402-8

[21] Smida, M. A. Ben, Bolje, A., Ouerhani, A., Barhoumi, M., Mejri, H., Cafsi, M. El, & Fehri-Bedoui, R. (2014). Effects of Drying on the Biochemical Composition of Atherina boyeri from the Tunisian Coast. Food and Nutrition Sciences, 05(14), 1399–1407. https://doi.org/10.4236/fns.2014.514152

[22] Tavares, J., Martins, A., Fidalgo, L. G., Lima, V., Amaral, R. A., Pinto, C. A., Silva, A. M., & Saraiva, J. A. (2021). Fresh fish degradation and advances in preservation using physical emerging technologies. Foods, 10(4), 1–20. https://doi.org/10.3390/foods10040780

[23] Wang, X., McClements, D. J., Xu, Z., Meng, M., Qiu, C., Long, J., Jin, Z., & Chen, L. (2023). Recent advances in the optimization of the sensory attributes of fried foods: Appearance, flavor, and texture. Trends in Food Science and Technology, 138(March), 297–309. https://doi.org/10.1016/j.tifs.2023.06.012

[24] Lawless H.T., & Heymann H. (2010). Sensory Evaluation of Food: Principles and Practices Second Edition. London: Springer New York Dordrecht Heidelberg London. https://doi.org/10.1007/978-1-4419-6488-5

[25] Cardello A. V. (1995). Food quality: Relativity, context and consumer expectations. Food Quality and Preference, 06(3), 163-170. https://doi.org/10.1016/0950-3293(94)00039-X

[26] Jeong S., & Lee J. (2021). Effects of cultural background on consumer perception and acceptability of foods and drinks: a review of latest cross-cultural studies. Current Opinion in Food Science, 42, 248-256. https://doi.org/10.1016/j.cofs.2021.07.004

International Symposium on Food and Agro-Biodiversity 2024 (ISFA 2024) IOP Publishing
IOP Conf. Series: Earth and Environmental Science **1460** (2025) 012057 doi:10.1088/1755-1315/1460/1/012057

Factors affecting diversification of food consumption in poorer households in Samigaluh, Kulonprogo, Yogyakarta

Latifah Dita Cahyani, Wiwit Rahayu and Umi Barokah

[1] Major Agribusiness, Faculty of Agriculture, Sebelas Maret University, Surakarta, Indonesia

*umibarokah.uns@gmail.com

Abstract. Food diversification is an action to increase the variety of food types with the aim of obtaining more balanced nutrition from various nutritional sources. One of the obstacles in diversifying food consumption is limited income. Based on data from BPS in 2022, the percentage of poorer people in Kulonprogo Regency was 16.39%, while in Samigaluh District, it was 66.65%. This research aims to analyze the amount and diversification of food consumption of poorer households in Samigaluh, Kulonprogo and the influencing factors. This research uses primary data obtained from 70 respondents. The data analysis methods include energy consumption levels, desirable dietary patterns, and multiple linear regression. The research results show that household energy consumption is 5.160,6 kcal/day or 1.753,6 kcal/capita/day, with an energy consumption level of 81,9%. Diversification of food consumption on poorer households in Samigaluh is still lacking with a desirable dietary pattern score of 75,6. Number of household members have a negative effect on food consumption diversification, while expenditure on food consumption, nutritional knowledge, and perceptions of the benefits of the food consumption diversification program have a positive effect on food consumption diversification. Income and Education has no effect on food consumption diversification.

1. Background

Food is an essential basic human need. Based on Law No. 18 of 2012, food includes biological resources such as agricultural products, plantations, forestry, fisheries, livestock, waters, and water, both processed and unprocessed, that are used for human consumption. This includes food additives, food raw materials, and other materials used in the process of preparing, processing, or manufacturing food or beverages. Adequate food fulfillment will support the formation of competent and highly competitive Human Resources (HR), which in turn will realize national food security and reflect the success of development (Decree of the Minister of Health of the Republic of Indonesia in 2015; Ashari et al., 2012).

High poverty conditions, especially in the Special Region Province of Yogyakarta, hinder people's access to nutritious food. According to BPS data in 2022, the Special Region of Yogyakarta Province had the highest poverty rate in Java Island, with a percentage of 11.04%. Kulonprogo Regency, one of the districts in the province, had a high poverty rate, with Samigaluh District in 4th place with a percentage of 66.65% (BPS, 2022; Kulon Progo in Numbers, 2023).

Household food consumption habits are greatly influenced by the poverty level. Consumption of indiverse meals can result in health problems such as malnutrition and stunting due to insufficient nutritional intake. Ideally, a healthy diet includes 15% protein, 20% fat, and 65% carbohydrates (Irianto & Kusno, 2007). Food diversification, which is increasing the variety of food types consumed, is an important step to achieve food security. This can be measured through the desirable dietary pattern, which reflects the balance of energy and nutrients of various food groups (FAO-RAPA, 1989; Food Security Agency, Ministry of Agriculture, 2015).

Kulonprogo Regency has diverse local food potential, including 49 types of commodities such as peanuts and cassava (Rijanta et al., 2018). However, the implementation of food diversification is not optimal, the main obstacles are low public knowledge, food culture that is difficult to change, and slow food innovation. The tradition of prioritizing rice as a staple food hinders food diversification (Ariani, 2010). The Kulonprogo government has carried out various initiatives to support food independence, such as the "local food excitement" activity and the food independence program through field school activities for integrated crop management, conducting food processing and yard utilization training, as well as the sustainability of health activities at the integrated service post. These efforts are expected to be in accordance with food diversification to create qualified human resources (Satyarini et al., 2018).

2. Methods

2.1 Sampling Method, Location, Population and Technique

This study uses a descriptive analytical method, which is a research method by interpreting and interpreting the data obtained. This research is located in Samigaluh District, Kulonprogo, Yogyakarta because it is an area that has many poorer people and has a fairly diverse local food and there is a promotion related to food security where it can affect people's consumption patterns and food diversification. Samigaluh District is the poorest area in the Kulon Progo, Samigaluh is in fourth place poorest area with 29.824 people classified as poor (BPS, 2023).

The population in this study are poorer households in Samigaluh District. Based on data in 2022, there were three villages with the highest percentage of poorer people in Samigaluh District, namely Purwoharjo, Sidoharjo, and Kebonharjo with a total of 8,490 people (Department of Social Affairs of Kulon Progo Regency, 2022). The number of samples used was 70 poorer households. The number of samples in each village is determined by the proportional random sampling technique which is calculated with the formula:

$$ni = \frac{nk}{N}n$$

Information:

ni = Number of samples per village
nk = Number of poorer people from each village
N = Number of population
n = Number of samples taken

Table 1. Number of Selected Samples in Samigaluh District

No	Urban Village	Number of Poorer people	Data Poorer people from Urban Village	Selected Sample
1.	Purwoharjo	2.902	1.880	24
2.	Sidoharjo	3.908	120	32
3.	Kebonharjo	1.680	25	14
	Total	**8.490**	**2.025**	**70**

*Social Service of Kulonprogo Regency, 2022

2.2 Types and Techniques of Collection, as well as Data Analysis

The types of data used are primary data and secondary data. Primary data was obtained through data collection techniques with interviews, observations, 24-hour recalls, and recording. The primary data used were in the form of identity, income, expenses, number of family members, education, nutritional knowledge, and food consumed by poorer households in Samigaluh. Secondary data refers to information obtained from relevant agencies. The secondary data used in this study are poverty data in Java, poverty data in Kulonprogo, poverty data in Samigaluh, and energy consumption data in Kulonprogo Regency.

The data analysis method used to determine the amount of energy consumption is to calculate the energy consumption of each food consumed with the formula:

$$Gej = \left(\frac{BPj}{100}\right) x \left(\frac{Bddj}{100}\right) x\ KGej$$

Information:

Gej = Amount of energy from food j (kcal)
Cpm = Weight of food ingredients consumed (grams)
Bddj = Edible food ingredient j (percent)
KGej = Energy content from food 100 grams of food j(kcal)

The data analysis used to determine the diversification of food consumption is by calculating the desirable dietary pattern score. The calculation of the desirable dietary pattern score is carried out in accordance with the guidelines of the 2015 Food Security Agency as follows:

a. Grouping the food consumed into 9 food groups according to desirable dietary pattern standards.
b. The food consumed is converted to the same unit (grams/day)
c. Calculating the energy consumption of each food group
d. Calculating the energy contribution of each food group
e. Calculating the energy contribution of each food group to the Recommended Dietary Allowances
f. Calculating the actual score
g. Calculating desirable dietary pattern score

The data analysis method to determine the influence of socioeconomic factors on the diversification of consumption of poorer households in Samigaluh District, Kulonprogo, Yogyakarta was carried out by multiple linear regression test with the following equations:

$$Y = \alpha+\beta_1X_1+ \beta_2X_2+ \beta_3X_3+ \beta_4X_4+ \beta_5X_5+ \beta_6X_6+e$$

Information:

Y = Diversification of household food consumption (Desirable dietary pattern score)
X1 = Revenue (Rupiah/month)

X2 = Expenditure on food consumption (Rupiah/month)
X3 = Number of family members (People)
X4 = Education (Years)
X5 = Nutrition knowledge (score 1 – 10)
X6 = Perception of food consumption diversification programs (if "Feeling the role of the government" is worth 1 and if "Not feeling the role of the government" is worth 0)
e = *error*

3. Results and discussion

Kulonprogo Regency is one of the districts in the Special Region of Yogyakarta Province. Kulonprogo Regency has 12 sub-districts with an area of 586.28 km^2 which are located between 7° 38'42" – 7° 59'3" South Latitude and between 110° 1'37" – 110° 16'26" East Longitude. The population in Samigaluh District is dominated by the productive age population (15-64 years) with a total of 18,511 people, more detailed data regarding the respondents' age can be found in Table 2.

Tabel 2. Characteristics of Respondents by Age

No	Age	People	Percentage (%)
1.	0 – 14	0	0
2.	15 – 64	55	79
3.	>65	15	21
	Total	70	100
	Average	52	

*Primary data, 2024

The average number of respondent household members is 3-4 people with a percentage of 46% and the number of family members of 1-2 people is also high, which is 43%. Most of these respondents have formal education in the elementary school of primary school, which is as much as 44%. Household income refers to the income received by all members of the household. Most of the respondents' income comes from farming activities. Below is the distribution of household income for respondents in this study.

Tabel 3. Household Income of the Poorer People in Samigaluh District, Kulonprogo Regency 2024

No	Monthly income	People	Percentage (%)
1.	100.000 – 1.500.000	36	51
2.	1.500.001 – 2.500.000	17	24
3.	2.500.001 – 3.500.000	13	19
4.	>3.500.001	4	6
	Average	Rp1.670.556	

*Primary data, 2024

The average household income of respondents is IDR 1,670,556 which indicates that the average household income is classified as moderate based on income classification according to the Central Statistics Agency, but many households have low income, which is 51%.

3.1 Amount of Food Consumption of Poorer households

The amount of food consumption in a household reflects the intake of food and beverages consumed by all family members to meet their nutritional needs. This consumption was measured for 24 hours and converted in energy units (kcal) using the Nutrisurvey 2007 application. Based on the amount of household energy consumption of respondents, the level of household energy consumption can be known. Energy intake describes the percentage of energy consumption compared to the Recommended Dietary Allowances. Data related to energy consumption and Energy intake of poorer households in Samigaluh District can be seen in Table 5.

Table 4. Average Energy Consumption and Energy intake of Poorer households in Samigaluh District

Information	Average
Individual Energy Consumption (kcal/person/day)	1.753,6
Household Energy Consumption (kcal/day)	5.160,6
Individual Energy Adequacy Rate (kcal/person/day)	2.140,6
Household Energy Adequacy Rate (kcal/day)	6.299,4
Individual Energy Consumption Rate (%)	81,9
Household Energy Consumption Rate (%)	81,9

*Primary data, 2024

Based on Table 4. the average energy consumption of poorer households in Samigaluh District is 5.160,6 kcal/day while the energy consumption of individuals is 1.753,6 kcal/person/day. The average household Energy Adequacy Rate is 6.299,4 kcal/day, while the average Individual Energy Adequacy Rate is 1.753,6 kcal/person/day. Based on this data, it can be concluded that the energy consumption of the people of Samigaluh District is still below the recommended levels stipulated in Permenkes (Regulation of the Minister of Health) Number 28 in 2019. Energy intake of individuals and households is 81.9%. The consumption of poorer households in Samigaluh is dominated by the consumption of grain food, so households are highly dependent on staple foods, such as rice and corn. The distribution of Energy intake of poorer households in Samigaluh District can be seen in Table 5.

Table 5. Distribution of Energy intake Categories for Poorer households in Samigaluh District 2024

Energy Intake Category	Number of Households	Presented (%)
Severe Deficit (<70%)	12	17
Moderate Deficit (70 – 79%)	22	31
Light Deficit (80 – 89%)	16	23
Normal (90 – 119%)	19	27
More (≥120%)	1	1

*Primary data, 2024

Based on Table 5. It can be known the distribution of the energy intake category in respondents' households. The majority of households have energy intake in the deficit category, either a deficit of care, a moderate deficit, or a severe deficit with a percentage of 23%, 31%, and 17%, respectively, which means that poorer households have a energy intake in the deficit category of 71%. Meanwhile, poorer households that have a normal energy intake category are 31%.

The amount of energy consumed is related to the availability of household food. Food is a basic need for humans, both individually and as a group, because it is essential for growth. Food and nutrition play a crucial role in improving the quality of human life. Therefore, food must always be available in good quantity and quality. The availability of adequate food at the household level is very important for the improvement of family nutrition (Hartina *et al.*, 2020). The distribution of food availability for poorer households in Samigaluh District can be seen in Table 6.

Table 6. Distribution of Sources of Food Availability for Poorer households in Samigaluh District, Kulonprogo Regency 2024

Food	Source of Availability					
	OP	Since	Buy	OP, Give, Buy	OP, Give	OP, Buy
Grains	-	-	-	20	30	20
Tubers	39	-	31	-	-	-
Animal Food	-	-	27	-	-	43
Oils and Fats	-	-	70	-	-	-
Oily fruits/seeds	43	-	17	-	-	-
Legumes	-	-	69	-	-	1
Yolk	-	-	70	-	-	-
Vegetables and Fruits	45	-	-	-	-	25
Others	-	-	70	-	-	-

*Primary data, 2024
OP = goods produced independently by a household, often through agricultural activities, home gardening, or small-scale crafts and manufacturing
Since = resources through financial or material aid provided by government programs or non-governmental organizations (NGOs)
Buy = acquiring resources through financial transactions in a market or store
Give = receiving gifts from neighbors, friends, or family members represents a form of social exchange deeply embedded in communal and kinship networks.

The availability of household food in Samigaluh mostly comes from a combination of self-production and purchases. This is because the production carried out is only a small amount and there are production products that are sold to meet other needs. Grains are mostly obtained through self-production and aid, while oil and fats and sugars are entirely from purchase. Tubers such as cassava are mostly grown by themselves, while animal food is more often obtained from a combination of self-production and purchase.

Poorer households in Samigaluh District are highly dependent on rice as their primary energy source. Meanwhile, food production is insufficient and needs to be improved to meet their daily needs. Water limitations and the application of modern agricultural technology are obstacles in achieving optimal food security.

Diversification of household food consumption is an effort to consume a variety of foods from the availability of available food so as to obtain balanced nutrition. The calculation of food consumption diversification is carried out by calculating the desirable dietary pattern. The consumption of food and drink consumed by respondents for 24 hours based on the results of interviews with the 24-hour recall method. If the food consumed by respondents covers all food groups with the right amount, the desirable dietary pattern score will be high, which means that the diversification of people's food consumption is also high. The average data on the proportion of energy donations and the desirable dietary pattern of households in Samigaluh District in 2024 can be seen in Table 7.

Table 7. Average Proportion of Energy Contributions and desirable dietary pattern of Poorer households in Samigaluh District 2024

Food Group	Energy (kcal)	%		Weight	Shoes				+/-
		Current	AKE		Current	AKE	Maxim be	Desirable Dietary Pattern	
Grains	1.039,5	59,5	49,7	0,5	29,8	24,8	25	24,8	+
Tubers	79,3	4,5	3,8	0,5	2,3	1,9	2,5	1,9	-
Animal Food	60,6	3,5	2,9	2	6,9	5,8	24	5,8	-
Oils and Fats	44,8	2,6	2,1	0,5	1,3	1,1	5	1,1	-
Oily Fruits/Seeds	22,7	1,3	1,1	0,5	0,6	0,5	1	0,5	-
Legumes	254,3	14,5	12,1	2	29,1	24,2	10	10	+
Yolk	63,2	3,6	3,0	0,5	1,8	1,5	2,5	1,5	-
Vegetables and Fruits	146,8	8,4	7,0	5	42,0	35,0	30	30	+
Others	38,2	2,2	1,8	0	0,0	0,0	0	0	0
Sum	1.749,4	100,0	83,3		113,7	94,7	100	75,6	-

*Primary data, 2024

Based on Table 7. It can be known that the average score of the desirable dietary pattern of poorer households in Samigaluh District is 75.6, which means that it is still below the ideal desirable dietary pattern score. Ideal desirable dietary pattern score is 100. If the PPH score reaches 100, household food consumption is considered to be of high quality and diverse. The higher the score, the better the level of dietary diversity achieved (Ariani, 2010). Most food group consumption is still lacking. The food groups that have excess scores are rice, nuts, and vegetables and fruits, while the food groups that have low scores and need to be improved are tubers, animal foods, oils and fats, oily fruits/seeds, and sugar. However, the necessary increase must also be measured so as not to cause new problems due to excess food consumption. Excessive consumption of nuts because poorer households consume a lot of tofu and tempeh from soybeans as a source of protein because they cannot meet the needs of animal protein which has a higher price.

The low diversification of household food consumption in Samigaluh District is due to the fact that the local community consumes a lot of their own farming products in the form of rice, cassava, corn, and various vegetables. This makes people rarely consume animal based food such as meat, eggs, milk, and chicken that has a high weight in the calculation of desirable dietary pattern. This is also because the price of animal based food such as meat, eggs, milk, and chicken is more expensive so that people prefer to consume their crops. Many people do not pay attention to the nutrition consumed but only attach importance to consuming food and becoming full. This makes people consume more foods that contain carbohydrates, especially rice because most people think that they have not eaten if they have not eaten rice. People feel that consuming carbohydrates gives a feeling of fullness compared to consuming other foods, even though protein consumption provides a longer feeling of fullness, but the habits and stigmas of the community are still difficult to get rid of. This is in accordance with the statement that protein consumption helps satiety faster and has satiating properties than carbohydrates and fats, so protein source foods must always be on the daily menu (Cahyaningrum, 2015).

3.2 The Influence of Socio-Economic Factors on the Diversification of Food Consumption of Poorer households

Diversification of food consumption is an effort to consume a more diverse diet to meet nutritional needs. The measurement of food consumption diversification in this study is by calculating the desirable dietary pattern score. The difference in household income tax scores in Samigaluh District is influenced by several factors. These factors will affect the extent of the diversification of food consumption at the household level. The factors studied in this study are socio-cultural factors consisting of income, expenditure on food consumption, number of family members, education, understanding related to nutrition, and perception of food consumption diversification programs.

The analysis of the influence of socioeconomic factors on the diversification of food consumption was carried out using multiple linear regression analysis. The data used was obtained through direct interviews with poorer households in Samigaluh District. The form of function in the equation model of this research is as follows.

$Y = \alpha+\beta1X1+ \beta2X2+ \beta3X3+ \beta4X4+ \beta5X5+ \beta6X6+e$

$Y = 24{,}607 - 7{,}490E\text{-}6X1 + 1{,}774E\text{-}5X2 - 1{,}558X3 + 0{,}154X4 + 1{,}904X5 + 1{,}026X6$

Information:

And = Diversification of household food consumption (PPH Score)
X1 = Revenue (Rupiah/month)
X2 = Expenditure on food consumption (Rupiah/month)
X3 = Number of family members (People)
X4 = Education (Years)
X5 = Nutrition knowledge (score 1 – 10)
X6 = Perception of food consumption diversification programs (if "Feeling the role of the government" is worth 1 and if "Not feeling the role of the government" is worth 0)
and = *error*

Before the F test and t test were carried out to determine the influence of these factors on the diversification of food consumption, a classical assumption test was carried out. The following are the results of the classical assumption test. Normality test, the data to be used is distributed normally. The distribution of data on the resulting graph spreads around the diagonal line. Multicoloniarity Test, collinearity tolerance value of all variables was >0.1 and VIF value of all variables was <10, so it can be concluded that the data did not experience multicoloniality. Heterokedasticity Test, heterokedasticity test in this study used *a* sccaterplot. If the sccaterplot graph spreads randomly, it shows that there is no heterokedasticity in the regression model formed (Suliyanto, 2011). Scatterplot diagram it can be seen that the randomly spread points do not form a specific pattern and are scattered in the range of 0, so it can be concluded that the data does not experience heterokedasticity.

After the data met all the classical assumption tests, it was continued with multiple linear regression analysis consisting of R2 Test, F Test, and t Test. The following are the results of the multiple linear regression test in this study:

3.2.1 R^2 Test

The determination coefficient (R^2) test was carried out to find out how much independent variables could give rise to dependent variables. The determination coefficient is a value that measures how much the independent variable (x) contributes to the change of the dependent variable (y). In this study, an *Ajusted R Square* value of 0.628 was produced, which means that the independent variable can explain the dependent variable of 62.8%. These independent variables include income, expenditure on food consumption, number of family members, education, understanding related to nutrition, and perception of food consumption diversification programs. While the other 37.2% was influenced by other variables that were not studied in this study such as appetite and feeding by others.

3.2.2 Test F

The F test was carried out to find out whether the independent variables studied had a real effect on the dependent variables together. If the significance value F is <0.05, it can be interpreted that the independent variables together affect the dependent variables. In this study, the significance value of the data used in this study was 0.000 which means more than 0.05. Based on these results, it can be concluded that all independent variables together affect the dependent variables. Based on this, it can be interpreted that income, expenditure on food consumption, number of family members, education, nutritional knowledge, and perception of food consumption diversification programs together have a real effect on food consumption diversification.

3.2.3 Test t

The t-test was carried out to determine the influence of individual independent variables on the diversification of household food consumption in Samigaluh District. Independent variables that have a significance value of less than 0.05 are considered to have an effect on the diversification of household food consumption. The results of the t-test in this study can be seen in Table 8.

Table 8. Test Results t

Variable	B	t	Mr.
(Constant)	24,607		
Income	-7,490E-7	-0,853	0,397
Expenditure on Food Consumption	1,774E-5**	4,949	0,000
Number of Family Members	-1,558*	-2,544	0,013
Education	0,151	0,587	0,559
Understanding Nutrition	1,904**	5,065	0,000
Perception of food consumption diversification programs	1,026**	4,749	0,000

*Primary data, 2024
Information:
* = significant at a 95% confidence level
** = significant at a 99% confidence level

Based on Table 8. It can be seen that the variables of expenditure on food consumption, understanding related to nutrition, and perception of individual food consumption diversification programs have a real effect on the diversification of household food consumption in Samigaluh District. The income variable doesn't have a significant effect on the diversification of household food consumption. This is influenced by the fact that the income of the respondents is mostly in the low income group so that the increase in income will be used to increase rice consumption so that it does not have a significant effect on the Hope Food Pattern score. In low-income households, the increase in income actually increases rice consumption and reduces or replaces staple foods such as corn and cassava (Ariani and Ashari, 2003). The increase in income, which is still considered low, will boost energy consumption, but only for one food group. The Total Energy Consumption (TKE) will rise due to increased food consumption, but this does not happen evenly across all food groups. As a result, the desirable dietary pattern score will be low, indicating that food consumption diversification is not optimal. This occurs because households tend to prefer consuming rice, as it is considered more filling compared to other carbohydrate sources. Meanwhile, the increase in income is still insufficient to afford animal-based foods or other higher-quality foods, as the income remains in the low category.

The expenditure variable for food consumption has a significant effect on the diversification of food consumption. The increasing expenditure incurred by respondents on food consumption can be an indicator of high purchasing power, thereby increasing food diversification because more types of food are purchased. The higher the purchasing power of households, the more food diversification will be achieved (Hutagaol and Sinaga, 2022). The increase in expenditures for food consumption can lead to higher energy consumption. When households allocate more income for food consumption, they are likely to purchase larger quantities of food or foods with higher calorie content, which in turn increases their total energy consumption. An increase in food consumption expenditures not only boosts the quantity of food but also its quality. Therefore, an increase in food consumption expenditures influences the diversification of food consumption.

The variable of the number of family members has a significant effect on the diversification of food consumption but the correlation value is negative. This is because the number of family members who are more has more obstacles in meeting food needs such as more food needed and difficulty in dividing existing food so that all family members get food with good nutrition (Herawati, *et al.,* 2011). The larger the number of family members, the lower the food consumption per individual, as the available food must be shared among all members to ensure everyone gets enough. The distribution of food energy in households with more members can become more challenging. Households must ensure that each family member receives an adequate amount of calories. When food resources are limited, this may mean that not all family members get enough food energy, leading to energy deficiencies for some individuals, which could cause the total energy consumption to decrease. As a result, households may opt for lower-quality foods, allowing them to buy larger quantities and meet their energy consumption needs.

The educational variable don't have a significant effect on the diversification of household food consumption. The level of education don't have relationship with food consumption patterns so that education doesn't have a significant impact on the diversification of food consumption (Handayani and Yulistiyono, 2023). This happens because of many influencing factors such as economic limitations, food availability, habits, and access to nutrition-related information. This is consistent with the study by Dewanti et al. (2020), which states that although education broadens knowledge, without other factors such as food access and economic stability, achieving food consumption diversification becomes difficult.

The nutrition-related understanding variable has a significant effect on diversification of food consumption. In an effort to diversify food, a mother needs to have knowledge about food ingredients and nutrients, as well as understand how to serve and manage them (Mufida, *et al.,* 2020). This is because if housewives know about nutritional needs, it will increase their ability to make healthier choices with available food resources. The variable of perception of the food consumption diversification program has a significance value of 0.000 so that it can be interpreted that the variable of perception of the food consumption diversification program has a real effect on the diversification of food consumption.

The Perception variable for the food consumption diversification program has a significant effect on diversification of food consumption. The role of the government that has a real influence on public perception in terms of food diversification is the role as a communicator and the role as a motivator (Azhari, *et al.,* 2013). The role of the government in Samigaluh District has been to provide support for increasing food consumption diversification through socialization on food diversity and the provision of plant seedlings, thereby enhancing local food diversity. In addition, socialization on nutrition is also carried out through Posyandu (integrated health posts) and the Family Welfare Empowerment program. Through Posyandu and Family Welfare Empowerment program, the knowledge of household mothers regarding nutrition can be improved, which in turn helps to increase food consumption diversification.

4 Conclusion

The average household food consumption in Samigaluh District still has not reached the ideal level of energy consumption, which is 5,160.6 kcal/day or 1,753.6 kcal/person/day. This is due to some households that have not met the energy consumption recommendations. The majority of poorer households in Samigaluh District are included in the Energy intake deficit category, with an average Energy intake of 81.9%. Diversification of household food consumption has not reached the ideal figure. This is seen from the household income tax score of 75.6 below the ideal score of 100. The variable of expenditure variables for food consumption, understanding related to nutrition, and perception of food consumption diversification programs had a effect on the diversification of household food consumption in Samigaluh District. Meanwhile, income and education variables did not have a effect on the diversification of household food consumption in Samigaluh District. Income does not affect the expected dietary pattern because the respondents' income is relatively low, so any increase has not been sufficient to significantly improve food quality. Education does not have a significant effect because, without the availability of resources and specific knowledge about proper food consumption, it cannot make a meaningful impact.

References

[1] Ariani, M., Ashari. (2003). Arah, Kendala dan Pentingnya Diversifikasi Konsumsi Pangan di Indonesia. *Forum Penelitian Agro Ekonomi*, 21(2), 99 – 112.

[2] Ariani, M. (2010). Analisis Konsumsi Pangan Tingkat Masyarakat Mendukung Pencapaian Diversifikasi Pangan *Gizi Indon*, 33(1), 20-28.

[3] Ashari, Saptana, & Purwantini, T. B. (2012). Potensi dan Prospek Pemanfaatan Lahan Pekarangan untuk Mendukung Ketahanan Pangan. *Forum Penelitian Agro Ekonomi* 30(1), 13-30.

[4] Azhari, R., Muljono, P., & Tjitropranoto, P. (2013). Peran Penyuluh Dalam Peningkatan Diversifikasi Pangan Rumah Tangga. *Jurnal Agro Ekonomi*, 31(2), 181 – 198.

[5] Badan Ketahanan Pangan Kementerian Pertanian. (2015). *Panduan Penghitungan Pola Pangan Harapan (PPH)*. Jakarta: Badan Ketahanan Pangan Kementerian Pertanian.

[6] Badan Pusat Statistik. (2023). *Kabupaten Kulonprogo Dalam Angka Tahun 2023.* Kulonprogo: Badan Pusat Statistik.

[7] Badan Pusat Statistik. (2023). *Kecamatan Samigaluh Dalam Angka Tahun 2023.* Kulonprogo: Badan Pusat Statistik.

[8] Cahyaningrum, A. (2015). Lepitin Sebagai Indikator Obesitas. *Jurnal Kesehatan Prima* 9(1), 1364 – 1371.

[9]Dewanti, S., Rijanta, R., & Rofi, A. (2020) Keragaman Konsumsi Pangan Rumah Tangga di Provinsi Jawa Tengah. Jurnal Kawistara 3(10), 265 – 294.

[10] Handayani, S., & Yulistiyono, H. (2023). Pengaruh Pendapatan, Jumlah Anggota Keluarga Dan Pendidikan Terhadap Konsumsi Rumah Tangga Miskin Di Kabupaten Banyuwangi. *Neo-Bis*, 12(1), 32–47.

[11]Hartina, Laenggeng, A. H., & Nurjanah. (2020). Hubungan Pola Makan dan Ketersediaan Pangan Rumah Tangga dengan status Gizi remaja di Huntara asam III Kec. Ulujadi Kota Palu. *Jurnal Kolaboratif sains*, 3(4), 203 – 209.

[12] Herawati, T. et al. (2011). Ketahanan Pangan Keluarga Peserta Program Pemberdayaan Masyarakat di Pedesaan. *Jurnal Gizi dan Pangan*, 6(3), 208 – 216.

[13] Hutagaol, M. P. & Sinaga, R. (2022). The Influence of Food Income and Prices on Food Diversification in Java. *Saentific Journal of Reflection: Economic, Accounting, Management and Business*, 5(3), 702 – 715.

[14] Irianto, K., & Kusno, W. (2007). *Gizi dan Pola Hidup Sehat*. Bandung: Yrama Widya.

[15] Mufida, L., Sartono, A., & Mufnaetty. (2020). Pengetahuan Gizi dan Praktik Diversifikasi Makanan Keluarga Kelurahan Purworejo, Kecamatan Margoyoso, Pati. *Jurnal Gizi Unimus*, 9(2), 180 – 188.

[16] Republik Indonesia. (2012). *Pangan Lokal dalam Pembangunan Pedesaan: Dimensi Sosio-Spasial Pangan Lokal di Kabupaten Kulonprogo. Yogyakarta* Undang-undang Nomor 18 Tahun 2012 tentang Pangan. Sekretariat Negara. Jakarta.

[17] Rijanta et al. (2018). *Pangan Lokal dalam Pembangunan Pedesaan: Dimensi Sosio-Spasial Pangan Lokal di Kabupaten Kulonprogo.* Yogyakarta: Penerbit Pustaka Pelajar Yogyakarta.

[18] Satyarini, T. B., Widodo, & Mashithoh, H. (2018). Aspek Ketahanan Pangan dan Program Desa Mandiri Pangan (Studi Kasus Desa Pagerharjo, Kecamatan Samigaluh, Kabupaten Kulonprogo). *Seminar Nasional*. Universitas Udayana. Bali.

[19] Sihabudin, et al. (2021*). Ekonometrika Dasar Teori dan Praktik Berbasis SPSS*. Banyumas: CV. Pena Persada.

[20] Suliyanto. (2011). Ekonometrika Terapan. Yogyakarta: ANDI.

In silico approach to predict antidiabetic activity of bioactive peptide from *Leucaena leucocephala* seed during gastrointestinal digestion

Dwi Ishartani[1,2*], Danar Praseptiangga[1,2], Setyaningrum Ariviani[1,2] and Muhammad Zukhrufuz Zaman[1,2]

[1] Doctoral Program of Agricultural Science, Faculty of Agriculture, Universitas Sebelas Maret, Surakarta, Indonesia
[2] Department of Food Science and Technology, Universitas Sebelas Maret, Surakarta, Indonesia

*E-mail: dwiishartani@staff.uns.ac.id

Abstract. *Leucaena leucocephala* (leucaena) was an easy growing species with year-round flowering and fruiting and abundant seed production. Its mature seed contained 30.81%-50.37% protein that potential to be explored as bioactive peptide source, as well as nutrition source. Previous study showed that the seed extract has antidiabetic activity. However, study of bioactive peptide role on antidiabetic activity of leucaena was limited. *In silico* approach recently was widely used to predict the potential of bioactive peptide derived from certain protein source due to its supporting result to laboratory experimental approach. This study aimed to investigate the antidiabetic potential of bioactive peptide of leucaena's protein during gastrointestinal digestion. *In silico* digestion and toxicity prediction were performed using BIOPEP and ToxinPred online tools, respectively. After *in silico* digestion using pepsin, trypsin and chymotrypsin, the result showed that leucaena seed was a potential source of bioactive peptide with DPP-4 inhibitor, alpha glucosidase inhibitor, glucose absorption stimulation and antioxidative activities. The potential peptide with biological activity was predicted as safe.

Keywords: *Leucaena leucocephala*, in silico, antidiabetic, peptide, gastrointestinal

1. Introduction

Leucaena leucocephala was a tropical species belong to the family of Fabaceae. *Leucaena leucocephala* called by numerous names, such as leucaena, lead tree, white popinac in English, lamtoro in Indonesian, and klandingan or mlanding in Javanese. It was an invasive species that can grow easily due its year-round flowering and fruiting, hard seed coat, ability to resprout after cutting or fire attack, and abundant seed production. Its tree could be utilized as erosion control, shade, reclamation, nitrogen fixing, soil improver, ornamental, barrier support and intercropping. Its parts also have been used as variety of purposes, including pod, seed and leaf tip for food, leaves for fodder, flower that bloom throughout year for apiculture, wood for timber, fuel and fibre [1,2].

Leucaena leucocephala seed turned from green into dark brown and shining when mature, as well as the pod. Major constituents of mature seed dry matter were fibre, protein, oil and mineral [3]. It contained 5.44% oil, consisting of 21.4% oleic acid, 13.21% palmitic acid, 5.92% stearic acid and 175.47mg/100g α-tocopherol [4]. Protein content of the mature seed was reported range from 30.81% [5] to 50.37% which consist of globulin 43.5%, albumin 28.4%, and glutein 25% [6]. Mimosine [6] and trypsin inhibitor [7] also have been found in mature seed. This anti nutritional factor limited *Leucaena leucocephala* digestibility [8]. On the other hand, with its high content of protein, year-round fruiting and abundant seed production, leucaena was a promising source of food protein. Protein was an important food ingredient that responsible to the food properties, a nutritional ingredient that responsible to the body's growth, maintenance and source of energy and also a functional ingredient that attributed to the bioactive peptides.

Bioactive peptides could be produced from parent protein during gastrointestinal digestion using digestive enzymes and during food processing using enzymatic hydrolysis or chemical hydrolysis [9]. Various biological activities from bioactive peptides have been reported and several of them acted as metabolic enzymes inhibitor. Thus, bioactive peptide was potential to be developed as therapeutic agent in managing metabolic diseases [10,11].

Bioactive peptides derived from leucaena mature seed protein fraction and also from germinated leucaena flour showed biological activity, especially related to antihypertensive and antidiabetic. Enzymatic hydrolysis using alcalase, trypsin and alpha chymotrypsin of globulin and glutelin fraction showed *in vitro* antioxidant, Angiotensin Converting Enzymes (ACE) inhibitory and alpha amylase inhibitory activity [12]. Either peptide extract or peptide fraction (BM <3kDa) from pepsin-pancreatin hydrolysis of germinated leucaena flour had *in vitro* ACE inhibitory activity [13,14]. However, so far study on antidiabetic activity of leucaena seed using digestive enzyme has not been reported.

In silico or bioinformatic approaches now days were used to evaluate the potential of proteins as precursors of bioactive peptides. This computational method could manage, curate and interpret bioactive peptide information [15]. *In silico* approach prior to experimental studies was a good choice due to its time-saving and economically benefits. Experimental studies performed supporting results with *in silico* approach [16]. The aim of this study was to perform an *in silico* evaluation of *Leucaena leucocephala* proteins as potential precursors of biologically active peptides. An *in silico* enzymatic hydrolysis of the selected proteins was perform using digestive enzymes. The obtained peptides were evaluated according to their antidiabetic activities using *in silico* approach.

2. Materials and Methods

2.1 Protein sequence

Protein sequence of *Leucaena leucocephala* was chosen from UniProtKB/Swiss-Prot (reviewed) database(https://www.uniprot.org/uniprotkb?query=leucaena+leucocephala&facets=reviewed%3Atrue).

2.2 Hydrolysis simulation

Potential of biological activity, degree of hydrolysis and enzymatic hydrolysis was simulated using gastrointestinal digestion enzyme, including pepsin, trypsin and chymotrypsin using BIOPEP tool (https://biochemia.uwm.edu.pl/biopep/peptide_data.php).

2.3 Selection of antidiabetic peptide

Potential peptide with antidiabetic activity was selected using PeptideRanker tool (http://distilldeep.ucd.ie/PeptideRanker/).

2.4 Toxic prediction

Toxic prediction was perform using ToxinPred tool (https ://webs.iiitd.edu.in/ragha va/toxin pred/index .html).

3. Results and Discussion

3.1 Protein sequence

Protein sequence used in this research was chosen from UniProtKB/Swiss-Prot (reviewed) database that manually-annotated, i.e. trypsin inhibitor (ID P83036 with amino acid length 179, **Figure *1*** (a)) and ferredoxin (ID P00225 with amino acid length 96, **Figure *1*** (b)). Trypsin inhibitor was purified from *L. leucocephala* cotyledons [7] whether ferredoxin from the leaves [17].

a)

```
        10         20         30         40         50
QVLVDLDGDP LYNGMSYYIL PVARGKGGGL ELARTGSESC PRTVVQTRSE

        60         70         80         90        100
TSRGLPARLA SPYRILIGSN IPLTIEFQPQ KPYSCHGHSS RSLQWKVEKT

       110        120        130        140        150
QMVKIASSDE EQRLFGPFQI QPYRNHYKLV YCESESRNHH DDCRDLGISI

       160        170
DDQQNRLLVV KNGDPLVVQF AKANRGGDD
```

b)

```
        10         20         30         40         50
AFKVKLLTPD GPKEFECPDD VYILDQAEEL GIELPYSCRA GSCSSCAGKL

        60         70         80         90
VEGDLDQSDQ SFLDDEQIEE GWVLTCAAYP RSDVVIETHK EEELTG
```

Figure 1. Protein sequences of leucaena. a) Trypsin inhibitor (ID P83036); b) Ferredoxin (ID P00225)

3.2 Potential of biological activity and degree of hydrolysis

Both trypsin inhibitor and ferredoxin had potential as parent protein for bioactive peptide with antidiabetic activity as well as with antihipertensive action (Figure 2). Trypsin inhibitor was better source of peptide with alpha glucosidase inhibitor and stimulating glucose uptake activity, meanwhile ferredoxin was better source of peptide with antioxidative, DPP-IV inhibitor and ACE inhibitor activity.

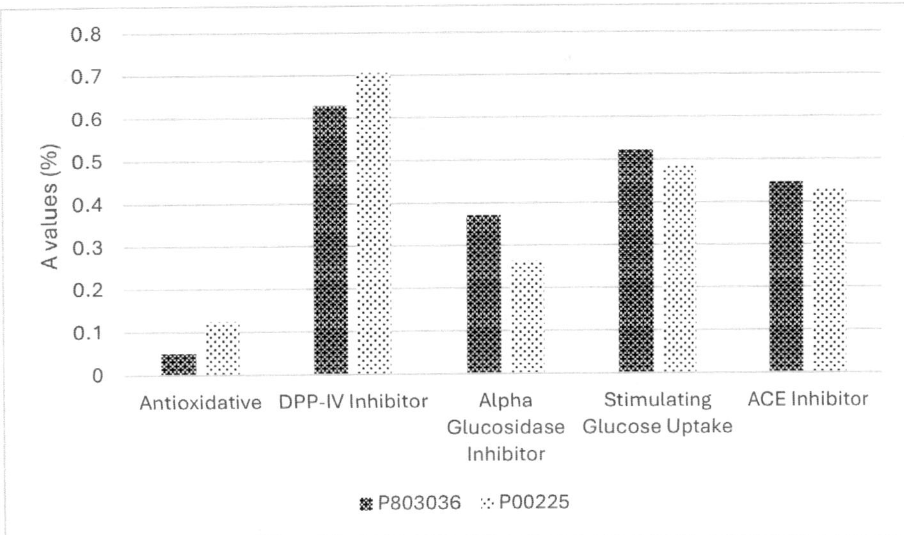

Figure 2. Biological activity potential of parent protein from *L. leucocephala*

Hydrolysis was simulated using gastrointestinal enzyme combination, i.e. pepsin pH 1.3 (later called pepsin 1.3), pepsin pH>2 (later called pepsin >2), trypsin, chymotrypsin A and chymotripsin C. Both trypsin inhibitor and ferredoxin hydrolysis using pepsin >2 generated higher degree of hydrolysis (more than 70%) than of using pepsin 1.3 (less than 50%), as can seen in Figure 3. When pepsin >2 combined with trypsin followed by chymotrypsin A or chymotrypsin C the degree of hydrolysis was increase from 70.52%, 73.68% and 77.89% respectively. This result was similar to the gastrointestinal digestion simulation of protein from jack bean (*C. ensiformis*) [18].

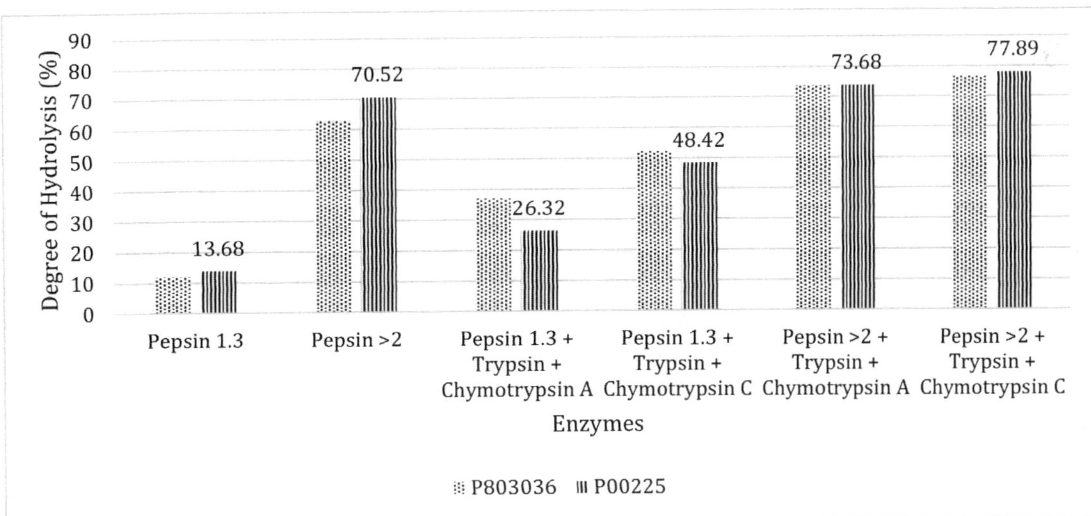

Figure 3. Degree of hydrolysis of leucaena's gastrointestinal digestion

3.3. Potential peptide with antidiabetic activity and its toxic prediction

As showed in Figure 4, gastrointestinal simulation of *L. leucocephala* produced peptides with antidiabetic activity, i.e. antioxidative, stimulating glucose uptake, dipeptidyl peptidase IV inhibitor and alpha glucosidase inhibitor activity. Antidiabetic activity of bioactive peptide was reported through several mechanism, such as inhibitory of carbohydrate digestion and glucose homeostasis management. Alpha amylase inhibitor and alpha glucosidase inhibitor worked to delay carbohydrate digestion. DPP-IV inhibitor, stimulating glucose uptake and antioxidative, worked to maintain glucose homeostasis [19–22].

Hydrolysis using combination of pepsin >2 + trypsin + chymotriypsin A or chymotrypsin C resulted greatest number and bioactivities both for trypsin inhibitor and ferredoxin. However, hydrolysis using pepsin 1.3 resulted least sequences number and least bioactivities (Figure 4 and Table 1). At pH 1.3 pepsin works less specific compare to at pH >2 [23].

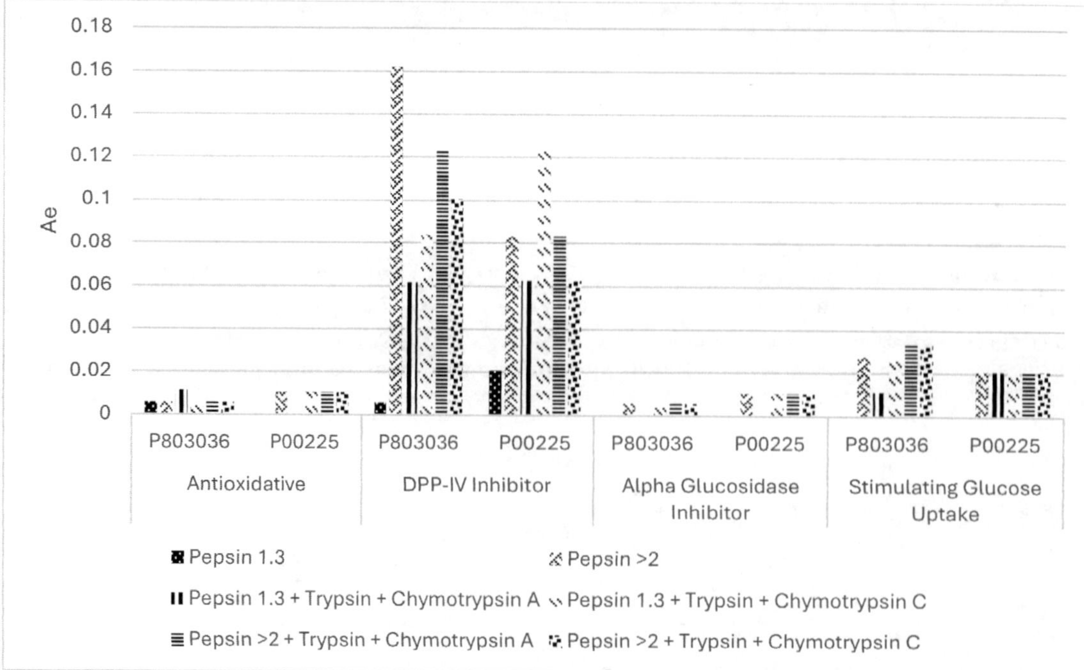

Figure 4. Potential of peptide with antidiabetic activity from leucaena protein hydrolyzed using digestive enzymes

Potential peptide with antidiabetic activity was predicted using PeptideRanker with threshold over 0.8 to reduce the false positive rate. However, peptide with threshold 0.5 was labelled as bioactive. Nine potential bioactive peptide with antidiabetic activity was selected from two protein parents using six digestive enzyme combinations. GL, PL, GP, QW, SF, AF, GPF, GF PF sequences were compiled from the lowest to the highest threshold, as shown in Table 2. DPP-IV inhibitor was the prominent activity related to antidiabetic, followed by antioxidative wich appeared only in GW sequence. GL, PL, GP, SF, AF, GW sequence also showed antihypertensive avtivity beside antidiabetic activity. All of the nine peptide sequences were predicted as non-toxic with net charge 0 and pI 5.88. The molecular weigh was ranged from 188.25 – 332.38 Da. Further analysis using *in vitro* and *in vivo* technique is necessary to prove this prediction since *in silico* approach is only a supporting tool to screen bioactive peptide but not to claim them [24].

Table 1. Peptide sequence from digestive hydrolysis simulation of leucaena protein*

Parent Protein	Enzymes**	Antioxidative	DPP-4 Inhibitor	Alpha glucosidase Inhibitor	Glucose uptake stimulating
Trypsin inhibitor	Pepsin 1.3	EL	GPF		
	Pepsin >2	VY	IL, PK, PY, SF, VE, VK, VL, VY	VE	VL, IL
	Pepsin 1.3 + Trypsin + Chymotrypsin A	EL, VY	SL, GL, GH, IL, PY, QW, SY, VK, VY, GPF		IL
	Pepsin 1.3 + Trypsin + Chymotrypsin C	VY	GP, IP, SL, GL, HY, IL, IQ, SY, TQ, TR, VE, VK, VL, VY	VE	VL, IL
	Pepsin >2 + Trypsin + Chymotrypsin A	VY, VE,	VA, PA, IA, SL, PL, IL, IQ, PF, PQ, PY, SY, VD, VE, VK, VL, VQ, VY		VL, IL
	Pepsin >2 + Trypsin + Chymotrypsin C	VY,	VA, IP, SP, IA, SL, HY, IL, IQ, SY, VD, VE, VK, VL, VQ, VY	VE	VL, IL
Ferredoxin	Pepsin 1.3		AF, TG		
	Pepsin >2	VY	IL, PK, PY, SF, VE, VK, VL, VY	VE	VL, IL
	Pepsin 1.3 + Trypsin + Chymotrypsin A		AF, IL, PY, TG, VK, VL		VL, IL
	Pepsin 1.3 + Trypsin + Chymotrypsin C	GW	TP, AE, AF, DQ, GW, IL, SF, TG, VE, VK, VL	VE	VL, IL
	Pepsin >2 + Trypsin + Chymotrypsin A	VY	IL, PK, PY, SF, VE, VK, VL, VY	VE	VL, IL
	Pepsin >2 + Trypsin + Chymotrypsin C	VY	IL, SF, VE, VK, VL, VY	VE	VL, IL

* Peptide sequence is presented in one-letter symbol based on the IUPAC recommendation
** Pepsin 1.3 works at pH 1.3, cleaves more specific; pepsin > 2 works at pH >2, cleaves less specific

Table 2. Selected peptide sequences with antidiabetic activity

Peptide	Threshold	Activity	Toxic Prediction	Charge	pI	Molecular Weight
GL	0.808777	DPP-IV inhibitor, ACE inhibitor	Non-toxin	0	5.88	188.25
PL	0.811148	DPP-IV inhibitor, ACE inhibitor	Non-toxin	0	5.88	228.31
GP	0.905487	DPP-IV inhibitor, ACE inhibitor	Non-toxin	0	5.88	172.20
QW	0.928524	DPP-IV inhibitor	Non-toxin	0	5.88	332.38
SF	0.948796	DPP-IV inhibitor, ACE inhibitor	Non-toxin	0	5.88	252.28
AF	0.973259	DPP-IV inhibitor, ACE inhibitor	Non-toxin	0	5.88	236.28
GPF	0.989324	DPP-IV inhibitor	Non-toxin	0	5.88	319.39
GW	0.993164	DPP-IV inhibitor, antioxidative, ACE inhibitor	Non-toxin	0	5.88	261.30
PF	0.993430	DPP-IV inhibitor	Non-toxin	0	5.88	262.32

4. Conclusion

From this *in silico* analysis we can predict that gastrointestinal hydrolysis of *Leucaena leucocephala* protein might produce bioactive peptide with antidiabetic activity. Nine peptide sequences were selected as potential bioactive peptide with dominant action as DPP-IV inhibitor. The nine sequences were predicted as non-toxic peptides. A further *in vitro* and *in vivo* analysis would be beneficial to confirm this *in silico* data.

Acknowledgments

This work was financially supported by research project of Non APBN UNS No: 194.2/UN27.22/PT.01.03/2024 from Universitas Sebelas Maret, Indonesia.

References

[1] Orwa C, Mutua A, Kindt R, Jamnadass R and Anthony S 2009 Leucaena leucocephala (Lam.) de Wit *Agroforestree Database: a tree reference and selection guide version 4.0* 1–9

[2] Parrotta J A 1992 Leucaena leucocephala (Lam.) de Wit L 1–8

[3] Adeneye J A 1979 A Note on The Nutrient and Mineral Compositin of Leucaena leucocephala in Western Nigeria *Anim Feed Sci Technol* **4** 221–5

[4] Nehdi I A, Sbihi H, Tan C P and Al-Resayes S I 2014 Leucaena leucocephala (Lam.) de Wit seed oil: Characterization and uses *Ind Crops Prod* **52** 582–7

[5] Ekpenyong T E 1986 Nutrient and amino acid composition of Leucaena leucocephala (Lam.) de Wit *Anim Feed Sci Technol* **15** 183–7

[6] Sethi P and Kulkarni P R 1993 Fractionation of Leucaena seed-kernel proteins based on their solubility characteristics *Food Chem* **48** 173–7

[7] Oliva M L V, Souza-Pinto J C, Batista I F C, Araujo M S, Silveira V F, Auerswald E A, Mentele R, Eckerskorn C, Sampaio M U and Sampaio C A M 2000 Leucaena leucocephala serine proteinase inhibitor: primary structure and action on blood coagulation, kinin release and rat paw edema 1 *Biochim Biophys Acta* **1477** 64–74

[8] Sethi P and Kulkarni P R 1993 *In vitro* protein digestibility of Leucaena leucocephala seed kernels and protein isolate *Food Chem* **46** 159–62

[9] Ulug S K, Jahandideh F and Wu J 2021 Novel technologies for the production of bioactive peptides *Trends Food Sci Technol* **108** 27–39

[10] Awosika T O and Aluko R E 2019 Inhibition of the *in vitro* activities of α-amylase, α-glucosidase and pancreatic lipase by yellow field pea (Pisum sativum L.) protein hydrolysates *Int J Food Sci Technol* **54** 2021–34

[11] Baco N, Oslan S N H, Shapawi R, Mohhtar R A M, Noordin W N M and Huda N 2022 Antibacterial activity of functional bioactive peptides derived from fish protein hydrolysate *IOP Conference Series: Earth and Environmental Science 967* vol 967 (IOP Publishing Ltd) pp 1–12

[12] Balderaz-Leon I, Baigts-Allende D and Cardador-Martínez A 2021 Antioxidant, angiotensin-converting enzyme, and α-amylase inhibitory activities of protein hydrolysat of Leucaena leucocephala seeds *CYTA – JOURNAL OF FOOD* **19** 349–359

[13] Fitriani A, Indrati R, Marsono Y and Supriyadi S 2022 Angiotensin-I-Converting Enzyme Inhibitory (ACE-I) Peptide from Germinated Lamtoro Gung (Leucaena leucocephala ssp. glabrata (Rose) S. Zarate) Flour *Sains Malays* **51** 3703–14

[14] Fitriani A, Indrati R, Marsono Y and Supriyadi S 2022 Impact of Gastrointestinal Digestion Simulation on the Formation of Angiotensin-I-Converting Enzyme Inhibitory (ACE-I) Peptides from Germinated Lamtoro Gung Flour *Foods* **11**

[15] Li-Chan E C Y 2015 Bioactive peptides and protein hydrolysates: Research trends and challenges for application as nutraceuticals and functional food ingredients *Curr Opin Food Sci* **1** 28–37

[16] Kartal C, Kaplan Türköz B and Otles S 2020 Prediction, identification and evaluation of bioactive peptides from tomato seed proteins using in silico approach *Journal of Food Measurement and Characterization* **14** 1865–83

[17] Benson A M and Yasunobu K T 1969 Non-heme Iron Proteins. X. The Amino Acid Sequences of Ferredoxins from Leucaena glauca *J Biol Chem* **244** 955–63

[18] Zulvana A H, Andriati N, Sri A and Widiastuti S 2019 In silico approach in evaluation of jack bean (Canavalia ensiformis) canavalin protein as precursors of bioactive peptides with dual antioxidant and angiotensin i-converting enzyme inhibitor *Materials Science Forum* vol 948 MSF (Trans Tech Publications Ltd) pp 85–94

[19] Jahandideh F and Wu J 2022 A review on mechanisms of action of bioactive peptides against glucose intolerance and insulin resistance *Food Science and Human Wellness* **11** 1441–54

[20] Patil S P, Goswami A, Kalia K and Kate A S 2020 Plant-Derived Bioactive Peptides: A Treatment to Cure Diabetes *Int J Pept Res Ther* **26** 955–68

[21] Nazarian-Samani Z, Sewell R D E, Lorigooini Z and Rafieian-Kopaei M 2018 Medicinal Plants with Multiple Effects on Diabetes Mellitus and Its Complications: a Systematic Review *Curr Diab Rep* **18**

[22] Ríos J L, Francini F and Schinella G R 2015 Natural Products for the Treatment of Type 2 Diabetes Mellitus *Planta Med* **81** 975–94

[23] Keil B 1992 *Specificity of Proteolysis* (Berlin: Springer-Verlag)

[24] Rivero-Pino F, Millan-Linares M C and Montserrat-de-la-Paz S 2023 Strengths and limitations of in silico tools to assess physicochemical properties, bioactivity, and bioavailability of food-derived peptides *Trends Food Sci Technol* **138** 433–40

Application of the Check All That Apply Method to Determine Sensory Profile of Commercial UHT Full-Cream Milk in Indonesia

Rafli Zulfa Kamil*, Heni Rizqiati, Nurwantoro, Nafisah Zainuba Hasan, Adam Permana Nur Saputra and Faiza Aisya Rizky

Food Technology Study Program, Department of Agriculture, Faculty of Animal and Agricultural Sciences, Universitas Diponegoro, Semarang 50275, Indonesia.

*E-mail: rafli.zulfa@live.undip.ac.id

Abstract. The quality of sensory attributes can influence consumer acceptance of UHT full-cream milk. The consumption of commercial dairy products has led to competition among manufacturers to produce milk that consumers consider ideal based on its sensory qualities. This study focused on identifying the sensory attributes and preference mapping of ideal products in full-cream UHT milk. Fifty untrained panelists who met the inclusion criteria were involved in sensory evaluation using the CATA method. Fifteen trained panelists determined the sensory attributes of the test products to be included in the research questionnaire through FGDs. Untrained panelists performed sensory evaluations by ticking the ideal sensory attributes, identifying the sensory attributes present in the products, and scoring their level of preference on a seven-point scale. The CATA method successfully profiled the ideal sensory attributes in full-cream UHT milk products, which were ivory white color, milky aroma, creamy taste, smooth texture, light texture, and aftertaste aroma. The product closest to the ideal is the one coded "173." This study recommends the identified sensory attributes and ideal products for UHT full-cream milk for product optimization and development.

Keywords: check all that apply, sensory profile, UHT full cream milk, consumer preference

1. Introduction

Milk is the most widely consumed food around the world. Milk commonly found in the market has typically undergone sterilization [1]. The sterilization method usually used is heating at high temperatures with the help of steam pressure. In contrast, the ultra-high temperature (UHT) method uses high temperatures quickly [2]. Increased milk consumption has led manufacturers to offer milk in various brands, flavors, and package sizes. The variety of dairy products on the market can make it difficult for consumers to choose milk that suits their preferences.

The production of full-cream UHT milk in Indonesia has increased significantly in recent years, increasing consumer awareness of the sensory quality of milk. However, this production process involves high-temperature heating that can damage the sensory quality. According to research by Andrés et al [3], the flavor of milk often falls short of expectations and is not preferred by most consumers. The imbalance between national milk supply and consumer demand encourages dairy producers to use skim and whey powder to increase total solids, affecting milk's viscosity and its less favorable taste. In addition, the Foot and Mouth Disease (FMD) outbreak in Indonesia recently, exacerbated this situation with a decline in national fresh milk production by 2022, thereby increasing fresh milk prices due to supply shortages.

Competition among producers of full-cream UHT milk is increasingly intense, prompting the need for strategies to attract consumer attention. Sensory evaluation can help identify ideal sensory attributes that match consumer preferences to increase sales [4]. Sensory evaluation methods usually involve trained panelists, which require time and money. Manufacturers need to be efficient in determining the sensory attributes that consumers prefer. Check All That Apply (CATA) is a practical solution to this problem which helps to identify the sensory attributes expected by consumers [5,6]. Unlike other sensory evaluation techniques, the CATA method is easier for participants to use, faster to conduct, reduces potential biases, provides broader insights, supports complex data analysis, and is highly flexible for different products and attributes. The identification of ideal sensory attributes can increase the producer's profit. Therefore, this study aims to determine the sensory attributes of commercial UHT full-cream milk in Indonesia by using the CATA method.

2. Materials and Methods

2.1 Analisis of Physicochemical Properties

Physicochemical analysis was conducted to determine product characteristics and their relationship with consumer acceptance. Physicochemical characteristics were observed for pH, Total Dissolved Solids (TDS), viscosity, and color. The pH value was analyzed by using a calibrated pH meter (pH 009i, China), total dissolved solids using refractometer (ATAGO, Japan), viscosity using a viscometer (Brookfield, USA), and color using color reader (CR-20, Konica Minolta, Japan).

2.2 Panelist Selection

The ISO 8586:2012 standard on sensory analysis selected panelists in this study, which provides guidelines for selecting, training, and monitoring experienced panelists. Panelists selection was also approved by the Health Research Ethics Commission, Faculty of Medicine, Diponegoro University No. 400/EC/KEPK/FK-UNDIP/VIII/2023. Untrained panelist candidates were required to complete a questionnaire via Google Form, including identity, consumption frequency, and health-related lactose intolerance. The selected panelists were from around Universitas Diponegoro, regularly consumed full-cream UHT at least once a week, were not lactose intolerant, were willing to taste the products, provided feedback, and were in good health during testing. Of the 82 candidates who applied for the panelist position, only 50 qualified based on the inclusion criteria. Among these 50 candidates, 41 were females, and 9 were males.

2.3 Focus group Discussion (FGD)

This study utilized FGDs to identify sensory attributes in a questionnaire completed by fifteen trained panelists. The brand with the most votes became the subject of the FGDs. The FGDs stages involved material presentation, product sensory evaluation, discussion, and endorsement of the sensory attributes resulting from the discussion. Based on the FGD results, the sensory attributes attached to the questionnaire and its description can be seen in Table 1.

Table 1. List of Sensory Attributes of UHT Full-Cream Milk.

Sensory Attributes	Definition
Yellow to dark color	yellows that are closer to darker colors, such as dark yellow or brown yellow
Ivory white color	intense yellowish-white color
White color	chalky white color
Wheylike color	pale yellow and transparent color
Milky aroma	typical milk odor, natural scent, and no aberrant odors
Thick aroma	thick and concentrated milk aroma
Sweet taste	taste on the tongue caused by sweet substances
Savory taste	a taste on the tongue caused by salty or savory substances
Creamy taste	creamy and thick taste on the tongue

Fishy taste	flavors on the tongue associated with animal products
Wheylike flavor	slightly sweet taste on the tongue and creamy taste
Buttery flavor	butter-like flavor that is savory and feels soft on the tongue
Smooth texture	the sensation of smooth and silky milk consistency when drinking
Greasy texture	the oily sensation when drinking
Light texture	the light sensation when drinking
after milky aroma	sensation of inhaled milk aroma after consuming the product
Astringent aftertaste	sensation of stiffness or hardness in the tongue and mouth after consuming the product
Cooked off-flavor	burnt flavor and aroma in cooked milk
Cowy off-flavor	beef flavor and aroma in the product

2.4 Test Product Preparation

This study This study used four brands of full-cream UHT with a maximum shelf life of less than one year. The test products were prepared by pouring 40 ml of UHT full-cream milk refrigerated for 24 hours into 90 ml plastic cups. To avoid physical and microbiological contamination, the product serving was done at 8 - 15°C, 10 minutes before sensory evaluation.

2.5 Sensory Evaluation Procedure

The sensory evaluation procedure using the CATA method refers to Hunaefi and Farhan [1] where untrained panelists are asked to tick the ideal sensory attributes that best describe the test product. Panelists are given a set of test products to evaluate one code first and return to tick the list of identified sensory attributes. Panelists were encouraged to neutralize their taste buds with mineral water and fresh bread provided after each product code change. Panelists were also asked to rate their level of preference on a seven-point scale (1 = strongly dislike, 2 = somewhat dislike, 3 = dislike, 4 = average, 5 = like, 6 = somewhat like, and 7 = strongly like).

2.6 Data Analysis

Data on the physicochemical properties of the products were analyzed using the SPSS version 26 application with the Analysis of Variance (ANOVA) test at the 5% significance level to determine differences in the value of physicochemical properties. ANOVA results were further tested using Duncan's Multiple Range Test (DMRT). CATA data was analyzed using the XLSTAT version 2021 application on the Sensory Data Analysis menu with the CATA Analysis method. Physicochemical property test data of UHT full-cream milk products, along with the level of panelist preference, were analyzed using the Spearman test to determine the level of correlation.

3. Results and Discussion

Analysis of Physicochemical Properties

3.1 pH Value

The pH value is a test that aims to measure the acidity or basicity of a liquid product. The pH analysis results in Table 2 show significant differences in all full-cream UHT milk products. The highest average pH value was found in product code 369, while the lowest was in product code 173. Generally, milk has a pH between 6.7 and 6.9 [2]. Low pH values can result in the degradation of casein micelles, thereby increasing the casein-calcium-phosphate complex [3]. Milk is amphoteric, which can be acidic or alkaline, influenced by compounds such as phosphoric acid, citric acid, amino acids, and carbon dioxide [4]. A pH above 6.7 indicates mastitis, while below 6.5 indicates colostrum [5].

IOP Conf. Series: Earth and Environmental Science **1460** (2025) 012059 doi:10.1088/1755-1315/1460/1/012059

Table 2. pH Values of Dairy Products.

UHT Full-Cream Milk	
Product Code	pH value
173	6.70 ± 0.01[a]
369	7.03 ± 0.06[d]
540	6.80 ± 0.01[b]
742	6.90 ± 0.01[c]

Note: Different lowercase superscripts indicate significant differences ($p<0.05$) based on ANOVA and Duncan's test. Data were obtained from three replicates and shown as mean ± standard deviation.

3.2 Viscosity

The A Brookfield viscometer was used to analyze the viscosity of UHT full-cream milk products. This viscosity analyzes the resistance of the liquid to rotation using a spindle.

Table 3. Viscosity of Dairy Products.

UHT Full-Cream Milk	
Product Code	Viscosity (cP)
173	3.17 ± 0.76[a]
369	6.67 ± 0.29[b]
540	3.00 ± 1.00[a]
742	4.33 ± 0.58[a]

Note: Different lowercase superscripts indicate significant differences ($p<0.05$) based on ANOVA and Duncan's test. Data were obtained from three replicates and shown as mean ± standard deviation.

Analytical results from Table 3 show a significant difference in the average viscosity of full-cream UHT milk in product code 369, while there was no significant difference between product codes 173, 540, and 742. The lowest viscosity was recorded in product code 540, while the highest was in product code 369. The high viscosity of milk can create a thick and creamy sensation when drunk.

The composition of milk has a significant effect on its viscosity. Products with low viscosity usually consist only of fresh milk, while those with high viscosity contain skim milk powder, whey powder, vegetable stabilizers, and other components in specific proportions. However, composition is not the only influencing factor. Storage duration of 30 - 90 days can increase viscosity due to age gelation and a decrease in pH [6]. High fat also makes milk thicker, while casein proteins tend to coagulate at high temperatures, thus increasing viscosity [7]. Fat degradation and fat globule coagulation can also increase viscosity. Sterile milk processing temperatures above 70°C can increase viscosity due to changes in globular protein structure and aggregation of globular whey proteins [8].

3.3 Total Dissolved Solids

TDS analysis measures the concentration of milk-soluble compounds, either naturally occurring or intentionally added during processing. The results are shown in Table 4.

Table 4. Total Dissolved Solids of Dairy Products.

UHT Full-Cream Milk	
Product Code	TDS (°Bx)
173	13.73 ± 0.06
369	13.97 ± 0.71
540	14.30 ± 0.44
742	14.53 ± 0.87

Note: Different lowercase superscripts indicate significant differences (p<0.05) based on ANOVA and Duncan's test. Data were obtained from three replicates and shown as mean ± standard deviation.

The results of the TDS analysis in Table 4 show no significant mean TDS value in all brands of full-cream UHT milk. The highest average TDS in full-cream UHT milk was recorded in product code 742, while the lowest was in product code 173. TDS is an indicator of the content of dissolved materials in the solution. The average TDS value of fresh milk is 11.44% in Total Solid (TS), which follows the SNI 3141.1.2011 standard [9]. Another study by Yusof et al. [10] mentioned that adding minerals to full-cream UHT and plain sterilized milk can affect TDS. In full-cream UHT milk with product code 742, the high TDS was caused by adding skim milk powder, whey powder, vegetable stabilizers, minerals, and vitamins. The addition of skim milk powder increased the TDS as the concentration of skim milk powder increased. High TDS indicates the addition or contamination of minerals during processing, while low TDS indicates excessive dilution during the processing of UHT full-cream milk.

3.4 Color

Color is a sensory quality parameter of milk that affects consumer preference. In this study, color analysis was conducted using a color reader to analyze L* (brightness), a* (redness), and b* (yellowness).

Table 5. Color of Dairy Products.

UHT Full-Cream Milk			
Product Code	Lightness (L*)	Redness (a*)	Yellowish (b*)
173	139.20 ± 1.30^{a}	-2.07 ± 0.06^{d}	33.37 ± 0.06^{b}
369	139.10 ± 0.46^{a}	-2.50 ± 0.02^{c}	34.03 ± 0.38^{c}
540	145.13 ± 0.06^{c}	-4.87 ± 0.06^{a}	29.40 ± 0.03^{a}
742	141.37 ± 0.06^{b}	-4.60 ± 0.10^{b}	35.50 ± 0.01^{d}

Note: Different lowercase superscripts indicate significant differences (p<0.05) based on ANOVA and Duncan's test. Data were obtained from three replicates and shown as mean ± standard deviation.

The results of the color analysis (Table 5) showed that the lightness of product codes 173 and 369 did not differ significantly from codes 540 and 742 in full-cream UHT milk. However, the redness and yellowness parameters showed significant differences in all these products. The highest lightness was found in code 540, while the lowest was in code 369. The highest redness was found in code 173, while the lowest was in code 540. Yellowness was highest in code 742 while the lowest in code 540.

Storage, fat content, processing, and packaging affect milk color. According to research Milovanovic et al. [11], temperature and storage time affect the lightness and yellowness of milk, while fat content also affects color change due to light scattering properties. Indirect heating methods in milk processing can alter its chemical components, causing discoloration [12]. Natural pigments such as carotenoids and riboflavin affect milk's yellow or orange color [13], [14]. The sterilization heat can also cause Maillard reactions that produce dark-colored substances [15].

IOP Conf. Series: Earth and Environmental Science **1460** (2025) 012059 doi:10.1088/1755-1315/1460/1/012059

3.5 Sensory Profile of Milk by CATA Method

The product profiles of full-cream UHT were evaluated using Cochran's Q test. The analysis results in Table 6 show significant differences in most sensory attributes of full-cream UHT milk, except wheylike color, cooked off-flavor, and cowy off-flavor. There was no significant difference between product codes 173 and 540 in yellow to dark color, but there was a significant difference between product codes 173 and 742.

Table 6. Cochran's Q Test Result Sensory Attributes of UHT Full-Cream Milk.

Sensory Attributes	P-value	Product code 173	Product code 369	Product code 540	Product code 742
Yellow to dark color	<0.00	0.00^{a}	0.18^{b}	0.00^{a}	0.28^{b}
Ivory white color	<0.00	0.80^{b}	0.76^{b}	0.38^{a}	0.62^{ab}
White color	<0.00	0.30^{a}	0.14^{a}	0.66^{b}	0.08^{a}
Wheylike color	0.22	0.00^{a}	0.02^{a}	0.06^{a}	0.02^{a}
Milky aroma	0.01	0.82^{b}	0.84^{b}	0.82^{b}	0.58^{a}
Thick aroma	0.00	0.28^{ab}	0.22^{a}	0.20^{a}	0.50^{b}
Sweet taste	<0.00	0.34^{ab}	0.64^{c}	0.18^{a}	0.50^{bc}
Savory taste	0.02	0.70^{b}	0.40^{a}	0.54^{ab}	0.60^{ab}
Creamy taste	0.02	0.54^{ab}	0.56^{b}	0.30^{a}	0.46^{ab}
Fishy taste	<0.00	0.20^{a}	0.16^{a}	0.62^{b}	0.54^{b}
Wheylike flavor	0.00	0.48^{ab}	0.72^{b}	0.38^{a}	0.60^{ab}
Buttery flavor	0.02	0.66^{ab}	0.42^{a}	0.70^{b}	0.66^{ab}
Smooth texture	0.00	0.54^{b}	0.62^{b}	0.26^{a}	0.56^{b}
Greasy texture	0.00	0.18^{ab}	0.14^{a}	0.38^{b}	0.38^{b}
Light texture	0.01	0.70^{b}	0.54^{ab}	0.62^{ab}	0.40^{a}
After milky aroma	<0.00	0.84^{b}	0.84^{b}	0.54^{a}	0.84^{b}
Astringent aftertaste	0.00	0.26^{a}	0.24^{a}	0.58^{b}	0.32^{a}
Cooked off-flavor	0.20	0.16^{a}	0.26^{a}	0.16^{a}	0.12^{a}
Cowy off-flavor	0.18	0.84^{a}	0.76^{a}	0.84^{a}	0.90^{a}

Note: Different lowercase superscripts in the same column indicate significant differences ($p<0.05$).

Cochran's Q test is a statistical method used to assess whether there are significant differences between groups in binary data [16]. The objective was to examine differences in the sensory attributes of UHT full-cream milk products. Results with a p=0.00 value showed highly significant differences, rejecting the null hypothesis and indicating variation in sensory attributes between the two product types. The insignificant results indicate that the panelists were unable to distinguish and identify the presence of these attributes in the products, which would lead to modeling on the symmetric plot being less accurate.

Correspondence analysis identifies the relationship between sensory attributes and the product being tested. The goal is to produce a symmetrical map depicting the sensory attributes and the ideal product. The ideal sensory attributes are determined before the panelists taste the product. Figure 1 displays the symmetry between the sensory attributes and the ideal product.

Based on Figure 1, the ideal sensory attributes of full-cream UHT milk include ivory white color, creamy taste, milky aroma, smooth texture, after milky aroma, and wheylike flavor. In contrast, less ideal attributes included white color and dark yellow color. The assessment of ideal attributes is based on the closeness of the position between the attribute plot and the ideal plot; the closer the position is, the more it is considered ideal. The full-cream UHT milk product considered ideal by panelists is product code 173, while the least ideal is product code 540.

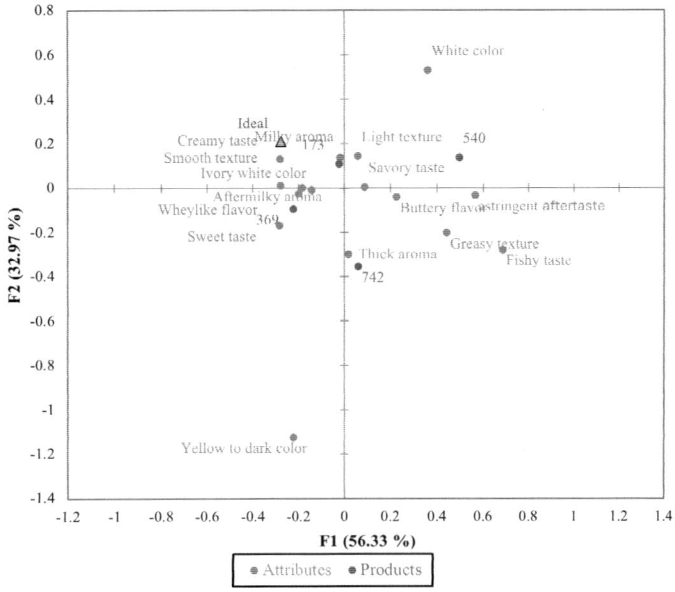

Figure 1. Symmetric Plot of Sensory Attributes and Ideal Product of UHT Full-Cream Milk

A liking plot describes panelists' preferences for the attributes of a product after tasting the test product, as can be seen in Figure 2. The proximity of the sensory attribute to the liking plot determines the preferred attribute; the closer the distance is, the more favorable it is. The preferred sensory attributes of full-cream UHT milk are after milky aroma, creamy taste, smooth texture, and ivory-white color. Less preferred attributes include greasy texture, astringent aftertaste, fishy taste, white color, and buttery flavor. Comparison between Figure 1 and 2 analysis results showed similarities between the ideal attributes and those preferred by the panelists, except for a milky aroma and wheylike flavor. This result indicates that panelists are not able to identify those two attributes in the tested UHT full-cream milk product, but they expect the presence of those two attributes.

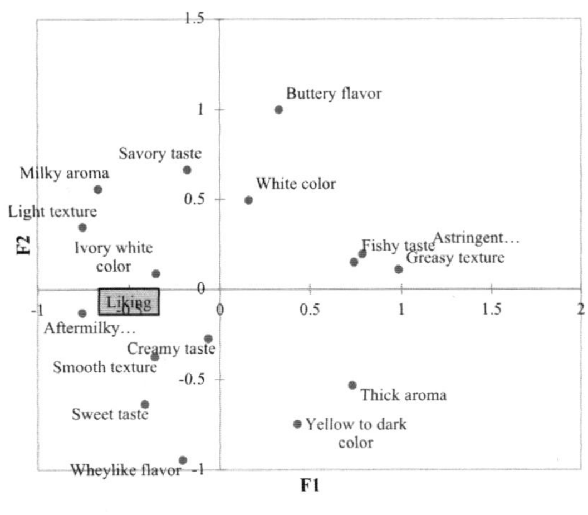

Figure 2. Correlation of Sensory Attributes with Preference for UHT Full-Cream Milk.

This study used The Spearman test to see the correlation between untrained panelists' preferences and the product's physicochemical properties. The results are presented in Table 7 with significance values (sig. 2-tailed) for each physicochemical parameter. The results showed no significant difference between the physicochemical properties and panelists' preferences, as the values exceeded the 5% significance level. A possible influencing factor is that untrained panelists assess preference for full-cream UHT milk based on its sensory qualities. This finding aligns with Mihafu et al. [17], who stated that the quality of a product's sensory attributes strongly influences consumer acceptance and preference.

Table 7. Correlation Results with Spearman Test on UHT Full-Cream Milk.

			Liking	pH value	Viscosity	L*	a*	b*	TDS
Spearman's rho	Liking	Correlation	1.00	-0.22	-0.31	-0.12	0.19	-0.43	0.11
		Sig. (2-tailed)		0.48	0.33	0.74	0.56	0.16	0.72
		N	200	12	12	12	12	12	12

The correlation coefficient represents the degree of relationship between untrained panelists' preferences and the product's physicochemical properties. In the case of full-cream UHT milk products, panelists' preferences are not significantly influenced by the physicochemical properties, so the correlation between the two is very weak. A correlation coefficient close to 1 indicates a strong relationship, while close to 0 indicates a weak relationship. Parameters such as pH, viscosity, brightness, and yellowness show a weak and opposite correlation with panelist preference, indicating that the higher the pH, viscosity, brightness, and yellowness values, the lower the consumer preference. On the other hand, redness and TDS showed a very weak but unidirectional correlation with panelists' preference, indicating that the higher the redness and TDS, the higher the consumer preference. In brief, there is no correlation between product's physicochemical characteristics and its consumer preference. This is in accordance with the statement of Apandi et al. [18] which stated that product selection is more influenced by attribute acceptance and overall liking for consumption.

The sensory attributes that impact panelists' preferences most are identified using penalty analysis. This analysis is widely used as a reference for new product development because the analysis results accurately determine the sensory attributes that match consumer preferences [19]. The penalty analysis in this study was grouped into must-have, nice-to-have, does not influence, does not harm, and must-not-have attributes as can be seen in Table 8.

Table 8. Summary of Sensory Attribute Penalty Analysis

Must have	Nice to have	Does not influence	Does not harm	Must not have
Ivory white color		Wheylike flavor	Yellow to dark color	Fishy taste
Milky aroma			White color	Greasy texture
Creamy taste			Thick aroma	Astringent aftertaste
Smooth texture			Sweet taste	
Light texture			Savory taste	
After milky aroma			Buttery flavor	

A penalty analysis of the sensory attributes of full-cream UHT milk identified several highly desirable attributes (must-haves), including ivory white color, milky aroma, creamy taste, smooth texture, light texture, and after milky aroma. These attributes significantly positively impact consumer preferences, and therefore need to be more emphasized during product improvement. On the other hand, there are must-not-have attributes, such as fishy taste, greasy texture, and astringent aftertaste, which mean that those attributes must not be encountered by the consumer when consuming UHT full-cream milk products. Attributes such as yellow to dark color, white color, thick aroma, and others were not significantly detrimental. The results of this penalty analysis are essential for manufacturers to develop products that match consumer preferences and increase the chances of success in a competitive market.

4. Conclusion

Full-cream UHT milk products should have sensory attributes of ivory white color, milky aroma, creamy taste, smooth texture, light texture, and aftertaste milky aroma. The full-cream UHT milk product considered as closest to the ideal by untrained panelists is product 173.

References

[1] D. Hunaefi and Z. M. Farhan, "Karakterisasi Sensori Cheese Tea dengan Metode Check All That Apply (CATA), Emotional Sensory Mapping (ESM), dan Ideal Profile Method (IPM)," *Jurnal Mutu Pangan : Indonesian Journal of Food Quality*, vol. 8, no. 1, pp. 1–9, Apr. 2021, doi: 10.29244/jmpi.2021.8.1.1.

[2] T. Ozcan, D. S. Horne, and J. A. Lucey, "Yogurt made from milk heated at different pH values," *J Dairy Sci*, vol. 98, no. 10, pp. 6749–6758, Oct. 2015, doi: 10.3168/jds.2015-9643.

[3] H. Sinaga, N. Bansal, and B. Bhandari, "Effects of milk pH alteration on casein micelle size and gelation properties of milk," *Int J Food Prop*, vol. 20, no. 1, pp. 179–197, Jan. 2017, doi: 10.1080/10942912.2016.1152480.

[4] M. N. Hidayat, "Meningkatkan Nilai Manfaat Susu dengan Penambahan Mikroba Probiotiak," *Teknosains*, vol. 11, no. 1, pp. 71–88, 2017.

[5] K. Sharun *et al.*, "Advances in therapeutic and managemental approaches of bovine mastitis: a comprehensive review," *Veterinary Quarterly*, vol. 41, no. 1, pp. 107–136, Jan. 2021, doi: 10.1080/01652176.2021.1882713.

[6] E.-S. I., A. Ibrahim, and A. S., "Effect of Milk Quality and Stabilizers on Some Physicochemical Properties of UHT-Milk," *Alexandria Journal of Food Science and Technology*, vol. 15, no. 1, pp. 49–64, Jun. 2018, doi: 10.21608/ajfs.2018.16348.

[7] K. Christensen and J. Lindberg, "Carbohydrate and Fluid Requirements for Endurance Runners," no. May, 2014.

[8] M. Quevedo, U. Jandt, U. Kulozik, H. P. Karbstein, and M. A. Emin, "Investigation on the influence of high protein concentrations on the thermal reaction behaviour of β-lactoglobulin by experimental and numerical analyses," *Int Dairy J*, vol. 97, pp. 99–110, Oct. 2019, doi: 10.1016/j.idairyj.2019.06.004.

[9] W. Wasito, "Perception and Adoption SNI 3141-1: 2011 Dairy Cows Family Farmers in Livestock Business Region Bogor District," *Jurnal Standardisasi* , vol. 19, no. 3, pp. 241–254, 2017.

[10] N. H. Yusof, N. Abdullah Sani, F. H. Anuar, M. S. Jamil, and S. I. Zubairi, "Rapid Microbial Detection Model System in UHT Milk Products using Poly(L-Lactic Acid) (PLLA) Thin Film," *Sains Malays*, vol. 47, no. 11, pp. 2677–2683, Nov. 2018, doi: 10.17576/jsm-2018-4711-10.

[11] B. Milovanovic *et al.*, "Colour assessment of milk and milk products using computer vision system and colorimeter," *Int Dairy J*, vol. 120, p. 105084, Sep. 2021, doi: 10.1016/j.idairyj.2021.105084.

[12] M. A. Karlsson *et al.*, "Changes in stability and shelf-life of ultra-high temperature treated milk during long term storage at different temperatures," *Heliyon*, vol. 5, no. 9, p. e02431, Sep. 2019, doi: 10.1016/j.heliyon.2019.e02431.

[13] S. Laverroux, F. Picard, D. Andueza, and B. Graulet, "Vitamin B2 concentration in cow milk: Quantification by a new UHPLC method and prediction by visible and near-infrared spectral analysis," *Food Chem*, vol. 342, p. 128310, Apr. 2021, doi: 10.1016/j.foodchem.2020.128310.

[14] R. Conboy Stephenson, R. P. Ross, and C. Stanton, "Carotenoids in Milk and the Potential for Dairy Based Functional Foods," *Foods*, vol. 10, no. 6, p. 1263, Jun. 2021, doi: 10.3390/foods10061263.

[15] V. Andrés, M.-J. Villanueva, and M.-D. Tenorio, "Influence of high pressure processing on microbial shelf life, sensory profile, soluble sugars, organic acids, and mineral content of milk- and soy-smoothies," *LWT*, vol. 65, pp. 98–105, Jan. 2016, doi: 10.1016/j.lwt.2015.07.066.

[16] F. Llobell, D. Giacalone, A. Labenne, and E. M. Qannari, "Assessment of the agreement and cluster analysis of the respondents in a CATA experiment," *Food Qual Prefer*, vol. 77, pp. 184–190, Oct. 2019, doi: 10.1016/j.foodqual.2019.05.017.

[17] F. D. Mihafu, J. Y. Issa, and M. W. Kamiyango, "Implication of Sensory Evaluation and Quality Assessment in Food Product Development: A Review," *Current Research in Nutrition and Food Science Journal*, vol. 8, no. 3, pp. 690–702, Dec. 2020, doi: 10.12944/CRNFSJ.8.3.03.

[18] I. Apandi, F. Restuhadi, and Yusmarini, "CONSUMERS PREFERENCE MAPPING ANALYSIS FOR THE SENSORY ATTRIBUTE OF SOYGHURT PRODUCT WITHIN THE STUDENT AGRICULTURAL FACULTY OF UNIVERSITY OF RIAU," *Jom Faperta*, vol. 3, no. 1, 2016.

[19] A. Gere, "Penalty analysis is an efficient tool for food product development: Case study with lemonades," *Journal on Processing and Energy in Agriculture*, vol. 26, no. 3–4, pp. 90–94, 2022, doi: 10.5937/jpea26-41654.

Spicy level determination on chili-based food products by electrochemical capsaicin detection

Arum Kurnia Sari[1*], Dian Muzdalifah[2,3], Sri Koerniati[4], Hasim Munawar[2], Media Y U[5], April Hari Wardhana[6], Ireng Darwati[7], Msartin Peacock[8]

[1] PT Mitra Asiatek Biosensor, Jakarta Selatan 12520, Indonesia
[2]Research Center for Chemistry, National Research and Innovation Agency, South Tangerang 15314, Indonesia
[3]Doctoral Program of Chemistry, Faculty of Mathematics and Natural Science, Gadjah Mada University, Yogyakarta 55281, Indonesia
[4]Research Center for Genetic Engineering, National Research and Innovation Agency, Cibinong 16911, Indonesia
[5]PT Galic Bina Mada, West Cikarang, Bekasi 17520, Indonesia
[6]Research Center for Veterinary Science, Research Organization for Health, National Research and Innovation Agency, Cibinong 16911, Indonesia
[7]Research Center for Estate Crops, Research Organization for Agriculture and Food, National Research and Innovation Agency, Cibinong 16911, Indonesia
[8]Zimmer and Peacock AS, Horten, Norway

*E-mail: arum@mitraasiatek.com

Abstract. Capsaicinoids are a group of chemicals that give chili peppers their characteristic pungency, with capsaicin being a significant contributor. Accurately measuring capsaicinoid content is crucial for assessing the quality of chili-based food products. Herein, the spicy level of chili-based food products was measured using an electrochemical detection tool, an Android-based FoodSense G4 (FsG4). In this study, the robust evaluation was examined to investigate the accuracy and precision of this commercial sensor. The oil- and water-based capsaicin standard solutions were diluted in buffer solutions with specific ratios and dropped onto the sensor chip connected to the FsG4 reader. The content of capsaicin was measured six times. An Android mobile phone and Djuli software were employed to record the Scoville Heat Unit (SHU) and cyclic voltammogram. As a result, the accuracy and precision were outstanding, 98–116% and 90–99%, respectively, from both standard solutions. The results showed acceptable repeatability, and the recoveries indicated good accuracy for this method. The ten local and five imported food products were tested without traditional laboratory equipment. The highest and lowest SHU were both from imported food products, 19130 and 377, respectively. Besides the spicy level data, the correlation between the percentage of chili and SHU was investigated, although further studies were needed. Even though the FsG4 tool needs improvement, this sensor could be helpful in obtaining capsaicin-related pungency values, making it highly suitable for commercial quality assessment of chili-based products.

Keywords: Capsaicin, Electrochemical Methods, Scoville Heat Unit

1. Introduction

Chili peppers are one of the most popular and primary ingredients in the world due to their unique spicy flavors and aromas in a variety of food products [1-3]. The primary source of pungency (spiciness) in chili peppers comes from a chemical group called capsaicinoids, which can be found in the capsicum family [2],[4-5]. Capsaicin ($C_{18}H_{27}NO_3$, trans-8-methyl-N-vanillyl-6-nonenamide) is the central part of capsaicinoids, with proportions of 69%, followed by dihydrocapsaicin (22%) and nordihydrocapsaicin (7%) [6-10]. Capsaicin is responsible for the hotness and spiciness of chili peppers and related products. It activates TRPV1 (transient potential receptor 1) sensory neurons on the tongue and mucous membranes, which stimulates the sense of heat and pain [11-13]. In the food industry, quantifying the amount of capsaicin is used to measure the spiciness level of commercial chili peppers and related products. However, scientific research and the pharmaceutical industry also depend on understanding the amount of capsaicin [11]. Therefore, it is highly desired to develop a simple, accurate, and cost-effective method for quantifying capsaicin [2].

Scoville Heat Units (SHU) are used as a scale to measure the spiciness level; a higher SHU indicates a higher concentration of capsaicin and a spiciness level [14]. Known as the 'Scoville Organoleptic Test,' Wilbur Lincoln Scoville created this scale as an industry standard technique for determining the amount of capsaicin in chili peppers and to assess their spiciness level [2],[15]. Capsaicin is extracted from chili peppers and diluted using a sugar and water solution until the tester is unable to detect any more spiciness. More dilutions indicate a higher spiciness level and, therefore, a higher concentration of capsaicin. However, because it depends on the testers, which can have different sensitivity levels to the level of spiciness, the accuracy of this method is questionable. The amount of oral heat receptors varies significantly between individuals, and sensory exhaustion is a significant concern. This happens when the palate becomes desensitized to capsaicin, usually after consuming a few samples within a brief period. When this approach is utilized, the findings differ significantly amongst laboratories [12],[16-17].

Many instrumental methods, such as spectrophotometry, high-performance liquid chromatography (HPLC), LC-mass spectrometry (LC-MS), gas chromatography (GC), and GC-mass spectrometry (GC-MS), are the most often methods used for detecting capsaicin [16],[18]. However, they need complex procedures, longer sample pretreatment times, longer response times, and expensive instruments. The need for other methods to measure capsaicin has increased as a result of this, as well as the subjectivity of the Scoville Organoleptic Test. It is essential to develop a method for measuring capsaicin that is simple, rapid, sensitive, specific, and cost-effective [4],[6-7],[19].

The electrochemical methods could be used as an alternative to determine capsaicin. Among the various advantages of the electrochemical methods are their simplicity, ease of use, rapid detection, sensitivity, reproducibility, cost-effectiveness, and suitability for on-site testing [2],[4],[7],[13],[19]. In related studies [2],[4],[6-7],[13-15],[17-21], electrochemical techniques were developed to determine capsaicinoid concentration. The most popular electrochemical technique for examining the electrochemical behavior of electroactive materials is cyclic voltammetry (CV) [4],[7].

The commercial FsG4 analyzes capsaicin solutions in the buffer on the chili sensor's (SPCE) surface. The advantages of using the screen-printed carbon electrode or SPCE include ease of use, portability, affordability, and mass production [7],[19]. In addition to being intended for on-site testing with minimal technical expertise, the commercial FsG4 is combined with a mobile phone application that provides the capsaicin concentration in Scoville Heat Units (SHU). It takes 30 seconds to obtain a capsaicin value corresponding to the level of spiciness. The capsaicin can be determined using cyclic voltammetry (CV). With cyclic voltammetry (CV), the amount of capsaicin can be determined. In cyclic voltammetry, the voltage is raised to a predetermined maximum and then lowered to the initial level. A voltammogram records the specific current response that arises as a result of the oxidation/reduction reaction throughout this voltage cycle. The voltammogram's data can be utilized to determine capsaicin concentrations. In this study, we, therefore, use a commercial FsG4 chili sensor to provide a simplified electrochemical capsaicin detection method for spicy level determination on chili-based food products using a commercial FsG4 chili sensor. The ten local and five imported chili-based food products were tested to determine their spicy level (SHU). They investigated the correlation between the percentage of chili and their spicy level (SHU).

2. Experimental

2.1 Chemicals

The oil- and water-based capsaicin standard solutions (Capsicum oleoresin 500,000 SHU) were supplied by PT Galic Bina Mada (Indonesia). Chili buffer was purchased from Zimmer & Peacock AS (Norway). The ten local and five imported chili-based food products were purchased from local supermarkets in Indonesia.

2.2 Instruments

The electrochemical measurements were performed using a potentiostat, FoodSense Gen 4 (FsG4), connected to a mobile phone via Bluetooth and controlled by the SenseitAll Android/IOS application (Norway). All measurements were conducted using a chili sensor (SPCE), a screen-printed three-electrode with capacitors. There are three electrodes in the integrated screen-printed electrode. The working electrode for the screen-printed electrodes was carbon coated with an analyte selective coating, whereas the reference and auxiliary electrodes were Ag/AgCl. Both potentiostat and chili sensors were purchased from Zimmer & Peacock AS.

2.3 Standard preparation

The procedure for preparing the standard and sample solution followed the FsG4 protocol from Zimmer Peacock, modified from previous studies [2],[14],[22],[30].

Standard preparation. 10 µl of oil-based capsaicin standard solution (Capsicum oleoresin 500,000 SHU) and 990 µl of buffer solution were added to microtubes (1 in 100 dilutions) and vortexed. The same procedure was also performed on the water-based capsaicin standard solution.

Sample preparation. 100 µl of chili-based food product sample and 900 µl of buffer solution were added to microtubes (1 in 10 dilution) and vortexed. The same procedure was also performed on the other chili-based food product sample. If the result of the application shows a capsaicin value that is "too high," consider using a higher dilution and selecting an alternative dilution factor (1:100; 1:1000; 1:10,000).

3. Results and Discussion

3.1 Electrochemical behavior of capsaicin measurement

In order to better understand the oxidation-reduction mechanism on the surface of the chili sensor, the electrochemical behavior of capsaicin determination was investigated using the CV, as shown in **Figure 1**. There is an oxidation peak (I) and a reduction peak (II) in the first scan (black curve). The irreversible oxidation of capsaicin is represented by Peak I. Capsaicin's phenolic hydroxyl group is oxidized to create a carbocation intermediate, which is then quickly hydrolyzed through the capsaicin's 2-methoxy group to produce o-benzoquinone. Peak II then shows the reduction of O-benzoquinone to catechol. However, the second scan (red curve) shows an oxidation peak (III) and a reduction peak (II) that is the same as the first scan, but with the initial peak (I) decreasing [2],[4],[12],[17],[20],[22]. Peak III is a result of catechol's reversible oxidation to o-benzoquinone. Since catechol was initially absent from the solution, Peak III only showed up in the second scan. Only after the reduction of o-benzoquinone in the first scan does catechol form. Due to the creation of catechol in Peak II, which oxidizes easier than capsaicin, and the decrease of capsaicin in the sample solution, Peak I for the second scan is smaller than that for the first scan. These three peaks are part of capsaicin's characteristic redox reactions [15],[23]. The results indicate that capsaicin formed a new species during the first scan, and that the new species continued to carry out electrochemical reactions [20],[24]. Furthermore, Peak I showed the highest sensitivity. Peak I was chosen to quantify the amount of capsaicin [4],[18].

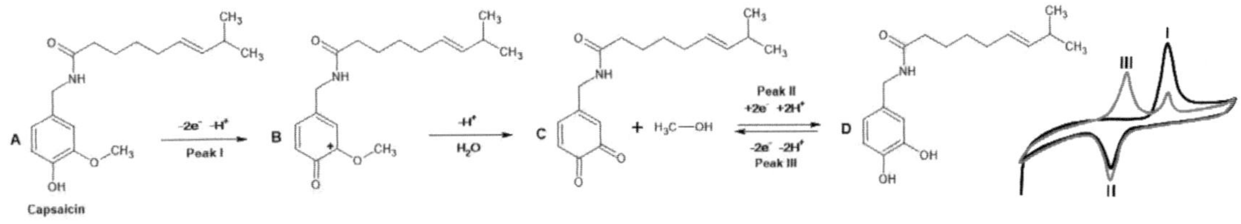

Figure 1. Electrochemical oxidation-reduction mechanism in the presence of capsaicin.

3.2 Accuracy and precision of the commercial sensor

A robust evaluation was conducted to investigate the accuracy and precision of the commercial sensor. The oil- and water-based capsaicin standard solutions are tested at 500,000 SHU, six times the measurements, to determine precision and accuracy, as shown in **Figure 2**. Precision shows how consistent and repeatable a measurement is, whereas accuracy shows how near a measurement is to the known concentration [25]. According to the calculations, the commercial sensor's accuracy and precision values for the two standard solutions are 90–99% and 98–116%, respectively, as shown in **Table 1**. The results presented satisfactory and excellent repeatability and accuracy of the commercial sensor.

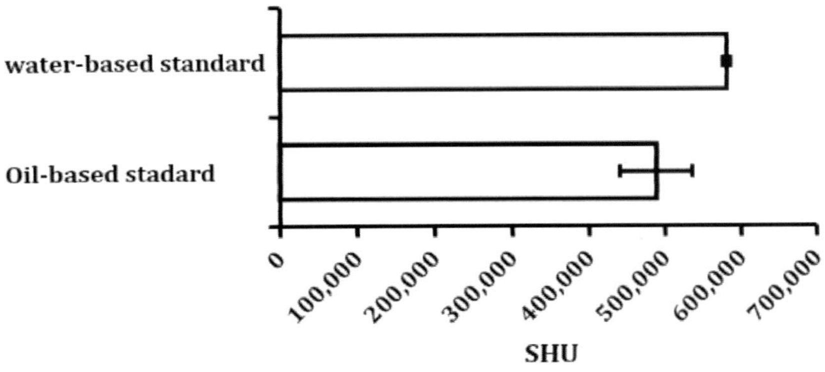

Figure 2. Accuracy of the FsG4 on detecting capsaicin concentration in oil- and water-based capsaicin standard solutions.

Table 1. Analytical parameter of capsaicin standard solutions.

Capsaicin standard solutions	Min	Max	Mean	STDEV	RSD	Accuracy	Precision
	SHU					%	
Oil-based	417,600	558,400	488,416	47,898	9.8	98	90
Water-based	576,700	587,400	581,550	4,645	0.8	116	99

3.3 Spicy levels of chili-based food products

A higher level of spiciness is indicated by a higher amount of capsaicin, which is measured by the Scoville scale. The Scoville Heat Unit, sometimes known as the SHU, is the unit of measurement for the Scoville scale [16],[24].

Fresh chili peppers, chili powders, chili sauces, and any other chili-related product can all have their level of spiciness measured by the commercial FsG4 chili sensor. The ten local (L) and five imported (I) chili-based food products were tested for quality assessment of the commercial product based on their spiciness level (SHU). Customers can use the SHU information of the products as a reference and make confident decisions about the spiciness level they can tolerate or like. The determination of SHU is carried out by mixing the chili-based food product sample with the buffer solution (1 in 10 dilutions), then measuring it using a commercial FsG4 chili sensor three times, as shown in **Figure 3**. Based on the result, the highest and lowest SHU were both from imported products: I5 with 19,130 SHU and I1 with 377 SHU. A range of chili peppers, which can be mild, medium, hot, or extremely hot, are used to make chili-based food products. However, there have not been many studies examining the worldwide spicy level of chili-based food products, especially in Indonesia. Therefore, further studies are needed on this matter.

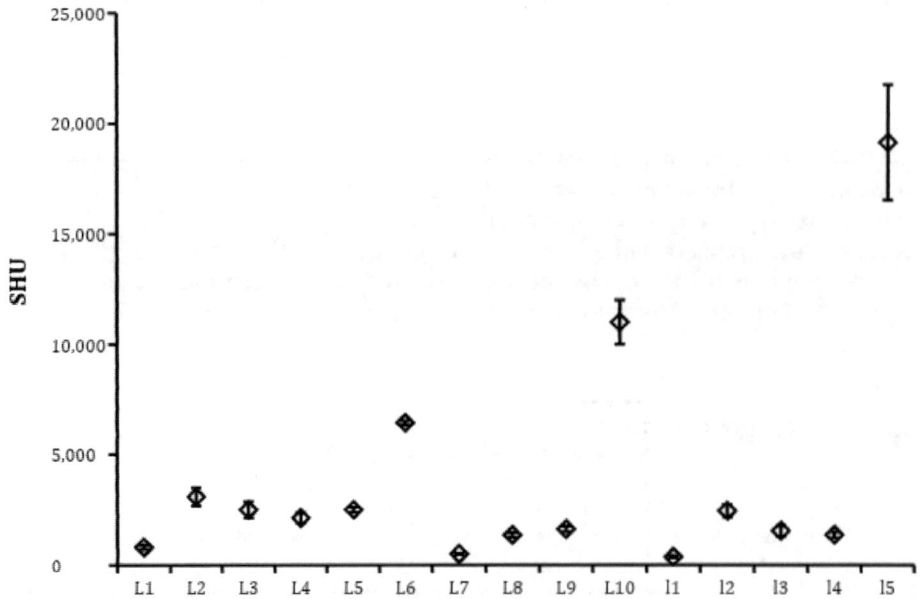

Figure 3. Spicy levels (SHU) of local and imported chili-based food products.

Note:
L = local chili-based food products
I = imported chili-based food products

3.4 Correlation between percentage of chili and spicy level on chili-based food products

Many recipes for chili-based food products exist; however, chili peppers of various types remain the primary component responsible for their spicy flavor [26]. The specific recipe varies by area. Milder chili peppers, like jalapeños and cayenne, have lower capsaicin contents; thus, milder chili peppers create milder chili-based food products. While hotter chili peppers, like Scotch Bonnet, Habanero, or Ghost pepper, have higher capsaicin contents and are utilized to manufacture spicier chili-based food products. Very highly spicy (>80,000 SHU), highly spicy (25,000–80,000 SHU), moderately spicy (3,000–25,000 SHU), mildly spicy (700–3,000 SHU), and non-spicy (0–700 SHU) are the five primary levels of spiciness that are categorized using SHU [22]. The Carolina Reaper held the previous world record of 1.64 million SHU. Still, on August 23, 2023, Guinness World Records recognized Pepper X as the hottest chili pepper in the world with a measurement of 2.69 million SHU (Guinness World Records). Demands in the food industry and food service may require an understanding of why people enjoy spicy food. Culture, exposure, gender, genetics, and personality are some of the many factors that influence the creation of a spicy taste preference. Exposure to chili peppers is influenced by culture; Mexico and several Asian nations have the highest consumption trends—people who eat more chili peppers like them more than people who don't. Therefore, repeated exposure can also improve liking, and early childhood exposure encourages the use of chili peppers as an adult [27].

The correlation between the percentage of chili and spicy level (SHU) was investigated using Pearson correlation as previously worked [28]. The Pearson correlation measures the linear relationship between the two variables. A value of -1 represents a negative linear correlation; a value of 0 means no connection and a value of +1 represents a positive correlation [29]. The r value (-0.8) showed the opposite correlation between chili percent and SHU from 12 samples (**Figure 4**). The amount of capsaicinoids in chili peppers or chili-based food products, of which capsaicin is the main contributor, is associated with the spicy level measured in SHU. It was expected that chili-based food products containing a high percentage of chili would have a high SHU value due to their high capsaicin content. Despite the fact that, based on the result, not all chili-based food products with a high percentage of chili had a high SHU value, further investigation was needed to obtain a comprehensive correlation.

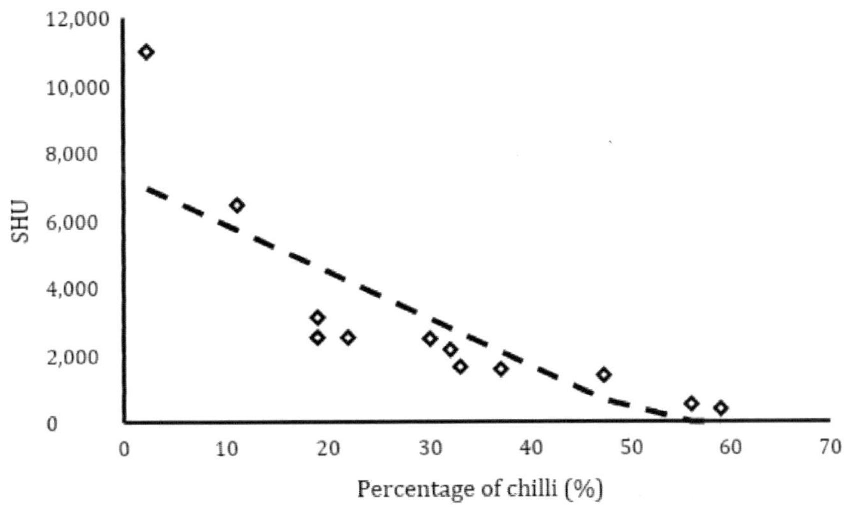

Figure 4. Correlation between percentage of chili and spicy level (SHU).

4. Conclusion

A commercial FsG4 chili sensor is a novel electrochemical detection tool for measuring the spiciness of chili-based food products in Scoville Heat Units (SHU) by measuring the presence of capsaicin. With the optimal conditions, the portable commercial sensor can be used to detect capsaicin directly through CV with good accuracy and precision. Future studies are necessary to find the effect of the matrix samples and other analytical parameters, such as recovery and uncertainty.

References

[1] L. Čižmek, I. N. Jovanović and Š. Komorsky-Lovrić. Quantitative determination of capsaicinoids in-ground hot pepper samples using voltammetry of microparticles. *Bulgarian Chemical Communications*, 49, pp. 233– 240 (2017).

[2] A. Soleh, K. Saisahas, K. Promsuwan, P. Thavarungkul, P. Kanatharana and W. Limbut. N-Doped Graphene Nanoplatelets for Direct Capsaicin Detection in Chili Pepper Samples. *Applied Nano Material*, (2019). 10.1021/acsanm.0c02079

[3] P. Popelka, P. Jevinová, K. Šmejkal and P. Roba. Determination of Capsaicin Content and Pungency Level of Different Fresh and Dried Chilli Peppers. *Folia Veterinaria*, 61, 2, (2017). 10.1515/fv-2017-0012

[4] W. Numphud, O. Chienthavorn and W. Siriwatcharapiboon. Cyclic voltammetric determination of capsaicin by using electrochemically deposited tin and reduced graphene oxide on screen-printed carbon electrodes. *ScienceAsia*, 46, 586–594, (2020). http://dx.doi.org/10.2306/scienceasia1513-1874.2020.076

[5] N. Yun, C. Lu, T. Sun, B. Xu, Y. Song, Z. Zong, K. Chen, G. Huang, X. Chen and Q. Gu. High sensitivity detection of capsaicin in red pepper oil based on reduced graphene oxide enhanced by β-cyclodextrin. *Food Analytical Methods*, 16, 318–329, (2023). https://doi.org/10.21203/rs.3.rs-1890129/v1

[6] E. D. de L. Zavala, L. M. T. Rodríguez, A. M. Rojas, V. H. T. Mendoza and A. E. L. González. Study of electrochemical determination of capsaicin and dihydrocapsaicin at carbon paste electrodes modified by β cyclodextrin. *Journal of Electroanalytical Chemistry*, (2018). 10.1016/j.jelechem.2018.02.055

[7] W. Lyu, M. Ding, Y. Zhou, M. Jiang, Y. Li, Y. Ding, Z. Zhang, X. Wei and X. Zhang. A Highly Sensitive Electrochemical Sensor for Capsaicinoids and Its Application in the Identification of Illegal Cooking Oil. *Biosensors*, 13, 863, (2023). https:// doi.org/10.3390/bios13090863

[8] E. P. Randviir, J. P. Metters, J. Stainton and C. E. Banks. Electrochemical impedance spectroscopy versus cyclic voltammetry for the electroanalytical sensing of capsaicin utilising screen printed carbon nanotube electrodes. *Analyst*, (2013). 10.1039/c3an00368j

[9] F. Wang, Y. Xue, L. Fu, Y. Wang, M. He, L. Zhao and X. Liao. Extraction, purification, bioactivity and pharmacological effects of capsaicin: a review. *Critical Reviews in Food Science and Nutrition*, (2021). 10.1080/10408398.2021.1884840

[10] H. Naskar, B. Ghatak, S. Biswas, B. Tudu, R. Bandyopadhay and P. Pramanik. Electrochemical Detection of Capsaicin Using Yttrium Oxide Nanoparticles Modified Graphite Paste Electrode (Y2O3/GPE). *IEEE International Symposium on Olfaction and Electronic Nose (ISOEN)*, 1-3, (2019). 10.1109/ISOEN.2019.8823183

[11] D. R. Singh, V. S. Banu and S. Singh. Insights into capsaicin distribution in Andaman chili peppers: A comprehensive regional analysis. *Annals of Phytomedicine*, 13, 1, 1149-1155, (2024). http://dx.doi.org/10.54085/ ap.2024.13.1.123

[12] B. B. Y. Lau, J. Panchompoo and L. Aldous. Extraction and electrochemical detection of capsaicin and ascorbic acid from fresh chili using ionic liquids. *New Journal of Chemistry*, (2014). 10.1039/c4nj01416b

[13] F. Zhong, Z. Liu, Y. Han and Y. Guo. Electrochemical Sensor for Sensitive Determination of Capsaicin Using Pd Decorated Reduced Graphene Oxide. *Electroanalysis*, 31, 1–8, (2019). 10.1002/elan.201900048

[14] S. Søpstad, K. Imenes and E. A. Johannessen. A hybrid electrochemical sensor platform for capsaicin determination is used using coarsely stepped cyclic square wave voltammetry. *Biosensors and Bioelectronic*, (2018). https://doi.org/10.1016/j.bios.2018.09.036

[15] P. Supchocksoonthorn, N. Thongsai, W. Wei, P. Gopalan and P. Paoprasert. Susceptible and stable sensor for the detection of capsaicin using electrocatalytic carbon dots grafted onto indium tin oxide. *Sensors and Actuators B Chemical*, (2020). https://doi.org/10.1016/j.snb.2020.129160

[16] M. G. Usman, M. Y. Rafii, M. R. Ismail, Md. A. Malek and M. A. Latif. Capsaicin and Dihydrocapsaicin Determination in Chili Pepper Genotypes Using Ultra-Fast Liquid Chromatography. *Molecules*, 19, 6474-6488, (2014). 10.3390/molecules19056474

[17] I. Jiménez, C. Pérez-Ràfols, N. Serrano, M. del Valle and J. M. Díaz-Cruz. Carbon-based electrodes are used for the voltammetric determination of capsaicin in spicy samples. *Microchemical Journal*, 191, 108757, (2023). https://doi.org/10.1016/j.microc.2023.108757

[18] W. Lyu, X. Zhang, Z. Zhang, X. Chen, Y. Zhou, H. Chen, H. Wang and M. Ding. A simple and sensitive electrochemical method for the determination of capsaicinoids in chili peppers. *Sensors and Actuators B*, (2019). https://doi.org/10.1016/j.snb.2019.02.104

[19] X. Fang and R. Duan. Highly Sensitive Capsaicin Electrochemical Sensor Based on Bimetallic Metal-Organic Framework Nanocage. *Frontiers in Chemistry*, 10, 822619, (2022). 10.3389/fchem.2022.822619

[20] P. B. Deroco, O. Fatibello-Filho, F. Arduini and D. Moscone. Electrochemical determination of capsaicin in pepper samples using sustainable paper-based screen-printed bulk modified with carbon black. *Electrochimica Acta*, (2020). https://doi.org/10.1016/j.electacta.2020.136628

[21] G. Ziyatdinova, E. Ziganshina, A. Shamsevalieva and H. Budnikov. Voltammetric determination of capsaicin using CeO2-surfactant/SWNT-modified electrode. *Arabian Journal of Chemistry*, 13, 1624-1632, (2020). https://doi.org/10.1016/j.arabjc.2017.12.019

[22] R. D. Crapnell and C. E. Banks. Electroanalytical overview: the pungency of chile and chili products determined via the sensing of capsaicinoids. *Analyst*, 146, 2769 (2021). 10.1039/d1an00086a

[23] S. A. Shafee, W. H. Danial, S. C. Perry, Z. I. Ali, M. A. M. Huri and A. S. M. Sabere. Qualitative and Quantitative Methods of Capsaicinoids: a Mini-Review. *Food Analytical Methods*, 15, 2424–2435, (2022). https://doi.org/10.1007/s12161-022-02306-2

[24] Q. Gu, C. Lu, K. Chen,X. Chen, P. Ma, Z. Wang and B. Xu. Electrochemical Determination of Capsaicinoids Content in Soy Sauce and Pot-Roast Meat Products Based on Glassy Carbon Electrode Modified with B-Cyclodextrin/Carboxylated Multi-Wall Carbon Nanotubes. *Foods*, 10, 1743, (2021). https://doi.org/10.3390/foods10081743

[25] R. Setiyono, M. I. H. L. Zein, P. Daud, A. Anggraeni, Y. W. Hartati and H. H. Bahti. A novel in-house built printed circuit board-ceria based electrochemical device for rapid detection of Epithelial Sodium Channel (ENaC), a hypertension biomarker. *Sensors and Actuators Reports*, 7, 100200, (2024). https://doi.org/10.1016/j.snr.2024.100200

[26] A. H. Prayitno, R. Meswari, and M. Diauddin. The study of chemical contents, daily values, and microbiology of chicken chili sauce. *Food Technology, Nutritions, and Culinary*, 3(1), 49–56, (2020). 10.20956/canrea.v3i1.257

[27] E. Siebert, S-Y. Lee and M. P. Prescott. Chili pepper preference development and its impact on dietary intake: A narrative review. Frontiers in Nutrition, 9, (2022). 10.3389/fnut.2022.1039207

[28] H. Munawar, A. H. M. Safaryan, A. De Girolamod, A. Garcia-Cruza, P. Marotee, K. Karima, V. Lippolisd, M. Pascaled and S. A. Piletskya. Determination of Fumonisin B1 in maize using molecularly imprinted polymer nanoparticles-based assay. *Food Chemistry,* 298, 125044, (2019). https://doi.org/10.1016/j.foodchem.2019.125044

[29] B. Williams, C. Halloin, W. Löbel, F. Finklea, E. Lipke, R. Zweigerdt and S. Cremaschi. Data-Driven Model Development for Cardiomyocyte Production Experimental Failure Prediction. *Computer Aided Chemical Engineering*, 48, 1639-1644, (2020). https://doi.org/10.1016/B978-0-12-823377-1.50274-3

[30] R. T. Kachoosangi, G. G. Wildgoose and R. G. Compton. Carbon nanotube-based electrochemical sensors for quantifying the 'heat' of chilli peppers: the adsorptive stripping voltammetric determination of capsaicin. *Analyst*, 133, 888–895, (2008). 10.1039/b803588a

Identification of Red Dragon Fruit Peel Flour Adulterated with Purple Sweet Potato Flour Using FTIR Spectroscopy Combined with Chemometrics

Toto Nugroho Triputra Maryana, Yuli Perwita Sari* and Chatarina Lilis Suryani

Study program of Agricultural Product Technology, Agroindustry Faculty, Universitas Mercu Buana Yogyakarta, Indonesia

*E-mail: yuli.perwita@mercubuana-yogya.ac.id

Abstract. The vibrant peel of the red dragon fruit, with its striking magenta hue and unique texture, is rich in antioxidants, making it a valuable addition to a healthy diet. It can be processed into flour, which simplifies storage and expands its applications. A significant advantage of this material is its ability to biodegrade alongside other substances. Therefore, this study aimed to identify red dragon fruit peel flour that has been adulterated with purple sweet potato flour using FTIR spectroscopy combined with chemometrics. This study used a model of adulteration, mixing red dragon fruit peel flour (RDFPL) with purple sweet potato flour (PSPF). The study examined adulteration rates ranging from 3 – 100%. A colorimeter can significantly identify color changes in RDFPL adulteration with PSPF ($p<0.05$). The presence of (C-H) bending alkene-specific bonds to starch was indicated by spectral peaks at wavenumber 862 cm^{-1}, which was specific indicator of RDFPL adulteration. Based on the results, the first derivative spectrum at wavenumbers 650–3603 cm^{-1} as the calibration and validation model for the adulteration rate prediction, with R^2 values of 0.9999 and 0.9880, respectively, and RMSEC and RMSEP values of 0.304 and 2.86, respectively. The results of the multivariate analysis using Principal Component Analysis (PCA) categorize the level of adulteration in red dragon fruit peel flour (RDFPL) into three groups: low, medium, and high, achieving an accuracy rate of 98.7%. Additionally, when combined with chemometrics, Fourier Transform Infrared (FTIR) spectroscopy successfully identified the degree of adulteration in red dragon fruit peel flour when mixed with purple sweet potato flour.

Keywords: adulteration, chemometrics, FTIR spectroscopy, red dragon fruit peel flour

1. Introduction

The peel of the red dragon fruit (*Hylocereus polyrhizus*) can be used as a natural food coloring due to its content of anthocyanin pigments. The extract contains 84.00 mg of anthocyanins per 100 grams (1), along with 150.46 mg of betacyanin, 0.95% protein, 0.10% ash, 6.20% carbohydrates, 10.8% pectin, and 0.10% fat (2). Additionally, the peel can be processed into flour, which expands its applications in the food industry. Rochmawati (2019) demonstrated the successful creation of cookies with high nutritional value using red dragon fruit peel flour. Incorporating dragon fruit peel flour into a cookie recipe, using a blend of 90% wheat flour and 10% dragon fruit peel flour, yields a distinct and nutritious treat. These cookies boast a water content of 8.06%, which helps maintain their moist texture. The nutritional profile reveals an ash content of 6.81%, indicative of mineral richness, and a protein content of 5.63%, contributing to their overall health benefits (3). Moreover, the cookies contain a noteworthy fat content of 27.03%, which enhances flavor and provides a satisfying mouthfeel. The carbohydrate content stands at 52.47%, offering a quick source of energy, while the impressive fiber content of 31.26% promotes digestive health and contributes to a feeling of fullness. When enjoying a serving size of 35 grams of these cookies, one can receive 4% of the daily protein needs, 15% of daily fat requirements, 6% of carbohydrates, a remarkable 37% of fiber, and 7% of total caloric intake. These figures highlight the potential of dragon fruit peel flour as a valuable ingredient, suggesting that it can be further explored and developed into a variety of innovative food products.

This flour can be used to produce a variety of items, such as delectable cakes, vibrant food coloring, and a range of dried noodles. Despite its many applications, this distinctive flour has not achieved widespread recognition, largely due to its relatively high price, which restricts consumer access. The manufacturing process for this flour demands precision to preserve its quality and nutritional benefits. Furthermore, the limited availability of essential raw materials raises concerns about the potential for adulteration in the marketplace. This highlights the need for vigilance to ensure that consumers receive authentic red dragon fruit peel flour (RDFPL) free from any unwanted additives. Common practices of adulteration include the addition or substitution of original ingredients with others (4–6).

Food adulteration, the intentional modification of food products to enhance their appearance or weight, can be effectively identified using various analytical techniques. One prominent method employed for this purpose is Fourier Transform Infrared (FTIR) spectroscopy. This technique is invaluable in detecting contaminants in food by providing detailed insights into the chemical composition of ingredients. FTIR spectroscopy operates by analyzing how different chemical molecules interact with electromagnetic radiation. It specifically focuses on the vibrations of functional groups—clusters of atoms that impart specific characteristics to a molecule—or polyatomic chemical bonds, which involve multiple atoms. When molecules absorb infrared light, they undergo vibrational transitions that produce unique spectral signatures. These signatures can be analyzed to detect the presence of specific compounds, enabling the identification of adulterated substances. Referred to as vibrational spectroscopy, FTIR is renowned for its non-destructive nature, meaning that the chemical structure of the food sample remains intact after analysis. This allows for additional testing or examination without damaging the sample. Moreover, FTIR provides rapid results, making it an essential tool in the food industry for ensuring product safety and quality control. By utilizing FTIR spectroscopy, food safety professionals can efficiently identify and address food adulteration concerns, ultimately safeguarding consumers and promoting public health (7,8). In previous research, FTIR combined with chemometrics has demonstrated the ability to accurately predict and differentiate corn flour adulteration from chickpea flour, even at substitution levels ranging from 1% to 90% (9).

Determining the authenticity of this product is crucial. This process helps prevent counterfeiting, ensures quality and nutritional content, and guarantees consumer safety. In this study, red dragon fruit peel flour was mixed with purple sweet potato flour due to their similar coloration. This study sought to investigate the presence of purple sweet potato flour as an adulterant in red dragon fruit peel flour. To achieve this, we employed Fourier-transform infrared (FTIR) spectroscopy in combination with advanced chemometric techniques. This approach allowed for a detailed analysis of the flour composition, enabling precise identification of any mixtures and offering valuable insights into the quality and authenticity of the flour samples.

2. Materials and Methods

2.1 Materials
Red dragon fruit peel *(Hylocereus polyrhizus)* and purple sweet potato *(Ipomoea batatas L.)* were obtained from Yogyakarta, salt (Refina), sodium metabisulfite, water, and 96% ethanol.

2.2 Methods
2.2.1 Sample preparation
The red dragon fruit peel flour was carried out according to (10) with modifications. The selected samples were washed, followed by reducing the size to approximately 3mm thickness, soaking in saltwater at 5g/l for 25 minutes, blanching for 1 minute, drying in a cabinet dryer (75 °C, 9 hours), and then grinding and sieving to 80 mesh.
The production of purple sweet potato flour involved removing the peel from the sweet potatoes, reducing the size to approximately 1mm thickness, soaking in sodium metabisulfite at 3g/l for 30 minutes, blanching for 1 minute, drying in a cabinet dryer at 60ºC for 7 hours, followed by grinding and sieving through an 80 mesh.
2.2.2 Adulteration model
The model for the adulteration of red dragon fruit peel flour (RDFPL) with purple sweet potato flour (PSPF) were created by mixing both flours in 10 g. The range of adulteration levels were 3, 5, 7, 10, 13, 15, 17, 20, 23, 25, 27, 30, 33, 35, 37, 40, 43, 45, 47, 50, and 53%. Pure RDFPL and PSPF at 100% were also prepared for FTIR spectroscopy analysis.
2.2.3 Color analysis
Pure RDFPL and pure PSPF, along with 21 samples of adulterated RDFPL, were analyzed using the NH300 Portable Colorimeter. Each sample was analyzed three times.
2.2.4 FTIR analysis
Pure RDFPL, pure PSPF, and 21 samples of counterfeit RDFPL were analyzed using FTIR spectroscopy (Thermo Scientific Bruker Alpha Platinum) was equipped with attenuated total reflectance (ATR) and operated with Omnic software. Measurement in the mid-infrared area (4000–650 cm^{-1}). A reading were carried out at a resolution of 8 cm^{-1} 32 times. Analysis three times on each sample.
2.2.5 Chemometrics analysis
Chemometric analysis was carried out using advanced techniques such as Partial Least Squares Regression (PLSR) and Principal Component Regression (PCR). This analysis was performed with the aid of TQ Analyst software, specifically version 9, to ensure accurate data interpretation and modeling (Thermo Fisher Scientific, Inc). Principal Component Analysis (PCA) were carried out using Metabolomic 6.0 software.

3. Results and Discussion

3.1 Color analysis
The analysis conducted with a colorimeter distinguishes three color components: brightness (L*), redness (a*), and yellowness (b*). The addition of varying levels of adulterated purple sweet potato flour (PSPF) to red dragon fruit peel flour (RDFPL) significantly affects the color characteristics, resulting in higher L* values and lower a* and b* values (see Figure 1). The changes in color are attributed to the different natural pigments present in RDFPL and PSPF, particularly the anthocyanins. The vibrant anthocyanins found in purple sweet potatoes are primarily derived from complex chemical structures known as mono- or diacetyl-3-(2-glucosyl) glucosyl-5-glucosyl peonidin and cyanidin derivatives (11). These compounds are responsible for the characteristic deep purple hue of the tuber, which is a result of their ability to absorb and reflect different wavelengths of light. Anthocyanins are a type of flavonoid that not only contribute to the color but also possess various health benefits, including antioxidant properties. The

presence of glucosyl groups in their structure enhances their stability and solubility, making them important for both nutritional value and aesthetic appeal in culinary applications. The anthocyanins in red dragon fruit peel come from cyanidin 3-glucoside, delphinidin 3-glucoside, and pelargonidin 3-glucoside (12).

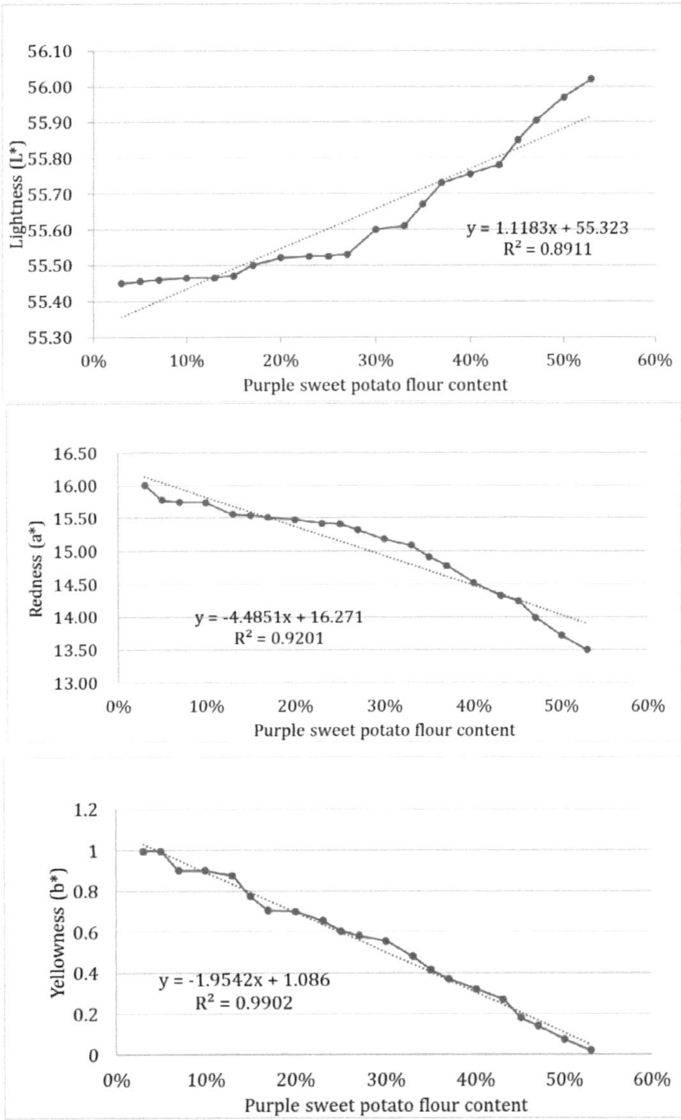

Figure 1. The changes of color parameters (L*, a*, and b*) in all the samples containing purple sweet potato flour in different contents.

3.2 FTIR spectrum analysis
The mid-infrared area (4000-650 cm^{-1}), used for FTIR spectrum analysis. The FTIR spectrum results can be analyzed and compared to a reference table of functional group characteristics, focusing on their absorption forms, positions, and intensities. The vibrations in the spectrum were stretching and bending.

Based on Figure 2 showed that the RDFPL and PSPF spectrum were similar. The medium absorption appeared at 3269 cm^{-1} from both RDFPL and PSPF samples, indicating the presence of O-H (hydroxyl) stretching vibrations. The absorption of 2919 cm^{-1} contains C-H stretching vibration absorption. The absorption point of 1727 cm^{-1} appears as the bending vibration absorption of C=O (carboxyl). The C=C (aromatic) stretching vibration absorption was found at 1612 cm^{-1}. At an absorption of 1375 cm^{-1}, the PSPF sample did not have any signal, while the RDFPL sample appeared absorbance. At 1375 cm^{-1}, stretching vibration absorption of CH_3 (methyl). There is a C-N (amine) bending vibration at 1315 cm^{-1} and 1244 cm^{-1}. At 1150 cm^{-1} was found an absorption point in the RDFPL spectrum, while in the PSPF spectrum, the absorption was unclear. The absorption at 1150 cm^{-1} was the bending vibration of O (ester). At the 1012 cm^{-1}, both samples have sharp absorption with C-O (alcohol) stretching vibrations. Absorption of 862 cm^{-1} with C-H bending vibration (alkene) there is a difference; this signal appeared in the PSPF sample only. C-H (aromatic) bending vibration absorption detected at the 759 cm^{-1} and 664 cm^{-1}.

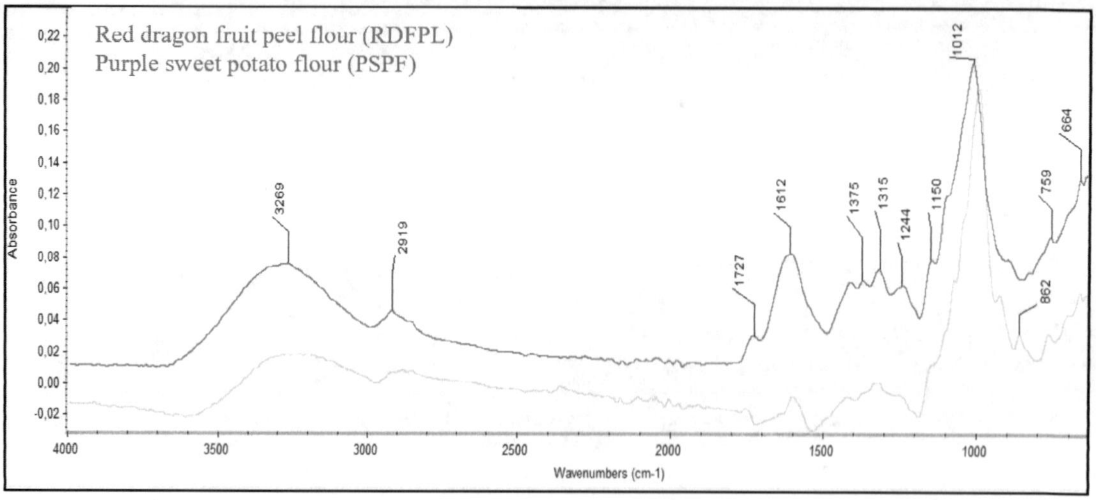

Figure 2. FTIR spectrum of RDFPL and PSPF

The anthocyanin content of the two samples can be observed by appearance of the hydroxyl group (O-H) at the absorption point of 3269 cm^{-1}, aromatic (C=C) at the 1612 cm^{-1}, and the alcohol group (C-O) at 1012 cm^{-1}. It was confirmed by the results of previous research which stated that the content of anthocyanin compounds can be detected by the appearance of hydroxyl O-H, aromatic C=C, and C-O alcohol groups in the FTIR spectrum (13).

The differences in the spectrum results of the two RDFPL and PSPF samples were marked at the absorption points of 1375 cm^{-1}, 1150 cm^{-1}, and 862 cm^{-1}. The absorption point of 1375 cm^{-1}, stretching vibration absorption of CH_3 (methyl) and 1150 cm^{-1} bending vibration of O (ester), which appeared in the RDFPL sample, was thought to be a group that forms pectin compounds. Previous research conducted stated that the constituent groups of pectin compounds in dragon fruit peel are characterized by -OH hydroxyl, -CH3, -C=O carboxyl, and -O- ester groups (14). The absorption point of 862 cm^{-1} with C-H (alkene) bending vibrations appearing in the PSPF spectrum was thought to be the presence of a starch-forming compound, namely glucose. A CH bending type of vibration characterizes the glucose group shown at a wave number of 875.62 cm-1 (15).

3.3 Multivariat Calibration

A multivariate calibration optimization process is used to obtain an optimal prediction model. Multivariate calibration used chemometrics with Principal Component Regression (PCR) and partial least squares regression (PLSR). The FTIR spectrum was optimized on a normal spectrum, first and second derivatives. The method that was ultimately chosen for this study is the one that demonstrated the most favorable performance metrics. Specifically, it exhibited the lowest values of root mean square error of calibration (RMSEC) and root mean square error of prediction (RMSEP). Additionally, this method achieved the highest coefficient of determination (R^2) among all the tested methods. These indicators collectively suggest that this approach provides a more accurate and reliable model for the data in question.

Table 1. Optimization of FTIR spectrum and multivariate calibration at different wavenumbers

Multivariate calibration	Wavenumber (cm^{-1})	Spectrum	Calibration		Prediction	
			RMSEC	R^2	RMSEP	R^2
PLSR	650-3603	Normal	5.98	0.9608	7.61	0.7549
		1st derivative	**0.304**	**0.9999**	**2.86**	**0.9880**
		2nd derivative	7.70	0.9341	5.06	0.9004
	651-811	Normal	6.70	0.9506	6.22	0.7048
		1st derivative	4.01	0.9826	8.65	0.8403
		2nd derivative	8.70	0.9151	2.78	0.9855
	883-1181	Normal	7.92	0.9302	8.64	0.8465
		1st derivative	3.40	0.9875	6.07	0.9657
		2nd derivative	2.30	0.9943	5.81	0.9901
	1187-1766	Normal	1.44	0.9978	3.17	0.9403
		1st derivative	0.967	0.9990	4.25	0.9632
		2nd derivative	2.61	0.9926	5.27	0.9104
	2833-3576	Normal	1.44	0.9978	3.17	0.9403
		1st derivative	1.22	0.9984	11.1	0.9008
		2nd derivative	5.83	0.9629	23.4	0.7032
PCR	650-3603	Normal	4.70	0.9750	5.99	0.9387
		1st derivative	6.68	0.9508	6.99	0.9518
		2nd derivative	7.99	0.9290	6.90	0.6348
	651-811	Normal	19.8	0.3979	24.5	0.7234
		1st derivative	18.2	0.5356	25.4	0.8324
		2nd derivative	20.0	0.3732	22.3	0.4645
	883-1181	Normal	7.98	0.9292	8.66	0.8452
		1st derivative	7.18	0.9430	8.48	0.9521
		2nd derivative	7.49	0.9379	9.88	0.9514
	1187-1766	Normal	4.56	0.9775	6.77	0.8477
		1st derivative	5.21	0.9704	5.30	0.9165
		2nd derivative	6.20	0.9578	6.94	0.9174
	2833-3576	Normal	11.0	0.8593	13.1	0.0506
		1st derivative	10.9	0.8633	9.77	0.9504
		2nd derivative	21.1	0.2077	21.9	0.8163

The calibration and validation model chosen was a PLSR first derivative spectrum with wavelengths of 650-3603 cm^{-1} with R^2 values of 0.9999 and 0.9880 respectively, the selected variables are marked in bold. (Table 1). The RMSEC and RMSEP values are 0.304 and 2.86, respectively. Low values of RMSEC and RMSEP and high R^2 value indicated that the model developed for RDFPL identification was entirely accurate. Figure 3 presents a detailed illustration of the relationship between actual concentration and predicted concentration as determined by the Partial Least Squares Regression (PLSR) technique. The actual concentration values are

IOP Conf. Series: Earth and Environmental Science **1460** (2025) 012061 doi:10.1088/1755-1315/1460/1/012061

plotted along the x-axis, while the predicted concentration values are shown on the y-axis. The analysis is based on the first derivative spectrum within the wavelength range of 3603 to 650 cm^{-1}. This figure highlights the effectiveness of the PLSR method in accurately predicting concentrations, and it allows for a visual assessment of how closely the predicted values align with the actual measurements.

Chemometric techniques are extensively employed in the analysis and classification of both pure and adulterated samples (5,8). These methods utilize quantitative data obtained from chemical analyses to develop models that can distinguish between authentic products and those

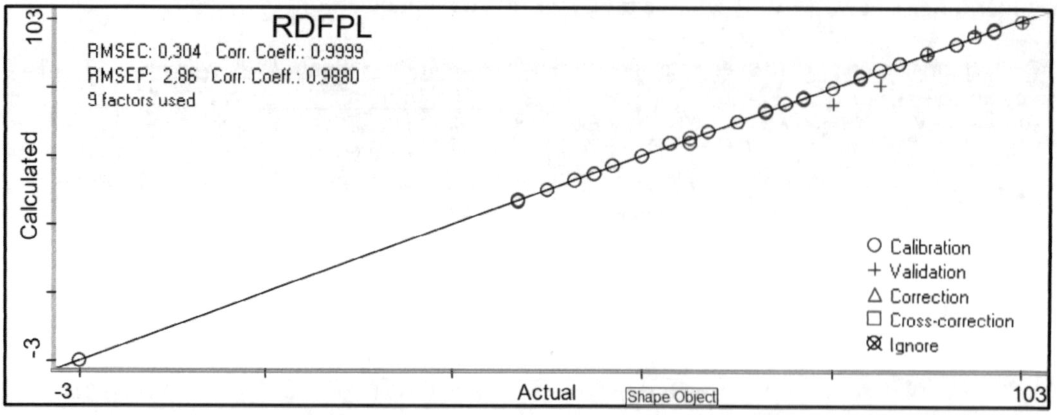

Figure 3. The correlation between actual concentration (x-axis) and predicted concentration (y-axis) of RDFPL adulterated with PSPF using the PLSR technique

that have been compromised. By applying various statistical and mathematical approaches, researchers can identify unique patterns and characteristics within the data, allowing for accurate differentiation and ensuring quality control in various industries. Classification of pure and adulterated samples can also use Principal component analysis (PCA). Figure 4 showed the score plot results, which group pure and adulterated RDFPL with a PC accuracy level (98.7%). The dark red color has a low level of RDFPL adulteration with a range of 3-20%; the green color has a medium level of counterfeiting in the range of 23-37%; the dark blue color has a high level of counterfeiting in the range of 40-53%, while the light blue and purple samples are pure RDFPL and PSPF. Therefore, FTIR, combined with chemometrics, was able to identify the degree of adulteration of red dragon fruit peel flour with purple sweet potato flour. PCA is able to group pure and adulterated samples by separating them according to the largest component of a sample and can group them according to the similarity of the samples.

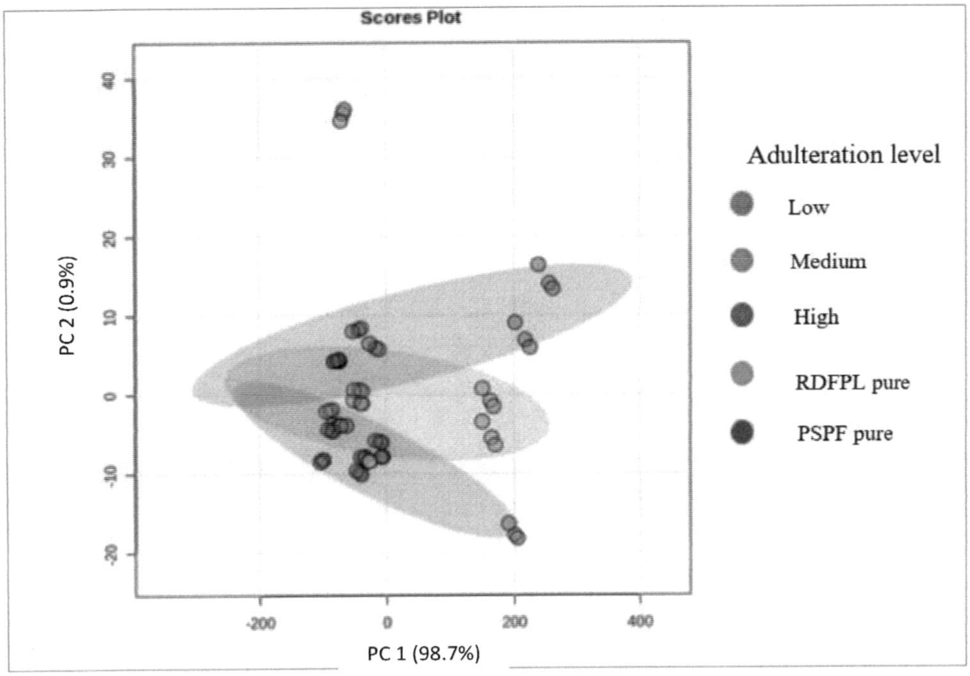

Figure 4. The score plot RDFPL adulterated with PSPF at different levels

4. Conclusion

The level of RDFPL adulteration with PSPF can be identified using a colorimeter and FTIR spectroscopy combined with chemometrics and PCA. Indications of RDFPL adulteration can be determined by a spectrum peak at wave number 862 cm^{-1}, which indicated the presence of (C-H) bending alkene-specific bonds in starch. The calibration and validation model chosen was a PLSR with first derivative spectrum with wavelengths of 650-3603 cm^{-1} with R^2 values of 0.9999 and 0.9880, respectively. The RMSEC and RMSEP values are 0.304 and 2.86, respectively. PCA can classify pure RDFPL samples and their fakes with a PC accuracy level (98.7%).

References

[1] Harni M, Anggraini T, Rini, Suliansyah I. Identifikasi kualitas warna buah naga *(Hylocerecus)* dengan ekstraksi menggunakan microwave-assisted extract (MAE). Jurnal Teknologi Pertanian Andalas. 2023;27(1):104–9.

[2] Jamilah B, Shu, Kharidah M, Dzulkifly MA, Noranizan A. Physico-chemical characteristics of red pitaya *(Hylocereus polyrhizus)* peel. Int Food Res J. 2011;18:279–86.

[3] Rochmawati N. Pemanfaatan kulit buah naga merah *(Hylocereus polyrhizus)* sebagai tepung untuk pembuatan cookies. Jurnal Pangan dan Agroindustri. 2019;7(3):19–24.

[4] Maruzy A, Budiarti M, Subositi D. Autentikasi *Centella asiatica* (L.) Urb. (Pegagan) dan Adulterannya Berdasarkan Karakter Makroskopis, Mikroskopis, dan Profil Kimia. Jurnal Kefarmasian Indonesia. 2020 Feb 24;19–30.

[5] Putri AR, Rohman A, Riyanto S, Setyaningsih W. Autentikasi Minyak Ikan Patin *(Pangasius micronemus)* menggunakan Metode Spektroskopi FTIR yang dikombinasikan dengan Kemometrika. Indonesian Journal of Chemometrics and Pharmaceutical Analysis. 2021;1(1):22–7.

[6] Putri AR, Aliaño-González MJ, Ferreiro M, Setyaningsih W, Rohman A, Riyanto S, et al. Development of a methodology based on headspace-gas chromatography-ion mobility spectrometry for the rapid detection and determination of patin fish oil adulterated with palm oil. Arabian Journal of Chemistry. 2020;13(10):7524–32.

[7] Dastjerdi AG, Akhgari M, Kamali A, Mousavi Z. Principal component analysis of synthetic adulterants in herbal supplements advertised as weight loss drugs. Complement Ther Clin Pract [Internet]. 2018 May 1;31:236–41. Available from: https://linkinghub.elsevier.com/retrieve/pii/S1744388117305704

[8] Irnawati I, Windarsih A, Indrianingsih AW, Apriyana W, Ratnawati YA, Hazairin Nadia LOM, et al. Rapid detection of tuna fish oil adulteration using FTIR-ATR spectroscopy and chemometrics for halal authentication. J Appl Pharm Sci [Internet]. 2023 Apr 1;13(4):231–9. Available from: https://japsonline.com/abstract.php?article_id=3893&sts=2

[9] Bala M, Sethi S, Sharma S, Mridula D, Kaur G. Prediction of maize flour adulteration in chickpea flour *(besan)* using near infrared spectroscopy. J Food Sci Technol. 2022 Aug 1;59(8):3130–8.

[10] Ho LH, Abdul Latif NW binti. Nutritional composition, physical properties, and sensory evaluation of cookies prepared from wheat flour and pitaya *(Hylocereus undatus)* peel flour blends. Cogent Food Agric. 2016;2(1):1–10.

[11] El Husna N, Novita M, Rohaya S. Kandungan antosianin dan aktivitas antioksidan ubi jalar ungu segar dan produk olahannya. AGRITECH. 2013;33(3):296–302.

[12] Saenjum C, Pattananandecha T, Nakagawa K. Antioxidative and Anti-Inflammatory Phytochemicals and Related Stable Paramagnetic Species in Different Parts of Dragon Fruit. Molecules. 2021 Jun 1;26(12).

[13] Nadia N, Mursal M, Jalil Z. Optical characterization of purple sweet potato *(Ipomoea batatas)* and carrot *(Daucus carota L.)* extracts for Dye Sensitized Solar Cell (DSSC) application. Journal of Aceh Physics Society. 2022 Aug 12;11(3):80–4.

[14] Nadir M, Latifah F, Meylinda P. Rendemen dan karakteristik pektin dari kulit nenas dan kulit buah naga dengan microwave assisted extraction (MAE). In: Prosiding Seminar Nasional Penelitian & Pengabdian Kepada Masyarakat. 2019. p. 124–8.

[15] Widjanarko SB, Nugroho A, Estiasih T. Functional interaction components of protein isolates and glucomannan in food bars by FTIR and SEM studies. African Journal of Food Science [Internet]. 2011;5(1):12–21. Available from: https://www.researchgate.net/publication/228690033

Encapsulation of *Lactobacillus plantarum* with Kepok Banana Flour through Microbiology and Physics Parameters

Yoyok Budi Pramono, Anang M. Legowo, Antonius Hintono, Bhakti Etza Setiani*, Aulia Zalfa Nabila, Nisfa Rama Kamila Handoko

Department of Agricultural Science, Universitas Diponegoro, Semarang, Indonesia

*E-mail: bhaktietzasetiani@lecturer.undip.ac.id

Abstract. Kepok banana flour contains prebiotic fiber in the form of inulin and fructooligosaccharides (FOS) which can stimulate the growth of *Lactobacillus plantarum* bacteria in the digestive tract. The microencapsulation technique is carried out to maintain the viability of probiotic bacteria from damage due to environmental conditions by packaging using coating materials in the form of skim milk and maltodextrin. This research aims to determine the effect of adding kepok banana flour with different concentrations on *L. plantarum* encapsulates. The research design used was a Completely Randomized Design (CRD) with 5 treatments and 4 replications. The treatment given a concentration of kepok banana flour of 0%, 2%, 4%, 6%, and 8%. The results obtained were a LAB viability value of 8.72 Log CFU/g-9.16 Log CFU/g, a pH value of 6.55-6.43, yield value of 23.39%-23.43%, color of 3.36-3.12, aroma of 2.80-2.88, texture of 3.08-2.92, and overall 3.12-3.36. The addition of 8% kepok banana flour can increase LAB viability, color and texture, but reduces the pH value, yield and aroma of *L. plantarum* encapsulates, and produces a well-encapsulated microcapsule morphology.

Keywords: freeze drying, kepok banana, LAB viability, pH, yield

1. Introduction

Bananas are easy to grow in countries located on the equator such as Indonesia, which has an elevated temperature of 28 - 35°C each year [1]. Kepok banana is one type varians of banana cultivar which are widely consumed in Indonesia [2]. Banana kepok is a climatic fruit that is easily damaged or has a short shelflife. Processing kepok banana into flour is one way to prevent banana shelflife [3]. Banana flour can be used as an alternative to wheat flour which is high nutrition and gluten free [4].

Kepok banana is a type of plantain banana that contains high starch content. The starch contained in plantain bananas includes resistant starch [5]. The resistant starch content of pisang kepok is 27.7%, so it can be classified as one of the fruits that is high in fiber [6]. Within 100 g of plantain banana, there are 60.7 g of carbohydrates, 536 kcal of energy, 3.57 g of protein, and 3.6 g of fiber [7]. In addition, plantain bananas are also high in fructooligosaccharide (FOS), which is beneficial for the digestive system [8]. Processing kepok banana into flour can be utilized as a source of prebiotics for probiotic bacteria because it has a high content of resistant starch and fructooligosaccharide (FOS) [9].

Lactobacillus plantarum as probiotic bacteria that can grow optimally at a temperature of 30 - 37°C and has the potential as a probiotic that is more resistant to acid than other types of bacteria [10]. Probiotics in food products must be protected from various things that can affect processing, namely by microencapsulation. This encapsulation is done by wrapping probiotic bacteria and using certain encapsulant materials that maintain the viability of probiotic bacteria from environmental conditions [11].

One method that can be used in the encapsulation process is freeze drying. The damage to probiotic bacteria cells by freeze drying can be minimized by the addition of coating materials such as skim milk and maltodextrin [12]. The coating material can produce a fairly high viability of Lactobacillus bacteria, besides the freeze-drying method that is maintained at the appropriate temperature and humidity can maintain the viability of Lactobacillus bacteria [13]. The growth of Lactobacillus bacteria is influenced by low pH (6.2 to 5.0). Low pH can cause Lactobacillus bacteria growth to be more optimal [14]. The processs on making banana flour utilizes citric acid to produce banana flour that tends to be acidic [15]. Citric acid soaking in flour making can result in a low flour pH [16]. Therefore, this study was conducted to make KBF functional by adding probiotic bacteria which is *Lactobacillus plantarum.*

2. Materials and Methods

The research was carried out at the Food Chemistry and Nutrition Laboratory and the Food Engineering and Agricultural Products, Faculty of Animal and Agriculture, Universitas Diponegoro in August – October 2023. The main ingredients used in the encapsulation of *L. plantarum* are KBF which comes from raw kepok bananas obtained homogeneously from one of the banana plantations in Semarang, pure isolate of *L. plantarum* strain ATCC 8014, skim milk indoprima, maltodextrin food grade DE 10-12, MRSB and MRSA media, pH 4 and 7 buffers, 0.85% physiological NaCl (8,5 gram NaCl on 100 ml aquades), and citric acid. Meanwhile, the tools used in this research were freezer dryer, autoclave, freezer, oven, cabinet dryer, pH meter, SEM tool, incubator, laminar air flow, colony counter, micropipette tip, micropipette, and petri dishes.

2.1 Research Design

Samples were prepared in a completely randomized design which was carried out in stages including: 1) Making KBF, 2) Making encapsulation of *L. plantarum*, 3) Comparative analysis of microbiology and physics parameters encapsulation of *L. plantarum*. The treatment given to *L. plantarum* encapsulates being the addition of KBF concentrations of 0%, 2%, 4%, 6%, and 8%.

2.2 Research Procedure

2.2.1 Making Kepok Banana Flour

The production of KBF has been adopted [15] with modifications. The steps for making banana kepok flour start with sorting and washing of kepok banana. The banana kepok is then peeled and sliced to a of ± 2 mm. The banana kepok was soaked with 1500 ml water and added citric acid according to the treatment (w/v) 0.8% for 3 minutes. The banana pieces are dried using a *cabinet dryer* at 60°C for 6 hours. The dried banana is crushed and sived thorough 80 mesh.

2.2.2 Making Encapsulation of *L. plantarum*

Preparation of *L. plantarum* encapsulate has been adopted from [17] which was carried out by incubation one ose of *L. plantarum* bacteria in MRSB, then incubation for 48 hours at a temperature of 37°C. Next, 5 ml of the harvested *L. plantarum* bacterial culture was added to 0.85% physiological NaCl. Then 4 ml of *L. plantarum* culture was mixed in 36 ml of 10% skim milk medium and KBF was added with concentrations of 0%, 2%, 4%, 6% and 8% and incubated for 2 hours at 37°C. The *L. plantarum* suspension that has been formed is added with 20% maltodextrin, then each sample in liquid state was taken to carry out the viability test of LAB bacteria using the pour plate dilution method and incubated for seven days to determine how much *L. plantarum* bacteria grew before being encapsulated with a freeze dryer and other samples frozen in a freezer at -24°C for 24 hours, then the frozen material is placed in a freeze dryer at a temperature of -50°C and a pressure of 0.040 mbar for 48 hours until a dry culture is formed *L. plantarum* bacteria (microcapsules).

2.2.3 Parameters Testing

Parameter assessment was carried out on *L. plantarum* encapsulation with the addition of 0% KBF as a control, KBF 2%, KBF 4%, KBF 4%, KBF 6%, and KBF 8%. LAB viability testing was carried out on [17] using pour plate method with 10^{-7} dilution. Testing of pH values was carried out at [18] using a pH meter. Yield testing was carried out in [19] with weight comparisons. Microcapsule morphology testing with a modification of the method [20] with 2000x magnification. Organoleptic testing using the scoring method.

2.3 Data Analysis

The data were analyzed using an application named SPSS for Windows 26.0. The data obtained from testing LAB viability, pH and yield were analyzed using the *Analysis of Variance* (ANOVA) parametric test with significance level ($p<0.05$) and when there is a significant difference in treatment will be continued by *Duncan's Multiple Range Test* (DMRT). The organoleptic testing using *Kruskal Wallis* analysis. If the significant level has a real effect ($p<0.05$), then proceed with the *Mann-Whitney test.* While resistant starch was statistically analyzed using Excel application and microcapsule morphology was descriptive.

3. Results and Discussion

3.1 Encapsulation of Lactobacillus plantarum with kepok banana flour microbiology characteristics

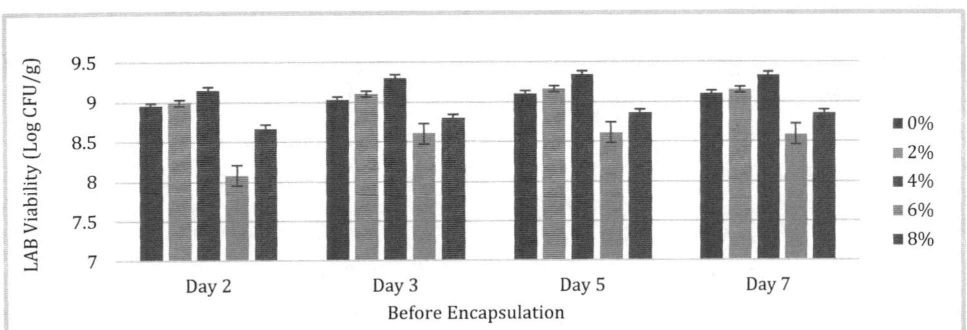

Figure 1. Microbiology characteristics before KBF encapsulation are shown with values presented as means with standard deviations. LAB viability value is carried out descriptively.

IOP Conf. Series: Earth and Environmental Science **1460** (2025) 012062 doi:10.1088/1755-1315/1460/1/012062

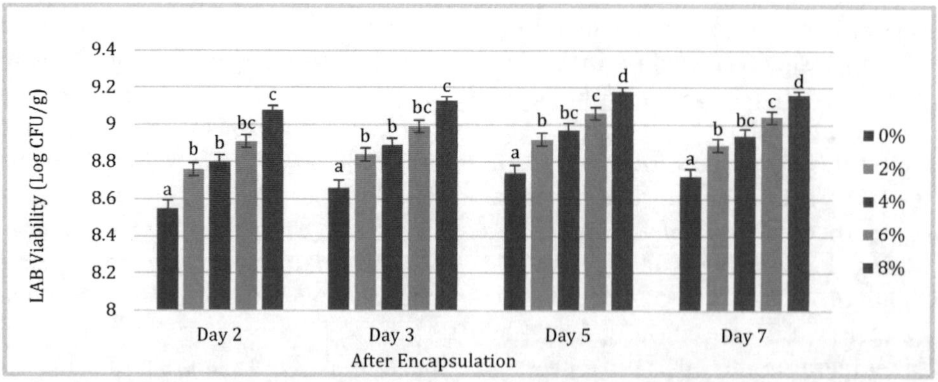

Figure 2. Microbiology characteristics after KBF encapsulation are shown with values presented as means with standard deviations. Different letters above the lines indicate statistically significant differences at $p<0.05$.

The microbiology characteristics of *L. plantarum* with KBF is the Lactic Acid Bacteria (LAB) viability as it describes on the Figure 1 and 2. Before *L. plantarum* was encapsulated, a LAB viability test was carried out without repetition to determine whether after encapsulation with freeze-drying *L. plantarum* continued to grow optimally.

Encapsulated *L. plantarum* with a concentration of 8% kepok banana flour had the highest viability value after encapsulation of 9.18 Log CFU/g which was on the 5th day of observation, while encapsulated *L. plantarum* with a concentration of 0% kepok banana flour had the lowest viability value of 8.55 Log CFU/g which was on the 2nd day of observation. LAB viability after encapsulation of *L. plantarum* at a concentration of 0% to 4% decreased in the number of colonies, whereas at concentration of 6% and 8% KBF *L. plantarum* experienced an increase in the number of colonies. The higher the addition of KBF, it indicates that the amount of fiber content contained in it is also high, so that when KBF is mixed with the coating material it can create a gel in the sample solution and can be more optimally utilized as a growth medium by bacteria probiotics [21].

The viability of probiotic bacteria can be measured in Log CFU/g, where a high Log CFU/g value indicates better viability. The minimum number of probiotic bacteria used for functional food is no less than 6 Log CFU/g [22]. Based on the data obtained, the viability of the probiotic before and after encapsulation shows that the number of cells is above 6 Log CFU/g, which means it meets the requirements as a probiotic. The exact number of live cells that trigger probiotic effects ranges from 6 to 9 Log CFU/g [23]. Thus, KBF can encapsulate *L. plantarum* optimally.

The longer the fermentation time, the total number of lactic acid bacteria increases. Lactic acid bacteria can reproduce by utilizing the nutrients in the *L. plantarum* encapsulate medium during the fermentation process to hydrolyze sugar into simpler components, namely CO2, H2O, and lactic acid, so that the number of bacterial cells increases [24]. KBF contains resistant starch which has the potential to act as a prebiotic which can protect probiotics. The probiotic bacteria *L. plantarum* can survive in the large intestine when contained in resistant starch granules, because resistant starch as a potential prebiotic cannot be digested in the small intestine [25].

3.2 Encapsulation of Lactobacillus plantarum with kepok banana flour physics characteristics

Physics characteristics of *L. plantarum* with KBF were yield and pH value as it describes on the Figure 3.

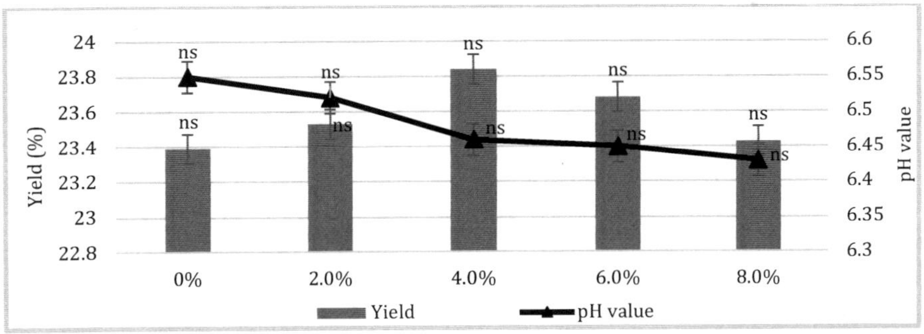

Figure 3. Physics characteristic of KBF encapsulation values are shown as averages with standard deviations. The same letters "ns" above the line indicate that there is no statistically significant difference between treatments at $p>0.05$.

The maltodextrin coating material has an influence on the yield value because maltodextrin has natural characteristics that produce sponge microcapsules with many remaining water molecules [26]. The freeze-drying method used resulted in very few encapsulate samples coming out of the container and many remaining in the container and the addition of maltodextrin which could increase the total solids of the material and increase the volume, so that the encapsulate yield value was high [27]. However, the maltodextrin concentration for each treatment was the same. The coating material maltodextrin and skim milk with the same concentration in each treatment resulted in yield values that were not much different.

Lactobacillus plantarum can thrive in the pH range of 4 to 6 [28]. The data obtained was that the pH of the *L. plantarum* encapsulate was in the range of pH 6, there was no significant difference because each *L. plantarum* treatment was added in the same percentage amount. The 8% KBF treatment showed pH results that tended to be lower compared to other treatments, this influenced increased bacterial growth. Encapsulation with a freeze dryer can increase the tolerance of *L. plantarum* to pH in acidic conditions [29]. The pH value is closely related to the number of bacteria produced in the encapsulated sample. The greater the number of probiotic bacterial cells, the more acidic the resulting encapsulate acidity value will be increased [30].

The pH value of *L. plantarum* encapsulates which becomes more acidic or decreases with the higher concentration of KBF indicates that the nutritional needs of probiotic bacteria have been met by the addition of KBF, so that the number of bacterial cells increases [31]. The pH value which tends to be acidic can help prevent the development of unwanted microorganisms so that it can increase the shelf life of *L. plantarum* encapsulates [32].

3.3 Microcapsule Morphology

Scanning Electron Microscope of encapsulated *L. plantarum* KBF concentration with the same magnification of 2000x as it describes on the Figure 4.

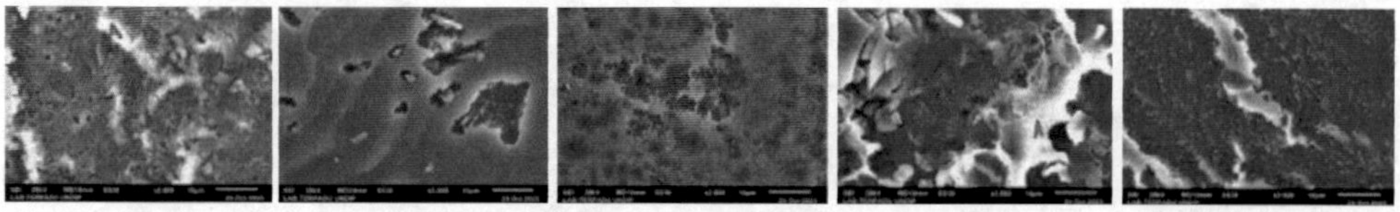

(a) (b) (c) (d) (e)

Figure 4. Is presented below in figure 4(a) the treatment with the addition of 0% KBF shows quite a lot of *L. plantarum*, figure 4(b) the treatment with the addition of 2% KBF shows *L. plantarum* growing slightly, figure 4(c) treatment with the addition of 4% KBF shows that *L. plantarum* is seen growing, figure 4(d) treatment with the addition of 6% KBF shows *L. plantarum* growing a lot, and figure 4(e) treatment with the addition of 8% KBF shows that *L. plantarum* grows more optimally than the previous treatment, which is indicated by a white image from the results of Scanning Electron Microscope (SEM) observations.

Note : Kepok Banana Flour (KBF)
A : *L. plantarum*
B : KBF, skim milk, maltodextrin

The morphology of *L. plantarum* microcapsules shows that the increasing use of KBF is able to function as a carrier medium for *L. plantarum*. It can be seen at the same magnification (2000x) that there are more *L. plantarum* attached to the surface of the KBF in the 8% KBF formulation. *L. plantarum* is shown in the SEM results to be bright in color, which shows that *L. plantarum* is under the KBF. KBF and other components in the form of skim milk and maltodextrin are darker in color, which means they are on the surface of *L. plantarum*. According to [33], observing the morphology of *L. plantarum* using SEM can be shown by the results of *L. plantarum* being rod-shaped and looking shiny white.

The SEM image of *L. plantarum* encapsulates (Figure 4) shows that the morphology of the microcapsules is similar with the surface shrinking due to the uniform drying process by freeze drying [34]. A smoother microcapsule surface and uniform structure was demonstrated with a concentration of 8% KBF, 20% maltodextrin, and 10% skim milk. The probiotic bacteria in the encapsulation results are located randomly on the surface of the microcapsule, this is due to the type of matrix with scattered cells and may be located on the surface of the coating material [26].

The aim of using a scanning electron microscope on *L. plantarum* encapsulates was to observe the probiotics coated with the coating material. Figure 1 shows that a thin layer of protective matrix is formed covering the cells with regular intercellular spacing and a dense structure [35]. From the SEM image, it can be that its morphology is irregular and heterogeneous. Coating ingredients for skim milk and maltodextrin with the same concentration for each treatment caused the size of the encapsulated *L. plantarum* microcapsules to tend to be uniform between treatments [36].

The shape of the microcapsules looks smooth and slightly wrinkled which can be caused by the high ratio of maltodextrin encapsulant compared to skim milk. Skim milk as a protein-based coating material which has a double emulsion system which can produce microcapsules with a shape that does not look wrinkled and smooth, but due to the presence of another encapsulant material, namely maltodextrin with a high concentration, the texture looks wrinkled [37].

Maltodextrin coating does not have good emulsion ability, so the microcapsule shape has wrinkles and size non-uniformity due to the high viscosity of the emulsion [38].

3.4 Organoleptic Properties

The organoleptic testing attributes of *L. plantarum* encapsulates assessed include color, aroma, texture, and overall Organoleptic encapsulation of L. *plantarum* with KBF as it describes on the Figure 5. The organoleptic testing was carried out with the help of 25 semi-trained.

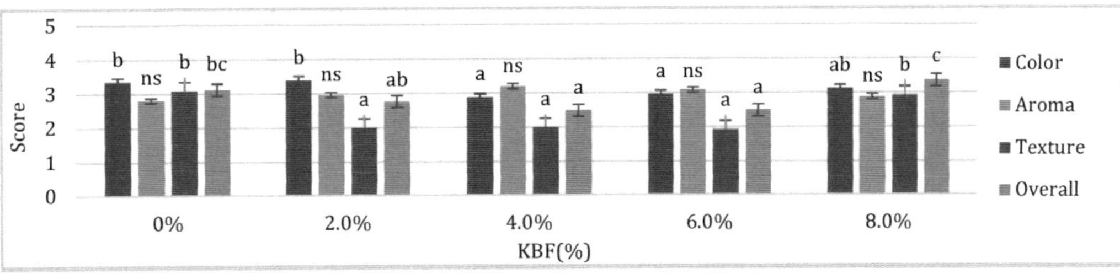

Figure 5. Organoleptic of KBF encapsulation values are shown as averages with standard deviations. Different letters above the line indicate that there are statistically significant differences between treatments in color, texture, and overall at $p<0.05$. While the "ns" letter above the line indicates that there is no statistically significant difference between treatments in aroma at $p>0.05$.

L. plantarum encapsulates have a white color, the color of each *L. plantarum* encapsulate treatment tends to fluctuate. The addition of high levels of maltodextrin resulted in the encapsulate color of *L. plantarum* increasing or being white and tending to be bright, this was due to the increasing number of total solids and white maltodextrin [39]. The color of *L. plantarum* encapsulates is also influenced by the color of the skim milk used. The research results of [40] the encapsulation results of red peppers have a red color which comes from the red pepper raw materials used. This can also be interpreted as the encapsulation results of *L. plantarum* are white, influenced by the color of the material used during encapsulation.

L. plantarum encapsulates tend not to have a typical banana aroma, this is because the concentration of the addition of KBF is not too high. Adding a low concentration of coating material can result in the aroma of the encapsulate still having a distinctive banana aroma because it cannot cover the strong characteristic aroma of kepok banana [36]. However, the addition of KBF also affects the resulting aroma. The small concentration of KBF in the *L. plantarum* encapsulation means that the aroma produced cannot be seen significantly between treatments [41]. Overall, the encapsulate produced was as expected, namely it did not have a typical banana aroma.

Encapsulation of *L. plantarum* with the addition of KBF using freeze drying can make the texture of the encapsulate less rough which is caused during the drying process carried out by freezing and low temperatures [42]. *L. plantarum* encapsulates have a rough to non-coarse texture which can be caused by the uneven grinding of the encapsulates between treatments. *L. plantarum* encapsulates produce a texture that is not rough due to the addition of maltodextrin which can interact with water so that it is able to maintain the texture of the *L. plantarum* encapsulates [43].

L. plantarum encapsulates have overall criteria that tend to fluctuate. However, with the addition of 8% the KBF had a high score and was liked by the panelists. The treatment with the highest score is the best treatment that can be received by the panelists [44]. Overall acceptance is an assessment of *L. plantarum* encapsulates on color, aroma, and texture parameters. These results indicate that the ratio of KBF, coating material, and *L. plantarum* at 8% KBF according to the expectations and criteria preferred by the panelists.

4. Conclusion

This research shows the effect of adding KBF on *L. plantarum* encapsulation. Our findings offer definitive evidence that this phenomenon at a KBF concretration of 8% is the best treatment with results in high LAB viability, low pH and yield values, microcapsule morphology on *L. plantarum* grows are optimally, non-grainy texture, white color, non-banana aroma and overall very favorable levels. This change was caused by the addition of higher KBF concentrations.

Acknowledgements
The authors would like to thank DIPA No. 4/UN7 F.5/HK/III/2023, Faculty of Animal and Agricultural Sciences, Diponegoro University, for their sponsorship and financial support

References

[1] C. Sugianti, T. Imaizumi, M. Thammawong, M. Tsuta, M. Nagata, and K. Nakano 2024, "Time–temperature tolerance of harvested green bananas exposed to high temperatures," *Sci Hortic*, vol. 329, doi: 10.1016/j.scienta.2024.112970.

[2] A. Setiawan, L. R. Dianti, N. E. Mayangsari, D. R. Widiana, and D. Dermawan 2024, "Removal of methylene blue using heterogeneous Fenton process with Fe impregnated kepok banana (Musa acuminate L.) peel activated carbon as catalyst," *Inorg Chem Commun*, vol. 152, doi: 10.1016/j.inoche.2023.110715.

[3] A. Campuzano, C. M. Rosell, and F. Cornejo 2018, "Physicochemical and nutritional characteristics of banana flour during ripening," *Food Chem*, vol. 256, pp. 11–17, doi: 10.1016/j.foodchem.2018.02.113.

[4] Y. T. Cheng *et al.* 2024, "Investigate the composition and physicochemical properties attributes of banana starch and flour during ripening," *Carbohydrate Polymer Technologies and Applications*, vol. 7, doi: 10.1016/j.carpta.2024.100446.

[5] C. V. Borges *et al.* 2020, "Nutritional value and antioxidant compounds during the ripening and after domestic cooking of bananas and plantains," *Food Research International*, vol. 132, doi: 10.1016/j.foodres.2020.109061.

[6] R. N. Arifah and A. Sofyan 2024, "Effect of Modified Banana Kepok (Musa paradisiaca L.) Starch Substitution on Resistant Starch, Protein, and Water Content in Steamed Brownies," MDPI AG, p. 24. doi: 10.3390/engproc2024063024.

[7] U.S. DEPARTEMENT OF AGRICULTURE, "BANANA PLANTAIN STRIPS, BANANA." Accessed: Jun. 03, 2024. [Online]. Available: https://fdc.nal.usda.gov/fdc-app.html#/food-details/2627618/nutrients

[8] O. M. Rahmawati *et al.* 2023, "Effect of unripe banana flour as a functional feed ingredient on growth performance, internal organ relative weight and carcass traits of broilers," *Vet Med Sci*, vol. 9, no. 2, pp. 851–859, doi: 10.1002/vms3.1070.

[9] P. Powthong, B. Jantrapanukorn, P. Suntornthiticharoen, and K. Laohaphatanalert 2020, "Study of prebiotic properties of selected banana species in Thailand," *J Food Sci Technol*, vol. 57, no. 7, pp. 2490–2500, doi: 10.1007/s13197-020-04284-x.

[10] J. Ilmu dan Teknologi Pangan *et al.* 2019, "The Viability of Lactobacillus plantarum FNCC-0027 in Apple Juice (Malus sylvestris Mill) with Different Varieties," vol. 8, no. 4, pp. 440–447.

[11] A. F. Massounga Bora, X. Li, Y. Zhu, and L. Du 2019, "Improved Viability of Microencapsulated Probiotics in a Freeze-Dried Banana Powder During Storage and Under Simulated Gastrointestinal Tract," *Probiotics Antimicrob Proteins*, vol. 11, no. 4, pp. 1330–1339, doi: 10.1007/s12602-018-9464-1.

[12] S. O. Oluwatosin, S. L. Tai, and M. A. Fagan-Endres 2022, "Sucrose, maltodextrin and inulin efficacy as cryoprotectant, preservative and prebiotic – towards a freeze dried Lactobacillus plantarum topical probiotic," *Biotechnology Reports*, vol. 33, doi: 10.1016/j.btre.2021.e00696.

[13] T. Hongpattarakere and S. Uraipan 2015, "Bifidogenic characteristic and protective effect of saba starch on survival of Lactobacillus plantarum CIF17AN2 during vacuum-drying and storage," *Carbohydr Polym*, vol. 117, pp. 230–235, doi: 10.1016/j.carbpol.2014.09.065.

[14] S. Jung and J. H. Lee 2020, "Characterization of transcriptional response of Lactobacillus plantarum under acidic conditions provides insight into bacterial adaptation in fermentative environments," *Sci Rep*, vol. 10, no. 1, doi: 10.1038/s41598-020-76171-6.

[15] A. Legowo, Y. Pramono, A. Hintono, B. E. Setiani, A. Nabila, and N. Handoko 2024, "Inhibition of Maillard Reaction of Kepok Banana Flour with Citric Acid," *Indonesian Journal of Agricultural Research*, vol. 7, no. 1, pp. 29–37, doi: 10.32734/injar.v7i1.15207.

[16] W. Fan, J. Zhao, and Q. Li 2023, "Effect of different food additives on the color protection of instant pumpkin flour," *Food Chemistry Advances*, vol. 3, doi: 10.1016/j.focha.2023.100413.

[17] S. D. Moody, I. I. Hanidah, I. L. Kayaputri, T. Rialita, E. Sukarminah, and R. A. P. Zakaria 2019, "Microencapsulation of Lactobacillus acidophilus with freeze drying method and application to synbiotic beverage of banana corm stone," *Int J Adv Sci Eng Inf Technol*, vol. 9, no. 2, pp. 532–537, doi: 10.18517/ijaseit.9.2.7903.

[18] L. I. Machado Vasconcelos *et al.* 2021, "Functional fermented sausages incorporated with microencapsulated Lactobacillus plantarum BG 112 in Acrycoat S100," *LWT*, vol. 148, doi: 10.1016/j.lwt.2021.111596.

[19] K. S. Yoha, J. A. Moses, and C. Anandharamakrishnan 2020, "Effect of encapsulation methods on the physicochemical properties and the stability of Lactobacillus plantarum (NCIM 2083) in synbiotic powders and in-vitro digestion conditions," *J Food Eng*, vol. 283, doi: 10.1016/j.jfoodeng.2020.110033.

[20] L. Chang *et al.* 2022, "Structural, physicochemical, antioxidant and in vitro digestibility properties of banana flours from different banana varieties (Musa spp.)," *Food Biosci*, vol. 47, doi: 10.1016/j.fbio.2022.101624.

[21] N. Azizah and A. Sofyan 2024, "The Effect of Modified Kepok Banana (Musa acuminata × balbisiana) Starch Substitution on the Fat, Dietary Fiber, and Resistant Starch Content of Product Cookies," MDPI AG, p. 7. doi: 10.3390/engproc2024063007.

[22] V. Y. Marinova, I. K. Rasheva, Y. K. Kizheva, Y. D. Dermenzhieva, and P. K. Hristova 2019, "Microbiological quality of probiotic dietary supplements," *Biotechnology and Biotechnological Equipment*, vol. 33, no. 1, pp. 834–841, doi: 10.1080/13102818.2019.1621208.

[23] A. Olivares, C. Soto, E. Caballero, and C. Altamirano 2019, "Survival of microencapsulated Lactobacillus casei (prepared by vibration technology) in fruit juice during cold storage," doi: 10.1016/j.

[24] Y. Wang *et al.* 2021, "Metabolism Characteristics of Lactic Acid Bacteria and the Expanding Applications in Food Industry," *Frontiers in Bioengineering and Biotechnology*, vol. 9. Frontiers Media S.A., doi: 10.3389/fbioe.2021.612285.

[25] W. Y. Koh, X. X. Lim, T. C. Tan, R. Kobun, and B. Rasti 2022, "Encapsulated Probiotics: Potential Techniques and Coating Materials for Non-Dairy Food Applications," *Applied Sciences (Switzerland)*, vol. 12, no. 19. MDPI, doi: 10.3390/app121910005.

[26] W. Sun *et al.* 2023, "Microencapsulation and Application of Probiotic Bacteria Lactiplantibacillus plantarum 299v Strain," *Microorganisms*, vol. 11, no. 4, doi: 10.3390/microorganisms11040947.

[27] F. Riany Eris, S. Kusumasari, V. Yoesepa Pamela, M. Rizal Febriansah, and R. Ahmad Riyanto, "INDONESIAN FOOD SCIENCE AND TECHNOLOGY JOURNAL (IFSTJ) Dextrose Equivalent (DE) Variation and Maltodextrin Concentration Effects in Yoghurt Powder Characteristics Using Foam-mat Drying," 2023.

[28] U. Nwachukwu, U. George-Okafor, U. Ozoani, and N. Ojiagu 2019, "Assessment of probiotic potentials of Lactobacillus plantarum CS and Micrococcus luteus CS from fermented milled corn-soybean waste-meal," *Sci Afr*, vol. 6, doi: 10.1016/j.sciaf.2019.e00183.

[29] W. F. Tee, R. Nazaruddin, Y. N. Tan, and M. K. Ayob 2014, "Effects of encapsulation on the viability of potential probiotic Lactobacillus plantarum exposed to high acidity condition and presence of bile salts," *Food Science and Technology International*, vol. 20, no. 6, pp. 399–404, doi: 10.1177/1082013213488775.

[30] A. A. Mendonça, W. de P. Pinto-Neto, G. A. da Paixão, D. da S. Santos, M. A. De Morais, and R. B. De Souza 2023, "Journey of the Probiotic Bacteria: Survival of the Fittest," *Microorganisms*, vol. 11, no. 1. MDPI, doi: 10.3390/microorganisms11010095.

[31] I. Handayani and N. Aini, "POTENTIAL ADDITION OF Lactobacillus casei AND FLOUR OF AMBON BANANA TO INHIBIT PATHOGENS AT YOGURT."

[32] M. S. Maia, M. M. Domingos, and J. F. B. de São José 2023, "Viability of Probiotic Microorganisms and the Effect of Their Addition to Fruit and Vegetable Juices," *Microorganisms*, vol. 11, no. 5. MDPI, doi: 10.3390/microorganisms11051335.

[33] S. Liu *et al.* 2019, "Evaluation and proteomic analysis of lead adsorption by lactic acid bacteria," *Int J Mol Sci*, vol. 20, no. 22, doi: 10.3390/ijms20225540.

[34] J. Baranauskaite-Ortasöz, B. Üner, G. Duman, and Ç. Taş 2022, "The microencapsulation of oregano extract by using different techniques: Spray-drying and freeze-drying techniques and their in vitro characterization," *Journal of Research in Pharmacy*, vol. 26, no. 1, pp. 102–111, doi: 10.29228/jrp.107.

[35] T. H. Nguyen, J. S. Kim, H. J. Kwon, and C. H. Kang 2023, "The Effect of a Glutathione (GSH)-Containing Cryo-Protectant on the Viability of Probiotic Cells Using a Freeze-Drying Process," *Fermentation*, vol. 9, no. 2, doi: 10.3390/fermentation9020187.

[36] F. Shoaei, A. Heshmati, R. Mahjub, A. D. Garmakhany, and M. Taheri 2022, "The assessment of microencapsulated Lactobacillus plantarum survivability in rose petal jam and the changes in physicochemical, textural and sensorial characteristics of the product during storage," *Sci Rep*, vol. 12, no. 1, doi: 10.1038/s41598-022-10224-w.

[37] T. T. Gao, J. X. Liu, X. Gao, G. Q. Zhang, and X. Z. Tang 2023, "Stability and Digestive Properties of a Dual-Protein Emulsion System Based on Soy Protein Isolate and Whey Protein Isolate," *Foods*, vol. 12, no. 11, doi: 10.3390/foods12112247.

[38] D. Ogrodowska, I. Z. Konopka, M. Tańska, W. Brandt, and B. Piłat 2022, "Effect of Maltodextrin Replacement by Selected Native Starches and Disaccharides on Physicochemical Properties of Pumpkin Oil Capsules Prepared by Spray-Drying," *Applied Sciences (Switzerland)*, vol. 12, no. 1, doi: 10.3390/app12010033.

[39] T. Shadordizadeh, E. Mahdian, and M. A. Hesarinejad 2023, "Application of encapsulated Indigofera tinctoria extract as a natural antioxidant and colorant in ice cream," *Food Sci Nutr*, vol. 11, no. 4, pp. 1940–1951, doi: 10.1002/fsn3.3228.

[40] J. Vulić *et al.* 2019, "Bioavailability and bioactivity of encapsulated phenolics and carotenoids isolated from red pepper waste," *Molecules*, vol. 24, no. 15, doi: 10.3390/molecules24152837.

[41] A. Abdelazez, M. Refaey, D. M. Mohamed, M. M. M. Refaey, and J. Niu, "0123456789) 1 3 Journal of Food Measurement and Characterization", doi: 10.13140/RG.2.2.19045.17126.

[42] D. Nowak and E. Jakubczyk 2020, "The freeze-drying of foods⇔the characteristic of the process course and the effect of its parameters on the physical properties of food materials," *Foods*, vol. 9, no. 10. MDPI AG, doi: 10.3390/foods9101488.

[43] U. S. Prabowo and P. Saraswati 2021, "Effect of maltodextrin concentration and drying temperature on the characteristics of watermelon (Citrullus vulgaris S.) albedo instant drink enriched with telang flower (Clitorea ternatea) extract," *Anjoro: International Journal of Agriculture and Business*, vol. 2, no. 2, pp. 50–57, doi: 10.31605/anjoro.v2i2.1058.

[44] M. Dimopoulou, P. Vareltzis, S. Floros, O. Androutsos, A. Bargiota, and O. Gortzi 2023, "Development of a Functional Acceptable Diabetic and Plant-Based Snack Bar Using Mushroom (Coprinus comatus) Powder," *Foods*, vol. 12, no. 14, doi: 10.3390/foods12142702.

AUTHOR INDEX